Linear Algebra

Fifth Edition

Seymour Lipschutz, PhD

Temple University

Marc Lars Lipson, PhD

University of Virginia

Schaum's Outline Series

**New York Chicago San Francisco Lisbon London
Madrid Mexico City Milan New Delhi San Juan
Seoul Singapore Sydney Toronto**

The McGraw·Hill Companies

SEYMOUR LIPSCHUTZ is on the faculty of Temple University and formally taught at the Polytechnic Institute of Brooklyn. He received his PhD in 1960 at Courant Institute of Mathematical Sciences of New York University. He is one of Schaum's most prolific authors. In particular, he has written, among others, *Beginning Linear Algebra*, *Probability*, *Discrete Mathematics*, *Set Theory*, *Finite Mathematics*, and *General Topology*.

MARC LARS LIPSON is on the faculty of the University of Virginia and formerly taught at the University of Georgia, he received his PhD in finance in 1994 from the University of Michigan. He is also the coauthor of *Discrete Mathematics* and *Probability* with Seymour Lipschutz.

5 6 7 8 9 10 QVS/QVS 20 19 18 17 16

ISBN 978-0-07-179456-5
MHID 0-07-179456-5

e-ISBN 978-0-07-179457-2 (basic e-book)
e-MHID 0-07-179457-3

e-ISBN 978-0-07-181388-4 (enhanced e-book)
e-MHID 0-07-181388-8

Library of Congress Control Number: 2012948438

This book is printed on acid-free paper.

Preface

Linear algebra has in recent years become an essential part of the mathematical background required by mathematicians and mathematics teachers, engineers, computer scientists, physicists, economists, and statisticians, among others. This requirement reflects the importance and wide applications of the subject matter.

This book is designed for use as a textbook for a formal course in linear algebra or as a supplement to all current standard texts. It aims to present an introduction to linear algebra which will be found helpful to all readers regardless of their fields of specification. More material has been included than can be covered in most first courses. This has been done to make the book more flexible, to provide a useful book of reference, and to stimulate further interest in the subject.

Each chapter begins with clear statements of pertinent definitions, principles, and theorems together with illustrative and other descriptive material. This is followed by graded sets of solved and supplementary problems. The solved problems serve to illustrate and amplify the theory, and to provide the repetition of basic principles so vital to effective learning. Numerous proofs, especially those of all essential theorems, are included among the solved problems. The supplementary problems serve as a complete review of the material of each chapter.

The first three chapters treat vectors in Euclidean space, matrix algebra, and systems of linear equations. These chapters provide the motivation and basic computational tools for the abstract investigations of vector spaces and linear mappings which follow. After chapters on inner product spaces and orthogonality and on determinants, there is a detailed discussion of eigenvalues and eigenvectors giving conditions for representing a linear operator by a diagonal matrix. This naturally leads to the study of various canonical forms, specifically, the triangular, Jordan, and rational canonical forms. Later chapters cover linear functions and the dual space V^*, and bilinear, quadratic, and Hermitian forms. The last chapter treats linear operators on inner product spaces.

The main changes in the fourth edition have been in the appendices. First of all, we have expanded Appendix A on the tensor and exterior products of vector spaces where we have now included proofs on the existence and uniqueness of such products. We also added appendices covering algebraic structures, including modules, and polynomials over a field. Appendix D, "Odds and Ends," includes the Moore–Penrose generalized inverse which appears in various applications, such as statistics. There are also many additional solved and supplementary problems.

Finally, we wish to thank the staff of the McGraw-Hill Schaum's Outline Series, especially Charles Wall, for their unfailing cooperation.

SEYMOUR LIPSCHUTZ
MARC LARS LIPSON

List of Symbols

$A = [a_{ij}]$, matrix, 27
$\bar{A} = [\bar{a}_{ij}]$, conjugate matrix, 38
$|A|$, determinant, 264, 268
A^*, adjoint, 377
A^H, conjugate transpose, 38
A^T, transpose, 33
A^+, Moore–Penrose inverse, 418
A_{ij}, minor, 269
$A(I,J)$, minor, 273
$A(V)$, linear operators, 174
adj A, adjoint (classical), 271
$A \sim B$, row equivalence, 72
$A \simeq B$, congruence, 360
C, complex numbers, 11
$\mathbf{C}^n$, complex n-space, 13
$C[a,b]$, continuous functions, 228
$C(f)$, companion matrix, 304
colsp (A), column space, 120
$d(u,v)$, distance, 5, 241
$\text{diag}(a_{11}, \ldots, a_{nn})$, diagonal matrix, 35
$\text{diag}(A_{11}, \ldots, A_{nn})$, block diagonal, 40
$\det(A)$, determinant, 268
$\dim V$, dimension, 124
$\{e_1, \ldots, e_n\}$, usual basis, 125
E_k, projections, 384
$f: A \to B$, mapping, 164
$F(X)$, function space, 114
$G \circ F$, composition, 173
$\text{Hom}(V, U)$, homomorphisms, 174
i, **j**, **k**, 9
I_n, identity matrix, 33
$Im\,F$, image, 169
$J(\lambda)$, Jordan block, 329
K, field of scalars, 112
Ker F, kernel, 169
$m(t)$, minimal polynomial, 303
$\mathbf{M}_{m,n}$, $m \times n$ matrices, 114
n-space, 5, 13, 227, 240
$\mathbf{P}(t)$, polynomials, 114
$\mathbf{P}_n(t)$, polynomials, 114
$\text{proj}(u, v)$, projection, 6, 234
$\text{proj}(u, V)$, projection, 235
Q, rational numbers, 11
R, real numbers, 1
$\mathbf{R}^n$, real n-space, 2
rowsp (A), row-space, 120
$S^\perp$, orthogonal complement, 231
sgn σ, sign, parity, 267
$\text{span}(S)$, linear span, 119
$\text{tr}(A)$, trace, 33
$[T]_S$, matrix representation, 195
T^*, adjoint, 377
T-invariant, 327
T^t, transpose, 351
$\|u\|$, norm, 5, 13, 227, 241
$[u]_S$, coordinate vector, 130
$u \cdot v$, dot product, 4, 13
$\langle u, v \rangle$, inner product, 226, 238
$u \times v$, cross product, 10
$u \otimes v$, tensor product, 396
$u \wedge v$, exterior product, 401
$u \oplus v$, direct sum, 129, 327
$V \cong U$, isomorphism, 132, 169
$V \otimes W$, tensor product, 396
V^*, dual space, 349
V^{**}, second dual space, 350
$\bigwedge^r V$, exterior product, 401
W^0, annihilator, 351
$\bar{z}$, complex conjugate, 12
$Z(v, T)$, T-cyclic subspace, 330
δ_{ij}, Kronecker delta, 37
$\Delta(t)$, characteristic polynomial, 294
λ, eigenvalue, 296
$\sum$, summation symbol, 29

Contents

Vectors in R^n and C^n, Spatial Vectors

1.1 Introduction

There are two ways to motivate the notion of a vector: one is by means of lists of numbers and subscripts, and the other is by means of certain objects in physics. We discuss these two ways below.

Here we assume the reader is familiar with the elementary properties of the field of real numbers, denoted by **R**. On the other hand, we will review properties of the field of complex numbers, denoted by **C**. In the context of vectors, the elements of our number fields are called *scalars*.

Although we will restrict ourselves in this chapter to vectors whose elements come from **R** and then from **C**, many of our operations also apply to vectors whose entries come from some arbitrary field K.

Lists of Numbers

Suppose the weights (in pounds) of eight students are listed as follows:

$$156, \quad 125, \quad 145, \quad 134, \quad 178, \quad 145, \quad 162, \quad 193$$

One can denote all the values in the list using only one symbol, say w, but with different subscripts; that is,

$$w_1, \quad w_2, \quad w_3, \quad w_4, \quad w_5, \quad w_6, \quad w_7, \quad w_8$$

Observe that each subscript denotes the position of the value in the list. For example,

$$w_1 = 156, \text{ the first number}, \quad w_2 = 125, \text{ the second number}, \quad \ldots$$

Such a list of values,

$$w = (w_1, w_2, w_3, \ldots, w_8)$$

is called a *linear array* or *vector*.

Vectors in Physics

Many physical quantities, such as temperature and speed, possess only "magnitude." These quantities can be represented by real numbers and are called *scalars*. On the other hand, there are also quantities, such as force and velocity, that possess both "magnitude" and "direction." These quantities, which can be represented by arrows having appropriate lengths and directions and emanating from some given reference point O, are called *vectors*.

Now we assume the reader is familiar with the space $\mathbf{R}^3$ where all the points in space are represented by ordered triples of real numbers. Suppose the origin of the axes in $\mathbf{R}^3$ is chosen as the reference point O for the vectors discussed above. Then every vector is uniquely determined by the coordinates of its endpoint, and vice versa.

There are two important operations, vector addition and scalar multiplication, associated with vectors in physics. The definition of these operations and the relationship between these operations and the endpoints of the vectors are as follows.

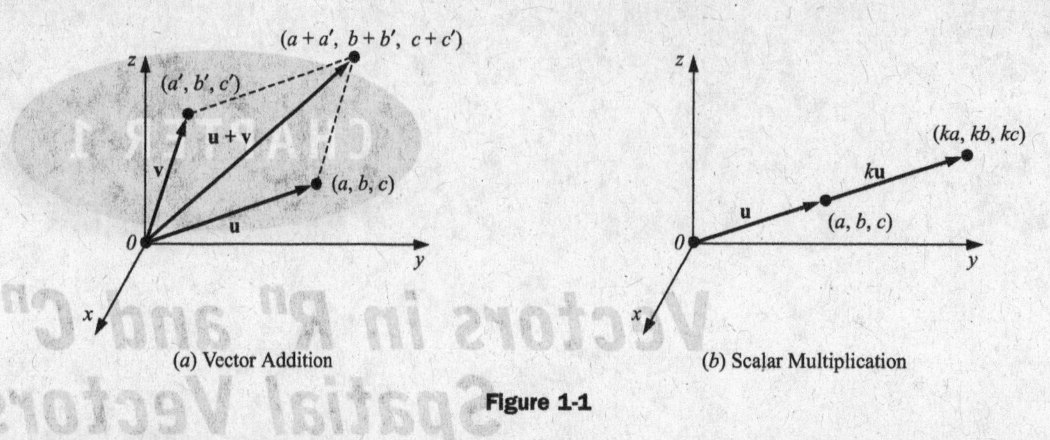

(a) Vector Addition

(b) Scalar Multiplication

Figure 1-1

(i) ***Vector Addition:*** The resultant $\mathbf{u}+\mathbf{v}$ of two vectors $\mathbf{u}$ and $\mathbf{v}$ is obtained by the *parallelogram law*; that is, $\mathbf{u}+\mathbf{v}$ is the diagonal of the parallelogram formed by $\mathbf{u}$ and $\mathbf{v}$. Furthermore, if (a,b,c) and (a',b',c') are the endpoints of the vectors $\mathbf{u}$ and $\mathbf{v}$, then $(a+a',\ b+b',\ c+c')$ is the endpoint of the vector $\mathbf{u}+\mathbf{v}$. These properties are pictured in Fig. 1-1(a).

(ii) ***Scalar Multiplication:*** The product $k\mathbf{u}$ of a vector $\mathbf{u}$ by a real number k is obtained by multiplying the magnitude of $\mathbf{u}$ by k and retaining the same direction if $k>0$ or the opposite direction if $k<0$. Also, if (a,b,c) is the endpoint of the vector $\mathbf{u}$, then (ka,kb,kc) is the endpoint of the vector $k\mathbf{u}$. These properties are pictured in Fig. 1-1(b).

Mathematically, we identify the vector $\mathbf{u}$ with its (a,b,c) and write $\mathbf{u}=(a,b,c)$. Moreover, we call the ordered triple (a,b,c) of real numbers a point or vector depending upon its interpretation. We generalize this notion and call an n-tuple $(a_1,a_2,\ldots,a_n)$ of real numbers a vector. However, special notation may be used for the vectors in $\mathbf{R}^3$ called *spatial* vectors (Section 1.6).

1.2 Vectors in R^n

The set of all n-tuples of real numbers, denoted by $\mathbf{R}^n$, is called *n-space*. A particular n-tuple in $\mathbf{R}^n$, say

$$u=(a_1,a_2,\ldots,a_n)$$

is called a *point* or *vector*. The numbers a_i are called the *coordinates*, *components*, *entries*, or *elements* of u. Moreover, when discussing the space $\mathbf{R}^n$, we use the term *scalar* for the elements of $\mathbf{R}$.

Two vectors, u and v, are *equal*, written $u=v$, if they have the same number of components and if the corresponding components are equal. Although the vectors $(1,2,3)$ and $(2,3,1)$ contain the same three numbers, these vectors are not equal because corresponding entries are not equal.

The vector $(0,0,\ldots,0)$ whose entries are all 0 is called the *zero vector* and is usually denoted by 0.

EXAMPLE 1.1

(a) The following are vectors:

$$(2,-5),\quad (7,9),\quad (0,0,0),\quad (3,4,5)$$

The first two vectors belong to $\mathbf{R}^2$, whereas the last two belong to $\mathbf{R}^3$. The third is the zero vector in $\mathbf{R}^3$.

(b) Find x,y,z such that $(x-y,\ x+y,\ z-1)=(4,2,3)$.

By definition of equality of vectors, corresponding entries must be equal. Thus,

$$x-y=4,\qquad x+y=2,\qquad z-1=3$$

Solving the above system of equations yields $x=3$, $y=-1$, $z=4$.

Column Vectors

Sometimes a vector in n-space $\mathbf{R}^n$ is written vertically rather than horizontally. Such a vector is called a *column vector*, and, in this context, the horizontally written vectors in Example 1.1 are called *row vectors*. For example, the following are column vectors with 2, 2, 3, and 3 components, respectively:

$$\begin{bmatrix} 1 \\ 2 \end{bmatrix}, \quad \begin{bmatrix} 3 \\ -4 \end{bmatrix}, \quad \begin{bmatrix} 1 \\ 5 \\ -6 \end{bmatrix}, \quad \begin{bmatrix} 1.5 \\ \frac{2}{3} \\ -15 \end{bmatrix}$$

We also note that any operation defined for row vectors is defined analogously for column vectors.

1.3 Vector Addition and Scalar Multiplication

Consider two vectors u and v in $\mathbf{R}^n$, say

$$u = (a_1, a_2, \ldots, a_n) \qquad \text{and} \qquad v = (b_1, b_2, \ldots, b_n)$$

Their *sum*, written $u + v$, is the vector obtained by adding corresponding components from u and v. That is,

$$u + v = (a_1 + b_1,\ a_2 + b_2,\ \ldots,\ a_n + b_n)$$

The *product*, of the vector u by a real number k, written ku, is the vector obtained by multiplying each component of u by k. That is,

$$ku = k(a_1, a_2, \ldots, a_n) = (ka_1, ka_2, \ldots, ka_n)$$

Observe that $u + v$ and ku are also vectors in $\mathbf{R}^n$. The sum of vectors with different numbers of components is not defined.

Negatives and subtraction are defined in $\mathbf{R}^n$ as follows:

$$-u = (-1)u \qquad \text{and} \qquad u - v = u + (-v)$$

The vector $-u$ is called the *negative* of u, and $u - v$ is called the *difference* of u and v.

Now suppose we are given vectors $u_1, u_2, \ldots, u_m$ in $\mathbf{R}^n$ and scalars $k_1, k_2, \ldots, k_m$ in $\mathbf{R}$. We can multiply the vectors by the corresponding scalars and then add the resultant scalar products to form the vector

$$v = k_1u_1 + k_2u_2 + k_3u_3 + \cdots + k_mu_m$$

Such a vector v is called a *linear combination* of the vectors $u_1, u_2, \ldots, u_m$.

EXAMPLE 1.2

(a) Let $u = (2, 4, -5)$ and $v = (1, -6, 9)$. Then

$$\begin{aligned} u + v &= (2 + 1,\ 4 + (-6),\ -5 + 9) = (3, -2, 4) \\ 7u &= (7(2), 7(4), 7(-5)) = (14, 28, -35) \\ -v &= (-1)(1, -6, 9) = (-1, 6, -9) \\ 3u - 5v &= (6, 12, -15) + (-5, 30, -45) = (1, 42, -60) \end{aligned}$$

(b) The zero vector $0 = (0, 0, \ldots, 0)$ in $\mathbf{R}^n$ is similar to the scalar 0 in that, for any vector $u = (a_1, a_2, \ldots, a_n)$.

$$u + 0 = (a_1 + 0,\ a_2 + 0,\ \ldots,\ a_n + 0) = (a_1, a_2, \ldots, a_n) = u$$

(c) Let $u = \begin{bmatrix} 2 \\ 3 \\ -4 \end{bmatrix}$ and $v = \begin{bmatrix} 3 \\ -1 \\ -2 \end{bmatrix}$. Then $2u - 3v = \begin{bmatrix} 4 \\ 6 \\ -8 \end{bmatrix} + \begin{bmatrix} -9 \\ 3 \\ 6 \end{bmatrix} = \begin{bmatrix} -5 \\ 9 \\ -2 \end{bmatrix}$.

Basic properties of vectors under the operations of vector addition and scalar multiplication are described in the following theorem.

THEOREM 1.1: For any vectors u, v, w in $\mathbf{R}^n$ and any scalars k, k' in $\mathbf{R}$,

(i) $(u+v)+w = u+(v+w)$, (v) $k(u+v) = ku+kv$,

(ii) $u+0 = u$, (vi) $(k+k')u = ku+k'u$,

(iii) $u+(-u) = 0$, (vii) (kk')u=k(k'u),

(iv) $u+v = v+u$, (viii) $1u = u$.

We postpone the proof of Theorem 1.1 until Chapter 2, where it appears in the context of matrices (Problem 2.3).

Suppose u and v are vectors in $\mathbf{R}^n$ for which $u = kv$ for some nonzero scalar k in $\mathbf{R}$. Then u is called a *multiple* of v. Also, u is said to be in the *same* or *opposite direction* as v according to whether $k > 0$ or $k < 0$.

1.4 Dot (Inner) Product

Consider arbitrary vectors u and v in $\mathbf{R}^n$; say,

$$u = (a_1, a_2, \ldots, a_n) \quad \text{and} \quad v = (b_1, b_2, \ldots, b_n)$$

The *dot product* or *inner product* of u and v is denoted and defined by

$$u \cdot v = a_1b_1 + a_2b_2 + \cdots + a_nb_n$$

That is, $u \cdot v$ is obtained by multiplying corresponding components and adding the resulting products. The vectors u and v are said to be *orthogonal* (or *perpendicular*) if their dot product is zero—that is, if $u \cdot v = 0$.

EXAMPLE 1.3

(a) Let $u = (1, -2, 3)$, $v = (4, 5, -1)$, $w = (2, 7, 4)$. Then,

$$u \cdot v = 1(4) - 2(5) + 3(-1) = 4 - 10 - 3 = -9$$
$$u \cdot w = 2 - 14 + 12 = 0, \qquad v \cdot w = 8 + 35 - 4 = 39$$

Thus, u and w are orthogonal.

(b) Let $u = \begin{bmatrix} 2 \\ 3 \\ -4 \end{bmatrix}$ and $v = \begin{bmatrix} 3 \\ -1 \\ -2 \end{bmatrix}$. Then $u \cdot v = 6 - 3 + 8 = 11$.

(c) Suppose $u = (1, 2, 3, 4)$ and $v = (6, k, -8, 2)$. Find k so that u and v are orthogonal.

First obtain $u \cdot v = 6 + 2k - 24 + 8 = -10 + 2k$. Then set $u \cdot v = 0$ and solve for k:

$$-10 + 2k = 0 \quad \text{or} \quad 2k = 10 \quad \text{or} \quad k = 5$$

Basic properties of the dot product in $\mathbf{R}^n$ (proved in Problem 1.13) follow.

THEOREM 1.2: For any vectors u, v, w in $\mathbf{R}^n$ and any scalar k in $\mathbf{R}$:

(i) $(u+v) \cdot w = u \cdot w + v \cdot w$, (iii) $u \cdot v = v \cdot u$,

(ii) $(ku) \cdot v = k(u \cdot v)$, (iv) $u \cdot u \geq 0$, and $u \cdot u = 0$ iff $u = 0$.

Note that (ii) says that we can "take k out" from the first position in an inner product. By (iii) and (ii),

$$u \cdot (kv) = (kv) \cdot u = k(v \cdot u) = k(u \cdot v)$$

That is, we can also "take k out" from the second position in an inner product.

The space $\mathbf{R}^n$ with the above operations of vector addition, scalar multiplication, and dot product is usually called *Euclidean n-space*.

Norm (Length) of a Vector

The *norm* or *length* of a vector u in $\mathbf{R}^n$, denoted by $\|u\|$, is defined to be the nonnegative square root of $u \cdot u$. In particular, if $u = (a_1, a_2, \ldots, a_n)$, then

$$\|u\| = \sqrt{u \cdot u} = \sqrt{a_1^2 + a_2^2 + \cdots + a_n^2}$$

That is, $\|u\|$ is the square root of the sum of the squares of the components of u. Thus, $\|u\| \geq 0$, and $\|u\| = 0$ if and only if $u = 0$.

A vector u is called a *unit* vector if $\|u\| = 1$ or, equivalently, if $u \cdot u = 1$. For any nonzero vector v in $\mathbf{R}^n$, the vector

$$\hat{v} = \frac{1}{\|v\|} v = \frac{v}{\|v\|}$$

is the unique unit vector in the same direction as v. The process of finding $\hat{v}$ from v is called *normalizing* v.

EXAMPLE 1.4

(a) Suppose $u = (1, -2, -4, 5, 3)$. To find $\|u\|$, we can first find $\|u\|^2 = u \cdot u$ by squaring each component of u and adding, as follows:

$$\|u\|^2 = 1^2 + (-2)^2 + (-4)^2 + 5^2 + 3^2 = 1 + 4 + 16 + 25 + 9 = 55$$

Then $\|u\| = \sqrt{55}$.

(b) Let $v = (1, -3, 4, 2)$ and $w = (\frac{1}{2}, -\frac{1}{6}, \frac{5}{6}, \frac{1}{6})$. Then

$$\|v\| = \sqrt{1 + 9 + 16 + 4} = \sqrt{30} \quad \text{and} \quad \|w\| = \sqrt{\frac{9}{36} + \frac{1}{36} + \frac{25}{36} + \frac{1}{36}} = \sqrt{\frac{36}{36}} = \sqrt{1} = 1$$

Thus w is a unit vector, but v is not a unit vector. However, we can normalize v as follows:

$$\hat{v} = \frac{v}{\|v\|} = \left(\frac{1}{\sqrt{30}}, \frac{-3}{\sqrt{30}}, \frac{4}{\sqrt{30}}, \frac{2}{\sqrt{30}}\right)$$

This is the unique unit vector in the same direction as v.

The following formula (proved in Problem 1.14) is known as the Schwarz inequality or Cauchy–Schwarz inequality. It is used in many branches of mathematics.

THEOREM 1.3 (Schwarz): For any vectors u, v in $\mathbf{R}^n$, $|u \cdot v| \leq \|u\|\|v\|$.

Using the above inequality, we also prove (Problem 1.15) the following result known as the "triangle inequality" or Minkowski's inequality.

THEOREM 1.4 (Minkowski): For any vectors u, v in $\mathbf{R}^n$, $\|u + v\| \leq \|u\| + \|v\|$.

Distance, Angles, Projections

The *distance* between vectors $u = (a_1, a_2, \ldots, a_n)$ and $v = (b_1, b_2, \ldots, b_n)$ in $\mathbf{R}^n$ is denoted and defined by

$$d(u, v) = \|u - v\| = \sqrt{(a_1 - b_1)^2 + (a_2 - b_2)^2 + \cdots + (a_n - b_n)^2}$$

One can show that this definition agrees with the usual notion of distance in the Euclidean plane $\mathbf{R}^2$ or space $\mathbf{R}^3$.

The *angle* θ between nonzero vectors u, v in $\mathbf{R}^n$ is defined by

$$\cos\theta = \frac{u \cdot v}{\|u\|\|v\|}$$

This definition is well defined, because, by the Schwarz inequality (Theorem 1.3),

$$-1 \le \frac{u \cdot v}{\|u\|\|v\|} \le 1$$

Note that if $u \cdot v = 0$, then $\theta = 90^\circ$ (or $\theta = \pi/2$). This then agrees with our previous definition of orthogonality.

The *projection* of a vector u onto a nonzero vector v is the vector denoted and defined by

$$\text{proj}(u, v) = \frac{u \cdot v}{\|v\|^2} v = \frac{u \cdot v}{v \cdot v} v$$

We show below that this agrees with the usual notion of vector projection in physics.

EXAMPLE 1.5

(a) Suppose $u = (1, -2, 3)$ and $v = (2, 4, 5)$. Then

$$d(u, v) = \sqrt{(1-2)^2 + (-2-4)^2 + (3-5)^2} = \sqrt{1 + 36 + 4} = \sqrt{41}$$

To find $\cos\theta$, where θ is the angle between u and v, we first find

$$u \cdot v = 2 - 8 + 15 = 9, \qquad \|u\|^2 = 1 + 4 + 9 = 14, \qquad \|v\|^2 = 4 + 16 + 25 = 45$$

Then

$$\cos\theta = \frac{u \cdot v}{\|u\|\|v\|} = \frac{9}{\sqrt{14}\sqrt{45}}$$

Also,

$$\text{proj}(u, v) = \frac{u \cdot v}{\|v\|^2} v = \frac{9}{45}(2, 4, 5) = \frac{1}{5}(2, 4, 5) = \left(\frac{2}{5}, \frac{4}{5}, 1\right)$$

(b) Consider the vectors u and v in Fig. 1-2(a) (with respective endpoints A and B). The (perpendicular) projection of u onto v is the vector u^* with magnitude

$$\|u^*\| = \|u\| \cos\theta = \|u\| \frac{u \cdot v}{\|u\|\|v\|} = \frac{u \cdot v}{\|v\|}$$

To obtain u^*, we multiply its magnitude by the unit vector in the direction of v, obtaining

$$u^* = \|u^*\| \frac{v}{\|v\|} = \frac{u \cdot v}{\|v\|} \frac{v}{\|v\|} = \frac{u \cdot v}{\|v\|^2} v$$

This is the same as the above definition of proj(u, v).

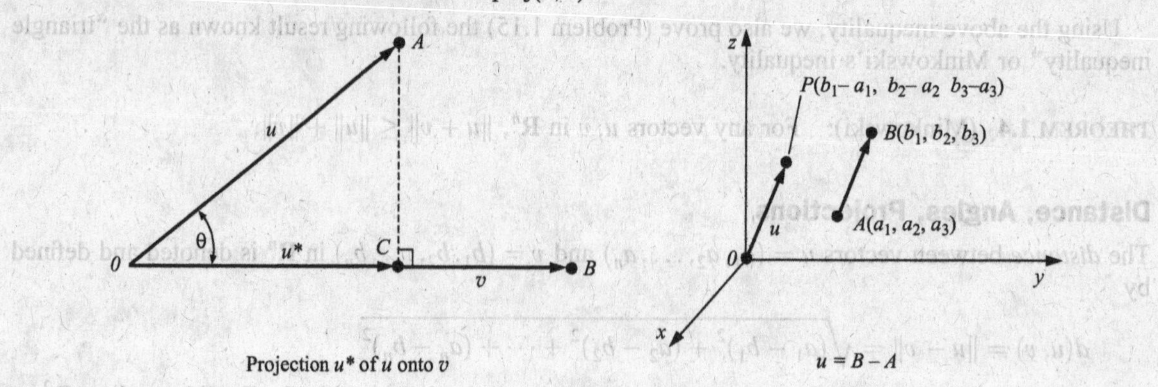

Figure 1-2

1.5 Located Vectors, Hyperplanes, Lines, Curves in R^n

This section distinguishes between an n-tuple $P(a_i) \equiv P(a_1, a_2, \ldots, a_n)$ viewed as a point in $\mathbf{R}^n$ and an n-tuple $u = [c_1, c_2, \ldots, c_n]$ viewed as a vector (arrow) from the origin O to the point $C(c_1, c_2, \ldots, c_n)$.

Located Vectors

Any pair of points $A(a_i)$ and $B(b_i)$ in $\mathbf{R}^n$ defines the *located vector* or *directed line segment* from A to B, written $\overrightarrow{AB}$. We identify $\overrightarrow{AB}$ with the vector

$$u = B - A = [b_1 - a_1, \; b_2 - a_2, \; \ldots, \; b_n - a_n]$$

because $\overrightarrow{AB}$ and u have the same magnitude and direction. This is pictured in Fig. 1-2(b) for the points $A(a_1, a_2, a_3)$ and $B(b_1, b_2, b_3)$ in $\mathbf{R}^3$ and the vector $u = B - A$ which has the endpoint $P(b_1 - a_1, \, b_2 - a_2, \, b_3 - a_3)$.

Hyperplanes

A *hyperplane H* in $\mathbf{R}^n$ is the set of points $(x_1, x_2, \ldots, x_n)$ that satisfy a linear equation

$$a_1x_1 + a_2x_2 + \cdots + a_nx_n = b$$

where the vector $u = [a_1, a_2, \ldots, a_n]$ of coefficients is not zero. Thus a hyperplane H in $\mathbf{R}^2$ is a line, and a hyperplane H in $\mathbf{R}^3$ is a plane. We show below, as pictured in Fig. 1-3(a) for $\mathbf{R}^3$, that u is orthogonal to any directed line segment $\overrightarrow{PQ}$, where $P(p_i)$ and $Q(q_i)$ are points in H. [For this reason, we say that u is *normal* to H and that H is *normal* to u.]

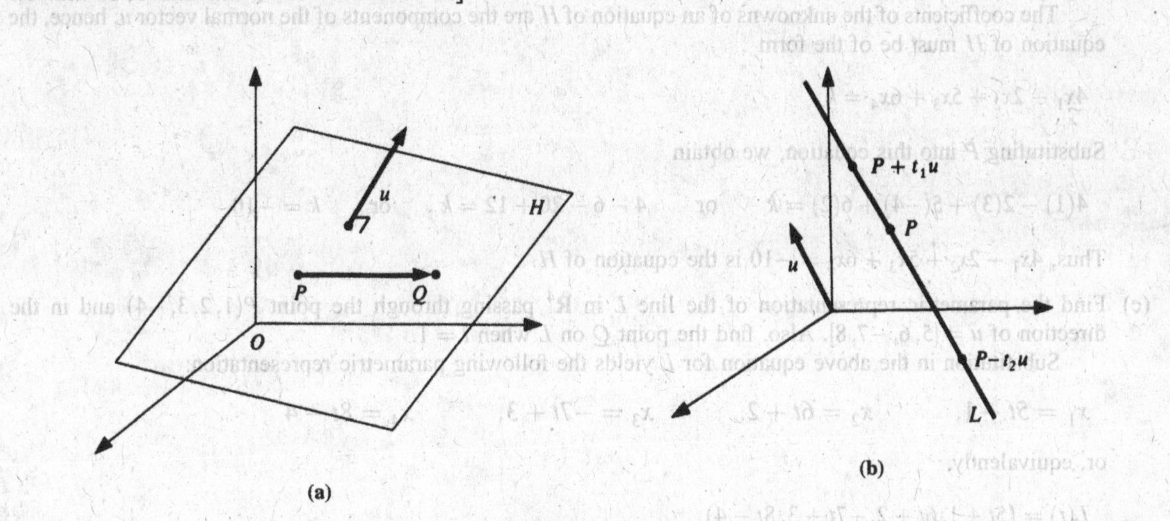

Figure 1-3

Because $P(p_i)$ and $Q(q_i)$ belong to H, they satisfy the above hyperplane equation—that is,

$$a_1p_1 + a_2p_2 + \cdots + a_np_n = b \quad \text{and} \quad a_1q_1 + a_2q_2 + \cdots + a_nq_n = b$$

Let $\quad v = \overrightarrow{PQ} = Q - P = [q_1 - p_1, q_2 - p_2, \ldots, q_n - p_n]$

Then

$$\begin{aligned} u \cdot v &= a_1(q_1 - p_1) + a_2(q_2 - p_2) + \cdots + a_n(q_n - p_n) \\ &= (a_1q_1 + a_2q_2 + \cdots + a_nq_n) - (a_1p_1 + a_2p_2 + \cdots + a_np_n) = b - b = 0 \end{aligned}$$

Thus $v = \overrightarrow{PQ}$ is orthogonal to u, as claimed.

Lines in R^n

The *line* L in $\mathbf{R}^n$ passing through the point $P(b_1, b_2, \ldots, b_n)$ and in the direction of a nonzero vector $u = [a_1, a_2, \ldots, a_n]$ consists of the points $X(x_1, x_2, \ldots, x_n)$ that satisfy

$$X = P + tu \qquad \text{or} \qquad \begin{cases} x_1 = a_1 t + b_1 \\ x_2 = a_2 t + b_2 \\ \ldots\ldots\ldots\ldots\ldots\ldots \\ x_n = a_n t + b_n \end{cases} \qquad \text{or } L(t) = (a_i t + b_i)$$

where the *parameter* t takes on all real values. Such a line L in $\mathbf{R}^3$ is pictured in Fig. 1-3(b).

EXAMPLE 1.6

(a) Let H be the plane in $\mathbf{R}^3$ corresponding to the linear equation $2x - 5y + 7z = 4$. Observe that $P(1, 1, 1)$ and $Q(5, 4, 2)$ are solutions of the equation. Thus P and Q and the directed line segment

$$v = \overrightarrow{PQ} = Q - P = [5 - 1,\ 4 - 1,\ 2 - 1] = [4, 3, 1]$$

lie on the plane H. The vector $u = [2, -5, 7]$ is normal to H, and, as expected,

$$u \cdot v = [2, -5, 7] \cdot [4, 3, 1] = 8 - 15 + 7 = 0$$

That is, u is orthogonal to v.

(b) Find an equation of the hyperplane H in $\mathbf{R}^4$ that passes through the point $P(1, 3, -4, 2)$ and is normal to the vector $u = [4, -2, 5, 6]$.

The coefficients of the unknowns of an equation of H are the components of the normal vector u; hence, the equation of H must be of the form

$$4x_1 - 2x_2 + 5x_3 + 6x_4 = k$$

Substituting P into this equation, we obtain

$$4(1) - 2(3) + 5(-4) + 6(2) = k \qquad \text{or} \qquad 4 - 6 - 20 + 12 = k \qquad \text{or} \qquad k = -10$$

Thus, $4x_1 - 2x_2 + 5x_3 + 6x_4 = -10$ is the equation of H.

(c) Find the parametric representation of the line L in $\mathbf{R}^4$ passing through the point $P(1, 2, 3, -4)$ and in the direction of $u = [5, 6, -7, 8]$. Also, find the point Q on L when $t = 1$.

Substitution in the above equation for L yields the following parametric representation:

$$x_1 = 5t + 1, \qquad x_2 = 6t + 2, \qquad x_3 = -7t + 3, \qquad x_4 = 8t - 4$$

or, equivalently,

$$L(t) = (5t + 1, 6t + 2, -7t + 3, 8t - 4)$$

Note that $t = 0$ yields the point P on L. Substitution of $t = 1$ yields the point $Q(6, 8, -4, 4)$ on L.

Curves in R^n

Let D be an interval (finite or infinite) on the real line $\mathbf{R}$. A continuous function $F: D \to \mathbf{R}^n$ is a *curve* in $\mathbf{R}^n$. Thus, to each point $t \in D$ there is assigned the following point in $\mathbf{R}^n$:

$$F(t) = [F_1(t), F_2(t), \ldots, F_n(t)]$$

Moreover, the derivative (if it exists) of $F(t)$ yields the vector

$$V(t) = \frac{dF(t)}{dt} = \left[\frac{dF_1(t)}{dt}, \frac{dF_2(t)}{dt}, \ldots, \frac{dF_n(t)}{dt}\right]$$

which is tangent to the curve. Normalizing $V(t)$ yields

$$\mathbf{T}(t) = \frac{V(t)}{\|V(t)\|}$$

Thus, $\mathbf{T}(t)$ is the unit tangent vector to the curve. (Unit vectors with geometrical significance are often presented in bold type.)

EXAMPLE 1.7 Consider the curve $F(t) = [\sin t, \cos t, t]$ in $\mathbf{R}^3$. Taking the derivative of $F(t)$ [or each component of $F(t)$] yields

$$V(t) = [\cos t, -\sin t, 1]$$

which is a vector tangent to the curve. We normalize $V(t)$. First we obtain

$$\|V(t)\|^2 = \cos^2 t + \sin^2 t + 1 = 1 + 1 = 2$$

Then the unit tangent vection $\mathbf{T}(t)$ to the curve follows:

$$\mathbf{T}(t) = \frac{V(t)}{\|V(t)\|} = \left[\frac{\cos t}{\sqrt{2}}, \frac{-\sin t}{\sqrt{2}}, \frac{1}{\sqrt{2}}\right]$$

1.6 Vectors in $\mathbf{R}^3$ (Spatial Vectors), ijk Notation

Vectors in $\mathbf{R}^3$, called *spatial vectors*, appear in many applications, especially in physics. In fact, a special notation is frequently used for such vectors as follows:

$\mathbf{i} = [1, 0, 0]$ denotes the unit vector in the x direction.
$\mathbf{j} = [0, 1, 0]$ denotes the unit vector in the y direction.
$\mathbf{k} = [0, 0, 1]$ denotes the unit vector in the z direction.

Then any vector $u = [a, b, c]$ in $\mathbf{R}^3$ can be expressed uniquely in the form

$$u = [a, b, c] = a\mathbf{i} + b\mathbf{j} + c\mathbf{k}$$

Because the vectors $\mathbf{i}, \mathbf{j}, \mathbf{k}$ are unit vectors and are mutually orthogonal, we obtain the following dot products:

$$\mathbf{i}\cdot\mathbf{i} = 1, \quad \mathbf{j}\cdot\mathbf{j} = 1, \quad \mathbf{k}\cdot\mathbf{k} = 1 \quad \text{and} \quad \mathbf{i}\cdot\mathbf{j} = 0, \quad \mathbf{i}\cdot\mathbf{k} = 0, \quad \mathbf{j}\cdot\mathbf{k} = 0$$

Furthermore, the vector operations discussed above may be expressed in the **ijk** notation as follows. Suppose

$$u = a_1\mathbf{i} + a_2\mathbf{j} + a_3\mathbf{k} \quad \text{and} \quad v = b_1\mathbf{i} + b_2\mathbf{j} + b_3\mathbf{k}$$

Then

$$u + v = (a_1 + b_1)\mathbf{i} + (a_2 + b_2)\mathbf{j} + (a_3 + b_3)\mathbf{k} \quad \text{and} \quad cu = ca_1\mathbf{i} + ca_2\mathbf{j} + ca_3\mathbf{k}$$

where c is a scalar. Also,

$$u \cdot v = a_1b_1 + a_2b_2 + a_3b_3 \quad \text{and} \quad \|u\| = \sqrt{u \cdot u} = \sqrt{a_1^2 + a_2^2 + a_3^2}$$

EXAMPLE 1.8 Suppose $u = 3\mathbf{i} + 5\mathbf{j} - 2\mathbf{k}$ and $v = 4\mathbf{i} - 8\mathbf{j} + 7\mathbf{k}$.

(a) To find $u + v$, add corresponding components, obtaining $u + v = 7\mathbf{i} - 3\mathbf{j} + 5\mathbf{k}$

(b) To find $3u - 2v$, first multiply by the scalars and then add:

$$3u - 2v = (9\mathbf{i} + 15\mathbf{j} - 6\mathbf{k}) + (-8\mathbf{i} + 16\mathbf{j} - 14\mathbf{k}) = \mathbf{i} + 31\mathbf{j} - 20\mathbf{k}$$

(c) To find $u \cdot v$, multiply corresponding components and then add:

$$u \cdot v = 12 - 40 - 14 = -42$$

(d) To find $\|u\|$, take the square root of the sum of the squares of the components:

$$\|u\| = \sqrt{9 + 25 + 4} = \sqrt{38}$$

Cross Product

There is a special operation for vectors u and v in $\mathbf{R}^3$ that is not defined in $\mathbf{R}^n$ for $n \neq 3$. This operation is called the *cross product* and is denoted by $u \times v$. One way to easily remember the formula for $u \times v$ is to use the determinant (of order two) and its negative, which are denoted and defined as follows:

$$\begin{vmatrix} a & b \\ c & d \end{vmatrix} = ad - bc \qquad \text{and} \qquad -\begin{vmatrix} a & b \\ c & d \end{vmatrix} = bc - ad$$

Here a and d are called the *diagonal* elements and b and c are the *nondiagonal* elements. Thus, the determinant is the product ad of the diagonal elements minus the product bc of the nondiagonal elements, but vice versa for the negative of the determinant.

Now suppose $u = a_1\mathbf{i} + a_2\mathbf{j} + a_3\mathbf{k}$ and $v = b_1\mathbf{i} + b_2\mathbf{j} + b_3\mathbf{k}$. Then

$$u \times v = (a_2b_3 - a_3b_2)\mathbf{i} + (a_3b_1 - a_1b_3)\mathbf{j} + (a_1b_2 - a_2b_1)\mathbf{k}$$

$$= \begin{vmatrix} a_1 & a_2 & a_3 \\ b_1 & b_2 & b_3 \end{vmatrix}\mathbf{i} - \begin{vmatrix} a_1 & a_2 & a_3 \\ b_1 & b_2 & b_3 \end{vmatrix}\mathbf{j} + \begin{vmatrix} a_1 & a_2 & a_3 \\ b_1 & b_2 & b_3 \end{vmatrix}\mathbf{k}$$

That is, the three components of $u \times v$ are obtained from the array

$$\begin{bmatrix} a_1 & a_2 & a_3 \\ b_1 & b_2 & b_3 \end{bmatrix}$$

(which contain the components of u above the component of v) as follows:

(1) Cover the first column and take the determinant.
(2) Cover the second column and take the negative of the determinant.
(3) Cover the third column and take the determinant.

Note that $u \times v$ is a vector; hence, $u \times v$ is also called the *vector product* or *outer product* of u and v.

EXAMPLE 1.9 Find $u \times v$ where: (a) $u = 4\mathbf{i} + 3\mathbf{j} + 6\mathbf{k}$, $v = 2\mathbf{i} + 5\mathbf{j} - 3\mathbf{k}$, (b) $u = [2, -1, 5]$, $v = [3, 7, 6]$.

(a) Use $\begin{bmatrix} 4 & 3 & 6 \\ 2 & 5 & -3 \end{bmatrix}$ to get $u \times v = (-9 - 30)\mathbf{i} + (12 + 12)\mathbf{j} + (20 - 6)\mathbf{k} = -39\mathbf{i} + 24\mathbf{j} + 14\mathbf{k}$

(b) Use $\begin{bmatrix} 2 & -1 & 5 \\ 3 & 7 & 6 \end{bmatrix}$ to get $u \times v = [-6 - 35, 15 - 12, 14 + 3] = [-41, 3, 17]$

Remark: The cross products of the vectors $\mathbf{i}, \mathbf{j}, \mathbf{k}$ are as follows:

$$\mathbf{i} \times \mathbf{j} = \mathbf{k}, \qquad \mathbf{j} \times \mathbf{k} = \mathbf{i}, \qquad \mathbf{k} \times \mathbf{i} = \mathbf{j}$$
$$\mathbf{j} \times \mathbf{i} = -\mathbf{k}, \qquad \mathbf{k} \times \mathbf{j} = -\mathbf{i}, \qquad \mathbf{i} \times \mathbf{k} = -\mathbf{j}$$

Thus, if we view the triple $(\mathbf{i}, \mathbf{j}, \mathbf{k})$ as a cyclic permutation, where $\mathbf{i}$ follows $\mathbf{k}$ and hence $\mathbf{k}$ precedes $\mathbf{i}$, then the product of two of them in the given direction is the third one, but the product of two of them in the opposite direction is the negative of the third one.

Two important properties of the cross product are contained in the following theorem.

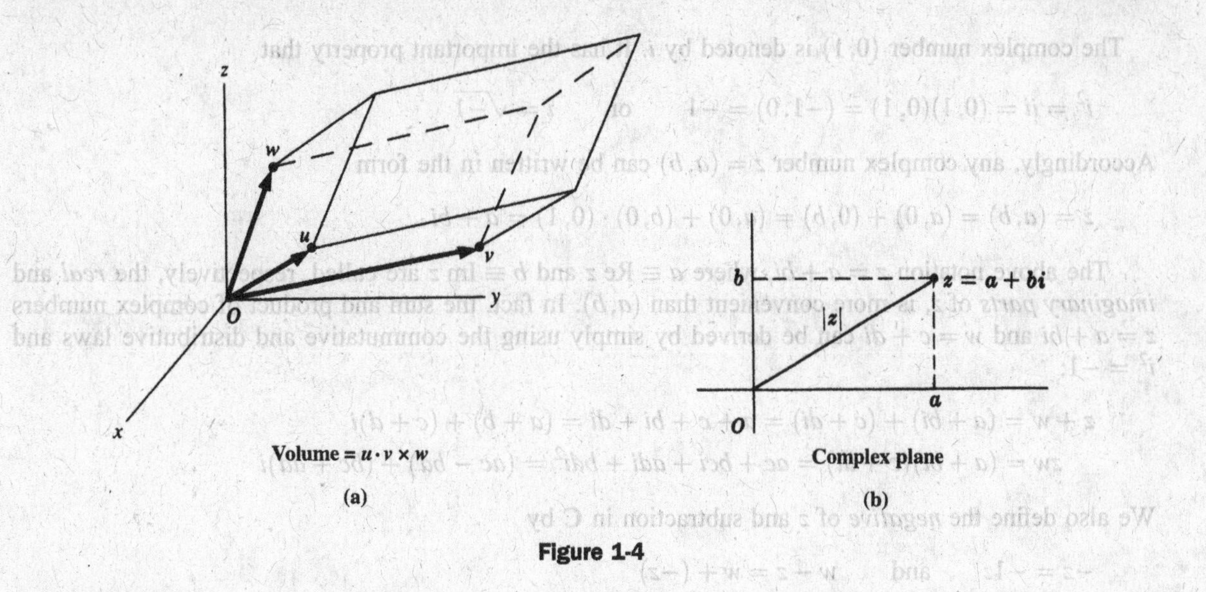

Figure 1-4

THEOREM 1.5: Let u, v, w be vectors in $\mathbf{R}^3$.

(a) The vector $u \times v$ is orthogonal to both u and v.

(b) The absolute value of the "triple product"

$$u \cdot v \times w$$

represents the volume of the parallelepiped formed by the vectors u, v, w. [See Fig. 1-4(a).]

We note that the vectors $u, v, u \times v$ form a right-handed system, and that the following formula gives the magnitude of $u \times v$:

$$\|u \times v\| = \|u\|\|v\| \sin\theta$$

where θ is the angle between u and v.

1.7 Complex Numbers

The set of complex numbers is denoted by **C**. Formally, a complex number is an ordered pair (a, b) of real numbers where equality, addition, and multiplication are defined as follows:

$$(a, b) = (c, d) \quad \text{if and only if } a = c \text{ and } b = d$$
$$(a, b) + (c, d) = (a + c,\ b + d)$$
$$(a, b) \cdot (c, d) = (ac - bd,\ ad + bc)$$

We identify the real number a with the complex number $(a, 0)$; that is,

$$a \leftrightarrow (a, 0)$$

This is possible because the operations of addition and multiplication of real numbers are preserved under the correspondence; that is,

$$(a, 0) + (b, 0) = (a + b,\ 0) \qquad \text{and} \qquad (a, 0) \cdot (b, 0) = (ab, 0)$$

Thus we view **R** as a subset of **C**, and replace $(a, 0)$ by a whenever convenient and possible.

We note that the set **C** of complex numbers with the above operations of addition and multiplication is a *field* of numbers, like the set **R** of real numbers and the set **Q** of *rational numbers*.

The complex number $(0, 1)$ is denoted by i. It has the important property that

$$i^2 = ii = (0, 1)(0, 1) = (-1, 0) = -1 \qquad \text{or} \qquad i = \sqrt{-1}$$

Accordingly, any complex number $z = (a, b)$ can be written in the form

$$z = (a, b) = (a, 0) + (0, b) = (a, 0) + (b, 0) \cdot (0, 1) = a + bi$$

The above notation $z = a + bi$, where $a \equiv \text{Re}\, z$ and $b \equiv \text{Im}\, z$ are called, respectively, the *real* and *imaginary parts* of z, is more convenient than (a, b). In fact, the sum and product of complex numbers $z = a + bi$ and $w = c + di$ can be derived by simply using the commutative and distributive laws and $i^2 = -1$:

$$z + w = (a + bi) + (c + di) = a + c + bi + di = (a + b) + (c + d)i$$
$$zw = (a + bi)(c + di) = ac + bci + adi + bdi^2 = (ac - bd) + (bc + ad)i$$

We also define the *negative* of z and subtraction in **C** by

$$-z = -1z \qquad \text{and} \qquad w - z = w + (-z)$$

Warning: The letter i representing $\sqrt{-1}$ has no relationship whatsoever to the vector $\mathbf{i} = [1, 0, 0]$ in Section 1.6.

Complex Conjugate, Absolute Value

Consider a complex number $z = a + bi$. The *conjugate* of z is denoted and defined by

$$\bar{z} = \overline{a + bi} = a - bi$$

Then $z\bar{z} = (a + bi)(a - bi) = a^2 - b^2i^2 = a^2 + b^2$. Note that z is real if and only if $\bar{z} = z$.

The *absolute value* of z, denoted by $|z|$, is defined to be the nonnegative square root of $z\bar{z}$. Namely,

$$|z| = \sqrt{z\bar{z}} = \sqrt{a^2 + b^2}$$

Note that $|z|$ is equal to the norm of the vector (a, b) in $\mathbf{R}^2$.

Suppose $z \neq 0$. Then the inverse z^{-1} of z and division in **C** of w by z are given, respectively, by

$$z^{-1} = \frac{\bar{z}}{z\bar{z}} = \frac{a}{a^2 + b^2} - \frac{b}{a^2 + b^2}i \qquad \text{and} \qquad \frac{w}{z} = \frac{w\bar{z}}{z\bar{z}} = wz^{-1}$$

EXAMPLE 1.10 Suppose $z = 2 + 3i$ and $w = 5 - 2i$. Then

$$z + w = (2 + 3i) + (5 - 2i) = 2 + 5 + 3i - 2i = 7 + i$$
$$zw = (2 + 3i)(5 - 2i) = 10 + 15i - 4i - 6i^2 = 16 + 11i$$
$$\bar{z} = \overline{2 + 3i} = 2 - 3i \qquad \text{and} \qquad \bar{w} = \overline{5 - 2i} = 5 + 2i$$
$$\frac{w}{z} = \frac{5 - 2i}{2 + 3i} = \frac{(5 - 2i)(2 - 3i)}{(2 + 3i)(2 - 3i)} = \frac{4 - 19i}{13} = \frac{4}{13} - \frac{19}{13}i$$
$$|z| = \sqrt{4 + 9} = \sqrt{13} \qquad \text{and} \qquad |w| = \sqrt{25 + 4} = \sqrt{29}$$

Complex Plane

Recall that the real numbers **R** can be represented by points on a line. Analogously, the complex numbers **C** can be represented by points in the plane. Specifically, we let the point (a, b) in the plane represent the complex number $a + bi$ as shown in Fig. 1-4(b). In such a case, $|z|$ is the distance from the origin O to the point z. The plane with this representation is called the *complex plane*, just like the line representing **R** is called the *real line*.

1.8 Vectors in C^n

The set of all n-tuples of complex numbers, denoted by $\mathbf{C}^n$, is called *complex* n*-space*. Just as in the real case, the elements of $\mathbf{C}^n$ are called *points* or *vectors*, the elements of $\mathbf{C}$ are called *scalars*, and vector addition in $\mathbf{C}^n$ and scalar multiplication on $\mathbf{C}^n$ are given by

$$[z_1, z_2, \ldots, z_n] + [w_1, w_2, \ldots, w_n] = [z_1 + w_1,\ z_2 + w_2,\ \ldots,\ z_n + w_n]$$
$$z[z_1, z_2, \ldots, z_n] = [zz_1, zz_2, \ldots, zz_n]$$

where the z_i, w_i, and z belong to $\mathbf{C}$.

EXAMPLE 1.11 Consider vectors $u = [2 + 3i,\ 4 - i,\ 3]$ and $v = [3 - 2i,\ 5i,\ 4 - 6i]$ in $\mathbf{C}^3$. Then

$$u + v = [2 + 3i,\ 4 - i,\ 3] + [3 - 2i,\ 5i,\ 4 - 6i] = [5 + i,\ 4 + 4i,\ 7 - 6i]$$
$$(5 - 2i)u = [(5 - 2i)(2 + 3i),\ (5 - 2i)(4 - i),\ (5 - 2i)(3)] = [16 + 11i,\ 18 - 13i,\ 15 - 6i]$$

Dot (Inner) Product in C^n

Consider vectors $u = [z_1, z_2, \ldots, z_n]$ and $v = [w_1, w_2, \ldots, w_n]$ in $\mathbf{C}^n$. The *dot* or *inner product* of u and v is denoted and defined by

$$u \cdot v = z_1\bar{w}_1 + z_2\bar{w}_2 + \cdots + z_n\bar{w}_n$$

This definition reduces to the real case because $\bar{w}_i = w_i$ when w_i is real. The norm of u is defined by

$$\|u\| = \sqrt{u \cdot u} = \sqrt{z_1\bar{z}_1 + z_2\bar{z}_2 + \cdots + z_n\bar{z}_n} = \sqrt{|z_1|^2 + |z_2|^2 + \cdots + |v_n|^2}$$

We emphasize that $u \cdot u$ and so $\|u\|$ are real and positive when $u \neq 0$ and 0 when $u = 0$.

EXAMPLE 1.12 Consider vectors $u = [2 + 3i,\ 4 - i,\ 3 + 5i]$ and $v = [3 - 4i,\ 5i,\ 4 - 2i]$ in $\mathbf{C}^3$. Then

$$\begin{aligned} u \cdot v &= (2 + 3i)(\overline{3 - 4i}) + (4 - i)(\overline{5i}) + (3 + 5i)(\overline{4 - 2i}) \\ &= (2 + 3i)(3 + 4i) + (4 - i)(-5i) + (3 + 5i)(4 + 2i) \\ &= (-6 + 13i) + (-5 - 20i) + (2 + 26i) = -9 + 19i \end{aligned}$$
$$u \cdot u = |2 + 3i|^2 + |4 - i|^2 + |3 + 5i|^2 = 4 + 9 + 16 + 1 + 9 + 25 = 64$$
$$\|u\| = \sqrt{64} = 8$$

The space $\mathbf{C}^n$ with the above operations of vector addition, scalar multiplication, and dot product, is called *complex Euclidean* n*-space*. Theorem 1.2 for $\mathbf{R}^n$ also holds for $\mathbf{C}^n$ if we replace $u \cdot v = v \cdot u$ by

$$u \cdot v = \overline{u \cdot v}$$

On the other hand, the Schwarz inequality (Theorem 1.3) and Minkowski's inequality (Theorem 1.4) are true for $\mathbf{C}^n$ with no changes.

SOLVED PROBLEMS

Vectors in R^n

1.1. Determine which of the following vectors are equal:

$$u_1 = (1, 2, 3), \quad u_2 = (2, 3, 1), \quad u_3 = (1, 3, 2), \quad u_4 = (2, 3, 1)$$

Vectors are equal only when corresponding entries are equal; hence, only $u_2 = u_4$.

1.2. Let $u = (2, -7, 1)$, $v = (-3, 0, 4)$, $w = (0, 5, -8)$. Find:

(a) $3u - 4v$,
(b) $2u + 3v - 5w$.

First perform the scalar multiplication and then the vector addition.

(a) $3u - 4v = 3(2, -7, 1) - 4(-3, 0, 4) = (6, -21, 3) + (12, 0, -16) = (18, -21, -13)$
(b) $2u + 3v - 5w = (4, -14, 2) + (-9, 0, 12) + (0, -25, 40) = (-5, -39, 54)$

1.3. Let $u = \begin{bmatrix} 5 \\ 3 \\ -4 \end{bmatrix}, v = \begin{bmatrix} -1 \\ 5 \\ 2 \end{bmatrix}, w = \begin{bmatrix} 3 \\ -1 \\ -2 \end{bmatrix}$. Find:

(a) $5u - 2v$,
(b) $-2u + 4v - 3w$.

First perform the scalar multiplication and then the vector addition:

(a) $5u - 2v = 5\begin{bmatrix} 5 \\ 3 \\ -4 \end{bmatrix} - 2\begin{bmatrix} -1 \\ 5 \\ 2 \end{bmatrix} = \begin{bmatrix} 25 \\ 15 \\ -20 \end{bmatrix} + \begin{bmatrix} 2 \\ -10 \\ -4 \end{bmatrix} = \begin{bmatrix} 27 \\ 5 \\ -24 \end{bmatrix}$

(b) $-2u + 4v - 3w = \begin{bmatrix} -10 \\ -6 \\ 8 \end{bmatrix} + \begin{bmatrix} -4 \\ 20 \\ 8 \end{bmatrix} + \begin{bmatrix} -9 \\ 3 \\ 6 \end{bmatrix} = \begin{bmatrix} -23 \\ 17 \\ 22 \end{bmatrix}$

1.4. Find x and y, where: (a) $(x, 3) = (2, x + y)$, (b) $(4, y) = x(2, 3)$.

(a) Because the vectors are equal, set the corresponding entries equal to each other, yielding

$$x = 2, \qquad 3 = x + y$$

Solve the linear equations, obtaining $x = 2, y = 1$.

(b) First multiply by the scalar x to obtain $(4, y) = (2x, 3x)$. Then set corresponding entries equal to each other to obtain

$$4 = 2x, \qquad y = 3x$$

Solve the equations to yield $x = 2, y = 6$.

1.5. Write the vector $v = (1, -2, 5)$ as a linear combination of the vectors $u_1 = (1, 1, 1)$, $u_2 = (1, 2, 3)$, $u_3 = (2, -1, 1)$.

We want to express v in the form $v = xu_1 + yu_2 + zu_3$ with x, y, z as yet unknown. First we have

$$\begin{bmatrix} 1 \\ -2 \\ 5 \end{bmatrix} = x\begin{bmatrix} 1 \\ 1 \\ 1 \end{bmatrix} + y\begin{bmatrix} 1 \\ 2 \\ 3 \end{bmatrix} + z\begin{bmatrix} 2 \\ -1 \\ 1 \end{bmatrix} = \begin{bmatrix} x + y + 2z \\ x + 2y - z \\ x + 3y + z \end{bmatrix}$$

(It is more convenient to write vectors as columns than as rows when forming linear combinations.) Set corresponding entries equal to each other to obtain

$$\begin{aligned} x + y + 2z &= 1 \\ x + 2y - z &= -2 \\ x + 3y + z &= 5 \end{aligned} \qquad \text{or} \qquad \begin{aligned} x + y + 2z &= 1 \\ y - 3z &= -3 \\ 2y - z &= 4 \end{aligned} \qquad \text{or} \qquad \begin{aligned} x + y + 2z &= 1 \\ y - 3z &= -3 \\ 5z &= 10 \end{aligned}$$

This unique solution of the triangular system is $x = -6, y = 3, z = 2$. Thus, $v = -6u_1 + 3u_2 + 2u_3$.

1.6. Write $v = (2, -5, 3)$ as a linear combination of

$$u_1 = (1, -3, 2), u_2 = (2, -4, -1), u_3 = (1, -5, 7).$$

Find the equivalent system of linear equations and then solve. First,

$$\begin{bmatrix} 2 \\ -5 \\ 3 \end{bmatrix} = x\begin{bmatrix} 1 \\ -3 \\ 2 \end{bmatrix} + y\begin{bmatrix} 2 \\ -4 \\ -1 \end{bmatrix} + z\begin{bmatrix} 1 \\ -5 \\ 7 \end{bmatrix} = \begin{bmatrix} x + 2y + z \\ -3x - 4y - 5z \\ 2x - y + 7z \end{bmatrix}$$

Set the corresponding entries equal to each other to obtain

$$\begin{aligned} x + 2y + z &= 2 \\ -3x - 4y - 5z &= -5 \\ 2x - y + 7z &= 3 \end{aligned} \quad \text{or} \quad \begin{aligned} x + 2y + z &= 2 \\ 2y - 2z &= 1 \\ -5y + 5z &= -1 \end{aligned} \quad \text{or} \quad \begin{aligned} x + 2y + z &= 2 \\ 2y - 2z &= 1 \\ 0 &= 3 \end{aligned}$$

The third equation, $0x + 0y + 0z = 3$, indicates that the system has no solution. Thus, v cannot be written as a linear combination of the vectors u_1, u_2, u_3.

Dot (Inner) Product, Orthogonality, Norm in R^n

1.7. Find $u \cdot v$ where:

(a) $u = (2, -5, 6)$ and $v = (8, 2, -3)$,

(b) $u = (4, 2, -3, 5, -1)$ and $v = (2, 6, -1, -4, 8)$.

Multiply the corresponding components and add:

(a) $u \cdot v = 2(8) - 5(2) + 6(-3) = 16 - 10 - 18 = -12$

(b) $u \cdot v = 8 + 12 + 3 - 20 - 8 = -5$

1.8. Let $u = (5, 4, 1)$, $v = (3, -4, 1)$, $w = (1, -2, 3)$. Which pair of vectors, if any, are perpendicular (orthogonal)?

Find the dot product of each pair of vectors:

$$u \cdot v = 15 - 16 + 1 = 0, \qquad v \cdot w = 3 + 8 + 3 = 14, \qquad u \cdot w = 5 - 8 + 3 = 0$$

Thus, u and v are orthogonal, u and w are orthogonal, but v and w are not.

1.9. Find k so that u and v are orthogonal, where:

(a) $u = (1, k, -3)$ and $v = (2, -5, 4)$,

(b) $u = (2, 3k, -4, 1, 5)$ and $v = (6, -1, 3, 7, 2k)$.

Compute $u \cdot v$, set $u \cdot v$ equal to 0, and then solve for k:

(a) $u \cdot v = 1(2) + k(-5) - 3(4) = -5k - 10$. Then $-5k - 10 = 0$, or $k = -2$.

(b) $u \cdot v = 12 - 3k - 12 + 7 + 10k = 7k + 7$. Then $7k + 7 = 0$, or $k = -1$.

1.10. Find $\|u\|$, where: (a) $u = (3, -12, -4)$, (b) $u = (2, -3, 8, -7)$.

First find $\|u\|^2 = u \cdot u$ by squaring the entries and adding. Then $\|u\| = \sqrt{\|u\|^2}$.

(a) $\|u\|^2 = (3)^2 + (-12)^2 + (-4)^2 = 9 + 144 + 16 = 169$. Then $\|u\| = \sqrt{169} = 13$.

(b) $\|u\|^2 = 4 + 9 + 64 + 49 = 126$. Then $\|u\| = \sqrt{126}$.

1.11. Recall that *normalizing* a nonzero vector v means finding the unique unit vector $\hat{v}$ in the same direction as v, where

$$\hat{v} = \frac{1}{\|v\|} v$$

Normalize: (a) $u = (3, -4)$, (b) $v = (4, -2, -3, 8)$, (c) $w = (\frac{1}{2}, \frac{2}{3}, -\frac{1}{4})$.

(a) First find $\|u\| = \sqrt{9+16} = \sqrt{25} = 5$. Then divide each entry of u by 5, obtaining $\hat{u} = (\frac{3}{5}, -\frac{4}{5})$.

(b) Here $\|v\| = \sqrt{16+4+9+64} = \sqrt{93}$. Then

$$\hat{v} = \left(\frac{4}{\sqrt{93}}, \frac{-2}{\sqrt{93}}, \frac{-3}{\sqrt{93}}, \frac{8}{\sqrt{93}}\right)$$

(c) Note that w and any positive multiple of w will have the same normalized form. Hence, first multiply w by 12 to "clear fractions"—that is, first find $w' = 12w = (6, 8, -3)$. Then

$$\|w'\| = \sqrt{36+64+9} = \sqrt{109} \text{ and } \hat{w} = \widehat{w'} = \left(\frac{6}{\sqrt{109}}, \frac{8}{\sqrt{109}}, \frac{-3}{\sqrt{109}}\right)$$

1.12. Let $u = (1, -3, 4)$ and $v = (3, 4, 7)$. Find:

(a) $\cos\theta$, where θ is the angle between u and v;
(b) $\text{proj}(u, v)$, the projection of u onto v;
(c) $d(u, v)$, the distance between u and v.

First find $u \cdot v = 3 - 12 + 28 = 19$, $\|u\|^2 = 1 + 9 + 16 = 26$, $\|v\|^2 = 9 + 16 + 49 = 74$. Then

(a) $\cos\theta = \dfrac{u \cdot v}{\|u\|\|v\|} = \dfrac{19}{\sqrt{26}\sqrt{74}}$,

(b) $\text{proj}(u, v) = \dfrac{u \cdot v}{\|v\|^2} v = \dfrac{19}{74}(3, 4, 7) = \left(\dfrac{57}{74}, \dfrac{76}{74}, \dfrac{133}{74}\right) = \left(\dfrac{57}{74}, \dfrac{38}{37}, \dfrac{133}{74}\right)$,

(c) $d(u, v) = \|u - v\| = \|(-2, -7 - 3)\| = \sqrt{4 + 49 + 9} = \sqrt{62}$.

1.13. Prove Theorem 1.2: For any u, v, w in $\mathbf{R}^n$ and k in $\mathbf{R}$:

(i) $(u + v) \cdot w = u \cdot w + v \cdot w$, (ii) $(ku) \cdot v = k(u \cdot v)$, (iii) $u \cdot v = v \cdot u$,
(iv) $u \cdot u \geq 0$, and $u \cdot u = 0$ iff $u = 0$.

Let $u = (u_1, u_2, \ldots, u_n)$, $v = (v_1, v_2, \ldots, v_n)$, $w = (w_1, w_2, \ldots, w_n)$.

(i) Because $u + v = (u_1 + v_1,\ u_2 + v_2,\ \ldots,\ u_n + v_n)$,

$$\begin{aligned}(u + v) \cdot w &= (u_1 + v_1)w_1 + (u_2 + v_2)w_2 + \cdots + (u_n + v_n)w_n \\ &= u_1w_1 + v_1w_1 + u_2w_2 + \cdots + u_nw_n + v_nw_n \\ &= (u_1w_1 + u_2w_2 + \cdots + u_nw_n) + (v_1w_1 + v_2w_2 + \cdots + v_nw_n) \\ &= u \cdot w + v \cdot w\end{aligned}$$

(ii) Because $ku = (ku_1, ku_2, \ldots, ku_n)$,

$$(ku) \cdot v = ku_1v_1 + ku_2v_2 + \cdots + ku_nv_n = k(u_1v_1 + u_2v_2 + \cdots + u_nv_n) = k(u \cdot v)$$

(iii) $u \cdot v = u_1v_1 + u_2v_2 + \cdots + u_nv_n = v_1u_1 + v_2u_2 + \cdots + v_nu_n = v \cdot u$

(iv) Because u_i^2 is nonnegative for each i, and because the sum of nonnegative real numbers is nonnegative,

$$u \cdot u = u_1^2 + u_2^2 + \cdots + u_n^2 \geq 0$$

Furthermore, $u \cdot u = 0$ iff $u_i = 0$ for each i, that is, iff $u = 0$.

1.14. Prove Theorem 1.3 (Schwarz): $|u \cdot v| \le \|u\|\|v\|$.

For any real number t, and using Theorem 1.2, we have

$$0 \le (tu+v)\cdot(tu+v) = t^2(u\cdot u) + 2t(u\cdot v) + (v\cdot v) = \|u\|^2 t^2 + 2(u\cdot v)t + \|v\|^2$$

Let $a = \|u\|^2$, $b = 2(u\cdot v)$, $c = \|v\|^2$. Then, for every value of t, $at^2 + bt + c \ge 0$. This means that the quadratic polynomial cannot have two real roots. This implies that the discriminant $D = b^2 - 4ac \le 0$ or, equivalently, $b^2 \le 4ac$. Thus,

$$4(u\cdot v)^2 \le 4\|u\|^2\|v\|^2$$

Dividing by 4 gives us our result.

1.15. Prove Theorem 1.4 (Minkowski): $\|u+v\| \le \|u\| + \|v\|$.

By the Schwarz inequality and other properties of the dot product,

$$\|u+v\|^2 = (u+v)\cdot(u+v) = (u\cdot u) + 2(u\cdot v) + (v\cdot v) \le \|u\|^2 + 2\|u\|\|v\| + \|v\|^2 = (\|u\| + \|v\|)^2$$

Taking the square root of both sides yields the desired inequality.

Points, Lines, Hyperplanes in R^n

Here we distinguish between an n-tuple $P(a_1, a_2, \ldots, a_n)$ viewed as a point in $\mathbf{R}^n$ and an n-tuple $u = [c_1, c_2, \ldots, c_n]$ viewed as a vector (arrow) from the origin O to the point $C(c_1, c_2, \ldots, c_n)$.

1.16. Find the vector u identified with the directed line segment $\overrightarrow{PQ}$ for the points:

(a) $P(1,-2,4)$ and $Q(6,1,-5)$ in $\mathbf{R}^3$, (b) $P(2,3,-6,5)$ and $Q(7,1,4,-8)$ in $\mathbf{R}^4$.

(a) $u = \overrightarrow{PQ} = Q - P = [6-1,\ 1-(-2),\ -5-4] = [5,3,-9]$

(b) $u = \overrightarrow{PQ} = Q - P = [7-2,\ 1-3,\ 4+6,\ -8-5] = [5,-2,10,-13]$

1.17. Find an equation of the hyperplane H in $\mathbf{R}^4$ that passes through $P(3,-4,1,-2)$ and is normal to $u = [2,5,-6,-3]$.

The coefficients of the unknowns of an equation of H are the components of the normal vector u. Thus, an equation of H is of the form $2x_1 + 5x_2 - 6x_3 - 3x_4 = k$. Substitute P into this equation to obtain $k = -26$. Thus, an equation of H is $2x_1 + 5x_2 - 6x_3 - 3x_4 = -26$.

1.18. Find an equation of the plane H in $\mathbf{R}^3$ that contains $P(1,-3,-4)$ and is parallel to the plane H' determined by the equation $3x - 6y + 5z = 2$.

The planes H and H' are parallel if and only if their normal directions are parallel or antiparallel (opposite direction). Hence, an equation of H is of the form $3x - 6y + 5z = k$. Substitute P into this equation to obtain $k = 1$. Then an equation of H is $3x - 6y + 5z = 1$.

1.19. Find a parametric representation of the line L in $\mathbf{R}^4$ passing through $P(4,-2,3,1)$ in the direction of $u = [2,5,-7,8]$.

Here L consists of the points $X(x_i)$ that satisfy

$$X = P + tu \quad \text{or} \quad x_i = a_i t + b_i \quad \text{or} \quad L(t) = (a_i t + b_i)$$

where the parameter t takes on all real values. Thus we obtain

$$x_1 = 4 + 2t,\ x_2 = -2 + 2t,\ x_3 = 3 - 7t,\ x_4 = 1 + 8t \quad \text{or} \quad L(t) = (4+2t,\ -2+2t,\ 3-7t,\ 1+8t)$$

1.20. Let C be the curve $F(t) = (t^2,\ 3t-2,\ t^3,\ t^2+5)$ in $\mathbf{R}^4$, where $0 \le t \le 4$.

(a) Find the point P on C corresponding to $t = 2$.

(b) Find the initial point Q and terminal point Q' of C.

(c) Find the unit tangent vector $\mathbf{T}$ to the curve C when $t = 2$.

(a) Substitute $t = 2$ into $F(t)$ to get $P = f(2) = (4, 4, 8, 9)$.

(b) The parameter t ranges from $t = 0$ to $t = 4$. Hence, $Q = f(0) = (0, -2, 0, 5)$ and $Q' = F(4) = (16, 10, 64, 21)$.

(c) Take the derivative of $F(t)$—that is, of each component of $F(t)$—to obtain a vector V that is tangent to the curve:

$$V(t) = \frac{dF(t)}{dt} = [2t, 3, 3t^2, 2t]$$

Now find V when $t = 2$; that is, substitute $t = 2$ in the equation for $V(t)$ to obtain $V = V(2) = [4, 3, 12, 4]$. Then normalize V to obtain the desired unit tangent vector $\mathbf{T}$. We have

$$\|V\| = \sqrt{16+9+144+16} = \sqrt{185} \quad \text{and} \quad \mathbf{T} = \left[\frac{4}{\sqrt{185}}, \frac{3}{\sqrt{185}}, \frac{12}{\sqrt{185}}, \frac{4}{\sqrt{185}}\right]$$

Spatial Vectors (Vectors in $\mathbf{R}^3$), ijk Notation, Cross Product

1.21. Let $u = 2\mathbf{i} - 3\mathbf{j} + 4\mathbf{k}$, $v = 3\mathbf{i} + \mathbf{j} - 2\mathbf{k}$, $w = \mathbf{i} + 5\mathbf{j} + 3\mathbf{k}$. Find:

(a) $u + v$, (b) $2u - 3v + 4w$, (c) $u \cdot v$ and $u \cdot w$, (d) $\|u\|$ and $\|v\|$.

Treat the coefficients of **i**, **j**, **k** just like the components of a vector in $\mathbf{R}^3$.

(a) Add corresponding coefficients to get $u + v = 5\mathbf{i} - 2\mathbf{j} - 2\mathbf{k}$.

(b) First perform the scalar multiplication and then the vector addition:

$$\begin{aligned} 2u - 3v + 4w &= (4\mathbf{i} - 6\mathbf{j} + 8\mathbf{k}) + (-9\mathbf{i} - 3\mathbf{j} + 6\mathbf{k}) + (4\mathbf{i} + 20\mathbf{j} + 12\mathbf{k}) \\ &= -\mathbf{i} + 11\mathbf{j} + 26\mathbf{k} \end{aligned}$$

(c) Multiply corresponding coefficients and then add:

$$u \cdot v = 6 - 3 - 8 = -5 \quad \text{and} \quad u \cdot w = 2 - 15 + 12 = -1$$

(d) The norm is the square root of the sum of the squares of the coefficients:

$$\|u\| = \sqrt{4+9+16} = \sqrt{29} \quad \text{and} \quad \|v\| = \sqrt{9+1+4} = \sqrt{14}$$

1.22. Find the (parametric) equation of the line L:

(a) through the points $P(1, 3, 2)$ and $Q(2, 5, -6)$;

(b) containing the point $P(1, -2, 4)$ and perpendicular to the plane H given by the equation $3x + 5y + 7z = 15$.

(a) First find $v = \overrightarrow{PQ} = Q - P = [1, 2, -8] = \mathbf{i} + 2\mathbf{j} - 8\mathbf{k}$. Then

$$L(t) = (t+1,\ 2t+3,\ -8t+2) = (t+1)\mathbf{i} + (2t+3)\mathbf{j} + (-8t+2)\mathbf{k}$$

(b) Because L is perpendicular to H, the line L is in the same direction as the normal vector $\mathbf{N} = 3\mathbf{i} + 5\mathbf{j} + 7\mathbf{k}$ to H. Thus,

$$L(t) = (3t+1,\ 5t-2,\ 7t+4) = (3t+1)\mathbf{i} + (5t-2)\mathbf{j} + (7t+4)\mathbf{k}$$

1.23. Let S be the surface $xy^2 + 2yz = 16$ in $\mathbf{R}^3$.

(a) Find the normal vector $\mathbf{N}(x, y, z)$ to the surface S.

(b) Find the tangent plane H to S at the point $P(1, 2, 3)$.

(a) The formula for the normal vector to a surface $F(x,y,z)=0$ is

$$\mathbf{N}(x,y,z)=F_x\mathbf{i}+F_y\mathbf{j}+F_z\mathbf{k}$$

where F_x, F_y, F_z are the partial derivatives. Using $F(x,y,z)=xy^2+2yz-16$, we obtain

$$F_x=y^2, \qquad F_y=2xy+2z, \qquad F_z=2y$$

Thus, $\mathbf{N}(x,y,z)=y^2\mathbf{i}+(2xy+2z)\mathbf{j}+2y\mathbf{k}$.

(b) The normal to the surface S at the point P is

$$\mathbf{N}(P)=\mathbf{N}(1,2,3)=4\mathbf{i}+10\mathbf{j}+4\mathbf{k}$$

Hence, $\mathbf{N}=2\mathbf{i}+5\mathbf{j}+2\mathbf{k}$ is also normal to S at P. Thus an equation of H has the form $2x+5y+2z=c$. Substitute P in this equation to obtain $c=18$. Thus the tangent plane H to S at P is $2x+5y+2z=18$.

1.24. Evaluate the following determinants and negative of determinants of order two:

(a) (i) $\begin{vmatrix}3&4\\5&9\end{vmatrix}$, (ii) $\begin{vmatrix}2&-1\\4&3\end{vmatrix}$, (iii) $\begin{vmatrix}4&-5\\3&-2\end{vmatrix}$

(b) (i) $-\begin{vmatrix}3&6\\4&2\end{vmatrix}$, (ii) $-\begin{vmatrix}7&-5\\3&2\end{vmatrix}$, (iii) $-\begin{vmatrix}4&-1\\8&-3\end{vmatrix}$

Use $\begin{vmatrix}a&b\\c&d\end{vmatrix}=ad-bc$ and $-\begin{vmatrix}a&b\\c&d\end{vmatrix}=bc-ad$. Thus,

(a) (i) $27-20=7$, (ii) $6+4=10$, (iii) $-8+15=7$.

(b) (i) $24-6=18$, (ii) $-15-14=-29$, (iii) $-8+12=4$.

1.25. Let $u=2\mathbf{i}-3\mathbf{j}+4\mathbf{k}$, $v=3\mathbf{i}+\mathbf{j}-2\mathbf{k}$, $w=\mathbf{i}+5\mathbf{j}+3\mathbf{k}$.
Find: (a) $u\times v$, (b) $u\times w$

(a) Use $\begin{bmatrix}2&-3&4\\3&1&-2\end{bmatrix}$ to get $u\times v=(6-4)\mathbf{i}+(12+4)\mathbf{j}+(2+9)\mathbf{k}=2\mathbf{i}+16\mathbf{j}+11\mathbf{k}$.

(b) Use $\begin{bmatrix}2&-3&4\\1&5&3\end{bmatrix}$ to get $u\times w=(-9-20)\mathbf{i}+(4-6)\mathbf{j}+(10+3)\mathbf{k}=-29\mathbf{i}-2\mathbf{j}+13\mathbf{k}$.

1.26. Find $u\times v$, where: (a) $u=(1,2,3)$, $v=(4,5,6)$; (b) $u=(-4,7,3)$, $v=(6,-5,2)$.

(a) Use $\begin{bmatrix}1&2&3\\4&5&6\end{bmatrix}$ to get $u\times v=[12-15,\ 12-6,\ 5-8]=[-3,6,-3]$.

(b) Use $\begin{bmatrix}-4&7&3\\6&-5&2\end{bmatrix}$ to get $u\times v=[14+15,\ 18+8,\ 20-42]=[29,26,-22]$.

1.27. Find a unit vector u orthogonal to $v=[1,3,4]$ and $w=[2,-6,-5]$.

First find $v\times w$, which is orthogonal to v and w.

The array $\begin{bmatrix}1&3&4\\2&-6&-5\end{bmatrix}$ gives $v\times w=[-15+24,\ 8+5,\ -6-61]=[9,13,-12]$.

Normalize $v\times w$ to get $u=[9/\sqrt{394},\ 13/\sqrt{394},\ -12/\sqrt{394}]$.

1.28. Let $u=(a_1,a_2,a_3)$ and $v=(b_1,b_2,b_3)$ so $u\times v=(a_2b_3-a_3b_2,\ a_3b_1-a_1b_3,\ a_1b_2-a_2b_1)$.
Prove:

(a) $u\times v$ is orthogonal to u and v [Theorem 1.5(a)].

(b) $\|u\times v\|^2=(u\cdot u)(v\cdot v)-(u\cdot v)^2$ (Lagrange's identity).

(a) We have

$$u \cdot (u \times v) = a_1(a_2b_3 - a_3b_2) + a_2(a_3b_1 - a_1b_3) + a_3(a_1b_2 - a_2b_1)$$
$$= a_1a_2b_3 - a_1a_3b_2 + a_2a_3b_1 - a_1a_2b_3 + a_1a_3b_2 - a_2a_3b_1 = 0$$

Thus, $u \times v$ is orthogonal to u. Similarly, $u \times v$ is orthogonal to v.

(b) We have

$$\|u \times v\|^2 = (a_2b_3 - a_3b_2)^2 + (a_3b_1 - a_1b_3)^2 + (a_1b_2 - a_2b_1)^2 \quad (1)$$
$$(u \cdot u)(v \cdot v) - (u \cdot v)^2 = (a_1^2 + a_2^2 + a_3^2)(b_1^2 + b_2^2 + b_3^2) - (a_1b_1 + a_2b_2 + a_3b_3)^2 \quad (2)$$

Expansion of the right-hand sides of (1) and (2) establishes the identity.

Complex Numbers, Vectors in C^n

1.29. Suppose $z = 5 + 3i$ and $w = 2 - 4i$. Find: (a) $z + w$, (b) $z - w$, (c) zw.

Use the ordinary rules of algebra together with $i^2 = -1$ to obtain a result in the standard form $a + bi$.

(a) $z + w = (5 + 3i) + (2 - 4i) = 7 - i$

(b) $z - w = (5 + 3i) - (2 - 4i) = 5 + 3i - 2 + 4i = 3 + 7i$

(c) $zw = (5 + 3i)(2 - 4i) = 10 - 14i - 12i^2 = 10 - 14i + 12 = 22 - 14i$

1.30. Simplify: (a) $(5 + 3i)(2 - 7i)$, (b) $(4 - 3i)^2$, (c) $(1 + 2i)^3$.

(a) $(5 + 3i)(2 - 7i) = 10 + 6i - 35i - 21i^2 = 31 - 29i$

(b) $(4 - 3i)^2 = 16 - 24i + 9i^2 = 7 - 24i$

(c) $(1 + 2i)^3 = 1 + 6i + 12i^2 + 8i^3 = 1 + 6i - 12 - 8i = -11 - 2i$

1.31. Simplify: (a) i^0, i^3, i^4, (b) i^5, i^6, i^7, i^8, (c) $i^{39}, i^{174}, i^{252}, i^{317}$.

(a) $i^0 = 1, \quad i^3 = i^2(i) = (-1)(i) = -i, \quad i^4 = (i^2)(i^2) = (-1)(-1) = 1$

(b) $i^5 = (i^4)(i) = (1)(i) = i, \quad i^6 = (i^4)(i^2) = (1)(i^2) = i^2 = -1, \quad i^7 = i^3 = -i, \quad i^8 = i^4 = 1$

(c) Using $i^4 = 1$ and $i^n = i^{4q+r} = (i^4)^q i^r = 1^q i^r = i^r$, divide the exponent n by 4 to obtain the remainder r:

$$i^{39} = i^{4(9)+3} = (i^4)^9 i^3 = 1^9 i^3 = i^3 = -i, \qquad i^{174} = i^2 = -1, \qquad i^{252} = i^0 = 1, \qquad i^{317} = i^1 = i$$

1.32. Find the complex conjugate of each of the following:
(a) $6 + 4i$, $7 - 5i$, $4 + i$, $-3 - i$, (b) 6, -3, $4i$, $-9i$.

(a) $\overline{6 + 4i} = 6 - 4i$, $\overline{7 - 5i} = 7 + 5i$, $\overline{4 + i} = 4 - i$, $\overline{-3 - i} = -3 + i$

(b) $\overline{6} = 6$, $\overline{-3} = -3$, $\overline{4i} = -4i$, $\overline{-9i} = 9i$

(Note that the conjugate of a real number is the original number, but the conjugate of a pure imaginary number is the negative of the original number.)

1.33. Find $z\bar{z}$ and $|z|$ when $z = 3 + 4i$.

For $z = a + bi$, use $z\bar{z} = a^2 + b^2$ and $z = \sqrt{z\bar{z}} = \sqrt{a^2 + b^2}$.

$$z\bar{z} = 9 + 16 = 25, \qquad |z| = \sqrt{25} = 5$$

1.34. Simplify $\dfrac{2 - 7i}{5 + 3i}$.

To simplify a fraction z/w of complex numbers, multiply both numerator and denominator by $\bar{w}$, the conjugate of the denominator:

$$\frac{2 - 7i}{5 + 3i} = \frac{(2 - 7i)(5 - 3i)}{(5 + 3i)(5 - 3i)} = \frac{-11 - 41i}{34} = -\frac{11}{34} - \frac{41}{34}i$$

1.35. Prove: For any complex numbers $z, w \in \mathbf{C}$, (i) $\overline{z+w} = \bar{z} + \bar{w}$, (ii) $\overline{zw} = \bar{z}\bar{w}$, (iii) $\bar{\bar{z}} = z$.

Suppose $z = a + bi$ and $w = c + di$ where $a, b, c, d \in \mathbf{R}$.

$$\text{(i)}\quad \overline{z+w} = \overline{(a+bi)+(c+di)} = \overline{(a+c)+(b+d)i} = (a+c)-(b+d)i = a+c-bi-di = (a-bi)+(c-di) = \bar{z}+\bar{w}$$

$$\text{(ii)}\quad \overline{zw} = \overline{(a+bi)(c+di)} = \overline{(ac-bd)+(ad+bc)i} = (ac-bd)-(ad+bc)i = (a-bi)(c-di) = \bar{z}\bar{w}$$

$$\text{(iii)}\quad \bar{\bar{z}} = \overline{\overline{a+bi}} = \overline{a-bi} = a-(-b)i = a+bi = z$$

1.36. Prove: For any complex numbers $z, w \in \mathbf{C}$, $|zw| = |z||w|$.

By (ii) of Problem 1.35,

$$|zw|^2 = (zw)(\overline{zw}) = (zw)(\bar{z}\bar{w}) = (z\bar{z})(w\bar{w}) = |z|^2|w|^2$$

The square root of both sides gives us the desired result.

1.37. Prove: For any complex numbers $z, w \in \mathbf{C}$, $|z+w| \le |z| + |w|$.

Suppose $z = a + bi$ and $w = c + di$ where $a, b, c, d \in \mathbf{R}$. Consider the vectors $u = (a, b)$ and $v = (c, d)$ in $\mathbf{R}^2$. Note that

$$|z| = \sqrt{a^2+b^2} = \|u\|, \qquad |w| = \sqrt{c^2+d^2} = \|v\|$$

and

$$|z+w| = |(a+c)+(b+d)i| = \sqrt{(a+c)^2+(b+d)^2} = \|(a+c, b+d)\| = \|u+v\|$$

By Minkowski's inequality (Problem 1.15), $\|u+v\| \le \|u\| + \|v\|$, and so

$$|z+w| = \|u+v\| \le \|u\| + \|v\| = |z| + |w|$$

1.38. Find the dot products $u \cdot v$ and $v \cdot u$ where: (a) $u = (1-2i,\ 3+i)$, $v = (4+2i,\ 5-6i)$, (b) $u = (3-2i,\ 4i,\ 1+6i)$, $v = (5+i,\ 2-3i,\ 7+2i)$.

Recall that conjugates of the second vector appear in the dot product

$$(z_1, \ldots, z_n) \cdot (w_1, \ldots, w_n) = z_1\bar{w}_1 + \cdots + z_n\bar{w}_n$$

(a)
$$u \cdot v = (1-2i)\overline{(4+2i)} + (3+i)\overline{(5-6i)} = (1-2i)(4-2i) + (3+i)(5+6i) = -10i + 9 + 23i = 9 + 13i$$
$$v \cdot u = (4+2i)\overline{(1-2i)} + (5-6i)\overline{(3+i)} = (4+2i)(1+2i) + (5-6i)(3-i) = 10i + 9 - 23i = 9 - 13i$$

(b)
$$u \cdot v = (3-2i)\overline{(5+i)} + (4i)\overline{(2-3i)} + (1+6i)\overline{(7+2i)} = (3-2i)(5-i) + (4i)(2+3i) + (1+6i)(7-2i) = 20 + 35i$$
$$v \cdot u = (5+i)\overline{(3-2i)} + (2-3i)\overline{(4i)} + (7+2i)\overline{(1+6i)} = (5+i)(3+2i) + (2-3i)(-4i) + (7+2i)(1-6i) = 20 - 35i$$

In both cases, $v \cdot u = \overline{u \cdot v}$. This holds true in general, as seen in Problem 1.40.

1.39. Let $u = (7-2i,\ 2+5i)$ and $v = (1+i,\ -3-6i)$. Find:
(a) $u+v$, (b) $2iu$, (c) $(3-i)v$, (d) $u \cdot v$, (e) $\|u\|$ and $\|v\|$.

(a) $u + v = (7-2i+1+i,\ 2+5i-3-6i) = (8-i,\ -1-i)$

(b) $2iu = (14i-4i^2,\ 4i+10i^2) = (4+14i,\ -10+4i)$

(c) $(3-i)v = (3+3i-i-i^2,\ -9-18i+3i+6i^2) = (4+2i,\ -15-15i)$

(d) $u \cdot v = (7-2i)(\overline{1+i}) + (2+5i)(\overline{-3-6i})$
$= (7-2i)(1-i) + (2+5i)(-3+6i) = 5-9i-36-3i = -31-12i$

(e) $\|u\| = \sqrt{7^2 + (-2)^2 + 2^2 + 5^2} = \sqrt{82}$ and $\|v\| = \sqrt{1^2 + 1^2 + (-3)^2 + (-6)^2} = \sqrt{47}$

1.40. Prove: For any vectors $u, v \in \mathbf{C}^n$ and any scalar $z \in \mathbf{C}$, (i) $u \cdot v = \overline{v \cdot u}$, (ii) $(zu) \cdot v = z(u \cdot v)$, (iii) $u \cdot (zv) = \bar{z}(u \cdot v)$.

Suppose $u = (z_1, z_2, \ldots, z_n)$ and $v = (w_1, w_2, \ldots, w_n)$.

(i) Using the properties of the conjugate,

$$\overline{v \cdot u} = \overline{w_1\bar{z}_1 + w_2\bar{z}_2 + \cdots + w_n\bar{z}_n} = \overline{w_1\bar{z}_1} + \overline{w_2\bar{z}_2} + \cdots + \overline{w_n\bar{z}_n}$$
$$= \bar{w}_1 z_1 + \bar{w}_2 z_2 + \cdots + \bar{w}_n z_n = z_1\bar{w}_1 + z_2\bar{w}_2 + \cdots + z_n\bar{w}_n = u \cdot v$$

(ii) Because $zu = (zz_1, zz_2, \ldots, zz_n)$,

$$(zu) \cdot v = zz_1\bar{w}_1 + zz_2\bar{w}_2 + \cdots + zz_n\bar{w}_n = z(z_1\bar{w}_1 + z_2\bar{w}_2 + \cdots + z_n\bar{w}_n) = z(u \cdot v)$$

(Compare with Theorem 1.2 on vectors in $\mathbf{R}^n$.)

(iii) Using (i) and (ii),

$$u \cdot (zv) = \overline{(zv) \cdot u} = \overline{z(v \cdot u)} = \bar{z}(\overline{v \cdot u}) = \bar{z}(u \cdot v)$$

SUPPLEMENTARY PROBLEMS

Vectors in R^n

1.41. Let $u = (1, -2, 4)$, $v = (3, 5, 1)$, $w = (2, 1, -3)$. Find:

(a) $3u - 2v$; (b) $5u + 3v - 4w$; (c) $u \cdot v$, $u \cdot w$, $v \cdot w$; (d) $\|u\|$, $\|v\|$, $\|w\|$;
(e) $\cos\theta$, where θ is the angle between u and v; (f) $d(u, v)$; (g) $\text{proj}(u, v)$.

1.42. Repeat Problem 1.41 for vectors $u = \begin{bmatrix} 1 \\ 3 \\ -4 \end{bmatrix}$, $v = \begin{bmatrix} 2 \\ 1 \\ 5 \end{bmatrix}$, $w = \begin{bmatrix} 3 \\ -2 \\ 6 \end{bmatrix}$.

1.43. Let $u = (2, -5, 4, 6, -3)$ and $v = (5, -2, 1, -7, -4)$. Find:

(a) $4u - 3v$; (b) $5u + 2v$; (c) $u \cdot v$; (d) $\|u\|$ and $\|v\|$; (e) $\text{proj}(u, v)$; (f) $d(u, v)$.

1.44. Normalize each vector:

(a) $u = (5, -7)$; (b) $v = (1, 2, -2, 4)$; (c) $w = \left(\frac{1}{2}, -\frac{1}{3}, \frac{3}{4}\right)$.

1.45. Let $u = (1, 2, -2)$, $v = (3, -12, 4)$, and $k = -3$.

(a) Find $\|u\|$, $\|v\|$, $\|u + v\|$, $\|ku\|$.
(b) Verify that $\|ku\| = |k|\|u\|$ and $\|u + v\| \le \|u\| + \|v\|$.

1.46. Find x and y where:

(a) $(x, y + 1) = (y - 2, 6)$; (b) $x(2, y) = y(1, -2)$.

1.47. Find x, y, z where $(x, y + 1, y + z) = (2x + y, 4, 3z)$.

1.48. Write $v = (2,5)$ as a linear combination of u_1 and u_2, where:

(a) $u_1 = (1,2)$ and $u_2 = (3,5)$;
(b) $u_1 = (3,-4)$ and $u_2 = (2,-3)$.

1.49. Write $v = \begin{bmatrix} 9 \\ -3 \\ 16 \end{bmatrix}$ as a linear combination of $u_1 = \begin{bmatrix} 1 \\ 2 \\ 3 \end{bmatrix}$, $u_2 = \begin{bmatrix} 2 \\ 5 \\ -1 \end{bmatrix}$, $u_3 = \begin{bmatrix} 4 \\ -2 \\ 3 \end{bmatrix}$.

1.50. Find k so that u and v are orthogonal, where:

(a) $u = (3,k,-2)$, $v = (6,-4,-3)$;
(b) $u = (5,k,-4,2)$, $v = (1,-3,2,2k)$;
(c) $u = (1,\ 7,\ k+2,\ -2)$, $v = (3,k,-3,k)$.

Located Vectors, Hyperplanes, Lines in R^n

1.51. Find the vector v identified with the directed line segment $\overrightarrow{PQ}$ for the points:

(a) $P(2,3,-7)$ and $Q(1,-6,-5)$ in $\mathbf{R}^3$;
(b) $P(1,-8,-4,6)$ and $Q(3,-5,2,-4)$ in $\mathbf{R}^4$.

1.52. Find an equation of the hyperplane H in $\mathbf{R}^4$ that:

(a) contains $P(1,2,-3,2)$ and is normal to $u = [2,3,-5,6]$;
(b) contains $P(3,-1,2,5)$ and is parallel to $2x_1 - 3x_2 + 5x_3 - 7x_4 = 4$.

1.53. Find a parametric representation of the line in $\mathbf{R}^4$ that:

(a) passes through the points $P(1,2,1,2)$ and $Q(3,-5,7,-9)$;
(b) passes through $P(1,1,3,3)$ and is perpendicular to the hyperplane $2x_1 + 4x_2 + 6x_3 - 8x_4 = 5$.

Spatial Vectors (Vectors in R^3), ijk Notation

1.54. Given $u = 3\mathbf{i} - 4\mathbf{j} + 2\mathbf{k}$, $v = 2\mathbf{i} + 5\mathbf{j} - 3\mathbf{k}$, $w = 4\mathbf{i} + 7\mathbf{j} + 2\mathbf{k}$. Find:

(a) $2u - 3v$; (b) $3u + 4v - 2w$; (c) $u \cdot v$, $u \cdot w$, $v \cdot w$; (d) $\|u\|$, $\|v\|$, $\|w\|$.

1.55. Find the equation of the plane H:

(a) with normal $\mathbf{N} = 3\mathbf{i} - 4\mathbf{j} + 5\mathbf{k}$ and containing the point $P(1,2,-3)$;
(b) parallel to $4x + 3y - 2z = 11$ and containing the point $Q(2,-1,3)$.

1.56. Find the (parametric) equation of the line L:

(a) through the point $P(2,5,-3)$ and in the direction of $v = 4\mathbf{i} - 5\mathbf{j} + 7\mathbf{k}$;
(b) perpendicular to the plane $2x - 3y + 7z = 4$ and containing $P(1,-5,7)$.

1.57. Consider the following curve C in $\mathbf{R}^3$ where $0 \le t \le 5$:

$$F(t) = t^3\mathbf{i} - t^2\mathbf{j} + (2t-3)\mathbf{k}$$

(a) Find the point P on C corresponding to $t = 2$.
(b) Find the initial point Q and the terminal point Q'.
(c) Find the unit tangent vector $\mathbf{T}$ to the curve C when $t = 2$.

1.58. Consider a moving body B whose position at time t is given by $R(t) = t^2\mathbf{i} + t^3\mathbf{j} + 2t\mathbf{k}$. [Then $V(t) = dR(t)/dt$ and $A(t) = dV(t)/dt$ denote, respectively, the velocity and acceleration of B.] When $t = 1$, find for the body B:

(a) position; (b) velocity v; (c) speed s; (d) acceleration a.

1.59. Find a normal vector **N** and the tangent plane H to each surface at the given point:

(a) surface $x^2y + 3yz = 20$ and point $P(1, 3, 2)$;
(b) surface $x^2 + 3y^2 - 5z^2 = 160$ and point $P(3, -2, 1)$.

Cross Product

1.60. Evaluate the following determinants and negative of determinants of order two:

(a) $\begin{vmatrix} 2 & 5 \\ 3 & 6 \end{vmatrix}$, $\begin{vmatrix} 3 & -6 \\ 1 & -4 \end{vmatrix}$, $\begin{vmatrix} -4 & -2 \\ 7 & -3 \end{vmatrix}$

(b) $-\begin{vmatrix} 6 & 4 \\ 7 & 5 \end{vmatrix}$, $-\begin{vmatrix} 1 & -3 \\ 2 & 4 \end{vmatrix}$, $-\begin{vmatrix} 8 & -3 \\ -6 & -2 \end{vmatrix}$

1.61. Given $u = 3\mathbf{i} - 4\mathbf{j} + 2\mathbf{k}$, $v = 2\mathbf{i} + 5\mathbf{j} - 3\mathbf{k}$, $w = 4\mathbf{i} + 7\mathbf{j} + 2\mathbf{k}$, find:

(a) $u \times v$, (b) $u \times w$, (c) $v \times w$.

1.62. Given $u = [2, 1, 3]$, $v = [4, -1, 2]$, $w = [1, 1, 5]$, find:

(a) $u \times v$, (b) $u \times w$, (c) $v \times w$.

1.63. Find the volume V of the parallelopiped formed by the vectors u, v, w appearing in:

(a) Problem 1.61 (b) Problem 1.62.

1.64. Find a unit vector u orthogonal to:

(a) $v = [1, 2, 3]$ and $w = [1, -1, 2]$;
(b) $v = 3\mathbf{i} - \mathbf{j} + 2\mathbf{k}$ and $w = 4\mathbf{i} - 2\mathbf{j} - \mathbf{k}$.

1.65. Prove the following properties of the cross product:

(a) $u \times v = -(v \times u)$
(b) $u \times u = 0$ for any vector u
(c) $(ku) \times v = k(u \times v) = u \times (kv)$
(d) $u \times (v + w) = (u \times v) + (u \times w)$
(e) $(v + w) \times u = (v \times u) + (w \times u)$
(f) $(u \times v) \times w = (u \cdot w)v - (v \cdot w)u$

Complex Numbers

1.66. Simplify:

(a) $(4 - 7i)(9 + 2i)$; (b) $(3 - 5i)^2$; (c) $\dfrac{1}{4 - 7i}$; (d) $\dfrac{9 + 2i}{3 - 5i}$; (e) $(1 - i)^3$.

1.67. Simplify: (a) $\dfrac{1}{2i}$; (b) $\dfrac{2 + 3i}{7 - 3i}$; (c) i^{15}, i^{25}, i^{34}; (d) $\left(\dfrac{1}{3 - i}\right)^2$.

1.68. Let $z = 2 - 5i$ and $w = 7 + 3i$. Find:

(a) $v + w$; (b) zw; (c) z/w; (d) $\bar{z}, \bar{w}$; (e) $|z|, |w|$.

1.69. Show that for complex numbers z and w:

(a) $\text{Re}\, z = \frac{1}{2}(z + \bar{z})$, (b) $\text{Im}\, z = \frac{1}{2}(z - \bar{z})$, (c) $zw = 0$ implies $z = 0$ or $w = 0$.

Vectors in C^n

1.70. Let $u = (1 + 7i,\ 2 - 6i)$ and $v = (5 - 2i,\ 3 - 4i)$. Find:

(a) $u + v$ (b) $(3 + i)u$ (c) $2iu + (4 + 7i)v$ (d) $u \cdot v$ (e) $\|u\|$ and $\|v\|$.

1.71. Prove: For any vectors u, v, w in $\mathbf{C}^n$:

(a) $(u+v)\cdot w = u\cdot w + v\cdot w$, (b) $w\cdot(u+v) = w\cdot u + w\cdot v$.

1.72. Prove that the norm in $\mathbf{C}^n$ satisfies the following laws:

$[N_1]$ For any vector u, $\|u\| \geq 0$; and $\|u\| = 0$ if and only if $u = 0$.
$[N_2]$ For any vector u and complex number z, $\|zu\| = |z|\|u\|$.
$[N_3]$ For any vectors u and v, $\|u+v\| \leq \|u\| + \|v\|$.

ANSWERS TO SUPPLEMENTARY PROBLEMS

1.41. (a) $(-3,-16,10)$; (b) $(6,1,35)$; (c) $-3,-12,8$; (d) $\sqrt{21}, \sqrt{35}, \sqrt{14}$;
(e) $-3/\sqrt{21}\sqrt{35}$; (f) $\sqrt{62}$; (g) $-\frac{3}{35}(3,5,1) = (-\frac{9}{35}, -\frac{15}{35}, -\frac{3}{35})$

1.42. (Column vectors) (a) $(-1,7,-22)$; (b) $(-1,26,-29)$; (c) $-15,-27,34$;
(d) $\sqrt{26}, \sqrt{30}$; (e) $-15/(\sqrt{26}\sqrt{30})$; (f) $\sqrt{86}$; (g) $-\frac{15}{30}v = (-1, -\frac{1}{2}, -\frac{5}{2})$

1.43. (a) $(-7,-14,13,45,0)$; (b) $(20,-29,22,16,-23)$; (c) -6; (d) $\sqrt{90}, \sqrt{95}$;
(e) $-\frac{6}{95}v$; (f) $\sqrt{197}$

1.44. (a) $(5/\sqrt{74}, -9/\sqrt{74})$; (b) $(\frac{1}{5}, \frac{2}{5}, -\frac{2}{5}, \frac{4}{5})$; (c) $(6/\sqrt{133}, -4/\sqrt{133}, 9/\sqrt{133})$

1.45. (a) 3, 13, $\sqrt{120}$, 9

1.46. (a) $x=3$, $y=5$; (b) $x=0$, $y=0$, and $x=-2$, $y=-4$

1.47. $x=-3$, $y=3$, $z=\frac{3}{2}$

1.48. (a) $v = 5u_1 - u_2$; (b) $v = 16u_1 - 23u_2$

1.49. $v = 3u_1 - u_2 + 2u_3$

1.50. (a) 6; (b) 3; (c) $\frac{3}{2}$

1.51. (a) $v = [-1,-9,2]$; (b) $[2,3,6,-10]$

1.52. (a) $2x_1 + 3x_2 - 5x_3 + 6x_4 = 35$; (b) $2x_1 - 3x_2 + 5x_3 - 7x_4 = -16$

1.53. (a) $[2t+1, -7t+2, 6t+1, -11t+2]$; (b) $[2t+1, 4t+1, 6t+3, -8t+3]$

1.54. (a) $-23\mathbf{j} + 13\mathbf{k}$; (b) $9\mathbf{i} - 6\mathbf{j} - 10\mathbf{k}$; (c) $-20,-12,37$; (d) $\sqrt{29}, \sqrt{38}, \sqrt{69}$

1.55. (a) $3x - 4y + 5z = -20$; (b) $4x + 3y - 2z = -1$

1.56. (a) $[4t+2, -5t+5, 7t-3]$; (b) $[2t+1, -3t-5, 7t+7]$

1.57. (a) $P = F(2) = 8\mathbf{i} - 4\mathbf{j} + \mathbf{k}$; (b) $Q = F(0) = -3\mathbf{k}$, $Q' = F(5) = 125\mathbf{i} - 25\mathbf{j} + 7\mathbf{k}$;
(c) $\mathbf{T} = (6\mathbf{i} - 2\mathbf{j} + \mathbf{k})/\sqrt{41}$

1.58. (a) $\mathbf{i} + \mathbf{j} + 2\mathbf{k}$; (b) $2\mathbf{i} + 3\mathbf{j} + 2\mathbf{k}$; (c) $\sqrt{17}$; (d) $2\mathbf{i} + 6\mathbf{j}$

1.59. (a) $\mathbf{N} = 6\mathbf{i} + 7\mathbf{j} + 9\mathbf{k}$, $6x + 7y + 9z = 45$; (b) $\mathbf{N} = 6\mathbf{i} - 12\mathbf{j} - 10\mathbf{k}$, $3x - 6y - 5z = 16$

1.60. (a) $-3, -6, 26$; (b) $-2, -10, 34$

1.61. (a) $2\mathbf{i} + 13\mathbf{j} + 23\mathbf{k}$; (b) $-22\mathbf{i} + 2\mathbf{j} + 37\mathbf{k}$; (c) $31\mathbf{i} - 16\mathbf{j} - 6\mathbf{k}$

1.62. (a) $[5, 8, -6]$; (b) $[2, -7, 1]$; (c) $[-7, -18, 5]$

1.63. (a) 145; (b) 17

1.64. (a) $(7, 1, -3)/\sqrt{59}$; (b) $(5\mathbf{i} + 11\mathbf{j} - 2\mathbf{k})/\sqrt{150}$

1.66. (a) $50 - 55i$; (b) $-16 - 30i$; (c) $\frac{1}{65}(4 + 7i)$; (d) $\frac{1}{2}(1 + 3i)$; (e) $-2 - 2i$

1.67. (a) $-\frac{1}{2}i$; (b) $\frac{1}{58}(5 + 27i)$; (c) $-i, i, -1$; (d) $\frac{1}{50}(4 + 3i)$

1.68. (a) $9 - 2i$; (b) $29 - 29i$; (c) $\frac{1}{58}(-1 - 41i)$; (d) $2 + 5i,\ 7 - 3i$; (e) $\sqrt{29}, \sqrt{58}$

1.69. (c) *Hint:* If $zw = 0$, then $|zw| = |z||w| = |0| = 0$

1.70. (a) $(6 + 5i,\ 5 - 10i)$; (b) $(-4 + 22i,\ 12 - 16i)$; (c) $(20 + 29i,\ 52 + 9i)$;
(d) $21 + 27i$; (e) $\sqrt{90}, \sqrt{54}$

CHAPTER 2

Algebra of Matrices

2.1 Introduction

This chapter investigates matrices and algebraic operations defined on them. These matrices may be viewed as rectangular arrays of elements where each entry depends on two subscripts (as compared with vectors, where each entry depended on only one subscript). Systems of linear equations and their solutions (Chapter 3) may be efficiently investigated using the language of matrices. Furthermore, certain abstract objects introduced in later chapters, such as "change of basis," "linear transformations," and "quadratic forms," can be represented by these matrices (rectangular arrays). On the other hand, the abstract treatment of linear algebra presented later on will give us new insight into the structure of these matrices.

The entries in our matrices will come from some arbitrary, but fixed, field K. The elements of K are called *numbers* or *scalars*. Nothing essential is lost if the reader assumes that K is the real field $\mathbf{R}$.

2.2 Matrices

A *matrix A over a field K* or, simply, a *matrix A* (when K is implicit) is a rectangular array of scalars usually presented in the following form:

$$A = \begin{bmatrix} a_{11} & a_{12} & \dots & a_{1n} \\ a_{21} & a_{22} & \dots & a_{2n} \\ \dots & \dots & \dots & \dots \\ a_{m1} & a_{m2} & \dots & a_{mn} \end{bmatrix}$$

The *rows* of such a matrix A are the m horizontal lists of scalars:

$$(a_{11}, a_{12}, \dots, a_{1n}), \quad (a_{21}, a_{22}, \dots, a_{2n}), \quad \dots, \quad (a_{m1}, a_{m2}, \dots, a_{mn})$$

and the *columns* of A are the n vertical lists of scalars:

$$\begin{bmatrix} a_{11} \\ a_{21} \\ \dots \\ a_{m1} \end{bmatrix}, \quad \begin{bmatrix} a_{12} \\ a_{22} \\ \dots \\ a_{m2} \end{bmatrix}, \quad \dots, \quad \begin{bmatrix} a_{1n} \\ a_{2n} \\ \dots \\ a_{mn} \end{bmatrix}$$

Note that the element a_{ij}, called the *ij-entry* or *ij-element*, appears in row i and column j. We frequently denote such a matrix by simply writing $A = [a_{ij}]$.

A matrix with m rows and n columns is called an *m by n* matrix, written $m \times n$. The pair of numbers m and n is called the *size* of the matrix. Two matrices A and B are *equal*, written $A = B$, if they have the same size and if corresponding elements are equal. Thus, the equality of two $m \times n$ matrices is equivalent to a system of mn equalities, one for each corresponding pair of elements.

A matrix with only one row is called a *row matrix* or *row vector*, and a matrix with only one column is called a *column matrix* or *column vector*. A matrix whose entries are all zero is called a *zero matrix* and will usually be denoted by 0.

Matrices whose entries are all real numbers are called *real matrices* and are said to be *matrices over* **R**. Analogously, matrices whose entries are all complex numbers are called *complex matrices* and are said to be *matrices over* **C**. This text will be mainly concerned with such real and complex matrices.

EXAMPLE 2.1

(a) The rectangular array $A = \begin{bmatrix} 1 & -4 & 5 \\ 0 & 3 & -2 \end{bmatrix}$ is a 2×3 matrix. Its rows are $(1, -4, 5)$ and $(0, 3, -2)$, and its columns are

$$\begin{bmatrix} 1 \\ 0 \end{bmatrix}, \quad \begin{bmatrix} -4 \\ 3 \end{bmatrix}, \quad \begin{bmatrix} 5 \\ -2 \end{bmatrix}$$

(b) The 2×4 zero matrix is the matrix $0 = \begin{bmatrix} 0 & 0 & 0 & 0 \\ 0 & 0 & 0 & 0 \end{bmatrix}$.

(c) Find x, y, z, t such that

$$\begin{bmatrix} x+y & 2z+t \\ x-y & z-t \end{bmatrix} = \begin{bmatrix} 3 & 7 \\ 1 & 5 \end{bmatrix}$$

By definition of equality of matrices, the four corresponding entries must be equal. Thus,

$$x+y=3, \qquad x-y=1, \qquad 2z+t=7, \qquad z-t=5$$

Solving the above system of equations yields $x = 2$, $y = 1$, $z = 4$, $t = -1$.

2.3 Matrix Addition and Scalar Multiplication

Let $A = [a_{ij}]$ and $B = [b_{ij}]$ be two matrices with the same size, say $m \times n$ matrices. The *sum* of A and B, written $A + B$, is the matrix obtained by adding corresponding elements from A and B. That is,

$$A + B = \begin{bmatrix} a_{11}+b_{11} & a_{12}+b_{12} & \dots & a_{1n}+b_{1n} \\ a_{21}+b_{21} & a_{22}+b_{22} & \dots & a_{2n}+b_{2n} \\ \dots & \dots & \dots & \dots \\ a_{m1}+b_{m1} & a_{m2}+b_{m2} & \dots & a_{mn}+b_{mn} \end{bmatrix}$$

The *product* of the matrix A by a scalar k, written $k \cdot A$ or simply kA, is the matrix obtained by multiplying each element of A by k. That is,

$$kA = \begin{bmatrix} ka_{11} & ka_{12} & \dots & ka_{1n} \\ ka_{21} & ka_{22} & \dots & ka_{2n} \\ \dots & \dots & \dots & \dots \\ ka_{m1} & ka_{m2} & \dots & ka_{mn} \end{bmatrix}$$

Observe that $A + B$ and kA are also $m \times n$ matrices. We also define

$$-A = (-1)A \qquad \text{and} \qquad A - B = A + (-B)$$

The matrix $-A$ is called the *negative* of the matrix A, and the matrix $A - B$ is called the *difference* of A and B. The sum of matrices with different sizes is not defined.

EXAMPLE 2.2 Let $A = \begin{bmatrix} 1 & -2 & 3 \\ 0 & 4 & 5 \end{bmatrix}$ and $B = \begin{bmatrix} 4 & 6 & 8 \\ 1 & -3 & -7 \end{bmatrix}$. Then

$$A + B = \begin{bmatrix} 1+4 & -2+6 & 3+8 \\ 0+1 & 4+(-3) & 5+(-7) \end{bmatrix} = \begin{bmatrix} 5 & 4 & 11 \\ 1 & 1 & -2 \end{bmatrix}$$

$$3A = \begin{bmatrix} 3(1) & 3(-2) & 3(3) \\ 3(0) & 3(4) & 3(5) \end{bmatrix} = \begin{bmatrix} 3 & -6 & 9 \\ 0 & 12 & 15 \end{bmatrix}$$

$$2A - 3B = \begin{bmatrix} 2 & -4 & 6 \\ 0 & 8 & 10 \end{bmatrix} + \begin{bmatrix} -12 & -18 & -24 \\ -3 & 9 & 21 \end{bmatrix} = \begin{bmatrix} -10 & -22 & -18 \\ -3 & 17 & 31 \end{bmatrix}$$

The matrix $2A - 3B$ is called a *linear combination* of A and B.

Basic properties of matrices under the operations of matrix addition and scalar multiplication follow.

THEOREM 2.1: Consider any matrices A, B, C (with the same size) and any scalars k and k'. Then

(i) $(A + B) + C = A + (B + C)$, (v) $k(A + B) = kA + kB$,

(ii) $A + 0 = 0 + A = A$, (vi) $(k + k')A = kA + k'A$,

(iii) $A + (-A) = (-A) + A = 0$, (vii) $(kk')A = k(k'A)$,

(iv) $A + B = B + A$, (viii) $1 \cdot A = A$.

Note first that the 0 in (ii) and (iii) refers to the zero matrix. Also, by (i) and (iv), any sum of matrices

$$A_1 + A_2 + \cdots + A_n$$

requires no parentheses, and the sum does not depend on the order of the matrices. Furthermore, using (vi) and (viii), we also have

$$A + A = 2A, \qquad A + A + A = 3A, \qquad \ldots$$

and so on.

The proof of Theorem 2.1 reduces to showing that the ij-entries on both sides of each matrix equation are equal. (See Problem 2.3.)

Observe the similarity between Theorem 2.1 for matrices and Theorem 1.1 for vectors. In fact, the above operations for matrices may be viewed as generalizations of the corresponding operations for vectors.

2.4 Summation Symbol

Before we define matrix multiplication, it will be instructive to first introduce the *summation symbol* Σ (the Greek capital letter sigma).

Suppose $f(k)$ is an algebraic expression involving the letter k. Then the expression

$$\sum_{k=1}^{n} f(k) \qquad \text{or equivalently} \qquad \Sigma_{k=1}^{n} f(k)$$

has the following meaning. First we set $k = 1$ in $f(k)$, obtaining

$$f(1)$$

Then we set $k = 2$ in $f(k)$, obtaining $f(2)$, and add this to $f(1)$, obtaining

$$f(1) + f(2)$$

Then we set $k = 3$ in $f(k)$, obtaining $f(3)$, and add this to the previous sum, obtaining

$$f(1) + f(2) + f(3)$$

We continue this process until we obtain the sum

$$f(1) + f(2) + \cdots + f(n)$$

Observe that at each step we increase the value of k by 1 until we reach n. The letter k is called the *index*, and 1 and n are called, respectively, the *lower* and *upper* limits. Other letters frequently used as indices are i and j.

We also generalize our definition by allowing the sum to range from any integer n_1 to any integer n_2. That is, we define

$$\sum_{k=n_1}^{n_2} f(k) = f(n_1) + f(n_1 + 1) + f(n_1 + 2) + \cdots + f(n_2)$$

EXAMPLE 2.3

(a) $\sum_{k=1}^{5} x_k = x_1 + x_2 + x_3 + x_4 + x_5$ and $\sum_{i=1}^{n} a_i b_i = a_1 b_1 + a_2 b_2 + \cdots + a_n b_n$

(b) $\sum_{j=2}^{5} j^2 = 2^2 + 3^2 + 4^2 + 5^2 = 54$ and $\sum_{i=0}^{n} a_i x^i = a_0 + a_1 x + a_2 x^2 + \cdots + a_n x^n$

(c) $\sum_{k=1}^{p} a_{ik} b_{kj} = a_{i1} b_{1j} + a_{i2} b_{2j} + a_{i3} b_{3j} + \cdots + a_{ip} b_{pj}$

2.5 Matrix Multiplication

The product of matrices A and B, written AB, is somewhat complicated. For this reason, we first begin with a special case.

The product AB of a row matrix $A = [a_i]$ and a column matrix $B = [b_i]$ with the same number of elements is defined to be the scalar (or 1×1 matrix) obtained by multiplying corresponding entries and adding; that is,

$$AB = [a_1, a_2, \ldots, a_n] \begin{bmatrix} b_1 \\ b_2 \\ \cdots \\ b_n \end{bmatrix} = a_1 b_1 + a_2 b_2 + \cdots + a_n b_n = \sum_{k=1}^{n} a_k b_k$$

We emphasize that AB is a scalar (or a 1×1 matrix). The product AB is not defined when A and B have different numbers of elements.

EXAMPLE 2.4

(a) $[7, -4, 5] \begin{bmatrix} 3 \\ 2 \\ -1 \end{bmatrix} = 7(3) + (-4)(2) + 5(-1) = 21 - 8 - 5 = 8$

(b) $[6, -1, 8, 3] \begin{bmatrix} 4 \\ -9 \\ -2 \\ 5 \end{bmatrix} = 24 + 9 - 16 + 15 = 32$

We are now ready to define matrix multiplication in general.

DEFINITION: Suppose $A = [a_{ik}]$ and $B = [b_{kj}]$ are matrices such that the number of columns of A is equal to the number of rows of B; say, A is an $m \times p$ matrix and B is a $p \times n$ matrix. Then the product AB is the $m \times n$ matrix whose ij-entry is obtained by multiplying the ith row of A by the jth column of B. That is,

$$\begin{bmatrix} a_{11} & \dots & a_{1p} \\ \cdot & \dots & \cdot \\ a_{i1} & \dots & a_{ip} \\ \cdot & \dots & \cdot \\ a_{m1} & \dots & a_{mp} \end{bmatrix} \begin{bmatrix} b_{11} & \dots & b_{1j} & \dots & b_{1n} \\ \cdot & \dots & \cdot & \dots & \cdot \\ \cdot & \dots & \cdot & \dots & \cdot \\ \cdot & \dots & \cdot & \dots & \cdot \\ b_{p1} & \dots & b_{pj} & \dots & b_{pn} \end{bmatrix} = \begin{bmatrix} c_{11} & \dots & c_{1n} \\ \cdot & \dots & \cdot \\ \cdot & c_{ij} & \cdot \\ \cdot & \dots & \cdot \\ c_{m1} & \dots & c_{mn} \end{bmatrix}$$

where $\quad c_{ij} = a_{i1}b_{1j} + a_{i2}b_{2j} + \cdots + a_{ip}b_{pj} = \sum_{k=1}^{p} a_{ik}b_{kj}$

The product AB is not defined if A is an $m \times p$ matrix and B is a $q \times n$ matrix, where $p \neq q$.

EXAMPLE 2.5

(a) Find AB where $A = \begin{bmatrix} 1 & 3 \\ 2 & -1 \end{bmatrix}$ and $B = \begin{bmatrix} 2 & 0 & -4 \\ 5 & -2 & 6 \end{bmatrix}$.

Because A is 2×2 and B is 2×3, the product AB is defined and AB is a 2×3 matrix. To obtain the first row of the product matrix AB, multiply the first row $[1, 3]$ of A by each column of B,

$$\begin{bmatrix} 2 \\ 5 \end{bmatrix}, \quad \begin{bmatrix} 0 \\ -2 \end{bmatrix}, \quad \begin{bmatrix} -4 \\ 6 \end{bmatrix}$$

respectively. That is,

$$AB = \begin{bmatrix} 2+15 & 0-6 & -4+18 \\ & & \end{bmatrix} = \begin{bmatrix} 17 & -6 & 14 \\ & & \end{bmatrix}$$

To obtain the second row of AB, multiply the second row $[2, -1]$ of A by each column of B. Thus,

$$AB = \begin{bmatrix} 17 & -6 & 14 \\ 4-5 & 0+2 & -8-6 \end{bmatrix} = \begin{bmatrix} 17 & -6 & 14 \\ -1 & 2 & -14 \end{bmatrix}$$

(b) Suppose $A = \begin{bmatrix} 1 & 2 \\ 3 & 4 \end{bmatrix}$ and $B = \begin{bmatrix} 5 & 6 \\ 0 & -2 \end{bmatrix}$. Then

$$AB = \begin{bmatrix} 5+0 & 6-4 \\ 15+0 & 18-8 \end{bmatrix} = \begin{bmatrix} 5 & 2 \\ 15 & 10 \end{bmatrix} \quad \text{and} \quad BA = \begin{bmatrix} 5+18 & 10+24 \\ 0-6 & 0-8 \end{bmatrix} = \begin{bmatrix} 23 & 34 \\ -6 & -8 \end{bmatrix}$$

The above example shows that matrix multiplication is not commutative—that is, in general, $AB \neq BA$. However, matrix multiplication does satisfy the following properties.

THEOREM 2.2: Let A, B, C be matrices. Then, whenever the products and sums are defined,

(i) $(AB)C = A(BC)$ (associative law),

(ii) $A(B + C) = AB + AC$ (left distributive law),

(iii) $(B + C)A = BA + CA$ (right distributive law),

(iv) $k(AB) = (kA)B = A(kB)$, where k is a scalar.

We note that $0A = 0$ and $B0 = 0$, where 0 is the zero matrix.

2.6 Transpose of a Matrix

The *transpose* of a matrix A, written A^T, is the matrix obtained by writing the columns of A, in order, as rows. For example,

$$\begin{bmatrix} 1 & 2 & 3 \\ 4 & 5 & 6 \end{bmatrix}^T = \begin{bmatrix} 1 & 4 \\ 2 & 5 \\ 3 & 6 \end{bmatrix} \quad \text{and} \quad [1, -3, -5]^T = \begin{bmatrix} 1 \\ -3 \\ -5 \end{bmatrix}$$

In other words, if $A = [a_{ij}]$ is an $m \times n$ matrix, then $A^T = [b_{ij}]$ is the $n \times m$ matrix where $b_{ij} = a_{ji}$.

Observe that the tranpose of a row vector is a column vector. Similarly, the transpose of a column vector is a row vector.

The next theorem lists basic properties of the transpose operation.

THEOREM 2.3: Let A and B be matrices and let k be a scalar. Then, whenever the sum and product are defined,

$$\text{(i)}\ (A+B)^T = A^T + B^T, \qquad \text{(iii)}\ (kA)^T = kA^T,$$
$$\text{(ii)}\ (A^T)^T = A, \qquad \text{(iv)}\ (AB)^T = B^T A^T.$$

We emphasize that, by (iv), the transpose of a product is the product of the transposes, but in the reverse order.

2.7 Square Matrices

A *square matrix* is a matrix with the same number of rows as columns. An $n \times n$ square matrix is said to be of *order* n and is sometimes called an *n-square matrix*.

Recall that not every two matrices can be added or multiplied. However, if we only consider square matrices of some given order n, then this inconvenience disappears. Specifically, the operations of addition, multiplication, scalar multiplication, and transpose can be performed on any $n \times n$ matrices, and the result is again an $n \times n$ matrix.

EXAMPLE 2.6 The following are square matrices of order 3:

$$A = \begin{bmatrix} 1 & 2 & 3 \\ -4 & -4 & -4 \\ 5 & 6 & 7 \end{bmatrix} \quad \text{and} \quad B = \begin{bmatrix} 2 & -5 & 1 \\ 0 & 3 & -2 \\ 1 & 2 & -4 \end{bmatrix}$$

The following are also matrices of order 3:

$$A+B = \begin{bmatrix} 3 & -3 & 4 \\ -4 & -1 & -6 \\ 6 & 8 & 3 \end{bmatrix}, \qquad 2A = \begin{bmatrix} 2 & 4 & 6 \\ -8 & -8 & -8 \\ 10 & 12 & 14 \end{bmatrix}, \qquad A^T = \begin{bmatrix} 1 & -4 & 5 \\ 2 & -4 & 6 \\ 3 & -4 & 7 \end{bmatrix}$$

$$AB = \begin{bmatrix} 5 & 7 & -15 \\ -12 & 0 & 20 \\ 17 & 7 & -35 \end{bmatrix}, \qquad BA = \begin{bmatrix} 27 & 30 & 33 \\ -22 & -24 & -26 \\ -27 & -30 & -33 \end{bmatrix}$$

Diagonal and Trace

Let $A = [a_{ij}]$ be an n-square matrix. The *diagonal* or *main diagonal* of A consists of the elements with the same subscripts—that is,

$$a_{11}, \quad a_{22}, \quad a_{33}, \quad \ldots, \quad a_{nn}$$

The *trace* of A, written $\operatorname{tr}(A)$, is the sum of the diagonal elements. Namely,

$$\operatorname{tr}(A) = a_{11} + a_{22} + a_{33} + \cdots + a_{nn}$$

The following theorem applies.

THEOREM 2.4: Suppose $A = [a_{ij}]$ and $B = [b_{ij}]$ are n-square matrices and k is a scalar. Then

(i) $\operatorname{tr}(A+B) = \operatorname{tr}(A) + \operatorname{tr}(B)$, (iii) $\operatorname{tr}(A^T) = \operatorname{tr}(A)$,

(ii) $\operatorname{tr}(kA) = k\operatorname{tr}(A)$, (iv) $\operatorname{tr}(AB) = \operatorname{tr}(BA)$.

EXAMPLE 2.7 Let A and B be the matrices A and B in Example 2.6. Then

$$\text{diagonal of } A = \{1, -4, 7\} \quad \text{and} \quad \operatorname{tr}(A) = 1 - 4 + 7 = 4$$
$$\text{diagonal of } B = \{2, 3, -4\} \quad \text{and} \quad \operatorname{tr}(B) = 2 + 3 - 4 = 1$$

Moreover,

$$\operatorname{tr}(A+B) = 3 - 1 + 3 = 5, \quad \operatorname{tr}(2A) = 2 - 8 + 14 = 8, \quad \operatorname{tr}(A^T) = 1 - 4 + 7 = 4$$
$$\operatorname{tr}(AB) = 5 + 0 - 35 = -30, \quad \operatorname{tr}(BA) = 27 - 24 - 33 = -30$$

As expected from Theorem 2.4,

$$\operatorname{tr}(A+B) = \operatorname{tr}(A) + \operatorname{tr}(B), \quad \operatorname{tr}(A^T) = \operatorname{tr}(A), \quad \operatorname{tr}(2A) = 2\operatorname{tr}(A)$$

Furthermore, although $AB \neq BA$, the traces are equal.

Identity Matrix, Scalar Matrices

The n-square *identity* or *unit* matrix, denoted by I_n, or simply I, is the n-square matrix with 1's on the diagonal and 0's elsewhere. The identity matrix I is similar to the scalar 1 in that, for any n-square matrix A,

$$AI = IA = A$$

More generally, if B is an $m \times n$ matrix, then $BI_n = I_m B = B$.

For any scalar k, the matrix kI that contains k's on the diagonal and 0's elsewhere is called the *scalar matrix* corresponding to the scalar k. Observe that

$$(kI)A = k(IA) = kA$$

That is, multiplying a matrix A by the scalar matrix kI is equivalent to multiplying A by the scalar k.

EXAMPLE 2.8 The following are the identity matrices of orders 3 and 4 and the corresponding scalar matrices for $k = 5$:

$$\begin{bmatrix} 1 & 0 & 0 \\ 0 & 1 & 0 \\ 0 & 0 & 1 \end{bmatrix}, \quad \begin{bmatrix} 1 & & & \\ & 1 & & \\ & & 1 & \\ & & & 1 \end{bmatrix}, \quad \begin{bmatrix} 5 & 0 & 0 \\ 0 & 5 & 0 \\ 0 & 0 & 5 \end{bmatrix}, \quad \begin{bmatrix} 5 & & & \\ & 5 & & \\ & & 5 & \\ & & & 5 \end{bmatrix}$$

Remark 1: It is common practice to omit blocks or patterns of 0's when there is no ambiguity, as in the above second and fourth matrices.

Remark 2: The *Kronecker delta function* δ_{ij} is defined by

$$\delta_{ij} = \begin{cases} 0 & \text{if } i \neq j \\ 1 & \text{if } i = j \end{cases}$$

Thus, the identity matrix may be defined by $I = [\delta_{ij}]$.

2.8 Powers of Matrices, Polynomials in Matrices

Let A be an n-square matrix over a field K. *Powers* of A are defined as follows:

$$A^2 = AA, \quad A^3 = A^2A, \quad \ldots, \quad A^{n+1} = A^nA, \quad \ldots, \quad \text{and} \quad A^0 = I$$

Polynomials in the matrix A are also defined. Specifically, for any polynomial

$$f(x) = a_0 + a_1x + a_2x^2 + \cdots + a_nx^n$$

where the a_i are scalars in K, $f(A)$ is defined to be the following matrix:

$$f(A) = a_0I + a_1A + a_2A^2 + \cdots + a_nA^n$$

[Note that $f(A)$ is obtained from $f(x)$ by substituting the matrix A for the variable x and substituting the scalar matrix a_0I for the scalar a_0.] If $f(A)$ is the zero matrix, then A is called a *zero* or *root* of $f(x)$.

EXAMPLE 2.9 Suppose $A = \begin{bmatrix} 1 & 2 \\ 3 & -4 \end{bmatrix}$. Then

$$A^2 = \begin{bmatrix} 1 & 2 \\ 3 & -4 \end{bmatrix}\begin{bmatrix} 1 & 2 \\ 3 & -4 \end{bmatrix} = \begin{bmatrix} 7 & -6 \\ -9 & 22 \end{bmatrix} \quad \text{and} \quad A^3 = A^2A = \begin{bmatrix} 7 & -6 \\ -9 & 22 \end{bmatrix}\begin{bmatrix} 1 & 2 \\ 3 & -4 \end{bmatrix} = \begin{bmatrix} -11 & 38 \\ 57 & -106 \end{bmatrix}$$

Suppose $f(x) = 2x^2 - 3x + 5$ and $g(x) = x^2 + 3x - 10$. Then

$$f(A) = 2\begin{bmatrix} 7 & -6 \\ -9 & 22 \end{bmatrix} - 3\begin{bmatrix} 1 & 2 \\ 3 & -4 \end{bmatrix} + 5\begin{bmatrix} 1 & 0 \\ 0 & 1 \end{bmatrix} = \begin{bmatrix} 16 & -18 \\ -27 & 61 \end{bmatrix}$$

$$g(A) = \begin{bmatrix} 7 & -6 \\ -9 & 22 \end{bmatrix} + 3\begin{bmatrix} 1 & 2 \\ 3 & -4 \end{bmatrix} - 10\begin{bmatrix} 1 & 0 \\ 0 & 1 \end{bmatrix} = \begin{bmatrix} 0 & 0 \\ 0 & 0 \end{bmatrix}$$

Thus, A is a zero of the polynomial $g(x)$.

2.9 Invertible (Nonsingular) Matrices

A square matrix A is said to be *invertible* or *nonsingular* if there exists a matrix B such that

$$AB = BA = I$$

where I is the identity matrix. Such a matrix B is unique. That is, if $AB_1 = B_1A = I$ and $AB_2 = B_2A = I$, then

$$B_1 = B_1I = B_1(AB_2) = (B_1A)B_2 = IB_2 = B_2$$

We call such a matrix B the *inverse* of A and denote it by A^{-1}. Observe that the above relation is symmetric; that is, if B is the inverse of A, then A is the inverse of B.

EXAMPLE 2.10 Suppose that $A = \begin{bmatrix} 2 & 5 \\ 1 & 3 \end{bmatrix}$ and $B = \begin{bmatrix} 3 & -5 \\ -1 & 2 \end{bmatrix}$. Then

$$AB = \begin{bmatrix} 6-5 & -10+10 \\ 3-3 & -5+6 \end{bmatrix} = \begin{bmatrix} 1 & 0 \\ 0 & 1 \end{bmatrix} \quad \text{and} \quad BA = \begin{bmatrix} 6-5 & 15-15 \\ -2+2 & -5+6 \end{bmatrix} = \begin{bmatrix} 1 & 0 \\ 0 & 1 \end{bmatrix}$$

Thus, A and B are inverses.

It is known (Theorem 3.18) that $AB = I$ if and only if $BA = I$. Thus, it is necessary to test only one product to determine whether or not two given matrices are inverses. (See Problem 2.17.)

Now suppose A and B are invertible. Then AB is invertible and $(AB)^{-1} = B^{-1}A^{-1}$. More generally, if $A_1, A_2, \ldots, A_k$ are invertible, then their product is invertible and

$$(A_1A_2 \ldots A_k)^{-1} = A_k^{-1} \ldots A_2^{-1}A_1^{-1}$$

the product of the inverses in the reverse order.

Inverse of a 2 × 2 Matrix

Let A be an arbitrary 2×2 matrix, say $A = \begin{bmatrix} a & b \\ c & d \end{bmatrix}$. We want to derive a formula for A^{-1}, the inverse of A. Specifically, we seek $2^2 = 4$ scalars, say x_1, y_1, x_2, y_2, such that

$$\begin{bmatrix} a & b \\ c & d \end{bmatrix}\begin{bmatrix} x_1 & x_2 \\ y_1 & y_2 \end{bmatrix} = \begin{bmatrix} 1 & 0 \\ 0 & 1 \end{bmatrix} \quad \text{or} \quad \begin{bmatrix} ax_1 + by_1 & ax_2 + by_2 \\ cx_1 + dy_1 & cx_2 + dy_2 \end{bmatrix} = \begin{bmatrix} 1 & 0 \\ 0 & 1 \end{bmatrix}$$

Setting the four entries equal to the corresponding entries in the identity matrix yields four equations, which can be partitioned into two 2×2 systems as follows:

$$\begin{aligned} ax_1 + by_1 &= 1, \qquad & ax_2 + by_2 &= 0 \\ cx_1 + dy_1 &= 0, \qquad & cx_2 + dy_2 &= 1 \end{aligned}$$

Suppose we let $|A| = ad - bc$ (called the *determinant* of A). Assuming $|A| \neq 0$, we can solve uniquely for the above unknowns x_1, y_1, x_2, y_2, obtaining

$$x_1 = \frac{d}{|A|}, \qquad y_1 = \frac{-c}{|A|}, \qquad x_2 = \frac{-b}{|A|}, \qquad y_2 = \frac{a}{|A|}$$

Accordingly,

$$A^{-1} = \begin{bmatrix} a & b \\ c & d \end{bmatrix}^{-1} = \begin{bmatrix} d/|A| & -b/|A| \\ -c/|A| & a/|A| \end{bmatrix} = \frac{1}{|A|}\begin{bmatrix} d & -b \\ -c & a \end{bmatrix}$$

In other words, when $|A| \neq 0$, the inverse of a 2×2 matrix A may be obtained from A as follows:

(1) Interchange the two elements on the diagonal.

(2) Take the negatives of the other two elements.

(3) Multiply the resulting matrix by $1/|A|$ or, equivalently, divide each element by $|A|$.

In case $|A| = 0$, the matrix A is not invertible.

EXAMPLE 2.11 Find the inverse of $A = \begin{bmatrix} 2 & 3 \\ 4 & 5 \end{bmatrix}$ and $B = \begin{bmatrix} 1 & 3 \\ 2 & 6 \end{bmatrix}$.

First evaluate $|A| = 2(5) - 3(4) = 10 - 12 = -2$. Because $|A| \neq 0$, the matrix A is invertible and

$$A^{-1} = \frac{1}{-2}\begin{bmatrix} 5 & -3 \\ -4 & 2 \end{bmatrix} = \begin{bmatrix} -\frac{5}{2} & \frac{3}{2} \\ 2 & -1 \end{bmatrix}$$

Now evaluate $|B| = 1(6) - 3(2) = 6 - 6 = 0$. Because $|B| = 0$, the matrix B has no inverse.

Remark: The above property that a matrix is invertible if and only if A has a nonzero determinant is true for square matrices of any order. (See Chapter 8.)

Inverse of an $n \times n$ Matrix

Suppose A is an arbitrary n-square matrix. Finding its inverse A^{-1} reduces, as above, to finding the solution of a collection of $n \times n$ systems of linear equations. The solution of such systems and an efficient way of solving such a collection of systems is treated in Chapter 3.

2.10 Special Types of Square Matrices

This section describes a number of special kinds of square matrices.

Diagonal and Triangular Matrices

A square matrix $D = [d_{ij}]$ is *diagonal* if its nondiagonal entries are all zero. Such a matrix is sometimes denoted by

$$D = \text{diag}(d_{11}, d_{22}, \ldots, d_{nn})$$

where some or all the d_{ii} may be zero. For example,

$$\begin{bmatrix} 3 & 0 & 0 \\ 0 & -7 & 0 \\ 0 & 0 & 2 \end{bmatrix}, \qquad \begin{bmatrix} 4 & 0 \\ 0 & -5 \end{bmatrix}, \qquad \begin{bmatrix} 6 & & & \\ & 0 & & \\ & & -9 & \\ & & & 8 \end{bmatrix}$$

are diagonal matrices, which may be represented, respectively, by

$$\text{diag}(3, -7, 2), \qquad \text{diag}(4, -5), \qquad \text{diag}(6, 0, -9, 8)$$

(Observe that patterns of 0's in the third matrix have been omitted.)

A square matrix $A = [a_{ij}]$ is *upper triangular* or simply *triangular* if all entries below the (main) diagonal are equal to 0—that is, if $a_{ij} = 0$ for $i > j$. Generic upper triangular matrices of orders 2, 3, 4 are as follows:

$$\begin{bmatrix} a_{11} & a_{12} \\ 0 & a_{22} \end{bmatrix}, \qquad \begin{bmatrix} b_{11} & b_{12} & b_{13} \\ & b_{22} & b_{23} \\ & & b_{33} \end{bmatrix}, \qquad \begin{bmatrix} c_{11} & c_{12} & c_{13} & c_{14} \\ & c_{22} & c_{23} & c_{24} \\ & & c_{33} & c_{34} \\ & & & c_{44} \end{bmatrix}$$

(As with diagonal matrices, it is common practice to omit patterns of 0's.)

The following theorem applies.

THEOREM 2.5: Suppose $A = [a_{ij}]$ and $B = [b_{ij}]$ are $n \times n$ (upper) triangular matrices. Then

(i) $A + B$, kA, AB are triangular with respective diagonals:

$$(a_{11} + b_{11}, \ \ldots, \ a_{nn} + b_{nn}), \qquad (ka_{11}, \ \ldots, \ ka_{nn}), \qquad (a_{11}b_{11}, \ \ldots, \ a_{nn}b_{nn})$$

(ii) For any polynomial $f(x)$, the matrix $f(A)$ is triangular with diagonal

$$(f(a_{11}), f(a_{22}), \ldots, f(a_{nn}))$$

(iii) A is invertible if and only if each diagonal element $a_{ii} \neq 0$, and when A^{-1} exists it is also triangular.

A *lower triangular matrix* is a square matrix whose entries above the diagonal are all zero. We note that Theorem 2.5 is true if we replace "triangular" by either "lower triangular" or "diagonal."

Remark: A nonempty collection A of matrices is called an *algebra* (of matrices) if A is closed under the operations of matrix addition, scalar multiplication, and matrix multiplication. Clearly, the square matrices with a given order form an algebra of matrices, but so do the scalar, diagonal, triangular, and lower triangular matrices.

Special Real Square Matrices: Symmetric, Orthogonal, Normal [Optional until Chapter 12]

Suppose now A is a square matrix with real entries—that is, a real square matrix. The relationship between A and its transpose A^T yields important kinds of matrices.

(a) Symmetric Matrices

A matrix A is *symmetric* if $A^T = A$. Equivalently, $A = [a_{ij}]$ is symmetric if *symmetric elements* (mirror elements with respect to the diagonal) are equal—that is, if each $a_{ij} = a_{ji}$.

A matrix A is *skew-symmetric* if $A^T = -A$ or, equivalently, if each $a_{ij} = -a_{ji}$. Clearly, the diagonal elements of such a matrix must be zero, because $a_{ii} = -a_{ii}$ implies $a_{ii} = 0$.

(Note that a matrix A must be square if $A^T = A$ or $A^T = -A$.)

EXAMPLE 2.12 Let $A = \begin{bmatrix} 2 & -3 & 5 \\ -3 & 6 & 7 \\ 5 & 7 & -8 \end{bmatrix}, B = \begin{bmatrix} 0 & 3 & -4 \\ -3 & 0 & 5 \\ 4 & -5 & 0 \end{bmatrix}, C = \begin{bmatrix} 1 & 0 & 0 \\ 0 & 0 & 1 \end{bmatrix}$.

(a) By inspection, the symmetric elements in A are equal, or $A^T = A$. Thus, A is symmetric.

(b) The diagonal elements of B are 0 and symmetric elements are negatives of each other, or $B^T = -B$. Thus, B is skew-symmetric.

(c) Because C is not square, C is neither symmetric nor skew-symmetric.

(b) Orthogonal Matrices

A real matrix A is *orthogonal* if $A^T = A^{-1}$—that is, if $AA^T = A^TA = I$. Thus, A must necessarily be square and invertible.

EXAMPLE 2.13 Let $A = \begin{bmatrix} \frac{1}{9} & \frac{8}{9} & -\frac{4}{9} \\ \frac{4}{9} & -\frac{4}{9} & -\frac{7}{9} \\ \frac{8}{9} & \frac{1}{9} & \frac{4}{9} \end{bmatrix}$. Multiplying A by A^T yields I; that is, $AA^T = I$. This means $A^TA = I$, as well. Thus, $A^T = A^{-1}$; that is, A is orthogonal.

Now suppose A is a real orthogonal 3×3 matrix with rows

$$u_1 = (a_1, a_2, a_3), \qquad u_2 = (b_1, b_2, b_3), \qquad u_3 = (c_1, c_2, c_3)$$

Because A is orthogonal, we must have $AA^T = I$. Namely,

$$AA^T = \begin{bmatrix} a_1 & a_2 & a_3 \\ b_1 & b_2 & b_3 \\ c_1 & c_2 & c_3 \end{bmatrix} \begin{bmatrix} a_1 & b_1 & c_1 \\ a_2 & b_2 & c_2 \\ a_3 & b_3 & c_3 \end{bmatrix} = \begin{bmatrix} 1 & 0 & 0 \\ 0 & 1 & 0 \\ 0 & 0 & 1 \end{bmatrix} = I$$

Multiplying A by A^T and setting each entry equal to the corresponding entry in I yields the following nine equations:

$$\begin{aligned} a_1^2 + a_2^2 + a_3^2 &= 1, & a_1b_1 + a_2b_2 + a_3b_3 &= 0, & a_1c_1 + a_2c_2 + a_3c_3 &= 0 \\ b_1a_1 + b_2a_2 + b_3a_3 &= 0, & b_1^2 + b_2^2 + b_3^2 &= 1, & b_1c_1 + b_2c_2 + b_3c_3 &= 0 \\ c_1a_1 + c_2a_2 + c_3a_3 &= 0, & c_1b_1 + c_2b_2 + c_3b_3 &= 0, & c_1^2 + c_2^2 + c_3^2 &= 1 \end{aligned}$$

Accordingly, $u_1 \cdot u_1 = 1$, $u_2 \cdot u_2 = 1$, $u_3 \cdot u_3 = 1$, and $u_i \cdot u_j = 0$ for $i \neq j$. Thus, the rows u_1, u_2, u_3 are unit vectors and are orthogonal to each other.

Generally speaking, vectors $u_1, u_2, \ldots, u_m$ in $\mathbf{R}^n$ are said to form an *orthonormal set* of vectors if the vectors are unit vectors and are orthogonal to each other; that is,

$$u_i \cdot u_j = \begin{cases} 0 & \text{if } i \neq j \\ 1 & \text{if } i = j \end{cases}$$

In other words, $u_i \cdot u_j = \delta_{ij}$ where δ_{ij} is the Kronecker delta function.

We have shown that the condition $AA^T = I$ implies that the rows of A form an orthonormal set of vectors. The condition $A^TA = I$ similarly implies that the columns of A also form an orthonormal set of vectors. Furthermore, because each step is reversible, the converse is true.

The above results for 3×3 matrices are true in general. That is, the following theorem holds.

THEOREM 2.6: Let A be a real matrix. Then the following are equivalent:
(a) A is orthogonal.
(b) The rows of A form an orthonormal set.
(c) The columns of A form an orthonormal set.

For $n = 2$, we have the following result (proved in Problem 2.28).

THEOREM 2.7: Let A be a real 2×2 orthogonal matrix. Then, for some real number θ,

$$A = \begin{bmatrix} \cos\theta & \sin\theta \\ -\sin\theta & \cos\theta \end{bmatrix} \quad \text{or} \quad A = \begin{bmatrix} \cos\theta & \sin\theta \\ \sin\theta & -\cos\theta \end{bmatrix}$$

(c) Normal Matrices

A real matrix A is *normal* if it *commutes* with its transpose A^T—that is, if $AA^T = A^TA$. If A is symmetric, orthogonal, or skew-symmetric, then A is normal. There are also other normal matrices.

EXAMPLE 2.14 Let $A = \begin{bmatrix} 6 & -3 \\ 3 & 6 \end{bmatrix}$. Then

$$AA^T = \begin{bmatrix} 6 & -3 \\ 3 & 6 \end{bmatrix}\begin{bmatrix} 6 & 3 \\ -3 & 6 \end{bmatrix} = \begin{bmatrix} 45 & 0 \\ 0 & 45 \end{bmatrix} \quad \text{and} \quad A^TA = \begin{bmatrix} 6 & 3 \\ -3 & 6 \end{bmatrix}\begin{bmatrix} 6 & -3 \\ 3 & 6 \end{bmatrix} = \begin{bmatrix} 45 & 0 \\ 0 & 45 \end{bmatrix}$$

Because $AA^T = A^TA$, the matrix A is normal.

2.11 Complex Matrices

Let A be a complex matrix—that is, a matrix with complex entries. Recall (Section 1.7) that if $z = a + bi$ is a complex number, then $\bar{z} = a - bi$ is its conjugate. The *conjugate* of a complex matrix A, written $\bar{A}$, is the matrix obtained from A by taking the conjugate of each entry in A. That is, if $A = [a_{ij}]$, then $\bar{A} = [b_{ij}]$, where $b_{ij} = \bar{a}_{ij}$. (We denote this fact by writing $\bar{A} = [\bar{a}_{ij}]$.)

The two operations of transpose and conjugation commute for any complex matrix A, and the special notation A^H is used for the conjugate transpose of A. That is,

$$A^H = (\bar{A})^T = (\overline{A^T})$$

Note that if A is real, then $A^H = A^T$. [Some texts use A^* instead of A^H.]

EXAMPLE 2.15 Let $A = \begin{bmatrix} 2+8i & 5-3i & 4-7i \\ 6i & 1-4i & 3+2i \end{bmatrix}$. Then $A^H = \begin{bmatrix} 2-8i & -6i \\ 5+3i & 1+4i \\ 4+7i & 3-2i \end{bmatrix}$.

Special Complex Matrices: Hermitian, Unitary, Normal [Optional until Chapter 12]

Consider a complex matrix A. The relationship between A and its conjugate transpose A^H yields important kinds of complex matrices (which are analogous to the kinds of real matrices described above).

A complex matrix A is said to be *Hermitian* or *skew-Hermitian* according as to whether

$$A^H = A \quad \text{or} \quad A^H = -A.$$

Clearly, $A = [a_{ij}]$ is Hermitian if and only if symmetric elements are conjugate—that is, if each $a_{ij} = \bar{a}_{ji}$—in which case each diagonal element a_{ii} must be real. Similarly, if A is skew-symmetric, then each diagonal element $a_{ii} = 0$. (Note that A must be square if $A^H = A$ or $A^H = -A$.)

A complex matrix A is *unitary* if $A^HA^{-1} = A^{-1}A^H = I$—that is, if

$$A^H = A^{-1}.$$

Thus, A must necessarily be square and invertible. We note that a complex matrix A is unitary if and only if its rows (columns) form an orthonormal set relative to the dot product of complex vectors.

A complex matrix A is said to be *normal* if it commutes with A^H—that is, if

$$AA^H = A^HA$$

(Thus, A must be a square matrix.) This definition reduces to that for real matrices when A is real.

EXAMPLE 2.16 Consider the following complex matrices:

$$A = \begin{bmatrix} 3 & 1-2i & 4+7i \\ 1+2i & -4 & -2i \\ 4-7i & 2i & 5 \end{bmatrix} \qquad B = \frac{1}{2}\begin{bmatrix} 1 & -i & -1+i \\ i & 1 & 1+i \\ 1+i & -1+i & 0 \end{bmatrix} \qquad C = \begin{bmatrix} 2+3i & 1 \\ i & 1+2i \end{bmatrix}$$

(a) By inspection, the diagonal elements of A are real, and the symmetric elements $1-2i$ and $1+2i$ are conjugate, $4+7i$ and $4-7i$ are conjugate, and $-2i$ and $2i$ are conjugate. Thus, A is Hermitian.

(b) Multiplying B by B^H yields I; that is, $BB^H = I$. This implies $B^H B = I$, as well. Thus, $B^H = B^{-1}$, which means B is unitary.

(c) To show C is normal, we evaluate CC^H and $C^H C$:

$$CC^H = \begin{bmatrix} 2+3i & 1 \\ i & 1+2i \end{bmatrix}\begin{bmatrix} 2-3i & -i \\ 1 & 1-2i \end{bmatrix} = \begin{bmatrix} 14 & 4-4i \\ 4+4i & 6 \end{bmatrix}$$

and similarly $C^H C = \begin{bmatrix} 14 & 4-4i \\ 4+4i & 6 \end{bmatrix}$. Because $CC^H = C^H C$, the complex matrix C is normal.

We note that when a matrix A is real, Hermitian is the same as symmetric, and unitary is the same as orthogonal.

2.12 Block Matrices

Using a system of horizontal and vertical (dashed) lines, we can partition a matrix A into submatrices called *blocks* (or *cells*) of A. Clearly a given matrix may be divided into blocks in different ways. For example,

$$\left[\begin{array}{cc|cc|c} 1 & -2 & 0 & 1 & 3 \\ 2 & 3 & 5 & 7 & -2 \\ \hline 3 & 1 & 4 & 5 & 9 \\ 4 & 6 & -3 & 1 & 8 \end{array}\right], \quad \left[\begin{array}{cc|ccc} 1 & -2 & 0 & 1 & 3 \\ \hline 2 & 3 & 5 & 7 & -2 \\ 3 & 1 & 4 & 5 & 9 \\ \hline 4 & 6 & -3 & 1 & 8 \end{array}\right], \quad \left[\begin{array}{ccc|cc} 1 & -2 & 0 & 1 & 3 \\ 2 & 3 & 5 & 7 & -2 \\ \hline 3 & 1 & 4 & 5 & 9 \\ 4 & 6 & -3 & 1 & 8 \end{array}\right]$$

The convenience of the partition of matrices, say A and B, into blocks is that the result of operations on A and B can be obtained by carrying out the computation with the blocks, just as if they were the actual elements of the matrices. This is illustrated below, where the notation $A = [A_{ij}]$ will be used for a block matrix A with blocks A_{ij}.

Suppose that $A = [A_{ij}]$ and $B = [B_{ij}]$ are block matrices with the same numbers of row and column blocks, and suppose that corresponding blocks have the same size. Then adding the corresponding blocks of A and B also adds the corresponding elements of A and B, and multiplying each block of A by a scalar k multiplies each element of A by k. Thus,

$$A + B = \begin{bmatrix} A_{11}+B_{11} & A_{12}+B_{12} & \ldots & A_{1n}+B_{1n} \\ A_{21}+B_{21} & A_{22}+B_{22} & \ldots & A_{2n}+B_{2n} \\ \ldots & \ldots & \ldots & \ldots \\ A_{m1}+B_{m1} & A_{m2}+B_{m2} & \ldots & A_{mn}+B_{mn} \end{bmatrix}$$

and

$$kA = \begin{bmatrix} kA_{11} & kA_{12} & \ldots & kA_{1n} \\ kA_{21} & kA_{22} & \ldots & kA_{2n} \\ \ldots & \ldots & \ldots & \ldots \\ kA_{m1} & kA_{m2} & \ldots & kA_{mn} \end{bmatrix}$$

The case of matrix multiplication is less obvious, but still true. That is, suppose that $U = [U_{ik}]$ and $V = [V_{kj}]$ are block matrices such that the number of columns of each block U_{ik} is equal to the number of rows of each block V_{kj}. (Thus, each product $U_{ik}V_{kj}$ is defined.) Then

$$UV = \begin{bmatrix} W_{11} & W_{12} & \cdots & W_{1n} \\ W_{21} & W_{22} & \cdots & W_{2n} \\ \cdots & \cdots & \cdots & \cdots \\ W_{m1} & W_{m2} & \cdots & W_{mn} \end{bmatrix}, \qquad \text{where} \quad W_{ij} = U_{i1}V_{1j} + U_{i2}V_{2j} + \cdots + U_{ip}V_{pj}$$

The proof of the above formula for UV is straightforward but detailed and lengthy. It is left as an exercise (Problem 2.85).

Square Block Matrices

Let M be a block matrix. Then M is called a *square block matrix* if

(i) M is a square matrix.
(ii) The blocks form a square matrix.
(iii) The diagonal blocks are also square matrices.

The latter two conditions will occur if and only if there are the same number of horizontal and vertical lines and they are placed symmetrically.

Consider the following two block matrices:

$$A = \left[\begin{array}{cc|cc|c} 1 & 2 & 3 & 4 & 5 \\ 1 & 1 & 1 & 1 & 1 \\ \hline 9 & 8 & 7 & 6 & 5 \\ \hline 4 & 4 & 4 & 4 & 4 \\ 3 & 5 & 3 & 5 & 3 \end{array}\right] \quad \text{and} \quad B = \left[\begin{array}{cc|cc|c} 1 & 2 & 3 & 4 & 5 \\ 1 & 1 & 1 & 1 & 1 \\ \hline 9 & 8 & 7 & 6 & 5 \\ 4 & 4 & 4 & 4 & 4 \\ \hline 3 & 5 & 3 & 5 & 3 \end{array}\right]$$

The block matrix A is not a square block matrix, because the second and third diagonal blocks are not square. On the other hand, the block matrix B is a square block matrix.

Block Diagonal Matrices

Let $M = [A_{ij}]$ be a square block matrix such that the nondiagonal blocks are all zero matrices; that is, $A_{ij} = 0$ when $i \neq j$. Then M is called a *block diagonal matrix*. We sometimes denote such a block diagonal matrix by writing

$$M = \operatorname{diag}(A_{11}, A_{22}, \ldots, A_{rr}) \quad \text{or} \quad M = A_{11} \oplus A_{22} \oplus \cdots \oplus A_{rr}$$

The importance of block diagonal matrices is that the algebra of the block matrix is frequently reduced to the algebra of the individual blocks. Specifically, suppose $f(x)$ is a polynomial and M is the above block diagonal matrix. Then $f(M)$ is a block diagonal matrix, and

$$f(M) = \operatorname{diag}(f(A_{11}), f(A_{22}), \ldots, f(A_{rr}))$$

Also, M is invertible if and only if each A_{ii} is invertible, and, in such a case, M^{-1} is a block diagonal matrix, and

$$M^{-1} = \operatorname{diag}(A_{11}^{-1}, A_{22}^{-1}, \ldots, A_{rr}^{-1})$$

Analogously, a square block matrix is called a *block upper triangular matrix* if the blocks below the diagonal are zero matrices and a *block lower triangular matrix* if the blocks above the diagonal are zero matrices.

EXAMPLE 2.17 Determine which of the following square block matrices are upper diagonal, lower diagonal, or diagonal:

$$A = \left[\begin{array}{cc|c} 1 & 2 & 0 \\ 3 & 4 & 5 \\ \hline 0 & 0 & 6 \end{array}\right], \quad B = \left[\begin{array}{c|cc|c} 1 & 0 & 0 & 0 \\ \hline 2 & 3 & 4 & 0 \\ 5 & 0 & 6 & 0 \\ \hline 0 & 7 & 8 & 9 \end{array}\right], \quad C = \left[\begin{array}{c|cc} 1 & 0 & 0 \\ \hline 0 & 2 & 3 \\ 0 & 4 & 5 \end{array}\right], \quad D = \left[\begin{array}{cc|c} 1 & 2 & 0 \\ 3 & 4 & 5 \\ \hline 0 & 6 & 7 \end{array}\right]$$

(a) A is upper triangular because the block below the diagonal is a zero block.

(b) B is lower triangular because all blocks above the diagonal are zero blocks.

(c) C is diagonal because the blocks above and below the diagonal are zero blocks.

(d) D is neither upper triangular nor lower triangular. Also, no other partitioning of D will make it into either a block upper triangular matrix or a block lower triangular matrix.

SOLVED PROBLEMS

Matrix Addition and Scalar Multiplication

2.1 Given $A = \begin{bmatrix} 1 & -2 & 3 \\ 4 & 5 & -6 \end{bmatrix}$ and $B = \begin{bmatrix} 3 & 0 & 2 \\ -7 & 1 & 8 \end{bmatrix}$, find:

(a) $A + B$, (b) $2A - 3B$.

(a) Add the corresponding elements:

$$A + B = \begin{bmatrix} 1+3 & -2+0 & 3+2 \\ 4-7 & 5+1 & -6+8 \end{bmatrix} = \begin{bmatrix} 4 & -2 & 5 \\ -3 & 6 & 2 \end{bmatrix}$$

(b) First perform the scalar multiplication and then a matrix addition:

$$2A - 3B = \begin{bmatrix} 2 & -4 & 6 \\ 8 & 10 & -12 \end{bmatrix} + \begin{bmatrix} -9 & 0 & -6 \\ 21 & -3 & -24 \end{bmatrix} = \begin{bmatrix} -7 & -4 & 0 \\ 29 & 7 & -36 \end{bmatrix}$$

(Note that we multiply B by -3 and then add, rather than multiplying B by 3 and subtracting. This usually prevents errors.)

2.2. Find x, y, z, t where $3\begin{bmatrix} x & y \\ z & t \end{bmatrix} = \begin{bmatrix} x & 6 \\ -1 & 2t \end{bmatrix} + \begin{bmatrix} 4 & x+y \\ z+t & 3 \end{bmatrix}$.

Write each side as a single equation:

$$\begin{bmatrix} 3x & 3y \\ 3z & 3t \end{bmatrix} = \begin{bmatrix} x+4 & x+y+6 \\ z+t-1 & 2t+3 \end{bmatrix}$$

Set corresponding entries equal to each other to obtain the following system of four equations:

$$3x = x+4, \qquad 3y = x+y+6, \qquad 3z = z+t-1, \qquad 3t = 2t+3$$

or $\quad 2x = 4, \qquad 2y = 6+x, \qquad 2z = t-1, \qquad t = 3$

The solution is $x = 2$, $y = 4$, $z = 1$, $t = 3$.

2.3. Prove Theorem 2.1 (i) and (v): (i) $(A+B)+C = A+(B+C)$, (v) $k(A+B) = kA + kB$.

Suppose $A = [a_{ij}]$, $B = [b_{ij}]$, $C = [c_{ij}]$. The proof reduces to showing that corresponding ij-entries in each side of each matrix equation are equal. [We prove only (i) and (v), because the other parts of Theorem 2.1 are proved similarly.]

(i) The ij-entry of $A+B$ is $a_{ij}+b_{ij}$; hence, the ij-entry of $(A+B)+C$ is $(a_{ij}+b_{ij})+c_{ij}$. On the other hand, the ij-entry of $B+C$ is $b_{ij}+c_{ij}$; hence, the ij-entry of $A+(B+C)$ is $a_{ij}+(b_{ij}+c_{ij})$. However, for scalars in K,

$$(a_{ij}+b_{ij})+c_{ij}=a_{ij}+(b_{ij}+c_{ij})$$

Thus, $(A+B)+C$ and $A+(B+C)$ have identical ij-entries. Therefore, $(A+B)+C=A+(B+C)$.

(v) The ij-entry of $A+B$ is $a_{ij}+b_{ij}$; hence, $k(a_{ij}+b_{ij})$ is the ij-entry of $k(A+B)$. On the other hand, the ij-entries of kA and kB are ka_{ij} and kb_{ij}, respectively. Thus, $ka_{ij}+kb_{ij}$ is the ij-entry of $kA+kB$. However, for scalars in K,

$$k(a_{ij}+b_{ij})=ka_{ij}+kb_{ij}$$

Thus, $k(A+B)$ and $kA+kB$ have identical ij-entries. Therefore, $k(A+B)=kA+kB$.

Matrix Multiplication

2.4. Calculate: (a) $[8,-4,5]\begin{bmatrix}3\\2\\-1\end{bmatrix}$, (b) $[6,-1,7,5]\begin{bmatrix}4\\-9\\-3\\2\end{bmatrix}$, (c) $[3,8,-2,4]\begin{bmatrix}5\\-1\\6\end{bmatrix}$

(a) Multiply the corresponding entries and add:

$$[8,-4,5]\begin{bmatrix}3\\2\\-1\end{bmatrix}=8(3)+(-4)(2)+5(-1)=24-8-5=11$$

(b) Multiply the corresponding entries and add:

$$[6,-1,7,5]\begin{bmatrix}4\\-9\\-3\\2\end{bmatrix}=24+9-21+10=22$$

(c) The product is not defined when the row matrix and the column matrix have different numbers of elements.

2.5. Let $(r\times s)$ denote an $r\times s$ matrix. Find the sizes of those matrix products that are defined:

(a) $(2\times 3)(3\times 4)$, (c) $(1\times 2)(3\times 1)$, (e) $(4\times 4)(3\times 3)$
(b) $(4\times 1)(1\times 2)$, (d) $(5\times 2)(2\times 3)$, (f) $(2\times 2)(2\times 4)$

In each case, the product is defined if the inner numbers are equal, and then the product will have the size of the outer numbers in the given order.

(a) 2×4, (c) not defined, (e) not defined
(b) 4×2, (d) 5×3, (f) 2×4

2.6. Let $A=\begin{bmatrix}1&3\\2&-1\end{bmatrix}$ and $B=\begin{bmatrix}2&0&-4\\3&-2&6\end{bmatrix}$. Find: (a) AB, (b) BA.

(a) Because A is a 2×2 matrix and B a 2×3 matrix, the product AB is defined and is a 2×3 matrix. To obtain the entries in the first row of AB, multiply the first row $[1,3]$ of A by the columns $\begin{bmatrix}2\\3\end{bmatrix},\begin{bmatrix}0\\-2\end{bmatrix},\begin{bmatrix}-4\\6\end{bmatrix}$ of B, respectively, as follows:

$$AB=\begin{bmatrix}1&3\\2&-1\end{bmatrix}\begin{bmatrix}2&0&-4\\3&-2&6\end{bmatrix}=\begin{bmatrix}2+9&0-6&-4+18\end{bmatrix}=\begin{bmatrix}11&-6&14\end{bmatrix}$$

To obtain the entries in the second row of AB, multiply the second row $[2, -1]$ of A by the columns of B:

$$AB = \begin{bmatrix} 1 & 3 \\ 2 & -1 \end{bmatrix}\begin{bmatrix} 2 & 0 & -4 \\ 3 & -2 & 6 \end{bmatrix} = \begin{bmatrix} 11 & -6 & 14 \\ 4-3 & 0+2 & -8-6 \end{bmatrix}$$

Thus,

$$AB = \begin{bmatrix} 11 & -6 & 14 \\ 1 & 2 & -14 \end{bmatrix}.$$

(b) The size of B is 2×3 and that of A is 2×2. The inner numbers 3 and 2 are not equal; hence, the product BA is not defined.

2.7. Find AB, where $A = \begin{bmatrix} 2 & 3 & -1 \\ 4 & -2 & 5 \end{bmatrix}$ and $B = \begin{bmatrix} 2 & -1 & 0 & 6 \\ 1 & 3 & -5 & 1 \\ 4 & 1 & -2 & 2 \end{bmatrix}$.

Because A is a 2×3 matrix and B a 3×4 matrix, the product AB is defined and is a 2×4 matrix. Multiply the rows of A by the columns of B to obtain

$$AB = \begin{bmatrix} 4+3-4 & -2+9-1 & 0-15+2 & 12+3-2 \\ 8-2+20 & -4-6+5 & 0+10-10 & 24-2+10 \end{bmatrix} = \begin{bmatrix} 3 & 6 & -13 & 13 \\ 26 & -5 & 0 & 32 \end{bmatrix}.$$

2.8. Find: (a) $\begin{bmatrix} 1 & 6 \\ -3 & 5 \end{bmatrix}\begin{bmatrix} 2 \\ -7 \end{bmatrix}$, (b) $\begin{bmatrix} 2 \\ -7 \end{bmatrix}\begin{bmatrix} 1 & 6 \\ -3 & 5 \end{bmatrix}$, (c) $[2, -7]\begin{bmatrix} 1 & 6 \\ -3 & 5 \end{bmatrix}$.

(a) The first factor is 2×2 and the second is 2×1, so the product is defined as a 2×1 matrix:

$$\begin{bmatrix} 1 & 6 \\ -3 & 5 \end{bmatrix}\begin{bmatrix} 2 \\ -7 \end{bmatrix} = \begin{bmatrix} 2-42 \\ -6-35 \end{bmatrix} = \begin{bmatrix} -40 \\ -41 \end{bmatrix}$$

(b) The product is not defined, because the first factor is 2×1 and the second factor is 2×2.

(c) The first factor is 1×2 and the second factor is 2×2, so the product is defined as a 1×2 (row) matrix:

$$[2, -7]\begin{bmatrix} 1 & 6 \\ -3 & 5 \end{bmatrix} = [2+21, \ 12-35] = [23, -23]$$

2.9. Clearly, $0A = 0$ and $A0 = 0$, where the 0's are *zero matrices* (with possibly different sizes). Find matrices A and B with no zero entries such that $AB = 0$.

Let $A = \begin{bmatrix} 1 & 2 \\ 2 & 4 \end{bmatrix}$ and $B = \begin{bmatrix} 6 & 2 \\ -3 & -1 \end{bmatrix}$. Then $AB = \begin{bmatrix} 0 & 0 \\ 0 & 0 \end{bmatrix}$.

2.10. Prove Theorem 2.2(i): $(AB)C = A(BC)$.

Let $A = [a_{ij}]$, $B = [b_{jk}]$, $C = [c_{kl}]$, and let $AB = S = [s_{ik}]$, $BC = T = [t_{jl}]$. Then

$$s_{ik} = \sum_{j=1}^{m} a_{ij}b_{jk} \quad \text{and} \quad t_{jl} = \sum_{k=1}^{n} b_{jk}c_{kl}$$

Multiplying $S = AB$ by C, the il-entry of $(AB)C$ is

$$s_{i1}c_{1l} + s_{i2}c_{2l} + \cdots + s_{in}c_{nl} = \sum_{k=1}^{n} s_{ik}c_{kl} = \sum_{k=1}^{n}\sum_{j=1}^{m} (a_{ij}b_{jk})c_{kl}$$

On the other hand, multiplying A by $T = BC$, the il-entry of $A(BC)$ is

$$a_{i1}t_{1l} + a_{i2}t_{2l} + \cdots + a_{im}t_{ml} = \sum_{j=1}^{m} a_{ij}t_{jl} = \sum_{j=1}^{m}\sum_{k=1}^{n} a_{ij}(b_{jk}c_{kl})$$

The above sums are equal; that is, corresponding elements in $(AB)C$ and $A(BC)$ are equal. Thus, $(AB)C = A(BC)$.

2.11. Prove Theorem 2.2(ii): $A(B+C) = AB + AC$.

Let $A = [a_{ij}]$, $B = [b_{jk}]$, $C = [c_{jk}]$, and let $D = B + C = [d_{jk}]$, $E = AB = [e_{ik}]$, $F = AC = [f_{ik}]$. Then

$$d_{jk} = b_{jk} + c_{jk}, \qquad e_{ik} = \sum_{j=1}^{m} a_{ij}b_{jk}, \qquad f_{ik} = \sum_{j=1}^{m} a_{ij}c_{jk}$$

Thus, the ik-entry of the matrix $AB + AC$ is

$$e_{ik} + f_{ik} = \sum_{j=1}^{m} a_{ij}b_{jk} + \sum_{j=1}^{m} a_{ij}c_{jk} = \sum_{j=1}^{m} a_{ij}(b_{jk} + c_{jk})$$

On the other hand, the ik-entry of the matrix $AD = A(B+C)$ is

$$a_{i1}d_{1k} + a_{i2}d_{2k} + \cdots + a_{im}d_{mk} = \sum_{j=1}^{m} a_{ij}d_{jk} = \sum_{j=1}^{m} a_{ij}(b_{jk} + c_{jk})$$

Thus, $A(B+C) = AB + AC$, because the corresponding elements are equal.

Transpose

2.12. Find the transpose of each matrix:

$$A = \begin{bmatrix} 1 & -2 & 3 \\ 7 & 8 & -9 \end{bmatrix}, \qquad B = \begin{bmatrix} 1 & 2 & 3 \\ 2 & 4 & 5 \\ 3 & 5 & 6 \end{bmatrix}, \qquad C = [1, -3, 5, -7], \qquad D = \begin{bmatrix} 2 \\ -4 \\ 6 \end{bmatrix}$$

Rewrite the rows of each matrix as columns to obtain the transpose of the matrix:

$$A^T = \begin{bmatrix} 1 & 7 \\ -2 & 8 \\ 3 & -9 \end{bmatrix}, \qquad B^T = \begin{bmatrix} 1 & 2 & 3 \\ 2 & 4 & 5 \\ 3 & 5 & 6 \end{bmatrix}, \qquad C^T = \begin{bmatrix} 1 \\ -3 \\ 5 \\ -7 \end{bmatrix}, \qquad D^T = [2, -4, 6]$$

(Note that $B^T = B$; such a matrix is said to be *symmetric*. Note also that the transpose of the row vector C is a column vector, and the transpose of the column vector D is a row vector.)

2.13. Prove Theorem 2.3(iv): $(AB)^T = B^T A^T$.

Let $A = [a_{ik}]$ and $B = [b_{kj}]$. Then the ij-entry of AB is

$$a_{i1}b_{1j} + a_{i2}b_{2j} + \cdots + a_{im}b_{mj}$$

This is the ji-entry (reverse order) of $(AB)^T$. Now column j of B becomes row j of B^T, and row i of A becomes column i of A^T. Thus, the ij-entry of $B^T A^T$ is

$$[b_{1j}, b_{2j}, \ldots, b_{mj}][a_{i1}, a_{i2}, \ldots, a_{im}]^T = b_{1j}a_{i1} + b_{2j}a_{i2} + \cdots + b_{mj}a_{im}$$

Thus, $(AB)^T = B^T A^T$ on because the corresponding entries are equal.

Square Matrices

2.14. Find the diagonal and trace of each matrix:

$$\text{(a)}\ A = \begin{bmatrix} 1 & 3 & 6 \\ 2 & -5 & 8 \\ 4 & -2 & 9 \end{bmatrix}, \qquad \text{(b)}\ B = \begin{bmatrix} 2 & 4 & 8 \\ 3 & -7 & 9 \\ -5 & 0 & 2 \end{bmatrix}, \qquad \text{(c)}\ C = \begin{bmatrix} 1 & 2 & -3 \\ 4 & -5 & 6 \end{bmatrix}.$$

(a) The diagonal of A consists of the elements from the upper left corner of A to the lower right corner of A or, in other words, the elements a_{11}, a_{22}, a_{33}. Thus, the diagonal of A consists of the numbers 1, −5, and 9. The trace of A is the sum of the diagonal elements. Thus,

$$\mathrm{tr}(A) = 1 - 5 + 9 = 5$$

(b) The diagonal of B consists of the numbers 2, −7, and 2. Hence,

$$\mathrm{tr}(B) = 2 - 7 + 2 = -3$$

(c) The diagonal and trace are only defined for square matrices.

2.15. Let $A = \begin{bmatrix} 1 & 2 \\ 4 & -3 \end{bmatrix}$, and let $f(x) = 2x^3 - 4x + 5$ and $g(x) = x^2 + 2x + 11$. Find

(a) A^2, (b) A^3, (c) $f(A)$, (d) $g(A)$.

(a) $A^2 = AA = \begin{bmatrix} 1 & 2 \\ 4 & -3 \end{bmatrix}\begin{bmatrix} 1 & 2 \\ 4 & -3 \end{bmatrix} = \begin{bmatrix} 1+8 & 2-6 \\ 4-12 & 8+9 \end{bmatrix} = \begin{bmatrix} 9 & -4 \\ -8 & 17 \end{bmatrix}$

(b) $A^3 = AA^2 = \begin{bmatrix} 1 & 2 \\ 4 & -3 \end{bmatrix}\begin{bmatrix} 9 & -4 \\ -8 & 17 \end{bmatrix} = \begin{bmatrix} 9-16 & -4+34 \\ 36+24 & -16-51 \end{bmatrix} = \begin{bmatrix} -7 & 30 \\ 60 & -67 \end{bmatrix}$

(c) First substitute A for x and $5I$ for the constant in $f(x)$, obtaining

$$f(A) = 2A^3 - 4A + 5I = 2\begin{bmatrix} -7 & 30 \\ 60 & -67 \end{bmatrix} - 4\begin{bmatrix} 1 & 2 \\ 4 & -3 \end{bmatrix} + 5\begin{bmatrix} 1 & 0 \\ 0 & 1 \end{bmatrix}$$

Now perform the scalar multiplication and then the matrix addition:

$$f(A) = \begin{bmatrix} -14 & 60 \\ 120 & -134 \end{bmatrix} + \begin{bmatrix} -4 & -8 \\ -16 & 12 \end{bmatrix} + \begin{bmatrix} 5 & 0 \\ 0 & 5 \end{bmatrix} = \begin{bmatrix} -13 & 52 \\ 104 & -117 \end{bmatrix}$$

(d) Substitute A for x and $11I$ for the constant in $g(x)$, and then calculate as follows:

$$g(A) = A^2 + 2A - 11I = \begin{bmatrix} 9 & -4 \\ -8 & 17 \end{bmatrix} + 2\begin{bmatrix} 1 & 2 \\ 4 & -3 \end{bmatrix} - 11\begin{bmatrix} 1 & 0 \\ 0 & 1 \end{bmatrix}$$

$$= \begin{bmatrix} 9 & -4 \\ -8 & 17 \end{bmatrix} + \begin{bmatrix} 2 & 4 \\ 8 & -6 \end{bmatrix} + \begin{bmatrix} -11 & 0 \\ 0 & -11 \end{bmatrix} = \begin{bmatrix} 0 & 0 \\ 0 & 0 \end{bmatrix}$$

Because $g(A)$ is the zero matrix, A is a root of the polynomial $g(x)$.

2.16. Let $A = \begin{bmatrix} 1 & 3 \\ 4 & -3 \end{bmatrix}$. (a) Find a nonzero column vector $u = \begin{bmatrix} x \\ y \end{bmatrix}$ such that $Au = 3u$.

(b) Describe all such vectors.

(a) First set up the matrix equation $Au = 3u$, and then write each side as a single matrix (column vector) as follows:

$$\begin{bmatrix} 1 & 3 \\ 4 & -3 \end{bmatrix}\begin{bmatrix} x \\ y \end{bmatrix} = 3\begin{bmatrix} x \\ y \end{bmatrix}, \quad \text{and then} \quad \begin{bmatrix} x+3y \\ 4x-3y \end{bmatrix} = \begin{bmatrix} 3x \\ 3y \end{bmatrix}$$

Set the corresponding elements equal to each other to obtain a system of equations:

$$\begin{aligned} x + 3y &= 3x \\ 4x - 3y &= 3y \end{aligned} \quad \text{or} \quad \begin{aligned} 2x - 3y &= 0 \\ 4x - 6y &= 0 \end{aligned} \quad \text{or} \quad 2x - 3y = 0$$

The system reduces to one nondegenerate linear equation in two unknowns, and so has an infinite number of solutions. To obtain a nonzero solution, let, say, $y = 2$; then $x = 3$. Thus, $u = (3, 2)^T$ is a desired nonzero vector.

(b) To find the general solution, set $y = a$, where a is a parameter. Substitute $y = a$ into $2x - 3y = 0$ to obtain $x = \frac{3}{2}a$. Thus, $u = (\frac{3}{2}a, a)^T$ represents all such solutions.

Invertible Matrices, Inverses

2.17. Show that $A = \begin{bmatrix} 1 & 0 & 2 \\ 2 & -1 & 3 \\ 4 & 1 & 8 \end{bmatrix}$ and $B = \begin{bmatrix} -11 & 2 & 2 \\ -4 & 0 & 1 \\ 6 & -1 & -1 \end{bmatrix}$ are inverses.

Compute the product AB, obtaining

$$AB = \begin{bmatrix} -11+0+12 & 2+0-2 & 2+0-2 \\ -22+4+18 & 4+0-3 & 4-1-3 \\ -44-4+48 & 8+0-8 & 8+1-8 \end{bmatrix} = \begin{bmatrix} 1 & 0 & 0 \\ 0 & 1 & 0 \\ 0 & 0 & 1 \end{bmatrix} = I$$

Because $AB = I$, we can conclude (Theorem 3.18) that $BA = I$. Accordingly, A and B are inverses.

2.18. Find the inverse, if possible, of each matrix:

(a) $A = \begin{bmatrix} 5 & 3 \\ 4 & 2 \end{bmatrix}$, (b) $B = \begin{bmatrix} 2 & -3 \\ 1 & 3 \end{bmatrix}$, (c) $\begin{bmatrix} -2 & 6 \\ 3 & -9 \end{bmatrix}$.

Use the formula for the inverse of a 2×2 matrix appearing in Section 2.9.

(a) First find $|A| = 5(2) - 3(4) = 10 - 12 = -2$. Next interchange the diagonal elements, take the negatives of the nondiagonal elements, and multiply by $1/|A|$:

$$A^{-1} = -\frac{1}{2}\begin{bmatrix} 2 & -3 \\ -4 & 5 \end{bmatrix} = \begin{bmatrix} -1 & \frac{3}{2} \\ 2 & -\frac{5}{2} \end{bmatrix}$$

(b) First find $|B| = 2(3) - (-3)(1) = 6 + 3 = 9$. Next interchange the diagonal elements, take the negatives of the nondiagonal elements, and multiply by $1/|B|$:

$$B^{-1} = \frac{1}{9}\begin{bmatrix} 3 & 3 \\ -1 & 2 \end{bmatrix} = \begin{bmatrix} \frac{1}{3} & \frac{1}{3} \\ -\frac{1}{9} & \frac{2}{9} \end{bmatrix}$$

(c) First find $|C| = -2(-9) - 6(3) = 18 - 18 = 0$. Because $|C| = 0$, C has no inverse.

2.19. Let $A = \begin{bmatrix} 1 & 1 & 1 \\ 0 & 1 & 2 \\ 1 & 2 & 4 \end{bmatrix}$. Find $A^{-1} = \begin{bmatrix} x_1 & x_2 & x_3 \\ y_1 & y_2 & y_3 \\ z_1 & z_2 & z_3 \end{bmatrix}$.

Multiplying A by A^{-1} and setting the nine entries equal to the nine entries of the identity matrix I yields the following three systems of three equations in three of the unknowns:

$$\begin{aligned} x_1 + y_1 + z_1 &= 1 \\ y_1 + 2z_1 &= 0 \\ x_1 + 2y_1 + 4z_1 &= 0 \end{aligned} \qquad \begin{aligned} x_2 + y_2 + z_2 &= 0 \\ y_2 + 2z_2 &= 1 \\ x_2 + 2y_2 + 4z_2 &= 0 \end{aligned} \qquad \begin{aligned} x_3 + y_3 + z_3 &= 0 \\ y_3 + 2z_3 &= 0 \\ x_3 + 2y_3 + 4z_3 &= 1 \end{aligned}$$

[Note that A is the coefficient matrix for all three systems.]
Solving the three systems for the nine unknowns yields

$$x_1 = 0, \quad y_1 = 2, \quad z_1 = -1; \qquad x_2 = -2, \quad y_2 = 3, \quad z_2 = -1; \qquad x_3 = 1, \quad y_3 = -2, \quad z_3 = 1$$

Thus,
$$A^{-1} = \begin{bmatrix} 0 & -2 & 1 \\ 2 & 3 & -2 \\ -1 & -1 & 1 \end{bmatrix}$$

(**Remark:** Chapter 3 gives an efficient way to solve the three systems.)

2.20. Let A and B be invertible matrices (with the same size). Show that AB is also invertible and $(AB)^{-1} = B^{-1}A^{-1}$. [Thus, by induction, $(A_1A_2 \ldots A_m)^{-1} = A_m^{-1} \ldots A_2^{-1}A_1^{-1}$.]

Using the associativity of matrix multiplication, we get

$$\begin{aligned} (AB)(B^{-1}A^{-1}) &= A(BB^{-1})A^{-1} = AIA^{-1} = AA^{-1} = I \\ (B^{-1}A^{-1})(AB) &= B^{-1}(A^{-1}A)B = A^{-1}IB = B^{-1}B = I \end{aligned}$$

Thus, $(AB)^{-1} = B^{-1}A^{-1}$.

Diagonal and Triangular Matrices

2.21. Write out the diagonal matrices $A = \text{diag}(4, -3, 7)$, $B = \text{diag}(2, -6)$, $C = \text{diag}(3, -8, 0, 5)$.

Put the given scalars on the diagonal and 0's elsewhere:

$$A = \begin{bmatrix} 4 & 0 & 0 \\ 0 & -3 & 0 \\ 0 & 0 & 7 \end{bmatrix}, \qquad B = \begin{bmatrix} 2 & 0 \\ 0 & -6 \end{bmatrix}, \qquad C = \begin{bmatrix} 3 & & & \\ & -8 & & \\ & & 0 & \\ & & & 5 \end{bmatrix}$$

2.22. Let $A = \text{diag}(2, 3, 5)$ and $B = \text{diag}(7, 0, -4)$. Find

(a) AB, A^2, B^2; (b) $f(A)$, where $f(x) = x^2 + 3x - 2$; (c) A^{-1} and B^{-1}.

(a) The product matrix AB is a diagonal matrix obtained by multiplying corresponding diagonal entries; hence,

$$AB = \text{diag}(2(7),\ 3(0),\ 5(-4)) = \text{diag}(14, 0, -20)$$

Thus, the squares A^2 and B^2 are obtained by squaring each diagonal entry; hence,

$$A^2 = \text{diag}(2^2, 3^2, 5^2) = \text{diag}(4, 9, 25) \qquad \text{and} \qquad B^2 = \text{diag}(49, 0, 16)$$

(b) $f(A)$ is a diagonal matrix obtained by evaluating $f(x)$ at each diagonal entry. We have

$$f(2) = 4 + 6 - 2 = 8, \qquad f(3) = 9 + 9 - 2 = 16, \qquad f(5) = 25 + 15 - 2 = 38$$

Thus, $f(A) = \text{diag}(8, 16, 38)$.

(c) The inverse of a diagonal matrix is a diagonal matrix obtained by taking the inverse (reciprocal) of each diagonal entry. Thus, $A^{-1} = \text{diag}(\frac{1}{2}, \frac{1}{3}, \frac{1}{5})$, but B has no inverse because there is a 0 on the diagonal.

2.23. Find a 2×2 matrix A such that A^2 is diagonal but not A.

Let $A = \begin{bmatrix} 1 & 2 \\ 3 & -1 \end{bmatrix}$. Then $A^2 = \begin{bmatrix} 7 & 0 \\ 0 & 7 \end{bmatrix}$, which is diagonal.

2.24. Find an upper triangular matrix A such that $A^3 = \begin{bmatrix} 8 & -57 \\ 0 & 27 \end{bmatrix}$.

Set $A = \begin{bmatrix} x & y \\ 0 & z \end{bmatrix}$. Then $x^3 = 8$, so $x = 2$; and $z^3 = 27$, so $z = 3$. Next calculate A^3 using $x = 2$ and $y = 3$:

$$A^2 = \begin{bmatrix} 2 & y \\ 0 & 3 \end{bmatrix}\begin{bmatrix} 2 & y \\ 0 & 3 \end{bmatrix} = \begin{bmatrix} 4 & 5y \\ 0 & 9 \end{bmatrix} \qquad \text{and} \qquad A^3 = \begin{bmatrix} 2 & y \\ 0 & 3 \end{bmatrix}\begin{bmatrix} 4 & 5y \\ 0 & 9 \end{bmatrix} = \begin{bmatrix} 8 & 19y \\ 0 & 27 \end{bmatrix}$$

Thus, $19y = -57$, or $y = -3$. Accordingly, $A = \begin{bmatrix} 2 & -3 \\ 0 & 3 \end{bmatrix}$.

2.25. Let $A = [a_{ij}]$ and $B = [b_{ij}]$ be upper triangular matrices. Prove that AB is upper triangular with diagonal $a_{11}b_{11}, a_{22}b_{22}, \ldots, a_{nn}b_{nn}$.

Let $AB = [c_{ij}]$. Then $c_{ij} = \sum_{k=1}^{n} a_{ik}b_{kj}$ and $c_{ii} = \sum_{k=1}^{n} a_{ik}b_{ki}$. Suppose $i > j$. Then, for any k, either $i > k$ or $k > j$, so that either $a_{ik} = 0$ or $b_{kj} = 0$. Thus, $c_{ij} = 0$, and AB is upper triangular. Suppose $i = j$. Then, for $k < i$, we have $a_{ik} = 0$; and, for $k > i$, we have $b_{ki} = 0$. Hence, $c_{ii} = a_{ii}b_{ii}$, as claimed. [This proves one part of Theorem 2.5(i); the statements for $A + B$ and kA are left as exercises.]

Special Real Matrices: Symmetric and Orthogonal

2.26. Determine whether or not each of the following matrices is *symmetric*—that is, $A^T = A$—or *skew-symmetric*—that is, $A^T = -A$:

$$\text{(a)}\quad A = \begin{bmatrix} 5 & -7 & 1 \\ -7 & 8 & 2 \\ 1 & 2 & -4 \end{bmatrix}, \qquad \text{(b)}\quad B = \begin{bmatrix} 0 & 4 & -3 \\ -4 & 0 & 5 \\ 3 & -5 & 0 \end{bmatrix}, \qquad \text{(c)}\quad C = \begin{bmatrix} 0 & 0 & 0 \\ 0 & 0 & 0 \end{bmatrix}$$

(a) By inspection, the symmetric elements (mirror images in the diagonal) are -7 and -7, 1 and 1, 2 and 2. Thus, A is symmetric, because symmetric elements are equal.

(b) By inspection, the diagonal elements are all 0, and the symmetric elements, 4 and -4, -3 and 3, and 5 and -5, are negatives of each other. Hence, B is skew-symmetric.

(c) Because C is not square, C is neither symmetric nor skew-symmetric.

2.27. Suppose $B = \begin{bmatrix} 4 & x+2 \\ 2x-3 & x+1 \end{bmatrix}$ is symmetric. Find x and B.

Set the symmetric elements $x+2$ and $2x-3$ equal to each other, obtaining $2x-3 = x+2$ or $x = 5$. Hence, $B = \begin{bmatrix} 4 & 7 \\ 7 & 6 \end{bmatrix}$.

2.28. Let A be an arbitrary 2×2 (real) orthogonal matrix.

(a) Prove: If (a, b) is the first row of A, then $a^2 + b^2 = 1$ and

$$A = \begin{bmatrix} a & b \\ -b & a \end{bmatrix} \quad \text{or} \quad A = \begin{bmatrix} a & b \\ b & -a \end{bmatrix}.$$

(b) Prove Theorem 2.7: For some real number θ,

$$A = \begin{bmatrix} \cos\theta & \sin\theta \\ -\sin\theta & \cos\theta \end{bmatrix} \quad \text{or} \quad A = \begin{bmatrix} \cos\theta & \sin\theta \\ \sin\theta & -\cos\theta \end{bmatrix}$$

(a) Suppose (x, y) is the second row of A. Because the rows of A form an orthonormal set, we get

$$a^2 + b^2 = 1, \qquad x^2 + y^2 = 1, \qquad ax + by = 0$$

Similarly, the columns form an orthogonal set, so

$$a^2 + x^2 = 1, \qquad b^2 + y^2 = 1, \qquad ab + xy = 0$$

Therefore, $x^2 = 1 - a^2 = b^2$, whence $x = \pm b$.

Case (i): $x = b$. Then $b(a + y) = 0$, so $y = -a$.

Case (ii): $x = -b$. Then $b(y - a) = 0$, so $y = a$.

This means, as claimed,

$$A = \begin{bmatrix} a & b \\ -b & a \end{bmatrix} \quad \text{or} \quad A = \begin{bmatrix} a & b \\ b & -a \end{bmatrix}$$

(b) Because $a^2 + b^2 = 1$, we have $-1 \le a \le 1$. Let $a = \cos\theta$. Then $b^2 = 1 - \cos^2\theta$, so $b = \sin\theta$. This proves the theorem.

2.29. Find a 2×2 orthogonal matrix A whose first row is a (positive) multiple of $(3, 4)$.

Normalize $(3, 4)$ to get $(\frac{3}{5}, \frac{4}{5})$. Then, by Problem 2.28,

$$A = \begin{bmatrix} \frac{3}{5} & \frac{4}{5} \\ -\frac{4}{5} & \frac{3}{5} \end{bmatrix} \quad \text{or} \quad A = \begin{bmatrix} \frac{3}{5} & \frac{4}{5} \\ \frac{4}{5} & -\frac{3}{5} \end{bmatrix}.$$

2.30. Find a 3×3 orthogonal matrix P whose first two rows are multiples of $u_1 = (1, 1, 1)$ and $u_2 = (0, -1, 1)$, respectively. (Note that, as required, u_1 and u_2 are orthogonal.)

First find a nonzero vector u_3 orthogonal to u_1 and u_2; say (cross product) $u_3 = u_1 \times u_2 = (2, -1, -1)$. Let A be the matrix whose rows are u_1, u_2, u_3; and let P be the matrix obtained from A by normalizing the rows of A. Thus,

$$A = \begin{bmatrix} 1 & 1 & 1 \\ 0 & -1 & 1 \\ 2 & -1 & -1 \end{bmatrix} \quad \text{and} \quad P = \begin{bmatrix} 1/\sqrt{3} & 1/\sqrt{3} & 1/\sqrt{3} \\ 0 & -1/\sqrt{2} & 1/\sqrt{2} \\ 2/\sqrt{6} & -1/\sqrt{6} & -1/\sqrt{6} \end{bmatrix}$$

Complex Matrices: Hermitian and Unitary Matrices

2.31. Find A^H where (a) $A = \begin{bmatrix} 3-5i & 2+4i \\ 6+7i & 1+8i \end{bmatrix}$, (b) $A = \begin{bmatrix} 2-3i & 5+8i \\ -4 & 3-7i \\ -6-i & 5i \end{bmatrix}$

Recall that $A^H = \bar{A}^T$, the conjugate tranpose of A. Thus,

(a) $A^H = \begin{bmatrix} 3+5i & 6-7i \\ 2-4i & 1-8i \end{bmatrix}$, (b) $A^H = \begin{bmatrix} 2+3i & -4 & -6+i \\ 5-8i & 3+7i & -5i \end{bmatrix}$

2.32. Show that $A = \begin{bmatrix} \frac{1}{3} - \frac{2}{3}i & \frac{2}{3}i \\ -\frac{2}{3}i & -\frac{1}{3} - \frac{2}{3}i \end{bmatrix}$ is unitary.

The rows of A form an orthonormal set:

$$\left(\tfrac{1}{3} - \tfrac{2}{3}i, \tfrac{2}{3}i\right) \cdot \left(\tfrac{1}{3} - \tfrac{2}{3}i, \tfrac{2}{3}i\right) = \left(\tfrac{1}{9} + \tfrac{4}{9}\right) + \tfrac{4}{9} = 1$$

$$\left(\tfrac{1}{3} - \tfrac{2}{3}i, \tfrac{2}{3}i\right) \cdot \left(-\tfrac{2}{3}i, -\tfrac{1}{3} - \tfrac{2}{3}i\right) = \left(\tfrac{2}{9}i + \tfrac{4}{9}\right) + \left(-\tfrac{2}{9}i - \tfrac{4}{9}\right) = 0$$

$$\left(-\tfrac{2}{3}i, -\tfrac{1}{3} - \tfrac{2}{3}i\right) \cdot \left(-\tfrac{2}{3}i, -\tfrac{1}{3} - \tfrac{2}{3}i\right) = \tfrac{4}{9} + \left(\tfrac{1}{9} + \tfrac{4}{9}\right) = 1$$

Thus, A is unitary.

2.33. Prove the complex analogue of Theorem 2.6: Let A be a complex matrix. Then the following are equivalent: (i) A is unitary. (ii) The rows of A form an orthonormal set. (iii) The columns of A form an orthonormal set.

(The proof is almost identical to the proof on page 37 for the case when A is a 3×3 real matrix.)

First recall that the vectors $u_1, u_2, \ldots, u_n$ in $\mathbf{C}^n$ form an orthonormal set if they are unit vectors and are orthogonal to each other, where the dot product in $\mathbf{C}^n$ is defined by

$$(a_1, a_2, \ldots, a_n) \cdot (b_1, b_2, \ldots, b_n) = a_1\bar{b}_1 + a_2\bar{b}_2 + \cdots + a_n\bar{b}_n$$

Suppose A is unitary, and $R_1, R_2, \ldots, R_n$ are its rows. Then $\bar{R}_1^T, \bar{R}_2^T, \ldots, \bar{R}_n^T$ are the columns of A^H. Let $AA^H = [c_{ij}]$. By matrix multiplication, $c_{ij} = R_i\bar{R}_j^T = R_i \cdot R_j$. Because A is unitary, we have $AA^H = I$. Multiplying A by A^H and setting each entry c_{ij} equal to the corresponding entry in I yields the following n^2 equations:

$$R_1 \cdot R_1 = 1, \quad R_2 \cdot R_2 = 1, \quad \ldots, \quad R_n \cdot R_n = 1, \qquad \text{and} \qquad R_i \cdot R_j = 0, \quad \text{for } i \neq j$$

Thus, the rows of A are unit vectors and are orthogonal to each other; hence, they form an orthonormal set of vectors. The condition $A^T A = I$ similarly shows that the columns of A also form an orthonormal set of vectors. Furthermore, because each step is reversible, the converse is true. This proves the theorem.

Block Matrices

2.34. Consider the following block matrices (which are partitions of the same matrix):

(a) $\left[\begin{array}{cc|cc|c} 1 & -2 & 0 & 1 & 3 \\ 2 & 3 & 5 & 7 & -2 \\ \hline 3 & 1 & 4 & 5 & 9 \end{array}\right]$, (b) $\left[\begin{array}{ccc|cc} 1 & -2 & 0 & 1 & 3 \\ \hline 2 & 3 & 5 & 7 & -2 \\ \hline 3 & 1 & 4 & 5 & 9 \end{array}\right]$

Find the size of each block matrix and also the size of each block.

(a) The block matrix has two rows of matrices and three columns of matrices; hence, its size is 2×3. The block sizes are 2×2, 2×2, and 2×1 for the first row; and 1×2, 1×2, and 1×1 for the second row.

(b) The size of the block matrix is 3×2; and the block sizes are 1×3 and 1×2 for each of the three rows.

2.35. Compute AB using block multiplication, where

$$A = \left[\begin{array}{cc|c} 1 & 2 & 1 \\ 3 & 4 & 0 \\ \hline 0 & 0 & 2 \end{array}\right] \quad \text{and} \quad B = \left[\begin{array}{ccc|c} 1 & 2 & 3 & 1 \\ 4 & 5 & 6 & 1 \\ \hline 0 & 0 & 0 & 1 \end{array}\right]$$

Here $A = \begin{bmatrix} E & F \\ 0_{1\times 2} & G \end{bmatrix}$ and $B = \begin{bmatrix} R & S \\ 0_{1\times 3} & T \end{bmatrix}$, where E, F, G, R, S, T are the given blocks, and $0_{1\times 2}$ and $0_{1\times 3}$ are zero matrices of the indicated sites. Hence,

$$AB = \begin{bmatrix} ER & ES + FT \\ 0_{1\times 3} & GT \end{bmatrix} = \begin{bmatrix} \begin{bmatrix} 9 & 12 & 15 \\ 19 & 26 & 33 \end{bmatrix} & \begin{bmatrix} 3 \\ 7 \end{bmatrix} + \begin{bmatrix} 1 \\ 0 \end{bmatrix} \\ [0 \quad 0 \quad 0] & 2 \end{bmatrix} = \begin{bmatrix} 9 & 12 & 15 & 4 \\ 19 & 26 & 33 & 7 \\ 0 & 0 & 0 & 2 \end{bmatrix}$$

2.36. Let $M = \text{diag}(A, B, C)$, where $A = \begin{bmatrix} 1 & 2 \\ 3 & 4 \end{bmatrix}$, $B = [5]$, $C = \begin{bmatrix} 1 & 3 \\ 5 & 7 \end{bmatrix}$. Find M^2.

Because M is block diagonal, square each block:

$$A^2 = \begin{bmatrix} 7 & 10 \\ 15 & 22 \end{bmatrix}, \qquad B^2 = [25], \qquad C^2 = \begin{bmatrix} 16 & 24 \\ 40 & 64 \end{bmatrix},$$

so

$$M^2 = \left[\begin{array}{cc|c|cc} 7 & 10 & & & \\ 15 & 22 & & & \\ \hline & & 25 & & \\ \hline & & & 16 & 24 \\ & & & 40 & 64 \end{array}\right]$$

Miscellaneous Problem

2.37. Let $f(x)$ and $g(x)$ be polynomials and let A be a square matrix. Prove

(a) $(f+g)(A) = f(A) + g(A)$,

(b) $(f \cdot g)(A) = f(A)g(A)$,

(c) $f(A)g(A) = g(A)f(A)$.

Suppose $f(x) = \sum_{i=1}^{r} a_i x^i$ and $g(x) = \sum_{j=1}^{s} b_j x^j$.

(a) We can assume $r = s = n$ by adding powers of x with 0 as their coefficients. Then

$$f(x) + g(x) = \sum_{i=1}^{n} (a_i + b_i)x^i$$

Hence,

$$(f+g)(A) = \sum_{i=1}^{n} (a_i + b_i)A^i = \sum_{i=1}^{n} a_i A^i + \sum_{i=1}^{n} b_i A^i = f(A) + g(A)$$

(b) We have $f(x)g(x) = \sum_{i,j} a_i b_j x^{i+j}$. Then

$$f(A)g(A) = \left(\sum_i a_i A^i\right)\left(\sum_j b_j A^j\right) = \sum_{i,j} a_i b_j A^{i+j} = (fg)(A)$$

(c) Using $f(x)g(x) = g(x)f(x)$, we have

$$f(A)g(A) = (fg)(A) = (gf)(A) = g(A)f(A)$$

SUPPLEMENTARY PROBLEMS

Algebra of Matrices

Problems 2.38–2.41 refer to the following matrices:

$$A = \begin{bmatrix} 1 & 2 \\ 3 & -4 \end{bmatrix}, \quad B = \begin{bmatrix} 5 & 0 \\ -6 & 7 \end{bmatrix}, \quad C = \begin{bmatrix} 1 & -3 & 4 \\ 2 & 6 & -5 \end{bmatrix}, \quad D = \begin{bmatrix} 3 & 7 & -1 \\ 4 & -8 & 9 \end{bmatrix}$$

2.38. Find (a) $5A - 2B$, (b) $2A + 3B$, (c) $2C - 3D$.

2.39. Find (a) AB and $(AB)C$, (b) BC and $A(BC)$. [Note that $(AB)C = A(BC)$.]

2.40. Find (a) A^2 and A^3, (b) AD and BD, (c) CD.

2.41. Find (a) A^T, (b) B^T, (c) $(AB)^T$, (d) $A^T B^T$. [Note that $A^T B^T \neq (AB)^T$.]

Problems 2.42 and 2.43 refer to the following matrices:

$$A = \begin{bmatrix} 1 & -1 & 2 \\ 0 & 3 & 4 \end{bmatrix}, \quad B = \begin{bmatrix} 4 & 0 & -3 \\ -1 & -2 & 3 \end{bmatrix}, \quad C = \begin{bmatrix} 2 & -3 & 0 & 1 \\ 5 & -1 & -4 & 2 \\ -1 & 0 & 0 & 3 \end{bmatrix}, \quad D = \begin{bmatrix} 2 \\ -1 \\ 3 \end{bmatrix}.$$

2.42. Find (a) $3A - 4B$, (b) AC, (c) BC, (d) AD, (e) BD, (*f*) CD.

2.43. Find (a) A^T, (b) $A^T B$, (c) $A^T C$.

2.44. Let $A = \begin{bmatrix} 1 & 2 \\ 3 & 6 \end{bmatrix}$. Find a 2×3 matrix B with distinct nonzero entries such that $AB = 0$.

2.45 Let $e_1 = [1, 0, 0]$, $e_2 = [0, 1, 0]$, $e_3 = [0, 0, 1]$, and $A = \begin{bmatrix} a_1 & a_2 & a_3 & a_4 \\ b_1 & b_2 & b_3 & b_4 \\ c_1 & c_2 & c_3 & c_4 \end{bmatrix}$. Find $e_1 A$, $e_2 A$, $e_3 A$.

2.46. Let $e_i = [0, \ldots, 0, 1, 0, \ldots, 0]$, where 1 is the ith entry. Show

(a) $e_i A = A_i$, ith row of A.

(b) $Be_j^T = B^j$, jth column of B.

(c) If $e_i A = e_i B$, for each i, then $A = B$.

(d) If $Ae_j^T = Be_j^T$, for each j, then $A = B$.

2.47. Prove Theorem 2.2(iii) and (iv): (iii) $(B + C)A = BA + CA$, (iv) $k(AB) = (kA)B = A(kB)$.

2.48. Prove Theorem 2.3: (i) $(A + B)^T = A^T + B^T$, (ii) $(A^T)^T = A$, (iii) $(kA)^T = kA^T$.

2.49. Show (a) If A has a zero row, then AB has a zero row. (b) If B has a zero column, then AB has a zero column.

Square Matrices, Inverses

2.50. Find the diagonal and trace of each of the following matrices:

(a) $A = \begin{bmatrix} 2 & -5 & 8 \\ 3 & -6 & -7 \\ 4 & 0 & -1 \end{bmatrix}$, (b) $B = \begin{bmatrix} 1 & 3 & -4 \\ 6 & 1 & 7 \\ 2 & -5 & -1 \end{bmatrix}$, (c) $C = \begin{bmatrix} 4 & 3 & -6 \\ 2 & -5 & 0 \end{bmatrix}$

Problems 2.51–2.53 refer to $A = \begin{bmatrix} 2 & -5 \\ 3 & 1 \end{bmatrix}$, $B = \begin{bmatrix} 4 & -2 \\ 1 & -6 \end{bmatrix}$, $C = \begin{bmatrix} 6 & -4 \\ 3 & -2 \end{bmatrix}$.

2.51. Find (a) A^2 and A^3, (b) $f(A)$ and $g(A)$, where

$$f(x) = x^3 - 2x^2 - 5, \qquad g(x) = x^2 - 3x + 17.$$

2.52. Find (a) B^2 and B^3, (b) $f(B)$ and $g(B)$, where

$$f(x) = x^2 + 2x - 22, \qquad g(x) = x^2 - 3x - 6.$$

2.53. Find a nonzero column vector u such that $Cu = 4u$.

2.54. Find the inverse of each of the following matrices (if it exists):

$$A = \begin{bmatrix} 7 & 4 \\ 5 & 3 \end{bmatrix}, \qquad B = \begin{bmatrix} 2 & 3 \\ 4 & 5 \end{bmatrix}, \qquad C = \begin{bmatrix} 4 & -6 \\ -2 & 3 \end{bmatrix}, \qquad D = \begin{bmatrix} 5 & -2 \\ 6 & -3 \end{bmatrix}$$

2.55. Find the inverses of $A = \begin{bmatrix} 1 & 1 & 2 \\ 1 & 2 & 5 \\ 1 & 3 & 7 \end{bmatrix}$ and $B = \begin{bmatrix} 1 & -1 & 1 \\ 0 & 1 & -1 \\ 1 & 3 & -2 \end{bmatrix}$. [*Hint:* See Problem 2.19.]

2.56. Suppose A is invertible. Show that if $AB = AC$, then $B = C$. Give an example of a nonzero matrix A such that $AB = AC$ but $B \neq C$.

2.57. Find 2×2 invertible matrices A and B such that $A + B \neq 0$ and $A + B$ is not invertible.

2.58. Show (a) A is invertible if and only if A^T is invertible. (b) The operations of inversion and transpose commute; that is, $(A^T)^{-1} = (A^{-1})^T$. (c) If A has a zero row or zero column, then A is not invertible.

Diagonal and triangular matrices

2.59. Let $A = \text{diag}(1, 2, -3)$ and $B = \text{diag}(2, -5, 0)$. Find

(a) AB, A^2, B^2; (b) $f(A)$, where $f(x) = x^2 + 4x - 3$; (c) A^{-1} and B^{-1}.

2.60. Let $A = \begin{bmatrix} 1 & 2 \\ 0 & 1 \end{bmatrix}$ and $B = \begin{bmatrix} 1 & 1 & 0 \\ 0 & 1 & 1 \\ 0 & 0 & 1 \end{bmatrix}$. (a) Find A^n. (b) Find B^n.

2.61. Find all real triangular matrices A such that $A^2 = B$, where (a) $B = \begin{bmatrix} 4 & 21 \\ 0 & 25 \end{bmatrix}$, (b) $B = \begin{bmatrix} 1 & 4 \\ 0 & -9 \end{bmatrix}$.

2.62. Let $A = \begin{bmatrix} 5 & 2 \\ 0 & k \end{bmatrix}$. Find all numbers k for which A is a root of the polynomial:

(a) $f(x) = x^2 - 7x + 10$, (b) $g(x) = x^2 - 25$, (c) $h(x) = x^2 - 4$.

2.63. Let $B = \begin{bmatrix} 1 & 0 \\ 26 & 27 \end{bmatrix}$. Find a matrix A such that $A^3 = B$.

2.64. Let $B = \begin{bmatrix} 1 & 8 & 5 \\ 0 & 9 & 5 \\ 0 & 0 & 4 \end{bmatrix}$. Find a triangular matrix A with positive diagonal entries such that $A^2 = B$.

2.65. Using only the elements 0 and 1, find the number of 3×3 matrices that are (a) diagonal, (b) upper triangular, (c) nonsingular and upper triangular. Generalize to $n \times n$ matrices.

2.66. Let $D_k = kI$, the scalar matrix belonging to the scalar k. Show

(a) $D_k A = kA$, (b) $BD_k = kB$, (c) $D_k + D_{k'} = D_{k+k'}$, (d) $D_k D_{k'} = D_{kk'}$

2.67. Suppose $AB = C$, where A and C are upper triangular.

(a) Find 2×2 nonzero matrices A, B, C, where B is not upper triangular.

(b) Suppose A is also invertible. Show that B must also be upper triangular.

Special Types of Real Matrices

2.68. Find x, y, z such that A is symmetric, where

(a) $A = \begin{bmatrix} 2 & x & 3 \\ 4 & 5 & y \\ z & 1 & 7 \end{bmatrix}$, (b) $A = \begin{bmatrix} 7 & -6 & 2x \\ y & z & -2 \\ x & -2 & 5 \end{bmatrix}$.

2.69. Suppose A is a square matrix. Show (a) $A + A^T$ is symmetric, (b) $A - A^T$ is skew-symmetric, (c) $A = B + C$, where B is symmetric and C is skew-symmetric.

2.70. Write $A = \begin{bmatrix} 4 & 5 \\ 1 & 3 \end{bmatrix}$ as the sum of a symmetric matrix B and a skew-symmetric matrix C.

2.71. Suppose A and B are symmetric. Show that the following are also symmetric:
(a) $A + B$; (b) kA, for any scalar k; (c) A^2;
(d) A^n, for $n > 0$; (e) $f(A)$, for any polynomial $f(x)$.

2.72. Find a 2×2 orthogonal matrix P whose first row is a multiple of
(a) $(3, -4)$, (b) $(1, 2)$.

2.73. Find a 3×3 orthogonal matrix P whose first two rows are multiples of
(a) $(1, 2, 3)$ and $(0, -2, 3)$, (b) $(1, 3, 1)$ and $(1, 0, -1)$.

2.74. Suppose A and B are orthogonal matrices. Show that A^T, A^{-1}, AB are also orthogonal.

2.75. Which of the following matrices are normal? $A = \begin{bmatrix} 3 & -4 \\ 4 & 3 \end{bmatrix}$, $B = \begin{bmatrix} 1 & -2 \\ 2 & 3 \end{bmatrix}$, $C = \begin{bmatrix} 1 & 1 & 1 \\ 0 & 1 & 1 \\ 0 & 0 & 1 \end{bmatrix}$.

Complex Matrices

2.76. Find real numbers x, y, z such that A is Hermitian, where $A = \begin{bmatrix} 3 & x + 2i & yi \\ 3 - 2i & 0 & 1 + zi \\ yi & 1 - xi & -1 \end{bmatrix}$.

2.77. Suppose A is a complex matrix. Show that AA^H and A^HA are Hermitian.

2.78. Let A be a square matrix. Show that (a) $A + A^H$ is Hermitian, (b) $A - A^H$ is skew-Hermitian, (c) $A = B + C$, where B is Hermitian and C is skew-Hermitian.

2.79. Determine which of the following matrices are unitary:

$$A = \begin{bmatrix} i/2 & -\sqrt{3}/2 \\ \sqrt{3}/2 & -i/2 \end{bmatrix}, \qquad B = \frac{1}{2}\begin{bmatrix} 1+i & 1-i \\ 1-i & 1+i \end{bmatrix}, \qquad C = \frac{1}{2}\begin{bmatrix} 1 & -i & -1+i \\ i & 1 & 1+i \\ 1+i & -1+i & 0 \end{bmatrix}$$

2.80. Suppose A and B are unitary. Show that A^H, A^{-1}, AB are unitary.

2.81. Determine which of the following matrices are normal: $A = \begin{bmatrix} 3+4i & 1 \\ i & 2+3i \end{bmatrix}$ and $B = \begin{bmatrix} 1 & 0 \\ 1-i & i \end{bmatrix}$.

Block Matrices

2.82. Let $U = \left[\begin{array}{cc|ccc} 1 & 2 & 0 & 0 & 0 \\ 3 & 4 & 0 & 0 & 0 \\ \hline 0 & 0 & 5 & 1 & 2 \\ 0 & 0 & 3 & 4 & 1 \end{array}\right]$ and $V = \left[\begin{array}{cc|cc} 3 & -2 & 0 & 0 \\ 2 & 4 & 0 & 0 \\ \hline 0 & 0 & 1 & 2 \\ 0 & 0 & 2 & -3 \\ 0 & 0 & -4 & 1 \end{array}\right]$.

(a) Find UV using block multiplication. (b) Are U and V block diagonal matrices?
(c) Is UV block diagonal?

2.83. Partition each of the following matrices so that it becomes a square block matrix with as many diagonal blocks as possible:

$$A = \begin{bmatrix} 1 & 0 & 0 \\ 0 & 0 & 2 \\ 0 & 0 & 3 \end{bmatrix}, \qquad B = \begin{bmatrix} 1 & 2 & 0 & 0 & 0 \\ 3 & 0 & 0 & 0 & 0 \\ 0 & 0 & 4 & 0 & 0 \\ 0 & 0 & 5 & 0 & 0 \\ 0 & 0 & 0 & 0 & 6 \end{bmatrix}, \qquad C = \begin{bmatrix} 0 & 1 & 0 \\ 0 & 0 & 0 \\ 2 & 0 & 0 \end{bmatrix}$$

2.84. Find M^2 and M^3 for (a) $M = \left[\begin{array}{c|cc|c} 2 & 0 & 0 & 0 \\ \hline 0 & 1 & 4 & 0 \\ 0 & 2 & 1 & 0 \\ \hline 0 & 0 & 0 & 3 \end{array}\right]$, (b) $M = \left[\begin{array}{cc|cc} 1 & 1 & 0 & 0 \\ 2 & 3 & 0 & 0 \\ \hline 0 & 0 & 1 & 2 \\ 0 & 0 & 4 & 5 \end{array}\right]$.

2.85. For each matrix M in Problem 2.84, find $f(M)$ where $f(x) = x^2 + 4x - 5$.

2.86. Suppose $U = [U_{ik}]$ and $V = [V_{kj}]$ are block matrices for which UV is defined and the number of columns of each block U_{ik} is equal to the number of rows of each block V_{kj}. Show that $UV = [W_{ij}]$, where $W_{ij} = \sum_k U_{ik} V_{kj}$.

2.87. Suppose M and N are block diagonal matrices where corresponding blocks have the same size, say $M = \text{diag}(A_i)$ and $N = \text{diag}(B_i)$. Show

(i) $M + N = \text{diag}(A_i + B_i)$, (iii) $MN = \text{diag}(A_i B_i)$,
(ii) $kM = \text{diag}(kA_i)$, (iv) $f(M) = \text{diag}(f(A_i))$ for any polynomial $f(x)$.

ANSWERS TO SUPPLEMENTARY PROBLEMS

Notation: $A = [R_1;\ R_2;\ \ldots]$ denotes a matrix A with rows $R_1, R_2, \ldots$

2.38. (a) $[-5, 10;\ 27, -34]$, (b) $[17, 4;\ -12, 13]$, (c) $[-7, -27, 11;\ -8, 36, -37]$

2.39. (a) $[-7, 14;\ 39, -28]$, $[21, 105, -98;\ -17, -285, 296]$
(b) $[5, -15, 20;\ 8, 60, -59]$, $[21, 105, -98;\ -17, -285, 296]$

2.40. (a) $[7, -6;\ -9, 22]$, $[-11, 38;\ 57, -106]$;
(b) $[11, -9, 17;\ -7, 53, -39]$, $[15, 35, -5;\ 10, -98, 69]$; (c) not defined

2.41. (a) $[1, 3;\ 2, -4]$, (b) $[5, -6;\ 0, 7]$, (c) $[-7, 39;\ 14, -28]$, (d) $[5, 15;\ 10, -40]$

2.42. (a) $[-13, -3, 18;\ 4, 17, 0]$, (b) $[-5, -2, 4, 5;\ 11, -3, -12, 18]$,
(c) $[11, -12, 0, -5;\ -15, 5, 8, 4]$, (d) $[9;\ 9]$, (e) $[-1;\ 9]$, (f) not defined

2.43. (a) $[1, 0;\ \ -1, 3;\ \ 2, 4]$, (b) $[4, 0, -3;\ \ -7, -6, 12;\ \ 4, -8, 6]$, (c) not defined

2.44. $[2, 4, 6;\ \ -1, -2, -3]$

2.45. $[a_1, a_2, a_3, a_4]$, $[b_1, b_2, b_3, b_4]$, $[c_1, c_2, c_3, c_4]$

2.50. (a) $2, -6, -1, \operatorname{tr}(A) = -5$, (b) $1, 1, -1, \operatorname{tr}(B) = 1$, (c) not defined

2.51. (a) $[-11, -15;\ \ 9, -14]$, $[-67, 40;\ \ -24, -59]$, (b) $[-50, 70;\ \ -42, -36]$, $g(A) = 0$

2.52. (a) $[14, 4;\ \ -2, 34]$, $[60, -52;\ \ 26, -200]$, (b) $f(B) = 0$, $[-4, 10;\ \ -5, 46]$

2.53. $u = [2a, a]^T$

2.54. $[3, -4;\ \ -5, 7]$, $[-\frac{5}{2}, \frac{3}{2};\ \ 2, -1]$, not defined, $[1, -\frac{2}{3};\ \ 2, -\frac{5}{3}]$

2.55. $[1, 1, -1;\ \ 2, -5, 3;\ \ -1, 2, -1]$, $[1, 1, 0;\ \ -1, -3, 1;\ \ -1, -4, 1]$

2.56. $A = [1, 2;\ \ 1, 2]$, $B = [0, 0;\ \ 1, 1]$, $C = [2, 2;\ \ 0, 0]$

2.57. $A = [1, 2;\ \ 0, 3]$; $B = [4, 3;\ \ 3, 0]$

2.58. (c) *Hint:* Use Problem 2.48

2.59. (a) $AB = \operatorname{diag}(2, -10, 0)$, $A^2 = \operatorname{diag}(1, 4, 9)$, $B^2 = \operatorname{diag}(4, 25, 0)$;
(b) $f(A) = \operatorname{diag}(2, 9, -6)$; (c) $A^{-1} = \operatorname{diag}(1, \frac{1}{2}, -\frac{1}{3})$, B^{-1} does not exist

2.60. (a) $[1, 2n;\ \ 0, 1]$, (b) $[1, n, \frac{1}{2}n(n-1);\ \ 0, 1, n;\ \ 0, 0, 1]$

2.61. (a) $[2, 3;\ \ 0, 5]$, $[-2, -3;\ \ 0, -5]$, $[2, -7;\ \ 0, -5]$, $[-2, 7;\ \ 0, 5]$, (b) none

2.62. (a) $k = 2$, (b) $k = -5$, (c) none

2.63. $[1, 0;\ \ 2, 3]$

2.64. $[1, 2, 1;\ \ 0, 3, 1;\ \ 0, 0, 2]$

2.65. All entries below the diagonal must be 0 to be upper triangular, and all diagonal entries must be 1 to be nonsingular.
(a) $8\ (2^n)$, (b) $2^6\ (2^{n(n+1)/2})$, (c) $2^3\ (2^{n(n-1)/2})$

2.67. (a) $A = [1, 1;\ \ 0, 0]$, $B = [1, 2;\ \ 3, 4]$, $C = [4, 6;\ \ 0, 0]$

2.68. (a) $x = 4, y = 1, z = 3$; (b) $x = 0, y = -6$, z any real number

2.69. (c) *Hint:* Let $B = \frac{1}{2}(A + A^T)$ and $C = \frac{1}{2}(A - A^T)$.

2.70. $B = [4, 3;\ \ 3, 3]$, $C = [0, 2;\ \ -2, 0]$

2.72. (a) $[\frac{3}{5}, -\frac{4}{5};\ \ \frac{4}{5}, \frac{3}{5}]$, (b) $[1/\sqrt{5}, 2/\sqrt{5};\ \ 2/\sqrt{5}, -1/\sqrt{5}]$

2.73. (a) $[1/\sqrt{14}, 2/\sqrt{14}, 3/\sqrt{14};\ \ 0, -2/\sqrt{13}, 3/\sqrt{13};\ \ 12/\sqrt{157}, -3/\sqrt{157}, -2/\sqrt{157}]$
(b) $[1/\sqrt{11}, 3/\sqrt{11}, 1/\sqrt{11};\ \ 1/\sqrt{2}, 0, -1/\sqrt{2};\ \ 3/\sqrt{22}, -2/\sqrt{22}, 3/\sqrt{22}]$

2.75. A, C

2.76. $x = 3, y = 0, z = 3$

2.78. (c) *Hint:* Let $B = \frac{1}{2}(A + A^H)$ and $C = \frac{1}{2}(A - A^H)$.

2.79. A, B, C

2.81. A

2.82. (a) $UV = \text{diag}([7,6;\ \ 17,10];\ [-1,9;\ \ 7,-5])$; (b) no; (c) yes

2.83. A: line between first and second rows (columns);
B: line between second and third rows (columns) and between fourth and fifth rows (columns);
C: C itself—no further partitioning of C is possible.

2.84. (a) $M^2 = \text{diag}([4],\ [9,8;\ \ 4,9],\ \ [9])$,
$M^3 = \text{diag}([8],\ [25,44;\ \ 22,25],\ [27])$
(b) $M^2 = \text{diag}([3,4;\ \ 8,11],\ \ [9,12;\ \ 24,33])$
$M^3 = \text{diag}([11,15;\ \ 30,41],\ \ [57,78;\ \ 156,213])$

2.85. (a) $\text{diag}([7],\ \ [8,24;\ \ 12,8],\ [16])$, (b) $\text{diag}([2,8;\ \ 16,181],\ \ [8,20;\ \ 40,48])$

CHAPTER 3

Systems of Linear Equations

3.1 Introduction

Systems of linear equations play an important and motivating role in the subject of linear algebra. In fact, many problems in linear algebra reduce to finding the solution of a system of linear equations. Thus, the techniques introduced in this chapter will be applicable to abstract ideas introduced later. On the other hand, some of the abstract results will give us new insights into the structure and properties of systems of linear equations.

All our systems of linear equations involve scalars as both coefficients and constants, and such scalars may come from any number field K. There is almost no loss in generality if the reader assumes that all our scalars are real numbers—that is, that they come from the real field $\mathbf{R}$.

3.2 Basic Definitions, Solutions

This section gives basic definitions connected with the solutions of systems of linear equations. The actual algorithms for finding such solutions will be treated later.

Linear Equation and Solutions

A *linear equation* in unknowns $x_1, x_2, \ldots, x_n$ is an equation that can be put in the *standard form*

$$a_1x_1 + a_2x_2 + \cdots + a_nx_n = b \tag{3.1}$$

where $a_1, a_2, \ldots, a_n$, and b are constants. The constant a_k is called the *coefficient* of x_k, and b is called the *constant term* of the equation.

A solution of the linear equation (3.1) is a list of values for the unknowns or, equivalently, a vector u in K^n, say

$$x_1 = k_1, \quad x_2 = k_2, \quad \ldots, \quad x_n = k_n \quad \text{or} \quad u = (k_1, k_2, \ldots, k_n)$$

such that the following statement (obtained by substituting k_i for x_i in the equation) is true:

$$a_1k_1 + a_2k_2 + \cdots + a_nk_n = b$$

In such a case we say that u *satisfies* the equation.

Remark: Equation (3.1) implicitly assumes there is an ordering of the unknowns. In order to avoid subscripts, we will usually use x, y for two unknowns; x, y, z for three unknowns; and x, y, z, t for four unknowns; they will be ordered as shown.

EXAMPLE 3.1 Consider the following linear equation in three unknowns x, y, z:

$$x + 2y - 3z = 6$$

We note that $x = 5, y = 2, z = 1$, or, equivalently, the vector $u = (5, 2, 1)$ is a solution of the equation. That is,

$$5 + 2(2) - 3(1) = 6 \quad \text{or} \quad 5 + 4 - 3 = 6 \quad \text{or} \quad 6 = 6$$

On the other hand, $w = (1, 2, 3)$ is not a solution, because on substitution, we do not get a true statement:

$$1 + 2(2) - 3(3) = 6 \quad \text{or} \quad 1 + 4 - 9 = 6 \quad \text{or} \quad -4 = 6$$

System of Linear Equations

A system of linear equations is a list of linear equations with the same unknowns. In particular, a system of m linear equations $L_1, L_2, \ldots, L_m$ in n unknowns $x_1, x_2, \ldots, x_n$ can be put in the *standard form*

$$\begin{aligned} a_{11}x_1 + a_{12}x_2 + \cdots + a_{1n}x_n &= b_1 \\ a_{21}x_1 + a_{22}x_2 + \cdots + a_{2n}x_n &= b_2 \\ \cdots\cdots\cdots\cdots\cdots\cdots\cdots\cdots\cdots\cdots \\ a_{m1}x_1 + a_{m2}x_2 + \cdots + a_{mn}x_n &= b_m \end{aligned} \tag{3.2}$$

where the a_{ij} and b_i are constants. The number a_{ij} is the *coefficient* of the unknown x_j in the equation L_i, and the number b_i is the *constant* of the equation L_i.

The system (3.2) is called an $m \times n$ (read: m by n) system. It is called a *square system* if $m = n$—that is, if the number m of equations is equal to the number n of unknowns.

The system (3.2) is said to be *homogeneous* if all the constant terms are zero—that is, if $b_1 = 0$, $b_2 = 0, \ldots, b_m = 0$. Otherwise the system is said to be *nonhomogeneous*.

A *solution* (or a *particular solution*) of the system (3.2) is a list of values for the unknowns or, equivalently, a vector u in K^n, which is a solution of each of the equations in the system. The set of all solutions of the system is called the *solution set* or the *general solution* of the system.

EXAMPLE 3.2 Consider the following system of linear equations:

$$\begin{aligned} x_1 + x_2 + 4x_3 + 3x_4 &= 5 \\ 2x_1 + 3x_2 + x_3 - 2x_4 &= 1 \\ x_1 + 2x_2 - 5x_3 + 4x_4 &= 3 \end{aligned}$$

It is a 3×4 system because it has three equations in four unknowns. Determine whether (a) $u = (-8, 6, 1, 1)$ and (b) $v = (-10, 5, 1, 2)$ are solutions of the system.

(a) Substitute the values of u in each equation, obtaining

$$\begin{aligned} -8 + 6 + 4(1) + 3(1) = 5 \quad &\text{or} \quad -8 + 6 + 4 + 3 = 5 \quad \text{or} \quad 5 = 5 \\ 2(-8) + 3(6) + 1 - 2(1) = 1 \quad &\text{or} \quad -16 + 18 + 1 - 2 = 1 \quad \text{or} \quad 1 = 1 \\ -8 + 2(6) - 5(1) + 4(1) = 3 \quad &\text{or} \quad -8 + 12 - 5 + 4 = 3 \quad \text{or} \quad 3 = 3 \end{aligned}$$

Yes, u is a solution of the system because it is a solution of each equation.

(b) Substitute the values of v into each successive equation, obtaining

$$\begin{aligned} -10 + 5 + 4(1) + 3(2) = 5 \quad &\text{or} \quad -10 + 5 + 4 + 6 = 5 \quad \text{or} \quad 5 = 5 \\ 2(-10) + 3(5) + 1 - 2(2) = 1 \quad &\text{or} \quad -20 + 15 + 1 - 4 = 1 \quad \text{or} \quad -8 = 1 \end{aligned}$$

No, v is not a solution of the system, because it is not a solution of the second equation. (We do not need to substitute v into the third equation.)

The system (3.2) of linear equations is said to be *consistent* if it has one or more solutions, and it is said to be *inconsistent* if it has no solution. If the field K of scalars is infinite, such as when K is the real field $\mathbf{R}$ or the complex field $\mathbf{C}$, then we have the following important result.

THEOREM 3.1: Suppose the field K is infinite. Then any system $\mathscr{L}$ of linear equations has (i) a unique solution, (ii) no solution, or (iii) an infinite number of solutions.

This situation is pictured in Fig. 3-1. The three cases have a geometrical description when the system $\mathscr{L}$ consists of two equations in two unknowns (Section 3.4).

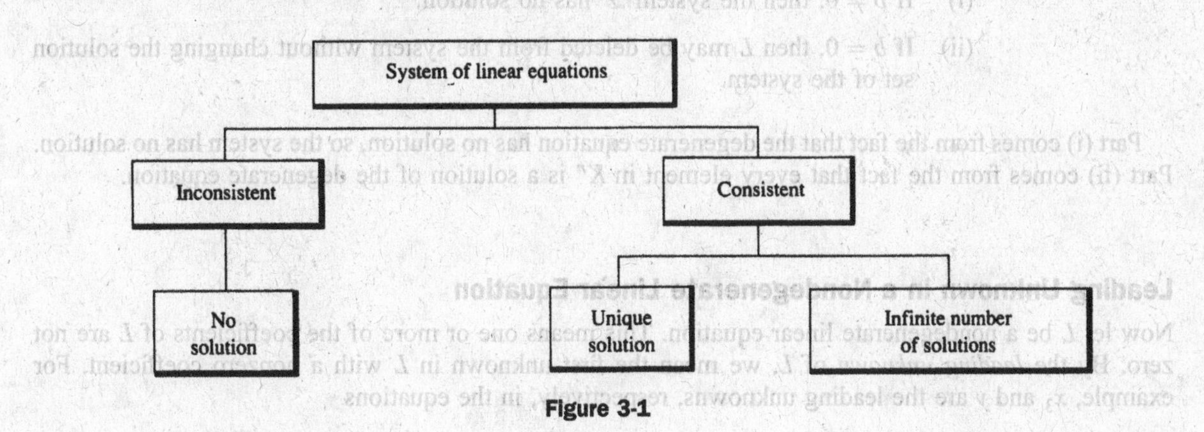

Figure 3-1

Augmented and Coefficient Matrices of a System

Consider again the general system (3.2) of m equations in n unknowns. Such a system has associated with it the following two matrices:

$$M = \begin{bmatrix} a_{11} & a_{12} & \dots & a_{1n} & b_1 \\ a_{21} & a_{22} & \dots & a_{2n} & b_2 \\ \dots & \dots & \dots & \dots & \dots \\ a_{m1} & a_{m2} & \dots & a_{mn} & b_n \end{bmatrix} \quad \text{and} \quad A = \begin{bmatrix} a_{11} & a_{12} & \dots & a_{1n} \\ a_{21} & a_{22} & \dots & a_{2n} \\ \dots & \dots & \dots & \dots \\ a_{m1} & a_{m2} & \dots & a_{mn} \end{bmatrix}$$

The first matrix M is called the *augmented matrix* of the system, and the second matrix A is called the *coefficient matrix*.

The coefficient matrix A is simply the matrix of coefficients, which is the augmented matrix M without the last column of constants. Some texts write $M = [A, B]$ to emphasize the two parts of M, where B denotes the column vector of constants. The augmented matrix M and the coefficient matrix A of the system in Example 3.2 are as follows:

$$M = \begin{bmatrix} 1 & 1 & 4 & 3 & 5 \\ 2 & 3 & 1 & -2 & 1 \\ 1 & 2 & -5 & 4 & 3 \end{bmatrix} \quad \text{and} \quad A = \begin{bmatrix} 1 & 1 & 4 & 3 \\ 2 & 3 & 1 & -2 \\ 1 & 2 & -5 & 4 \end{bmatrix}$$

As expected, A consists of all the columns of M except the last, which is the column of constants.

Clearly, a system of linear equations is completely determined by its augmented matrix M, and vice versa. Specifically, each row of M corresponds to an equation of the system, and each column of M corresponds to the coefficients of an unknown, except for the last column, which corresponds to the constants of the system.

Degenerate Linear Equations

A linear equation is said to be *degenerate* if all the coefficients are zero—that is, if it has the form

$$0x_1 + 0x_2 + \cdots + 0x_n = b \tag{3.3}$$

The solution of such an equation depends only on the value of the constant b. Specifically,

(i) If $b \neq 0$, then the equation has no solution.
(ii) If $b = 0$, then every vector $u = (k_1, k_2, \ldots, k_n)$ in K^n is a solution.

The following theorem applies.

THEOREM 3.2: Let $\mathscr{L}$ be a system of linear equations that contains a degenerate equation L, say with constant b.

(i) If $b \neq 0$, then the system $\mathscr{L}$ has no solution.

(ii) If $b = 0$, then L may be deleted from the system without changing the solution set of the system.

Part (i) comes from the fact that the degenerate equation has no solution, so the system has no solution. Part (ii) comes from the fact that every element in K^n is a solution of the degenerate equation.

Leading Unknown in a Nondegenerate Linear Equation

Now let L be a nondegenerate linear equation. This means one or more of the coefficients of L are not zero. By the *leading unknown* of L, we mean the first unknown in L with a nonzero coefficient. For example, x_3 and y are the leading unknowns, respectively, in the equations

$$0x_1 + 0x_2 + 5x_3 + 6x_4 + 0x_5 + 8x_6 = 7 \quad \text{and} \quad 0x + 2y - 4z = 5$$

We frequently omit terms with zero coefficients, so the above equations would be written as

$$5x_3 + 6x_4 + 8x_6 = 7 \quad \text{and} \quad 2y - 4z = 5$$

In such a case, the leading unknown appears first.

3.3 Equivalent Systems, Elementary Operations

Consider the system (3.2) of m linear equations in n unknowns. Let L be the linear equation obtained by multiplying the m equations by constants $c_1, c_2, \ldots, c_m$, respectively, and then adding the resulting equations. Specifically, let L be the following linear equation:

$$(c_1a_{11} + \cdots + c_ma_{m1})x_1 + \cdots + (c_1a_{1n} + \cdots + c_ma_{mn})x_n = c_1b_1 + \cdots + c_mb_m$$

Then L is called a *linear combination* of the equations in the system. One can easily show (Problem 3.43) that any solution of the system (3.2) is also a solution of the linear combination L.

EXAMPLE 3.3 Let L_1, L_2, L_3 denote, respectively, the three equations in Example 3.2. Let L be the equation obtained by multiplying L_1, L_2, L_3 by $3, -2, 4$, respectively, and then adding. Namely,

$$\begin{array}{rl} 3L_1: & 3x_1 + 3x_2 + 12x_3 + 9x_4 = 15 \\ -2L_2: & -4x_1 - 6x_2 - 2x_3 + 4x_4 = -2 \\ 4L_1: & 4x_1 + 8x_2 - 20x_3 + 16x_4 = 12 \\ \hline \text{(Sum) } L: & 3x_1 + 5x_2 - 10x_3 + 29x_4 = 25 \end{array}$$

Then L is a linear combination of L_1, L_2, L_3. As expected, the solution $u = (-8, 6, 1, 1)$ of the system is also a solution of L. That is, substituting u in L, we obtain a true statement:

$$3(-8) + 5(6) - 10(1) + 29(1) = 25 \quad \text{or} \quad -24 + 30 - 10 + 29 = 25 \quad \text{or} \quad 9 = 9$$

The following theorem holds.

THEOREM 3.3: Two systems of linear equations have the same solutions if and only if each equation in each system is a linear combination of the equations in the other system.

Two systems of linear equations are said to be *equivalent* if they have the same solutions. The next subsection shows one way to obtain equivalent systems of linear equations.

Elementary Operations

The following operations on a system of linear equations $L_1, L_2, \ldots, L_m$ are called *elementary operations*.

[E_1] Interchange two of the equations. We indicate that the equations L_i and L_j are interchanged by writing:

$$\text{"Interchange } L_i \text{ and } L_j\text{"} \quad \text{or} \quad \text{"}L_i \longleftrightarrow L_j\text{"}$$

[E_2] Replace an equation by a nonzero multiple of itself. We indicate that equation L_i is replaced by kL_i (where $k \neq 0$) by writing

$$\text{"Replace } L_i \text{ by } kL_i\text{"} \quad \text{or} \quad \text{"}kL_i \to L_i\text{"}$$

[E_3] Replace an equation by the sum of a multiple of another equation and itself. We indicate that equation L_j is replaced by the sum of kL_i and L_j by writing

$$\text{"Replace } L_j \text{ by } kL_i + L_j\text{"} \quad \text{or} \quad \text{"}kL_i + L_j \to L_j\text{"}$$

The arrow $\to$ in [E_2] and [E_3] may be read as "replaces."

The main property of the above elementary operations is contained in the following theorem (proved in Problem 3.45).

THEOREM 3.4: Suppose a system of $\mathcal{M}$ of linear equations is obtained from a system $\mathcal{L}$ of linear equations by a finite sequence of elementary operations. Then $\mathcal{M}$ and $\mathcal{L}$ have the same solutions.

Remark: Sometimes (say to avoid fractions when all the given scalars are integers) we may apply [E_2] and [E_3] in one step; that is, we may apply the following operation:

[E] Replace equation L_j by the sum of kL_i and $k'L_j$ (where $k' \neq 0$), written

$$\text{"Replace } L_j \text{ by } kL_i + k'L_j\text{"} \quad \text{or} \quad \text{"}kL_i + k'L_j \to L_j\text{"}$$

We emphasize that in operations [E_3] and [E], only equation L_j is changed.

Gaussian elimination, our main method for finding the solution of a given system of linear equations, consists of using the above operations to transform a given system into an equivalent system whose solution can be easily obtained.

The details of Gaussian elimination are discussed in subsequent sections.

3.4 Small Square Systems of Linear Equations

This section considers the special case of one equation in one unknown, and two equations in two unknowns. These simple systems are treated separately because their solution sets can be described geometrically, and their properties motivate the general case.

Linear Equation in One Unknown

The following simple basic result is proved in Problem 3.5.

THEOREM 3.5: Consider the linear equation $ax = b$.

(i) If $a \neq 0$, then $x = b/a$ is a unique solution of $ax = b$.

(ii) If $a = 0$, but $b \neq 0$, then $ax = b$ has no solution.

(iii) If $a = 0$ and $b = 0$, then every scalar k is a solution of $ax = b$.

EXAMPLE 3.4 Solve (a) $4x - 1 = x + 6$, (b) $2x - 5 - x = x + 3$, (c) $4 + x - 3 = 2x + 1 - x$.

(a) Rewrite the equation in standard form obtaining $3x = 7$. Then $x = \frac{7}{3}$ is the unique solution [Theorem 3.5(i)].

(b) Rewrite the equation in standard form, obtaining $0x = 8$. The equation has no solution [Theorem 3.5(ii)].

(c) Rewrite the equation in standard form, obtaining $0x = 0$. Then every scalar k is a solution [Theorem 3.5(iii)].

System of Two Linear Equations in Two Unknowns (2×2 System)

Consider a system of two nondegenerate linear equations in two unknowns x and y, which can be put in the standard form

$$\begin{aligned} A_1x + B_1y &= C_1 \\ A_2x + B_2y &= C_2 \end{aligned} \tag{3.4}$$

Because the equations are nondegenerate, A_1 and B_1 are not both zero, and A_2 and B_2 are not both zero.

The general solution of the system (3.4) belongs to one of three types as indicated in Fig. 3-1. If $\mathbf{R}$ is the field of scalars, then the graph of each equation is a line in the plane $\mathbf{R}^2$ and the three types may be described geometrically as pictured in Fig. 3-2. Specifically,

(1) *The system has exactly one solution.*
Here the two lines intersect in one point [Fig. 3-2(a)]. This occurs when the lines have distinct slopes or, equivalently, when the coefficients of x and y are not proportional:

$$\frac{A_1}{A_2} \neq \frac{B_1}{B_2} \qquad \text{or, equivalently,} \qquad A_1B_2 - A_2B_1 \neq 0$$

For example, in Fig. 3-2(a), $1/3 \neq -1/2$.

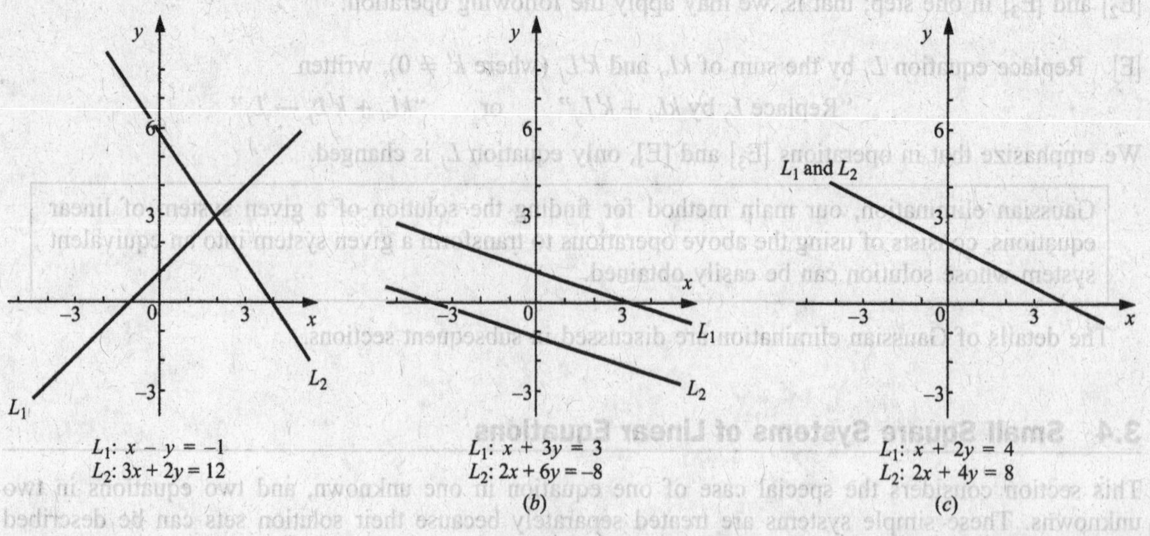

Figure 3-2

(2) *The system has no solution.*
Here the two lines are parallel [Fig. 3-2(b)]. This occurs when the lines have the same slopes but different y intercepts, or when

$$\frac{A_1}{A_2} = \frac{B_1}{B_2} \neq \frac{C_1}{C_2}$$

For example, in Fig. 3-2(*b*), $1/2 = 3/6 \neq -3/8$.

(3) *The system has an infinite number of solutions.*
Here the two lines coincide [Fig. 3-2(c)]. This occurs when the lines have the same slopes and same y intercepts, or when the coefficients and constants are proportional,

$$\frac{A_1}{A_2} = \frac{B_1}{B_2} = \frac{C_1}{C_2}$$

For example, in Fig. 3-2(c), $1/2 = 2/4 = 4/8$.

Remark: The following expression and its value is called a *determinant of order two*:

$$\begin{vmatrix} A_1 & B_1 \\ A_2 & B_2 \end{vmatrix} = A_1B_2 - A_2B_1$$

Determinants will be studied in Chapter 8. Thus, the system (3.4) has a unique solution if and only if the determinant of its coefficients is not zero. (We show later that this statement is true for any square system of linear equations.)

Elimination Algorithm

The solution to system (3.4) can be obtained by the process of elimination, whereby we reduce the system to a single equation in only one unknown. Assuming the system has a unique solution, this elimination algorithm has two parts.

ALGORITHM 3.1: The input consists of two nondegenerate linear equations L_1 and L_2 in two unknowns with a unique solution.

Part A. (Forward Elimination) Multiply each equation by a constant so that the resulting coefficients of one unknown are negatives of each other, and then add the two equations to obtain a new equation L that has only one unknown.

Part B. (Back-Substitution) Solve for the unknown in the new equation L (which contains only one unknown), substitute this value of the unknown into one of the original equations, and then solve to obtain the value of the other unknown.

Part A of Algorithm 3.1 can be applied to any system even if the system does not have a unique solution. In such a case, the new equation L will be degenerate and Part B will not apply.

EXAMPLE 3.5 (Unique Case). Solve the system

$$\begin{aligned} L_1&: \ 2x - 3y = -8 \\ L_2&: \ 3x + 4y = \ \ 5 \end{aligned}$$

The unknown x is eliminated from the equations by forming the new equation $L = -3L_1 + 2L_2$. That is, we multiply L_1 by -3 and L_2 by 2 and add the resulting equations as follows:

$$\begin{aligned} -3L_1&: \quad -6x + 9y = 24 \\ 2L_2&: \quad \ \ 6x + 8y = 10 \\ \hline \text{Addition}&: \qquad\qquad 17y = 34 \end{aligned}$$

We now solve the new equation for y, obtaining $y = 2$. We substitute $y = 2$ into one of the original equations, say L_1, and solve for the other unknown x, obtaining

$$2x - 3(2) = -8 \quad \text{or} \quad 2x - 6 = 8 \quad \text{or} \quad 2x = -2 \quad \text{or} \quad x = -1$$

Thus, $x = -1$, $y = 2$, or the pair $u = (-1, 2)$ is the unique solution of the system. The unique solution is expected, because $2/3 \neq -3/4$. [Geometrically, the lines corresponding to the equations intersect at the point $(-1, 2)$.]

EXAMPLE 3.6 (Nonunique Cases)

(a) Solve the system

$$L_1: \quad x - 3y = 4$$
$$L_2: \quad -2x + 6y = 5$$

We eliminated x from the equations by multiplying L_1 by 2 and adding it to L_2—that is, by forming the new equation $L = 2L_1 + L_2$. This yields the degenerate equation

$$0x + 0y = 13$$

which has a nonzero constant $b = 13$. Thus, this equation and the system have no solution. This is expected, because $1/(-2) = -3/6 \neq 4/5$. (Geometrically, the lines corresponding to the equations are parallel.)

(b) Solve the system

$$L_1: \quad x - 3y = 4$$
$$L_2: \quad -2x + 6y = -8$$

We eliminated x from the equations by multiplying L_1 by 2 and adding it to L_2—that is, by forming the new equation $L = 2L_1 + L_2$. This yields the degenerate equation

$$0x + 0y = 0$$

where the constant term is also zero. Thus, the system has an infinite number of solutions, which correspond to the solutions of either equation. This is expected, because $1/(-2) = -3/6 = 4/(-8)$. (Geometrically, the lines corresponding to the equations coincide.)

To find the general solution, let $y = a$, and substitute into L_1 to obtain

$$x - 3a = 4 \quad \text{or} \quad x = 3a + 4$$

Thus, the general solution of the system is

$$x = 3a + 4, y = a \quad \text{or} \quad u = (3a + 4, \ a)$$

where a (called a *parameter*) is any scalar.

3.5 Systems in Triangular and Echelon Forms

The main method for solving systems of linear equations, Gaussian elimination, is treated in Section 3.6. Here we consider two simple types of systems of linear equations: systems in triangular form and the more general systems in echelon form.

Triangular Form

Consider the following system of linear equations, which is in *triangular form*:

$$\begin{aligned} 2x_1 - 3x_2 + 5x_3 - 2x_4 &= 9 \\ 5x_2 - \ x_3 + 3x_4 &= 1 \\ 7x_3 - \ x_4 &= 3 \\ 2x_4 &= 8 \end{aligned}$$

That is, the first unknown x_1 is the leading unknown in the first equation, the second unknown x_2 is the leading unknown in the second equation, and so on. Thus, in particular, the system is square and each leading unknown is *directly* to the right of the leading unknown in the preceding equation.

Such a triangular system always has a unique solution, which may be obtained by *back-substitution*. That is,

(1) First solve the last equation for the last unknown to get $x_4 = 4$.

(2) Then substitute this value $x_4 = 4$ in the next-to-last equation, and solve for the next-to-last unknown x_3 as follows:

$$7x_3 - 4 = 3 \quad \text{or} \quad 7x_3 = 7 \quad \text{or} \quad x_3 = 1$$

(3) Now substitute $x_3 = 1$ and $x_4 = 4$ in the second equation, and solve for the second unknown x_2 as follows:

$$5x_2 - 1 + 12 = 1 \quad \text{or} \quad 5x_2 + 11 = 1 \quad \text{or} \quad 5x_2 = -10 \quad \text{or} \quad x_2 = -2$$

(4) Finally, substitute $x_2 = -2$, $x_3 = 1$, $x_4 = 4$ in the first equation, and solve for the first unknown x_1 as follows:

$$2x_1 + 6 + 5 - 8 = 9 \quad \text{or} \quad 2x_1 + 3 = 9 \quad \text{or} \quad 2x_1 = 6 \quad \text{or} \quad x_1 = 3$$

Thus, $x_1 = 3$, $x_2 = -2$, $x_3 = 1$, $x_4 = 4$, or, equivalently, the vector $u = (3, -2, 1, 4)$ is the unique solution of the system.

Remark: There is an alternative form for back-substitution (which will be used when solving a system using the matrix format). Namely, after first finding the value of the last unknown, we substitute this value for the last unknown in all the preceding equations before solving for the next-to-last unknown. This yields a triangular system with one less equation and one less unknown. For example, in the above triangular system, we substitute $x_4 = 4$ in all the preceding equations to obtain the triangular system

$$\begin{aligned} 2x_1 - 3x_2 + 5x_3 &= 17 \\ 5x_2 - x_3 &= -1 \\ 7x_3 &= 7 \end{aligned}$$

We then repeat the process using the new last equation. And so on.

Echelon Form, Pivot and Free Variables

The following system of linear equations is said to be in *echelon form*:

$$\begin{aligned} 2x_1 + 6x_2 - x_3 + 4x_4 - 2x_5 &= 15 \\ x_3 + 2x_4 + 2x_5 &= 5 \\ 3x_4 - 9x_5 &= 6 \end{aligned}$$

That is, no equation is degenerate and the leading unknown in each equation other than the first is to the right of the leading unknown in the preceding equation. The leading unknowns in the system, x_1, x_3, x_4, are called *pivot* variables, and the other unknowns, x_2 and x_5, are called *free* variables.

Generally speaking, an *echelon system* or a *system in echelon form* has the following form:

$$\begin{aligned} a_{11}x_1 + a_{12}x_2 + a_{13}x_3 + a_{14}x_4 + \cdots + a_{1n}x_n &= b_1 \\ a_{2j_2}x_{j_2} + a_{2,j_2+1}x_{j_2+1} + \cdots + a_{2n}x_n &= b_2 \\ \cdots\cdots\cdots\cdots\cdots\cdots\cdots\cdots\cdots\cdots & \\ a_{rj_r}x_{j_r} + \cdots + a_{rn}x_n &= b_r \end{aligned} \tag{3.5}$$

where $1 < j_2 < \cdots < j_r$ and $a_{11}, a_{2j_2}, \ldots, a_{rj_r}$ are not zero. The *pivot* variables are $x_1, x_{j_2}, \ldots, x_{j_r}$. Note that $r \leq n$.

The solution set of any echelon system is described in the following theorem (proved in Problem 3.10).

THEOREM 3.6: Consider a system of linear equations in echelon form, say with r equations in n unknowns. There are two cases:

(i) $r = n$. That is, there are as many equations as unknowns (triangular form). Then the system has a unique solution.

(ii) $r < n$. That is, there are more unknowns than equations. Then we can arbitrarily assign values to the $n - r$ free variables and solve uniquely for the r pivot variables, obtaining a solution of the system.

Suppose an echelon system contains more unknowns than equations. Assuming the field K is infinite, the system has an infinite number of solutions, because each of the $n - r$ free variables may be assigned any scalar.

The general solution of a system with free variables may be described in either of two equivalent ways, which we illustrate using the above echelon system where there are $r = 3$ equations and $n = 5$ unknowns. One description is called the "Parametric Form" of the solution, and the other description is called the "Free-Variable Form."

Parametric Form

Assign arbitrary values, called *parameters*, to the free variables x_2 and x_5, say $x_2 = a$ and $x_5 = b$, and then use back-substitution to obtain values for the pivot variables x_1, x_3, x_5 in terms of the parameters a and b. Specifically,

(1) Substitute $x_5 = b$ in the last equation, and solve for x_4:

$$3x_4 - 9b = 6 \quad \text{or} \quad 3x_4 = 6 + 9b \quad \text{or} \quad x_4 = 2 + 3b$$

(2) Substitute $x_4 = 2 + 3b$ and $x_5 = b$ into the second equation, and solve for x_3:

$$x_3 + 2(2 + 3b) + 2b = 5 \quad \text{or} \quad x_3 + 4 + 8b = 5 \quad \text{or} \quad x_3 = 1 - 8b$$

(3) Substitute $x_2 = a$, $x_3 = 1 - 8b$, $x_4 = 2 + 3b$, $x_5 = b$ into the first equation, and solve for x_1:

$$2x_1 + 6a - (1 - 8b) + 4(2 + 3b) - 2b = 15 \quad \text{or} \quad x_1 = 4 - 3a - 9b$$

Accordingly, the general solution in *parametric form* is

$$x_1 = 4 - 3a - 9b, \qquad x_2 = a, \qquad x_3 = 1 - 8b, \qquad x_4 = 2 + 3b, \qquad x_5 = b$$

or, equivalently, $v = (4 - 3a - 9b,\ a,\ 1 - 8b,\ 2 + 3b,\ b)$ where a and b are arbitrary numbers.

Free-Variable Form

Use back-substitution to solve for the pivot variables x_1, x_3, x_4 directly in terms of the free variables x_2 and x_5. That is, the last equation gives $x_4 = 2 + 3x_5$. Substitution in the second equation yields $x_3 = 1 - 8x_5$, and then substitution in the first equation yields $x_1 = 4 - 3x_2 - 9x_5$. Accordingly,

$$x_1 = 4 - 3x_2 - 9x_5, \quad x_2 = \text{free variable}, \quad x_3 = 1 - 8x_5, \quad x_4 = 2 + 3x_5, \quad x_5 = \text{free variable}$$

or, equivalently,

$$v = (4 - 3x_2 - 9x_5,\ x_2,\ 1 - 8x_5,\ 2 + 3x_5,\ x_5)$$

is the *free-variable form* for the general solution of the system.

We emphasize that there is no difference between the above two forms of the general solution, and the use of one or the other to represent the general solution is simply a matter of taste.

Remark: A particular solution of the above system can be found by assigning any values to the free variables and then solving for the pivot variables by back-substitution. For example, setting $x_2 = 1$ and $x_5 = 1$, we obtain

$$x_4 = 2 + 3 = 5, \qquad x_3 = 1 - 8 = -7, \qquad x_1 = 4 - 3 - 9 = -8$$

Thus, $u = (-8, 1, 7, 5, 1)$ is the particular solution corresponding to $x_2 = 1$ and $x_5 = 1$.

3.6 Gaussian Elimination

The main method for solving the general system (3.2) of linear equations is called *Gaussian elimination*. It essentially consists of two parts:

Part A. (Forward Elimination) Step-by-step reduction of the system yielding either a degenerate equation with no solution (which indicates the system has no solution) or an equivalent simpler system in triangular or echelon form.

Part B. (Backward Elimination) Step-by-step back-substitution to find the solution of the simpler system.

Part B has already been investigated in Section 3.4. Accordingly, we need only give the algorithm for Part A, which is as follows.

ALGORITHM 3.2 for (Part A): Input: The $m \times n$ system (3.2) of linear equations.

ELIMINATION STEP: Find the first unknown in the system with a nonzero coefficient (which now must be x_1).

(a) Arrange so that $a_{11} \neq 0$. That is, if necessary, interchange equations so that the first unknown x_1 appears with a nonzero coefficient in the first equation.

(b) Use a_{11} *as a pivot* to eliminate x_1 from all equations except the first equation. That is, for $i > 1$:

(1) Set $m = -a_{i1}/a_{11}$; (2) Replace L_i by $mL_1 + L_i$

The system now has the following form:

$$\begin{aligned} a_{11}x_1 + a_{12}x_2 + a_{13}x_3 + \cdots + a_{1n}x_n &= b_1 \\ a_{2j_2}x_{j_2} + \cdots + a_{2n}x_n &= b_2 \\ \cdots\cdots\cdots\cdots\cdots\cdots\cdots\cdots& \\ a_{mj_2}x_{j_2} + \cdots + a_{mn}x_n &= b_n \end{aligned}$$

where x_1 does not appear in any equation except the first, $a_{11} \neq 0$, and x_{j_2} denotes the first unknown with a nonzero coefficient in any equation other than the first.

(c) Examine each new equation L.

(1) If L has the form $0x_1 + 0x_2 + \cdots + 0x_n = b$ with $b \neq 0$, then

STOP

The system is *inconsistent* and has no solution.

(2) If L has the form $0x_1 + 0x_2 + \cdots + 0x_n = 0$ or if L is a multiple of another equation, then delete L from the system.

RECURSION STEP: Repeat the Elimination Step with each new "smaller" subsystem formed by all the equations excluding the first equation.

OUTPUT: Finally, the system is reduced to triangular or echelon form, or a degenerate equation with no solution is obtained indicating an inconsistent system.

The next remarks refer to the Elimination Step in Algorithm 3.2.

(1) The following number m in (b) is called the *multiplier*:

$$m = -\frac{a_{i1}}{a_{11}} = -\frac{\text{coefficient to be deleted}}{\text{pivot}}$$

(2) One could alternatively apply the following operation in (b):

Replace L_i by $-a_{i1}L_1 + a_{11}L_i$

This would avoid fractions if all the scalars were originally integers.

Gaussian Elimination Example

Here we illustrate in detail Gaussian elimination using the following system of linear equations:

$$\begin{aligned} L_1:&\quad x - 3y - 2z = 6 \\ L_2:&\quad 2x - 4y - 3z = 8 \\ L_3:&\quad -3x + 6y + 8z = -5 \end{aligned}$$

Part A. We use the coefficient 1 of x in the first equation L_1 as the pivot in order to eliminate x from the second equation L_2 and from the third equation L_3. This is accomplished as follows:

(1) Multiply L_1 by the multiplier $m = -2$ and add it to L_2; that is, "Replace L_2 by $-2L_1 + L_2$."

(2) Multiply L_1 by the multiplier $m = 3$ and add it to L_3; that is, "Replace L_3 by $3L_1 + L_3$."

These steps yield

$$\begin{aligned} (-2)L_1:&\quad -2x + 6y + 4z = -12 \\ L_2:&\quad 2x - 4y - 3z = 8 \\ \hline \text{New } L_2:&\quad 2y + z = -4 \end{aligned} \qquad \begin{aligned} 3L_1:&\quad 3x - 9y - 6z = 18 \\ L_3:&\quad -3x + 6y + 8z = -5 \\ \hline \text{New } L_3:&\quad -3y + 2z = 13 \end{aligned}$$

Thus, the original system is replaced by the following system:

$$\begin{aligned} L_1:&\quad x - 3y - 2z = 6 \\ L_2:&\quad 2y + z = -4 \\ L_3:&\quad -3y + 2z = 13 \end{aligned}$$

(Note that the equations L_2 and L_3 form a subsystem with one less equation and one less unknown than the original system.)

Next we use the coefficient 2 of y in the (new) second equation L_2 as the pivot in order to eliminate y from the (new) third equation L_3. This is accomplished as follows:

(3) Multiply L_2 by the multiplier $m = \frac{3}{2}$ and add it to L_3; that is, "Replace L_3 by $\frac{3}{2}L_2 + L_3$." (Alternately, "Replace L_3 by $3L_2 + 2L_3$," which will avoid fractions.)

This step yields

$$\begin{aligned} \tfrac{3}{2}L_2:&\quad 3y + \tfrac{3}{2}z = -6 \\ L_3:&\quad -3y + 2z = 13 \\ \hline \text{New } L_3:&\quad \tfrac{7}{2}z = 7 \end{aligned} \qquad \text{or} \qquad \begin{aligned} 3L_2:&\quad 6y + 3z = -12 \\ 2L_3:&\quad -6y + 4z = 26 \\ \hline \text{New } L_3:&\quad 7z = 14 \end{aligned}$$

Thus, our system is replaced by the following system:

$$\begin{aligned} L_1:&\quad x - 3y - 2z = 6 \\ L_2:&\quad 2y + z = -4 \\ L_3:&\quad 7z = 14 \quad (\text{or } \tfrac{7}{2}z = 7) \end{aligned}$$

The system is now in triangular form, so Part A is completed.

Part B. The values for the unknowns are obtained in reverse order, z, y, x, by back-substitution. Specifically,

(1) Solve for z in L_3 to get $z = 2$.

(2) Substitute $z = 2$ in L_2, and solve for y to get $y = -3$.

(3) Substitute $y = -3$ and $z = 2$ in L_1, and solve for x to get $x = 1$.

Thus, the solution of the triangular system and hence the original system is as follows:

$$x = 1, \quad y = -3, \quad z = 2 \quad \text{or, equivalently,} \quad u = (1, -3, 2).$$

Condensed Format

The Gaussian elimination algorithm involves rewriting systems of linear equations. Sometimes we can avoid excessive recopying of some of the equations by adopting a "condensed format." This format for the solution of the above system follows:

Number	Equation	Operation
(1)	$x - 3y - 2z = 6$	
(2)	$2x - 4y - 3z = 8$	
(3)	$-3x + 6y + 8z = -5$	
(2′)	$2y + z = -4$	Replace L_2 by $-2L_1 + L_2$
(3′)	$-3y + 2z = 13$	Replace L_3 by $3L_1 + L_3$
(3″)	$7z = 14$	Replace L_3 by $3L_2 + 2L_3$

That is, first we write down the number of each of the original equations. As we apply the Gaussian elimination algorithm to the system, we only write down the new equations, and we label each new equation using the same number as the original corresponding equation, but with an added prime. (After each new equation, we will indicate, for instructional purposes, the elementary operation that yielded the new equation.)

The system in triangular form consists of equations (1), (2′), and (3″), the numbers with the largest number of primes. Applying back-substitution to these equations again yields $x = 1, y = -3, z = 2$.

Remark: If two equations need to be interchanged, say to obtain a nonzero coefficient as a pivot, then this is easily accomplished in the format by simply renumbering the two equations rather than changing their positions.

EXAMPLE 3.7 Solve the following system:

$$\begin{aligned} x + 2y - 3z &= 1 \\ 2x + 5y - 8z &= 4 \\ 3x + 8y - 13z &= 7 \end{aligned}$$

We solve the system by Gaussian elimination.

Part A. (Forward Elimination) We use the coefficient 1 of x in the first equation L_1 as the pivot in order to eliminate x from the second equation L_2 and from the third equation L_3. This is accomplished as follows:

(1) Multiply L_1 by the multiplier $m = -2$ and add it to L_2; that is, "Replace L_2 by $-2L_1 + L_2$."
(2) Multiply L_1 by the multiplier $m = -3$ and add it to L_3; that is, "Replace L_3 by $-3L_1 + L_3$."

The two steps yield

$$\begin{aligned} x + 2y - 3z &= 1 \\ y - 2z &= 2 \\ 2y - 4z &= 4 \end{aligned} \qquad \text{or} \qquad \begin{aligned} x + 2y - 3z &= 1 \\ y - 2z &= 2 \end{aligned}$$

(The third equation is deleted, because it is a multiple of the second equation.) The system is now in echelon form with free variable z.

Part B. (Backward Elimination) To obtain the general solution, let the free variable $z = a$, and solve for x and y by back-substitution. Substitute $z = a$ in the second equation to obtain $y = 2 + 2a$. Then substitute $z = a$ and $y = 2 + 2a$ into the first equation to obtain

$$x + 2(2 + 2a) - 3a = 1 \qquad \text{or} \qquad x + 4 + 4a - 3a = 1 \qquad \text{or} \qquad x = -3 - a$$

Thus, the following is the general solution where a is a parameter:

$$x = -3 - a, \quad y = 2 + 2a, \quad z = a \qquad \text{or} \qquad u = (-3 - a,\ 2 + 2a,\ a)$$

EXAMPLE 3.8 Solve the following system:

$$\begin{aligned} x_1 + 3x_2 - 2x_3 + 5x_4 &= 4 \\ 2x_1 + 8x_2 - x_3 + 9x_4 &= 9 \\ 3x_1 + 5x_2 - 12x_3 + 17x_4 &= 7 \end{aligned}$$

We use Gaussian elimination.

Part A. (Forward Elimination) We use the coefficient 1 of x_1 in the first equation L_1 as the pivot in order to eliminate x_1 from the second equation L_2 and from the third equation L_3. This is accomplished by the following operations:

(1) "Replace L_2 by $-2L_1 + L_2$" and (2) "Replace L_3 by $-3L_1 + L_3$"

These yield:

$$\begin{aligned} x_1 + 3x_2 - 2x_3 + 5x_4 &= 4 \\ 2x_2 + 3x_3 - x_4 &= 1 \\ -4x_2 - 6x_3 + 2x_4 &= -5 \end{aligned}$$

We now use the coefficient 2 of x_2 in the second equation L_2 as the pivot and the multiplier $m = 2$ in order to eliminate x_2 from the third equation L_3. This is accomplished by the operation "Replace L_3 by $2L_2 + L_3$," which then yields the degenerate equation

$$0x_1 + 0x_2 + 0x_3 + 0x_4 = -3$$

This equation and, hence, the original system have no solution:

DO NOT CONTINUE

Remark 1: As in the above examples, Part A of Gaussian elimination tells us whether or not the system has a solution—that is, whether or not the system is consistent. Accordingly, Part B need never be applied when a system has no solution.

Remark 2: If a system of linear equations has more than four unknowns and four equations, then it may be more convenient to use the matrix format for solving the system. This matrix format is discussed later.

3.7 Echelon Matrices, Row Canonical Form, Row Equivalence

One way to solve a system of linear equations is by working with its augmented matrix M rather than the system itself. This section introduces the necessary matrix concepts for such a discussion. These concepts, such as echelon matrices and elementary row operations, are also of independent interest.

Echelon Matrices

A matrix A is called an *echelon matrix*, or is said to be in *echelon form*, if the following two conditions hold (where a *leading nonzero element* of a row of A is the first nonzero element in the row):

(1) All zero rows, if any, are at the bottom of the matrix.

(2) Each leading nonzero entry in a row is to the right of the leading nonzero entry in the preceding row.

That is, $A = [a_{ij}]$ is an echelon matrix if there exist nonzero entries

$$a_{1j_1}, a_{2j_2}, \ldots, a_{rj_r}, \qquad \text{where } j_1 < j_2 < \cdots < j_r$$

with the property that

$$a_{ij} = 0 \qquad \text{for} \qquad \begin{cases} \text{(i) } i \le r, \quad j < j_i \\ \text{(ii) } i > r \end{cases}$$

The entries $a_{1j_1}, a_{2j_2}, \ldots, a_{rj_r}$, which are the leading nonzero elements in their respective rows, are called the *pivots* of the echelon matrix.

EXAMPLE 3.9 The following is an echelon matrix whose pivots have been circled:

$$A = \begin{bmatrix} 0 & \textcircled{2} & 3 & 4 & 5 & 9 & 0 & 7 \\ 0 & 0 & 0 & \textcircled{3} & 4 & 1 & 2 & 5 \\ 0 & 0 & 0 & 0 & 0 & \textcircled{5} & 7 & 2 \\ 0 & 0 & 0 & 0 & 0 & 0 & \textcircled{8} & 6 \\ 0 & 0 & 0 & 0 & 0 & 0 & 0 & 0 \end{bmatrix}$$

Observe that the pivots are in columns C_2, C_4, C_6, C_7, and each is to the right of the one above. Using the above notation, the pivots are

$$a_{1j_1} = 2, \qquad a_{2j_2} = 3, \qquad a_{3j_3} = 5, \qquad a_{4j_4} = 8$$

where $j_1 = 2, \; j_2 = 4, \; j_3 = 6, \; j_4 = 7$. Here $r = 4$.

Row Canonical Form

A matrix A is said to be in *row canonical form* (or *row-reduced echelon form*) if it is an echelon matrix—that is, if it satisfies the above properties (1) and (2), and if it satisfies the following additional two properties:

(3) Each pivot (leading nonzero entry) is equal to 1.

(4) Each pivot is the only nonzero entry in its column.

The major difference between an echelon matrix and a matrix in row canonical form is that in an echelon matrix there must be zeros below the pivots [Properties (1) and (2)], but in a matrix in row canonical form, each pivot must also equal 1 [Property (3)] and there must also be zeros above the pivots [Property (4)].

The zero matrix 0 of any size and the identity matrix I of any size are important special examples of matrices in row canonical form.

EXAMPLE 3.10

The following are echelon matrices whose pivots have been circled:

$$\begin{bmatrix} \textcircled{2} & 3 & 2 & 0 & 4 & 5 & -6 \\ 0 & 0 & 0 & \textcircled{1} & -3 & 2 & 0 \\ 0 & 0 & 0 & 0 & 0 & \textcircled{6} & 2 \\ 0 & 0 & 0 & 0 & 0 & 0 & 0 \end{bmatrix}, \quad \begin{bmatrix} \textcircled{1} & 2 & 3 \\ 0 & 0 & \textcircled{1} \\ 0 & 0 & 0 \end{bmatrix}, \quad \begin{bmatrix} 0 & \textcircled{1} & 3 & 0 & 0 & 4 \\ 0 & 0 & 0 & \textcircled{1} & 0 & -3 \\ 0 & 0 & 0 & 0 & \textcircled{1} & 2 \end{bmatrix}$$

The third matrix is also an example of a matrix in row canonical form. The second matrix is not in row canonical form, because it does not satisfy property (4); that is, there is a nonzero entry above the second pivot in the third column. The first matrix is not in row canonical form, because it satisfies neither property (3) nor property (4); that is, some pivots are not equal to 1 and there are nonzero entries above the pivots.

Elementary Row Operations

Suppose A is a matrix with rows $R_1, R_2, \ldots, R_m$. The following operations on A are called *elementary row operations*.

[E$_1$] (Row Interchange): Interchange rows R_i and R_j. This may be written as

$$\text{"Interchange } R_i \text{ and } R_j\text{"} \quad \text{or} \quad \text{"}R_i \longleftrightarrow R_j\text{"}$$

[E$_2$] (Row Scaling): Replace row R_i by a nonzero multiple kR_i of itself. This may be written as

$$\text{"Replace } R_i \text{ by } kR_i \ (k \neq 0)\text{"} \quad \text{or} \quad \text{"}kR_i \to R_i\text{"}$$

[E$_3$] (Row Addition): Replace row R_j by the sum of a multiple kR_i of a row R_i and itself. This may be written as

$$\text{"Replace } R_j \text{ by } kR_i + R_j\text{"} \quad \text{or} \quad \text{"}kR_i + R_j \to R_j\text{"}$$

The arrow $\to$ in E_2 and E_3 may be read as "replaces."

Sometimes (say to avoid fractions when all the given scalars are integers) we may apply [E$_2$] and [E$_3$] in one step; that is, we may apply the following operation:

[E] Replace R_j by the sum of a multiple kR_i of a row R_i and a nonzero multiple $k'R_j$ of itself. This may be written as

$$\text{"Replace } R_j \text{ by } kR_i + k'R_j \ (k' \neq 0)\text{"} \quad \text{or} \quad \text{"}kR_i + k'R_j \to R_j\text{"}$$

We emphasize that in operations [E$_3$] and [E] only row R_j is changed.

Row Equivalence, Rank of a Matrix

A matrix A is said to be *row equivalent* to a matrix B, written

$$A \sim B$$

if B can be obtained from A by a sequence of elementary row operations. In the case that B is also an echelon matrix, B is called an *echelon form* of A.

The following are two basic results on row equivalence.

THEOREM 3.7: Suppose $A = [a_{ij}]$ and $B = [b_{ij}]$ are row equivalent echelon matrices with respective pivot entries

$$a_{1j_1}, a_{2j_2}, \ldots a_{rj_r} \quad \text{and} \quad b_{1k_1}, b_{2k_2}, \ldots b_{sk_s}$$

Then A and B have the same number of nonzero rows—that is, $r = s$—and the pivot entries are in the same positions—that is, $j_1 = k_1, \ j_2 = k_2, \ \ldots, \ j_r = k_r$.

THEOREM 3.8: Every matrix A is row equivalent to a unique matrix in row canonical form.

The proofs of the above theorems will be postponed to Chapter 4. The unique matrix in Theorem 3.8 is called the *row canonical form* of A.

Using the above theorems, we can now give our first definition of the rank of a matrix.

DEFINITION: The *rank* of a matrix A, written rank(A), is equal to the number of pivots in an echelon form of A.

> The rank is a very important property of a matrix and, depending on the context in which the matrix is used, it will be defined in many different ways. Of course, all the definitions lead to the same number.

The next section gives the matrix format of Gaussian elimination, which finds an echelon form of any matrix A (and hence the rank of A), and also finds the row canonical form of A.

One can show that row equivalence is an *equivalence relation*. That is,

(1) $A \sim A$ for any matrix A.
(2) If $A \sim B$, then $B \sim A$.
(3) If $A \sim B$ and $B \sim C$, then $A \sim C$.

Property (2) comes from the fact that each elementary row operation has an inverse operation of the same type. Namely,

(i) "Interchange R_i and R_j" is its own inverse.
(ii) "Replace R_i by kR_i" and "Replace R_i by $(1/k)R_i$" are inverses.
(iii) "Replace R_j by $kR_i + R_j$" and "Replace R_j by $-kR_i + R_j$" are inverses.

There is a similar result for operation [E] (Problem 3.73).

3.8 Gaussian Elimination, Matrix Formulation

This section gives two matrix algorithms that accomplish the following:

(1) Algorithm 3.3 transforms any matrix A into an echelon form.
(2) Algorithm 3.4 transforms the echelon matrix into its row canonical form.

These algorithms, which use the elementary row operations, are simply restatements of *Gaussian elimination* as applied to matrices rather than to linear equations. (The term "row reduce" or simply "reduce" will mean to transform a matrix by the elementary row operations.)

ALGORITHM 3.3 (Forward Elimination): The input is any matrix A. (The algorithm puts 0's below each pivot, working from the "top-down.") The output is an echelon form of A.

Step 1. Find the first column with a nonzero entry. Let j_1 denote this column.

(a) Arrange so that $a_{1j_1} \neq 0$. That is, if necessary, interchange rows so that a nonzero entry appears in the first row in column j_1.
(b) Use a_{1j_1} as a pivot to obtain 0's below a_{1j_1}.

Specifically, for $i > 1$:

(1) Set $m = -a_{ij_1}/a_{1j_1}$; (2) Replace R_i by $mR_1 + R_i$

[That is, apply the operation $-(a_{ij_1}/a_{1j_1})R_1 + R_i \to R_i$.]

Step 2. Repeat Step 1 with the submatrix formed by all the rows excluding the first row. Here we let j_2 denote the first column in the subsystem with a nonzero entry. Hence, at the end of Step 2, we have $a_{2j_2} \neq 0$.

Steps 3 to *r*. Continue the above process until a submatrix has only zero rows.

We emphasize that at the end of the algorithm, the pivots will be

$$a_{1j_1}, a_{2j_2}, \ldots, a_{rj_r}$$

where r denotes the number of nonzero rows in the final echelon matrix.

Remark 1: The following number m in Step 1(b) is called the *multiplier*:

$$m = -\frac{a_{ij_1}}{a_{1j_1}} = -\frac{\text{entry to be deleted}}{\text{pivot}}$$

Remark 2: One could replace the operation in Step 1(b) by the following which would avoid fractions if all the scalars were originally integers.

Replace R_i by $-a_{ij_1}R_1 + a_{1j_1}R_i$.

ALGORITHM 3.4 (Backward Elimination): The input is a matrix $A = [a_{ij}]$ in echelon form with pivot entries

$$a_{1j_1}, \quad a_{2j_2}, \quad \ldots, \quad a_{rj_r}$$

The output is the row canonical form of A.

Step 1. (a) (Use row scaling so the last pivot equals 1.) Multiply the last nonzero row R_r by $1/a_{rj_r}$.

(b) (Use $a_{rj_r} = 1$ to obtain 0's above the pivot.) For $i = r-1, \quad r-2, \quad \ldots, \quad 2, \quad 1$:

(1) Set $m = -a_{ij_r}$; (2) Replace R_i by $mR_r + R_i$

(That is, apply the operations $-a_{ij_r}R_r + R_i \to R_i$.)

Steps 2 to $r-1$. Repeat Step 1 for rows $R_{r-1}, R_{r-2}, \ldots, R_2$.

Step r. (Use row scaling so the first pivot equals 1.) Multiply R_1 by $1/a_{1j_1}$.

There is an alternative form of Algorithm 3.4, which we describe here in words. The formal description of this algorithm is left to the reader as a supplementary problem.

ALTERNATIVE ALGORITHM 3.4 Puts 0's above the pivots row by row from the bottom up (rather than column by column from right to left).

The alternative algorithm, when applied to an augmented matrix M of a system of linear equations, is essentially the same as solving for the pivot unknowns one after the other from the bottom up.

Remark: We emphasize that Gaussian elimination is a two-stage process. Specifically,

Stage A (Algorithm 3.3). Puts 0's below each pivot, working from the top row R_1 down.

Stage B (Algorithm 3.4). Puts 0's above each pivot, working from the bottom row R_r up.

There is another algorithm, called *Gauss–Jordan*, that also row reduces a matrix to its row canonical form. The difference is that Gauss–Jordan puts 0's both below and above each pivot as it works its way from the top row R_1 down. Although Gauss–Jordan may be easier to state and understand, it is much less efficient than the two-stage Gaussian elimination algorithm.

EXAMPLE 3.11 Consider the matrix $A = \begin{bmatrix} 1 & 2 & -3 & 1 & 2 \\ 2 & 4 & -4 & 6 & 10 \\ 3 & 6 & -6 & 9 & 13 \end{bmatrix}$.

(a) Use Algorithm 3.3 to reduce A to an echelon form.

(b) Use Algorithm 3.4 to further reduce A to its row canonical form.

(a) First use $a_{11} = 1$ as a pivot to obtain 0's below a_{11}; that is, apply the operations "Replace R_2 by $-2R_1 + R_2$" and "Replace R_3 by $-3R_1 + R_3$." Then use $a_{23} = 2$ as a pivot to obtain 0 below a_{23}; that is, apply the operation "Replace R_3 by $-\frac{3}{2}R_2 + R_3$." This yields

$$A \sim \begin{bmatrix} 1 & 2 & -3 & 1 & 2 \\ 0 & 0 & 2 & 4 & 6 \\ 0 & 0 & 3 & 6 & 7 \end{bmatrix} \sim \begin{bmatrix} 1 & 2 & -3 & 1 & 2 \\ 0 & 0 & 2 & 4 & 6 \\ 0 & 0 & 0 & 0 & -2 \end{bmatrix}$$

The matrix is now in echelon form.

(b) Multiply R_3 by $-\frac{1}{2}$ so the pivot entry $a_{35} = 1$, and then use $a_{35} = 1$ as a pivot to obtain 0's above it by the operations "Replace R_2 by $-6R_3 + R_2$" and then "Replace R_1 by $-2R_3 + R_1$." This yields

$$A \sim \begin{bmatrix} 1 & 2 & -3 & 1 & 2 \\ 0 & 0 & 2 & 4 & 6 \\ 0 & 0 & 0 & 0 & 1 \end{bmatrix} \sim \begin{bmatrix} 1 & 2 & -3 & 1 & 0 \\ 0 & 0 & 2 & 4 & 0 \\ 0 & 0 & 0 & 0 & 1 \end{bmatrix}.$$

Multiply R_2 by $\frac{1}{2}$ so the pivot entry $a_{23} = 1$, and then use $a_{23} = 1$ as a pivot to obtain 0's above it by the operation "Replace R_1 by $3R_2 + R_1$." This yields

$$A \sim \begin{bmatrix} 1 & 2 & -3 & 1 & 0 \\ 0 & 0 & 1 & 2 & 0 \\ 0 & 0 & 0 & 0 & 1 \end{bmatrix} \sim \begin{bmatrix} 1 & 2 & 0 & 7 & 0 \\ 0 & 0 & 1 & 2 & 0 \\ 0 & 0 & 0 & 0 & 1 \end{bmatrix}.$$

The last matrix is the row canonical form of A.

Application to Systems of Linear Equations

One way to solve a system of linear equations is by working with its augmented matrix M rather than the equations themselves. Specifically, we reduce M to echelon form (which tells us whether the system has a solution), and then further reduce M to its row canonical form (which essentially gives the solution of the original system of linear equations). The justification for this process comes from the following facts:

(1) Any elementary row operation on the augmented matrix M of the system is equivalent to applying the corresponding operation on the system itself.

(2) The system has a solution if and only if the echelon form of the augmented matrix M does not have a row of the form $(0, 0, \ldots, 0, b)$ with $b \neq 0$.

(3) In the row canonical form of the augmented matrix M (excluding zero rows), the coefficient of each basic variable is a pivot entry equal to 1, and it is the only nonzero entry in its respective column; hence, the free-variable form of the solution of the system of linear equations is obtained by simply transferring the free variables to the other side.

This process is illustrated below.

EXAMPLE 3.12 Solve each of the following systems:

$$\begin{array}{ccc} \begin{aligned} x_1 + x_2 - 2x_3 + 4x_4 &= 5 \\ 2x_1 + 2x_2 - 3x_3 + x_4 &= 3 \\ 3x_1 + 3x_2 - 4x_3 - 2x_4 &= 1 \end{aligned} & \begin{aligned} x_1 + x_2 - 2x_3 + 3x_4 &= 4 \\ 2x_1 + 3x_2 + 3x_3 - x_4 &= 3 \\ 5x_1 + 7x_2 + 4x_3 + x_4 &= 5 \end{aligned} & \begin{aligned} x + 2y + z &= 3 \\ 2x + 5y - z &= -4 \\ 3x - 2y - z &= 5 \end{aligned} \\ \text{(a)} & \text{(b)} & \text{(c)} \end{array}$$

(a) Reduce its augmented matrix M to echelon form and then to row canonical form as follows:

$$M = \begin{bmatrix} 1 & 1 & -2 & 4 & 5 \\ 2 & 2 & -3 & 1 & 3 \\ 3 & 3 & -4 & -2 & 1 \end{bmatrix} \sim \begin{bmatrix} 1 & 1 & -2 & 4 & 5 \\ 0 & 0 & 1 & -7 & -7 \\ 0 & 0 & 2 & -14 & -14 \end{bmatrix} \sim \begin{bmatrix} 1 & 1 & 0 & -10 & -9 \\ 0 & 0 & 1 & -7 & -7 \\ 0 & 0 & 0 & 0 & 0 \end{bmatrix}$$

Rewrite the row canonical form in terms of a system of linear equations to obtain the free variable form of the solution. That is,

$$\begin{aligned} x_1 + x_2 - 10x_4 &= -9 \\ x_3 - 7x_4 &= -7 \end{aligned} \quad \text{or} \quad \begin{aligned} x_1 &= -9 - x_2 + 10x_4 \\ x_3 &= -7 + 7x_4 \end{aligned}$$

(The zero row is omitted in the solution.) Observe that x_1 and x_3 are the pivot variables, and x_2 and x_4 are the free variables.

(b) First reduce its augmented matrix M to echelon form as follows:

$$M = \begin{bmatrix} 1 & 1 & -2 & 3 & 4 \\ 2 & 3 & 3 & -1 & 3 \\ 5 & 7 & 4 & 1 & 5 \end{bmatrix} \sim \begin{bmatrix} 1 & 1 & -2 & 3 & 4 \\ 0 & 1 & 7 & -7 & -5 \\ 0 & 2 & 14 & -14 & -15 \end{bmatrix} \sim \begin{bmatrix} 1 & 1 & -2 & 3 & 4 \\ 0 & 1 & 7 & -7 & -5 \\ 0 & 0 & 0 & 0 & -5 \end{bmatrix}$$

There is no need to continue to find the row canonical form of M, because the echelon form already tells us that the system has no solution. Specifically, the third row of the echelon matrix corresponds to the degenerate equation

$$0x_1 + 0x_2 + 0x_3 + 0x_4 = -5$$

which has no solution. Thus, the system has no solution.

(c) Reduce its augmented matrix M to echelon form and then to row canonical form as follows:

$$M = \begin{bmatrix} 1 & 2 & 1 & 3 \\ 2 & 5 & -1 & -4 \\ 3 & -2 & -1 & 5 \end{bmatrix} \sim \begin{bmatrix} 1 & 2 & 1 & 3 \\ 0 & 1 & -3 & -10 \\ 0 & -8 & -4 & -4 \end{bmatrix} \sim \begin{bmatrix} 1 & 2 & 1 & 3 \\ 0 & 1 & -3 & -10 \\ 0 & 0 & -28 & -84 \end{bmatrix}$$

$$\sim \begin{bmatrix} 1 & 2 & 1 & 3 \\ 0 & 1 & -3 & -10 \\ 0 & 0 & 1 & 3 \end{bmatrix} \sim \begin{bmatrix} 1 & 2 & 0 & 0 \\ 0 & 1 & 0 & -1 \\ 0 & 0 & 1 & 3 \end{bmatrix} \sim \begin{bmatrix} 1 & 0 & 0 & 2 \\ 0 & 1 & 0 & -1 \\ 0 & 0 & 1 & 3 \end{bmatrix}$$

Thus, the system has the unique solution $x = 2, y = -1, z = 3$, or, equivalently, the vector $u = (2, -1, 3)$. We note that the echelon form of M already indicated that the solution was unique, because it corresponded to a triangular system.

Application to Existence and Uniqueness Theorems

This subsection gives theoretical conditions for the existence and uniqueness of a solution of a system of linear equations using the notion of the rank of a matrix.

THEOREM 3.9: Consider a system of linear equations in n unknowns with augmented matrix $M = [A, B]$. Then,

(a) The system has a solution if and only if $\text{rank}(A) = \text{rank}(M)$.

(b) The solution is unique if and only if $\text{rank}(A) = \text{rank}(M) = n$.

Proof of (a). The system has a solution if and only if an echelon form of $M = [A, B]$ does not have a row of the form

$$(0, 0, \ldots, 0, b), \quad \text{with} \quad b \neq 0$$

If an echelon form of M does have such a row, then b is a pivot of M but not of A, and hence, $\text{rank}(M) > \text{rank}(A)$. Otherwise, the echelon forms of A and M have the same pivots, and hence, $\text{rank}(A) = \text{rank}(M)$. This proves (a).

Proof of (b). The system has a unique solution if and only if an echelon form has no free variable. This means there is a pivot for each unknown. Accordingly, $n = \text{rank}(A) = \text{rank}(M)$. This proves (b).

The above proof uses the fact (Problem 3.74) that an echelon form of the augmented matrix $M = [A, B]$ also automatically yields an echelon form of A.

3.9 Matrix Equation of a System of Linear Equations

The general system (3.2) of m linear equations in n unknowns is equivalent to the matrix equation

$$\begin{bmatrix} a_{11} & a_{12} & \cdots & a_{1n} \\ a_{21} & a_{22} & \cdots & a_{2n} \\ \cdots & \cdots & \cdots & \cdots \\ a_{m1} & a_{m2} & \cdots & a_{mn} \end{bmatrix} \begin{bmatrix} x_1 \\ x_2 \\ x_3 \\ \cdots \\ x_n \end{bmatrix} = \begin{bmatrix} b_1 \\ b_2 \\ \cdots \\ b_m \end{bmatrix} \quad \text{or} \quad AX = B$$

where $A = [a_{ij}]$ is the coefficient matrix, $X = [x_j]$ is the column vector of unknowns, and $B = [b_i]$ is the column vector of constants. (Some texts write $Ax = b$ rather than $AX = B$, in order to emphasize that x and b are simply column vectors.)

The statement that the system of linear equations and the matrix equation are equivalent means that any vector solution of the system is a solution of the matrix equation, and vice versa.

EXAMPLE 3.13 The following system of linear equations and matrix equation are equivalent:

$$\begin{aligned} x_1 + 2x_2 - 4x_3 + 7x_4 &= 4 \\ 3x_1 - 5x_2 + 6x_3 - 8x_4 &= 8 \\ 4x_1 - 3x_2 - 2x_3 + 6x_4 &= 11 \end{aligned} \quad \text{and} \quad \begin{bmatrix} 1 & 2 & -4 & 7 \\ 3 & -5 & 6 & -8 \\ 4 & -3 & -2 & 6 \end{bmatrix} \begin{bmatrix} x_1 \\ x_2 \\ x_3 \\ x_4 \end{bmatrix} = \begin{bmatrix} 4 \\ 8 \\ 11 \end{bmatrix}$$

We note that $x_1 = 3, \ x_2 = 1, \ x_3 = 2, \ x_4 = 1$, or, in other words, the vector $u = [3, 1, 2, 1]$ is a solution of the system. Thus, the (column) vector u is also a solution of the matrix equation.

The matrix form $AX = B$ of a system of linear equations is notationally very convenient when discussing and proving properties of systems of linear equations. This is illustrated with our first theorem (described in Fig. 3-1), which we restate for easy reference.

THEOREM 3.10: Suppose the field K is infinite. Then the system $AX = B$ has: (a) a unique solution, (b) no solution, or (c) an infinite number of solutions.

Proof. It suffices to show that if $AX = B$ has more than one solution, then it has infinitely many. Suppose u and v are distinct solutions of $AX = B$; that is, $Au = B$ and $Av = B$. Then, for any $k \in K$,

$$A[u + k(u - v)] = Au + k(Au - Av) = B + k(B - B) = B$$

Thus, for each $k \in K$, the vector $u + k(u - v)$ is a solution of $AX = B$. Because all such solutions are distinct (Problem 3.47), $AX = B$ has an infinite number of solutions.

Observe that the above theorem is true when K is the real field $\mathbf{R}$ (or the complex field $\mathbf{C}$). Section 3.3 shows that the theorem has a geometrical description when the system consists of two equations in two unknowns, where each equation represents a line in $\mathbf{R}^2$. The theorem also has a geometrical description when the system consists of three nondegenerate equations in three unknowns, where the three equations correspond to planes H_1, H_2, H_3 in $\mathbf{R}^3$. That is,

(a) *Unique solution:* Here the three planes intersect in exactly one point.

(b) *No solution:* Here the planes may intersect pairwise but with no common point of intersection, or two of the planes may be parallel.

(c) *Infinite number of solutions:* Here the three planes may intersect in a line (one free variable), or they may coincide (two free variables).

These three cases are pictured in Fig. 3-3.

Matrix Equation of a Square System of Linear Equations

A system $AX = B$ of linear equations is square if and only if the matrix A of coefficients is square. In such a case, we have the following important result.

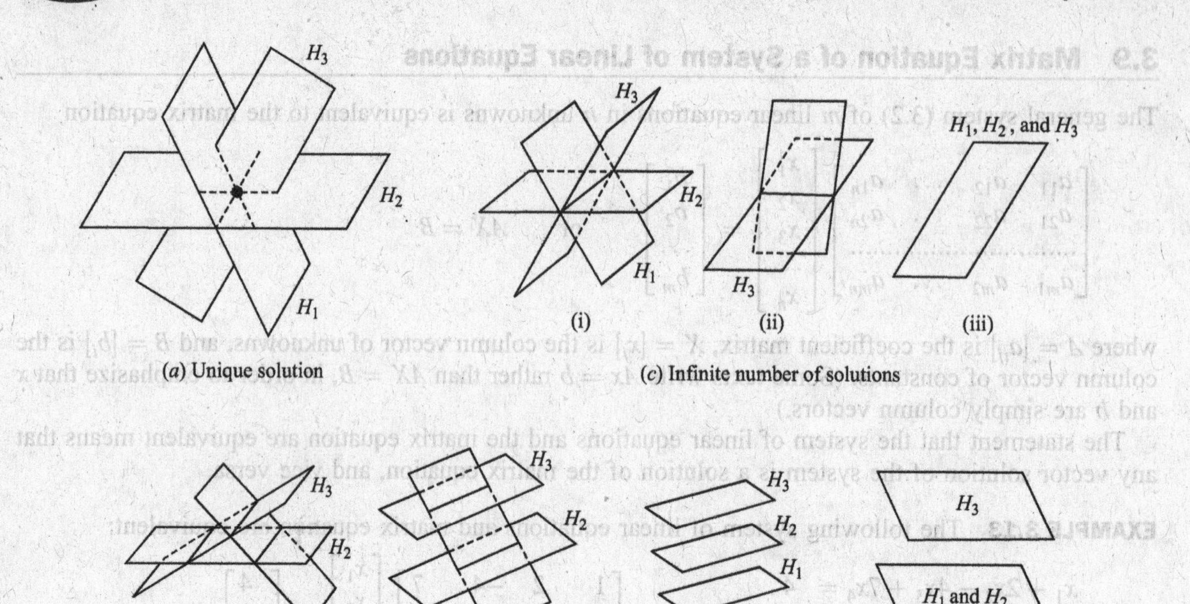

Figure 3-3

THEOREM 3.10: A square system $AX = B$ of linear equations has a unique solution if and only if the matrix A is invertible. In such a case, $A^{-1}B$ is the unique solution of the system.

We only prove here that if A is invertible, then $A^{-1}B$ is a unique solution. If A is invertible, then

$$A(A^{-1}B) = (AA^{-1})B = IB = B$$

and hence, $A^{-1}B$ is a solution. Now suppose v is any solution, so $Av = B$. Then

$$v = Iv = (A^{-1}A)v = A^{-1}(Av) = A^{-1}B$$

Thus, the solution $A^{-1}B$ is unique.

EXAMPLE 3.14 Consider the following system of linear equations, whose coefficient matrix A and inverse A^{-1} are also given:

$$\begin{aligned} x + 2y + \ 3z &= 1 \\ x + 3y + \ 6z &= 3, \\ 2x + 6y + 13z &= 5 \end{aligned} \qquad A = \begin{bmatrix} 1 & 2 & 3 \\ 1 & 3 & 6 \\ 2 & 6 & 13 \end{bmatrix}, \qquad A^{-1} = \begin{bmatrix} 3 & -8 & 3 \\ -1 & 7 & -3 \\ 0 & -2 & 1 \end{bmatrix}$$

By Theorem 3.10, the unique solution of the system is

$$A^{-1}B = \begin{bmatrix} 3 & -8 & 3 \\ -1 & 7 & -3 \\ 0 & -2 & 1 \end{bmatrix} \begin{bmatrix} 1 \\ 3 \\ 5 \end{bmatrix} = \begin{bmatrix} -6 \\ 5 \\ -1 \end{bmatrix}$$

That is, $x = -6$, $y = 5$, $z = -1$.

Remark: We emphasize that Theorem 3.10 does not usually help us to find the solution of a square system. That is, finding the inverse of a coefficient matrix A is not usually any easier than solving the system directly. Thus, unless we are given the inverse of a coefficient matrix A, as in Example 3.14, we usually solve a square system by Gaussian elimination (or some iterative method whose discussion lies beyond the scope of this text).

3.10 Systems of Linear Equations and Linear Combinations of Vectors

The general system (3.2) of linear equations may be rewritten as the following vector equation:

$$x_1 \begin{bmatrix} a_{11} \\ a_{21} \\ \cdots \\ a_{m1} \end{bmatrix} + x_2 \begin{bmatrix} a_{12} \\ a_{22} \\ \cdots \\ a_{m2} \end{bmatrix} + \cdots + x_n \begin{bmatrix} a_{1n} \\ a_{2n} \\ \cdots \\ a_{mn} \end{bmatrix} = \begin{bmatrix} b_1 \\ b_2 \\ \cdots \\ b_m \end{bmatrix}$$

Recall that a vector v in K^n is said to be a *linear combination* of vectors $u_1, u_2, \ldots, u_m$ in K^n if there exist scalars $a_1, a_2, \ldots, a_m$ in K such that

$$v = a_1 u_1 + a_2 u_2 + \cdots + a_m u_m$$

Accordingly, the general system (3.2) of linear equations and the above equivalent vector equation have a solution if and only if the column vector of constants is a linear combination of the columns of the coefficient matrix. We state this observation formally.

THEOREM 3.11: A system $AX = B$ of linear equations has a solution if and only if B is a linear combination of the columns of the coefficient matrix A.

Thus, the answer to the problem of expressing a given vector v in K^n as a linear combination of vectors $u_1, u_2, \ldots, u_m$ in K^n reduces to solving a system of linear equations.

Linear Combination Example

Suppose we want to write the vector $v = (1, -2, 5)$ as a linear combination of the vectors

$$u_1 = (1, 1, 1), \qquad u_2 = (1, 2, 3), \qquad u_3 = (2, -1, 1)$$

First we write $v = xu_1 + yu_2 + zu_3$ with unknowns x, y, z, and then we find the equivalent system of linear equations which we solve. Specifically, we first write

$$\begin{bmatrix} 1 \\ -2 \\ 5 \end{bmatrix} = x \begin{bmatrix} 1 \\ 1 \\ 1 \end{bmatrix} + y \begin{bmatrix} 1 \\ 2 \\ 3 \end{bmatrix} + z \begin{bmatrix} 2 \\ -1 \\ 1 \end{bmatrix} \qquad (*)$$

Then

$$\begin{bmatrix} 1 \\ -2 \\ 5 \end{bmatrix} = \begin{bmatrix} x \\ x \\ x \end{bmatrix} + \begin{bmatrix} y \\ 2y \\ 3y \end{bmatrix} + \begin{bmatrix} 2z \\ -z \\ z \end{bmatrix} = \begin{bmatrix} x + y + 2z \\ x + 2y - z \\ x + 3y + z \end{bmatrix}$$

Setting corresponding entries equal to each other yields the following equivalent system:

$$\begin{aligned} x + y + 2z &= 1 \\ x + 2y - z &= -2 \\ x + 3y + z &= 5 \end{aligned} \qquad (**)$$

For notational convenience, we have written the vectors in $\mathbf{R}^n$ as columns, because it is then easier to find the equivalent system of linear equations. In fact, one can easily go from the vector equation (*) directly to the system (**).

Now we solve the equivalent system of linear equations by reducing the system to echelon form. This yields

$$\begin{aligned} x + y + 2z &= 1 \\ y - 3z &= -3 \\ 2y - z &= 4 \end{aligned} \qquad \text{and then} \qquad \begin{aligned} x + y + 2z &= 1 \\ y - 3z &= -3 \\ 5z &= 10 \end{aligned}$$

Back-substitution yields the solution $x = -6$, $y = 3$, $z = 2$. Thus, $v = -6u_1 + 3u_2 + 2u_3$.

EXAMPLE 3.15

(a) Write the vector $v = (4, 9, 19)$ as a linear combination of

$$u_1 = (1, -2, 3), \qquad u_2 = (3, -7, 10), \qquad u_3 = (2, 1, 9).$$

Find the equivalent system of linear equations by writing $v = xu_1 + yu_2 + zu_3$, and reduce the system to an echelon form. We have

$$\begin{aligned} x + 3y + 2z &= 4 \\ -2x - 7y + z &= 9 \\ 3x + 10y + 9z &= 19 \end{aligned} \quad \text{or} \quad \begin{aligned} x + 3y + 2z &= 4 \\ -y + 5z &= 17 \\ y + 3z &= 7 \end{aligned} \quad \text{or} \quad \begin{aligned} x + 3y + 2z &= 4 \\ -y + 5z &= 17 \\ 8z &= 24 \end{aligned}$$

Back-substitution yields the solution $x = 4$, $y = -2$, $z = 3$. Thus, v is a linear combination of u_1, u_2, u_3. Specifically, $v = 4u_1 - 2u_2 + 3u_3$.

(b) Write the vector $v = (2, 3, -5)$ as a linear combination of

$$u_1 = (1, 2, -3), \qquad u_2 = (2, 3, -4), \qquad u_3 = (1, 3, -5)$$

Find the equivalent system of linear equations by writing $v = xu_1 + yu_2 + zu_3$, and reduce the system to an echelon form. We have

$$\begin{aligned} x + 2y + z &= 2 \\ 2x + 3y + 3z &= 3 \\ -3x - 4y - 5z &= -5 \end{aligned} \quad \text{or} \quad \begin{aligned} x + 2y + z &= 2 \\ -y + z &= -1 \\ 2y - 2z &= 1 \end{aligned} \quad \text{or} \quad \begin{aligned} x + 2y + z &= 2 \\ -5y + 5z &= -1 \\ 0 &= 3 \end{aligned}$$

The system has no solution. Thus, it is impossible to write v as a linear combination of u_1, u_2, u_3.

Linear Combinations of Orthogonal Vectors, Fourier Coefficients

Recall first (Section 1.4) that the dot (inner) product $u \cdot v$ of vectors $u = (a_1, \ldots, a_n)$ and $v = (b_1, \ldots, b_n)$ in $\mathbf{R}^n$ is defined by

$$u \cdot v = a_1b_1 + a_2b_2 + \cdots + a_nb_n$$

Furthermore, vectors u and v are said to be *orthogonal* if their dot product $u \cdot v = 0$.

Suppose that $u_1, u_2, \ldots, u_n$ in $\mathbf{R}^n$ are n nonzero pairwise orthogonal vectors. This means

$$\text{(i)} \quad u_i \cdot u_j = 0 \quad \text{for } i \neq j \qquad \text{and} \qquad \text{(ii)} \quad u_i \cdot u_i \neq 0 \quad \text{for each } i$$

Then, for any vector v in $\mathbf{R}^n$, there is an easy way to write v as a linear combination of $u_1, u_2, \ldots, u_n$, which is illustrated in the next example.

EXAMPLE 3.16 Consider the following three vectors in $\mathbf{R}^3$:

$$u_1 = (1, 1, 1), \qquad u_2 = (1, -3, 2), \qquad u_3 = (5, -1, -4)$$

These vectors are pairwise orthogonal; that is,

$$u_1 \cdot u_2 = 1 - 3 + 2 = 0, \qquad u_1 \cdot u_3 = 5 - 1 - 4 = 0, \qquad u_2 \cdot u_3 = 5 + 3 - 8 = 0$$

Suppose we want to write $v = (4, 14, -9)$ as a linear combination of u_1, u_2, u_3.

Method 1. Find the equivalent system of linear equations as in Example 3.14 and then solve, obtaining $v = 3u_1 - 4u_2 + u_3$.

Method 2. (This method uses the fact that the vectors u_1, u_2, u_3 are mutually orthogonal, and hence, the arithmetic is much simpler.) Set v as a linear combination of u_1, u_2, u_3 using unknown scalars x, y, z as follows:

$$(4, 14, -9) = x(1, 1, 1) + y(1, -3, 2) + z(5, -1, -4) \qquad (*)$$

Take the dot product of (*) with respect to u_1 to get

$$(4,14,-9)\cdot(1,1,1) = x(1,1,1)\cdot(1,1,1) \quad \text{or} \quad 9 = 3x \quad \text{or} \quad x = 3$$

(The last two terms drop out, because u_1 is orthogonal to u_2 and to u_3.) Next take the dot product of (*) with respect to u_2 to obtain

$$(4,14,-9)\cdot(1,-3,2) = y(1,-3,2)\cdot(1,-3,2) \quad \text{or} \quad -56 = 14y \quad \text{or} \quad y = -4$$

Finally, take the dot product of (*) with respect to u_3 to get

$$(4,14,-9)\cdot(5,-1,-4) = z(5,-1,-4)\cdot(5,-1,-4) \quad \text{or} \quad 42 = 42z \quad \text{or} \quad z = 1$$

Thus, $v = 3u_1 - 4u_2 + u_3$.

The procedure in Method 2 in Example 3.16 is valid in general. Namely,

THEOREM 3.12: Suppose $u_1, u_2, \ldots, u_n$ are nonzero mutually orthogonal vectors in $\mathbf{R}^n$. Then, for any vector v in $\mathbf{R}^n$,

$$v = \frac{v\cdot u_1}{u_1\cdot u_1}u_1 + \frac{v\cdot u_2}{u_2\cdot u_2}u_2 + \cdots + \frac{v\cdot u_n}{u_n\cdot u_n}u_n$$

We emphasize that there must be n such orthogonal vectors u_i in $\mathbf{R}^n$ for the formula to be used. Note also that each $u_i\cdot u_i \neq 0$, because each u_i is a nonzero vector.

Remark: The following scalar k_i (appearing in Theorem 3.12) is called the *Fourier coefficient* of v with respect to u_i:

$$k_i = \frac{v\cdot u_i}{u_i\cdot u_i} = \frac{v\cdot u_i}{\|u_i\|^2}$$

It is analogous to a coefficient in the celebrated Fourier series of a function.

3.11 Homogeneous Systems of Linear Equations

A system of linear equations is said to be *homogeneous* if all the constant terms are zero. Thus, a homogeneous system has the form $AX = 0$. Clearly, such a system always has the zero vector $0 = (0, 0, \ldots, 0)$ as a solution, called the *zero* or *trivial* solution. Accordingly, we are usually interested in whether or not the system has a nonzero solution.

Because a homogeneous system $AX = 0$ has at least the zero solution, it can always be put in an echelon form, say

$$\begin{array}{r}
a_{11}x_1 + a_{12}x_2 + a_{13}x_3 + a_{14}x_4 + \cdots + a_{1n}x_n = 0 \\
a_{2j_2}x_{j_2} + a_{2,j_2+1}x_{j_2+1} + \cdots + a_{2n}x_n = 0 \\
\cdots\cdots\cdots\cdots\cdots\cdots\cdots\cdots\cdots\cdots \\
a_{rj_r}x_{j_r} + \cdots + a_{rn}x_n = 0
\end{array}$$

Here r denotes the number of equations in echelon form and n denotes the number of unknowns. Thus, the echelon system has $n - r$ free variables.

The question of nonzero solutions reduces to the following two cases:

(i) $r = n$. The system has only the zero solution.

(ii) $r < n$. The system has a nonzero solution.

Accordingly, if we begin with fewer equations than unknowns, then, in echelon form, $r < n$, and the system has a nonzero solution. This proves the following important result.

THEOREM 3.13: A homogeneous system $AX = 0$ with more unknowns than equations has a nonzero solution.

EXAMPLE 3.17 Determine whether or not each of the following homogeneous systems has a nonzero solution:

$$\begin{array}{ccc} x+y-z=0 & x+y-z=0 & x_1+2x_2-3x_3+4x_4=0 \\ 2x-3y+z=0 & 2x+4y-z=0 & 2x_1-3x_2+5x_3-7x_4=0 \\ x-4y+2z=0 & 3x+2y+2z=0 & 5x_1+6x_2-9x_3+8x_4=0 \\ \text{(a)} & \text{(b)} & \text{(c)} \end{array}$$

(a) Reduce the system to echelon form as follows:

$$\begin{aligned} x+y-z&=0 \\ -5y+3z&=0 \\ -5y+3z&=0 \end{aligned} \qquad \text{and then} \qquad \begin{aligned} x+y-z&=0 \\ -5y+3z&=0 \end{aligned}$$

The system has a nonzero solution, because there are only two equations in the three unknowns in echelon form. Here z is a free variable. Let us, say, set $z=5$. Then, by back-substitution, $y=3$ and $x=2$. Thus, the vector $u=(2,3,5)$ is a particular nonzero solution.

(b) Reduce the system to echelon form as follows:

$$\begin{aligned} x+y-z&=0 \\ 2y+z&=0 \\ -y+5z&=0 \end{aligned} \qquad \text{and then} \qquad \begin{aligned} x+y-z&=0 \\ 2y+z&=0 \\ 11z&=0 \end{aligned}$$

In echelon form, there are three equations in three unknowns. Thus, the system has only the zero solution.

(c) The system must have a nonzero solution (Theorem 3.13), because there are four unknowns but only three equations. (Here we do not need to reduce the system to echelon form.)

Basis for the General Solution of a Homogeneous System

Let W denote the general solution of a homogeneous system $AX=0$. A list of nonzero solution vectors $u_1, u_2, \ldots, u_s$ of the system is said to be a *basis* for W if each solution vector $w \in W$ can be expressed uniquely as a linear combination of the vectors $u_1, u_2, \ldots, u_s$; that is, there exist unique scalars $a_1, a_2, \ldots, a_s$ such that

$$w = a_1u_1 + a_2u_2 + \cdots + a_su_s$$

The number s of such basis vectors is equal to the number of free variables. This number s is called the *dimension* of W, written as $\dim W = s$. When $W=\{0\}$—that is, the system has only the zero solution—we define $\dim W = 0$.

The following theorem, proved in Chapter 5, page 171, tells us how to find such a basis.

THEOREM 3.14: Let W be the general solution of a homogeneous system $AX=0$, and suppose that the echelon form of the homogeneous system has s free variables. Let $u_1, u_2, \ldots, u_s$ be the solutions obtained by setting one of the free variables equal to 1 (or any nonzero constant) and the remaining free variables equal to 0. Then $\dim W = s$, and the vectors $u_1, u_2, \ldots, u_s$ form a basis of W.

We emphasize that the general solution W may have many bases, and that Theorem 3.12 only gives us one such basis.

EXAMPLE 3.18 Find the dimension and a basis for the general solution W of the homogeneous system

$$\begin{aligned} x_1 + 2x_2 - 3x_3 + 2x_4 - 4x_5 &= 0 \\ 2x_1 + 4x_2 - 5x_3 + x_4 - 6x_5 &= 0 \\ 5x_1 + 10x_2 - 13x_3 + 4x_4 - 16x_5 &= 0 \end{aligned}$$

First reduce the system to echelon form. Apply the following operations:

"Replace L_2 by $-2L_1 + L_2$" and "Replace L_3 by $-5L_1 + L_3$" and then "Replace L_3 by $-2L_2 + L_3$"

These operations yield

$$\begin{aligned} x_1 + 2x_2 - 3x_3 + 2x_4 - 4x_5 &= 0 \\ x_3 - 3x_4 + 2x_5 &= 0 \\ 2x_3 - 6x_4 + 4x_5 &= 0 \end{aligned} \quad \text{and} \quad \begin{aligned} x_1 + 2x_2 - 3x_3 + 2x_4 - 4x_5 &= 0 \\ x_3 - 3x_4 + 2x_5 &= 0 \end{aligned}$$

The system in echelon form has three free variables, x_2, x_4, x_5; hence, $\dim W = 3$. Three solution vectors that form a basis for W are obtained as follows:

(1) Set $x_2 = 1$, $x_4 = 0$, $x_5 = 0$. Back-substitution yields the solution $u_1 = (-2, 1, 0, 0, 0)$.
(2) Set $x_2 = 0$, $x_4 = 1$, $x_5 = 0$. Back-substitution yields the solution $u_2 = (7, 0, 3, 1, 0)$.
(3) Set $x_2 = 0$, $x_4 = 0$, $x_5 = 1$. Back-substitution yields the solution $u_3 = (-2, 0, -2, 0, 1)$.

The vectors $u_1 = (-2, 1, 0, 0, 0)$, $u_2 = (7, 0, 3, 1, 0)$, $u_3 = (-2, 0, -2, 0, 1)$ form a basis for W.

Remark: Any solution of the system in Example 3.18 can be written in the form

$$\begin{aligned} au_1 + bu_2 + cu_3 &= a(-2, 1, 0, 0, 0) + b(7, 0, 3, 1, 0) + c(-2, 0, -2, 0, 1) \\ &= (-2a + 7b - 2c, \quad a, \quad 3b - 2c, \quad b, \quad c) \end{aligned}$$

or

$$x_1 = -2a + 7b - 2c, \quad x_2 = a, \quad x_3 = 3b - 2c, \quad x_4 = b, \quad x_5 = c$$

where a, b, c are arbitrary constants. Observe that this representation is nothing more than the parametric form of the general solution under the choice of parameters $x_2 = a$, $x_4 = b$, $x_5 = c$.

Nonhomogeneous and Associated Homogeneous Systems

Let $AX = B$ be a nonhomogeneous system of linear equations. Then $AX = 0$ is called the *associated homogeneous system*. For example,

$$\begin{aligned} x + 2y - 4z &= 7 \\ 3x - 5y + 6z &= 8 \end{aligned} \quad \text{and} \quad \begin{aligned} x + 2y - 4z &= 0 \\ 3x - 5y + 6z &= 0 \end{aligned}$$

show a nonhomogeneous system and its associated homogeneous system.

The relationship between the solution U of a nonhomogeneous system $AX = B$ and the solution W of its associated homogeneous system $AX = 0$ is contained in the following theorem.

THEOREM 3.15: Let v_0 be a particular solution of $AX = B$ and let W be the general solution of $AX = 0$. Then the following is the general solution of $AX = B$:

$$U = v_0 + W = \{v_0 + w : w \in W\}$$

That is, $U = v_0 + W$ is obtained by adding v_0 to each element in W. We note that this theorem has a geometrical interpretation in $\mathbf{R}^3$. Specifically, suppose W is a line through the origin O. Then, as pictured in Fig. 3-4, $U = v_0 + W$ is the line parallel to W obtained by adding v_0 to each element of W. Similarly, whenever W is a plane through the origin O, then $U = v_0 + W$ is a plane parallel to W.

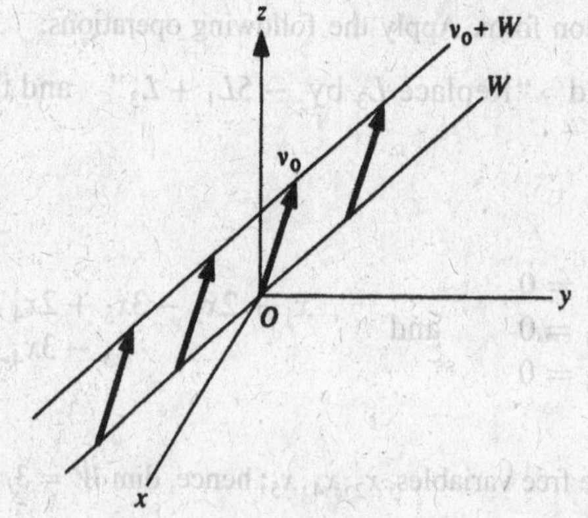

Figure 3-4

3.12 Elementary Matrices

Let e denote an elementary row operation and let $e(A)$ denote the results of applying the operation e to a matrix A. Now let E be the matrix obtained by applying e to the identity matrix I; that is,

$$E = e(I)$$

Then E is called the *elementary matrix* corresponding to the elementary row operation e. Note that E is always a square matrix.

EXAMPLE 3.19 Consider the following three elementary row operations:

(1) Interchange R_2 and R_3. (2) Replace R_2 by $-6R_2$. (3) Replace R_3 by $-4R_1 + R_3$.

The 3×3 elementary matrices corresponding to the above elementary row operations are as follows:

$$E_1 = \begin{bmatrix} 1 & 0 & 0 \\ 0 & 0 & 1 \\ 0 & 1 & 0 \end{bmatrix}, \qquad E_2 = \begin{bmatrix} 1 & 0 & 0 \\ 0 & -6 & 0 \\ 0 & 0 & 1 \end{bmatrix}, \qquad E_3 = \begin{bmatrix} 1 & 0 & 0 \\ 0 & 1 & 0 \\ -4 & 0 & 1 \end{bmatrix}$$

The following theorem, proved in Problem 3.34, holds.

THEOREM 3.16: Let e be an elementary row operation and let E be the corresponding $m \times m$ elementary matrix. Then

$$e(A) = EA$$

where A is any $m \times n$ matrix.

In other words, the result of applying an elementary row operation e to a matrix A can be obtained by premultiplying A by the corresponding elementary matrix E.

Now suppose e' is the inverse of an elementary row operation e, and let E' and E be the corresponding matrices. We note (Problem 3.33) that E is invertible and E' is its inverse. This means, in particular, that any product

$$P = E_k \ldots E_2 E_1$$

of elementary matrices is invertible.

Applications of Elementary Matrices

Using Theorem 3.16, we are able to prove (Problem 3.35) the following important properties of matrices.

THEOREM 3.17: Let A be a square matrix. Then the following are equivalent:

(a) A is invertible (nonsingular).

(b) A is row equivalent to the identity matrix I.

(c) A is a product of elementary matrices.

Recall that square matrices A and B are inverses if $AB = BA = I$. The next theorem (proved in Problem 3.36) demonstrates that we need only show that one of the products is true, say $AB = I$, to prove that matrices are inverses.

THEOREM 3.18: Suppose $AB = I$. Then $BA = I$, and hence, $B = A^{-1}$.

Row equivalence can also be defined in terms of matrix multiplication. Specifically, we will prove (Problem 3.37) the following.

THEOREM 3.19: B is row equivalent to A if and only if there exists a nonsingular matrix P such that $B = PA$.

Application to Finding the Inverse of an $n \times n$ Matrix

The following algorithm finds the inverse of a matrix.

ALGORITHM 3.5: The input is a square matrix A. The output is the inverse of A or that the inverse does not exist.

Step 1. Form the $n \times 2n$ (block) matrix $M = [A, I]$, where A is the left half of M and the identity matrix I is the right half of M.

Step 2. Row reduce M to echelon form. If the process generates a zero row in the A half of M, then

STOP

A has no inverse. (Otherwise A is in triangular form.)

Step 3. Further row reduce M to its row canonical form

$$M \sim [I, B]$$

where the identity matrix I has replaced A in the left half of M.

Step 4. Set $A^{-1} = B$, the matrix that is now in the right half of M.

The justification for the above algorithm is as follows. Suppose A is invertible and, say, the sequence of elementary row operations $e_1, e_2, \ldots, e_q$ applied to $M = [A, I]$ reduces the left half of M, which is A, to the identity matrix I. Let E_i be the elementary matrix corresponding to the operation e_i. Then, by applying Theorem 3.16. we get

$$E_q \ldots E_2E_1A = I \quad \text{or} \quad (E_q \ldots E_2E_1I)A = I, \quad \text{so} \quad A^{-1} = E_q \ldots E_2E_1I$$

That is, A^{-1} can be obtained by applying the elementary row operations $e_1, e_2, \ldots, e_q$ to the identity matrix I, which appears in the right half of M. Thus, $B = A^{-1}$, as claimed.

EXAMPLE 3.20

Find the inverse of the matrix $A = \begin{bmatrix} 1 & 0 & 2 \\ 2 & -1 & 3 \\ 4 & 1 & 8 \end{bmatrix}$.

First form the (block) matrix $M = [A, I]$ and row reduce M to an echelon form:

$$M = \begin{bmatrix} 1 & 0 & 2 & 1 & 0 & 0 \\ 2 & -1 & 3 & 0 & 1 & 0 \\ 4 & 1 & 8 & 0 & 0 & 1 \end{bmatrix} \sim \begin{bmatrix} 1 & 0 & 2 & 1 & 0 & 0 \\ 0 & -1 & -1 & -2 & 1 & 0 \\ 0 & 1 & 0 & -4 & 0 & 1 \end{bmatrix} \sim \begin{bmatrix} 1 & 0 & 2 & 1 & 0 & 0 \\ 0 & -1 & -1 & -2 & 1 & 0 \\ 0 & 0 & -1 & -6 & 1 & 1 \end{bmatrix}$$

In echelon form, the left half of M is in triangular form; hence, A has an inverse. Next we further row reduce M to its row canonical form:

$$M \sim \begin{bmatrix} 1 & 0 & 0 & -11 & 2 & 2 \\ 0 & -1 & 0 & 4 & 0 & -1 \\ 0 & 0 & 1 & 6 & -1 & -1 \end{bmatrix} \sim \begin{bmatrix} 1 & 0 & 0 & -11 & 2 & 2 \\ 0 & 1 & 0 & -4 & 0 & 1 \\ 0 & 0 & 1 & 6 & -1 & -1 \end{bmatrix}$$

The identity matrix is now in the left half of the final matrix; hence, the right half is A^{-1}. In other words,

$$A^{-1} = \begin{bmatrix} -11 & 2 & 2 \\ -4 & 0 & 1 \\ 6 & -1 & -1 \end{bmatrix}$$

Elementary Column Operations

Now let A be a matrix with columns $C_1, C_2, \ldots, C_n$. The following operations on A, analogous to the elementary row operations, are called *elementary column operations*:

[F_1] (Column Interchange): Interchange columns C_i and C_j.
[F_2] (Column Scaling): Replace C_i by kC_i (where $k \neq 0$).
[F_3] (Column Addition): Replace C_j by $kC_i + C_j$.

We may indicate each of the column operations by writing, respectively,

$$(1)\ C_i \leftrightarrow C_j, \qquad (2)\ kC_i \rightarrow C_i, \qquad (3)\ (kC_i + C_j) \rightarrow C_j$$

Moreover, each column operation has an inverse operation of the same type, just like the corresponding row operation.

Now let f denote an elementary column operation, and let F be the matrix obtained by applying f to the identity matrix I; that is,

$$F = f(I)$$

Then F is called the *elementary matrix* corresponding to the elementary column operation f. Note that F is always a square matrix.

EXAMPLE 3.21

Consider the following elementary column operations:

(1) Interchange C_1 and C_3; (2) Replace C_3 by $-2C_3$; (3) Replace C_3 by $-3C_2 + C_3$

The corresponding three 3×3 elementary matrices are as follows:

$$F_1 = \begin{bmatrix} 0 & 0 & 1 \\ 0 & 1 & 0 \\ 1 & 0 & 0 \end{bmatrix}, \quad F_2 = \begin{bmatrix} 1 & 0 & 0 \\ 0 & 1 & 0 \\ 0 & 0 & -2 \end{bmatrix}, \quad F_3 = \begin{bmatrix} 1 & 0 & 0 \\ 0 & 1 & -3 \\ 0 & 0 & 1 \end{bmatrix}$$

The following theorem is analogous to Theorem 3.16 for the elementary row operations.

THEOREM 3.20: For any matrix A, $f(A) = AF$.

That is, the result of applying an elementary column operation f on a matrix A can be obtained by postmultiplying A by the corresponding elementary matrix F.

Matrix Equivalence

A matrix B is *equivalent* to a matrix A if B can be obtained from A by a sequence of row and column operations. Alternatively, B is equivalent to A, if there exist nonsingular matrices P and Q such that $B = PAQ$. Just like row equivalence, equivalence of matrices is an equivalence relation.

The main result of this subsection (proved in Problem 3.38) is as follows.

THEOREM 3.21: Every $m \times n$ matrix A is equivalent to a unique block matrix of the form

$$\begin{bmatrix} I_r & 0 \\ 0 & 0 \end{bmatrix}$$

where I_r is the r-square identity matrix.

The following definition applies.

DEFINITION: The nonnegative integer r in Theorem 3.21 is called the *rank* of A, written rank(A).

Note that this definition agrees with the previous definition of the rank of a matrix.

3.13 *LU* DECOMPOSITION

Suppose A is a nonsingular matrix that can be brought into (upper) triangular form U using only row-addition operations; that is, suppose A can be triangularized by the following algorithm, which we write using computer notation.

ALGORITHM 3.6: The input is a matrix A and the output is a triangular matrix U.

Step 1. Repeat for $i = 1, 2, \ldots, n-1$:

Step 2. Repeat for $j = i+1, i+2, \ldots, n$

(a) Set $m_{ij} := -a_{ij}/a_{ii}$

(b) Set $R_j := m_{ij}R_i + R_j$

[End of Step 2 inner loop.]

[End of Step 1 outer loop.]

The numbers m_{ij} are called ***multipliers***. Sometimes we keep track of these multipliers by means of the following lower triangular matrix L:

$$L = \begin{bmatrix} 1 & 0 & 0 & \ldots & 0 & 0 \\ -m_{21} & 1 & 0 & \ldots & 0 & 0 \\ -m_{31} & -m_{32} & 1 & \ldots & 0 & 0 \\ \ldots\ldots & \ldots\ldots & \ldots\ldots & \ldots\ldots & \ldots\ldots & \ldots \\ -m_{n1} & -m_{n2} & -m_{n3} & \ldots & -m_{n,n-1} & 1 \end{bmatrix}$$

That is, L has 1's on the diagonal, 0's above the diagonal, and the negative of the multiplier m_{ij} as its ij-entry below the diagonal.

The above matrix L and the triangular matrix U obtained in Algorithm 3.6 give us the classical LU factorization of such a matrix A. Namely,

THEOREM 3.22: Let A be a nonsingular matrix that can be brought into triangular form U using only row-addition operations. Then $A = LU$, where L is the above lower triangular matrix with 1's on the diagonal, and U is an upper triangular matrix with no 0's on the diagonal.

EXAMPLE 3.22 Suppose $A = \begin{bmatrix} 1 & 2 & -3 \\ -3 & -4 & 13 \\ 2 & 1 & -5 \end{bmatrix}$. We note that A may be reduced to triangular form by the operations

"Replace R_2 by $3R_1 + R_2$"; "Replace R_3 by $-2R_1 + R_3$"; and then "Replace R_3 by $\frac{3}{2}R_2 + R_3$"

That is,

$$A \sim \begin{bmatrix} 1 & 2 & -3 \\ 0 & 2 & 4 \\ 0 & -3 & 1 \end{bmatrix} \sim \begin{bmatrix} 1 & 2 & -3 \\ 0 & 2 & 4 \\ 0 & 0 & 7 \end{bmatrix}$$

This gives us the classical factorization $A = LU$, where

$$L = \begin{bmatrix} 1 & 0 & 0 \\ -3 & 1 & 0 \\ 2 & -\frac{3}{2} & 1 \end{bmatrix} \quad \text{and} \quad U = \begin{bmatrix} 1 & 2 & -3 \\ 0 & 2 & 4 \\ 0 & 0 & 7 \end{bmatrix}$$

We emphasize:

(1) The entries $-3, 2, -\frac{3}{2}$ in L are the negatives of the multipliers in the above elementary row operations.

(2) U is the triangular form of A.

Application to Systems of Linear Equations

Consider a computer algorithm M. Let $C(n)$ denote the running time of the algorithm as a function of the size n of the input data. [The function $C(n)$ is sometimes called the *time complexity* or simply the *complexity* of the algorithm M.] Frequently, $C(n)$ simply counts the number of multiplications and divisions executed by M, but does not count the number of additions and subtractions because they take much less time to execute.

Now consider a square system of linear equations $AX = B$, where

$$A = [a_{ij}], \qquad X = [x_1, \ldots, x_n]^T, \qquad B = [b_1, \ldots, b_n]^T$$

and suppose A has an LU factorization. Then the system can be brought into triangular form (in order to apply back-substitution) by applying Algorithm 3.6 to the augmented matrix $M = [A, B]$ of the system. The time complexity of Algorithm 3.6 and back-substitution are, respectively,

$$C(n) \approx \tfrac{1}{2}n^3 \qquad \text{and} \qquad C(n) \approx \tfrac{1}{2}n^2$$

where n is the number of equations.

On the other hand, suppose we already have the factorization $A = LU$. Then, to triangularize the system, we need only apply the row operations in the algorithm (retained by the matrix L) to the column vector B. In this case, the time complexity is

$$C(n) \approx \tfrac{1}{2}n^2$$

Of course, to obtain the factorization $A = LU$ requires the original algorithm where $C(n) \approx \frac{1}{2}n^3$. Thus, nothing may be gained by first finding the LU factorization when a single system is involved. However, there are situations, illustrated below, where the LU factorization is useful.

Suppose, for a given matrix A, we need to solve the system

$$AX = B$$

repeatedly for a sequence of different constant vectors, say $B_1, B_2, \ldots, B_k$. Also, suppose some of the B_i depend upon the solution of the system obtained while using preceding vectors B_j. In such a case, it is more efficient to first find the LU factorization of A, and then to use this factorization to solve the system for each new B.

EXAMPLE 3.23 Consider the following system of linear equations:

$$\begin{aligned} x + 2y + z &= k_1 \\ 2x + 3y + 3z &= k_2 \\ -3x + 10y + 2z &= k_3 \end{aligned} \quad \text{or} \quad AX = B, \quad \text{where} \quad A = \begin{bmatrix} 1 & 2 & 1 \\ 2 & 3 & 3 \\ -3 & 10 & 2 \end{bmatrix} \quad \text{and} \quad B = \begin{bmatrix} k_1 \\ k_2 \\ k_3 \end{bmatrix}$$

Suppose we want to solve the system three times where B is equal, say, to B_1, B_2, B_3. Furthermore, suppose $B_1 = [1, 1, 1]^T$, and suppose

$$B_{j+1} = B_j + X_j \quad (\text{for } j = 1, 2)$$

where X_j is the solution of $AX = B_j$. Here it is more efficient to first obtain the LU factorization of A and then use the LU factorization to solve the system for each of the B's. (This is done in Problem 3.42.)

SOLVED PROBLEMS

Linear Equations, Solutions, 2 × 2 Systems

3.1. Determine whether each of the following equations is linear:

(a) $5x + 7y - 8yz = 16$, (b) $x + \pi y + ez = \log 5$, (c) $3x + ky - 8z = 16$

(a) No, because the product yz of two unknowns is of second degree.

(b) Yes, because π, e, and $\log 5$ are constants.

(c) As it stands, there are four unknowns: x, y, z, k. Because of the term ky it is not a linear equation. However, assuming k is a constant, the equation is linear in the unknowns x, y, z.

3.2. Determine whether the following vectors are solutions of $x_1 + 2x_2 - 4x_3 + 3x_4 = 15$: (a) $u = (3, 2, 1, 4)$ and (b) $v = (1, 2, 4, 5)$.

(a) Substitute to obtain $3 + 2(2) - 4(1) + 3(4) = 15$, or $15 = 15$; yes, it is a solution.

(b) Substitute to obtain $1 + 2(2) - 4(4) + 3(5) = 15$, or $4 = 15$; no, it is not a solution.

3.3. Solve (a) $ex = \pi$, (b) $3x - 4 - x = 2x + 3$, (c) $7 + 2x - 4 = 3x + 3 - x$

(a) Because $e \neq 0$, multiply by $1/e$ to obtain $x = \pi/e$.

(b) Rewrite in standard form, obtaining $0x = 7$. The equation has no solution.

(c) Rewrite in standard form, obtaining $0x = 0$. Every scalar k is a solution.

3.4. Prove Theorem 3.4: Consider the equation $ax = b$.

(i) If $a \neq 0$, then $x = b/a$ is a unique solution of $ax = b$.

(ii) If $a = 0$ but $b \neq 0$, then $ax = b$ has no solution.

(iii) If $a = 0$ and $b = 0$, then every scalar k is a solution of $ax = b$.

Suppose $a \neq 0$. Then the scalar b/a exists. Substituting b/a in $ax = b$ yields $a(b/a) = b$, or $b = b$; hence, b/a is a solution. On the other hand, suppose x_0 is a solution to $ax = b$, so that $ax_0 = b$. Multiplying both sides by $1/a$ yields $x_0 = b/a$. Hence, b/a is the unique solution of $ax = b$. Thus, (i) is proved.

On the other hand, suppose $a = 0$. Then, for any scalar k, we have $ak = 0k = 0$. If $b \neq 0$, then $ak \neq b$. Accordingly, k is not a solution of $ax = b$, and so (ii) is proved. If $b = 0$, then $ak = b$. That is, any scalar k is a solution of $ax = b$, and so (iii) is proved.

3.5. Solve each of the following systems:

(a) $\begin{aligned} 2x - 5y &= 11 \\ 3x + 4y &= 5 \end{aligned}$ (b) $\begin{aligned} 2x - 3y &= 8 \\ -6x + 9y &= 6 \end{aligned}$ (c) $\begin{aligned} 2x - 3y &= 8 \\ -4x + 6y &= -16 \end{aligned}$

(a) Eliminate x from the equations by forming the new equation $L = -3L_1 + 2L_2$. This yields the equation

$$23y = -23, \quad \text{and so} \quad y = -1$$

Substitute $y = -1$ in one of the original equations, say L_1, to get

$$2x - 5(-1) = 11 \quad \text{or} \quad 2x + 5 = 11 \quad \text{or} \quad 2x = 6 \quad \text{or} \quad x = 3$$

Thus, $x = 3$, $y = -1$ or the pair $u = (3, -1)$ is the unique solution of the system.

(b) Eliminate x from the equations by forming the new equation $L = 3L_1 + L_2$. This yields the equation

$$0x + 0y = 30$$

This is a degenerate equation with a nonzero constant; hence, this equation and the system have no solution. (Geometrically, the lines corresponding to the equations are parallel.)

(c) Eliminate x from the equations by forming the new equation $L = 2L_1 + L_2$. This yields the equation

$$0x + 0y = 0$$

This is a degenerate equation where the constant term is also zero. Thus, the system has an infinite number of solutions, which correspond to the solution of either equation. (Geometrically, the lines corresponding to the equations coincide.)

To find the general solution, set $y = a$ and substitute in L_1 to obtain

$$2x - 3a = 8 \quad \text{or} \quad 2x = 3a + 8 \quad \text{or} \quad x = \tfrac{3}{2}a + 4$$

Thus, the general solution is

$$x = \tfrac{3}{2}a + 4, \quad y = a \quad \text{or} \quad u = \left(\tfrac{3}{2}a + 4, \ a\right)$$

where a is any scalar.

3.6. Consider the system

$$\begin{aligned} x + ay &= 4 \\ ax + 9y &= b \end{aligned}$$

(a) For which values of a does the system have a unique solution?

(b) Find those pairs of values (a, b) for which the system has more than one solution.

(a) Eliminate x from the equations by forming the new equation $L = -aL_1 + L_2$. This yields the equation

$$(9 - a^2)y = b - 4a \tag{1}$$

The system has a unique solution if and only if the coefficient of y in (1) is not zero—that is, if $9 - a^2 \neq 0$ or if $a \neq \pm 3$.

(b) The system has more than one solution if both sides of (1) are zero. The left-hand side is zero when $a = \pm 3$. When $a = 3$, the right-hand side is zero when $b - 12 = 0$ or $b = 12$. When $a = -3$, the right-hand side is zero when $b + 12 - 0$ or $b = -12$. Thus, $(3, 12)$ and $(-3, -12)$ are the pairs for which the system has more than one solution.

Systems in Triangular and Echelon Form

3.7. Determine the pivot and free variables in each of the following systems:

(a) $\begin{aligned} 2x_1 - 3x_2 - 6x_3 - 5x_4 + 2x_5 &= 7 \\ x_3 + 3x_4 - 7x_5 &= 6 \\ x_4 - 2x_5 &= 1 \end{aligned}$ (b) $\begin{aligned} 2x - 6y + 7z &= 1 \\ 4y + 3z &= 8 \\ 2z &= 4 \end{aligned}$ (c) $\begin{aligned} x + 2y - 3z &= 2 \\ 2x + 3y + z &= 4 \\ 3x + 4y + 5z &= 8 \end{aligned}$

(a) In echelon form, the leading unknowns are the pivot variables, and the others are the free variables. Here x_1, x_3, x_4 are the pivot variables, and x_2 and x_5 are the free variables.

(b) The leading unknowns are x, y, z, so they are the pivot variables. There are no free variables (as in any triangular system).

(c) The notion of pivot and free variables applies only to a system in echelon form.

3.8. Solve the triangular system in Problem 3.7(b).

Because it is a triangular system, solve by back-substitution.

(i) The last equation gives $z = 2$.

(ii) Substitute $z = 2$ in the second equation to get $4y + 6 = 8$ or $y = \frac{1}{2}$.

(iii) Substitute $z = 2$ and $y = \frac{1}{2}$ in the first equation to get

$$2x - 6\left(\frac{1}{2}\right) + 7(2) = 1 \quad \text{or} \quad 2x + 11 = 1 \quad \text{or} \quad x = -5$$

Thus, $x = -5,\ y = \frac{1}{2},\ z = 2$ or $u = (-5, \frac{1}{2}, 2)$ is the unique solution to the system.

3.9. Solve the echelon system in Problem 3.7(a).

Assign parameters to the free variables, say $x_2 = a$ and $x_5 = b$, and solve for the pivot variables by back-substitution.

(i) Substitute $x_5 = b$ in the last equation to get $x_4 - 2b = 1$ or $x_4 = 2b + 1$.

(ii) Substitute $x_5 = b$ and $x_4 = 2b + 1$ in the second equation to get

$$x_3 + 3(2b + 1) - 7b = 6 \quad \text{or} \quad x_3 - b + 3 = 6 \quad \text{or} \quad x_3 = b + 3$$

(iii) Substitute $x_5 = b,\ x_4 = 2b + 1,\ x_3 = b + 3,\ x_2 = a$ in the first equation to get

$$2x_1 - 3a - 6(b + 3) - 5(2b + 1) + 2b = 7 \quad \text{or} \quad 2x_1 - 3a - 14b - 23 = 7$$

$$\text{or} \quad x_1 = \tfrac{3}{2}a + 7b + 15$$

Thus,

$$x_1 = \frac{3}{2}a + 7b + 15, \quad x_2 = a, \quad x_3 = b + 3, \quad x_4 = 2b + 1, \quad x_5 = b$$

$$\text{or} \quad u = \left(\frac{3}{2}a + 7b + 15,\ \ a,\ \ b + 3,\ \ 2b + 1,\ \ b\right)$$

is the parametric form of the general solution.

Alternatively, solving for the pivot variable x_1, x_3, x_4 in terms of the free variables x_2 and x_5 yields the following free-variable form of the general solution:

$$x_1 = \frac{3}{2}x_2 + 7x_5 + 15, \qquad x_3 = x_5 + 3, \qquad x_4 = 2x_5 + 1$$

3.10. Prove Theorem 3.6. Consider the system (3.4) of linear equations in echelon form with r equations and n unknowns.

(i) If $r = n$, then the system has a unique solution.

(ii) If $r < n$, then we can arbitrarily assign values to the $n - r$ free variable and solve uniquely for the r pivot variables, obtaining a solution of the system.

(i) Suppose $r = n$. Then we have a square system $AX = B$ where the matrix A of coefficients is (upper) triangular with nonzero diagonal elements. Thus, A is invertible. By Theorem 3.10, the system has a unique solution.

(ii) Assigning values to the $n - r$ free variables yields a triangular system in the pivot variables, which, by (i), has a unique solution.

Gaussian Elimination

3.11. Solve each of the following systems:

$$\begin{array}{ccc} x+2y-4z=-4 & x+2y-3z=-1 & x+2y-3z=1 \\ 2x+5y-9z=-10 & -3x+y-2z=-7 & 2x+5y-8z=4 \\ 3x-2y+3z=11 & 5x+3y-4z=2 & 3x+8y-13z=7 \\ \text{(a)} & \text{(b)} & \text{(c)} \end{array}$$

Reduce each system to triangular or echelon form using Gaussian elimination:

(a) Apply "Replace L_2 by $-2L_1+L_2$" and "Replace L_3 by $-3L_1+L_3$" to eliminate x from the second and third equations, and then apply "Replace L_3 by $8L_2+L_3$" to eliminate y from the third equation. These operations yield

$$\begin{aligned} x+2y-4z&=-4 \\ y-z&=-2 \\ -8y+15z&=23 \end{aligned} \quad \text{and then} \quad \begin{aligned} x+2y-4z&=-4 \\ y-z&=-2 \\ 7z&=7 \end{aligned}$$

The system is in triangular form. Solve by back-substitution to obtain the unique solution $u=(2,-1,1)$.

(b) Eliminate x from the second and third equations by the operations "Replace L_2 by $3L_1+L_2$" and "Replace L_3 by $-5L_1+L_3$." This gives the equivalent system

$$\begin{aligned} x+2y-3z&=-1 \\ 7y-11z&=-10 \\ -7y+11z&=7 \end{aligned}$$

The operation "Replace L_3 by L_2+L_3" yields the following degenerate equation with a nonzero constant:

$$0x+0y+0z=-3$$

This equation and hence the system have no solution.

(c) Eliminate x from the second and third equations by the operations "Replace L_2 by $-2L_1+L_2$" and "Replace L_3 by $-3L_1+L_3$." This yields the new system

$$\begin{aligned} x+2y-3z&=1 \\ y-2z&=2 \\ 2y-4z&=4 \end{aligned} \quad \text{or} \quad \begin{aligned} x+2y-3z&=1 \\ y-2z&=2 \end{aligned}$$

(The third equation is deleted, because it is a multiple of the second equation.) The system is in echelon form with pivot variables x and y and free variable z.

To find the parametric form of the general solution, set $z=a$ and solve for x and y by back-substitution. Substitute $z=a$ in the second equation to get $y=2+2a$. Then substitute $z=a$ and $y=2+2a$ in the first equation to get

$$x+2(2+2a)-3a=1 \quad \text{or} \quad x+4+a=1 \quad \text{or} \quad x=-3-a$$

Thus, the general solution is

$$x=-3-a, \quad y=2+2a, \quad z=a \quad \text{or} \quad u=(-3-a, \quad 2+2a, \quad a)$$

where a is a parameter.

3.12. Solve each of the following systems:

$$\begin{array}{cc} x_1-3x_2+2x_3-x_4+2x_5=2 & x_1+2x_2-3x_3+4x_4=2 \\ 3x_1-9x_2+7x_3-x_4+3x_5=7 & 2x_1+5x_2-2x_3+x_4=1 \\ 2x_1-6x_2+7x_3+4x_4-5x_5=7 & 5x_1+12x_2-7x_3+6x_4=3 \\ \text{(a)} & \text{(b)} \end{array}$$

Reduce each system to echelon form using Gaussian elimination:

(a) Apply "Replace L_2 by $-3L_1 + L_2$" and "Replace L_3 by $-2L_1 + L_3$" to eliminate x from the second and third equations. This yields

$$\begin{aligned} x_1 - 3x_2 + 2x_3 - x_4 + 2x_5 &= 2 \\ x_3 + 2x_4 - 3x_5 &= 1 \\ 3x_3 + 6x_4 - 9x_5 &= 3 \end{aligned} \qquad \text{or} \qquad \begin{aligned} x_1 - 3x_2 + 2x_3 - x_4 + 2x_5 &= 2 \\ x_3 + 2x_4 - 3x_5 &= 1 \end{aligned}$$

(We delete L_3, because it is a multiple of L_2.) The system is in echelon form with pivot variables x_1 and x_3 and free variables x_2, x_4, x_5.

To find the parametric form of the general solution, set $x_2 = a$, $x_4 = b$, $x_5 = c$, where a, b, c are parameters. Back-substitution yields $x_3 = 1 - 2b + 3c$ and $x_1 = 3a + 5b - 8c$. The general solution is

$$x_1 = 3a + 5b - 8c, \ x_2 = a, \ x_3 = 1 - 2b + 3c, \ x_4 = b, \ x_5 = c$$

or, equivalently, $u = (3a + 5b - 8c, \ a, \ 1 - 2b + 3c, \ b, \ c)$.

(b) Eliminate x_1 from the second and third equations by the operations "Replace L_2 by $-2L_1 + L_2$" and "Replace L_3 by $-5L_1 + L_3$." This yields the system

$$\begin{aligned} x_1 + 2x_2 - 3x_3 + 4x_4 &= 2 \\ x_2 + 4x_3 - 7x_4 &= -3 \\ 2x_2 + 8x_3 - 14x_4 &= -7 \end{aligned}$$

The operation "Replace L_3 by $-2L_2 + L_3$" yields the degenerate equation $0 = -1$. Thus, the system has no solution (even though the system has more unknowns than equations).

3.13. Solve using the condensed format:

$$\begin{aligned} 2y + 3z &= 3 \\ x + y + z &= 4 \\ 4x + 8y - 3z &= 35 \end{aligned}$$

The condensed format follows:

	Number	Equation	Operation
(2)	(~~1~~)	$2y + 3z = 3$	$L_1 \leftrightarrow L_2$
(1)	(~~2~~)	$x + y + z = 4$	$L_1 \leftrightarrow L_2$
	(3)	$4x + 8y - 3z = 35$	
	(3′)	$4y - 7z = 19$	Replace L_3 by $-4L_1 + L_3$
	(3″)	$-13z = 13$	Replace L_3 by $-2L_2 + L_3$

Here (1), (2), and (3″) form a triangular system. (We emphasize that the interchange of L_1 and L_2 is accomplished by simply renumbering L_1 and L_2 as above.)

Using back-substitution with the triangular system yields $z = -1$ from L_3, $y = 3$ from L_2, and $x = 2$ from L_1. Thus, the unique solution of the system is $x = 2$, $y = 3$, $z = -1$ or the triple $u = (2, 3, -1)$.

3.14. Consider the system

$$\begin{aligned} x + 2y + z &= 3 \\ ay + 5z &= 10 \\ 2x + 7y + az &= b \end{aligned}$$

(a) Find those values of a for which the system has a unique solution.

(b) Find those pairs of values (a, b) for which the system has more than one solution.

Reduce the system to echelon form. That is, eliminate x from the third equation by the operation "Replace L_3 by $-2L_1 + L_3$" and then eliminate y from the third equation by the operation

"Replace L_3 by $-3L_2 + aL_3$." This yields

$$\begin{aligned} x + 2y \quad\quad + z &= 3 \\ ay \quad + 5z &= 10 \\ 3y + (a-2)z &= b - 6 \end{aligned} \qquad \text{and then} \qquad \begin{aligned} x + 2y + z &= 3 \\ ay + 5z &= 10 \\ (a^2 - 2a - 15)z &= ab - 6a - 30 \end{aligned}$$

Examine the last equation $(a^2 - 2a - 15)z = ab - 6a - 30$.

(a) The system has a unique solution if and only if the coefficient of z is not zero; that is, if

$$a^2 - 2a - 15 = (a-5)(a+3) \neq 0 \quad \text{or} \quad a \neq 5 \quad \text{and} \quad a \neq -3.$$

(b) The system has more than one solution if both sides are zero. The left-hand side is zero when $a = 5$ or $a = -3$. When $a = 5$, the right-hand side is zero when $5b - 60 = 0$, or $b = 12$. When $a = -3$, the right-hand side is zero when $-3b - 12 = 0$, or $b = -4$. Thus, $(5, 12)$ and $(-3, -4)$ are the pairs for which the system has more than one solution.

Echelon Matrices, Row Equivalence, Row Canonical Form

3.15. Row reduce each of the following matrices to echelon form:

$$\text{(a)}\quad A = \begin{bmatrix} 1 & 2 & -3 & 0 \\ 2 & 4 & -2 & 2 \\ 3 & 6 & -4 & 3 \end{bmatrix}, \qquad \text{(b)}\quad B = \begin{bmatrix} -4 & 1 & -6 \\ 1 & 2 & -5 \\ 6 & 3 & -4 \end{bmatrix}$$

(a) Use $a_{11} = 1$ as a pivot to obtain 0's below a_{11}; that is, apply the row operations "Replace R_2 by $-2R_1 + R_2$" and "Replace R_3 by $-3R_1 + R_3$." Then use $a_{23} = 4$ as a pivot to obtain a 0 below a_{23}; that is, apply the row operation "Replace R_3 by $-5R_2 + 4R_3$." These operations yield

$$A \sim \begin{bmatrix} 1 & 2 & -3 & 0 \\ 0 & 0 & 4 & 2 \\ 0 & 0 & 5 & 3 \end{bmatrix} \sim \begin{bmatrix} 1 & 2 & -3 & 0 \\ 0 & 0 & 4 & 2 \\ 0 & 0 & 0 & 2 \end{bmatrix}$$

The matrix is now in echelon form.

(b) Hand calculations are usually simpler if the pivot element equals 1. Therefore, first interchange R_1 and R_2. Next apply the operations "Replace R_2 by $4R_1 + R_2$" and "Replace R_3 by $-6R_1 + R_3$"; and then apply the operation "Replace R_3 by $R_2 + R_3$." These operations yield

$$B \sim \begin{bmatrix} 1 & 2 & -5 \\ -4 & 1 & -6 \\ 6 & 3 & -4 \end{bmatrix} \sim \begin{bmatrix} 1 & 2 & -5 \\ 0 & 9 & -26 \\ 0 & -9 & 26 \end{bmatrix} \sim \begin{bmatrix} 1 & 2 & -5 \\ 0 & 9 & -26 \\ 0 & 0 & 0 \end{bmatrix}$$

The matrix is now in echelon form.

3.16. Describe the *pivoting* row-reduction algorithm. Also describe the advantages, if any, of using this pivoting algorithm.

The row-reduction algorithm becomes a pivoting algorithm if the entry in column j of greatest absolute value is chosen as the pivot a_{1j_1} and if one uses the row operation

$$(-a_{ij_1}/a_{1j_1})R_1 + R_i \to R_i$$

The main advantage of the pivoting algorithm is that the above row operation involves division by the (current) pivot a_{1j_1}, and, on the computer, roundoff errors may be substantially reduced when one divides by a number as large in absolute value as possible.

3.17. Let $A = \begin{bmatrix} 2 & -2 & 2 & 1 \\ -3 & 6 & 0 & -1 \\ 1 & -7 & 10 & 2 \end{bmatrix}$. Reduce A to echelon form using the pivoting algorithm.

First interchange R_1 and R_2 so that -3 can be used as the pivot, and then apply the operations "Replace R_2 by $\frac{2}{3}R_1 + R_2$" and "Replace R_3 by $\frac{1}{3}R_1 + R_3$." These operations yield

$$A \sim \begin{bmatrix} -3 & 6 & 0 & -1 \\ 2 & -2 & 2 & 1 \\ 1 & -7 & 10 & 2 \end{bmatrix} \sim \begin{bmatrix} -3 & 6 & 0 & -1 \\ 0 & 2 & 2 & \frac{1}{3} \\ 0 & -5 & 10 & \frac{5}{3} \end{bmatrix}$$

Now interchange R_2 and R_3 so that -5 can be used as the pivot, and then apply the operation "Replace R_3 by $\frac{2}{5}R_2 + R_3$." We obtain

$$A \sim \begin{bmatrix} -3 & 6 & 0 & -1 \\ 0 & -5 & 10 & \frac{5}{3} \\ 0 & 2 & 2 & \frac{1}{3} \end{bmatrix} \sim \begin{bmatrix} -3 & 6 & 0 & -1 \\ 0 & -5 & 10 & \frac{5}{3} \\ 0 & 0 & 6 & 1 \end{bmatrix}$$

The matrix has been brought to echelon form using partial pivoting.

3.18. Reduce each of the following matrices to row canonical form:

(a) $A = \begin{bmatrix} 2 & 2 & -1 & 6 & 4 \\ 4 & 4 & 1 & 10 & 13 \\ 8 & 8 & -1 & 26 & 23 \end{bmatrix}$, (b) $B = \begin{bmatrix} 5 & -9 & 6 \\ 0 & 2 & 3 \\ 0 & 0 & 7 \end{bmatrix}$

(a) First reduce A to echelon form by applying the operations "Replace R_2 by $-2R_1 + R_2$" and "Replace R_3 by $-4R_1 + R_3$," and then applying the operation "Replace R_3 by $-R_2 + R_3$." These operations yield

$$A \sim \begin{bmatrix} 2 & 2 & -1 & 6 & 4 \\ 0 & 0 & 3 & -2 & 5 \\ 0 & 0 & 3 & 2 & 7 \end{bmatrix} \sim \begin{bmatrix} 2 & 2 & -1 & 6 & 4 \\ 0 & 0 & 3 & -2 & 5 \\ 0 & 0 & 0 & 4 & 2 \end{bmatrix}$$

Now use back-substitution on the echelon matrix to obtain the row canonical form of A. Specifically, first multiply R_3 by $\frac{1}{4}$ to obtain the pivot $a_{34} = 1$, and then apply the operations "Replace R_2 by $2R_3 + R_2$" and "Replace R_1 by $-6R_3 + R_1$." These operations yield

$$A \sim \begin{bmatrix} 2 & 2 & -1 & 6 & 4 \\ 0 & 0 & 3 & -2 & 5 \\ 0 & 0 & 0 & 1 & \frac{1}{2} \end{bmatrix} \sim \begin{bmatrix} 2 & 2 & -1 & 0 & 1 \\ 0 & 0 & 3 & 0 & 6 \\ 0 & 0 & 0 & 1 & \frac{1}{2} \end{bmatrix}$$

Now multiply R_2 by $\frac{1}{3}$, making the pivot $a_{23} = 1$, and then apply "Replace R_1 by $R_2 + R_1$," yielding

$$A \sim \begin{bmatrix} 2 & 2 & -1 & 0 & 1 \\ 0 & 0 & 1 & 0 & 2 \\ 0 & 0 & 0 & 1 & \frac{1}{2} \end{bmatrix} \sim \begin{bmatrix} 2 & 2 & 0 & 0 & 3 \\ 0 & 0 & 1 & 0 & 2 \\ 0 & 0 & 0 & 1 & \frac{1}{2} \end{bmatrix}$$

Finally, multiply R_1 by $\frac{1}{2}$, so the pivot $a_{11} = 1$. Thus, we obtain the following row canonical form of A:

$$A \sim \begin{bmatrix} 1 & 1 & 0 & 0 & \frac{3}{2} \\ 0 & 0 & 1 & 0 & 2 \\ 0 & 0 & 0 & 1 & \frac{1}{2} \end{bmatrix}$$

(b) Because B is in echelon form, use back-substitution to obtain

$$B \sim \begin{bmatrix} 5 & -9 & 6 \\ 0 & 2 & 3 \\ 0 & 0 & 1 \end{bmatrix} \sim \begin{bmatrix} 5 & -9 & 0 \\ 0 & 2 & 0 \\ 0 & 0 & 1 \end{bmatrix} \sim \begin{bmatrix} 5 & -9 & 0 \\ 0 & 1 & 0 \\ 0 & 0 & 1 \end{bmatrix} \sim \begin{bmatrix} 5 & 0 & 0 \\ 0 & 1 & 0 \\ 0 & 0 & 1 \end{bmatrix} \sim \begin{bmatrix} 1 & 0 & 0 \\ 0 & 1 & 0 \\ 0 & 0 & 1 \end{bmatrix}$$

The last matrix, which is the identity matrix I, is the row canonical form of B. (This is expected, because B is invertible, and so its row canonical form must be I.)

3.19. Describe the Gauss–Jordan elimination algorithm, which also row reduces an arbitrary matrix A to its row canonical form.

The Gauss–Jordan algorithm is similar in some ways to the Gaussian elimination algorithm, except that here each pivot is used to place 0's both below and above the pivot, not just below the pivot, before working with the next pivot. Also, one variation of the algorithm first *normalizes* each row—that is, obtains a unit pivot—before it is used to produce 0's in the other rows, rather than normalizing the rows at the end of the algorithm.

3.20. Let $A = \begin{bmatrix} 1 & -2 & 3 & 1 & 2 \\ 1 & 1 & 4 & -1 & 3 \\ 2 & 5 & 9 & -2 & 8 \end{bmatrix}$. Use Gauss–Jordan to find the row canonical form of A.

Use $a_{11} = 1$ as a pivot to obtain 0's below a_{11} by applying the operations "Replace R_2 by $-R_1 + R_2$" and "Replace R_3 by $-2R_1 + R_3$." This yields

$$A \sim \begin{bmatrix} 1 & -2 & 3 & 1 & 2 \\ 0 & 3 & 1 & -2 & 1 \\ 0 & 9 & 3 & -4 & 4 \end{bmatrix}$$

Multiply R_2 by $\frac{1}{3}$ to make the pivot $a_{22} = 1$, and then produce 0's below and above a_{22} by applying the operations "Replace R_3 by $-9R_2 + R_3$" and "Replace R_1 by $2R_2 + R_1$." These operations yield

$$A \sim \begin{bmatrix} 1 & -2 & 3 & 1 & 2 \\ 0 & 1 & \frac{1}{3} & -\frac{2}{3} & \frac{1}{3} \\ 0 & 9 & 3 & -4 & 4 \end{bmatrix} \sim \begin{bmatrix} 1 & 0 & \frac{11}{3} & -\frac{1}{3} & \frac{8}{3} \\ 0 & 1 & \frac{1}{3} & -\frac{2}{3} & \frac{1}{3} \\ 0 & 0 & 0 & 2 & 1 \end{bmatrix}$$

Finally, multiply R_3 by $\frac{1}{2}$ to make the pivot $a_{34} = 1$, and then produce 0's above a_{34} by applying the operations "Replace R_2 by $\frac{2}{3}R_3 + R_2$" and "Replace R_1 by $\frac{1}{3}R_3 + R_1$." These operations yield

$$A \sim \begin{bmatrix} 1 & 0 & \frac{11}{3} & -\frac{1}{3} & \frac{8}{3} \\ 0 & 1 & \frac{1}{3} & -\frac{2}{3} & \frac{1}{3} \\ 0 & 0 & 0 & 1 & \frac{1}{2} \end{bmatrix} \sim \begin{bmatrix} 1 & 0 & \frac{11}{3} & 0 & \frac{17}{6} \\ 0 & 1 & \frac{1}{3} & 0 & \frac{2}{3} \\ 0 & 0 & 0 & 1 & \frac{1}{2} \end{bmatrix}$$

which is the row canonical form of A.

Systems of Linear Equations in Matrix Form

3.21. Find the augmented matrix M and the coefficient matrix A of the following system:

$$\begin{aligned} x + 2y - 3z &= 4 \\ 3y - 4z + 7x &= 5 \\ 6z + 8x - 9y &= 1 \end{aligned}$$

First align the unknowns in the system, and then use the aligned system to obtain M and A. We have

$$\begin{aligned} x + 2y - 3z &= 4 \\ 7x + 3y - 4z &= 5; \\ 8x - 9y + 6z &= 1 \end{aligned} \quad \text{then} \quad M = \begin{bmatrix} 1 & 2 & -3 & 4 \\ 7 & 3 & -4 & 5 \\ 8 & -9 & 6 & 1 \end{bmatrix} \quad \text{and} \quad A = \begin{bmatrix} 1 & 2 & -3 \\ 7 & 3 & -4 \\ 8 & -9 & 6 \end{bmatrix}$$

3.22. Solve each of the following systems using its augmented matrix M:

$$\begin{aligned} x + 2y - z &= 3 \\ x + 3y + z &= 5 \\ 3x + 8y + 4z &= 17 \end{aligned} \qquad \begin{aligned} x - 2y + 4z &= 2 \\ 2x - 3y + 5z &= 3 \\ 3x - 4y + 6z &= 7 \end{aligned} \qquad \begin{aligned} x + y + 3z &= 1 \\ 2x + 3y - z &= 3 \\ 5x + 7y + z &= 7 \end{aligned}$$

(a) (b) (c)

(a) Reduce the augmented matrix M to echelon form as follows:

$$M = \begin{bmatrix} 1 & 2 & -1 & 3 \\ 1 & 3 & 1 & 5 \\ 3 & 8 & 4 & 17 \end{bmatrix} \sim \begin{bmatrix} 1 & 2 & -1 & 3 \\ 0 & 1 & 2 & 2 \\ 0 & 2 & 7 & 8 \end{bmatrix} \sim \begin{bmatrix} 1 & 2 & -1 & 3 \\ 0 & 1 & 2 & 2 \\ 0 & 0 & 3 & 4 \end{bmatrix}$$

Now write down the corresponding triangular system

$$\begin{aligned} x + 2y - z &= 3 \\ y + 2z &= 2 \\ 3z &= 4 \end{aligned}$$

and solve by back-substitution to obtain the unique solution

$$x = \tfrac{17}{3},\; y = -\tfrac{2}{3},\; z = \tfrac{4}{3} \qquad \text{or} \qquad u = \left(\tfrac{17}{3}, -\tfrac{2}{3}, \tfrac{4}{3}\right)$$

Alternately, reduce the echelon form of M to row canonical form, obtaining

$$M \sim \begin{bmatrix} 1 & 2 & -1 & 3 \\ 0 & 1 & 2 & 2 \\ 0 & 0 & 1 & \frac{4}{3} \end{bmatrix} \sim \begin{bmatrix} 1 & 2 & 0 & \frac{13}{3} \\ 0 & 1 & 0 & -\frac{2}{3} \\ 0 & 0 & 1 & \frac{4}{3} \end{bmatrix} \sim \begin{bmatrix} 1 & 0 & 0 & \frac{17}{3} \\ 0 & 1 & 0 & -\frac{2}{3} \\ 0 & 0 & 1 & \frac{4}{3} \end{bmatrix}$$

This also corresponds to the above solution.

(b) First reduce the augmented matrix M to echelon form as follows:

$$M = \begin{bmatrix} 1 & -2 & 4 & 2 \\ 2 & -3 & 5 & 3 \\ 3 & -4 & 6 & 7 \end{bmatrix} \sim \begin{bmatrix} 1 & -2 & 4 & 2 \\ 0 & 1 & -3 & -1 \\ 0 & 2 & -6 & 1 \end{bmatrix} \sim \begin{bmatrix} 1 & -2 & 4 & 2 \\ 0 & 1 & -3 & -1 \\ 0 & 0 & 0 & 3 \end{bmatrix}$$

The third row corresponds to the degenerate equation $0x + 0y + 0z = 3$, which has no solution. Thus, "DO NOT CONTINUE." The original system also has no solution. (Note that the echelon form indicates whether or not the system has a solution.)

(c) Reduce the augmented matrix M to echelon form and then to row canonical form:

$$M = \begin{bmatrix} 1 & 1 & 3 & 1 \\ 2 & 3 & -1 & 3 \\ 5 & 7 & 1 & 7 \end{bmatrix} \sim \begin{bmatrix} 1 & 1 & 3 & 1 \\ 0 & 1 & -7 & 1 \\ 0 & 2 & -14 & 2 \end{bmatrix} \sim \begin{bmatrix} 1 & 0 & 10 & 0 \\ 0 & 1 & -7 & 1 \end{bmatrix}$$

(The third row of the second matrix is deleted, because it is a multiple of the second row and will result in a zero row.) Write down the system corresponding to the row canonical form of M and then transfer the free variables to the other side to obtain the free-variable form of the solution:

$$\begin{aligned} x + 10z &= 0 \\ y - 7z &= 1 \end{aligned} \qquad \text{and} \qquad \begin{aligned} x &= -10z \\ y &= 1 + 7z \end{aligned}$$

Here z is the only free variable. The parametric solution, using $z = a$, is as follows:

$$x = -10a,\; y = 1 + 7a,\; z = a \qquad \text{or} \qquad u = (-10a,\; 1 + 7a,\; a)$$

3.23. Solve the following system using its augmented matrix M:

$$\begin{aligned} x_1 + 2x_2 - 3x_3 - 2x_4 + 4x_5 &= 1 \\ 2x_1 + 5x_2 - 8x_3 - x_4 + 6x_5 &= 4 \\ x_1 + 4x_2 - 7x_3 + 5x_4 + 2x_5 &= 8 \end{aligned}$$

Reduce the augmented matrix M to echelon form and then to row canonical form:

$$M = \begin{bmatrix} 1 & 2 & -3 & -2 & 4 & 1 \\ 2 & 5 & -8 & -1 & 6 & 4 \\ 1 & 4 & -7 & 5 & 2 & 8 \end{bmatrix} \sim \begin{bmatrix} 1 & 2 & -3 & -2 & 4 & 1 \\ 0 & 1 & -2 & 3 & -2 & 2 \\ 0 & 2 & -4 & 7 & -2 & 7 \end{bmatrix} \sim \begin{bmatrix} 1 & 2 & -3 & -2 & 4 & 1 \\ 0 & 1 & -2 & 3 & -2 & 2 \\ 0 & 0 & 0 & 1 & 2 & 3 \end{bmatrix}$$

$$\sim \begin{bmatrix} 1 & 2 & -3 & 0 & 8 & 7 \\ 0 & 1 & -2 & 0 & -8 & -7 \\ 0 & 0 & 0 & 1 & 2 & 3 \end{bmatrix} \sim \begin{bmatrix} 1 & 0 & 1 & 0 & 24 & 21 \\ 0 & 1 & -2 & 0 & -8 & -7 \\ 0 & 0 & 0 & 1 & 2 & 3 \end{bmatrix}$$

Write down the system corresponding to the row canonical form of M and then transfer the free variables to the other side to obtain the free-variable form of the solution:

$$\begin{aligned} x_1 + x_3 + 24x_5 &= 21 \\ x_2 - 2x_3 - 8x_5 &= -7 \\ x_4 + 2x_5 &= 3 \end{aligned} \qquad \text{and} \qquad \begin{aligned} x_1 &= 21 - x_3 - 24x_5 \\ x_2 &= -7 + 2x_3 + 8x_5 \\ x_4 &= 3 - 2x_5 \end{aligned}$$

Here x_1, x_2, x_4 are the pivot variables and x_3 and x_5 are the free variables. Recall that the parametric form of the solution can be obtained from the free-variable form of the solution by simply setting the free variables equal to parameters, say $x_3 = a$, $x_5 = b$. This process yields

$$x_1 = 21 - a - 24b, \quad x_2 = -7 + 2a + 8b, \quad x_3 = a, \quad x_4 = 3 - 2b, \quad x_5 = b$$

or

$$u = (21 - a - 24b, \; -7 + 2a + 8b, \; a, \; 3 - 2b, \; b)$$

which is another form of the solution.

Linear Combinations, Homogeneous Systems

3.24. Write v as a linear combination of u_1, u_2, u_3, where

(a) $v = (3, 10, 7)$ and $u_1 = (1, 3, -2), u_2 = (1, 4, 2), u_3 = (2, 8, 1)$;

(b) $v = (2, 7, 10)$ and $u_1 = (1, 2, 3),\ u_2 = (1, 3, 5),\ u_3 = (1, 5, 9)$;

(c) $v = (1, 5, 4)$ and $u_1 = (1, 3, -2),\ u_2 = (2, 7, -1),\ u_3 = (1, 6, 7)$.

Find the equivalent system of linear equations by writing $v = xu_1 + yu_2 + zu_3$. Alternatively, use the augmented matrix M of the equivalent system, where $M = [u_1, u_2, u_3, v]$. (Here u_1, u_2, u_3, v are the columns of M.)

(a) The vector equation $v = xu_1 + yu_2 + zu_3$ for the given vectors is as follows:

$$\begin{bmatrix} 3 \\ 10 \\ 7 \end{bmatrix} = x\begin{bmatrix} 1 \\ 3 \\ -2 \end{bmatrix} + y\begin{bmatrix} 1 \\ 4 \\ 2 \end{bmatrix} + z\begin{bmatrix} 2 \\ 8 \\ 1 \end{bmatrix} = \begin{bmatrix} x + y + 2z \\ 3x + 4y + 8z \\ -2x + 2y + z \end{bmatrix}$$

Form the equivalent system of linear equations by setting corresponding entries equal to each other, and then reduce the system to echelon form:

$$\begin{aligned} x + y + 2z &= 3 \\ 3x + 4y + 8z &= 10 \\ -2x + 2y + z &= 7 \end{aligned} \qquad \text{or} \qquad \begin{aligned} x + y + 2z &= 3 \\ y + 2z &= 1 \\ 4y + 5z &= 13 \end{aligned} \qquad \text{or} \qquad \begin{aligned} x + y + 2z &= 3 \\ y + 2z &= 1 \\ -3z &= 9 \end{aligned}$$

The system is in triangular form. Back-substitution yields the unique solution $x = 2$, $y = 7$, $z = -3$. Thus, $v = 2u_1 + 7u_2 - 3u_3$.

Alternatively, form the augmented matrix $M = [u_1, u_2, u_3, v]$ of the equivalent system, and reduce M to echelon form:

$$M = \begin{bmatrix} 1 & 1 & 2 & 3 \\ 3 & 4 & 8 & 10 \\ -2 & 2 & 1 & 7 \end{bmatrix} \sim \begin{bmatrix} 1 & 1 & 2 & 3 \\ 0 & 1 & 2 & 1 \\ 0 & 4 & 5 & 13 \end{bmatrix} \sim \begin{bmatrix} 1 & 1 & 2 & 3 \\ 0 & 1 & 2 & 1 \\ 0 & 0 & -3 & 9 \end{bmatrix}$$

The last matrix corresponds to a triangular system that has a unique solution. Back-substitution yields the solution $x = 2$, $y = 7$, $z = -3$. Thus, $v = 2u_1 + 7u_2 - 3u_3$.

(b) Form the augmented matrix $M = [u_1, u_2, u_3, v]$ of the equivalent system, and reduce M to the echelon form:

$$M = \begin{bmatrix} 1 & 1 & 1 & 2 \\ 2 & 3 & 5 & 7 \\ 3 & 5 & 9 & 10 \end{bmatrix} \sim \begin{bmatrix} 1 & 1 & 1 & 2 \\ 0 & 1 & 3 & 3 \\ 0 & 2 & 6 & 4 \end{bmatrix} \sim \begin{bmatrix} 1 & 1 & 1 & 2 \\ 0 & 1 & 3 & 3 \\ 0 & 0 & 0 & -2 \end{bmatrix}$$

The third row corresponds to the degenerate equation $0x + 0y + 0z = -2$, which has no solution. Thus, the system also has no solution, and v cannot be written as a linear combination of u_1, u_2, u_3.

(c) Form the augmented matrix $M = [u_1, u_2, u_3, v]$ of the equivalent system, and reduce M to echelon form:

$$M = \begin{bmatrix} 1 & 2 & 1 & 1 \\ 3 & 7 & 6 & 5 \\ -2 & -1 & 7 & 4 \end{bmatrix} \sim \begin{bmatrix} 1 & 2 & 1 & 1 \\ 0 & 1 & 3 & 2 \\ 0 & 3 & 9 & 6 \end{bmatrix} \sim \begin{bmatrix} 1 & 2 & 1 & 1 \\ 0 & 1 & 3 & 2 \\ 0 & 0 & 0 & 0 \end{bmatrix}$$

The last matrix corresponds to the following system with free variable z:

$$x + 2y + z = 1$$
$$y + 3z = 2$$

Thus, v can be written as a linear combination of u_1, u_2, u_3 in many ways. For example, let the free variable $z = 1$, and, by back-substitution, we get $y = -2$ and $x = 2$. Thus, $v = 2u_1 - 2u_2 + u_3$.

3.25. Let $u_1 = (1, 2, 4)$, $u_2 = (2, -3, 1)$, $u_3 = (2, 1, -1)$ in $\mathbf{R}^3$. Show that u_1, u_2, u_3 are orthogonal, and write v as a linear combination of u_1, u_2, u_3, where (a) $v = (7, 16, 6)$, (b) $v = (3, 5, 2)$.

Take the dot product of pairs of vectors to get

$$u_1 \cdot u_2 = 2 - 6 + 4 = 0, \quad u_1 \cdot u_3 = 2 + 2 - 4 = 0, \quad u_2 \cdot u_3 = 4 - 3 - 1 = 0$$

Thus, the three vectors in $\mathbf{R}^3$ are orthogonal, and hence Fourier coefficients can be used. That is, $v = xu_1 + yu_2 + zu_3$, where

$$x = \frac{v \cdot u_1}{u_1 \cdot u_1}, \qquad y = \frac{v \cdot u_2}{u_2 \cdot u_2}, \qquad z = \frac{v \cdot u_3}{u_3 \cdot u_3}$$

(a) We have

$$x = \frac{7 + 32 + 24}{1 + 4 + 16} = \frac{63}{21} = 3, \qquad y = \frac{14 - 48 + 6}{4 + 9 + 1} = \frac{-28}{14} = -2, \qquad z = \frac{14 + 16 - 6}{4 + 1 + 1} = \frac{24}{6} = 4$$

Thus, $v = 3u_1 - 2u_2 + 4u_3$.

(b) We have

$$x = \frac{3 + 10 + 8}{1 + 4 + 16} = \frac{21}{21} = 1, \qquad y = \frac{6 - 15 + 2}{4 + 9 + 1} = \frac{-7}{14} = -\frac{1}{2}, \qquad z = \frac{6 + 5 - 2}{4 + 1 + 1} = \frac{9}{6} = \frac{3}{2}$$

Thus, $v = u_1 - \frac{1}{2}u_2 + \frac{3}{2}u_3$.

3.26. Find the dimension and a basis for the general solution W of each of the following homogeneous systems:

$$\begin{aligned} 2x_1 + 4x_2 - 5x_3 + 3x_4 &= 0 \\ 3x_1 + 6x_2 - 7x_3 + 4x_4 &= 0 \\ 5x_1 + 10x_2 - 11x_3 + 6x_4 &= 0 \end{aligned} \qquad \begin{aligned} x - 2y - 3z &= 0 \\ 2x + y + 3z &= 0 \\ 3x - 4y - 2z &= 0 \end{aligned}$$

(a) (b)

(a) Reduce the system to echelon form using the operations "Replace L_2 by $-3L_1 + 2L_2$," "Replace L_3 by $-5L_1 + 2L_3$," and then "Replace L_3 by $-2L_2 + L_3$." These operations yield

$$\begin{aligned} 2x_1 + 4x_2 - 5x_3 + 3x_4 &= 0 \\ x_3 - x_4 &= 0 \\ 3x_3 - 3x_4 &= 0 \end{aligned} \qquad \text{and} \qquad \begin{aligned} 2x_1 + 4x_2 - 5x_3 + 3x_4 &= 0 \\ x_3 - x_4 &= 0 \end{aligned}$$

The system in echelon form has two free variables, x_2 and x_4, so $\dim W = 2$. A basis $[u_1, u_2]$ for W may be obtained as follows:

(1) Set $x_2 = 1, x_4 = 0$. Back-substitution yields $x_3 = 0$, and then $x_1 = -2$. Thus, $u_1 = (-2, 1, 0, 0)$.
(2) Set $x_2 = 0, x_4 = 1$. Back-substitution yields $x_3 = 1$, and then $x_1 = 1$. Thus, $u_2 = (1, 0, 1, 1)$.

(b) Reduce the system to echelon form, obtaining

$$\begin{aligned} x - 2y - 3z &= 0 \\ 5y + 9z &= 0 \\ 2y + 7z &= 0 \end{aligned} \qquad \text{and} \qquad \begin{aligned} x - 2y - 3z &= 0 \\ 5y + 9z &= 0 \\ 17z &= 0 \end{aligned}$$

There are no free variables (the system is in triangular form). Hence, $\dim W = 0$, and W has no basis. Specifically, W consists only of the zero solution; that is, $W = \{0\}$.

3.27. Find the dimension and a basis for the general solution W of the following homogeneous system using matrix notation:

$$\begin{aligned} x_1 + 2x_2 + 3x_3 - 2x_4 + 4x_5 &= 0 \\ 2x_1 + 4x_2 + 8x_3 + x_4 + 9x_5 &= 0 \\ 3x_1 + 6x_2 + 13x_3 + 4x_4 + 14x_5 &= 0 \end{aligned}$$

Show how the basis gives the parametric form of the general solution of the system.

When a system is homogeneous, we represent the system by its coefficient matrix A rather than by its augmented matrix M, because the last column of the augmented matrix M is a zero column, and it will remain a zero column during any row-reduction process.

Reduce the coefficient matrix A to echelon form, obtaining

$$A = \begin{bmatrix} 1 & 2 & 3 & -2 & 4 \\ 2 & 4 & 8 & 1 & 9 \\ 3 & 6 & 13 & 4 & 14 \end{bmatrix} \sim \begin{bmatrix} 1 & 2 & 3 & -2 & 4 \\ 0 & 0 & 2 & 5 & 1 \\ 0 & 0 & 4 & 10 & 2 \end{bmatrix} \sim \begin{bmatrix} 1 & 2 & 3 & -2 & 4 \\ 0 & 0 & 2 & 5 & 1 \end{bmatrix}$$

(The third row of the second matrix is deleted, because it is a multiple of the second row and will result in a zero row.) We can now proceed in one of two ways.

(a) Write down the corresponding homogeneous system in echelon form:

$$\begin{aligned} x_1 + 2x_2 + 3x_3 - 2x_4 + 4x_5 &= 0 \\ 2x_3 + 5x_4 + \ x_5 &= 0 \end{aligned}$$

The system in echelon form has three free variables, x_2, x_4, x_5, so $\dim W = 3$. A basis $[u_1, u_2, u_3]$ for W may be obtained as follows:

(1) Set $x_2 = 1, x_4 = 0, x_5 = 0$. Back-substitution yields $x_3 = 0$, and then $x_1 = -2$. Thus,
$$u_1 = (-2, 1, 0, 0, 0).$$

(2) Set $x_2 = 0, x_4 = 1, x_5 = 0$. Back-substitution yields $x_3 = -\frac{5}{2}$, and then $x_1 = \frac{19}{2}$. Thus,
$$u_2 = (\tfrac{19}{2}, 0, -\tfrac{5}{2}, 1, 0).$$

(3) Set $x_2 = 0, x_4 = 0, x_5 = 1$. Back-substitution yields $x_3 = -\frac{1}{2}$, and then $x_1 = -\frac{5}{2}$. Thus,
$$u_3 = (-\tfrac{5}{2}, 0, -\tfrac{1}{2}, 0, 1).$$

[One could avoid fractions in the basis by choosing $x_4 = 2$ in (2) and $x_5 = 2$ in (3), which yields multiples of u_2 and u_3.] The parametric form of the general solution is obtained from the following linear combination of the basis vectors using parameters a, b, c:

$$au_1 + bu_2 + cu_3 = (-2a + \tfrac{19}{2}b - \tfrac{5}{2}c,\ a,\ -\tfrac{5}{2}b - \tfrac{1}{2}c,\ b,\ c)$$

(b) Reduce the echelon form of A to row canonical form:

$$A \sim \begin{bmatrix} 1 & 2 & 3 & -2 & 4 \\ 0 & 0 & 1 & \frac{5}{2} & \frac{1}{2} \end{bmatrix} \sim \begin{bmatrix} 1 & 2 & 3 & -\frac{19}{2} & \frac{5}{2} \\ 0 & 0 & 1 & \frac{5}{2} & \frac{1}{2} \end{bmatrix}$$

Write down the corresponding free-variable solution:

$$x_1 = -2x_2 + \frac{19}{2}x_4 - \frac{5}{2}x_5$$

$$x_3 = -\frac{5}{2}x_4 - \frac{1}{2}x_5$$

Using these equations for the pivot variables x_1 and x_3, repeat the above process to obtain a basis $[u_1, u_2, u_3]$ for W. That is, set $x_2 = 1, x_4 = 0, x_5 = 0$ to get u_1; set $x_2 = 0, x_4 = 1, x_5 = 0$ to get u_2; and set $x_2 = 0$, $x_4 = 0, x_5 = 1$ to get u_3.

3.28. Prove Theorem 3.15. Let v_0 be a particular solution of $AX = B$, and let W be the general solution of $AX = 0$. Then $U = v_0 + W = \{v_0 + w : w \in W\}$ is the general solution of $AX = B$.

Let w be a solution of $AX = 0$. Then

$$A(v_0 + w) = Av_0 + Aw = B + 0 = B$$

Thus, the sum $v_0 + w$ is a solution of $AX = B$. On the other hand, suppose v is also a solution of $AX = B$. Then

$$A(v - v_0) = Av - Av_0 = B - B = 0$$

Therefore, $v - v_0$ belongs to W. Because $v = v_0 + (v - v_0)$, we find that any solution of $AX = B$ can be obtained by adding a solution of $AX = 0$ to a solution of $AX = B$. Thus, the theorem is proved.

Elementary Matrices, Applications

3.29. Let e_1, e_2, e_3 denote, respectively, the elementary row operations

"Interchange rows R_1 and R_2," "Replace R_3 by $7R_3$," "Replace R_2 by $-3R_1 + R_2$"

Find the corresponding three-square elementary matrices E_1, E_2, E_3. Apply each operation to the 3×3 identity matrix I_3 to obtain

$$E_1 = \begin{bmatrix} 0 & 1 & 0 \\ 1 & 0 & 0 \\ 0 & 0 & 1 \end{bmatrix}, \qquad E_2 = \begin{bmatrix} 1 & 0 & 0 \\ 0 & 1 & 0 \\ 0 & 0 & 7 \end{bmatrix}, \qquad E_3 = \begin{bmatrix} 1 & 0 & 0 \\ -3 & 1 & 0 \\ 0 & 0 & 1 \end{bmatrix}$$

3.30. Consider the elementary row operations in Problem 3.29.

(a) Describe the inverse operations $e_1^{-1}, e_2^{-1}, e_3^{-1}$.

(b) Find the corresponding three-square elementary matrices E_1', E_2', E_3'.

(c) What is the relationship between the matrices E_1', E_2', E_3' and the matrices E_1, E_2, E_3?

(a) The inverses of e_1, e_2, e_3 are, respectively,

"Interchange rows R_1 and R_2," "Replace R_3 by $\frac{1}{7}R_3$," "Replace R_2 by $3R_1 + R_2$."

(b) Apply each inverse operation to the 3×3 identity matrix I_3 to obtain

$$E_1' = \begin{bmatrix} 0 & 1 & 0 \\ 1 & 0 & 0 \\ 0 & 0 & 1 \end{bmatrix}, \qquad E_2' = \begin{bmatrix} 1 & 0 & 0 \\ 0 & 1 & 0 \\ 0 & 0 & \frac{1}{7} \end{bmatrix}, \qquad E_3' = \begin{bmatrix} 1 & 0 & 0 \\ 3 & 1 & 0 \\ 0 & 0 & 1 \end{bmatrix}$$

(c) The matrices E_1', E_2', E_3' are, respectively, the inverses of the matrices E_1, E_2, E_3.

3.31. Write each of the following matrices as a product of elementary matrices:

$$\text{(a)}\ A = \begin{bmatrix} 1 & -3 \\ -2 & 4 \end{bmatrix}, \qquad \text{(b)}\ B = \begin{bmatrix} 1 & 2 & 3 \\ 0 & 1 & 4 \\ 0 & 0 & 1 \end{bmatrix}, \qquad \text{(c)}\ C = \begin{bmatrix} 1 & 1 & 2 \\ 2 & 3 & 8 \\ -3 & -1 & 2 \end{bmatrix}$$

The following three steps write a matrix M as a product of elementary matrices:

Step 1. Row reduce M to the identity matrix I, keeping track of the elementary row operations.

Step 2. Write down the inverse row operations.

Step 3. Write M as the product of the elementary matrices corresponding to the inverse operations. This gives the desired result.

If a zero row appears in Step 1, then M is not row equivalent to the identity matrix I, and M cannot be written as a product of elementary matrices.

(a) (1) We have

$$A = \begin{bmatrix} 1 & -3 \\ -2 & 4 \end{bmatrix} \sim \begin{bmatrix} 1 & -3 \\ 0 & -2 \end{bmatrix} \sim \begin{bmatrix} 1 & -3 \\ 0 & 1 \end{bmatrix} \sim \begin{bmatrix} 1 & 0 \\ 0 & 1 \end{bmatrix} = I$$

where the row operations are, respectively,

"Replace R_2 by $2R_1 + R_2$," "Replace R_2 by $-\frac{1}{2}R_2$," "Replace R_1 by $3R_2 + R_1$"

(2) Inverse operations:

"Replace R_2 by $-2R_1 + R_2$," "Replace R_2 by $-2R_2$," "Replace R_1 by $-3R_2 + R_1$"

(3) $A = \begin{bmatrix} 1 & 0 \\ -2 & 1 \end{bmatrix} \begin{bmatrix} 1 & 0 \\ 0 & -2 \end{bmatrix} \begin{bmatrix} 1 & -3 \\ 0 & 1 \end{bmatrix}$

(b) (1) We have

$$B = \begin{bmatrix} 1 & 2 & 3 \\ 0 & 1 & 4 \\ 0 & 0 & 1 \end{bmatrix} \sim \begin{bmatrix} 1 & 2 & 0 \\ 0 & 1 & 0 \\ 0 & 0 & 1 \end{bmatrix} \sim \begin{bmatrix} 1 & 0 & 0 \\ 0 & 1 & 0 \\ 0 & 0 & 1 \end{bmatrix} = I$$

where the row operations are, respectively,

"Replace R_2 by $-4R_3 + R_2$," "Replace R_1 by $-3R_3 + R_1$," "Replace R_1 by $-2R_2 + R_1$"

(2) Inverse operations:

"Replace R_2 by $4R_3 + R_2$," "Replace R_1 by $3R_3 + R_1$," "Replace R_1 by $2R_2 + R_1$"

(3) $B = \begin{bmatrix} 1 & 0 & 0 \\ 0 & 1 & 4 \\ 0 & 0 & 1 \end{bmatrix} \begin{bmatrix} 1 & 0 & 3 \\ 0 & 1 & 0 \\ 0 & 0 & 1 \end{bmatrix} \begin{bmatrix} 1 & 2 & 0 \\ 0 & 1 & 0 \\ 0 & 0 & 1 \end{bmatrix}$

(c) (1) First row reduce C to echelon form. We have

$$C = \begin{bmatrix} 1 & 1 & 2 \\ 2 & 3 & 8 \\ -3 & -1 & 2 \end{bmatrix} \sim \begin{bmatrix} 1 & 1 & 2 \\ 0 & 1 & 4 \\ 0 & 2 & 8 \end{bmatrix} \sim \begin{bmatrix} 1 & 1 & 2 \\ 0 & 1 & 4 \\ 0 & 0 & 0 \end{bmatrix}$$

In echelon form, C has a zero row. "STOP." The matrix C cannot be row reduced to the identity matrix I, and C cannot be written as a product of elementary matrices. (We note, in particular, that C has no inverse.)

3.32. Find the inverse of (a) $A = \begin{bmatrix} 1 & 2 & -4 \\ -1 & -1 & 5 \\ 2 & 7 & -3 \end{bmatrix}$, (b) $B = \begin{bmatrix} 1 & 3 & -4 \\ 1 & 5 & -1 \\ 3 & 13 & -6 \end{bmatrix}$.

(a) Form the matrix $M = [A, I]$ and row reduce M to echelon form:

$$M = \left[\begin{array}{ccc|ccc} 1 & 2 & -4 & 1 & 0 & 0 \\ -1 & -1 & 5 & 0 & 1 & 0 \\ 2 & 7 & -3 & 0 & 0 & 1 \end{array}\right] \sim \left[\begin{array}{ccc|ccc} 1 & 2 & -4 & 1 & 0 & 0 \\ 0 & 1 & 1 & 1 & 1 & 0 \\ 0 & 3 & 5 & -2 & 0 & 1 \end{array}\right]$$

$$\sim \left[\begin{array}{ccc|ccc} 1 & 2 & -4 & 1 & 0 & 0 \\ 0 & 1 & 1 & 1 & 1 & 0 \\ 0 & 0 & 2 & -5 & -3 & 1 \end{array}\right]$$

In echelon form, the left half of M is in triangular form; hence, A has an inverse. Further reduce M to row canonical form:

$$M \sim \left[\begin{array}{ccc|ccc} 1 & 2 & 0 & -9 & -6 & 2 \\ 0 & 1 & 0 & \frac{7}{2} & \frac{5}{2} & -\frac{1}{2} \\ 0 & 0 & 1 & -\frac{5}{2} & -\frac{3}{2} & \frac{1}{2} \end{array}\right] \sim \left[\begin{array}{ccc|ccc} 1 & 0 & 0 & -16 & -11 & 3 \\ 0 & 1 & 0 & \frac{7}{2} & \frac{5}{2} & -\frac{1}{2} \\ 0 & 0 & 1 & -\frac{5}{2} & -\frac{3}{2} & \frac{1}{2} \end{array}\right]$$

The final matrix has the form $[I, A^{-1}]$; that is, A^{-1} is the right half of the last matrix. Thus,

$$A^{-1} = \begin{bmatrix} -16 & -11 & 3 \\ \frac{7}{2} & \frac{5}{2} & -\frac{1}{2} \\ -\frac{5}{2} & -\frac{3}{2} & \frac{1}{2} \end{bmatrix}$$

(b) Form the matrix $M = [B, I]$ and row reduce M to echelon form:

$$M = \left[\begin{array}{ccc|ccc} 1 & 3 & -4 & 1 & 0 & 0 \\ 1 & 5 & -1 & 0 & 1 & 0 \\ 3 & 13 & -6 & 0 & 0 & 1 \end{array}\right] \sim \left[\begin{array}{ccc|ccc} 1 & 3 & -4 & 1 & 0 & 0 \\ 0 & 2 & 3 & -1 & 1 & 0 \\ 0 & 4 & 6 & -3 & 0 & 1 \end{array}\right] \sim \left[\begin{array}{ccc|ccc} 1 & 3 & -4 & 1 & 0 & 0 \\ 0 & 2 & 3 & -1 & 1 & 0 \\ 0 & 0 & 0 & -1 & -2 & 1 \end{array}\right]$$

In echelon form, M has a zero row in its left half; that is, B is not row reducible to triangular form. Accordingly, B has no inverse.

3.33. Show that every elementary matrix E is invertible, and its inverse is an elementary matrix.

Let E be the elementary matrix corresponding to the elementary operation e; that is, $e(I) = E$. Let e' be the inverse operation of e and let E' be the corresponding elementary matrix; that is, $e'(I) = E'$. Then

$$I = e'(e(I)) = e'(E) = E'E \quad \text{and} \quad I = e(e'(I)) = e(E') = EE'$$

Therefore, E' is the inverse of E.

3.34. Prove Theorem 3.16: Let e be an elementary row operation and let E be the corresponding m-square elementary matrix; that is, $E = e(I)$. Then $e(A) = EA$, where A is any $m \times n$ matrix.

Let R_i be the row i of A; we denote this by writing $A = [R_1, \ldots, R_m]$. If B is a matrix for which AB is defined then $AB = [R_1B, \ldots, R_mB]$. We also let

$$e_i = (0, \ldots, 0, \hat{1}, 0, \ldots, 0), \qquad \hat{} = i$$

Here $\hat{} = i$ means 1 is the ith entry. One can show (Problem 2.45) that $e_iA = R_i$. We also note that $I = [e_1, e_2, \ldots, e_m]$ is the identity matrix.

(i) Let e be the elementary row operation "Interchange rows R_i and R_j." Then, for $\hat{} = i$ and $\hat{\hat{}} = j$,

$$E = e(I) = [e_1, \ldots, \widehat{e_j}, \ldots, \widehat{\widehat{e_i}}, \ldots, e_m]$$

and

$$e(A) = [R_1, \ldots, \widehat{R_j}, \ldots, \widehat{\widehat{R_i}}, \ldots, R_m]$$

Thus,

$$EA = [e_1A, \ldots, \widehat{e_jA}, \ldots, \widehat{\widehat{e_iA}}, \ldots, e_mA] = [R_1, \ldots, \widehat{R_j}, \ldots, \widehat{\widehat{R_i}}, \ldots, R_m] = e(A)$$

(ii) Let e be the elementary row operation "Replace R_i by kR_i $(k \neq 0)$." Then, for $\hat{} = i$,

$$E = e(I) = [e_1, \ldots, \widehat{ke_i}, \ldots, e_m]$$

and

$$e(A) = [R_1, \ldots, \widehat{kR_i}, \ldots, R_m]$$

Thus,

$$EA = [e_1A, \ldots, \widehat{ke_iA}, \ldots, e_mA] = [R_1, \ldots, \widehat{kR_i}, \ldots, R_m] = e(A)$$

(iii) Let e be the elementary row operation "Replace R_i by $kR_j + R_i$." Then, for $\hat{} = i$,

$$E = e(I) = [e_1, \ \ldots, \ \widehat{ke_j + e_i}, \ \ldots, \ e_m]$$

and

$$e(A) = [R_1, \ \ldots, \ \widehat{kR_j + R_i}, \ \ldots, \ R_m]$$

Using $(ke_j + e_i)A = k(e_jA) + e_iA = kR_j + R_i$, we have

$$\begin{aligned} EA &= [e_1A, \quad \ldots, \quad (ke_j + e_i)A, \quad \ldots, \quad e_mA] \\ &= [R_1, \quad \ldots, \quad \widehat{kR_j + R_i}, \quad \ldots, \quad R_m] = e(A) \end{aligned}$$

3.35. Prove Theorem 3.17: Let A be a square matrix. Then the following are equivalent:

(a) A is invertible (nonsingular).

(b) A is row equivalent to the identity matrix I.

(c) A is a product of elementary matrices.

Suppose A is invertible and suppose A is row equivalent to matrix B in row canonical form. Then there exist elementary matrices $E_1, E_2, \ldots, E_s$ such that $E_s \ldots E_2E_1A = B$. Because A is invertible and each elementary matrix is invertible, B is also invertible. But if $B \neq I$, then B has a zero row; whence B is not invertible. Thus, $B = I$, and (a) implies (b).

If (b) holds, then there exist elementary matrices $E_1, E_2, \ldots, E_s$ such that $E_s \ldots E_2E_1A = I$. Hence, $A = (E_s \ldots E_2E_1)^{-1} = E_1^{-1}E_2^{-1} \ldots E_s^{-1}$. But the E_i^{-1} are also elementary matrices. Thus (b) implies (c).

If (c) holds, then $A = E_1E_2 \ldots E_s$. The E_i are invertible matrices; hence, their product A is also invertible. Thus, (c) implies (a). Accordingly, the theorem is proved.

3.36. Prove Theorem 3.18: If $AB = I$, then $BA = I$, and hence $B = A^{-1}$.

Suppose A is not invertible. Then A is not row equivalent to the identity matrix I, and so A is row equivalent to a matrix with a zero row. In other words, there exist elementary matrices $E_1, \ldots, E_s$ such that $E_s \ldots E_2E_1A$ has a zero row. Hence, $E_s \ldots E_2E_1AB = E_s \ldots E_2E_1$, an invertible matrix, also has a zero row. But invertible matrices cannot have zero rows; hence A is invertible, with inverse A^{-1}. Then also,

$$B = IB = (A^{-1}A)B = A^{-1}(AB) = A^{-1}I = A^{-1}$$

3.37. Prove Theorem 3.19: B is row equivalent to A (written $B \sim A$) if and only if there exists a nonsingular matrix P such that $B = PA$.

If $B \sim A$, then $B = e_s(\ldots(e_2(e_1(A)))\ldots) = E_s \ldots E_2E_1A = PA$ where $P = E_s \ldots E_2E_1$ is nonsingular. Conversely, suppose $B = PA$, where P is nonsingular. By Theorem 3.17, P is a product of elementary matrices, and so B can be obtained from A by a sequence of elementary row operations; that is, $B \sim A$. Thus, the theorem is proved.

3.38. Prove Theorem 3.21: Every $m \times n$ matrix A is equivalent to a unique block matrix of the form $\begin{bmatrix} I_r & 0 \\ 0 & 0 \end{bmatrix}$, where I_r is the $r \times r$ identity matrix.

The proof is constructive, in the form of an algorithm.

Step 1. Row reduce A to row canonical form, with leading nonzero entries $a_{1j_1}, a_{2j_2}, \ldots, a_{rj_r}$.

Step 2. Interchange C_1 and C_{1j_1}, interchange C_2 and $C_{2j_2}, \ldots,$ and interchange C_r and C_{jr}. This gives a matrix in the form $\left[\begin{array}{c|c} I_r & B \\ \hline 0 & 0 \end{array}\right]$, with leading nonzero entries $a_{11}, a_{22}, \ldots, a_{rr}$.

Step 3. Use column operations, with the a_{ii} as pivots, to replace each entry in B with a zero; that is, for $i = 1, 2, \ldots, r$ and $j = r+1, r+2, \ldots, n$, apply the operation $-b_{ij}C_i + C_j \to C_j$.

The final matrix has the desired form $\left[\begin{array}{c|c} I_r & 0 \\ \hline 0 & 0 \end{array}\right]$.

Lu Factorization

3.39. Find the LU factorization of (a) $A = \begin{bmatrix} 1 & -3 & 5 \\ 2 & -4 & 7 \\ -1 & -2 & 1 \end{bmatrix}$, (b) $B = \begin{bmatrix} 1 & 4 & -3 \\ 2 & 8 & 1 \\ -5 & -9 & 7 \end{bmatrix}$.

(a) Reduce A to triangular form by the following operations:

"Replace R_2 by $-2R_1 + R_2$," "Replace R_3 by $R_1 + R_3$," and then
"Replace R_3 by $\frac{5}{2}R_2 + R_3$"

These operations yield the following, where the triangular form is U:

$$A \sim \begin{bmatrix} 1 & -3 & 5 \\ 0 & 2 & -3 \\ 0 & -5 & 6 \end{bmatrix} \sim \begin{bmatrix} 1 & -3 & 5 \\ 0 & 2 & -3 \\ 0 & 0 & -\frac{3}{2} \end{bmatrix} = U \quad \text{and} \quad L = \begin{bmatrix} 1 & 0 & 0 \\ 2 & 1 & 0 \\ -1 & -\frac{5}{2} & 1 \end{bmatrix}$$

The entries $2, -1, -\frac{5}{2}$ in L are the negatives of the multipliers $-2, 1, \frac{5}{2}$ in the above row operations. (As a check, multiply L and U to verify $A = LU$.)

(b) Reduce B to triangular form by first applying the operations "Replace R_2 by $-2R_1 + R_2$" and "Replace R_3 by $5R_1 + R_3$." These operations yield

$$B \sim \begin{bmatrix} 1 & 4 & -3 \\ 0 & 0 & 7 \\ 0 & 11 & -8 \end{bmatrix}.$$

Observe that the second diagonal entry is 0. Thus, B cannot be brought into triangular form without row interchange operations. Accordingly, B is not LU-factorable. (There does exist a PLU factorization of such a matrix B, where P is a permutation matrix, but such a factorization lies beyond the scope of this text.)

3.40. Find the LDU factorization of the matrix A in Problem 3.39.

The $A = LDU$ factorization refers to the situation where L is a lower triangular matrix with 1's on the diagonal (as in the LU factorization of A), D is a diagonal matrix, and U is an upper triangular matrix with 1's on the diagonal. Thus, simply factor out the diagonal entries in the matrix U in the above LU factorization of A to obtain D and L. That is,

$$L = \begin{bmatrix} 1 & 0 & 0 \\ 2 & 1 & 0 \\ -1 & -\frac{5}{2} & 1 \end{bmatrix}, \quad D = \begin{bmatrix} 1 & 0 & 0 \\ 0 & 2 & 0 \\ 0 & 0 & -\frac{3}{2} \end{bmatrix}, \quad U = \begin{bmatrix} 1 & -3 & 5 \\ 0 & 1 & -3 \\ 0 & 0 & 1 \end{bmatrix}$$

3.41. Find the LU factorization of the matrix $A = \begin{bmatrix} 1 & 2 & 1 \\ 2 & 3 & 3 \\ -3 & -10 & 2 \end{bmatrix}$.

Reduce A to triangular form by the following operations:

(1) "Replace R_2 by $-2R_1 + R_2$," (2) "Replace R_3 by $3R_1 + R_3$," (3) "Replace R_3 by $-4R_2 + R_3$"

These operations yield the following, where the triangular form is U:

$$A \sim \begin{bmatrix} 1 & 2 & 1 \\ 0 & -1 & 1 \\ 0 & -4 & 5 \end{bmatrix} \sim \begin{bmatrix} 1 & 2 & 1 \\ 0 & -1 & 1 \\ 0 & 0 & 1 \end{bmatrix} = U \quad \text{and} \quad L = \begin{bmatrix} 1 & 0 & 0 \\ 2 & 1 & 0 \\ -3 & 4 & 1 \end{bmatrix}$$

The entries $2, -3, 4$ in L are the negatives of the multipliers $-2, 3, -4$ in the above row operations. (As a check, multiply L and U to verify $A = LU$.)

3.42. Let A be the matrix in Problem 3.41. Find X_1, X_2, X_3, where X_i is the solution of $AX = B_i$ for (a) $B_1 = (1, 1, 1)$, (b) $B_2 = B_1 + X_1$, (c) $B_3 = B_2 + X_2$.

(a) Find $L^{-1}B_1$ by applying the row operations (1), (2), and then (3) in Problem 3.41 to B_1:

$$B_1 = \begin{bmatrix} 1 \\ 1 \\ 1 \end{bmatrix} \xrightarrow{(1) \text{ and } (2)} \begin{bmatrix} 1 \\ -1 \\ 4 \end{bmatrix} \xrightarrow{(3)} \begin{bmatrix} 1 \\ -1 \\ 8 \end{bmatrix}$$

Solve $UX = B$ for $B = (1, -1, 8)$ by back-substitution to obtain $X_1 = (-25, 9, 8)$.

(b) First find $B_2 = B_1 + X_1 = (1, 1, 1) + (-25, 9, 8) = (-24, 10, 9)$. Then as above

$$B_2 = [-24, 10, 9]^T \xrightarrow{(1) \text{ and } (2)} [-24, 58, -63]^T \xrightarrow{(3)} [-24, 58, -295]^T$$

Solve $UX = B$ for $B = (-24, 58, -295)$ by back-substitution to obtain $X_2 = (943, -353, -295)$.

(c) First find $B_3 = B_2 + X_2 = (-24, 10, 9) + (943, -353, -295) = (919, -343, -286)$. Then, as above

$$B_3 = [943, -353, -295]^T \xrightarrow{(1) \text{ and } (2)} [919, -2181, 2671]^T \xrightarrow{(3)} [919, -2181, 11\,395]^T$$

Solve $UX = B$ for $B = (919, -2181, 11\,395)$ by back-substitution to obtain

$$X_3 = (-37\,628, 13\,576, 11\,395).$$

Miscellaneous Problems

3.43. Let L be a linear combination of the m equations in n unknowns in the system (3.2). Say L is the equation

$$(c_1a_{11} + \cdots + c_ma_{m1})x_1 + \cdots + (c_1a_{1n} + \cdots + c_ma_{mn})x_n = c_1b_1 + \cdots + c_mb_m \qquad (1)$$

Show that any solution of the system (3.2) is also a solution of L.

Let $u = (k_1, \ldots, k_n)$ be a solution of (3.2). Then

$$a_{i1}k_1 + a_{i2}k_2 + \cdots + a_{in}k_n = b_i \qquad (i = 1, 2, \ldots, m) \qquad (2)$$

Substituting u in the left-hand side of (1) and using (2), we get

$$\begin{aligned}(c_1a_{11} + \cdots + c_ma_{m1})k_1 &+ \cdots + (c_1a_{1n} + \cdots + c_ma_{mn})k_n \\ &= c_1(a_{11}k_1 + \cdots + a_{1n}k_n) + \cdots + c_m(a_{m1}k_1 + \cdots + a_{mn}k_n) \\ &= c_1b_1 + \cdots + c_mb_m\end{aligned}$$

This is the right-hand side of (1); hence, u is a solution of (1).

3.44. Suppose a system $\mathscr{M}$ of linear equations is obtained from a system $\mathscr{L}$ by applying an elementary operation (page 64). Show that $\mathscr{M}$ and $\mathscr{L}$ have the same solutions.

Each equation L in $\mathscr{M}$ is a linear combination of equations in $\mathscr{L}$. Hence, by Problem 3.43, any solution of $\mathscr{L}$ will also be a solution of $\mathscr{M}$. On the other hand, each elementary operation has an inverse elementary operation, so $\mathscr{L}$ can be obtained from $\mathscr{M}$ by an elementary operation. This means that any solution of $\mathscr{M}$ is a solution of $\mathscr{L}$. Thus, $\mathscr{L}$ and $\mathscr{M}$ have the same solutions.

3.45. Prove Theorem 3.4: Suppose a system $\mathscr{M}$ of linear equations is obtained from a system $\mathscr{L}$ by a sequence of elementary operations. Then $\mathscr{M}$ and $\mathscr{L}$ have the same solutions.

Each step of the sequence does not change the solution set (Problem 3.44). Thus, the original system $\mathscr{L}$ and the final system $\mathscr{M}$ (and any system in between) have the same solutions.

3.46. A system $\mathscr{L}$ of linear equations is said to be *consistent* if no linear combination of its equations is a degenerate equation L with a nonzero constant. Show that $\mathscr{L}$ is consistent if and only if $\mathscr{L}$ is reducible to echelon form.

Suppose $\mathscr{L}$ is reducible to echelon form. Then $\mathscr{L}$ has a solution, which must also be a solution of every linear combination of its equations. Thus, L, which has no solution, cannot be a linear combination of the equations in $\mathscr{L}$. Thus, $\mathscr{L}$ is consistent.

On the other hand, suppose $\mathscr{L}$ is not reducible to echelon form. Then, in the reduction process, it must yield a degenerate equation L with a nonzero constant, which is a linear combination of the equations in $\mathscr{L}$. Therefore, $\mathscr{L}$ is not consistent; that is, $\mathscr{L}$ is inconsistent.

3.47. Suppose u and v are distinct vectors. Show that, for distinct scalars k, the vectors $u + k(u - v)$ are distinct.

Suppose $u + k_1(u - v) = u + k_2(u - v)$. We need only show that $k_1 = k_2$. We have

$$k_1(u - v) = k_2(u - v), \qquad \text{and so} \qquad (k_1 - k_2)(u - v) = 0$$

Because u and v are distinct, $u - v \neq 0$. Hence, $k_1 - k_2 = 0$, and so $k_1 = k_2$.

3.48. Suppose AB is defined. Prove

(a) Suppose A has a zero row. Then AB has a zero row.

(b) Suppose B has a zero column. Then AB has a zero column.

(a) Let R_i be the zero row of A, and $C_1, \ldots, C_n$ the columns of B. Then the ith row of AB is

$$(R_iC_1, R_iC_2, \ldots, R_iC_n) = (0, 0, 0, \ldots, 0)$$

(b) B^T has a zero row, and so $B^TA^T = (AB)^T$ has a zero row. Hence, AB has a zero column.

SUPPLEMENTARY PROBLEMS

Linear Equations, 2 × 2 Systems

3.49. Determine whether each of the following systems is linear:

(a) $3x - 4y + 2yz = 8$, (b) $ex + 3y = \pi$, (c) $2x - 3y + kz = 4$

3.50. Solve (a) $\pi x = 2$, (b) $3x + 2 = 5x + 7 - 2x$, (c) $6x + 2 - 4x = 5 + 2x - 3$

3.51. Solve each of the following systems:

(a) $\begin{aligned} 2x + 3y &= 1 \\ 5x + 7y &= 3 \end{aligned}$ (b) $\begin{aligned} 4x - 2y &= 5 \\ -6x + 3y &= 1 \end{aligned}$ (c) $\begin{aligned} 2x - 4 &= 3y \\ 5y - x &= 5 \end{aligned}$ (d) $\begin{aligned} 2x - 4y &= 10 \\ 3x - 6y &= 15 \end{aligned}$

3.52. Consider each of the following systems in unknowns x and y:

(a) $\begin{aligned} x - ay &= 1 \\ ax - 4y &= b \end{aligned}$ (b) $\begin{aligned} ax + 3y &= 2 \\ 12x + ay &= b \end{aligned}$ (c) $\begin{aligned} x + ay &= 3 \\ 2x + 5y &= b \end{aligned}$

For which values of a does each system have a unique solution, and for which pairs of values (a, b) does each system have more than one solution?

General Systems of Linear Equations

3.53. Solve

(a) $\begin{aligned} x + y + 2z &= 4 \\ 2x + 3y + 6z &= 10 \\ 3x + 6y + 10z &= 17 \end{aligned}$ (b) $\begin{aligned} x - 2y + 3z &= 2 \\ 2x - 3y + 8z &= 7 \\ 3x - 4y + 13z &= 8 \end{aligned}$ (c) $\begin{aligned} x + 2y + 3z &= 3 \\ 2x + 3y + 8z &= 4 \\ 5x + 8y + 19z &= 11 \end{aligned}$

3.54. Solve

(a) $\begin{aligned} x - 2y &= 5 \\ 2x + 3y &= 3 \\ 3x + 2y &= 7 \end{aligned}$ (b) $\begin{aligned} x + 2y - 3z + 2t &= 2 \\ 2x + 5y - 8z + 6t &= 5 \\ 3x + 4y - 5z + 2t &= 4 \end{aligned}$ (c) $\begin{aligned} x + 2y + 4z - 5t &= 3 \\ 3x - y + 5z + 2t &= 4 \\ 5x - 4y + 6z + 9t &= 2 \end{aligned}$

3.55. Solve

(a) $\begin{aligned} 2x - y - 4z &= 2 \\ 4x - 2y - 6z &= 5 \\ 6x - 3y - 8z &= 8 \end{aligned}$ (b) $\begin{aligned} x + 2y - z + 3t &= 3 \\ 2x + 4y + 4z + 3t &= 9 \\ 3x + 6y - z + 8t &= 10 \end{aligned}$

3.56. Consider each of the following systems in unknowns x, y, z:

(a) $\begin{aligned} x - 2y &= 1 \\ x - y + az &= 2 \\ ay + 9z &= b \end{aligned}$ (b) $\begin{aligned} x + 2y + 2z &= 1 \\ x + ay + 3z &= 3 \\ x + 11y + az &= b \end{aligned}$ (c) $\begin{aligned} x + y + az &= 1 \\ x + ay + z &= 4 \\ ax + y + z &= b \end{aligned}$

For which values of a does the system have a unique solution, and for which pairs of values (a, b) does the system have more than one solution? The value of b does not have any effect on whether the system has a unique solution. Why?

Linear Combinations, Homogeneous Systems

3.57. Write v as a linear combination of u_1, u_2, u_3, where

(a) $v = (4, -9, 2)$, $u_1 = (1, 2, -1)$, $u_2 = (1, 4, 2)$, $u_3 = (1, -3, 2)$;
(b) $v = (1, 3, 2)$, $u_1 = (1, 2, 1)$, $u_2 = (2, 6, 5)$, $u_3 = (1, 7, 8)$;
(c) $v = (1, 4, 6)$, $u_1 = (1, 1, 2)$, $u_2 = (2, 3, 5)$, $u_3 = (3, 5, 8)$.

3.58. Let $u_1 = (1, 1, 2)$, $u_2 = (1, 3, -2)$, $u_3 = (4, -2, -1)$ in $\mathbf{R}^3$. Show that u_1, u_2, u_3 are orthogonal, and write v as a linear combination of u_1, u_2, u_3, where (a) $v = (5, -5, 9)$, (b) $v = (1, -3, 3)$, (c) $v = (1, 1, 1)$. (*Hint:* Use Fourier coefficients.)

3.59. Find the dimension and a basis of the general solution W of each of the following homogeneous systems:

(a) $\begin{aligned} x - y + 2z &= 0 \\ 2x + y + z &= 0 \\ 5x + y + 4z &= 0 \end{aligned}$ (b) $\begin{aligned} x + 2y - 3z &= 0 \\ 2x + 5y + 2z &= 0 \\ 3x - y - 4z &= 0 \end{aligned}$ (c) $\begin{aligned} x + 2y + 3z + t &= 0 \\ 2x + 4y + 7z + 4t &= 0 \\ 3x + 6y + 10z + 5t &= 0 \end{aligned}$

3.60. Find the dimension and a basis of the general solution W of each of the following systems:

(a) $\begin{aligned} x_1 + 3x_2 + 2x_3 - x_4 - x_5 &= 0 \\ 2x_1 + 6x_2 + 5x_3 + x_4 - x_5 &= 0 \\ 5x_1 + 15x_2 + 12x_3 + x_4 - 3x_5 &= 0 \end{aligned}$ (b) $\begin{aligned} 2x_1 - 4x_2 + 3x_3 - x_4 + 2x_5 &= 0 \\ 3x_1 - 6x_2 + 5x_3 - 2x_4 + 4x_5 &= 0 \\ 5x_1 - 10x_2 + 7x_3 - 3x_4 + 18x_5 &= 0 \end{aligned}$

Echelon Matrices, Row Canonical Form

3.61. Reduce each of the following matrices to echelon form and then to row canonical form:

(a) $\begin{bmatrix} 1 & 1 & 2 \\ 2 & 4 & 9 \\ 1 & 5 & 12 \end{bmatrix}$, (b) $\begin{bmatrix} 1 & 2 & -1 & 2 & 1 \\ 2 & 4 & 1 & -2 & 5 \\ 3 & 6 & 3 & -7 & 7 \end{bmatrix}$, (c) $\begin{bmatrix} 2 & 4 & 2 & -2 & 5 & 1 \\ 3 & 6 & 2 & 2 & 0 & 4 \\ 4 & 8 & 2 & 6 & -5 & 7 \end{bmatrix}$

3.62. Reduce each of the following matrices to echelon form and then to row canonical form:

(a) $\begin{bmatrix} 1 & 2 & 1 & 2 & 1 & 2 \\ 2 & 4 & 3 & 5 & 5 & 7 \\ 3 & 6 & 4 & 9 & 10 & 11 \\ 1 & 2 & 4 & 3 & 6 & 9 \end{bmatrix}$, (b) $\begin{bmatrix} 0 & 1 & 2 & 3 \\ 0 & 3 & 8 & 12 \\ 0 & 0 & 4 & 6 \\ 0 & 2 & 7 & 10 \end{bmatrix}$, (c) $\begin{bmatrix} 1 & 3 & 1 & 3 \\ 2 & 8 & 5 & 10 \\ 1 & 7 & 7 & 11 \\ 3 & 11 & 7 & 15 \end{bmatrix}$

3.63. Using only 0's and 1's, list all possible 2×2 matrices in row canonical form.

3.64. Using only 0's and 1's, find the number n of possible 3×3 matrices in row canonical form.

Elementary Matrices, Applications

3.65. Let e_1, e_2, e_3 denote, respectively, the following elementary row operations:

"Interchange R_2 and R_3," "Replace R_2 by $3R_2$," "Replace R_1 by $2R_3 + R_1$"

(a) Find the corresponding elementary matrices E_1, E_2, E_3.
(b) Find the inverse operations $e_1^{-1}, e_2^{-1}, e_3^{-1}$; their corresponding elementary matrices E_1', E_2', E_3'; and the relationship between them and E_1, E_2, E_3.
(c) Describe the corresponding elementary column operations f_1, f_2, f_3.
(d) Find elementary matrices F_1, F_2, F_3 corresponding to f_1, f_2, f_3, and the relationship between them and E_1, E_2, E_3.

3.66. Express each of the following matrices as a product of elementary matrices:

$$A = \begin{bmatrix} 1 & 2 \\ 3 & 4 \end{bmatrix}, \quad B = \begin{bmatrix} 3 & -6 \\ -2 & 4 \end{bmatrix}, \quad C = \begin{bmatrix} 2 & 6 \\ -3 & -7 \end{bmatrix}, \quad D = \begin{bmatrix} 1 & 2 & 0 \\ 0 & 1 & 3 \\ 3 & 8 & 7 \end{bmatrix}$$

3.67. Find the inverse of each of the following matrices (if it exists):

$$A = \begin{bmatrix} 1 & -2 & -1 \\ 2 & -3 & 1 \\ 3 & -4 & 4 \end{bmatrix}, \quad B = \begin{bmatrix} 1 & 2 & 3 \\ 2 & 6 & 1 \\ 3 & 10 & -1 \end{bmatrix}, \quad C = \begin{bmatrix} 1 & 3 & -2 \\ 2 & 8 & -3 \\ 1 & 7 & 1 \end{bmatrix}, \quad D = \begin{bmatrix} 2 & 1 & -1 \\ 5 & 2 & -3 \\ 0 & 2 & 1 \end{bmatrix}$$

3.68. Find the inverse of each of the following $n \times n$ matrices:

(a) A has 1's on the diagonal and *superdiagonal* (entries directly above the diagonal) and 0's elsewhere.
(b) B has 1's on and above the diagonal, and 0's below the diagonal.

Lu Factorization

3.69. Find the LU factorization of each of the following matrices:

(a) $\begin{bmatrix} 1 & -1 & -1 \\ 3 & -4 & -2 \\ 2 & -3 & -2 \end{bmatrix}$, (b) $\begin{bmatrix} 1 & 3 & -1 \\ 2 & 5 & 1 \\ 3 & 4 & 2 \end{bmatrix}$, (c) $\begin{bmatrix} 2 & 3 & 6 \\ 4 & 7 & 9 \\ 3 & 5 & 4 \end{bmatrix}$, (d) $\begin{bmatrix} 1 & 2 & 3 \\ 2 & 4 & 7 \\ 3 & 7 & 10 \end{bmatrix}$

3.70. Let A be the matrix in Problem 3.69(a). Find X_1, X_2, X_3, X_4, where

(a) X_1 is the solution of $AX = B_1$, where $B_1 = (1, 1, 1)^T$.
(b) For $k > 1$, X_k is the solution of $AX = B_k$, where $B_k = B_{k-1} + X_{k-1}$.

3.71. Let B be the matrix in Problem 3.69(b). Find the LDU factorization of B.

Miscellaneous Problems

3.72. Consider the following systems in unknowns x and y:

$$\text{(a)} \quad \begin{matrix} ax + by = 1 \\ cx + dy = 0 \end{matrix} \qquad \text{(b)} \quad \begin{matrix} ax + by = 0 \\ cx + dy = 1 \end{matrix}$$

Suppose $D = ad - bc \neq 0$. Show that each system has the unique solution:

(a) $x = d/D, y = -c/D$, (b) $x = -b/D, y = a/D$.

3.73. Find the inverse of the row operation "Replace R_i by $kR_j + k'R_i$ $(k' \neq 0)$."

3.74. Prove that deleting the last column of an echelon form (respectively, the row canonical form) of an augmented matrix $M = [A, B]$ yields an echelon form (respectively, the row canonical form) of A.

3.75. Let e be an elementary row operation and E its elementary matrix, and let f be the corresponding elementary column operation and F its elementary matrix. Prove

(a) $f(A) = (e(A^T))^T$, (b) $F = E^T$, (c) $f(A) = AF$.

3.76. Matrix A is *equivalent* to matrix B, written $A \approx B$, if there exist nonsingular matrices P and Q such that $B = PAQ$. Prove that $\approx$ is an *equivalence* relation; that is,

(a) $A \approx A$, (b) If $A \approx B$, then $B \approx A$, (c) If $A \approx B$ and $B \approx C$, then $A \approx C$.

ANSWERS TO SUPPLEMENTARY PROBLEMS

Notation: $A = [R_1;\ \ R_2;\ \ \ldots]$ denotes the matrix A with rows $R_1, R_2, \ldots$. The elements in each row are separated by commas (which may be omitted with single digits), the rows are separated by semicolons, and 0 denotes a zero row. For example,

$$A = [1,2,3,4;\quad 5,-6,7,-8;\quad 0] = \begin{bmatrix} 1 & 2 & 3 & 4 \\ 5 & -6 & 7 & -8 \\ 0 & 0 & 0 & 0 \end{bmatrix}$$

3.49. (a) no, (b) yes, (c) linear in x, y, z, not linear in x, y, z, k

3.50. (a) $x = 2/\pi$, (b) no solution, (c) every scalar k is a solution

3.51. (a) $(2, -1)$, (b) no solution, (c) $(5, 2)$, (d) $(5 + 2a,\ a)$

3.52. (a) $a \neq \pm 2$, $(2, 2)$, $(-2, -2)$, (b) $a \neq \pm 6$, $(6, 4)$, $(-6, -4)$, (c) $a \neq \frac{5}{2}$, $(\frac{5}{2}, 6)$

3.53. (a) $(2, 1, \frac{1}{2})$, (b) no solution, (c) $u = (-7a - 1,\ 2a + 2,\ a)$.

3.54. (a) $(3, -1)$, (b) $u = (-a + 2b,\ 1 + 2a - 2b,\ a,\ b)$, (c) no solution

3.55. (a) $u = (\frac{1}{2}a + 2,\ \ a,\ \ \frac{1}{2})$, (b) $u = (\frac{1}{2}(7 - 5b - 4a),\ \ a,\ \ \frac{1}{2}(1 + b),\ \ b)$

3.56. (a) $a \neq \pm 3$, $(3, 3)$, $(-3, -3)$, (b) $a \neq 5$ and $a \neq -1$, $(5, 7)$, $(-1, -5)$,
(c) $a \neq 1$ and $a \neq -2$, $(-2, 5)$

3.57. (a) $2, -1, 3$, (b) $6, -3, 1$, (c) not possible

3.58. (a) $3, -2, 1$, (b) $\frac{2}{3}, -1, \frac{1}{3}$, (c) $\frac{2}{3}, \frac{1}{7}, \frac{1}{21}$

3.59. (a) $\dim W = 1$, $u_1 = (-1, 1, 1)$, (b) $\dim W = 0$, no basis,
(c) $\dim W = 2$, $u_1 = (-2, 1, 0, 0)$, $u_2 = (5, 0, -2, 1)$

3.60. (a) $\dim W = 3$, $u_1 = (-3, 1, 0, 0, 0)$, $u_2 = (7, 0, -3, 1, 0)$, $u_3 = (3, 0, -1, 0, 1)$,
(b) $\dim W = 2$, $u_1 = (2, 1, 0, 0, 0)$, $u_2 = (5, 0, -5, -3, 1)$

3.61. (a) $[1, 0, -\frac{1}{2};\ \ 0, 1, \frac{5}{2};\ \ 0]$, (b) $[1, 2, 0, 0, 2;\ \ 0, 0, 1, 0, 5;\ \ 0, 0, 0, 1, 2]$,
(c) $[1, 2, 0, 4, -5, 3;\ \ 0, 0, 1, -5, \frac{15}{2}, -\frac{5}{2};\ \ 0]$

3.62. (a) $[1, 2, 0, 0, -4, -2;\ \ 0, 0, 1, 0, 1, 2;\ \ 0, 0, 0, 1, 2, 1;\ \ 0]$,
(b) $[0, 1, 0, 0;\ \ 0, 0, 1, 0;\ \ 0, 0, 0, 1;\ \ 0]$, (c) $[1, 0, 0, 4;\ \ 0, 1, 0, -1;\ \ 0, 0, 1, 2;\ \ 0]$

3.63. 5: $[1, 0;\ \ 0, 1]$, $[1, 1;\ \ 0, 0]$, $[1, 0;\ \ 0, 0]$, $[0, 1;\ \ 0, 0]$, 0

3.64. 16

3.65. (a) $[1, 0, 0;\ \ 0, 0, 1;\ \ 0, 1, 0]$, $[1, 0, 0;\ \ 0, 3, 0;\ \ 0, 0, 1]$, $[1, 0, 2;\ \ 0, 1, 0;\ \ 0, 0, 1]$,
(b) $R_2 \leftrightarrow R_3$; $\frac{1}{3}R_2 \to R_2$; $-2R_3 + R_1 \to R_1$; each $E_i' = E_i^{-1}$,
(c) $C_2 \leftrightarrow C_3, 3C_2 \to C_2, 2C_3 + C_1 \to C_1$, (d) each $F_i = E_i^T$.

3.66. $A = [1, 0;\ \ 3, 1][1, 0;\ \ 0, -2][1, 2;\ \ 0, 1]$, B is not invertible,
$C = [1, 0;\ \ -\frac{3}{2}, 1][1, 0;\ \ 0, 2][1, 6;\ \ 0, 1][2, 0;\ \ 0, 1]$,
$D = [100;\ \ 010;\ \ 301][100;\ \ 010;\ \ 021][100;\ \ 013;\ \ 001][120;\ \ 010;\ \ 001]$

3.67. $A^{-1} = [-8, 12, -5;\ \ -5, 7, -3;\ \ 1, -2, 1]$, B has no inverse,
$C^{-1} = [\frac{29}{2}, -\frac{17}{2}, \frac{7}{2};\ \ -\frac{5}{2}, \frac{3}{2}, -\frac{1}{2};\ \ 3, -2, 1]$, $D^{-1} = [8, -3, -1;\ \ -5, 2, 1;\ \ 10, -4, -1]$

3.68. $A^{-1} = [1, -1, 1, -1, \ldots; \quad 0, 1, -1, 1, -1, \ldots; \quad 0, 0, 1, -1, 1, -1, 1, \ldots; \quad \ldots; \quad \ldots; \quad 0, \ldots 0, 1]$
B^{-1} has 1's on diagonal, -1's on superdiagonal, and 0's elsewhere.

3.69. (a) $[100; \quad 310; \quad 211][1, -1, -1; \quad 0, -1, 1; \quad 0, 0, -1]$,
(b) $[100; \quad 210; \quad 351][1, 3, -1; \quad 0, -1, 3; \quad 0, 0, -10]$,
(c) $[100; \quad 210; \quad \frac{3}{2}, \frac{1}{2}; 1][2, 3, 6; \quad 0, 1, -3; \quad 0, 0, -\frac{7}{2}]$,
(d) There is no LU decomposition.

3.70. $X_1 = [1, 1, -1]^T, \quad B_2 = [2, 2, 0]^T, \quad X_2 = [6, 4, 0]^T, \quad B_3 = [8, 6, 0]^T, \quad X_3 = [22, 16, -2]^T,$
$B_4 = [30, 22, -2]^T, \quad X_4 = [86, 62, -6]^T$

3.71. $B = [100; \quad 210; \quad 351] \operatorname{diag}(1, -1, -10) [1, 3, -1; \quad 0, 1, 3; \quad 0, 0, 1]$

3.73. Replace R_i by $-kR_j + (1/k')R_i$.

3.75. (c) $f(A) = (e(A^T))^T = (EA^T)^T = (A^T)^T E^T = AF$

3.76. (a) $A = IAI$. (b) If $A = PBQ$, then $B = P^{-1}AQ^{-1}$.
(c) If $A = PBQ$ and $B = P'CQ'$, then $A = (PP')C(Q'Q)$.

CHAPTER 4

Vector Spaces

4.1 Introduction

This chapter introduces the underlying structure of linear algebra, that of a finite-dimensional vector space. The definition of a vector space V, whose elements are called *vectors*, involves an arbitrary field K, whose elements are called *scalars*. The following notation will be used (unless otherwise stated or implied):

V	the given vector space
u, v, w	vectors in V
K	the given number field
$a, b, c,$ or k	scalars in K

Almost nothing essential is lost if the reader assumes that K is the real field $\mathbf{R}$ or the complex field $\mathbf{C}$.

The reader might suspect that the real line $\mathbf{R}$ has "dimension" one, the cartesian plane $\mathbf{R}^2$ has "dimension" two, and the space $\mathbf{R}^3$ has "dimension" three. This chapter formalizes the notion of "dimension," and this definition will agree with the reader's intuition.

Throughout this text, we will use the following set notation:

$a \in A$	Element a belongs to set A
$a, b \in A$	Elements a and b belong to A
$\forall x \in A$	For every x in A
$\exists x \in A$	There exists an x in A
$A \subseteq B$	A is a subset of B
$A \cap B$	Intersection of A and B
$A \cup B$	Union of A and B
$\emptyset$	Empty set

4.2 Vector Spaces

The following defines the notion of a vector space V where K is the field of scalars.

DEFINITION: Let V be a nonempty set with two operations:

(i) ***Vector Addition:*** This assigns to any $u, v \in V$ a *sum* $u + v$ in V.

(ii) ***Scalar Multiplication:*** This assigns to any $u \in V$, $k \in K$ a *product* $ku \in V$.

Then V is called a *vector space* (over the field K) if the following axioms hold for any vectors $u, v, w \in V$:

[A$_1$] $(u+v)+w=u+(v+w)$
[A$_2$] There is a vector in V, denoted by 0 and called the *zero vector*, such that, for any $u \in V$,
$$u+0=0+u=u$$
[A$_3$] For each $u \in V$, there is a vector in V, denoted by $-u$, and called the *negative* of u, such that
$$u+(-u)=(-u)+u=0.$$
[A$_4$] $u+v=v+u$.
[M$_1$] $k(u+v)=ku+kv$, for any scalar $k \in K$.
[M$_2$] $(a+b)u=au+bu$, for any scalars $a, b \in K$.
[M$_3$] $(ab)u=a(bu)$, for any scalars $a, b \in K$.
[M$_4$] $1u=u$, for the unit scalar $1 \in K$.

The above axioms naturally split into two sets (as indicated by the labeling of the axioms). The first four are concerned only with the additive structure of V and can be summarized by saying V is a *commutative group* under addition. This means

(a) Any sum $v_1+v_2+\cdots+v_m$ of vectors requires no parentheses and does not depend on the order of the summands.
(b) The zero vector 0 is unique, and the negative $-u$ of a vector u is unique.
(c) (Cancellation Law) If $u+w=v+w$, then $u=v$.

Also, *subtraction* in V is defined by $u-v=u+(-v)$, where $-v$ is the unique negative of v.

On the other hand, the remaining four axioms are concerned with the "action" of the field K of scalars on the vector space V. Using these additional axioms, we prove (Problem 4.2) the following simple properties of a vector space.

THEOREM 4.1: Let V be a vector space over a field K.

(i) For any scalar $k \in K$ and $0 \in V$, $k0=0$.
(ii) For $0 \in K$ and any vector $u \in V$, $0u=0$.
(iii) If $ku=0$, where $k \in K$ and $u \in V$, then $k=0$ or $u=0$.
(iv) For any $k \in K$ and any $u \in V$, $(-k)u=k(-u)=-ku$.

4.3 Examples of Vector Spaces

This section lists important examples of vector spaces that will be used throughout the text.

Space K^n

Let K be an arbitrary field. The notation K^n is frequently used to denote the set of all n-tuples of elements in K. Here K^n is a vector space over K using the following operations:

(i) ***Vector Addition:*** $(a_1, a_2, \ldots, a_n)+(b_1, b_2, \ldots, b_n)=(a_1+b_1,\ a_2+b_2, \ldots,\ a_n+b_n)$
(ii) ***Scalar Multiplication:*** $k(a_1, a_2, \ldots, a_n)=(ka_1, ka_2, \ldots, ka_n)$

The zero vector in K^n is the n-tuple of zeros,
$$0=(0,0,\ldots,0)$$
and the negative of a vector is defined by
$$-(a_1, a_2, \ldots, a_n)=(-a_1, -a_2, \ldots, -a_n)$$

Observe that these are the same as the operations defined for $\mathbf{R}^n$ in Chapter 1. The proof that K^n is a vector space is identical to the proof of Theorem 1.1, which we now regard as stating that $\mathbf{R}^n$ with the operations defined there is a vector space over $\mathbf{R}$.

Polynomial Space $\mathbf{P}(t)$

Let $\mathbf{P}(t)$ denote the set of all polynomials of the form

$$p(t) = a_0 + a_1 t + a_2 t^2 + \cdots + a_s t^s \qquad (s = 1, 2, \ldots)$$

where the coefficients a_i belong to a field K. Then $\mathbf{P}(t)$ is a vector space over K using the following operations:

(i) ***Vector Addition:*** Here $p(t) + q(t)$ in $\mathbf{P}(t)$ is the usual operation of addition of polynomials.

(ii) ***Scalar Multiplication:*** Here $kp(t)$ in $\mathbf{P}(t)$ is the usual operation of the product of a scalar k and a polynomial $p(t)$.

The zero polynomial 0 is the zero vector in $\mathbf{P}(t)$.

Polynomial Space $\mathbf{P}_n(t)$

Let $\mathbf{P}_n(t)$ denote the set of all polynomials $p(t)$ over a field K, where the degree of $p(t)$ is less than or equal to n; that is,

$$p(t) = a_0 + a_1 t + a_2 t^2 + \cdots + a_s t^s$$

where $s \leq n$. Then $\mathbf{P}_n(t)$ is a vector space over K with respect to the usual operations of addition of polynomials and of multiplication of a polynomial by a constant (just like the vector space $\mathbf{P}(t)$ above). We include the zero polynomial 0 as an element of $\mathbf{P}_n(t)$, even though its degree is undefined.

Matrix Space $\mathbf{M}_{m,n}$

The notation $\mathbf{M}_{m,n}$, or simply $\mathbf{M}$, will be used to denote the set of all $m \times n$ matrices with entries in a field K. Then $\mathbf{M}_{m,n}$ is a vector space over K with respect to the usual operations of matrix addition and scalar multiplication of matrices, as indicated by Theorem 2.1.

Function Space $F(X)$

Let X be a nonempty set and let K be an arbitrary field. Let $F(X)$ denote the set of all functions of X into K. [Note that $F(X)$ is nonempty, because X is nonempty.] Then $F(X)$ is a vector space over K with respect to the following operations:

(i) ***Vector Addition:*** The sum of two functions f and g in $F(X)$ is the function $f + g$ in $F(X)$ defined by

$$(f + g)(x) = f(x) + g(x) \qquad \forall x \in X$$

(ii) ***Scalar Multiplication:*** The product of a scalar $k \in K$ and a function f in $F(X)$ is the function kf in $F(X)$ defined by

$$(kf)(x) = kf(x) \qquad \forall x \in X$$

The zero vector in $F(X)$ is the zero function $\mathbf{0}$, which maps every $x \in X$ into the zero element $0 \in K$;

$$\mathbf{0}(x) = 0 \qquad \forall x \in X$$

Also, for any function f in $F(X)$, negative of f is the function $-f$ in $F(X)$ defined by

$$(-f)(x) = -f(x) \qquad \forall x \in X$$

Fields and Subfields

Suppose a field E is an extension of a field K; that is, suppose E is a field that contains K as a subfield. Then E may be viewed as a vector space over K using the following operations:

(i) ***Vector Addition:*** Here $u + v$ in E is the usual addition in E.

(ii) ***Scalar Multiplication:*** Here ku in E, where $k \in K$ and $u \in E$, is the usual product of k and u as elements of E.

That is, the eight axioms of a vector space are satisfied by E and its subfield K with respect to the above two operations.

4.4 Linear Combinations, Spanning Sets

Let V be a vector space over a field K. A vector v in V is a *linear combination* of vectors $u_1, u_2, \ldots, u_m$ in V if there exist scalars $a_1, a_2, \ldots, a_m$ in K such that

$$v = a_1u_1 + a_2u_2 + \cdots + a_mu_m$$

Alternatively, v is a linear combination of $u_1, u_2, \ldots, u_m$ if there is a solution to the vector equation

$$v = x_1u_1 + x_2u_2 + \cdots + x_mu_m$$

where $x_1, x_2, \ldots, x_m$ are unknown scalars.

EXAMPLE 4.1 (Linear Combinations in $\mathbf{R}^n$) Suppose we want to express $v = (3, 7, -4)$ in $\mathbf{R}^3$ as a linear combination of the vectors

$$u_1 = (1, 2, 3), \qquad u_2 = (2, 3, 7), \qquad u_3 = (3, 5, 6)$$

We seek scalars x, y, z such that $v = xu_1 + yu_2 + zu_3$; that is,

$$\begin{bmatrix} 3 \\ 7 \\ -4 \end{bmatrix} = x\begin{bmatrix} 1 \\ 2 \\ 3 \end{bmatrix} + y\begin{bmatrix} 2 \\ 3 \\ 7 \end{bmatrix} + z\begin{bmatrix} 3 \\ 5 \\ 6 \end{bmatrix} \quad \text{or} \quad \begin{aligned} x + 2y + 3z &= 3 \\ 2x + 3y + 5z &= 7 \\ 3x + 7y + 6z &= -4 \end{aligned}$$

(For notational convenience, we have written the vectors in $\mathbf{R}^3$ as columns, because it is then easier to find the equivalent system of linear equations.) Reducing the system to echelon form yields

$$\begin{aligned} x + 2y + 3z &= 3 \\ -y - z &= 1 \\ y - 3z &= -13 \end{aligned} \quad \text{and then} \quad \begin{aligned} x + 2y + 3z &= 3 \\ -y - z &= 1 \\ -4z &= -12 \end{aligned}$$

Back-substitution yields the solution $x = 2$, $y = -4$, $z = 3$. Thus, $v = 2u_1 - 4u_2 + 3u_3$.

Remark: Generally speaking, the question of expressing a given vector v in K^n as a linear combination of vectors $u_1, u_2, \ldots, u_m$ in K^n is equivalent to solving a system $AX = B$ of linear equations, where v is the column B of constants, and the u's are the columns of the coefficient matrix A. Such a system may have a unique solution (as above), many solutions, or no solution. The last case—no solution—means that v cannot be written as a linear combination of the u's.

EXAMPLE 4.2 (Linear combinations in $\mathbf{P}(t)$) Suppose we want to express the polynomial $v = 3t^2 + 5t - 5$ as a linear combination of the polynomials

$$p_1 = t^2 + 2t + 1, \qquad p_2 = 2t^2 + 5t + 4, \qquad p_3 = t^2 + 3t + 6$$

We seek scalars x, y, z such that $v = xp_1 + yp_2 + zp_3$; that is,

$$3t^2 + 5t - 5 = x(t^2 + 2t + 1) + y(2t^2 + 5t + 4) + z(t^2 + 3t + 6) \tag{*}$$

There are two ways to proceed from here.

(1) Expand the right-hand side of (*) obtaining:

$$\begin{aligned} 3t^2 + 5t - 5 &= xt^2 + 2xt + x + 2yt^2 + 5yt + 4y + zt^2 + 3zt + 6z \\ &= (x + 2y + z)t^2 + (2x + 5y + 3z)t + (x + 4y + 6z) \end{aligned}$$

Set coefficients of the same powers of t equal to each other, and reduce the system to echelon form:

$$\begin{aligned} x + 2y + z &= 3 \\ 2x + 5y + 3z &= 5 \\ x + 4y + 6z &= -5 \end{aligned} \quad \text{or} \quad \begin{aligned} x + 2y + z &= 3 \\ y + z &= -1 \\ 2y + 5z &= -8 \end{aligned} \quad \text{or} \quad \begin{aligned} x + 2y + z &= 3 \\ y + z &= -1 \\ 3z &= -6 \end{aligned}$$

The system is in triangular form and has a solution. Back-substitution yields the solution $x = 3, y = 1, z = -2$. Thus,

$$v = 3p_1 + p_2 - 2p_3$$

(2) The equation (*) is actually an identity in the variable t; that is, the equation holds for any value of t. We can obtain three equations in the unknowns x, y, z by setting t equal to any three values. For example,

Set $t = 0$ in (1) to obtain: $x + 4y + 6z = -5$

Set $t = 1$ in (1) to obtain: $4x + 11y + 10z = 3$

Set $t = -1$ in (1) to obtain: $y + 4z = -7$

Reducing this system to echelon form and solving by back-substitution again yields the solution $x = 3, y = 1, z = -2$. Thus (again), $v = 3p_1 + p_2 - 2p_3$.

Spanning Sets

Let V be a vector space over K. Vectors $u_1, u_2, \ldots, u_m$ in V are said to *span* V or to form a *spanning set* of V if every v in V is a linear combination of the vectors $u_1, u_2, \ldots, u_m$—that is, if there exist scalars $a_1, a_2, \ldots, a_m$ in K such that

$$v = a_1u_1 + a_2u_2 + \cdots + a_mu_m$$

The following remarks follow directly from the definition.

Remark 1: Suppose $u_1, u_2, \ldots, u_m$ span V. Then, for any vector w, the set $w, u_1, u_2, \ldots, u_m$ also spans V.

Remark 2: Suppose $u_1, u_2, \ldots, u_m$ span V and suppose u_k is a linear combination of some of the other u's. Then the u's without u_k also span V.

Remark 3: Suppose $u_1, u_2, \ldots, u_m$ span V and suppose one of the u's is the zero vector. Then the u's without the zero vector also span V.

EXAMPLE 4.3 Consider the vector space $V = \mathbf{R}^3$.

(a) We claim that the following vectors form a spanning set of $\mathbf{R}^3$:

$$e_1 = (1, 0, 0), \qquad e_2 = (0, 1, 0), \qquad e_3 = (0, 0, 1)$$

Specifically, if $v = (a, b, c)$ is any vector in $\mathbf{R}^3$, then

$$v = ae_1 + be_2 + ce_3$$

For example, $v = (5, -6, 2) = 5e_1 - 6e_2 + 2e_3$.

(b) We claim that the following vectors also form a spanning set of $\mathbf{R}^3$:

$$w_1 = (1, 1, 1), \qquad w_2 = (1, 1, 0), \qquad w_3 = (1, 0, 0)$$

Specifically, if $v = (a, b, c)$ is any vector in $\mathbf{R}^3$, then (Problem 4.62)

$$v = (a, b, c) = cw_1 + (b - c)w_2 + (a - b)w_3$$

For example, $v = (5, -6, 2) = 2w_1 - 8w_2 + 11w_3$.

(c) One can show (Problem 3.24) that $v = (2, 7, 8)$ cannot be written as a linear combination of the vectors

$$u_1 = (1, 2, 3), \qquad u_2 = (1, 3, 5), \qquad u_3 = (1, 5, 9)$$

Accordingly, u_1, u_2, u_3 do not span $\mathbf{R}^3$.

EXAMPLE 4.4 Consider the vector space $V = \mathbf{P}_n(t)$ consisting of all polynomials of degree $\leq n$.

(a) Clearly every polynomial in $\mathbf{P}_n(t)$ can be expressed as a linear combination of the $n+1$ polynomials

$$1, \quad t, \quad t^2, \quad t^3, \quad \ldots, \quad t^n$$

Thus, these powers of t (where $1 = t^0$) form a spanning set for $\mathbf{P}_n(t)$.

(b) One can also show that, for any scalar c, the following $n+1$ powers of $t - c$,

$$1, \quad t-c, \quad (t-c)^2, \quad (t-c)^3, \quad \ldots, \quad (t-c)^n$$

(where $(t-c)^0 = 1$), also form a spanning set for $\mathbf{P}_n(t)$.

EXAMPLE 4.5 Consider the vector space $\mathbf{M} = \mathbf{M}_{2,2}$ consisting of all 2×2 matrices, and consider the following four matrices in $\mathbf{M}$:

$$E_{11} = \begin{bmatrix} 1 & 0 \\ 0 & 0 \end{bmatrix}, \qquad E_{12} = \begin{bmatrix} 0 & 1 \\ 0 & 0 \end{bmatrix}, \qquad E_{21} = \begin{bmatrix} 0 & 0 \\ 1 & 0 \end{bmatrix}, \qquad E_{22} = \begin{bmatrix} 0 & 0 \\ 0 & 1 \end{bmatrix}$$

Then clearly any matrix A in $\mathbf{M}$ can be written as a linear combination of the four matrices. For example,

$$A = \begin{bmatrix} 5 & -6 \\ 7 & 8 \end{bmatrix} = 5E_{11} - 6E_{12} + 7E_{21} + 8E_{22}$$

Accordingly, the four matrices $E_{11}, E_{12}, E_{21}, E_{22}$ span $\mathbf{M}$.

4.5 Subspaces

This section introduces the important notion of a subspace.

DEFINITION: Let V be a vector space over a field K and let W be a subset of V. Then W is a *subspace* of V if W is itself a vector space over K with respect to the operations of vector addition and scalar multiplication on V.

The way in which one shows that any set W is a vector space is to show that W satisfies the eight axioms of a vector space. However, if W is a subset of a vector space V, then some of the axioms automatically hold in W, because they already hold in V. Simple criteria for identifying subspaces follow.

THEOREM 4.2: Suppose W is a subset of a vector space V. Then W is a subspace of V if the following two conditions hold:

(a) The zero vector 0 belongs to W.

(b) For every $u, v \in W, k \in K$: (i) The sum $u + v \in W$. (ii) The multiple $ku \in W$.

Property (i) in (b) states that W is *closed under vector addition*, and property (ii) in (b) states that W is *closed under scalar multiplication*. Both properties may be combined into the following equivalent single statement:

(b') For every $u, v \in W, a, b \in K$, the linear combination $au + bv \in W$.

Now let V be any vector space. Then V automatically contains two subspaces: the set $\{0\}$ consisting of the zero vector alone and the whole space V itself. These are sometimes called the *trivial* subspaces of V. Examples of nontrivial subspaces follow.

EXAMPLE 4.6 Consider the vector space $V = \mathbf{R}^3$.

(a) Let U consist of all vectors in $\mathbf{R}^3$ whose entries are equal; that is,

$$U = \{(a, b, c) : a = b = c\}$$

For example, $(1, 1, 1)$, $(-3, -3, -3)$, $(7, 7, 7)$, $(-2, -2, -2)$ are vectors in U. Geometrically, U is the line through the origin O and the point $(1, 1, 1)$ as shown in Fig. 4-1(a). Clearly $0 = (0, 0, 0)$ belongs to U, because all entries in 0 are equal. Further, suppose u and v are arbitrary vectors in U, say, $u = (a, a, a)$ and $v = (b, b, b)$. Then, for any scalar $k \in \mathbf{R}$, the following are also vectors in U:

$$u + v = (a + b,\ a + b,\ a + b) \quad \text{and} \quad ku = (ka,\ ka,\ ka)$$

Thus, U is a subspace of $\mathbf{R}^3$.

(b) Let W be any plane in $\mathbf{R}^3$ passing through the origin, as pictured in Fig. 4-1(b). Then $0 = (0, 0, 0)$ belongs to W, because we assumed W passes through, the origin O. Further, suppose u and v are vectors in W. Then u and v may be viewed as arrows in the plane W emanating from the origin O, as in Fig. 4-1(b). The sum $u + v$ and any multiple ku of u also lie in the plane W. Thus, W is a subspace of $\mathbf{R}^3$.

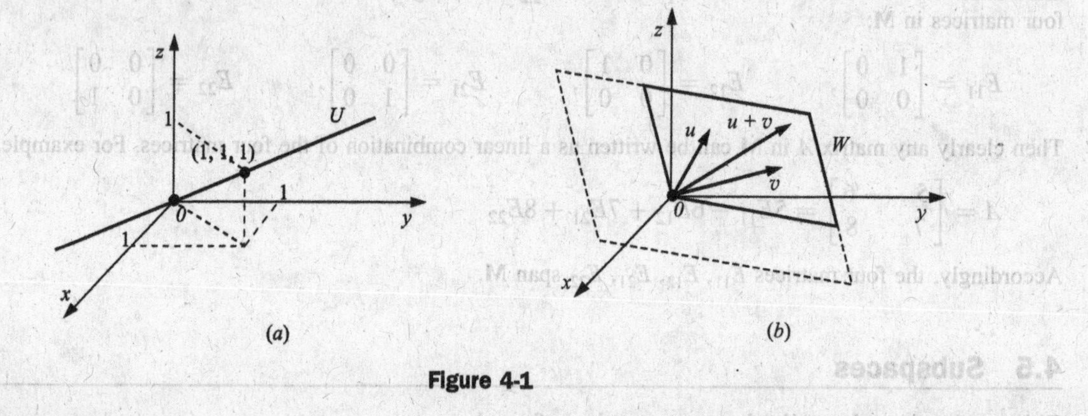

Figure 4-1

EXAMPLE 4.7

(a) Let $V = \mathbf{M}_{n,n}$, the vector space of $n \times n$ matrices. Let W_1 be the subset of all (upper) triangular matrices and let W_2 be the subset of all symmetric matrices. Then W_1 is a subspace of V, because W_1 contains the zero matrix 0 and W_1 is closed under matrix addition and scalar multiplication; that is, the sum and scalar multiple of such triangular matrices are also triangular. Similarly, W_2 is a subspace of V.

(b) Let $V = \mathbf{P}(t)$, the vector space $\mathbf{P}(t)$ of polynomials. Then the space $\mathbf{P}_n(t)$ of polynomials of degree at most n may be viewed as a subspace of $\mathbf{P}(t)$. Let $\mathbf{Q}(t)$ be the collection of polynomials with only even powers of t. For example, the following are polynomials in $\mathbf{Q}(t)$:

$$p_1 = 3 + 4t^2 - 5t^6 \quad \text{and} \quad p_2 = 6 - 7t^4 + 9t^6 + 3t^{12}$$

(We assume that any constant $k = kt^0$ is an even power of t.) Then $\mathbf{Q}(t)$ is a subspace of $\mathbf{P}(t)$.

(c) Let V be the vector space of real-valued functions. Then the collection W_1 of continuous functions and the collection W_2 of differentiable functions are subspaces of V.

Intersection of Subspaces

Let U and W be subspaces of a vector space V. We show that the intersection $U \cap W$ is also a subspace of V. Clearly, $0 \in U$ and $0 \in W$, because U and W are subspaces; whence $0 \in U \cap W$. Now suppose u and v belong to the intersection $U \cap W$. Then $u, v \in U$ and $u, v \in W$. Further, because U and W are subspaces, for any scalars $a, b \in K$,

$$au + bv \in U \quad \text{and} \quad au + bv \in W$$

Thus, $au + bv \in U \cap W$. Therefore, $U \cap W$ is a subspace of V.

The above result generalizes as follows.

THEOREM 4.3: The intersection of any number of subspaces of a vector space V is a subspace of V.

Solution Space of a Homogeneous System

Consider a system $AX = B$ of linear equations in n unknowns. Then every solution u may be viewed as a vector in K^n. Thus, the solution set of such a system is a subset of K^n. Now suppose the system is homogeneous; that is, suppose the system has the form $AX = 0$. Let W be its solution set. Because $A0 = 0$, the zero vector $0 \in W$. Moreover, suppose u and v belong to W. Then u and v are solutions of $AX = 0$, or, in other words, $Au = 0$ and $Av = 0$. Therefore, for any scalars a and b, we have

$$A(au + bv) = aAu + bAv = a0 + b0 = 0 + 0 = 0$$

Thus, $au + bv$ belongs to W, because it is a solution of $AX = 0$. Accordingly, W is a subspace of K^n.
We state the above result formally.

THEOREM 4.4: The solution set W of a homogeneous system $AX = 0$ in n unknowns is a subspace of K^n.

We emphasize that the solution set of a nonhomogeneous system $AX = B$ is not a subspace of K^n. In fact, the zero vector 0 does not belong to its solution set.

4.6 Linear Spans, Row Space of a Matrix

Suppose $u_1, u_2, \ldots, u_m$ are any vectors in a vector space V. Recall (Section 4.4) that any vector of the form $a_1u_1 + a_2u_2 + \cdots + a_mu_m$, where the a_i are scalars, is called a *linear combination* of $u_1, u_2, \ldots, u_m$. The collection of all such linear combinations, denoted by

$$\text{span}(u_1, u_2, \ldots, u_m) \quad \text{or} \quad \text{span}(u_i)$$

is called the *linear span* of $u_1, u_2, \ldots, u_m$.

Clearly the zero vector 0 belongs to $\text{span}(u_i)$, because

$$0 = 0u_1 + 0u_2 + \cdots + 0u_m$$

Furthermore, suppose v and v' belong to $\text{span}(u_i)$, say,

$$v = a_1u_1 + a_2u_2 + \cdots + a_mu_m \quad \text{and} \quad v' = b_1u_1 + b_2u_2 + \cdots + b_mu_m$$

Then,

$$v + v' = (a_1 + b_1)u_1 + (a_2 + b_2)u_2 + \cdots + (a_m + b_m)u_m$$

and, for any scalar $k \in K$,

$$kv = ka_1u_1 + ka_2u_2 + \cdots + ka_mu_m$$

Thus, $v + v'$ and kv also belong to $\text{span}(u_i)$. Accordingly, $\text{span}(u_i)$ is a subspace of V.
More generally, for any subset S of V, $\text{span}(S)$ consists of all linear combinations of vectors in S or, when $S = \phi$, $\text{span}(S) = \{0\}$. Thus, in particular, S is a spanning set (Section 4.4) of $\text{span}(S)$.
The following theorem, which was partially proved above, holds.

THEOREM 4.5: Let S be a subset of a vector space V.

(i) Then $\text{span}(S)$ is a subspace of V that contains S.
(ii) If W is a subspace of V containing S, then $\text{span}(S) \subseteq W$.

Condition (ii) in theorem 4.5 may be interpreted as saying that $\text{span}(S)$ is the "smallest" subspace of V containing S.

EXAMPLE 4.8 Consider the vector space $V = \mathbf{R}^3$.

(a) Let u be any nonzero vector in $\mathbf{R}^3$. Then $\text{span}(u)$ consists of all scalar multiples of u. Geometrically, $\text{span}(u)$ is the line through the origin O and the endpoint of u, as shown in Fig. 4-2(a).

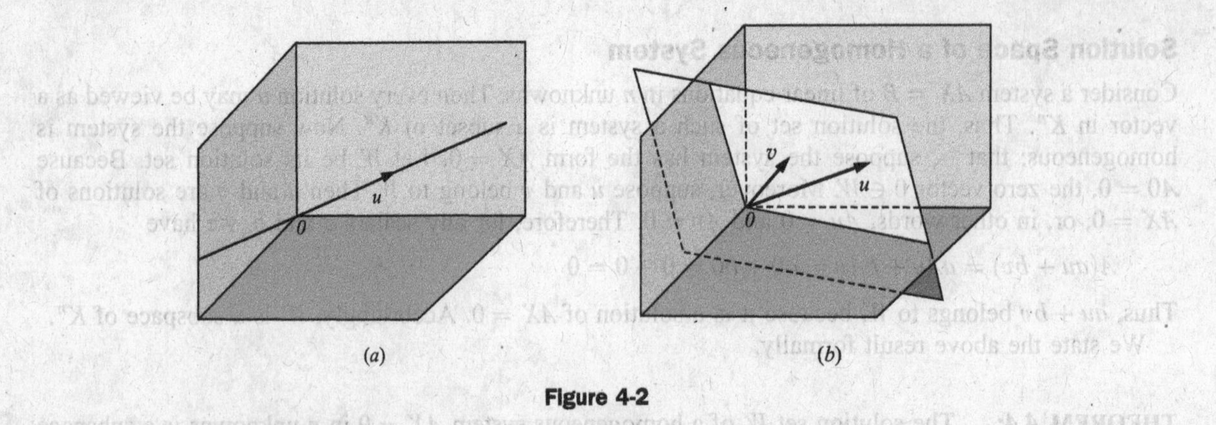

Figure 4-2

(b) Let u and v be vectors in $\mathbf{R}^3$ that are not multiples of each other. Then span(u, v) is the plane through the origin O and the endpoints of u and v as shown in Fig. 4-2(b).

(c) Consider the vectors $e_1 = (1,0,0)$, $e_2 = (0,1,0)$, $e_3 = (0,0,1)$ in $\mathbf{R}^3$. Recall [Example 4.1(a)] that every vector in $\mathbf{R}^3$ is a linear combination of e_1, e_2, e_3. That is, e_1, e_2, e_3 form a spanning set of $\mathbf{R}^3$. Accordingly, span$(e_1, e_2, e_3) = \mathbf{R}^3$.

Row Space of a Matrix

Let $A = [a_{ij}]$ be an arbitrary $m \times n$ matrix over a field K. The rows of A,

$$R_1 = (a_{11}, a_{12}, \ldots, a_{1n}), \qquad R_2 = (a_{21}, a_{22}, \ldots, a_{2n}), \qquad \ldots, \qquad R_m = (a_{m1}, a_{m2}, \ldots, a_{mn})$$

may be viewed as vectors in K^n; hence, they span a subspace of K^n called the *row space* of A and denoted by rowsp(A). That is,

$$\text{rowsp}(A) = \text{span}(R_1, R_2, \ldots, R_m)$$

Analagously, the columns of A may be viewed as vectors in K^m called the *column space* of A and denoted by colsp(A). Observe that colsp(A) = rowsp(A^T).

Recall that matrices A and B are row equivalent, written $A \sim B$, if B can be obtained from A by a sequence of elementary row operations. Now suppose M is the matrix obtained by applying one of the following elementary row operations on a matrix A:

(1) Interchange R_i and R_j, (2) Replace R_i by kR_i, (3) Replace R_j by $kR_i + R_j$

Then each row of M is a row of A or a linear combination of rows of A. Hence, the row space of M is contained in the row space of A. On the other hand, we can apply the inverse elementary row operation on M to obtain A; hence, the row space of A is contained in the row space of M. Accordingly, A and M have the same row space. This will be true each time we apply an elementary row operation. Thus, we have proved the following theorem.

THEOREM 4.6: Row equivalent matrices have the same row space.

We are now able to prove (Problems 4.45–4.47) basic results on row equivalence (which first appeared as Theorems 3.7 and 3.8 in Chapter 3).

THEOREM 4.7: Suppose $A = [a_{ij}]$ and $B = [b_{ij}]$ are row equivalent echelon matrices with respective pivot entries

$$a_{1j_1}, a_{2j_2}, \ldots, a_{rj_r} \quad \text{and} \quad b_{1k_1}, b_{2k_2}, \ldots, b_{sk_s}$$

Then A and B have the same number of nonzero rows—that is, $r = s$—and their pivot entries are in the same positions—that is, $j_1 = k_1, j_2 = k_2, \ldots, j_r = k_r$.

THEOREM 4.8: Suppose A and B are row canonical matrices. Then A and B have the same row space if and only if they have the same nonzero rows.

COROLLARY 4.9: Every matrix A is row equivalent to a unique matrix in row canonical form.

We apply the above results in the next example.

EXAMPLE 4.9 Consider the following two sets of vectors in $\mathbf{R}^4$:

$$u_1 = (1,2,-1,3), \quad u_2 = (2,4,1,-2), \quad u_3 = (3,6,3,-7)$$
$$w_1 = (1,2,-4,11), \quad w_2 = (2,4,-5,14)$$

Let $U = \text{span}(u_i)$ and $W = \text{span}(w_i)$. There are two ways to show that $U = W$.

(a) Show that each u_i is a linear combination of w_1 and w_2, and show that each w_i is a linear combination of u_1, u_2, u_3. Observe that we have to show that six systems of linear equations are consistent.

(b) Form the matrix A whose rows are u_1, u_2, u_3 and row reduce A to row canonical form, and form the matrix B whose rows are w_1 and w_2 and row reduce B to row canonical form:

$$A = \begin{bmatrix} 1 & 2 & -1 & 3 \\ 2 & 4 & 1 & -2 \\ 3 & 6 & 3 & -7 \end{bmatrix} \sim \begin{bmatrix} 1 & 2 & -1 & 3 \\ 0 & 0 & 3 & -8 \\ 0 & 0 & 6 & -16 \end{bmatrix} \sim \begin{bmatrix} 1 & 2 & 0 & \frac{1}{3} \\ 0 & 0 & 1 & -\frac{8}{3} \\ 0 & 0 & 0 & 0 \end{bmatrix}$$

$$B = \begin{bmatrix} 1 & 2 & -4 & 11 \\ 2 & 4 & -5 & 14 \end{bmatrix} \sim \begin{bmatrix} 1 & 2 & -4 & 11 \\ 0 & 0 & 3 & -8 \end{bmatrix} \sim \begin{bmatrix} 1 & 2 & 0 & \frac{1}{3} \\ 0 & 0 & 1 & -\frac{8}{3} \end{bmatrix}$$

Because the nonzero rows of the matrices in row canonical form are identical, the row spaces of A and B are equal. Therefore, $U = W$.

Clearly, the method in (b) is more efficient than the method in (a).

4.7 Linear Dependence and Independence

Let V be a vector space over a field K. The following defines the notion of linear dependence and independence of vectors over K. (One usually suppresses mentioning K when the field is understood.) This concept plays an essential role in the theory of linear algebra and in mathematics in general.

DEFINITION: We say that the vectors $v_1, v_2, \ldots, v_m$ in V are *linearly dependent* if there exist scalars $a_1, a_2, \ldots, a_m$ in K, not all of them 0, such that

$$a_1 v_1 + a_2 v_2 + \cdots + a_m v_m = 0$$

Otherwise, we say that the vectors are *linearly independent.*

The above definition may be restated as follows. Consider the vector equation

$$x_1 v_1 + x_2 v_2 + \cdots + x_m v_m = 0 \qquad (*)$$

where the x's are unknown scalars. This equation always has the *zero solution* $x_1 = 0$, $x_2 = 0, \ldots, x_m = 0$. Suppose this is the only solution; that is, suppose we can show:

$$x_1 v_1 + x_2 v_2 + \cdots + x_m v_m = 0 \quad \text{implies} \quad x_1 = 0, \quad x_2 = 0, \quad \ldots, \quad x_m = 0$$

Then the vectors $v_1, v_2, \ldots, v_m$ are linearly independent, On the other hand, suppose the equation (*) has a nonzero solution; then the vectors are linearly dependent.

A set $S = \{v_1, v_2, \ldots, v_m\}$ of vectors in V is linearly dependent or independent according to whether the vectors $v_1, v_2, \ldots, v_m$ are linearly dependent or independent.

An infinite set S of vectors is linearly dependent or independent according to whether there do or do not exist vectors $v_1, v_2, \ldots, v_k$ in S that are linearly dependent.

Warning: The set $S = \{v_1, v_2, \ldots, v_m\}$ above represents a *list* or, in other words, a finite sequence of vectors where the vectors are ordered and repetition is permitted.

The following remarks follow directly from the above definition.

Remark 1: Suppose 0 is one of the vectors $v_1, v_2, \ldots, v_m$, say $v_1 = 0$. Then the vectors must be linearly dependent, because we have the following linear combination where the coefficient of $v_1 \neq 0$:

$$1v_1 + 0v_2 + \cdots + 0v_m = 1 \cdot 0 + 0 + \cdots + 0 = 0$$

Remark 2: Suppose v is a nonzero vector. Then v, by itself, is linearly independent, because

$$kv = 0, \qquad v \neq 0 \qquad \text{implies} \qquad k = 0$$

Remark 3: Suppose two of the vectors $v_1, v_2, \ldots, v_m$ are equal or one is a scalar multiple of the other, say $v_1 = kv_2$. Then the vectors must be linearly dependent, because we have the following linear combination where the coefficient of $v_1 \neq 0$:

$$v_1 - kv_2 + 0v_3 + \cdots + 0v_m = 0$$

Remark 4: Two vectors v_1 and v_2 are linearly dependent if and only if one of them is a multiple of the other.

Remark 5: If the set $\{v_1, \ldots, v_m\}$ is linearly independent, then any rearrangement of the vectors $\{v_{i_1}, v_{i_2}, \ldots, v_{i_m}\}$ is also linearly independent.

Remark 6: If a set S of vectors is linearly independent, then any subset of S is linearly independent. Alternatively, if S contains a linearly dependent subset, then S is linearly dependent.

EXAMPLE 4.10

(a) Let $u = (1, 1, 0)$, $v = (1, 3, 2)$, $w = (4, 9, 5)$. Then u, v, w are linearly dependent, because

$$3u + 5v - 2w = 3(1, 1, 0) + 5(1, 3, 2) - 2(4, 9, 5) = (0, 0, 0) = 0$$

(b) We show that the vectors $u = (1, 2, 3)$, $v = (2, 5, 7)$, $w = (1, 3, 5)$ are linearly independent. We form the vector equation $xu + yv + zw = 0$, where x, y, z are unknown scalars. This yields

$$x\begin{bmatrix}1\\2\\3\end{bmatrix} + y\begin{bmatrix}2\\5\\7\end{bmatrix} + z\begin{bmatrix}1\\3\\5\end{bmatrix} = \begin{bmatrix}0\\0\\0\end{bmatrix} \quad \text{or} \quad \begin{aligned} x + 2y + z &= 0 \\ 2x + 5y + 3z &= 0 \\ 3x + 7y + 5z &= 0 \end{aligned} \quad \text{or} \quad \begin{aligned} x + 2y + z &= 0 \\ y + z &= 0 \\ 2z &= 0 \end{aligned}$$

Back-substitution yields $x = 0$, $y = 0$, $z = 0$. We have shown that

$$xu + yv + zw = 0 \qquad \text{implies} \qquad x = 0, \quad y = 0, \quad z = 0$$

Accordingly, u, v, w are linearly independent.

(c) Let V be the vector space of functions from **R** into **R**. We show that the functions $f(t) = \sin t$, $g(t) = e^t$, $h(t) = t^2$ are linearly independent. We form the vector (function) equation $xf + yg + zh = 0$, where x, y, z are unknown scalars. This function equation means that, for every value of t,

$$x \sin t + ye^t + zt^2 = 0$$

Thus, in this equation, we choose appropriate values of t to easily get $x = 0$, $y = 0$, $z = 0$. For example,

(i)	Substitute $t = 0$	to obtain $x(0) + y(1) + z(0) = 0$	or	$y = 0$
(ii)	Substitute $t = \pi$	to obtain $x(0) + 0(e^{\pi}) + z(\pi^2) = 0$	or	$z = 0$
(iii)	Substitute $t = \pi/2$	to obtain $x(1) + 0(e^{\pi/2}) + 0(\pi^2/4) = 0$	or	$x = 0$

We have shown

$$xf + yg + zf = 0 \qquad \text{implies} \qquad x = 0, \quad y = 0, \quad z = 0$$

Accordingly, u, v, w are linearly independent.

Linear Dependence in R^3

Linear dependence in the vector space $V = \mathbf{R}^3$ can be described geometrically as follows:

(a) Any two vectors u and v in $\mathbf{R}^3$ are linearly dependent if and only if they lie on the same line through the origin O, as shown in Fig. 4-3(a).

(b) Any three vectors u, v, w in $\mathbf{R}^3$ are linearly dependent if and only if they lie on the same plane through the origin O, as shown in Fig. 4-3(b).

Later, we will be able to show that any four or more vectors in $\mathbf{R}^3$ are automatically linearly dependent.

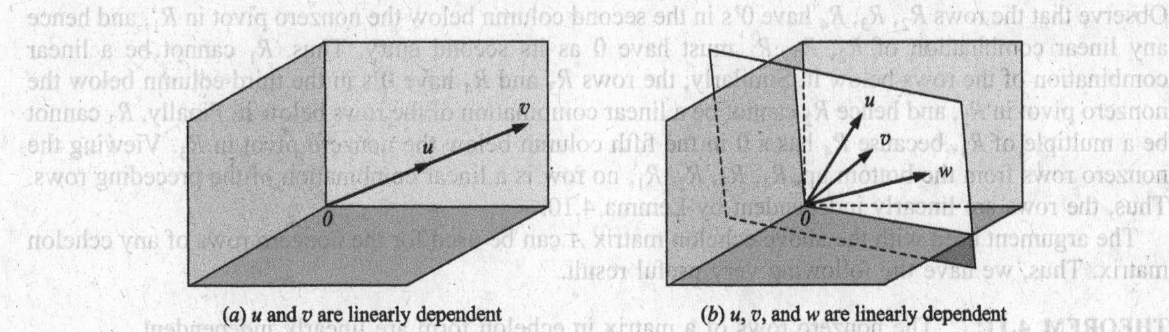

(a) u and v are linearly dependent

(b) u, v, and w are linearly dependent

Figure 4-3

Linear Dependence and Linear Combinations

The notions of linear dependence and linear combinations are closely related. Specifically, for more than one vector, we show that the vectors $v_1, v_2, \ldots, v_m$ are linearly dependent if and only if one of them is a linear combination of the others.

Suppose, say, v_i is a linear combination of the others,

$$v_i = a_1 v_1 + \cdots + a_{i-1} v_{i-1} + a_{i+1} v_{i+1} + \cdots + a_m v_m$$

Then by adding $-v_i$ to both sides, we obtain

$$a_1 v_1 + \cdots + a_{i-1} v_{i-1} - v_i + a_{i+1} v_{i+1} + \cdots + a_m v_m = 0$$

where the coefficient of v_i is not 0. Hence, the vectors are linearly dependent. Conversely, suppose the vectors are linearly dependent, say,

$$b_1 v_1 + \cdots + b_j v_j + \cdots + b_m v_m = 0, \qquad \text{where} \qquad b_j \neq 0$$

Then we can solve for v_j obtaining

$$v_j = b_j^{-1} b_1 v_1 - \cdots - b_j^{-1} b_{j-1} v_{j-1} - b_j^{-1} b_{j+1} v_{j+1} - \cdots - b_j^{-1} b_m v_m$$

and so v_j is a linear combination of the other vectors.

We now state a slightly stronger statement than the one above. This result has many important consequences.

LEMMA 4.10: Suppose two or more nonzero vectors $v_1, v_2, \ldots, v_m$ are linearly dependent. Then one of the vectors is a linear combination of the preceding vectors; that is, there exists $k > 1$ such that

$$v_k = c_1 v_1 + c_2 v_2 + \cdots + c_{k-1} v_{k-1}$$

Linear Dependence and Echelon Matrices

Consider the following echelon matrix A, whose pivots have been circled:

$$A = \begin{bmatrix} 0 & \textcircled{2} & 3 & 4 & 5 & 6 & 7 \\ 0 & 0 & \textcircled{4} & 3 & 2 & 3 & 4 \\ 0 & 0 & 0 & 0 & \textcircled{7} & 8 & 9 \\ 0 & 0 & 0 & 0 & 0 & \textcircled{6} & 7 \\ 0 & 0 & 0 & 0 & 0 & 0 & 0 \end{bmatrix}$$

Observe that the rows R_2, R_3, R_4 have 0's in the second column below the nonzero pivot in R_1, and hence any linear combination of R_2, R_3, R_4 must have 0 as its second entry. Thus, R_1 cannot be a linear combination of the rows below it. Similarly, the rows R_3 and R_4 have 0's in the third column below the nonzero pivot in R_2, and hence R_2 cannot be a linear combination of the rows below it. Finally, R_3 cannot be a multiple of R_4, because R_4 has a 0 in the fifth column below the nonzero pivot in R_3. Viewing the nonzero rows from the bottom up, R_4, R_3, R_2, R_1, no row is a linear combination of the preceding rows. Thus, the rows are linearly independent by Lemma 4.10.

The argument used with the above echelon matrix A can be used for the nonzero rows of any echelon matrix. Thus, we have the following very useful result.

THEOREM 4.11: The nonzero rows of a matrix in echelon form are linearly independent.

4.8 Basis and Dimension

First we state two equivalent ways to define a basis of a vector space V. (The equivalence is proved in Problem 4.28.)

DEFINITION A: A set $S = \{u_1, u_2, \ldots, u_n\}$ of vectors is a *basis* of V if it has the following two properties: (1) S is linearly independent. (2) S spans V.

DEFINITION B: A set $S = \{u_1, u_2, \ldots, u_n\}$ of vectors is a *basis* of V if every $v \in V$ can be written uniquely as a linear combination of the basis vectors.

The following is a fundamental result in linear algebra.

THEOREM 4.12: Let V be a vector space such that one basis has m elements and another basis has n elements. Then $m = n$.

A vector space V is said to be of *finite dimension n* or *n-dimensional*, written

$$\dim V = n$$

if V has a basis with n elements. Theorem 4.12 tells us that all bases of V have the same number of elements, so this definition is well defined.

The vector space $\{0\}$ is defined to have dimension 0.

Suppose a vector space V does not have a finite basis. Then V is said to be of *infinite dimension* or to be *infinite-dimensional*.

The above fundamental Theorem 4.12 is a consequence of the following "replacement lemma" (proved in Problem 4.35).

LEMMA 4.13: Suppose $\{v_1, v_2, \ldots, v_n\}$ spans V, and suppose $\{w_1, w_2, \ldots, w_m\}$ is linearly independent. Then $m \le n$, and V is spanned by a set of the form

$$\{w_1, w_2, \ldots, w_m, v_{i_1}, v_{i_2}, \ldots, v_{i_{n-m}}\}$$

Thus, in particular, $n+1$ or more vectors in V are linearly dependent.

Observe in the above lemma that we have replaced m of the vectors in the spanning set of V by the m independent vectors and still retained a spanning set.

Examples of Bases

This subsection presents important examples of bases of some of the main vector spaces appearing in this text.

(a) Vector space K^n: Consider the following n vectors in K^n:

$$e_1 = (1, 0, 0, 0, \ldots, 0, 0), \; e_2 = (0, 1, 0, 0, \ldots, 0, 0), \; \ldots, \; e_n = (0, 0, 0, 0, \ldots, 0, 1)$$

These vectors are linearly independent. (For example, they form a matrix in echelon form.) Furthermore, any vector $u = (a_1, a_2, \ldots, a_n)$ in K^n can be written as a linear combination of the above vectors. Specifically,

$$v = a_1e_1 + a_2e_2 + \cdots + a_ne_n$$

Accordingly, the vectors form a basis of K^n called the *usual* or *standard* basis of K^n. Thus (as one might expect), K^n has dimension n. In particular, any other basis of K^n has n elements.

(b) Vector space $\mathbf{M} = \mathbf{M}_{r,s}$ of all $r \times s$ matrices: The following six matrices form a basis of the vector space $\mathbf{M}_{2,3}$ of all 2×3 matrices over K:

$$\begin{bmatrix} 1 & 0 & 0 \\ 0 & 0 & 0 \end{bmatrix}, \begin{bmatrix} 0 & 1 & 0 \\ 0 & 0 & 0 \end{bmatrix}, \begin{bmatrix} 0 & 0 & 1 \\ 0 & 0 & 0 \end{bmatrix}, \begin{bmatrix} 0 & 0 & 0 \\ 1 & 0 & 0 \end{bmatrix}, \begin{bmatrix} 0 & 0 & 0 \\ 0 & 1 & 0 \end{bmatrix}, \begin{bmatrix} 0 & 0 & 0 \\ 0 & 0 & 1 \end{bmatrix}$$

More generally, in the vector space $\mathbf{M} = \mathbf{M}_{r,s}$ of all $r \times s$ matrices, let E_{ij} be the matrix with ij-entry 1 and 0's elsewhere. Then all such matrices form a basis of $\mathbf{M}_{r,s}$ called the *usual* or *standard* basis of $\mathbf{M}_{r,s}$. Accordingly, $\dim \mathbf{M}_{r,s} = rs$.

(c) Vector space $\mathbf{P}_n(t)$ of all polynomials of degree $\leq n$: The set $S = \{1, t, t^2, t^3, \ldots, t^n\}$ of $n + 1$ polynomials is a basis of $\mathbf{P}_n(t)$. Specifically, any polynomial $f(t)$ of degree $\leq n$ can be expessed as a linear combination of these powers of t, and one can show that these polynomials are linearly independent. Therefore, $\dim \mathbf{P}_n(t) = n + 1$.

(d) Vector space $\mathbf{P}(t)$ of all polynomials: Consider any finite set $S = \{f_1(t), f_2(t), \ldots, f_m(t)\}$ of polynomials in $\mathbf{P}(t)$, and let m denote the largest of the degrees of the polynomials. Then any polynomial $g(t)$ of degree exceeding m cannot be expressed as a linear combination of the elements of S. Thus, S cannot be a basis of $\mathbf{P}(t)$. This means that the dimension of $\mathbf{P}(t)$ is infinite. We note that the infinite set $S' = \{1, t, t^2, t^3, \ldots\}$, consisting of all the powers of t, spans $\mathbf{P}(t)$ and is linearly independent. Accordingly, S' is an infinite basis of $\mathbf{P}(t)$.

Theorems on Bases

The following three theorems (proved in Problems 4.37, 4.38, and 4.39) will be used frequently.

THEOREM 4.14: Let V be a vector space of finite dimension n. Then:

(i) Any $n + 1$ or more vectors in V are linearly dependent.

(ii) Any linearly independent set $S = \{u_1, u_2, \ldots, u_n\}$ with n elements is a basis of V.

(iii) Any spanning set $T = \{v_1, v_2, \ldots, v_n\}$ of V with n elements is a basis of V.

THEOREM 4.15: Suppose S spans a vector space V. Then:

(i) Any maximum number of linearly independent vectors in S form a basis of V.

(ii) Suppose one deletes from S every vector that is a linear combination of preceding vectors in S. Then the remaining vectors form a basis of V.

THEOREM 4.16: Let V be a vector space of finite dimension and let $S = \{u_1, u_2, \ldots, u_r\}$ be a set of linearly independent vectors in V. Then S is part of a basis of V; that is, S may be extended to a basis of V.

EXAMPLE 4.11

(a) The following four vectors in $\mathbf{R}^4$ form a matrix in echelon form:

$$(1,1,1,1),\ (0,1,1,1),\ (0,0,1,1),\ (0,0,0,1)$$

Thus, the vectors are linearly independent, and, because $\dim \mathbf{R}^4 = 4$, the four vectors form a basis of $\mathbf{R}^4$.

(b) The following $n+1$ polynomials in $\mathbf{P}_n(t)$ are of increasing degree:

$$1,\ t-1,\ (t-1)^2,\ \ldots,\ (t-1)^n$$

Therefore, no polynomial is a linear combination of preceding polynomials; hence, the polynomials are linear independent. Furthermore, they form a basis of $\mathbf{P}_n(t)$, because $\dim \mathbf{P}_n(t) = n+1$.

(c) Consider any four vectors in $\mathbf{R}^3$, say

$$(257, -132, 58), \quad (43, 0, -17), \quad (521, -317, 94), \quad (328, -512, -731)$$

By Theorem 4.14(i), the four vectors must be linearly dependent, because they come from the three-dimensional vector space $\mathbf{R}^3$.

Dimension and Subspaces

The following theorem (proved in Problem 4.40) gives the basic relationship between the dimension of a vector space and the dimension of a subspace.

THEOREM 4.17: Let W be a subspace of an n-dimensional vector space V. Then $\dim W \leq n$. In particular, if $\dim W = n$, then $W = V$.

EXAMPLE 4.12 Let W be a subspace of the real space $\mathbf{R}^3$. Note that $\dim \mathbf{R}^3 = 3$. Theorem 4.17 tells us that the dimension of W can only be 0, 1, 2, or 3. The following cases apply:

(a) If $\dim W = 0$, then $W = \{0\}$, a point.

(b) If $\dim W = 1$, then W is a line through the origin 0.

(c) If $\dim W = 2$, then W is a plane through the origin 0.

(d) If $\dim W = 3$, then W is the entire space $\mathbf{R}^3$.

4.9 Application to Matrices, Rank of a Matrix

Let A be any $m \times n$ matrix over a field K. Recall that the rows of A may be viewed as vectors in K^n and that the row space of A, written rowsp(A), is the subspace of K^n spanned by the rows of A. The following definition applies.

DEFINITION: The *rank* of a matrix A, written rank(A), is equal to the maximum number of linearly independent rows of A or, equivalently, the dimension of the row space of A.

Recall, on the other hand, that the columns of an $m \times n$ matrix A may be viewed as vectors in K^m and that the column space of A, written colsp(A), is the subspace of K^m spanned by the columns of A. Although m may not be equal to n—that is, the rows and columns of A may belong to different vector spaces—we have the following fundamental result.

THEOREM 4.18: The maximum number of linearly independent rows of any matrix A is equal to the maximum number of linearly independent columns of A. Thus, the dimension of the row space of A is equal to the dimension of the column space of A.

Accordingly, one could restate the above definition of the rank of A using columns instead of rows.

Basis-Finding Problems

This subsection shows how an echelon form of any matrix A gives us the solution to certain problems about A itself. Specifically, let A and B be the following matrices, where the echelon matrix B (whose pivots are circled) is an echelon form of A:

$$A = \begin{bmatrix} 1 & 2 & 1 & 3 & 1 & 2 \\ 2 & 5 & 5 & 6 & 4 & 5 \\ 3 & 7 & 6 & 11 & 6 & 9 \\ 1 & 5 & 10 & 8 & 9 & 9 \\ 2 & 6 & 8 & 11 & 9 & 12 \end{bmatrix} \quad \text{and} \quad B = \begin{bmatrix} \textcircled{1} & 2 & 1 & 3 & 1 & 2 \\ 0 & \textcircled{1} & 3 & 1 & 2 & 1 \\ 0 & 0 & 0 & \textcircled{1} & 1 & 2 \\ 0 & 0 & 0 & 0 & 0 & 0 \\ 0 & 0 & 0 & 0 & 0 & 0 \end{bmatrix}$$

We solve the following four problems about the matrix A, where $C_1, C_2, \ldots, C_6$ denote its columns:

(a) Find a basis of the row space of A.

(b) Find each column C_k of A that is a linear combination of preceding columns of A.

(c) Find a basis of the column space of A.

(d) Find the rank of A.

(a) We are given that A and B are row equivalent, so they have the same row space. Moreover, B is in echelon form, so its nonzero rows are linearly independent and hence form a basis of the row space of B. Thus, they also form a basis of the row space of A. That is,

$$\text{basis of rowsp}(A): \qquad (1,2,1,3,1,2), \qquad (0,1,3,1,2,1), \qquad (0,0,0,1,1,2)$$

(b) Let $M_k = [C_1, C_2, \ldots, C_k]$, the submatrix of A consisting of the first k columns of A. Then M_{k-1} and M_k are, respectively, the coefficient matrix and augmented matrix of the vector equation

$$x_1C_1 + x_2C_2 + \cdots + x_{k-1}C_{k-1} = C_k$$

Theorem 3.9 tells us that the system has a solution, or, equivalently, C_k is a linear combination of the preceding columns of A if and only if $\text{rank}(M_k) = \text{rank}(M_{k-1})$, where $\text{rank}(M_k)$ means the number of pivots in an echelon form of M_k. Now the first k column of the echelon matrix B is also an echelon form of M_k. Accordingly,

$$\text{rank}(M_2) = \text{rank}(M_3) = 2 \qquad \text{and} \qquad \text{rank}(M_4) = \text{rank}(M_5) = \text{rank}(M_6) = 3$$

Thus, C_3, C_5, C_6 are each a linear combination of the preceding columns of A.

(c) The fact that the remaining columns C_1, C_2, C_4 are not linear combinations of their respective preceding columns also tells us that they are linearly independent. Thus, they form a basis of the column space of A. That is,

$$\text{basis of colsp}(A): \qquad [1,2,3,1,2]^T, \qquad [2,5,7,5,6]^T, \qquad [3,6,11,8,11]^T$$

Observe that C_1, C_2, C_4 may also be characterized as those columns of A that contain the pivots in any echelon form of A.

(d) Here we see that three possible definitions of the rank of A yield the same value.

(i) There are three pivots in B, which is an echelon form of A.

(ii) The three pivots in B correspond to the nonzero rows of B, which form a basis of the row space of A.

(iii) The three pivots in B correspond to the columns of A, which form a basis of the column space of A.

Thus, $\text{rank}(A) = 3$.

Application to Finding a Basis for $W = \text{span}(u_1, u_2, \ldots, u_r)$

Frequently, we are given a list $S = \{u_1, u_2, \ldots, u_r\}$ of vectors in K^n and we want to find a basis for the subspace W of K^n spanned by the given vectors—that is, a basis of

$$W = \text{span}(S) = \text{span}(u_1, u_2, \ldots, u_r)$$

The following two algorithms, which are essentially described in the above subsection, find such a basis (and hence the dimension) of W.

Algorithm 4.1 (Row space algorithm)

Step 1. Form the matrix M whose *rows* are the given vectors.

Step 2. Row reduce M to echelon form.

Step 3. Output the nonzero rows of the echelon matrix.

Sometimes we want to find a basis that only comes from the original given vectors. The next algorithm accomplishes this task.

Algorithm 4.2 (Casting-out algorithm)

Step 1. Form the matrix M whose *columns* are the given vectors.

Step 2. Row reduce M to echelon form.

Step 3. For each column C_k in the echelon matrix without a pivot, delete (cast out) the vector u_k from the list S of given vectors.

Step 4. Output the remaining vectors in S (which correspond to columns with pivots).

We emphasize that in the first algorithm we form a matrix whose rows are the given vectors, whereas in the second algorithm we form a matrix whose columns are the given vectors.

EXAMPLE 4.13 Let W be the subspace of $\mathbf{R}^5$ spanned by the following vectors:

$$u_1 = (1, 2, 1, 3, 2), \quad u_2 = (1, 3, 3, 5, 3), \quad u_3 = (3, 8, 7, 13, 8)$$
$$u_4 = (1, 4, 6, 9, 7), \quad u_5 = (5, 13, 13, 25, 19)$$

Find a basis of W consisting of the original given vectors, and find $\dim W$.

Form the matrix M whose columns are the given vectors, and reduce M to echelon form:

$$M = \begin{bmatrix} 1 & 1 & 3 & 1 & 5 \\ 2 & 3 & 8 & 4 & 13 \\ 1 & 3 & 7 & 6 & 13 \\ 3 & 5 & 13 & 9 & 25 \\ 2 & 3 & 8 & 7 & 19 \end{bmatrix} \sim \begin{bmatrix} 1 & 1 & 3 & 1 & 5 \\ 0 & 1 & 2 & 2 & 3 \\ 0 & 0 & 0 & 1 & 2 \\ 0 & 0 & 0 & 0 & 0 \\ 0 & 0 & 0 & 0 & 0 \end{bmatrix}$$

The pivots in the echelon matrix appear in columns C_1, C_2, C_4. Accordingly, we "cast out" the vectors u_3 and u_5 from the original five vectors. The remaining vectors u_1, u_2, u_4, which correspond to the columns in the echelon matrix with pivots, form a basis of W. Thus, in particular, $\dim W = 3$.

Remark: The justification of the casting-out algorithm is essentially described above, but we repeat it again here for emphasis. The fact that column C_3 in the echelon matrix in Example 4.13 does not have a pivot means that the vector equation

$$xu_1 + yu_2 = u_3$$

has a solution, and hence u_3 is a linear combination of u_1 and u_2. Similarly, the fact that C_5 does not have a pivot means that u_5 is a linear combination of the preceding vectors. We have deleted each vector in the original spanning set that is a linear combination of preceding vectors. Thus, the remaining vectors are linearly independent and form a basis of W.

Application to Homogeneous Systems of Linear Equations

Consider again a homogeneous system $AX = 0$ of linear equations over K with n unknowns. By Theorem 4.4, the solution set W of such a system is a subspace of K^n, and hence W has a dimension. The following theorem, whose proof is postponed until Chapter 5, holds.

THEOREM 4.19: The dimension of the solution space W of a homogeneous system $AX = 0$ is $n - r$, where n is the number of unknowns and r is the rank of the coefficient matrix A.

In the case where the system $AX = 0$ is in echelon form, it has precisely $n - r$ free variables, say $x_{i_1}, x_{i_2}, \ldots, x_{i_{n-r}}$. Let v_j be the solution obtained by setting $x_{i_j} = 1$ (or any nonzero constant) and the remaining free variables equal to 0. We show (Problem 4.50) that the solutions $v_1, v_2, \ldots, v_{n-r}$ are linearly independent; hence, they form a basis of the solution space W.

We have already used the above process to find a basis of the solution space W of a homogeneous system $AX = 0$ in Section 3.11. Problem 4.48 gives three other examples.

4.10 Sums and Direct Sums

Let U and W be subsets of a vector space V. The sum of U and W, written $U + W$, consists of all sums $u + w$ where $u \in U$ and $w \in W$. That is,

$$U + W = \{v : v = u + w, \text{ where } u \in U \text{ and } w \in W\}$$

Now suppose U and W are subspaces of V. Then one can easily show (Problem 4.53) that $U + W$ is a subspace of V. Recall that $U \cap W$ is also a subspace of V. The following theorem (proved in Problem 4.58) relates the dimensions of these subspaces.

THEOREM 4.20: Suppose U and W are finite-dimensional subspaces of a vector space V. Then $U + W$ has finite dimension and

$$\dim(U + W) = \dim U + \dim W - \dim(U \cap W)$$

EXAMPLE 4.14 Let $V = \mathbf{M}_{2,2}$, the vector space of 2×2 matrices. Let U consist of those matrices whose second row is zero, and let W consist of those matrices whose second column is zero. Then

$$U = \left\{ \begin{bmatrix} a & b \\ 0 & 0 \end{bmatrix} \right\}, \quad W = \left\{ \begin{bmatrix} a & 0 \\ c & 0 \end{bmatrix} \right\} \quad \text{and} \quad U + W = \left\{ \begin{bmatrix} a & b \\ c & 0 \end{bmatrix} \right\}, \quad U \cap W = \left\{ \begin{bmatrix} a & 0 \\ 0 & 0 \end{bmatrix} \right\}$$

That is, $U + W$ consists of those matrices whose lower right entry is 0, and $U \cap W$ consists of those matrices whose second row and second column are zero. Note that $\dim U = 2$, $\dim W = 2$, $\dim(U \cap W) = 1$. Also, $\dim(U + W) = 3$, which is expected from Theorem 4.20. That is,

$$\dim(U + W) = \dim U + \dim V - \dim(U \cap W) = 2 + 2 - 1 = 3$$

Direct Sums

The vector space V is said to be the *direct sum* of its subspaces U and W, denoted by

$$V = U \oplus W$$

if every $v \in V$ can be written in one and only one way as $v = u + w$ where $u \in U$ and $w \in W$.

The following theorem (proved in Problem 4.59) characterizes such a decomposition.

THEOREM 4.21: The vector space V is the direct sum of its subspaces U and W if and only if: (i) $V = U + W$, (ii) $U \cap W = \{0\}$.

EXAMPLE 4.15 Consider the vector space $V = \mathbf{R}^3$.

(a) Let U be the xy-plane and let W be the yz-plane; that is,

$$U = \{(a, b, 0) : a, b \in \mathbf{R}\} \quad \text{and} \quad W = \{(0, b, c) : b, c \in \mathbf{R}\}$$

Then $\mathbf{R}^3 = U + W$, because every vector in $\mathbf{R}^3$ is the sum of a vector in U and a vector in W. However, $\mathbf{R}^3$ is not the direct sum of U and W, because such sums are not unique. For example,

$$(3, 5, 7) = (3, 1, 0) + (0, 4, 7) \quad \text{and also} \quad (3, 5, 7) = (3, -4, 0) + (0, 9, 7)$$

(b) Let U be the xy-plane and let W be the z-axis; that is,

$$U = \{(a, b, 0) : a, b \in \mathbf{R}\} \quad \text{and} \quad W = \{(0, 0, c) : c \in \mathbf{R}\}$$

Now any vector $(a, b, c) \in \mathbf{R}^3$ can be written as the sum of a vector in U and a vector in V in one and only one way:

$$(a, b, c) = (a, b, 0) + (0, 0, c)$$

Accordingly, $\mathbf{R}^3$ is the direct sum of U and W; that is, $\mathbf{R}^3 = U \oplus W$.

General Direct Sums

The notion of a direct sum is extended to more than one factor in the obvious way. That is, V is the *direct sum* of subspaces $W_1, W_2, \ldots, W_r$, written

$$V = W_1 \oplus W_2 \oplus \cdots \oplus W_r$$

if every vector $v \in V$ can be written in one and only one way as

$$v = w_1 + w_2 + \cdots + w_r$$

where $w_1 \in W_1, w_2 \in W_2, \ldots, w_r \in W_r$.

The following theorems hold.

THEOREM 4.22: Suppose $V = W_1 \oplus W_2 \oplus \cdots \oplus W_r$. Also, for each k, suppose S_k is a linearly independent subset of W_k. Then

(a) The union $S = \bigcup_k S_k$ is linearly independent in V.

(b) If each S_k is a basis of W_k, then $\bigcup_k S_k$ is a basis of V.

(c) $\dim V = \dim W_1 + \dim W_2 + \cdots + \dim W_r$.

THEOREM 4.23: Suppose $V = W_1 + W_2 + \cdots + W_r$ and $\dim V = \sum_k \dim W_k$. Then

$$V = W_1 \oplus W_2 \oplus \cdots \oplus W_r.$$

4.11 Coordinates

Let V be an n-dimensional vector space over K with basis $S = \{u_1, u_2, \ldots, u_n\}$. Then any vector $v \in V$ can be expressed uniquely as a linear combination of the basis vectors in S, say

$$v = a_1 u_1 + a_2 u_2 + \cdots + a_n u_n$$

These n scalars $a_1, a_2, \ldots, a_n$ are called the *coordinates* of v relative to the basis S, and they form a vector $[a_1, a_2, \ldots, a_n]$ in K^n called the *coordinate vector* of v relative to S. We denote this vector by $[v]_S$, or simply $[v]$, when S is understood. Thus,

$$[v]_S = [a_1, a_2, \ldots, a_n]$$

For notational convenience, brackets $[\ldots]$, rather than parentheses $(\ldots)$, are used to denote the coordinate vector.

Remark: The above n scalars $a_1, a_2, \ldots, a_n$ also form the *coordinate column vector* $[a_1, a_2, \ldots, a_n]^T$ of v relative to S. The choice of the column vector rather than the row vector to represent v depends on the context in which it is used. The use of such column vectors will become clear later in Chapter 6.

EXAMPLE 4.16 Consider the vector space $\mathbf{P}_2(t)$ of polynomials of degree ≤ 2. The polynomials

$$p_1 = t + 1, \qquad p_2 = t - 1, \qquad p_3 = (t-1)^2 = t^2 - 2t + 1$$

form a basis S of $\mathbf{P}_2(t)$. The coordinate vector $[v]$ of $v = 2t^2 - 5t + 9$ relative to S is obtained as follows.

Set $v = xp_1 + yp_2 + zp_3$ using unknown scalars x, y, z, and simplify:

$$\begin{aligned} 2t^2 - 5t + 9 &= x(t+1) + y(t-1) + z(t^2 - 2t + 1) \\ &= xt + x + yt - y + zt^2 - 2zt + z \\ &= zt^2 + (x + y - 2z)t + (x - y + z) \end{aligned}$$

Then set the coefficients of the same powers of t equal to each other to obtain the system

$$z = 2, \qquad x + y - 2z = -5, \qquad x - y + z = 9$$

The solution of the system is $x = 3$, $y = -4$, $z = 2$. Thus,

$$v = 3p_1 - 4p_2 + 2p_3, \quad \text{and hence,} \quad [v] = [3, -4, 2]$$

EXAMPLE 4.17 Consider real space $\mathbf{R}^3$. The following vectors form a basis S of $\mathbf{R}^3$:

$$u_1 = (1, -1, 0), \qquad u_2 = (1, 1, 0), \qquad u_3 = (0, 1, 1)$$

The coordinates of $v = (5, 3, 4)$ relative to the basis S are obtained as follows.

Set $v = xv_1 + yv_2 + zv_3$; that is, set v as a linear combination of the basis vectors using unknown scalars x, y, z. This yields

$$\begin{bmatrix} 5 \\ 3 \\ 4 \end{bmatrix} = x\begin{bmatrix} 1 \\ -1 \\ 0 \end{bmatrix} + y\begin{bmatrix} 1 \\ 1 \\ 0 \end{bmatrix} + z\begin{bmatrix} 0 \\ 1 \\ 1 \end{bmatrix}$$

The equivalent system of linear equations is as follows:

$$x + y = 5, \qquad -x + y + z = 3, \qquad z = 4$$

The solution of the system is $x = 3$, $y = 2$, $z = 4$. Thus,

$$v = 3u_1 + 2u_2 + 4u_3, \quad \text{and so} \quad [v]_s = [3, 2, 4]$$

Remark 1: There is a geometrical interpretation of the coordinates of a vector v relative to a basis S for the real space $\mathbf{R}^n$, which we illustrate using the basis S of $\mathbf{R}^3$ in Example 4.17. First consider the space $\mathbf{R}^3$ with the usual x, y, z axes. Then the basis vectors determine a new coordinate system of $\mathbf{R}^3$, say with x', y', z' axes, as shown in Fig. 4-4. That is,

(1) The x'-axis is in the direction of u_1 with unit length $\|u_1\|$.
(2) The y'-axis is in the direction of u_2 with unit length $\|u_2\|$.
(3) The z'-axis is in the direction of u_3 with unit length $\|u_3\|$.

Then each vector $v = (a, b, c)$ or, equivalently, the point $P(a, b, c)$ in $\mathbf{R}^3$ will have new coordinates with respect to the new x', y', z' axes. These new coordinates are precisely $[v]_S$, the coordinates of v with respect to the basis S. Thus, as shown in Example 4.17, the coordinates of the point $P(5, 3, 4)$ with the new axes form the vector $[3, 2, 4]$.

Remark 2: Consider the usual basis $E = \{e_1, e_2, \ldots, e_n\}$ of K^n defined by

$$e_1 = (1, 0, 0, \ldots, 0, 0), \qquad e_2 = (0, 1, 0, \ldots, 0, 0), \qquad \ldots, \qquad e_n = (0, 0, 0, \ldots, 0, 1)$$

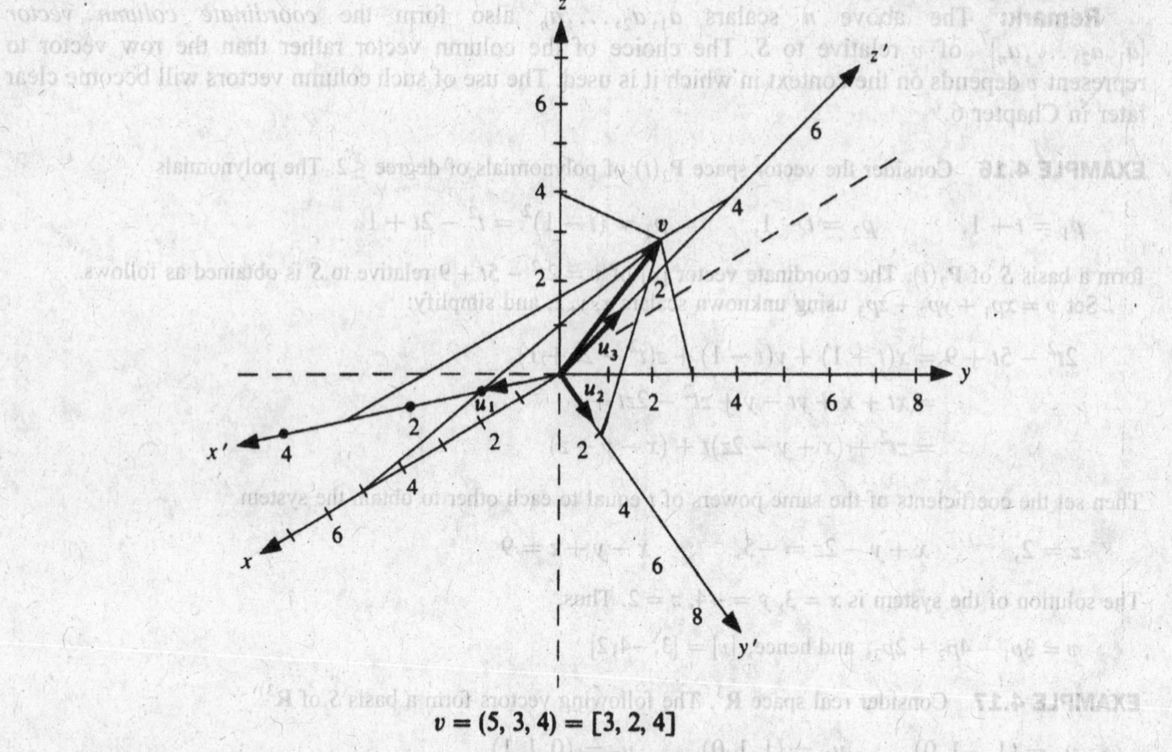

$v = (5, 3, 4) = [3, 2, 4]$

Figure 4-4

Let $v = (a_1, a_2, \ldots, a_n)$ be any vector in K^n. Then one can easily show that

$$v = a_1e_1 + a_2e_2 + \cdots + a_ne_n, \qquad \text{and so} \qquad [v]_E = [a_1, a_2, \ldots, a_n]$$

That is, the coordinate vector $[v]_E$ of any vector v relative to the usual basis E of K^n is identical to the original vector v.

Isomorphism of V and K^n

Let V be a vector space of dimension n over K, and suppose $S = \{u_1, u_2, \ldots, u_n\}$ is a basis of V. Then each vector $v \in V$ corresponds to a unique n-tuple $[v]_S$ in K^n. On the other hand, each n-tuple $[c_1, c_2, \ldots, c_n]$ in K^n corresponds to a unique vector $c_1u_1 + c_2u_2 + \cdots + c_nu_n$ in V. Thus, the basis S induces a one-to-one correspondence between V and K^n. Furthermore, suppose

$$v = a_1u_1 + a_2u_2 + \cdots + a_nu_n \qquad \text{and} \qquad w = b_1u_1 + b_2u_2 + \cdots + b_nu_n$$

Then

$$v + w = (a_1 + b_1)u_1 + (a_2 + b_2)u_2 + \cdots + (a_n + b_n)u_n$$
$$kv = (ka_1)u_1 + (ka_2)u_2 + \cdots + (ka_n)u_n$$

where k is a scalar. Accordingly,

$$[v + w]_S = [a_1 + b_1, \quad \ldots, \quad a_n + b_n] = [a_1, \ldots, a_n] + [b_1, \ldots, b_n] = [v]_S + [w]_S$$
$$[kv]_S = [ka_1, ka_2, \ldots, ka_n] = k[a_1, a_2, \ldots, a_n] = k[v]_S$$

Thus, the above one-to-one correspondence between V and K^n preserves the vector space operations of vector addition and scalar multiplication. We then say that V and K^n are isomorphic, written

$$V \cong K^n$$

We state this result formally.

THEOREM 4.24: Let V be an n-dimensional vector space over a field K. Then V and K^n a[re] isomorphic.

The next example gives a practical application of the above result.

EXAMPLE 4.18 Suppose we want to determine whether or not the following matrices in $V = \mathbf{M}_{2,3}$ are linearly dependent:

$$A = \begin{bmatrix} 1 & 2 & -3 \\ 4 & 0 & 1 \end{bmatrix}, \qquad B = \begin{bmatrix} 1 & 3 & -4 \\ 6 & 5 & 4 \end{bmatrix}, \qquad C = \begin{bmatrix} 3 & 8 & -11 \\ 16 & 10 & 9 \end{bmatrix}$$

The coordinate vectors of the matrices in the usual basis of $\mathbf{M}_{2,3}$ are as follows:

$$[A] = [1, 2, -3, 4, 0, 1], \qquad [B] = [1, 3, -4, 6, 5, 4], \qquad [C] = [3, 8, -11, 16, 10, 9]$$

Form the matrix M whose rows are the above coordinate vectors and reduce M to an echelon form:

$$M = \begin{bmatrix} 1 & 2 & -3 & 4 & 0 & 1 \\ 1 & 3 & -4 & 6 & 5 & 4 \\ 3 & 8 & -11 & 16 & 10 & 9 \end{bmatrix} \sim \begin{bmatrix} 1 & 2 & -3 & 4 & 0 & 1 \\ 0 & 1 & -1 & 2 & 5 & 3 \\ 0 & 2 & -2 & 4 & 10 & 6 \end{bmatrix} \sim \begin{bmatrix} 1 & 2 & -3 & 4 & 0 & 1 \\ 0 & 1 & -1 & 2 & 5 & 3 \\ 0 & 0 & 0 & 0 & 0 & 0 \end{bmatrix}$$

Because the echelon matrix has only two nonzero rows, the coordinate vectors $[A]$, $[B]$, $[C]$ span a subspace of dimension 2 and so are linearly dependent. Accordingly, the original matrices A, B, C are linearly dependent.

SOLVED PROBLEMS

Vector Spaces, Linear Combinations

4.1. Suppose u and v belong to a vector space V. Simplify each of the following expressions:

(a) $E_1 = 3(2u - 4v) + 5u + 7v$, (c) $E_3 = 2uv + 3(2u + 4v)$

(b) $E_2 = 3u - 6(3u - 5v) + 7u$, (d) $E_4 = 5u - \dfrac{3}{v} + 5u$

Multiply out and collect terms:

(a) $E_1 = 6u - 12v + 5u + 7v = 11u - 5v$

(b) $E_2 = 3u - 18u + 30v + 7u = -8u + 30v$

(c) E_3 is not defined because the product uv of vectors is not defined.

(d) E_4 is not defined because division by a vector is not defined.

4.2. Prove Theorem 4.1: Let V be a vector space over a field K.
(i) $k0 = 0$. (ii) $0u = 0$. (iii) If $ku = 0$, then $k = 0$ or $u = 0$. (iv) $(-k)u = k(-u) = -ku$.

(i) By Axiom $[A_2]$ with $u = 0$, we have $0 + 0 = 0$. Hence, by Axiom $[M_1]$, we have

$$k0 = k(0 + 0) = k0 + k0$$

Adding $-k0$ to both sides gives the desired result.

(ii) For scalars, $0 + 0 = 0$. Hence, by Axiom $[M_2]$, we have

$$0u = (0 + 0)u = 0u + 0u$$

Adding $-0u$ to both sides gives the desired result.

(iii) Suppose $ku = 0$ and $k \neq 0$. Then there exists a scalar k^{-1} such that $k^{-1}k = 1$. Thus,

$$u = 1u = (k^{-1}k)u = k^{-1}(ku) = k^{-1}0 = 0$$

(iv) Using $u + (-u) = 0$ and $k + (-k) = 0$ yields

$$0 = k0 = k[u + (-u)] = ku + k(-u) \qquad \text{and} \qquad 0 = 0u = [k + (-k)]u = ku + (-k)u$$

Adding $-ku$ to both sides of the first equation gives $-ku = k(-u)$, and adding $-ku$ to both sides of the second equation gives $-ku = (-k)u$. Thus, $(-k)u = k(-u) = -ku$.

4.3. Show that (a) $k(u-v) = ku - kv$, (b) $u + u = 2u$.

(a) Using the definition of subtraction, that $u - v = u + (-v)$, and Theorem 4.1(iv), that $k(-v) = -kv$, we have

$$k(u-v) = k[u+(-v)] = ku + k(-v) = ku + (-kv) = ku - kv$$

(b) Using Axiom [M_4] and then Axiom [M_2], we have

$$u + u = 1u + 1u = (1+1)u = 2u$$

4.4. Express $v = (1, -2, 5)$ in $\mathbf{R}^3$ as a linear combination of the vectors

$$u_1 = (1, 1, 1), \qquad u_2 = (1, 2, 3), \qquad u_3 = (2, -1, 1)$$

We seek scalars x, y, z, as yet unknown, such that $v = xu_1 + yu_2 + zu_3$. Thus, we require

$$\begin{bmatrix} 1 \\ -2 \\ 5 \end{bmatrix} = x\begin{bmatrix} 1 \\ 1 \\ 1 \end{bmatrix} + y\begin{bmatrix} 1 \\ 2 \\ 3 \end{bmatrix} + z\begin{bmatrix} 2 \\ -1 \\ 1 \end{bmatrix} \qquad \text{or} \qquad \begin{aligned} x + y + 2z &= 1 \\ x + 2y - z &= -2 \\ x + 3y + z &= 5 \end{aligned}$$

(For notational convenience, we write the vectors in $\mathbf{R}^3$ as columns, because it is then easier to find the equivalent system of linear equations.) Reducing the system to echelon form yields the triangular system

$$x + y + 2z = 1, \qquad y - 3z = -3, \qquad 5z = 10$$

The system is consistent and has a solution. Solving by back-substitution yields the solution $x = -6, y = 3, z = 2$. Thus, $v = -6u_1 + 3u_2 + 2u_3$.

Alternatively, write down the augmented matrix M of the equivalent system of linear equations, where u_1, u_2, u_3 are the first three columns of M and v is the last column, and then reduce M to echelon form:

$$M = \begin{bmatrix} 1 & 1 & 2 & 1 \\ 1 & 2 & -1 & -2 \\ 1 & 3 & 1 & 5 \end{bmatrix} \sim \begin{bmatrix} 1 & 1 & 2 & 1 \\ 0 & 1 & -3 & -3 \\ 0 & 2 & -1 & 4 \end{bmatrix} \sim \begin{bmatrix} 1 & 1 & 2 & 1 \\ 0 & 1 & -3 & -3 \\ 0 & 0 & 5 & 10 \end{bmatrix}$$

The last matrix corresponds to a triangular system, which has a solution. Solving the triangular system by back-substitution yields the solution $x = -6, y = 3, z = 2$. Thus, $v = -6u_1 + 3u_2 + 2u_3$.

4.5. Express $v = (2, -5, 3)$ in $\mathbf{R}^3$ as a linear combination of the vectors

$$u_1 = (1, -3, 2), \; u_2 = (2, -4, -1), \; u_3 = (1, -5, 7)$$

We seek scalars x, y, z, as yet unknown, such that $v = xu_1 + yu_2 + zu_3$. Thus, we require

$$\begin{bmatrix} 2 \\ -5 \\ 3 \end{bmatrix} = x\begin{bmatrix} 1 \\ -3 \\ 2 \end{bmatrix} + y\begin{bmatrix} 2 \\ -4 \\ -1 \end{bmatrix} + z\begin{bmatrix} 1 \\ -5 \\ 7 \end{bmatrix} \qquad \text{or} \qquad \begin{aligned} x + 2y + z &= 2 \\ -3x - 4y - 5z &= -5 \\ 2x - y + 7z &= 3 \end{aligned}$$

Reducing the system to echelon form yields the system

$$x + 2y + z = 2, \qquad 2y - 2z = 1, \qquad 0 = 3$$

The system is inconsistent and so has no solution. Thus, v cannot be written as a linear combination of u_1, u_2, u_3.

4.6. Express the polynomial $v = t^2 + 4t - 3$ in $\mathbf{P}(t)$ as a linear combination of the polynomials

$$p_1 = t^2 - 2t + 5, \qquad p_2 = 2t^2 - 3t, \qquad p_3 = t + 1$$

Set v as a linear combination of p_1, p_2, p_3 using unknowns x, y, z to obtain

$$t^2 + 4t - 3 = x(t^2 - 2t + 5) + y(2t^2 - 3t) + z(t + 1) \qquad (*)$$

We can proceed in two ways.

Method 1. Expand the right side of (*) and express it in terms of powers of t as follows:

$$t^2 + 4t - 3 = xt^2 - 2xt + 5x + 2yt^2 - 3yt + zt + z$$
$$= (x + 2y)t^2 + (-2x - 3y + z)t + (5x + 3z)$$

Set coefficients of the same powers of t equal to each other, and reduce the system to echelon form. This yields

$$\begin{aligned} x + 2y \quad\;\; &= 1 \\ -2x - 3y + z &= 4 \\ 5x \quad\;\; + 3z &= -3 \end{aligned} \qquad \text{or} \qquad \begin{aligned} x + 2y \quad\;\; &= 1 \\ y + z &= 6 \\ -10y + 3z &= -8 \end{aligned} \qquad \text{or} \qquad \begin{aligned} x + 2y \quad &= 1 \\ y + z &= 6 \\ 13z &= 52 \end{aligned}$$

The system is consistent and has a solution. Solving by back-substitution yields the solution $x = -3$, $y = 2$, $z = 4$. Thus, $v = -3p_1 + 2p_2 + 4p_2$.

Method 2. The equation (*) is an identity in t; that is, the equation holds for any value of t. Thus, we can set t equal to any numbers to obtain equations in the unknowns.

(a) Set $t = 0$ in (*) to obtain the equation $-3 = 5x + z$.

(b) Set $t = 1$ in (*) to obtain the equation $2 = 4x - y + 2z$.

(c) Set $t = -1$ in (*) to obtain the equation $-6 = 8x + 5y$.

Solve the system of the three equations to again obtain the solution $x = -3$, $y = 2$, $z = 4$. Thus, $v = -3p_1 + 2p_2 + 4p_3$.

4.7. Express M as a linear combination of the matrices A, B, C, where

$$M = \begin{bmatrix} 4 & 7 \\ 7 & 9 \end{bmatrix}, \quad \text{and} \quad A = \begin{bmatrix} 1 & 1 \\ 1 & 1 \end{bmatrix}, \quad B = \begin{bmatrix} 1 & 2 \\ 3 & 4 \end{bmatrix}, \quad C = \begin{bmatrix} 1 & 1 \\ 4 & 5 \end{bmatrix}$$

Set M as a linear combination of A, B, C using unknown scalars x, y, z; that is, set $M = xA + yB + zC$. This yields

$$\begin{bmatrix} 4 & 7 \\ 7 & 9 \end{bmatrix} = x\begin{bmatrix} 1 & 1 \\ 1 & 1 \end{bmatrix} + y\begin{bmatrix} 1 & 2 \\ 3 & 4 \end{bmatrix} + z\begin{bmatrix} 1 & 1 \\ 4 & 5 \end{bmatrix} = \begin{bmatrix} x + y + z & x + 2y + z \\ x + 3y + 4z & x + 4y + 5z \end{bmatrix}$$

Form the equivalent system of equations by setting corresponding entries equal to each other:

$$x + y + z = 4, \qquad x + 2y + z = 7, \qquad x + 3y + 4z = 7, \qquad x + 4y + 5z = 9$$

Reducing the system to echelon form yields

$$x + y + z = 4, \qquad y = 3, \qquad 3z = -3, \qquad 4z = -4$$

The last equation drops out. Solving the system by back-substitution yields $z = -1$, $y = 3$, $x = 2$. Thus, $M = 2A + 3B - C$.

Subspaces

4.8. Prove Theorem 4.2: W is a subspace of V if the following two conditions hold:

(a) $0 \in W$. (b) If $u, v \in W$, then $u + v$, $ku \in W$.

By (a), W is nonempty, and, by (b), the operations of vector addition and scalar multiplication are well defined for W. Axioms [A_1], [A_4], [M_1], [M_2], [M_3], [M_4] hold in W because the vectors in W belong to V. Thus, we need only show that [A_2] and [A_3] also hold in W. Now [A_2] holds because the zero vector in V belongs to W by (a). Finally, if $v \in W$, then $(-1)v = -v \in W$, and $v + (-v) = 0$. Thus [A_3] holds.

4.9. Let $V = \mathbf{R}^3$. Show that W is not a subspace of V, where

(a) $W = \{(a, b, c) : a \geq 0\}$, (b) $W = \{(a, b, c) : a^2 + b^2 + c^2 \leq 1\}$.

In each case, show that Theorem 4.2 does not hold.

(a) W consists of those vectors whose first entry is nonnegative. Thus, $v = (1, 2, 3)$ belongs to W. Let $k = -3$. Then $kv = (-3, -6, -9)$ does not belong to W, because -3 is negative. Thus, W is not a subspace of V.

(b) W consists of vectors whose length does not exceed 1. Hence, $u = (1, 0, 0)$ and $v = (0, 1, 0)$ belong to W, but $u + v = (1, 1, 0)$ does not belong to W, because $1^2 + 1^2 + 0^2 = 2 > 1$. Thus, W is not a subspace of V.

4.10. Let $V = \mathbf{P}(t)$, the vector space of real polynomials. Determine whether or not W is a subspace of V, where

(a) W consists of all polynomials with integral coefficients.

(b) W consists of all polynomials with degree ≥ 6 and the zero polynomial.

(c) W consists of all polynomials with only even powers of t.

(a) No, because scalar multiples of polynomials in W do not always belong to W. For example,

$$f(t) = 3 + 6t + 7t^2 \in W \quad \text{but} \quad \tfrac{1}{2}f(t) = \tfrac{3}{2} + 3t + \tfrac{7}{2}t^2 \notin W$$

(b and c) Yes. In each case, W contains the zero polynomial, and sums and scalar multiples of polynomials in W belong to W.

4.11. Let V be the vector space of functions $f : \mathbf{R} \to \mathbf{R}$. Show that W is a subspace of V, where

(a) $W = \{f(x) : f(1) = 0\}$, all functions whose value at 1 is 0.

(b) $W = \{f(x) : f(3) = f(1)\}$, all functions assigning the same value to 3 and 1.

(c) $W = \{f(t) : f(-x) = -f(x)\}$, all *odd functions*.

Let $\hat{0}$ denote the zero function, so $\hat{0}(x) = 0$ for every value of x.

(a) $\hat{0} \in W$, because $\hat{0}(1) = 0$. Suppose $f, g \in W$. Then $f(1) = 0$ and $g(1) = 0$. Also, for scalars a and b, we have

$$(af + bg)(1) = af(1) + bg(1) = a0 + b0 = 0$$

Thus, $af + bg \in W$, and hence W is a subspace.

(b) $\hat{0} \in W$, because $\hat{0}(3) = 0 = \hat{0}(1)$. Suppose $f, g \in W$. Then $f(3) = f(1)$ and $g(3) = g(1)$. Thus, for any scalars a and b, we have

$$(af + bg)(3) = af(3) + bg(3) = af(1) + bg(1) = (af + bg)(1)$$

Thus, $af + bg \in W$, and hence W is a subspace.

(c) $\hat{0} \in W$, because $\hat{0}(-x) = 0 = -0 = -\hat{0}(x)$. Suppose $f, g \in W$. Then $f(-x) = -f(x)$ and $g(-x) = -g(x)$. Also, for scalars a and b,

$$(af + bg)(-x) = af(-x) + bg(-x) = -af(x) - bg(x) = -(af + bg)(x)$$

Thus, $ab + gf \in W$, and hence W is a subspace of V.

4.12. Prove Theorem 4.3: The intersection of any number of subspaces of V is a subspace of V.

Let $\{W_i : i \in I\}$ be a collection of subspaces of V and let $W = \cap(W_i : i \in I)$. Because each W_i is a subspace of V, we have $0 \in W_i$, for every $i \in I$. Hence, $0 \in W$. Suppose $u, v \in W$. Then $u, v \in W_i$, for every $i \in I$. Because each W_i is a subspace, $au + bv \in W_i$, for every $i \in I$. Hence, $au + bv \in W$. Thus, W is a subspace of V.

Linear Spans

4.13. Show that the vectors $u_1 = (1, 1, 1)$, $u_2 = (1, 2, 3)$, $u_3 = (1, 5, 8)$ span $\mathbf{R}^3$.

We need to show that an arbitrary vector $v = (a, b, c)$ in $\mathbf{R}^3$ is a linear combination of u_1, u_2, u_3. Set $v = xu_1 + yu_2 + zu_3$; that is, set

$$(a, b, c) = x(1, 1, 1) + y(1, 2, 3) + z(1, 5, 8) = (x + y + z, \quad x + 2y + 5z, \quad x + 3y + 8z)$$

Form the equivalent system and reduce it to echelon form:

$$\begin{aligned} x+y+z&=a \\ x+2y+5z&=b \\ x+3y+8z&=c \end{aligned} \quad \text{or} \quad \begin{aligned} x+y+z&=a \\ y+4z&=b-a \\ 2y+7c&=c-a \end{aligned} \quad \text{or} \quad \begin{aligned} x+y+z&=a \\ y+4z&=b-a \\ -z&=c-2b+a \end{aligned}$$

The above system is in echelon form and is consistent; in fact,

$$x=-a+5b-3c, \quad y=3a-7b+4c, \quad z=a+2b-c$$

is a solution. Thus, u_1, u_2, u_3 span $\mathbf{R}^3$.

4.14. Find conditions on a, b, c so that $v=(a,b,c)$ in $\mathbf{R}^3$ belongs to $W=\text{span}(u_1,u_2,u_3)$, where

$$u_1=(1,2,0), \quad u_2=(-1,1,2), \quad u_3=(3,0,-4)$$

Set v as a linear combination of u_1, u_2, u_3 using unknowns x, y, z; that is, set $v=xu_1+yu_2+zu_3$. This yields

$$(a,b,c)=x(1,2,0)+y(-1,1,2)+z(3,0,-4)=(x-y+3z, \quad 2x+y, \quad 2y-4z)$$

Form the equivalent system of linear equations and reduce it to echelon form:

$$\begin{aligned} x-y+3z&=a \\ 2x+y\qquad&=b \\ 2y-4z&=c \end{aligned} \quad \text{or} \quad \begin{aligned} x-y+3z&=a \\ 3y-6z&=b-2a \\ 2y-4z&=c \end{aligned} \quad \text{or} \quad \begin{aligned} x-y+3z&=a \\ 3y-6z&=b-2a \\ 0&=4a-2b+3c \end{aligned}$$

The vector $v=(a,b,c)$ belongs to W if and only if the system is consistent, and it is consistent if and only if $4a-2b+3c=0$. Note, in particular, that u_1, u_2, u_3 do not span the whole space $\mathbf{R}^3$.

4.15. Show that the vector space $V=\mathbf{P}(t)$ of real polynomials cannot be spanned by a finite number of polynomials.

Any finite set S of polynomials contains a polynomial of maximum degree, say m. Then the linear span span(S) of S cannot contain a polynomial of degree greater than m. Thus, $\text{span}(S) \neq V$, for any finite set S.

4.16. Prove Theorem 4.5: Let S be a subset of V. (i) Then span(S) is a subspace of V containing S. (ii) If W is a subspace of V containing S, then $\text{span}(S) \subseteq W$.

(i) Suppose S is empty. By definition, $\text{span}(S)=\{0\}$. Hence $\text{span}(S)=\{0\}$ is a subspace of V and $S \subseteq \text{span}(S)$. Suppose S is not empty and $v \in S$. Then $v=1v \in \text{span}(S)$; hence, $S \subseteq \text{span}(S)$. Also $0=0v \in \text{span}(S)$. Now suppose $u, w \in \text{span}(S)$, say

$$u=a_1u_1+\cdots+a_ru_r=\sum_i a_iu_i \quad \text{and} \quad w=b_1w_1+\cdots+b_sw_s=\sum_j b_jw_j$$

where $u_i, w_j \in S$ and $a_i, b_j \in K$. Then

$$u+v=\sum_i a_iu_i+\sum_j b_jw_j \quad \text{and} \quad ku=k\left(\sum_i a_iu_i\right)=\sum_i ka_iu_i$$

belong to span(S) because each is a linear combination of vectors in S. Thus, span(S) is a subspace of V.

(ii) Suppose $u_1, u_2, \ldots, u_r \in S$. Then all the u_i belong to W. Thus, all multiples $a_1u_1, a_2u_2, \ldots, a_ru_r \in W$, and so the sum $a_1u_1+a_2u_2+\cdots+a_ru_r \in W$. That is, W contains all linear combinations of elements in S, or, in other words, $\text{span}(S) \subseteq W$, as claimed.

Linear Dependence

4.17. Determine whether or not u and v are linearly dependent, where

(a) $u=(1,2)$, $v=(3,-5)$, (c) $u=(1,2,-3)$, $v=(4,5,-6)$

(b) $u=(1,-3)$, $v=(-2,6)$, (d) $u=(2,4,-8)$, $v=(3,6,-12)$

Two vectors u and v are linearly dependent if and only if one is a multiple of the other.

(a) No. (b) Yes; for $v=-2u$. (c) No. (d) Yes, for $v=\frac{3}{2}u$.

4.18. Determine whether or not u and v are linearly dependent, where

(a) $u = 2t^2 + 4t - 3,\ v = 4t^2 + 8t - 6,$ (b) $u = 2t^2 - 3t + 4,\ v = 4t^2 - 3t + 2,$

(c) $u = \begin{bmatrix} 1 & 3 & -4 \\ 5 & 0 & -1 \end{bmatrix},\ v = \begin{bmatrix} -4 & -12 & 16 \\ -20 & 0 & 4 \end{bmatrix},$ (d) $u = \begin{bmatrix} 1 & 1 & 1 \\ 2 & 2 & 2 \end{bmatrix},\ v = \begin{bmatrix} 2 & 2 & 2 \\ 3 & 3 & 3 \end{bmatrix}$

Two vectors u and v are linearly dependent if and only if one is a multiple of the other.

(a) Yes; for $v = 2u$. (b) No. (c) Yes, for $v = -4u$. (d) No.

4.19. Determine whether or not the vectors $u = (1, 1, 2)$, $v = (2, 3, 1)$, $w = (4, 5, 5)$ in $\mathbf{R}^3$ are linearly dependent.

Method 1. Set a linear combination of u, v, w equal to the zero vector using unknowns x, y, z to obtain the equivalent homogeneous system of linear equations and then reduce the system to echelon form. This yields

$$x\begin{bmatrix} 1 \\ 1 \\ 1 \end{bmatrix} + y\begin{bmatrix} 2 \\ 3 \\ 1 \end{bmatrix} + z\begin{bmatrix} 4 \\ 5 \\ 5 \end{bmatrix} = \begin{bmatrix} 0 \\ 0 \\ 0 \end{bmatrix} \quad \text{or} \quad \begin{aligned} x + 2y + 4z &= 0 \\ x + 3y + 5z &= 0 \\ 2x + y + 5z &= 0 \end{aligned} \quad \text{or} \quad \begin{aligned} x + 2y + 4z &= 0 \\ y + z &= 0 \end{aligned}$$

The echelon system has only two nonzero equations in three unknowns; hence, it has a free variable and a nonzero solution. Thus, u, v, w are linearly dependent.

Method 2. Form the matrix A whose columns are u, v, w and reduce to echelon form:

$$A = \begin{bmatrix} 1 & 2 & 4 \\ 1 & 3 & 5 \\ 2 & 1 & 5 \end{bmatrix} \sim \begin{bmatrix} 1 & 2 & 4 \\ 0 & 1 & 1 \\ 0 & -3 & -3 \end{bmatrix} \sim \begin{bmatrix} 1 & 2 & 4 \\ 0 & 1 & 1 \\ 0 & 0 & 0 \end{bmatrix}$$

The third column does not have a pivot; hence, the third vector w is a linear combination of the first two vectors u and v. Thus, the vectors are linearly dependent. (Observe that the matrix A is also the coefficient matrix in Method 1. In other words, this method is essentially the same as the first method.)

Method 3. Form the matrix B whose rows are u, v, w, and reduce to echelon form:

$$B = \begin{bmatrix} 1 & 1 & 2 \\ 2 & 3 & 1 \\ 4 & 5 & 5 \end{bmatrix} \sim \begin{bmatrix} 0 & 1 & 2 \\ 0 & 1 & -3 \\ 0 & 1 & -3 \end{bmatrix} \sim \begin{bmatrix} 1 & 1 & 2 \\ 0 & 1 & -3 \\ 0 & 0 & 0 \end{bmatrix}$$

Because the echelon matrix has only two nonzero rows, the three vectors are linearly dependent. (The three given vectors span a space of dimension 2.)

4.20. Determine whether or not each of the following lists of vectors in $\mathbf{R}^3$ is linearly dependent:

(a) $u_1 = (1, 2, 5)$, $u_2 = (1, 3, 1)$, $u_3 = (2, 5, 7)$, $u_4 = (3, 1, 4)$,

(b) $u = (1, 2, 5)$, $v = (2, 5, 1)$, $w = (1, 5, 2)$,

(c) $u = (1, 2, 3)$, $v = (0, 0, 0)$, $w = (1, 5, 6)$.

(a) Yes, because any four vectors in $\mathbf{R}^3$ are linearly dependent.

(b) Use Method 2 above; that is, form the matrix A whose columns are the given vectors, and reduce the matrix to echelon form:

$$A = \begin{bmatrix} 1 & 2 & 1 \\ 2 & 5 & 5 \\ 5 & 1 & 2 \end{bmatrix} \sim \begin{bmatrix} 1 & 2 & 1 \\ 0 & 1 & 3 \\ 0 & -9 & -3 \end{bmatrix} \sim \begin{bmatrix} 1 & 2 & 1 \\ 0 & 1 & 3 \\ 0 & 0 & 24 \end{bmatrix}$$

Every column has a pivot entry; hence, no vector is a linear combination of the previous vectors. Thus, the vectors are linearly independent.

(c) Because $0 = (0, 0, 0)$ is one of the vectors, the vectors are linearly dependent.

4.21. Show that the functions $f(t) = \sin t$, $g(t) \cos t$, $h(t) = t$ from **R** into **R** are linearly independent.

Set a linear combination of the functions equal to the zero function **0** using unknown scalars x, y, z; that is, set $xf + yg + zh = \mathbf{0}$. Then show $x = 0, y = 0, z = 0$. We emphasize that $xf + yg + zh = \mathbf{0}$ means that, for every value of t, we have $xf(t) + yg(t) + zh(t) = 0$.

Thus, in the equation $x \sin t + y \cos t + zt = 0$:

(i) Set $t = 0$ to obtain $x(0) + y(1) + z(0) = 0$ or $y = 0$.
(ii) Set $t = \pi/2$ to obtain $x(1) + y(0) + z\pi/2 = 0$ or $x + \pi z/2 = 0$.
(iii) Set $t = \pi$ to obtain $x(0) + y(-1) + z(\pi) = 0$ or $-y + \pi z = 0$.

The three equations have only the zero solution; that is, $x = 0$, $y = 0$, $z = 0$. Thus, f, g, h are linearly independent.

4.22. Suppose the vectors u, v, w are linearly independent. Show that the vectors $u + v$, $u - v$, $u - 2v + w$ are also linearly independent.

Suppose $x(u + v) + y(u - v) + z(u - 2v + w) = 0$. Then

$$xu + xv + yu - yv + zu - 2zv + zw = 0$$

or

$$(x + y + z)u + (x - y - 2z)v + zw = 0$$

Because u, v, w are linearly independent, the coefficients in the above equation are each 0; hence,

$$x + y + z = 0, \qquad x - y - 2z = 0, \qquad z = 0$$

The only solution to the above homogeneous system is $x = 0, y = 0, z = 0$. Thus, $u + v$, $u - v$, $u - 2v + w$ are linearly independent.

4.23. Show that the vectors $u = (1 + i,\ 2i)$ and $w = (1,\ 1 + i)$ in $\mathbf{C}^2$ are linearly dependent over the complex field **C** but linearly independent over the real field **R**.

Recall that two vectors are linearly dependent (over a field K) if and only if one of them is a multiple of the other (by an element in K). Because

$$(1 + i)w = (1 + i)(1,\ 1 + i) = (1 + i,\ 2i) = u$$

u and w are linearly dependent over **C**. On the other hand, u and w are linearly independent over **R**, as no real multiple of w can equal u. Specifically, when k is real, the first component of $kw = (k,\ k + ki)$ must be real, and it can never equal the first component $1 + i$ of u, which is complex.

Basis and Dimension

4.24. Determine whether or not each of the following form a basis of $\mathbf{R}^3$:

(a) $(1, 1, 1)$, $(1, 0, 1)$; (c) $(1, 1, 1)$, $(1, 2, 3)$, $(2, -1, 1)$;

(b) $(1, 2, 3)$, $(1, 3, 5)$, $(1, 0, 1)$, $(2, 3, 0)$; (d) $(1, 1, 2)$, $(1, 2, 5)$, $(5, 3, 4)$.

(a and b) No, because a basis of $\mathbf{R}^3$ must contain exactly three elements because $\dim \mathbf{R}^3 = 3$.

(c) The three vectors form a basis if and only if they are linearly independent. Thus, form the matrix whose rows are the given vectors, and row reduce the matrix to echelon form:

$$\begin{bmatrix} 1 & 1 & 1 \\ 1 & 2 & 3 \\ 2 & -1 & 1 \end{bmatrix} \sim \begin{bmatrix} 1 & 1 & 1 \\ 0 & 1 & 2 \\ 0 & -3 & -1 \end{bmatrix} \sim \begin{bmatrix} 1 & 1 & 1 \\ 0 & 1 & 2 \\ 0 & 0 & 5 \end{bmatrix}$$

The echelon matrix has no zero rows; hence, the three vectors are linearly independent, and so they do form a basis of $\mathbf{R}^3$.

(d) Form the matrix whose rows are the given vectors, and row reduce the matrix to echelon form:

$$\begin{bmatrix} 1 & 1 & 2 \\ 1 & 2 & 5 \\ 5 & 3 & 4 \end{bmatrix} \sim \begin{bmatrix} 1 & 1 & 2 \\ 0 & 1 & 3 \\ 0 & -2 & -6 \end{bmatrix} \sim \begin{bmatrix} 1 & 1 & 2 \\ 0 & 1 & 3 \\ 0 & 0 & 0 \end{bmatrix}$$

The echelon matrix has a zero row; hence, the three vectors are linearly dependent, and so they do not form a basis of $\mathbf{R}^3$.

4.25. Determine whether $(1,1,1,1)$, $(1,2,3,2)$, $(2,5,6,4)$, $(2,6,8,5)$ form a basis of $\mathbf{R}^4$. If not, find the dimension of the subspace they span.

Form the matrix whose rows are the given vectors, and row reduce to echelon form:

$$B = \begin{bmatrix} 1 & 1 & 1 & 1 \\ 1 & 2 & 3 & 2 \\ 2 & 5 & 6 & 4 \\ 2 & 6 & 8 & 5 \end{bmatrix} \sim \begin{bmatrix} 1 & 1 & 1 & 1 \\ 0 & 1 & 2 & 1 \\ 0 & 3 & 4 & 2 \\ 0 & 4 & 6 & 3 \end{bmatrix} \sim \begin{bmatrix} 1 & 1 & 1 & 1 \\ 0 & 1 & 2 & 1 \\ 0 & 0 & -2 & -1 \\ 0 & 0 & -2 & -1 \end{bmatrix} \sim \begin{bmatrix} 1 & 1 & 1 & 1 \\ 0 & 1 & 2 & 1 \\ 0 & 0 & 2 & 1 \\ 0 & 0 & 0 & 0 \end{bmatrix}$$

The echelon matrix has a zero row. Hence, the four vectors are linearly dependent and do not form a basis of $\mathbf{R}^4$. Because the echelon matrix has three nonzero rows, the four vectors span a subspace of dimension 3.

4.26. Extend $\{u_1 = (1,1,1,1), u_2 = (2,2,3,4)\}$ to a basis of $\mathbf{R}^4$.

First form the matrix with rows u_1 and u_2, and reduce to echelon form:

$$\begin{bmatrix} 1 & 1 & 1 & 1 \\ 2 & 2 & 3 & 4 \end{bmatrix} \sim \begin{bmatrix} 1 & 1 & 1 & 1 \\ 0 & 0 & 1 & 2 \end{bmatrix}$$

Then $w_1 = (1,1,1,1)$ and $w_2 = (0,0,1,2)$ span the same set of vectors as spanned by u_1 and u_2. Let $u_3 = (0,1,0,0)$ and $u_4 = (0,0,0,1)$. Then w_1, u_3, w_2, u_4 form a matrix in echelon form. Thus, they are linearly independent, and they form a basis of $\mathbf{R}^4$. Hence, u_1, u_2, u_3, u_4 also form a basis of $\mathbf{R}^4$.

4.27. Consider the complex field $\mathbf{C}$, which contains the real field $\mathbf{R}$, which contains the rational field $\mathbf{Q}$. (Thus, $\mathbf{C}$ is a vector space over $\mathbf{R}$, and $\mathbf{R}$ is a vector space over $\mathbf{Q}$.)

(a) Show that $\{1, i\}$ is a basis of $\mathbf{C}$ over $\mathbf{R}$; hence, $\mathbf{C}$ is a vector space of dimension 2 over $\mathbf{R}$.

(b) Show that $\mathbf{R}$ is a vector space of infinite dimension over $\mathbf{Q}$.

(a) For any $v \in \mathbf{C}$, we have $v = a + bi = a(1) + b(i)$, where $a, b \in \mathbf{R}$. Hence, $\{1, i\}$ spans $\mathbf{C}$ over $\mathbf{R}$. Furthermore, if $x(1) + y(i) = 0$ or $x + yi = 0$, where $x, y \in \mathbf{R}$, then $x = 0$ and $y = 0$. Hence, $\{1, i\}$ is linearly independent over $\mathbf{R}$. Thus, $\{1, i\}$ is a basis for $\mathbf{C}$ over $\mathbf{R}$.

(b) It can be shown that π is a transcendental number; that is, π is not a root of any polynomial over $\mathbf{Q}$. Thus, for any n, the $n+1$ real numbers $1, \pi, \pi^2, \ldots, \pi^n$ are linearly independent over $\mathbf{Q}$. $\mathbf{R}$ cannot be of dimension n over $\mathbf{Q}$. Accordingly, $\mathbf{R}$ is of infinite dimension over $\mathbf{Q}$.

4.28. Suppose $S = \{u_1, u_2, \ldots, u_n\}$ is a subset of V. Show that the following Definitions A and B of a basis of V are equivalent:

(A) S is linearly independent and spans V.

(B) Every $v \in V$ is a unique linear combination of vectors in S.

Suppose (A) holds. Because S spans V, the vector v is a linear combination of the u_i, say

$$v = a_1u_1 + a_2u_2 + \cdots + a_nu_n \quad \text{and} \quad v = b_1u_1 + b_2u_2 + \cdots + b_nu_n$$

Subtracting, we get

$$0 = v - v = (a_1 - b_1)u_1 + (a_2 - b_2)u_2 + \cdots + (a_n - b_n)u_n$$

But the u_i are linearly independent. Hence, the coefficients in the above relation are each 0:

$$a_1 - b_1 = 0, \qquad a_2 - b_2 = 0, \qquad \ldots, \qquad a_n - b_n = 0$$

Therefore, $a_1 = b_1, a_2 = b_2, \ldots, a_n = b_n$. Hence, the representation of v as a linear combination of the u_i is unique. Thus, (A) implies (B).

Suppose (B) holds. Then S spans V. Suppose

$$0 = c_1u_1 + c_2u_2 + \cdots + c_nu_n$$

However, we do have

$$0 = 0u_1 + 0u_2 + \cdots + 0u_n$$

By hypothesis, the representation of 0 as a linear combination of the u_i is unique. Hence, each $c_i = 0$ and the u_i are linearly independent. Thus, (B) implies (A).

Dimension and Subspaces

4.29. Find a basis and dimension of the subspace W of $\mathbf{R}^3$ where

(a) $W = \{(a,b,c) : a+b+c=0\}$, (b) $W = \{(a,b,c) : (a=b=c)\}$

(a) Note that $W \neq \mathbf{R}^3$, because, for example, $(1,2,3) \notin W$. Thus, $\dim W < 3$. Note that $u_1 = (1,0,-1)$ and $u_2 = (0,1,-1)$ are two independent vectors in W. Thus, $\dim W = 2$, and so u_1 and u_2 form a basis of W.

(b) The vector $u = (1,1,1) \in W$. Any vector $w \in W$ has the form $w = (k,k,k)$. Hence, $w = ku$. Thus, u spans W and $\dim W = 1$.

4.30. Let W be the subspace of $\mathbf{R}^4$ spanned by the vectors

$$u_1 = (1,-2,5,-3), \qquad u_2 = (2,3,1,-4), \qquad u_3 = (3,8,-3,-5)$$

(a) Find a basis and dimension of W. (b) Extend the basis of W to a basis of $\mathbf{R}^4$.

(a) Apply Algorithm 4.1, the row space algorithm. Form the matrix whose rows are the given vectors, and reduce it to echelon form:

$$A = \begin{bmatrix} 1 & -2 & 5 & -3 \\ 2 & 3 & 1 & -4 \\ 3 & 8 & -3 & -5 \end{bmatrix} \sim \begin{bmatrix} 1 & -2 & 5 & -3 \\ 0 & 7 & -9 & 2 \\ 0 & 14 & -18 & 4 \end{bmatrix} \sim \begin{bmatrix} 1 & -2 & 5 & -3 \\ 0 & 7 & -9 & 2 \\ 0 & 0 & 0 & 0 \end{bmatrix}$$

The nonzero rows $(1,-2,5,-3)$ and $(0,7,-9,2)$ of the echelon matrix form a basis of the row space of A and hence of W. Thus, in particular, $\dim W = 2$.

(b) We seek four linearly independent vectors, which include the above two vectors. The four vectors $(1,-2,5,-3)$, $(0,7,-9,2)$, $(0,0,1,0)$, and $(0,0,0,1)$ are linearly independent (because they form an echelon matrix), and so they form a basis of $\mathbf{R}^4$, which is an extension of the basis of W.

4.31. Let W be the subspace of $\mathbf{R}^5$ spanned by $u_1 = (1,2,-1,3,4)$, $u_2 = (2,4,-2,6,8)$, $u_3 = (1,3,2,2,6)$, $u_4 = (1,4,5,1,8)$, $u_5 = (2,7,3,3,9)$. Find a subset of the vectors that form a basis of W.

Here we use Algorithm 4.2, the casting-out algorithm. Form the matrix M whose columns (not rows) are the given vectors, and reduce it to echelon form:

$$M = \begin{bmatrix} 1 & 2 & 1 & 1 & 2 \\ 2 & 4 & 3 & 4 & 7 \\ -1 & -2 & 2 & 5 & 3 \\ 3 & 6 & 2 & 1 & 3 \\ 4 & 8 & 6 & 8 & 9 \end{bmatrix} \sim \begin{bmatrix} 1 & 2 & 1 & 1 & 2 \\ 0 & 0 & 1 & 2 & 3 \\ 0 & 0 & 3 & 6 & 5 \\ 0 & 0 & -1 & -2 & -3 \\ 0 & 0 & 2 & 4 & 1 \end{bmatrix} \sim \begin{bmatrix} 1 & 2 & 1 & 1 & 2 \\ 0 & 0 & 1 & 2 & 3 \\ 0 & 0 & 0 & 0 & -4 \\ 0 & 0 & 0 & 0 & 0 \\ 0 & 0 & 0 & 0 & 0 \end{bmatrix}$$

The pivot positions are in columns C_1, C_3, C_5. Hence, the corresponding vectors u_1, u_3, u_5 form a basis of W, and $\dim W = 3$.

4.32. Let V be the vector space of 2×2 matrices over K. Let W be the subspace of symmetric matrices. Show that $\dim W = 3$, by finding a basis of W.

Recall that a matrix $A = [a_{ij}]$ is symmetric if $A^T = A$, or, equivalently, each $a_{ij} = a_{ji}$. Thus, $A = \begin{bmatrix} a & b \\ b & d \end{bmatrix}$ denotes an arbitrary 2×2 symmetric matrix. Setting (i) $a = 1$, $b = 0$, $d = 0$; (ii) $a = 0$, $b = 1$, $d = 0$; (iii) $a = 0$, $b = 0$, $d = 1$, we obtain the respective matrices:

$$E_1 = \begin{bmatrix} 1 & 0 \\ 0 & 0 \end{bmatrix}, \qquad E_2 = \begin{bmatrix} 0 & 1 \\ 1 & 0 \end{bmatrix}, \qquad E_3 = \begin{bmatrix} 0 & 0 \\ 0 & 1 \end{bmatrix}$$

We claim that $S = \{E_1, E_2, E_3\}$ is a basis of W; that is, (a) S spans W and (b) S is linearly independent.

(a) The above matrix $A = \begin{bmatrix} a & b \\ b & d \end{bmatrix} = aE_1 + bE_2 + dE_3$. Thus, S spans W.

(b) Suppose $xE_1 + yE_2 + zE_3 = 0$, where x, y, z are unknown scalars. That is, suppose

$$x\begin{bmatrix} 1 & 0 \\ 0 & 0 \end{bmatrix} + y\begin{bmatrix} 0 & 1 \\ 1 & 0 \end{bmatrix} + z\begin{bmatrix} 0 & 0 \\ 0 & 1 \end{bmatrix} = \begin{bmatrix} 0 & 0 \\ 0 & 0 \end{bmatrix} \qquad \text{or} \qquad \begin{bmatrix} x & y \\ y & z \end{bmatrix} = \begin{bmatrix} 0 & 0 \\ 0 & 0 \end{bmatrix}$$

Setting corresponding entries equal to each other yields $x = 0$, $y = 0$, $z = 0$. Thus, S is linearly independent. Therefore, S is a basis of W, as claimed.

Theorems on Linear Dependence, Basis, and Dimension

4.33. Prove Lemma 4.10: Suppose two or more nonzero vectors $v_1, v_2, \ldots, v_m$ are linearly dependent. Then one of them is a linear combination of the preceding vectors.

Because the v_i are linearly dependent, there exist scalars $a_1, \ldots, a_m$, not all 0, such that $a_1 v_1 + \cdots + a_m v_m = 0$. Let k be the largest integer such that $a_k \neq 0$. Then

$$a_1 v_1 + \cdots + a_k v_k + 0v_{k+1} + \cdots + 0v_m = 0 \qquad \text{or} \qquad a_1 v_1 + \cdots + a_k v_k = 0$$

Suppose $k = 1$; then $a_1 v_1 = 0$, $a_1 \neq 0$, and so $v_1 = 0$. But the v_i are nonzero vectors. Hence, $k > 1$ and

$$v_k = -a_k^{-1} a_1 v_1 - \cdots - a_k^{-1} a_{k-1} v_{k-1}$$

That is, v_k is a linear combination of the preceding vectors.

4.34. Suppose $S = \{v_1, v_2, \ldots, v_m\}$ spans a vector space V.

(a) If $w \in V$, then $\{w, v_1, \ldots, v_m\}$ is linearly dependent and spans V.

(b) If v_i is a linear combination of $v_1, \ldots, v_{i-1}$, then S without v_i spans V.

(a) The vector w is a linear combination of the v_i, because $\{v_i\}$ spans V. Accordingly, $\{w, v_1, \ldots, v_m\}$ is linearly dependent. Clearly, w with the v_i span V, as the v_i by themselves span V; that is, $\{w, v_1, \ldots, v_m\}$ spans V.

(b) Suppose $v_i = k_1 v_1 + \cdots + k_{i-1} v_{i-1}$. Let $u \in V$. Because $\{v_i\}$ spans V, u is a linear combination of the v_j's, say $u = a_1 v_1 + \cdots + a_m v_m$. Substituting for v_i, we obtain

$$\begin{aligned} u &= a_1 v_1 + \cdots + a_{i-1} v_{i-1} + a_i(k_1 v_1 + \cdots + k_{i-1} v_{i-1}) + a_{i+1} v_{i+1} + \cdots + a_m v_m \\ &= (a_1 + a_i k_1) v_1 + \cdots + (a_{i-1} + a_i k_{i-1}) v_{i-1} + a_{i+1} v_{i+1} + \cdots + a_m v_m \end{aligned}$$

Thus, $\{v_1, \ldots, v_{i-1}, v_{i+1}, \ldots, v_m\}$ spans V. In other words, we can delete v_i from the spanning set and still retain a spanning set.

4.35. Prove Lemma 4.13: Suppose $\{v_1, v_2, \ldots, v_n\}$ spans V, and suppose $\{w_1, w_2, \ldots, w_m\}$ is linearly independent. Then $m \leq n$, and V is spanned by a set of the form

$$\{w_1, w_2, \ldots, w_m, \ v_{i_1}, v_{i_2}, \ldots, v_{i_{n-m}}\}$$

Thus, any $n + 1$ or more vectors in V are linearly dependent.

It suffices to prove the lemma in the case that the v_i are all not 0. (Prove!) Because $\{v_i\}$ spans V, we have by Problem 4.34 that

$$\{w_1, v_1, \ldots, v_n\} \tag{1}$$

is linearly dependent and also spans V. By Lemma 4.10, one of the vectors in (1) is a linear combination of the preceding vectors. This vector cannot be w_1, so it must be one of the v's, say v_j. Thus by Problem 4.34, we can delete v_j from the spanning set (1) and obtain the spanning set

$$\{w_1, v_1, \ldots, v_{j-1}, \ v_{j+1}, \ldots, v_n\} \tag{2}$$

Now we repeat the argument with the vector w_2. That is, because (2) spans V, the set

$$\{w_1, w_2, v_1, \ldots, v_{j-1}, \ v_{j+1}, \ldots, v_n\} \tag{3}$$

is linearly dependent and also spans V. Again by Lemma 4.10, one of the vectors in (3) is a linear combination of the preceding vectors. We emphasize that this vector cannot be w_1 or w_2, because $\{w_1, \ldots, w_m\}$ is independent; hence, it must be one of the v's, say v_k. Thus, by Problem 4.34, we can delete v_k from the spanning set (3) and obtain the spanning set

$$\{w_1, w_2, v_1, \ldots, v_{j-1}, \ v_{j+1}, \ldots, v_{k-1}, \ v_{k+1}, \ldots, v_n\}$$

We repeat the argument with w_3, and so forth. At each step, we are able to add one of the w's and delete one of the v's in the spanning set. If $m \leq n$, then we finally obtain a spanning set of the required form:

$$\{w_1, \ldots, w_m, \ v_{i_1}, \ldots, v_{i_{n-m}}\}$$

Finally, we show that $m > n$ is not possible. Otherwise, after n of the above steps, we obtain the spanning set $\{w_1, \ldots, w_n\}$. This implies that w_{n+1} is a linear combination of $w_1, \ldots, w_n$, which contradicts the hypothesis that $\{w_i\}$ is linearly independent.

4.36. Prove Theorem 4.12: Every basis of a vector space V has the same number of elements.

Suppose $\{u_1, u_2, \ldots, u_n\}$ is a basis of V, and suppose $\{v_1, v_2, \ldots\}$ is another basis of V. Because $\{u_i\}$ spans V, the basis $\{v_1, v_2, \ldots\}$ must contain n or less vectors, or else it is linearly dependent by Problem 4.35—Lemma 4.13. On the other hand, if the basis $\{v_1, v_2, \ldots\}$ contains less than n elements, then $\{u_1, u_2, \ldots, u_n\}$ is linearly dependent by Problem 4.35. Thus, the basis $\{v_1, v_2, \ldots\}$ contains exactly n vectors, and so the theorem is true.

4.37. Prove Theorem 4.14: Let V be a vector space of finite dimension n. Then

(i) Any $n + 1$ or more vectors must be linearly dependent.

(ii) Any linearly independent set $S = \{u_1, u_2, \ldots u_n\}$ with n elements is a basis of V.

(iii) Any spanning set $T = \{v_1, v_2, \ldots, v_n\}$ of V with n elements is a basis of V.

Suppose $B = \{w_1, w_2, \ldots, w_n\}$ is a basis of V.

(i) Because B spans V, any $n + 1$ or more vectors are linearly dependent by Lemma 4.13.

(ii) By Lemma 4.13, elements from B can be adjoined to S to form a spanning set of V with n elements. Because S already has n elements, S itself is a spanning set of V. Thus, S is a basis of V.

(iii) Suppose T is linearly dependent. Then some v_i is a linear combination of the preceding vectors. By Problem 4.34, V is spanned by the vectors in T without v_i and there are $n - 1$ of them. By Lemma 4.13, the independent set B cannot have more than $n - 1$ elements. This contradicts the fact that B has n elements. Thus, T is linearly independent, and hence T is a basis of V.

4.38. Prove Theorem 4.15: Suppose S spans a vector space V. Then

(i) Any maximum number of linearly independent vectors in S form a basis of V.

(ii) Suppose one deletes from S every vector that is a linear combination of preceding vectors in S. Then the remaining vectors form a basis of V.

(i) Suppose $\{v_1, \ldots, v_m\}$ is a maximum linearly independent subset of S, and suppose $w \in S$. Accordingly, $\{v_1, \ldots, v_m, w\}$ is linearly dependent. No v_k can be a linear combination of preceding vectors.

Hence, w is a linear combination of the v_i. Thus, $w \in \text{span}(v_i)$, and hence $S \subseteq \text{span}(v_i)$. This leads to

$$V = \text{span}(S) \subseteq \text{span}(v_i) \subseteq V$$

Thus, $\{v_i\}$ spans V, and, as it is linearly independent, it is a basis of V.

(ii) The remaining vectors form a maximum linearly independent subset of S; hence, by (i), it is a basis of V.

4.39. Prove Theorem 4.16: Let V be a vector space of finite dimension and let $S = \{u_1, u_2, \ldots, u_r\}$ be a set of linearly independent vectors in V. Then S is part of a basis of V; that is, S may be extended to a basis of V.

Suppose $B = \{w_1, w_2, \ldots, w_n\}$ is a basis of V. Then B spans V, and hence V is spanned by

$$S \cup B = \{u_1, u_2, \ldots, u_r, w_1, w_2, \ldots, w_n\}$$

By Theorem 4.15, we can delete from $S \cup B$ each vector that is a linear combination of preceding vectors to obtain a basis B' for V. Because S is linearly independent, no u_k is a linear combination of preceding vectors. Thus, B' contains every vector in S, and S is part of the basis B' for V.

4.40. Prove Theorem 4.17: Let W be a subspace of an n-dimensional vector space V. Then $\dim W \leq n$. In particular, if $\dim W = n$, then $W = V$.

Because V is of dimension n, any $n+1$ or more vectors are linearly dependent. Furthermore, because a basis of W consists of linearly independent vectors, it cannot contain more than n elements. Accordingly, $\dim W \leq n$.

In particular, if $\{w_1, \ldots, w_n\}$ is a basis of W, then, because it is an independent set with n elements, it is also a basis of V. Thus, $W = V$ when $\dim W = n$.

Rank of a Matrix, Row and Column Spaces

4.41. Find the rank and basis of the row space of each of the following matrices:

$$\text{(a)}\quad A = \begin{bmatrix} 1 & 2 & 0 & -1 \\ 2 & 6 & -3 & -3 \\ 3 & 10 & -6 & -5 \end{bmatrix}, \qquad \text{(b)}\quad B = \begin{bmatrix} 1 & 3 & 1 & -2 & -3 \\ 1 & 4 & 3 & -1 & -4 \\ 2 & 3 & -4 & -7 & -3 \\ 3 & 8 & 1 & -7 & -8 \end{bmatrix}$$

(a) Row reduce A to echelon form:

$$A \sim \begin{bmatrix} 1 & 2 & 0 & -1 \\ 0 & 2 & -3 & -1 \\ 0 & 4 & -6 & -2 \end{bmatrix} \sim \begin{bmatrix} 1 & 2 & 0 & -1 \\ 0 & 2 & -3 & -1 \\ 0 & 0 & 0 & 0 \end{bmatrix}$$

The two nonzero rows $(1, 2, 0, -1)$ and $(0, 2, -3, -1)$ of the echelon form of A form a basis for rowsp(A). In particular, $\text{rank}(A) = 2$.

(b) Row reduce B to echelon form:

$$B \sim \begin{bmatrix} 1 & 3 & 1 & -2 & -3 \\ 0 & 1 & 2 & 1 & -1 \\ 0 & -3 & -6 & -3 & 3 \\ 0 & -1 & -2 & -1 & 1 \end{bmatrix} \sim \begin{bmatrix} 1 & 3 & 1 & -2 & -3 \\ 0 & 1 & 2 & 1 & -1 \\ 0 & 0 & 0 & 0 & 0 \\ 0 & 0 & 0 & 0 & 0 \end{bmatrix}$$

The two nonzero rows $(1, 3, 1, -2, -3)$ and $(0, 1, 2, 1, -1)$ of the echelon form of B form a basis for rowsp(B). In particular, $\text{rank}(B) = 2$.

4.42. Show that $U = W$, where U and W are the following subspaces of $\mathbf{R}^3$:

$$U = \text{span}(u_1, u_2, u_3) = \text{span}(1, 1, -1),\ (2, 3, -1),\ (3, 1, -5)\}$$
$$W = \text{span}(w_1, w_2, w_3) = \text{span}(1, -1, -3),\ (3, -2, -8),\ (2, 1, -3)\}$$

Form the matrix A whose rows are the u_i, and row reduce A to row canonical form:

$$A = \begin{bmatrix} 1 & 1 & -1 \\ 2 & 3 & -1 \\ 3 & 1 & -5 \end{bmatrix} \sim \begin{bmatrix} 1 & 1 & -1 \\ 0 & 1 & 1 \\ 0 & -2 & -2 \end{bmatrix} \sim \begin{bmatrix} 1 & 0 & -2 \\ 0 & 1 & 1 \\ 0 & 0 & 0 \end{bmatrix}$$

Next form the matrix B whose rows are the w_j, and row reduce B to row canonical form:

$$B = \begin{bmatrix} 1 & -1 & -3 \\ 3 & -2 & -8 \\ 2 & 1 & -3 \end{bmatrix} \sim \begin{bmatrix} 1 & -1 & -3 \\ 0 & 1 & 1 \\ 0 & 3 & 3 \end{bmatrix} \sim \begin{bmatrix} 1 & 0 & -2 \\ 0 & 1 & 1 \\ 0 & 0 & 0 \end{bmatrix}$$

Because A and B have the same row canonical form, the row spaces of A and B are equal, and so $U = W$.

4.43. Let $A = \begin{bmatrix} 1 & 2 & 1 & 2 & 3 & 1 \\ 2 & 4 & 3 & 7 & 7 & 4 \\ 1 & 2 & 2 & 5 & 5 & 6 \\ 3 & 6 & 6 & 15 & 14 & 15 \end{bmatrix}$.

(a) Find $\text{rank}(M_k)$, for $k = 1, 2, \ldots, 6$, where M_k is the submatrix of A consisting of the first k columns $C_1, C_2, \ldots, C_k$ of A.

(b) Which columns C_{k+1} are linear combinations of preceding columns $C_1, \ldots, C_k$?

(c) Find columns of A that form a basis for the column space of A.

(d) Express column C_4 as a linear combination of the columns in part (c).

(a) Row reduce A to echelon form:

$$A \sim \begin{bmatrix} 1 & 2 & 1 & 2 & 3 & 1 \\ 0 & 0 & 1 & 3 & 1 & 2 \\ 0 & 0 & 1 & 3 & 2 & 5 \\ 0 & 0 & 3 & 9 & 5 & 12 \end{bmatrix} \sim \begin{bmatrix} 1 & 2 & 1 & 2 & 3 & 1 \\ 0 & 0 & 1 & 3 & 1 & 2 \\ 0 & 0 & 0 & 0 & 1 & 3 \\ 0 & 0 & 0 & 0 & 0 & 0 \end{bmatrix}$$

Observe that this simultaneously reduces all the matrices M_k to echelon form; for example, the first four columns of the echelon form of A are an echelon form of M_4. We know that $\text{rank}(M_k)$ is equal to the number of pivots or, equivalently, the number of nonzero rows in an echelon form of M_k. Thus,

$$\text{rank}(M_1) = \text{rank}(M_2) = 1, \qquad \text{rank}(M_3) = \text{rank}(M_4) = 2$$
$$\text{rank}(M_5) = \text{rank}(M_6) = 3$$

(b) The vector equation $x_1C_1 + x_2C_2 + \cdots + x_kC_k = C_{k+1}$ yields the system with coefficient matrix M_k and augmented M_{k+1}. Thus, C_{k+1} is a linear combination of $C_1, \ldots, C_k$ if and only if $\text{rank}(M_k) = \text{rank}(M_{k+1})$ or, equivalently, if C_{k+1} does not contain a pivot. Thus, each of C_2, C_4, C_6 is a linear combination of preceding columns.

(c) In the echelon form of A, the pivots are in the first, third, and fifth columns. Thus, columns C_1, C_3, C_5 of A form a basis for the columns space of A. Alternatively, deleting columns C_2, C_4, C_6 from the spanning set of columns (they are linear combinations of other columns), we obtain, again, C_1, C_3, C_5.

(d) The echelon matrix tells us that C_4 is a linear combination of columns C_1 and C_3. The augmented matrix M of the vector equation $C_4 = xC_1 + yC_2$ consists of the columns C_1, C_3, C_4 of A which, when reduced to echelon form, yields the matrix (omitting zero rows)

$$\begin{bmatrix} 1 & 1 & 2 \\ 0 & 1 & 3 \end{bmatrix} \quad \text{or} \quad \begin{matrix} x + y = 2 \\ y = 3 \end{matrix} \quad \text{or} \quad x = -1, \quad y = 3$$

Thus, $C_4 = -C_1 + 3C_3 = -C_1 + 3C_3 + 0C_5$.

4.44. Suppose $u = (a_1, a_2, \ldots, a_n)$ is a linear combination of the rows $R_1, R_2, \ldots, R_m$ of a matrix $B = [b_{ij}]$, say $u = k_1R_1 + k_2R_2 + \cdots + k_mR_m$. Prove that

$$a_i = k_1b_{1i} + k_2b_{2i} + \cdots + k_mb_{mi}, \qquad i = 1, 2, \ldots, n$$

where $b_{1i}, b_{2i}, \ldots, b_{mi}$ are the entries in the ith column of B.

We are given that $u = k_1R_1 + k_2R_2 + \cdots + k_mR_m$. Hence,

$$\begin{aligned}(a_1, a_2, \ldots, a_n) &= k_1(b_{11}, \ldots, b_{1n}) + \cdots + k_m(b_{m1}, \ldots, b_{mn}) \\ &= (k_1b_{11} + \cdots + k_mb_{m1}, \ldots, k_1b_{1n} + \cdots + k_mb_{mn})\end{aligned}$$

Setting corresponding components equal to each other, we obtain the desired result.

4.45. Prove Theorem 4.7: Suppose $A = [a_{ij}]$ and $B = [b_{ij}]$ are row equivalent echelon matrices with respective pivot entries

$$a_{1j_1}, a_{2j_2}, \ldots, a_{rj_r} \quad \text{and} \quad b_{1k_1}, b_{2k_2}, \ldots, b_{sk_s}$$

(pictured in Fig. 4-5). Then A and B have the same number of nonzero rows—that is, $r = s$—and their pivot entries are in the same positions; that is, $j_1 = k_1, j_2 = k_2, \ldots, j_r = k_r$.

$$A = \begin{bmatrix} a_{1j_1} & * & * & * & * & * & * \\ & a_{2j_2} & * & * & * & * & \\ \cdots\cdots & \cdots\cdots & \cdots\cdots & \cdots\cdots & \cdots\cdots & \cdots\cdots & \\ & & & a_{rj_r} & * & * & \end{bmatrix}, \qquad B = \begin{bmatrix} b_{1k_1} & * & * & * & * & * \\ & b_{2k_2} & * & * & * & * \\ \cdots\cdots & \cdots\cdots & \cdots\cdots & \cdots\cdots & \cdots\cdots & \\ & & & b_{sk_s} & * & * \end{bmatrix}$$

Figure 4-5

Clearly $A = 0$ if and only if $B = 0$, and so we need only prove the theorem when $r \geq 1$ and $s \geq 1$. We first show that $j_1 = k_1$. Suppose $j_1 < k_1$. Then the j_1th column of B is zero. Because the first row R^* of A is in the row space of B, we have $R^* = c_1R_1 + c_1R_2 + \cdots + c_mR_m$, where the R_i are the rows of B. Because the j_1th column of B is zero, we have

$$a_{1j_1} = c_10 + c_20 + \cdots + c_m0 = 0$$

But this contradicts the fact that the pivot entry $a_{1j_1} \neq 0$. Hence, $j_1 \geq k_1$ and, similarly, $k_1 \geq j_1$. Thus $j_1 = k_1$.

Now let A' be the submatrix of A obtained by deleting the first row of A, and let B' be the submatrix of B obtained by deleting the first row of B. We prove that A' and B' have the same row space. The theorem will then follow by induction, because A' and B' are also echelon matrices.

Let $R = (a_1, a_2, \ldots, a_n)$ be any row of A' and let $R_1, \ldots, R_m$ be the rows of B. Because R is in the row space of B, there exist scalars $d_1, \ldots, d_m$ such that $R = d_1R_1 + d_2R_2 + \cdots + d_mR_m$. Because A is in echelon form and R is not the first row of A, the j_1th entry of R is zero: $a_i = 0$ for $i = j_1 = k_1$. Furthermore, because B is in echelon form, all the entries in the k_1th column of B are 0 except the first: $b_{1k_1} \neq 0$, but $b_{2k_1} = 0, \ldots, b_{mk_1} = 0$. Thus,

$$0 = a_{k_1} = d_1b_{1k_1} + d_20 + \cdots + d_m0 = d_1b_{1k_1}$$

Now $b_{1k_1} \neq 0$ and so $d_1 = 0$. Thus, R is a linear combination of $R_2, \ldots, R_m$ and so is in the row space of B'. Because R was any row of A', the row space of A' is contained in the row space of B'. Similarly, the row space of B' is contained in the row space of A'. Thus, A' and B' have the same row space, and so the theorem is proved.

4.46. Prove Theorem 4.8: Suppose A and B are row canonical matrices. Then A and B have the same row space if and only if they have the same nonzero rows.

Obviously, if A and B have the same nonzero rows, then they have the same row space. Thus we only have to prove the converse.

Suppose A and B have the same row space, and suppose $R \neq 0$ is the ith row of A. Then there exist scalars $c_1, \ldots, c_s$ such that

$$R = c_1R_1 + c_2R_2 + \cdots + c_sR_s \tag{1}$$

where the R_i are the nonzero rows of B. The theorem is proved if we show that $R = R_i$; that is, that $c_i = 1$ but $c_k = 0$ for $k \neq i$.

Let a_{ij}, be the pivot entry in R—that is, the first nonzero entry of R. By (1) and Problem 4.44,

$$a_{ij_i} = c_1 b_{1j_i} + c_2 b_{2j_i} + \cdots + c_s b_{sj_i} \quad (2)$$

But, by Problem 4.45, b_{ij_i} is a pivot entry of B, and, as B is row reduced, it is the only nonzero entry in the jth column of B. Thus, from (2), we obtain $a_{ij_i} = c_i b_{ij_i}$. However, $a_{ij_i} = 1$ and $b_{ij_i} = 1$, because A and B are row reduced; hence, $c_i = 1$.

Now suppose $k \neq i$, and b_{kj_k} is the pivot entry in R_k. By (1) and Problem 4.44,

$$a_{ij_k} = c_1 b_{1j_k} + c_2 b_{2j_k} + \cdots + c_s b_{sj_k} \quad (3)$$

Because B is row reduced, b_{kj_k} is the only nonzero entry in the jth column of B. Hence, by (3), $a_{ij_k} = c_k b_{kj_k}$. Furthermore, by Problem 4.45, a_{kj_k} is a pivot entry of A, and because A is row reduced, $a_{ij_k} = 0$. Thus, $c_k b_{kj_k} = 0$, and as $b_{kj_k} = 1$, $c_k = 0$. Accordingly $R = R_i$, and the theorem is proved.

4.47. Prove Corollary 4.9: Every matrix A is row equivalent to a unique matrix in row canonical form.

Suppose A is row equivalent to matrices A_1 and A_2, where A_1 and A_2 are in row canonical form. Then $\text{rowsp}(A) = \text{rowsp}(A_1)$ and $\text{rowsp}(A) = \text{rowsp}(A_2)$. Hence, $\text{rowsp}(A_1) = \text{rowsp}(A_2)$. Because A_1 and A_2 are in row canonical form, $A_1 = A_2$ by Theorem 4.8. Thus, the corollary is proved.

4.48. Suppose RB and AB are defined, where R is a row vector and A and B are matrices. Prove

(a) RB is a linear combination of the rows of B.

(b) The row space of AB is contained in the row space of B.

(c) The column space of AB is contained in the column space of A.

(d) If C is a column vector and AC is defined, then AC is a linear combination of the columns of A.

(e) $\text{rank}(AB) \leq \text{rank}(B)$ and $\text{rank}(AB) \leq \text{rank}(A)$.

(a) Suppose $R = (a_1, a_2, \ldots, a_m)$ and $B = [b_{ij}]$. Let $B_1, \ldots, B_m$ denote the rows of B and $B^1, \ldots, B^n$ its columns. Then

$$\begin{aligned} RB &= (RB^1, RB^2, \ldots, RB^n) \\ &= (a_1 b_{11} + a_2 b_{21} + \cdots + a_m b_{m1}, \quad \ldots, \quad a_1 b_{1n} + a_2 b_{2n} + \cdots + a_m b_{mn}) \\ &= a_1(b_{11}, b_{12}, \ldots, b_{1n}) + a_2(b_{21}, b_{22}, \ldots, b_{2n}) + \cdots + a_m(b_{m1}, b_{m2}, \ldots, b_{mn}) \\ &= a_1 B_1 + a_2 B_2 + \cdots + a_m B_m \end{aligned}$$

Thus, RB is a linear combination of the rows of B, as claimed.

(b) The rows of AB are $R_i B$, where R_i is the ith row of A. Thus, by part (a), each row of AB is in the row space of B. Thus, $\text{rowsp}(AB) \subseteq \text{rowsp}(B)$, as claimed.

(c) Using part (b), we have $\text{colsp}(AB) = \text{rowsp}(AB)^T = \text{rowsp}(B^T A^T) \subseteq \text{rowsp}(A^T) = \text{colsp}(A)$.

(d) Follows from (c) where C replaces B.

(e) The row space of AB is contained in the row space of B; hence, $\text{rank}(AB) \leq \text{rank}(B)$. Furthermore, the column space of AB is contained in the column space of A; hence, $\text{rank}(AB) \leq \text{rank}(A)$.

4.49. Let A be an n-square matrix. Show that A is invertible if and only if $\text{rank}(A) = n$.

Note that the rows of the n-square identity matrix I_n are linearly independent, because I_n is in echelon form; hence, $\text{rank}(I_n) = n$. Now if A is invertible, then A is row equivalent to I_n; hence, $\text{rank}(A) = n$. But if A is not invertible, then A is row equivalent to a matrix with a zero row; hence, $\text{rank}(A) < n$; that is, A is invertible if and only if $\text{rank}(A) = n$.

Applications to Linear Equations

4.50. Find the dimension and a basis of the solution space W of each homogeneous system:

$$\begin{array}{ccc} x+2y+2z-s+3t=0 & x+2y+z-2t=0 & x+y+2z=0 \\ x+2y+3z+s+t=0 & 2x+4y+4z-3t=0 & 2x+3y+3z=0 \\ 3x+6y+8z+s+5t=0 & 3x+6y+7z-4t=0 & x+3y+5z=0 \\ \text{(a)} & \text{(b)} & \text{(c)} \end{array}$$

(a) Reduce the system to echelon form:

$$\begin{array}{c} x+2y+2z-\ s+3t=0 \\ z+2s-2t=0 \\ 2z+4s-4t=0 \end{array} \quad \text{or} \quad \begin{array}{c} x+2y+2z-\ s+3t=0 \\ z+2s-2t=0 \end{array}$$

The system in echelon form has two (nonzero) equations in five unknowns. Hence, the system has $5-2=3$ free variables, which are y, s, t. Thus, $\dim W=3$. We obtain a basis for W:

(1) Set $y=1, s=0, t=0$ to obtain the solution $v_1=(-2,1,0,0,0)$.
(2) Set $y=0, s=1, t=0$ to obtain the solution $v_2=(5,0,-2,1,0)$.
(3) Set $y=0, s=0, t=1$ to obtain the solution $v_3=(-7,0,2,0,1)$.

The set $\{v_1, v_2, v_3\}$ is a basis of the solution space W.

(b) (Here we use the matrix format of our homogeneous system.) Reduce the coefficient matrix A to echelon form:

$$A=\begin{bmatrix} 1 & 2 & 1 & -2 \\ 2 & 4 & 4 & -3 \\ 3 & 6 & 7 & -4 \end{bmatrix} \sim \begin{bmatrix} 1 & 2 & 1 & -2 \\ 0 & 0 & 2 & 1 \\ 0 & 0 & 4 & 2 \end{bmatrix} \sim \begin{bmatrix} 1 & 2 & 1 & -2 \\ 0 & 0 & 2 & 1 \\ 0 & 0 & 0 & 0 \end{bmatrix}$$

This corresponds to the system

$$\begin{array}{c} x+2y+2z-2t=0 \\ 2z+\ t=0 \end{array}$$

The free variables are y and t, and $\dim W=2$.

(i) Set $y=1$, $z=0$ to obtain the solution $u_1=(-2,1,0,0)$.
(ii) Set $y=0$, $z=2$ to obtain the solution $u_2=(6,0,-1,2)$.

Then $\{u_1, u_2\}$ is a basis of W.

(c) Reduce the coefficient matrix A to echelon form:

$$A=\begin{bmatrix} 1 & 1 & 2 \\ 2 & 3 & 3 \\ 1 & 3 & 5 \end{bmatrix} \sim \begin{bmatrix} 1 & 1 & 2 \\ 0 & 1 & -1 \\ 0 & 2 & 3 \end{bmatrix} \sim \begin{bmatrix} 1 & 1 & 2 \\ 0 & 1 & -1 \\ 0 & 0 & 5 \end{bmatrix}$$

This corresponds to a triangular system with no free variables. Thus, 0 is the only solution; that is, $W=\{0\}$. Hence, $\dim W=0$.

4.51. Find a homogeneous system whose solution set W is spanned by

$$\{u_1, u_2, u_3\} = \{(1,-2,0,3), \quad (1,-1,-1,4), \quad (1,0,-2,5)\}$$

Let $v=(x,y,z,t)$. Then $v\in W$ if and only if v is a linear combination of the vectors u_1, u_2, u_3 that span W. Thus, form the matrix M whose first columns are u_1, u_2, u_3 and whose last column is v, and then row reduce M to echelon form. This yields

$$M=\begin{bmatrix} 1 & 1 & 1 & x \\ -2 & -1 & 0 & y \\ 0 & -1 & -2 & z \\ 3 & 4 & 5 & t \end{bmatrix} \sim \begin{bmatrix} 1 & 1 & 1 & x \\ 0 & 1 & 2 & 2x+y \\ 0 & -1 & -2 & z \\ 0 & 1 & 2 & -3x+t \end{bmatrix} \sim \begin{bmatrix} 1 & 1 & 1 & x \\ 0 & 1 & 2 & 2x+y \\ 0 & 0 & 0 & 2x+y+z \\ 0 & 0 & 0 & -5x-y+t \end{bmatrix}$$

Then v is a linear combination of u_1, u_2, u_3 if $\operatorname{rank}(M) = \operatorname{rank}(A)$, where A is the submatrix without column v. Thus, set the last two entries in the fourth column on the right equal to zero to obtain the required homogeneous system:

$$2x + y + z \quad = 0$$
$$5x + y \quad - t = 0$$

4.52. Let $x_{i_1}, x_{i_2}, \ldots, x_{i_k}$ be the free variables of a homogeneous system of linear equations with n unknowns. Let v_j be the solution for which $x_{i_j} = 1$, and all other free variables equal 0. Show that the solutions $v_1, v_2, \ldots, v_k$ are linearly independent.

Let A be the matrix whose rows are the v_i. We interchange column 1 and column i_1, then column 2 and column $i_2, \ldots,$ then column k and column i_k, and we obtain the $k \times n$ matrix

$$B = [I, C] = \begin{bmatrix} 1 & 0 & 0 & \ldots & 0 & 0 & c_{1,k+1} & \ldots & c_{1n} \\ 0 & 1 & 0 & \ldots & 0 & 0 & c_{2,k+1} & \ldots & c_{2n} \\ \ldots & \ldots & \ldots & \ldots & \ldots & \ldots & \ldots & \ldots & \ldots \\ 0 & 0 & 0 & \ldots & 0 & 1 & c_{k,k+1} & \ldots & c_{kn} \end{bmatrix}$$

The above matrix B is in echelon form, and so its rows are independent; hence, $\operatorname{rank}(B) = k$. Because A and B are column equivalent, they have the same rank—$\operatorname{rank}(A) = k$. But A has k rows; hence, these rows (i.e., the v_i) are linearly independent, as claimed.

Sums, Direct Sums, Intersections

4.53. Let U and W be subspaces of a vector space V. Show that

(a) $U + V$ is a subspace of V.

(b) U and W are contained in $U + W$.

(c) $U + W$ is the smallest subspace containing U and W; that is, $U + W = \operatorname{span}(U, W)$.

(d) $W + W = W$.

(a) Because U and W are subspaces, $0 \in U$ and $0 \in W$. Hence, $0 = 0 + 0$ belongs to $U + W$. Now suppose $v, v' \in U + W$. Then $v = u + w$ and $v' = u' + v'$, where $u, u' \in U$ and $w, w' \in W$. Then

$$av + bv' = (au + bu') + (aw + bw') \in U + W$$

Thus, $U + W$ is a subspace of V.

(b) Let $u \in U$. Because W is a subspace, $0 \in W$. Hence, $u = u + 0$ belongs to $U + W$. Thus, $U \subseteq U + W$. Similarly, $W \subseteq U + W$.

(c) Because $U + W$ is a subspace of V containing U and W, it must also contain the linear span of U and W. That is, $\operatorname{span}(U, W) \subseteq U + W$.

On the other hand, if $v \in U + W$, then $v = u + w = 1u + 1w$, where $u \in U$ and $w \in W$. Thus, v is a linear combination of elements in $U \cup W$, and so $v \in \operatorname{span}(U, W)$. Hence, $U + W \subseteq \operatorname{span}(U, W)$.

The two inclusion relations give the desired result.

(d) Because W is a subspace of V, we have that W is closed under vector addition; hence, $W + W \subseteq W$. By part (a), $W \subseteq W + W$. Hence, $W + W = W$.

4.54. Consider the following subspaces of $\mathbf{R}^5$:

$$U = \operatorname{span}(u_1, u_2, u_3) = \operatorname{span}\{(1, 3, -2, 2, 3), \quad (1, 4, -3, 4, 2), \quad (2, 3, -1, -2, 9)\}$$
$$W = \operatorname{span}(w_1, w_2, w_3) = \operatorname{span}\{(1, 3, 0, 2, 1), \quad (1, 5, -6, 6, 3), \quad (2, 5, 3, 2, 1)\}$$

Find a basis and the dimension of (a) $U + W$, (b) $U \cap W$.

(a) $U+W$ is the space spanned by all six vectors. Hence, form the matrix whose rows are the given six vectors, and then row reduce to echelon form:

$$\begin{bmatrix} 1 & 3 & -2 & 2 & 3 \\ 1 & 4 & -3 & 4 & 2 \\ 2 & 3 & -1 & -2 & 9 \\ 1 & 3 & 0 & 2 & 1 \\ 1 & 5 & -6 & 6 & 3 \\ 2 & 5 & 3 & 2 & 1 \end{bmatrix} \sim \begin{bmatrix} 1 & 3 & -2 & 2 & 3 \\ 0 & 1 & -1 & 2 & -1 \\ 0 & -3 & 3 & -6 & 3 \\ 0 & 0 & 2 & 0 & -2 \\ 0 & 2 & -4 & 4 & 0 \\ 0 & -1 & 7 & -2 & -5 \end{bmatrix} \sim \begin{bmatrix} 1 & 3 & -2 & 2 & 3 \\ 0 & 1 & -1 & 2 & -1 \\ 0 & 0 & 1 & 0 & -1 \\ 0 & 0 & 0 & 0 & 0 \\ 0 & 0 & 0 & 0 & 0 \\ 0 & 0 & 0 & 0 & 0 \end{bmatrix}$$

The following three nonzero rows of the echelon matrix form a basis of $U \cap W$:

$$(1,3,-2,2,2,3), \qquad (0,1,-1,2,-1), \qquad (0,0,1,0,-1)$$

Thus, $\dim(U+W)=3$.

(b) Let $v=(x,y,z,s,t)$ denote an arbitrary element in $\mathbf{R}^5$. First find, say as in Problem 4.49, homogeneous systems whose solution sets are U and W, respectively.

Let M be the matrix whose columns are the u_i and v, and reduce M to echelon form:

$$M=\begin{bmatrix} 1 & 1 & 2 & x \\ 3 & 4 & 3 & y \\ -2 & -3 & -1 & z \\ 2 & 4 & -2 & s \\ 3 & 2 & 9 & t \end{bmatrix} \sim \begin{bmatrix} 1 & 1 & 2 & x \\ 0 & 1 & -3 & -3x+y \\ 0 & 0 & 0 & -x+y+z \\ 0 & 0 & 0 & 4x-2y+s \\ 0 & 0 & 0 & -6x+y+t \end{bmatrix}$$

Set the last three entries in the last column equal to zero to obtain the following homogeneous system whose solution set is U:

$$-x+y+z=0, \qquad 4x-2y+s=0, \qquad -6x+y+t=0$$

Now let M' be the matrix whose columns are the w_i and v, and reduce M' to echelon form:

$$M'=\begin{bmatrix} 1 & 1 & 2 & x \\ 3 & 5 & 5 & y \\ 0 & -6 & 3 & z \\ 2 & 6 & 2 & s \\ 1 & 3 & 1 & t \end{bmatrix} \sim \begin{bmatrix} 1 & 1 & 2 & x \\ 0 & 2 & -1 & -3x+y \\ 0 & 0 & 0 & -9x+3y+z \\ 0 & 0 & 0 & 4x-2y+s \\ 0 & 0 & 0 & 2x-y+t \end{bmatrix}$$

Again set the last three entries in the last column equal to zero to obtain the following homogeneous system whose solution set is W:

$$-9+3+z=0, \qquad 4x-2y+s=0, \qquad 2x-y+t=0$$

Combine both of the above systems to obtain a homogeneous system, whose solution space is $U \cap W$, and reduce the system to echelon form, yielding

$$\begin{aligned} -x+y+\ z &= 0 \\ 2y+4z+\ s &= 0 \\ 8z+5s+2t &= 0 \\ s-2t &= 0 \end{aligned}$$

There is one free variable, which is t; hence, $\dim(U \cap W)=1$. Setting $t=2$, we obtain the solution $u=(1,4,-3,4,2)$, which forms our required basis of $U \cap W$.

4.55. Suppose U and W are distinct four-dimensional subspaces of a vector space V, where $\dim V=6$. Find the possible dimensions of $U \cap W$.

Because U and W are distinct, $U+W$ properly contains U and W; consequently, $\dim(U+W)>4$. But $\dim(U+W)$ cannot be greater than 6, as $\dim V=6$. Hence, we have two possibilities: (a) $\dim(U+W)=5$ or (b) $\dim(U+W)=6$. By Theorem 4.20,

$$\dim(U \cap W)=\dim U+\dim W-\dim(U+W)=8-\dim(U+W)$$

Thus (a) $\dim(U \cap W)=3$ or (b) $\dim(U \cap W)=2$.

4.56. Let U and W be the following subspaces of $\mathbf{R}^3$:

$$U = \{(a, b, c) : a = b = c\} \quad \text{and} \quad W = \{(0, b, c)\}$$

(Note that W is the yz-plane.) Show that $\mathbf{R}^3 = U \oplus W$.

First we show that $U \cap W = \{0\}$. Suppose $v = (a, b, c) \in U \cap W$. Then $a = b = c$ and $a = 0$. Hence, $a = 0$, $b = 0$, $c = 0$. Thus, $v = 0 = (0, 0, 0)$.

Next we show that $\mathbf{R}^3 = U + W$. For, if $v = (a, b, c) \in \mathbf{R}^3$, then

$$v = (a, a, a) + (0,\ b - a,\ c - a) \quad \text{where} \quad (a, a, a) \in U \quad \text{and} \quad (0,\ b - a,\ c - a) \in W$$

Both conditions $U \cap W = \{0\}$ and $U + W = \mathbf{R}^3$ imply that $\mathbf{R}^3 = U \oplus W$.

4.57. Suppose that U and W are subspaces of a vector space V and that $S = \{u_i\}$ spans U and $S' = \{w_j\}$ spans W. Show that $S \cup S'$ spans $U + W$. (Accordingly, by induction, if S_i spans W_i, for $i = 1, 2, \ldots, n$, then $S_1 \cup \ldots \cup S_n$ spans $W_1 + \cdots + W_n$.)

Let $v \in U + W$. Then $v = u + w$, where $u \in U$ and $w \in W$. Because S spans U, u is a linear combination of u_i, and as S' spans W, w is a linear combination of w_j; say

$$u = a_1 u_{i_1} + a_2 u_{i_2} + \cdots + a_r u_{i_r} \quad \text{and} \quad v = b_1 w_{j_1} + b_2 w_{j_2} + \cdots + b_s w_{j_s}$$

where $a_i, b_j \in K$. Then

$$v = u + w = a_1 u_{i_1} + a_2 u_{i_2} + \cdots + a_r u_{i_r} + b_1 w_{j_1} + b_2 w_{j_2} + \cdots + b_s w_{j_s}$$

Accordingly, $S \cup S' = \{u_i, w_j\}$ spans $U + W$.

4.58. Prove Theorem 4.20: Suppose U and V are finite-dimensional subspaces of a vector space V. Then $U + W$ has finite dimension and

$$\dim(U + W) = \dim U + \dim W - \dim(U \cap W)$$

Observe that $U \cap W$ is a subspace of both U and W. Suppose $\dim U = m$, $\dim W = n$, $\dim(U \cap W) = r$. Suppose $\{v_1, \ldots, v_r\}$ is a basis of $U \cap W$. By Theorem 4.16, we can extend $\{v_i\}$ to a basis of U and to a basis of W; say

$$\{v_1, \ldots, v_r, u_1, \ldots, u_{m-r}\} \quad \text{and} \quad \{v_1, \ldots, v_r, w_1, \ldots, w_{n-r}\}$$

are bases of U and W, respectively. Let

$$B = \{v_1, \ldots, v_r, u_1, \ldots, u_{m-r}, w_1, \ldots, w_{n-r}\}$$

Note that B has exactly $m + n - r$ elements. Thus, the theorem is proved if we can show that B is a basis of $U + W$. Because $\{v_i, u_j\}$ spans U and $\{v_i, w_k\}$ spans W, the union $B = \{v_i, u_j, w_k\}$ spans $U + W$. Thus, it suffices to show that B is independent.

Suppose

$$a_1 v_1 + \cdots + a_r v_r + b_1 u_1 + \cdots + b_{m-r} u_{m-r} + c_1 w_1 + \cdots + c_{n-r} w_{n-r} = 0 \tag{1}$$

where a_i, b_j, c_k are scalars. Let

$$v = a_1 v_1 + \cdots + a_r v_r + b_1 u_1 + \cdots + b_{m-r} u_{m-r} \tag{2}$$

By (1), we also have

$$v = -c_1 w_1 - \cdots - c_{n-r} w_{n-r} \tag{3}$$

Because $\{v_i, u_j\} \subseteq U$, $v \in U$ by (2); and as $\{w_k\} \subseteq W$, $v \in W$ by (3). Accordingly, $v \in U \cap W$. Now $\{v_i\}$ is a basis of $U \cap W$, and so there exist scalars $d_1, \ldots, d_r$ for which $v = d_1 v_1 + \cdots + d_r v_r$. Thus, by (3), we have

$$d_1 v_1 + \cdots + d_r v_r + c_1 w_1 + \cdots + c_{n-r} w_{n-r} = 0$$

But $\{v_i, w_k\}$ is a basis of W, and so is independent. Hence, the above equation forces $c_1 = 0, \ldots, c_{n-r} = 0$. Substituting this into (1), we obtain

$$a_1 v_1 + \cdots + a_r v_r + b_1 u_1 + \cdots + b_{m-r} u_{m-r} = 0$$

But $\{v_i, u_j\}$ is a basis of U, and so is independent. Hence, the above equation forces $a_1 = 0, \ldots, a_r = 0, b_1 = 0, \ldots, b_{m-r} = 0$.

Because (1) implies that the a_i, b_j, c_k are all 0, $B = \{v_i, u_j, w_k\}$ is independent, and the theorem is proved.

4.59. Prove Theorem 4.21: $V = U \oplus W$ if and only if (i) $V = U + W$, (ii) $U \cap W = \{0\}$.

Suppose $V = U \oplus W$. Then any $v \in V$ can be uniquely written in the form $v = u + w$, where $u \in U$ and $w \in W$. Thus, in particular, $V = U + W$. Now suppose $v \in U \cap W$. Then

$$(1) \quad v = v + 0, \text{ where } v \in U, \ 0 \in W, \quad (2) \quad v = 0 + v, \text{ where } 0 \in U, \ v \in W.$$

Thus, $v = 0 + 0 = 0$ and $U \cap W = \{0\}$.

On the other hand, suppose $V = U + W$ and $U \cap W = \{0\}$. Let $v \in V$. Because $V = U + W$, there exist $u \in U$ and $w \in W$ such that $v = u + w$. We need to show that such a sum is unique. Suppose also that $v = u' + w'$, where $u' \in U$ and $w' \in W$. Then

$$u + w = u' + w', \quad \text{and so} \quad u - u' = w' - w$$

But $u - u' \in U$ and $w' - w \in W$; hence, by $U \cap W = \{0\}$,

$$u - u' = 0, \quad w' - w = 0, \quad \text{and so} \quad u = u', \quad w = w'$$

Thus, such a sum for $v \in V$ is unique, and $V = U \oplus W$.

4.60. Prove Theorem 4.22 (for two factors): Suppose $V = U \oplus W$. Also, suppose $S = \{u_1, \ldots, u_m\}$ and $S' = \{w_1, \ldots, w_n\}$ are linearly independent subsets of U and W, respectively. Then

(a) The union $S \cup S'$ is linearly independent in V.

(b) If S and S' are bases of U and W, respectively, then $S \cup S'$ is a basis of V.

(c) $\dim V = \dim U + \dim W$.

(a) Suppose $a_1u_1 + \cdots + a_mu_m + b_1w_1 + \cdots + b_nw_n = 0$, where a_i, b_j are scalars. Then

$$(a_1u_1 + \cdots + a_mu_m) + (b_1w_1 + \cdots + b_nw_n) = 0 = 0 + 0$$

where $0, a_1u_1 + \cdots + a_mu_m \in U$ and $0, b_1w_1 + \cdots + b_nw_n \in W$. Because such a sum for 0 is unique, this leads to

$$a_1u_1 + \cdots + a_mu_m = 0 \quad \text{and} \quad b_1w_1 + \cdots + b_nw_n = 0$$

Because S_1 is linearly independent, each $a_i = 0$, and because S_2 is linearly independent, each $b_j = 0$. Thus, $S = S_1 \cup S_2$ is linearly independent.

(b) By part (a), $S = S_1 \cup S_2$ is linearly independent, and, by Problem 4.55, $S = S_1 \cup S_2$ spans $V = U + W$. Thus, $S = S_1 \cup S_2$ is a basis of V.

(c) This follows directly from part (b).

Coordinates

4.61. Relative to the basis $S = \{u_1, u_2\} = \{(1, 1), \ (2, 3)\}$ of $\mathbf{R}^2$, find the coordinate vector of v, where (a) $v = (4, -3)$, (b) $v = (a, b)$.

In each case, set

$$v = xu_1 + yu_2 = x(1, 1) + y(2, 3) = (x + 2y, \ x + 3y)$$

and then solve for x and y.

(a) We have

$$(4, -3) = (x + 2y, \ x + 3y) \quad \text{or} \quad \begin{aligned} x + 2y &= 4 \\ x + 3y &= -3 \end{aligned}$$

The solution is $x = 18$, $y = -7$. Hence, $[v] = [18, -7]$.

(b) We have

$$(a, b) = (x + 2y, \ x + 3y) \quad \text{or} \quad \begin{aligned} x + 2y &= a \\ x + 3y &= b \end{aligned}$$

The solution is $x = 3a - 2b$, $y = -a + b$. Hence, $[v] = [3a - 2b, \ \ a + b]$.

4.62. Find the coordinate vector of $v = (a, b, c)$ in $\mathbf{R}^3$ relative to

(a) the usual basis $E = \{(1,0,0),\ (0,1,0),\ (0,0,1)\}$,

(b) the basis $S = \{u_1, u_2, u_3\} = \{(1,1,1),\ (1,1,0),\ (1,0,0)\}$.

(a) Relative to the usual basis E, the coordinates of $[v]_E$ are the same as v. That is, $[v]_E = [a, b, c]$.

(b) Set v as a linear combination of u_1, u_2, u_3 using unknown scalars x, y, z. This yields

$$\begin{bmatrix} a \\ b \\ c \end{bmatrix} = x\begin{bmatrix} 1 \\ 1 \\ 1 \end{bmatrix} + y\begin{bmatrix} 1 \\ 1 \\ 0 \end{bmatrix} + z\begin{bmatrix} 1 \\ 0 \\ 0 \end{bmatrix} \qquad \text{or} \qquad \begin{aligned} x + y + z &= a \\ x + y \quad &= b \\ x \quad\quad &= c \end{aligned}$$

Solving the system yields $x = c$, $y = b - c$, $z = a - b$. Thus, $[v]_S = [c,\ b - c,\ a - b]$.

4.63. Consider the vector space $\mathbf{P}_3(t)$ of polynomials of degree ≤ 3.

(a) Show that $S = \{(t-1)^3,\ (t-1)^2,\ t - 1,\ 1\}$ is a basis of $\mathbf{P}_3(t)$.

(b) Find the coordinate vector $[v]$ of $v = 3t^3 - 4t^2 + 2t - 5$ relative to S.

(a) The degree of $(t-1)^k$ is k; writing the polynomials of S in reverse order, we see that no polynomial is a linear combination of preceding polynomials. Thus, the polynomials are linearly independent, and, because $\dim \mathbf{P}_3(t) = 4$, they form a basis of $\mathbf{P}_3(t)$.

(b) Set v as a linear combination of the basis vectors using unknown scalars x, y, z, s. We have

$$\begin{aligned} v = 3t^3 + 4t^2 + 2t - 5 &= x(t-1)^3 + y(t-1)^2 + z(t-1) + s(1) \\ &= x(t^3 - 3t^2 + 3t - 1) + y(t^2 - 2t + 1) + z(t - 1) + s(1) \\ &= xt^3 - 3xt^2 + 3xt - x + yt^2 - 2yt + y + zt - z + s \\ &= xt^3 + (-3x + y)t^2 + (3x - 2y + z)t + (-x + y - z + s) \end{aligned}$$

Then set coefficients of the same powers of t equal to each other to obtain

$$x = 3, \qquad -3x + y = 4, \qquad 3x - 2y + z = 2, \qquad -x + y - z + s = -5$$

Solving the system yields $x = 3$, $y = 13$, $z = 19$, $s = 4$. Thus, $[v] = [3, 13, 19, 4]$.

4.64. Find the coordinate vector of $A = \begin{bmatrix} 2 & 3 \\ 4 & -7 \end{bmatrix}$ in the real vector space $\mathbf{M} = \mathbf{M}_{2,2}$ relative to

(a) the basis $S = \left\{ \begin{bmatrix} 1 & 1 \\ 1 & 1 \end{bmatrix}, \begin{bmatrix} 1 & -1 \\ 1 & 0 \end{bmatrix}, \begin{bmatrix} 1 & -1 \\ 0 & 0 \end{bmatrix}, \begin{bmatrix} 1 & 0 \\ 0 & 0 \end{bmatrix} \right\}$,

(b) the usual basis $E = \left\{ \begin{bmatrix} 1 & 0 \\ 0 & 0 \end{bmatrix}, \begin{bmatrix} 0 & 1 \\ 0 & 0 \end{bmatrix}, \begin{bmatrix} 0 & 0 \\ 1 & 0 \end{bmatrix}, \begin{bmatrix} 0 & 0 \\ 0 & 1 \end{bmatrix} \right\}$

(a) Set A as a linear combination of the basis vectors using unknown scalars x, y, z, t as follows:

$$A = \begin{bmatrix} 2 & 3 \\ 4 & -7 \end{bmatrix} = x\begin{bmatrix} 1 & 1 \\ 1 & 1 \end{bmatrix} + y\begin{bmatrix} 1 & -1 \\ 1 & 0 \end{bmatrix} + z\begin{bmatrix} 1 & -1 \\ 0 & 0 \end{bmatrix} + t\begin{bmatrix} 1 & 0 \\ 0 & 0 \end{bmatrix} = \begin{bmatrix} x + y + z + t & x - y - z \\ x + y & x \end{bmatrix}$$

Set corresponding entries equal to each other to obtain the system

$$x + y + z + t = 2, \qquad x - y - z = 3, \qquad x + y = 4, \qquad x = -7$$

Solving the system yields $x = -7$, $y = 11$, $z = -21$, $t = 19$. Thus, $[A]_S = [-7, 11, -21, 19]$. (Note that the coordinate vector of A is a vector in $\mathbf{R}^4$, because $\dim \mathbf{M} = 4$.)

(b) Expressing A as a linear combination of the basis matrices yields

$$\begin{bmatrix} 2 & 3 \\ 4 & -7 \end{bmatrix} = x\begin{bmatrix} 1 & 0 \\ 0 & 0 \end{bmatrix} + y\begin{bmatrix} 0 & 1 \\ 0 & 0 \end{bmatrix} + z\begin{bmatrix} 0 & 0 \\ 1 & 0 \end{bmatrix} + t\begin{bmatrix} 0 & 0 \\ 0 & 1 \end{bmatrix} = \begin{bmatrix} x & y \\ z & t \end{bmatrix}$$

Thus, $x = 2$, $y = 3$, $z = 4$, $t = -7$. Hence, $[A] = [2, 3, 4, -7]$, whose components are the elements of A written row by row.

Remark: This result is true in general; that is, if A is any $m \times n$ matrix in $\mathbf{M} = \mathbf{M}_{m,n}$, then the coordinates of A relative to the usual basis of $\mathbf{M}$ are the elements of A written row by row.

4.65. In the space $\mathbf{M} = \mathbf{M}_{2,3}$, determine whether or not the following matrices are linearly dependent:

$$A = \begin{bmatrix} 1 & 2 & 3 \\ 4 & 0 & 5 \end{bmatrix}, \quad B = \begin{bmatrix} 2 & 4 & 7 \\ 10 & 1 & 13 \end{bmatrix}, \quad C = \begin{bmatrix} 1 & 2 & 5 \\ 8 & 2 & 11 \end{bmatrix}$$

If the matrices are linearly dependent, find the dimension and a basis of the subspace W of $\mathbf{M}$ spanned by the matrices.

The coordinate vectors of the above matrices relative to the usual basis of $\mathbf{M}$ are as follows:

$$[A] = [1, 2, 3, 4, 0, 5], \qquad [B] = [2, 4, 7, 10, 1, 13], \qquad [C] = [1, 2, 5, 8, 2, 11]$$

Form the matrix M whose rows are the above coordinate vectors, and reduce M to echelon form:

$$M = \begin{bmatrix} 1 & 2 & 3 & 4 & 0 & 5 \\ 2 & 4 & 7 & 10 & 1 & 13 \\ 1 & 2 & 5 & 8 & 2 & 11 \end{bmatrix} \sim \begin{bmatrix} 1 & 2 & 3 & 4 & 0 & 5 \\ 0 & 0 & 1 & 2 & 1 & 3 \\ 0 & 0 & 0 & 0 & 0 & 0 \end{bmatrix}$$

Because the echelon matrix has only two nonzero rows, the coordinate vectors $[A]$, $[B]$, $[C]$ span a space of dimension two, and so they are linearly dependent. Thus, A, B, C are linearly dependent. Furthermore, $\dim W = 2$, and the matrices

$$w_1 = \begin{bmatrix} 1 & 2 & 3 \\ 4 & 0 & 5 \end{bmatrix} \quad \text{and} \quad w_2 = \begin{bmatrix} 0 & 0 & 1 \\ 2 & 1 & 3 \end{bmatrix}$$

corresponding to the nonzero rows of the echelon matrix form a basis of W.

Miscellaneous Problems

4.66. Consider a finite sequence of vectors $S = \{v_1, v_2, \ldots, v_n\}$. Let T be the sequence of vectors obtained from S by one of the following "elementary operations": (i) interchange two vectors, (ii) multiply a vector by a nonzero scalar, (iii) add a multiple of one vector to another. Show that S and T span the same space W. Also show that T is independent if and only if S is independent.

Observe that, for each operation, the vectors in T are linear combinations of vectors in S. On the other hand, each operation has an inverse of the same type (Prove!); hence, the vectors in S are linear combinations of vectors in T. Thus S and T span the same space W. Also, T is independent if and only if $\dim W = n$, and this is true if and only if S is also independent.

4.67. Let $A = [a_{ij}]$ and $B = [b_{ij}]$ be row equivalent $m \times n$ matrices over a field K, and let $v_1, \ldots, v_n$ be any vectors in a vector space V over K. Let

$$\begin{aligned} u_1 &= a_{11}v_1 + a_{12}v_2 + \cdots + a_{1n}v_n \qquad & w_1 &= b_{11}v_1 + b_{12}v_2 + \cdots + b_{1n}v_n \\ u_2 &= a_{21}v_1 + a_{22}v_2 + \cdots + a_{2n}v_n \qquad & w_2 &= b_{21}v_1 + b_{22}v_2 + \cdots + b_{2n}v_n \\ &\cdots\cdots\cdots\cdots\cdots\cdots & &\cdots\cdots\cdots\cdots\cdots\cdots \\ u_m &= a_{m1}v_1 + a_{m2}v_2 + \cdots + a_{mn}v_n \qquad & w_m &= b_{m1}v_1 + b_{m2}v_2 + \cdots + b_{mn}v_n \end{aligned}$$

Show that $\{u_i\}$ and $\{w_i\}$ span the same space.

Applying an "elementary operation" of Problem 4.66 to $\{u_i\}$ is equivalent to applying an elementary row operation to the matrix A. Because A and B are row equivalent, B can be obtained from A by a sequence of elementary row operations; hence, $\{w_i\}$ can be obtained from $\{u_i\}$ by the corresponding sequence of operations. Accordingly, $\{u_i\}$ and $\{w_i\}$ span the same space.

4.68. Let $v_1, \ldots, v_n$ belong to a vector space V over K, and let $P = [a_{ij}]$ be an n-square matrix over K. Let

$$w_1 = a_{11}v_1 + a_{12}v_2 + \cdots + a_{1n}v_n, \qquad \ldots, \qquad w_n = a_{n1}v_1 + a_{n2}v_2 + \cdots + a_{nn}v_n$$

(a) Suppose P is invertible. Show that $\{w_i\}$ and $\{v_i\}$ span the same space; hence, $\{w_i\}$ is independent if and only if $\{v_i\}$ is independent.

(b) Suppose P is not invertible. Show that $\{w_i\}$ is dependent.

(c) Suppose $\{w_i\}$ is independent. Show that P is invertible.

(a) Because P is invertible, it is row equivalent to the identity matrix I. Hence, by Problem 4.67, $\{w_i\}$ and $\{v_i\}$ span the same space. Thus, one is independent if and only if the other is.

(b) Because P is not invertible, it is row equivalent to a matrix with a zero row. This means that $\{w_i\}$ spans a space that has a spanning set of less than n elements. Thus, $\{w_i\}$ is dependent.

(c) This is the contrapositive of the statement of (b), and so it follows from (b).

4.69. Find a homogeneous system whose solution space is spanned by

$$v_1 = (1,2,1,2,1), \quad v_2 = (1,3,2,5,3), \quad v_3 = (1,3,3,6,7)$$

First we seek the orthogonal complement to the v's, that is, the set of vectors $w = (a, b, c, d, e)$ orthogonal to v_1, v_2, v_3. Accordingly, we seek the solution to the system

$$\begin{aligned} a+2b+c+2d+e&=0 \\ a+3b+2c+5d+3e&=0 \\ a+3b+3c+6d+7e&=0 \end{aligned} \quad \text{or} \quad \begin{aligned} a+2b+c+2d+e&=0 \\ b+c+3d+2e&=0 \\ c+d+4e&=0 \end{aligned}$$

Here d and e are free variables. Setting (d, e) equal to $(1, 0)$ and then $(0, 1)$ yields the following two solutions of the system:

$$w_1 = (3,-2,-1,1,0) \quad \text{and} \quad w_2 = (5,2,-4,0,1)$$

Thus the homogeneous system follows:

$$3x-2y-z+s=0 \quad \text{and} \quad 5x+2y-4z+t=0$$

(Clearly the solution is not unique.)

4.70. Let K be a subfield of a field L, and let L be a subfield of a field E. (Thus, $K \subseteq L \subseteq E$, and K is a subfield of E.) Suppose E is of dimension n over L, and L is of dimension m over K. Show that E is of dimension mn over K.

Suppose $\{v_1, \ldots, v_n\}$ is a basis of E over L and $\{a_1, \ldots, a_m\}$ is a basis of L over K. We claim that $\{a_i v_j : i = 1, \ldots, m, j = 1, \ldots, n\}$ is a basis of E over K. Note that $\{a_i v_j\}$ contains mn elements.

Let w be any arbitrary element in E. Because $\{v_1, \ldots, v_n\}$ spans E over L, w is a linear combination of the v_i with coefficients in L:

$$w = b_1 v_1 + b_2 v_2 + \cdots + b_n v_n, \qquad b_i \in L \tag{1}$$

Because $\{a_1, \ldots, a_m\}$ spans L over K, each $b_i \in L$ is a linear combination of the a_j with coefficients in K:

$$\begin{aligned} b_1 &= k_{11}a_1 + k_{12}a_2 + \cdots + k_{1m}a_m \\ b_2 &= k_{21}a_1 + k_{22}a_2 + \cdots + k_{2m}a_m \\ &\cdots\cdots\cdots\cdots\cdots\cdots\cdots\cdots \\ b_n &= k_{n1}a_1 + k_{n2}a_2 + \cdots + k_{mn}a_m \end{aligned}$$

where $k_{ij} \in K$. Substituting in (1), we obtain

$$\begin{aligned} w &= (k_{11}a_1 + \cdots + k_{1m}a_m)v_1 + (k_{21}a_1 + \cdots + k_{2m}a_m)v_2 + \cdots + (k_{n1}a_1 + \cdots + k_{nm}a_m)v_n \\ &= k_{11}a_1v_1 + \cdots + k_{1m}a_mv_1 + k_{21}a_1v_2 + \cdots + k_{2m}a_mv_2 + \cdots + k_{n1}a_1v_n + \cdots + k_{nm}a_mv_n \\ &= \sum_{i,j} k_{ji}(a_iv_j) \end{aligned}$$

where $k_{ji} \in K$. Thus, w is a linear combination of the $a_i v_j$ with coefficients in K; hence, $\{a_i v_j\}$ spans E over K.

The proof is complete if we show that $\{a_i v_j\}$ is linearly independent over K. Suppose, for scalars $x_{ji} \in K$, we have $\sum_{i,j} x_{ji}(a_i v_j) = 0$; that is,

$$(x_{11}a_1v_1 + x_{12}a_2v_1 + \cdots + x_{1m}a_mv_1) + \cdots + (x_{n1}a_1v_n + x_{n2}a_2v_n + \cdots + x_{nm}a_mv_m) = 0$$

or

$$(x_{11}a_1 + x_{12}a_2 + \cdots + x_{1m}a_m)v_1 + \cdots + (x_{n1}a_1 + x_{n2}a_2 + \cdots + x_{nm}a_m)v_n = 0$$

Because $\{v_1, \ldots, v_n\}$ is linearly independent over L and the above coefficients of the v_i belong to L, each coefficient must be 0:

$$x_{11}a_1 + x_{12}a_2 + \cdots + x_{1m}a_m = 0, \quad \ldots, \quad x_{n1}a_1 + x_{n2}a_2 + \cdots + x_{nm}a_m = 0$$

But $\{a_1, \ldots, a_m\}$ is linearly independent over K; hence, because the $x_{ji} \in K$,

$$x_{11} = 0, \; x_{12} = 0, \; \ldots, \; x_{1m} = 0, \; \ldots, \; x_{n1} = 0, \; x_{n2} = 0, \; \ldots, \; x_{nm} = 0$$

Accordingly, $\{a_i v_j\}$ is linearly independent over K, and the theorem is proved.

SUPPLEMENTARY PROBLEMS

Vector Spaces

4.71. Suppose u and v belong to a vector space V. Simplify each of the following expressions:

(a) $E_1 = 4(5u - 6v) + 2(3u + v)$, (c) $E_3 = 6(3u + 2v) + 5u - 7v$,
(b) $E_2 = 5(2u - 3v) + 4(7v + 8)$, (d) $E_4 = 3(5u + 2/v)$.

4.72. Let V be the set of ordered pairs (a, b) of real numbers with addition in V and scalar multiplication on V defined by

$$(a, b) + (c, d) = (a + c, \quad b + d) \quad \text{and} \quad k(a, b) = (ka, 0)$$

Show that V satisfies all the axioms of a vector space except [M_4]—that is, except $1u = u$. Hence, [M_4] is not a consequence of the other axioms.

4.73. Show that Axiom [A_4] of a vector space V (that $u + v = v + u$) can be derived from the other axioms for V.

4.74. Let V be the set of ordered pairs (a, b) of real numbers. Show that V is not a vector space over $\mathbf{R}$ with addition and scalar multiplication defined by

(i) $(a, b) + (c, d) = (a + d, \; b + c)$ and $k(a, b) = (ka, kb)$,
(ii) $(a, b) + (c, d) = (a + c, \; b + d)$ and $k(a, b) = (a, b)$,
(iii) $(a, b) + (c, d) = (0, 0)$ and $k(a, b) = (ka, kb)$,
(iv) $(a, b) + (c, d) = (ac, bd)$ and $k(a, b) = (ka, kb)$.

4.75. Let V be the set of infinite sequences $(a_1, a_2, \ldots)$ in a field K. Show that V is a vector space over K with addition and scalar multiplication defined by

$$(a_1, a_2, \ldots) + (b_1, b_2, \ldots) = (a_1 + b_1, \; a_2 + b_2, \; \ldots) \quad \text{and} \quad k(a_1, a_2, \ldots) = (ka_1, ka_2, \ldots)$$

4.76. Let U and W be vector spaces over a field K. Let V be the set of ordered pairs (u, w) where $u \in U$ and $w \in W$. Show that V is a vector space over K with addition in V and scalar multiplication on V defined by

$$(u, w) + (u', w') = (u + u', \; w + w') \quad \text{and} \quad k(u, w) = (ku, kw)$$

(This space V is called the *external direct product* of U and W.)

Subspaces

4.77. Determine whether or not W is a subspace of $\mathbf{R}^3$ where W consists of all vectors (a, b, c) in $\mathbf{R}^3$ such that (a) $a = 3b$, (b) $a \le b \le c$, (c) $ab = 0$, (d) $a + b + c = 0$, (e) $b = a^2$, (*f*) $a = 2b = 3c$.

4.78. Let V be the vector space of n-square matrices over a field K. Show that W is a subspace of V if W consists of all matrices $A = [a_{ij}]$ that are

(a) symmetric ($A^T = A$ or $a_{ij} = a_{ji}$), (b) (upper) triangular, (c) diagonal, (d) scalar.

4.79. Let $AX = B$ be a nonhomogeneous system of linear equations in n unknowns; that is, $B \neq 0$. Show that the solution set is not a subspace of K^n.

4.80. Suppose U and W are subspaces of V for which $U \cup W$ is a subspace. Show that $U \subseteq W$ or $W \subseteq U$.

4.81. Let V be the vector space of all functions from the real field $\mathbf{R}$ into $\mathbf{R}$. Show that W is a subspace of V where W consists of all: (a) bounded functions, (b) even functions. [Recall that $f\colon \mathbf{R} \to \mathbf{R}$ is *bounded* if $\exists M \in \mathbf{R}$ such that $\forall x \in \mathbf{R}$, we have $|f(x)| \le M$; and $f(x)$ is *even* if $f(-x) = f(x), \forall x \in \mathbf{R}$.]

4.82. Let V be the vector space (Problem 4.75) of infinite sequences $(a_1, a_2, \ldots)$ in a field K. Show that W is a subspace of V if W consists of all sequences with (a) 0 as the first element, (b) only a finite number of nonzero elements.

Linear Combinations, Linear Spans

4.83. Consider the vectors $u = (1,2,3)$ and $v = (2,3,1)$ in $\mathbf{R}^3$.

(a) Write $w = (1,3,8)$ as a linear combination of u and v.
(b) Write $w = (2,4,5)$ as a linear combination of u and v.
(c) Find k so that $w = (1,k,4)$ is a linear combination of u and v.
(d) Find conditions on a, b, c so that $w = (a,b,c)$ is a linear combination of u and v.

4.84. Write the polynomial $f(t) = at^2 + bt + c$ as a linear combination of the polynomials $p_1 = (t-1)^2$, $p_2 = t - 1$, $p_3 = 1$. [Thus, p_1, p_2, p_3 span the space $\mathbf{P}_2(t)$ of polynomials of degree ≤ 2.]

4.85. Find one vector in $\mathbf{R}^3$ that spans the intersection of U and W where U is the xy-plane—that is, $U = \{(a,b,0)\}$—and W is the space spanned by the vectors $(1,1,1)$ and $(1,2,3)$.

4.86. Prove that span(S) is the intersection of all subspaces of V containing S.

4.87. Show that $\text{span}(S) = \text{span}(S \cup \{0\})$. That is, by joining or deleting the zero vector from a set, we do not change the space spanned by the set.

4.88. Show that (a) If $S \subseteq T$, then $\text{span}(S) \subseteq \text{span}(T)$. (b) $\text{span}[\text{span}(S)] = \text{span}(S)$.

Linear Dependence and Linear Independence

4.89. Determine whether the following vectors in $\mathbf{R}^4$ are linearly dependent or independent:

(a) $(1,2,-3,1)$, $(3,7,1,-2)$, $(1,3,7,-4)$; (b) $(1,3,1,-2)$, $(2,5,-1,3)$, $(1,3,7,-2)$.

4.90. Determine whether the following polynomials u, v, w in $\mathbf{P}(t)$ are linearly dependent or independent:

(a) $u = t^3 - 4t^2 + 3t + 3$, $v = t^3 + 2t^2 + 4t - 1$, $w = 2t^3 - t^2 - 3t + 5$;
(b) $u = t^3 - 5t^2 - 2t + 3$, $v = t^3 - 4t^2 - 3t + 4$, $w = 2t^3 - 17t^2 - 7t + 9$.

4.91. Show that the following functions f, g, h are linearly independent:

(a) $f(t) = e^t$, $g(t) = \sin t$, $h(t) = t^2$; (b) $f(t) = e^t$, $g(t) = e^{2t}$, $h(t) = t$.

4.92. Show that $u = (a,b)$ and $v = (c,d)$ in K^2 are linearly dependent if and only if $ad - bc = 0$.

4.93. Suppose u, v, w are linearly independent vectors. Prove that S is linearly independent where

(a) $S = \{u + v - 2w,\ u - v - w,\ u + w\}$; (b) $S = \{u + v - 3w,\ u + 3v - w,\ v + w\}$.

4.94. Suppose $\{u_1, \ldots, u_r, w_1, \ldots, w_s\}$ is a linearly independent subset of V. Show that

$$\text{span}(u_i) \cap \text{span}(w_j) = \{0\}$$

4.95. Suppose $v_1, v_2, \ldots, v_n$ are linearly independent. Prove that S is linearly independent where

(a) $S = \{a_1 v_1, a_2 v_2, \ldots, a_n v_n\}$ and each $a_i \neq 0$.
(b) $S = \{v_1, \ldots, v_{k-1}, w, v_{k+1}, \ldots, v_n\}$ and $w = \sum_i b_i v_i$ and $b_k \neq 0$.

4.96. Suppose $(a_{11}, \ldots, a_{1n})$, $(a_{21}, \ldots, a_{2n})$, $\ldots$, $(a_{m1}, \ldots, a_{mn})$ are linearly independent vectors in K^n, and suppose $v_1, v_2, \ldots, v_n$ are linearly independent vectors in a vector space V over K. Show that the following vectors are also linearly independent:

$$w_1 = a_{11}v_1 + \cdots + a_{1n}v_n, \quad w_2 = a_{21}v_1 + \cdots + a_{2n}v_n, \quad \ldots, \quad w_m = a_{m1}v_1 + \cdots + a_{mn}v_n$$

Basis and Dimension

4.97. Find a subset of u_1, u_2, u_3, u_4 that gives a basis for $W = \text{span}(u_i)$ of $\mathbf{R}^5$, where

(a) $u_1 = (1,1,1,2,3)$, $u_2 = (1,2,-1,-2,1)$, $u_3 = (3,5,-1,-2,5)$, $u_4 = (1,2,1,-1,4)$
(b) $u_1 = (1,-2,1,3,-1)$, $u_2 = (-2,4,-2,-6,2)$, $u_3 = (1,-3,1,2,1)$, $u_4 = (3,-7,3,8,-1)$
(c) $u_1 = (1,0,1,0,1)$, $u_2 = (1,1,2,1,0)$, $u_3 = (2,1,3,1,1)$, $u_4 = (1,2,1,1,1)$
(d) $u_1 = (1,0,1,1,1)$, $u_2 = (2,1,2,0,1)$, $u_3 = (1,1,2,3,4)$, $u_4 = (4,2,5,4,6)$

4.98. Consider the subspaces $U = \{(a,b,c,d) : b - 2c + d = 0\}$ and $W = \{(a,b,c,d) : a = d, b = 2c\}$ of $\mathbf{R}^4$. Find a basis and the dimension of (a) U, (b) W, (c) $U \cap W$.

4.99. Find a basis and the dimension of the solution space W of each of the following homogeneous systems:

$$\text{(a)} \quad \begin{aligned} x + 2y - 2z + 2s - t &= 0 \\ x + 2y - z + 3s - 2t &= 0 \\ 2x + 4y - 7z + s + t &= 0 \end{aligned} \qquad \text{(b)} \quad \begin{aligned} x + 2y - z + 3s - 4t &= 0 \\ 2x + 4y - 2z - s + 5t &= 0 \\ 2x + 4y - 2z + 4s - 2t &= 0 \end{aligned}$$

4.100. Find a homogeneous system whose solution space is spanned by the following sets of three vectors:

(a) $(1,-2,0,3,-1)$, $(2,-3,2,5,-3)$, $(1,-2,1,2,-2)$;
(b) $(1,1,2,1,1)$, $(1,2,1,4,3)$, $(3,5,4,9,7)$.

4.101. Determine whether each of the following is a basis of the vector space $\mathbf{P}_n(t)$:

(a) $\{1, \quad 1+t, \quad 1+t+t^2, \quad 1+t+t^2+t^3, \quad \ldots, \quad 1+t+t^2+\cdots+t^{n-1}+t^n\}$;
(b) $\{1+t, \quad t+t^2, \quad t^2+t^3, \quad \ldots, \quad t^{n-2}+t^{n-1}, \quad t^{n-1}+t^n\}$.

4.102. Find a basis and the dimension of the subspace W of $\mathbf{P}(t)$ spanned by

(a) $u = t^3 + 2t^2 - 2t + 1$, $v = t^3 + 3t^2 - 3t + 4$, $w = 2t^3 + 7t^2 - 7t + 11$,
(b) $u = t^3 + t^2 - 3t + 2$, $v = 2t^3 + t^2 + t - 4$, $w = 4t^3 + 3t^2 - 5t + 2$.

4.103. Find a basis and the dimension of the subspace W of $V = \mathbf{M}_{2,2}$ spanned by

$$A = \begin{bmatrix} 1 & -5 \\ -4 & 2 \end{bmatrix}, \quad B = \begin{bmatrix} 1 & 1 \\ -1 & 5 \end{bmatrix}, \quad C = \begin{bmatrix} 2 & -4 \\ -5 & 7 \end{bmatrix}, \quad D = \begin{bmatrix} 1 & -7 \\ -5 & 1 \end{bmatrix}$$

Rank of a Matrix, Row and Column Spaces

4.104. Find the rank of each of the following matrices:

$$\text{(a)} \begin{bmatrix} 1 & 3 & -2 & 5 & 4 \\ 1 & 4 & 1 & 3 & 5 \\ 1 & 4 & 2 & 4 & 3 \\ 2 & 7 & -3 & 6 & 13 \end{bmatrix}, \quad \text{(b)} \begin{bmatrix} 1 & 2 & -3 & -2 \\ 1 & 3 & -2 & 0 \\ 3 & 8 & -7 & -2 \\ 2 & 1 & -9 & -10 \end{bmatrix}, \quad \text{(c)} \begin{bmatrix} 1 & 1 & 2 \\ 4 & 5 & 5 \\ 5 & 8 & 1 \\ -1 & -2 & 2 \end{bmatrix}$$

4.105. For $k = 1, 2, \ldots, 5$, find the number n_k of linearly independent subsets consisting of k columns for each of the following matrices:

$$\text{(a)}\ A = \begin{bmatrix} 1 & 1 & 0 & 2 & 3 \\ 1 & 2 & 0 & 2 & 5 \\ 1 & 3 & 0 & 2 & 7 \end{bmatrix}, \quad \text{(b)}\ B = \begin{bmatrix} 1 & 2 & 1 & 0 & 2 \\ 1 & 2 & 3 & 0 & 4 \\ 1 & 1 & 5 & 0 & 6 \end{bmatrix}$$

4.106. Let (a) $A = \begin{bmatrix} 1 & 2 & 1 & 3 & 1 & 6 \\ 2 & 4 & 3 & 8 & 3 & 15 \\ 1 & 2 & 2 & 5 & 3 & 11 \\ 4 & 8 & 6 & 16 & 7 & 32 \end{bmatrix}$, (b) $B = \begin{bmatrix} 1 & 2 & 2 & 1 & 2 & 1 \\ 2 & 4 & 5 & 4 & 5 & 5 \\ 1 & 2 & 3 & 4 & 4 & 6 \\ 3 & 6 & 7 & 7 & 9 & 10 \end{bmatrix}$

For each matrix (where $C_1, \ldots, C_6$ denote its columns):

(i) Find its row canonical form M.
(ii) Find the columns that are linear combinations of preceding columns.
(iii) Find columns (excluding C_6) that form a basis for the column space.
(iv) Express C_6 as a linear combination of the basis vectors obtained in (iii).

4.107. Determine which of the following matrices have the same row space:

$$A = \begin{bmatrix} 1 & -2 & -1 \\ 3 & -4 & 5 \end{bmatrix}, \qquad B = \begin{bmatrix} 1 & -1 & 2 \\ 2 & 3 & -1 \end{bmatrix}, \qquad C = \begin{bmatrix} 1 & -1 & 3 \\ 2 & -1 & 10 \\ 3 & -5 & 1 \end{bmatrix}$$

4.108. Determine which of the following subspaces of $\mathbf{R}^3$ are identical:

$$U_1 = \text{span}[(1,1,-1),\ (2,3,-1),\ (3,1,-5)], \qquad U_2 = \text{span}[(1,-1,-3),\ (3,-2,-8),\ (2,1,-3)]$$
$$U_3 = \text{span}[(1,1,1),\ (1,-1,3),\ (3,-1,7)]$$

4.109. Determine which of the following subspaces of $\mathbf{R}^4$ are identical:

$$U_1 = \text{span}[(1,2,1,4),\ (2,4,1,5),\ (3,6,2,9)], \qquad U_2 = \text{span}[(1,2,1,2),\ (2,4,1,3)],$$
$$U_3 = \text{span}[(1,2,3,10),\ (2,4,3,11)]$$

4.110. Find a basis for (i) the row space and (ii) the column space of each matrix M:

(a) $M = \begin{bmatrix} 0 & 0 & 3 & 1 & 4 \\ 1 & 3 & 1 & 2 & 1 \\ 3 & 9 & 4 & 5 & 2 \\ 4 & 12 & 8 & 8 & 7 \end{bmatrix}$, (b) $M = \begin{bmatrix} 1 & 2 & 1 & 0 & 1 \\ 1 & 2 & 2 & 1 & 3 \\ 3 & 6 & 5 & 2 & 7 \\ 2 & 4 & 1 & -1 & 0 \end{bmatrix}$.

4.111. Show that if any row is deleted from a matrix in echelon (respectively, row canonical) form, then the resulting matrix is still in echelon (respectively, row canonical) form.

4.112. Let A and B be arbitrary $m \times n$ matrices. Show that $\text{rank}(A+B) \leq \text{rank}(A) + \text{rank}(B)$.

4.113. Let $r = \text{rank}(A+B)$. Find 2×2 matrices A and B such that

(a) $r < \text{rank}(A), \text{rank(B)}$; (b) $r = \text{rank}(A) = \text{rank}(B)$; (c) $r > \text{rank}(A), \text{rank(B)}$.

Sums, Direct Sums, Intersections

4.114. Suppose U and W are two-dimensional subspaces of K^3. Show that $U \cap W \neq \{0\}$.

4.115. Suppose U and W are subspaces of V such that $\dim U = 4$, $\dim W = 5$, and $\dim V = 7$. Find the possible dimensions of $U \cap W$.

4.116. Let U and W be subspaces of $\mathbf{R}^3$ for which $\dim U = 1$, $\dim W = 2$, and $U \not\subseteq W$. Show that $\mathbf{R}^3 = U \oplus W$.

4.117. Consider the following subspaces of $\mathbf{R}^5$:

$$U = \text{span}[(1,-1,-1,-2,0),\ \ (1,-2,-2,0,-3),\ \ (1,-1,-2,-2,1)]$$
$$W = \text{span}[(1,-2,-3,0,-2),\ \ (1,-1,-3,2,-4),\ \ (1,-1,-2,2,-5)]$$

(a) Find two homogeneous systems whose solution spaces are U and W, respectively.

(b) Find a basis and the dimension of $U \cap W$.

4.118. Let U_1, U_2, U_3 be the following subspaces of $\mathbf{R}^3$:

$$U_1 = \{(a,b,c) : a = c\}, \qquad U_2 = \{(a,b,c) : a+b+c = 0\}, \qquad U_3 = \{(0,0,c)\}$$

Show that (a) $\mathbf{R}^3 = U_1 + U_2$, (b) $\mathbf{R}^3 = U_2 + U_3$, (c) $\mathbf{R}^3 = U_1 + U_3$. When is the sum direct?

4.119. Suppose U, W_1, W_2 are subspaces of a vector space V. Show that

$$(U \cap W_1) + (U \cap W_2) \subseteq U \cap (W_1 + W_2)$$

Find subspaces of $\mathbf{R}^2$ for which equality does not hold.

4.120. Suppose $W_1, W_2, \ldots, W_r$ are subspaces of a vector space V. Show that

(a) $\text{span}(W_1, W_2, \ldots, W_r) = W_1 + W_2 + \cdots + W_r$.

(b) If S_i spans W_i for $i = 1, \ldots, r$, then $S_1 \cup S_2 \cup \cdots \cup S_r$ spans $W_1 + W_2 + \cdots + W_r$.

4.121. Suppose $V = U \oplus W$. Show that $\dim V = \dim U + \dim W$.

4.122. Let S and T be arbitrary nonempty subsets (not necessarily subspaces) of a vector space V and let k be a scalar. The sum $S + T$ and the scalar product kS are defined by

$$S + T = (u + v : u \in S, v \in T\}, \qquad kS = \{ku : u \in S\}$$

[We also write $w + S$ for $\{w\} + S$.] Let

$$S = \{(1,2),\ (2,3)\}, \qquad T = \{(1,4),\ (1,5),\ (2,5)\}, \qquad w = (1,1), \qquad k = 3$$

Find: (a) $S + T$, (b) $w + S$, (c) kS, (d) kT, (e) $kS + kT$, (f) $k(S + T)$.

4.123. Show that the above operations of $S + T$ and kS satisfy

(a) Commutative law: $S + T = T + S$.

(b) Associative law: $(S_1 + S_2) + S_3 = S_1 + (S_2 + S_3)$.

(c) Distributive law: $k(S + T) = kS + kT$.

(d) $S + \{0\} = \{0\} + S = S$ and $S + V = V + S = V$.

4.124. Let V be the vector space of n-square matrices. Let U be the subspace of upper triangular matrices, and let W be the subspace of lower triangular matrices. Find (a) $U \cap W$, (b) $U + W$.

4.125. Let V be the external direct sum of vector spaces U and W over a field K. (See Problem 4.76.) Let

$$\hat{U} = \{(u, 0) : u \in U\} \quad \text{and} \quad \hat{W} = \{(0, w) : w \in W\}$$

Show that (a) $\hat{U}$ and $\hat{W}$ are subspaces of V, (b) $V = \hat{U} \oplus \hat{W}$.

4.126. Suppose $V = U + W$. Let $\hat{V}$ be the external direct sum of U and W. Show that V is isomorphic to $\hat{V}$ under the correspondence $v = u + w \leftrightarrow (u, w)$.

4.127. Use induction to prove (a) Theorem 4.22, (b) Theorem 4.23.

Coordinates

4.128. The vectors $u_1 = (1, -2)$ and $u_2 = (4, -7)$ form a basis S of $\mathbf{R}^2$. Find the coordinate vector $[v]$ of v relative to S where (a) $v = (5, 3)$, (b) $v = (a, b)$.

4.129. The vectors $u_1 = (1, 2, 0)$, $u_2 = (1, 3, 2)$, $u_3 = (0, 1, 3)$ form a basis S of $\mathbf{R}^3$. Find the coordinate vector $[v]$ of v relative to S where (a) $v = (2, 7, -4)$, (b) $v = (a, b, c)$.

4.130. $S = \{t^3 + t^2, \quad t^2 + t, \quad t + 1, \quad 1\}$ is a basis of $\mathbf{P}_3(t)$. Find the coordinate vector $[v]$ of v relative to S where (a) $v = 2t^3 + t^2 - 4t + 2$, (b) $v = at^3 + bt^2 + ct + d$.

4.131. Let $V = \mathbf{M}_{2,2}$. Find the coordinate vector $[A]$ of A relative to S where

$$S = \left\{ \begin{bmatrix} 1 & 1 \\ 1 & 1 \end{bmatrix}, \begin{bmatrix} 1 & -1 \\ 1 & 0 \end{bmatrix}, \begin{bmatrix} 1 & 1 \\ 0 & 0 \end{bmatrix}, \begin{bmatrix} 1 & 0 \\ 0 & 0 \end{bmatrix} \right\} \text{ and (a) } A = \begin{bmatrix} 3 & -5 \\ 6 & 7 \end{bmatrix}, \text{ (b) } A = \begin{bmatrix} a & b \\ c & d \end{bmatrix}$$

4.132. Find the dimension and a basis of the subspace W of $\mathbf{P}_3(t)$ spanned by

$$u = t^3 + 2t^2 - 3t + 4, \qquad v = 2t^3 + 5t^2 - 4t + 7, \qquad w = t^3 + 4t^2 + t + 2$$

4.133. Find the dimension and a basis of the subspace W of $\mathbf{M} = \mathbf{M}_{2,3}$ spanned by

$$A = \begin{bmatrix} 1 & 2 & 1 \\ 3 & 1 & 2 \end{bmatrix}, \qquad B = \begin{bmatrix} 2 & 4 & 3 \\ 7 & 5 & 6 \end{bmatrix}, \qquad C = \begin{bmatrix} 1 & 2 & 3 \\ 5 & 7 & 6 \end{bmatrix}$$

Miscellaneous Problems

4.134. Answer true or false. If false, prove it with a counterexample.

(a) If u_1, u_2, u_3 span V, then $\dim V = 3$.
(b) If A is a 4×8 matrix, then any six columns are linearly dependent.
(c) If u_1, u_2, u_3 are linearly independent, then u_1, u_2, u_3, w are linearly dependent.
(d) If u_1, u_2, u_3, u_4 are linearly independent, then $\dim V \geq 4$.
(e) If u_1, u_2, u_3 span V, then w, u_1, u_2, u_3 span V.
(f) If u_1, u_2, u_3, u_4 are linearly independent, then u_1, u_2, u_3 are linearly independent.

4.135. Answer true or false. If false, prove it with a counterexample.

(a) If any column is deleted from a matrix in echelon form, then the resulting matrix is still in echelon form.
(b) If any column is deleted from a matrix in row canonical form, then the resulting matrix is still in row canonical form.
(c) If any column without a pivot is deleted from a matrix in row canonical form, then the resulting matrix is in row canonical form.

4.136. Determine the dimension of the vector space W of the following n-square matrices:

(a) symmetric matrices, (b) antisymmetric matrices,
(c) scalar matrices, (d) diagonal matrices.

4.137. Let $t_1, t_2, \ldots, t_n$ be symbols, and let K be any field. Let V be the following set of expressions where $a_i \in K$:

$$a_1 t_1 + a_2 t_2 + \cdots + a_n t_n$$

Define addition in V and scalar multiplication on V by

$$(a_1 t_1 + \cdots + a_n t_n) + (b_1 t_1 + \cdots + b_n t_n) = (a_1 + b_1) t_1 + \cdots + (a_n b_{nm}) t_n$$
$$k(a_1 t_1 + a_2 t_2 + \cdots + a_n t_n) = k a_1 t_1 + k a_2 t_2 + \cdots + k a_n t_n$$

Show that V is a vector space over K with the above operations. Also, show that $\{t_1, \ldots, t_n\}$ is a basis of V, where

$$t_j = 0t_1 + \cdots + 0t_{j-1} + 1t_j + 0t_{j+1} + \cdots + 0t_n$$

4.138. Suppose that $A_1, A_2, \ldots$ are linearly independent sets of vectors and that $A_1 \subseteq A_2 \subseteq \ldots$. Show that the union $A = A_1 \cup A_2 \cup \ldots$ is also linearly independent.

ANSWERS TO SUPPLEMENTARY PROBLEMS

[Some answers, such as bases, need not be unique.]

4.71. (a) $E_1 = 26u - 22v$; (b) The sum $7v + 8$ is not defined, so E_2 is not defined;
(c) $E_3 = 23u + 5v$; (d) Division by v is not defined, so E_4 is not defined.

4.77. (a) Yes; (b) No; e.g., $(1,2,3) \in W$ but $-2(1,2,3) \notin W$;
(c) No; e.g., $(1,0,0), (0,1,0) \in W$, but not their sum; (d) Yes;
(e) No; e.g., $(1,1,1) \in W$, but $2(1,1,1) \notin W$; (f) Yes

4.79. The zero vector 0 is not a solution.

4.83. (a) $w = 3u_1 - u_2$, (b) Impossible, (c) $k = \frac{11}{5}$, (d) $7a - 5b + c = 0$

4.84. Using $f = xp_1 + yp_2 + zp_3$, we get $x = a$, $y = 2a + b$, $z = a + b + c$

4.85. $v = (2,1,0)$

4.89. (a) Dependent, (b) Independent

4.90. (a) Independent, (b) Dependent

4.97. (a) u_1, u_2, u_4; (b) u_1, u_2, u_3; (c) u_1, u_2, u_4; (d) u_1, u_2, u_3

4.98. (a) $\dim U = 3$, (b) $\dim W = 2$, (c) $\dim(U \cap W) = 1$

4.99. (a) Basis: $\{(2,-1,0,0,0), \quad (4,0,1,-1,0), \quad (3,0,1,0,1)\}$; $\dim W = 3$;
(b) Basis: $\{(2,-1,0,0,0), \quad (1,0,1,0,0)\}$; $\dim W = 2$

4.100. (a) $5x + y - z - s = 0, \quad x + y - z - t = 0$;
(b) $3x - y - z = 0, \quad 2x - 3y + s = 0, \quad x - 2y + t = 0$

4.101. (a) Yes, (b) No, because $\dim \mathbf{P}_n(t) = n + 1$, but the set contains only n elements.

4.102. (a) $\dim W = 2$, (b) $\dim W = 3$

4.103. $\dim W = 2$

4.104. (a) 3, (b) 2, (c) 3

4.105. (a) $n_1 = 4, \quad n_2 = 5, \quad n_3 = n_4 = n_5 = 0$; (b) $n_1 = 4, \quad n_2 = 6, \quad n_3 = 3, \quad n_4 = n_5 = 0$

4.106. (a) (i) $M = [1,2,0,1,0,3; \quad 0,0,1,2,0,1; \quad 0,0,0,0,1,2; \quad 0]$;
(ii) C_2, C_4, C_6; (iii) C_1, C_3, C_5; (iv) $C_6 = 3C_1 + C_3 + 2C_5$.
(b) (i) $M = [1,2,0,0,3,1; \quad 0,0,1,0,-1,-1; \quad 0,0,0,1,1,2; \quad 0]$;
(ii) C_2, C_5, C_6; (iii) C_1, C_3, C_4; (iv) $C_6 = C_1 - C_3 + 2C_4$

4.107. A and C are row equivalent to $\begin{bmatrix} 1 & 0 & 7 \\ 0 & 1 & 4 \end{bmatrix}$, but not B

4.108. U_1 and U_2 are row equivalent to $\begin{bmatrix} 1 & 0 & -2 \\ 0 & 1 & 1 \end{bmatrix}$, but not U_3

4.109. U_1 and U_3 are row equivalent to $\begin{bmatrix} 1 & 2 & 0 & 1 \\ 0 & 0 & 1 & 3 \end{bmatrix}$, but not U_2

4.110. (a) (i) $(1,3,1,2,1), \ (0,0,1,-1,-1), \ (0,0,0,4,7)$; (ii) C_1, C_3, C_4;
(b) (i) $(1,2,1,0,1), \ (0,0,1,1,2)$; (ii) C_1, C_3

4.113. (a) $A = \begin{bmatrix} 1 & 1 \\ 0 & 0 \end{bmatrix}$, $B = \begin{bmatrix} -1 & -1 \\ 0 & 0 \end{bmatrix}$; (b) $A = \begin{bmatrix} 1 & 0 \\ 0 & 0 \end{bmatrix}$, $B = \begin{bmatrix} 0 & 2 \\ 0 & 0 \end{bmatrix}$;

(c) $A = \begin{bmatrix} 1 & 0 \\ 0 & 0 \end{bmatrix}$, $B = \begin{bmatrix} 0 & 0 \\ 0 & 1 \end{bmatrix}$

4.115. $\dim(U \cap W) = 2, 3$, or 4

4.117. (a) (i) $\begin{matrix} 3x + 4y - z - t = 0 \\ 4x + 2y + s = 0 \end{matrix}$ (ii) $\begin{matrix} 4x + 2y - s = 0 \\ 9x + 2y + z + t = 0 \end{matrix}$;

(b) Basis: $\{(1, -2, -5, 0, 0), \ (0, 0, 1, 0, -1)\}$; $\dim(U \cap W) = 2$

4.118. The sum is direct in (b) and (c).

4.119. In $\mathbf{R}^2$, let U, V, W be, respectively, the line $y = x$, the x-axis, the y-axis.

4.122. (a) $\{(2,6), \ (2,7), \ (3,7), \ (3,8), \ (4,8)\}$; (b) $\{(2,3), \ (3,4)\}$;
(c) $\{(3,6), \ (6,9)\}$; (d) $\{(3,12), \ (3,15), \ (6,15)\}$;
(e and f) $\{(6,18), \ (6,21), \ (9,21), \ (9,24), \ (12,24)\}$

4.124. (a) Diagonal matrices, (b) V

4.128. (a) $[-47, 13]$, (b) $[-7a - 4b, \ 2a + b]$

4.129. (a) $[-11, 13, -10]$, (b) $[c - 3b + 7a, \ -c + 3b - 6a, \ c - 2b + 4a]$

4.130. (a) $[2, -1, -2, 2]$, (b) $[a, \ b - c, \ c - b + a, \ d - c + b - a]$

4.131. (a) $[7, -1, -13, 10]$, (b) $[d, \ c - d, \ b + c - 2d, \ a - b - 2c + 2d]$

4.132. $\dim W = 2$; basis: $\{t^3 + 2t^2 - 3t + 4, \ t^2 + 2t - 1\}$

4.133. $\dim W = 2$; basis: $\{[1, 2, 1, 3, 1, 2], \ [0, 0, 1, 1, 3, 2]\}$

4.134. (a) False; $(1,1), (1,2), (2,1)$ span $\mathbf{R}^2$; (b) True;
(c) False; $(1,0,0,0), (0,1,0,0), (0,0,1,0), w = (0,0,0,1)$;
(d) True; (e) True; (f) True

4.135. (a) True; (b) False; e.g. delete C_2 from $\begin{bmatrix} 1 & 0 & 3 \\ 0 & 1 & 2 \end{bmatrix}$; (c) True

4.136. (a) $\frac{1}{2}n(n+1)$, (b) $\frac{1}{2}n(n-1)$, (c) n, (d) 1

CHAPTER 5

Linear Mappings

5.1 Introduction

The main subject matter of linear algebra is the study of linear mappings and their representation by means of matrices. This chapter introduces us to these linear maps and Chapter 6 shows how they can be represented by matrices. First, however, we begin with a study of mappings in general.

5.2 Mappings, Functions

Let A and B be arbitrary nonempty sets. Suppose to each element in $a \in A$ there is assigned a unique element of B; called the *image* of a. The collection f of such assignments is called a *mapping* (or map) from A into B, and it is denoted by

$$f : A \to B$$

The set A is called the *domain* of the mapping, and B is called the *target set*. We write $f(a)$, read "f of a," for the unique element of B that f assigns to $a \in A$.

One may also view a mapping $f : A \to B$ as a computer that, for each input value $a \in A$, produces a unique output $f(a) \in B$.

Remark: The term *function* is used synonymously with the word *mapping*, although some texts reserve the word "function" for a real-valued or complex-valued mapping.

Consider a mapping $f : A \to B$. If A' is any subset of A, then $f(A')$ denotes the set of images of elements of A'; and if B' is any subset of B, then $f^{-1}(B')$ denotes the set of elements of A, each of whose image lies in B. That is,

$$f(A') = \{f(a) : a \in A'\} \qquad \text{and} \qquad f^{-1}(B') = \{a \in A : f(a) \in B'\}$$

We call $f(A')$ the *image* of A' and $f^{-1}(B')$ the *inverse image* or *preimage* of B'. In particular, the set of all images (i.e., $f(A)$) is called the image or *range* of f.

To each mapping $f : A \to B$ there corresponds the subset of $A \times B$ given by $\{(a, f(a)) : a \in A\}$. We call this set the *graph* of f. Two mappings $f : A \to B$ and $g : A \to B$ are defined to be *equal*, written $f = g$, if $f(a) = g(a)$ for every $a \in A$—that is, if they have the same graph. Thus, we do not distinguish between a function and its graph. The negation of $f = g$ is written $f \neq g$ and is the statement:

$$\boxed{\text{There exists an } a \in A \text{ for which } f(a) \neq g(a).}$$

Sometimes the "barred" arrow $\mapsto$ is used to denote the image of an arbitrary element $x \in A$ under a mapping $f : A \to B$ by writing

$$x \mapsto f(x)$$

This is illustrated in the following example.

EXAMPLE 5.1

(a) Let $f: \mathbf{R} \to \mathbf{R}$ be the function that assigns to each real number x its square x^2. We can denote this function by writing

$$f(x) = x^2 \quad \text{or} \quad x \mapsto x^2$$

Here the image of -3 is 9, so we may write $f(-3) = 9$. However, $f^{-1}(9) = \{3, -3\}$. Also, $f(\mathbf{R}) = [0, \infty) = \{x : x \geq 0\}$ is the image of f.

(b) Let $A = \{a, b, c, d\}$ and $B = \{x, y, z, t\}$. Then the following defines a mapping $f: A \to B$:

$$f(a) = y,\ f(b) = x,\ f(c) = z,\ f(d) = y \quad \text{or} \quad f = \{(a, y),\ (b, x),\ (c, z),\ (d, y)\}$$

The first defines the mapping explicitly, and the second defines the mapping by its graph. Here,

$$f(\{a, b, d\}) = \{f(a), f(b), f(d)\} = \{y, x, y\} = \{x, y\}$$

Furthermore, $f(A) = \{x, y, z\}$ is the image of f.

EXAMPLE 5.2 Let V be the vector space of polynomials over $\mathbf{R}$, and let $p(t) = 3t^2 - 5t + 2$.

(a) The derivative defines a mapping $\mathbf{D}: V \to V$ where, for any polynomials $f(t)$, we have $\mathbf{D}(f) = df/dt$. Thus,

$$\mathbf{D}(p) = \mathbf{D}(3t^2 - 5t + 2) = 6t - 5$$

(b) The integral, say from 0 to 1, defines a mapping $\mathbf{J}: V \to \mathbf{R}$. That is, for any polynomial $f(t)$,

$$\mathbf{J}(f) = \int_0^1 f(t)\, dt, \quad \text{and so} \quad \mathbf{J}(p) = \int_0^1 (3t^2 - 5t + 2) = \tfrac{1}{2}$$

Observe that the mapping in (*b*) is from the vector space V into the scalar field $\mathbf{R}$, whereas the mapping in (*a*) is from the vector space V into itself.

Matrix Mappings

Let A be any $m \times n$ matrix over K. Then A determines a mapping $F_A: K^n \to K^m$ by

$$F_A(u) = Au$$

where the vectors in K^n and K^m are written as columns. For example, suppose

$$A = \begin{bmatrix} 1 & -4 & 5 \\ 2 & 3 & -6 \end{bmatrix} \quad \text{and} \quad u = \begin{bmatrix} 1 \\ 3 \\ -5 \end{bmatrix}$$

then

$$F_A(u) = Au = \begin{bmatrix} 1 & -4 & 5 \\ 2 & 3 & -6 \end{bmatrix} \begin{bmatrix} 1 \\ 3 \\ -5 \end{bmatrix} = \begin{bmatrix} -36 \\ 41 \end{bmatrix}$$

Remark: For notational convenience, we will frequently denote the mapping F_A by the letter A, the same symbol as used for the matrix.

Composition of Mappings

Consider two mappings $f: A \to B$ and $g: B \to C$, illustrated below:

$$A \xrightarrow{f} B \xrightarrow{g} C$$

The *composition* of f and g, denoted by $g \circ f$, is the mapping $g \circ f: A \to C$ defined by

$$(g \circ f)(a) \equiv g(f(a))$$

That is, first we apply f to $a \in A$, and then we apply g to $f(a) \in B$ to get $g(f(a)) \in C$. Viewing f and g as "computers," the composition means we first input $a \in A$ to get the output $f(a) \in B$ using f, and then we input $f(a)$ to get the output $g(f(a)) \in C$ using g.

Our first theorem tells us that the composition of mappings satisfies the associative law.

THEOREM 5.1: Let $f : A \to B$, $g : B \to C$, $h : C \to D$. Then
$$h \circ (g \circ f) = (h \circ g) \circ f$$

We prove this theorem here. Let $a \in A$. Then
$$(h \circ (g \circ f))(a) = h((g \circ f)(a)) = h(g(f(a)))$$
$$((h \circ g) \circ f)(a) = (h \circ g)(f(a)) = h(g(f(a)))$$

Thus, $(h \circ (g \circ f))(a) = ((h \circ g) \circ f)(a)$ for every $a \in A$, and so $h \circ (g \circ f) = (h \circ g) \circ f$.

One-to-One and Onto Mappings

We formally introduce some special types of mappings.

DEFINITION: A mapping $f : A \to B$ is said to be *one-to-one* (or 1-1 or *injective*) if different elements of A have distinct images; that is,

If $f(a) = f(a')$, then $a = a'$.

DEFINITION: A mapping $f : A \to B$ is said to be *onto* (or f maps A onto B or *surjective*) if every $b \in B$ is the image of at least one $a \in A$.

DEFINITION: A mapping $f : A \to B$ is said to be a *one-to-one correspondence* between A and B (or *bijective*) if f is both one-to-one and onto.

EXAMPLE 5.3 Let $f : \mathbf{R} \to \mathbf{R}$, $g : \mathbf{R} \to \mathbf{R}$, $h : \mathbf{R} \to \mathbf{R}$ be defined by
$$f(x) = 2^x, \qquad g(x) = x^3 - x, \qquad h(x) = x^2$$

The graphs of these functions are shown in Fig. 5-1. The function f is one-to-one. Geometrically, this means that each horizontal line does not contain more than one point of f. The function g is onto. Geometrically, this means that each horizontal line contains at least one point of g. The function h is neither one-to-one nor onto. For example, both 2 and -2 have the same image 4, and -16 has no preimage.

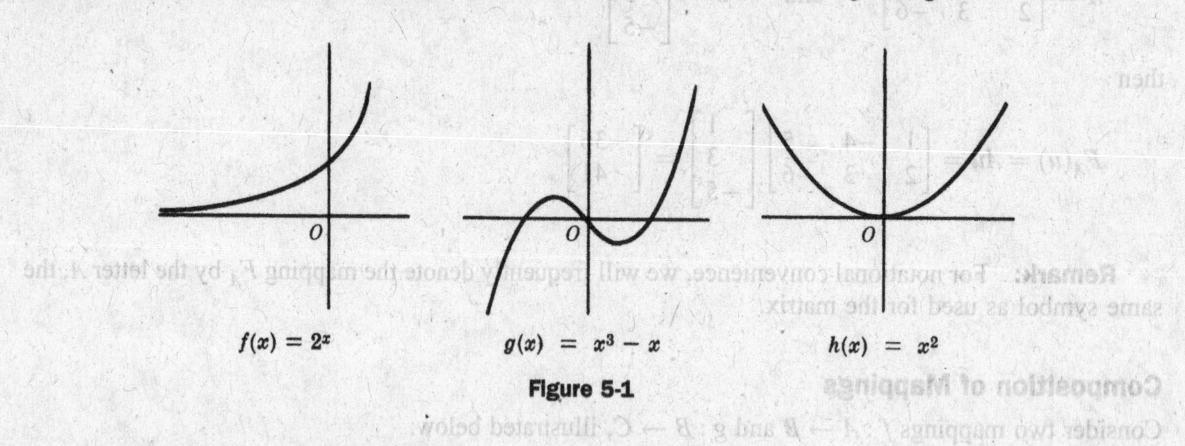

Figure 5-1

Identity and Inverse Mappings

Let A be any nonempty set. The mapping $f : A \to A$ defined by $f(a) = a$—that is, the function that assigns to each element in A itself—is called *identity mapping*. It is usually denoted by $\mathbf{1}_A$ or $\mathbf{1}$ or I. Thus, for any $a \in A$, we have $\mathbf{1}_A(a) = a$.

Now let $f : A \to B$. We call $g : B \to A$ the inverse of f, written f^{-1}, if

$$f \circ g = \mathbf{1}_B \quad \text{and} \quad g \circ f = \mathbf{1}_A$$

We emphasize that f has an inverse if and only if f is a one-to-one correspondence between A and B; that is, f is one-to-one and onto (Problem 5.7). Also, if $b \in B$, then $f^{-1}(b) = a$, where a is the unique element of A for which $f(a) = b$

5.3 Linear Mappings (Linear Transformations)

We begin with a definition.

DEFINITION: Let V and U be vector spaces over the same field K. A mapping $F : V \to U$ is called a *linear mapping* or *linear transformation* if it satisfies the following two conditions:

(1) For any vectors $v, w \in V$, $F(v + w) = F(v) + F(w)$.

(2) For any scalar k and vector $v \in V$, $F(kv) = kF(v)$.

Namely, $F : V \to U$ is linear if it "preserves" the two basic operations of a vector space, that of vector addition and that of scalar multiplication.

Substituting $k = 0$ into condition (2), we obtain $F(0) = 0$. Thus, every linear mapping takes the zero vector into the zero vector.

Now for any scalars $a, b \in K$ and any vector $v, w \in V$, we obtain

$$F(av + bw) = F(av) + F(bw) = aF(v) + bF(w)$$

More generally, for any scalars $a_i \in K$ and any vectors $v_i \in V$, we obtain the following basic property of linear mappings:

$$F(a_1 v_1 + a_2 v_2 + \cdots + a_m v_m) = a_1 F(v_1) + a_2 F(v_2) + \cdots + a_m F(v_m)$$

Remark 1: A linear mapping $F : V \to U$ is completely characterized by the condition

$$F(av + bw) = aF(v) + bF(w) \qquad (*)$$

and so this condition is sometimes used as its defintion.

Remark 2: The term *linear transformation* rather than *linear mapping* is frequently used for linear mappings of the form $F : \mathbf{R}^n \to \mathbf{R}^m$.

EXAMPLE 5.4

(a) Let $F : \mathbf{R}^3 \to \mathbf{R}^3$ be the "projection" mapping into the xy-plane; that is, F is the mapping defined by $F(x, y, z) = (x, y, 0)$. We show that F is linear. Let $v = (a, b, c)$ and $w = (a', b', c')$. Then

$$F(v + w) = F(a + a',\ b + b',\ c + c') = (a + a',\ b + b',\ 0)$$
$$= (a, b, 0) + (a', b', 0) = F(v) + F(w)$$

and, for any scalar k,

$$F(kv) = F(ka, kb, kc) = (ka, kb, 0) = k(a, b, 0) = kF(v)$$

Thus, F is linear.

(b) Let $G : \mathbf{R}^2 \to \mathbf{R}^2$ be the "translation" mapping defined by $G(x, y) = (x + 1,\ y + 2)$. [That is, G adds the vector $(1, 2)$ to any vector $v = (x, y)$ in $\mathbf{R}^2$.] Note that

$$G(0) = G(0, 0) = (1, 2) \neq 0$$

Thus, the zero vector is not mapped into the zero vector. Hence, G is not linear.

EXAMPLE 5.5 (Derivative and Integral Mappings) Consider the vector space $V = \mathbf{P}(t)$ of polynomials over the real field $\mathbf{R}$. Let $u(t)$ and $v(t)$ be any polynomials in V and let k be any scalar.

(a) Let $\mathbf{D}: V \to V$ be the derivative mapping. One proves in calculus that

$$\frac{d(u+v)}{dt} = \frac{du}{dt} + \frac{dv}{dt} \quad \text{and} \quad \frac{d(ku)}{dt} = k\frac{du}{dt}$$

That is, $\mathbf{D}(u+v) = \mathbf{D}(u) + \mathbf{D}(v)$ and $\mathbf{D}(ku) = k\mathbf{D}(u)$. Thus, the derivative mapping is linear.

(b) Let $\mathbf{J}: V \to \mathbf{R}$ be an integral mapping, say

$$\mathbf{J}(f(t)) = \int_0^1 f(t)\,dt$$

One also proves in calculus that,

$$\int_0^1 [u(t) + v(t)]dt = \int_0^1 u(t)\,dt + \int_0^1 v(t)\,dt$$

and

$$\int_0^1 ku(t)\,dt = k\int_0^1 u(t)\,dt$$

That is, $\mathbf{J}(u+v) = \mathbf{J}(u) + \mathbf{J}(v)$ and $\mathbf{J}(ku) = k\mathbf{J}(u)$. Thus, the integral mapping is linear.

EXAMPLE 5.6 (Zero and Identity Mappings)

(a) Let $F: V \to U$ be the mapping that assigns the zero vector $0 \in U$ to every vector $v \in V$. Then, for any vectors $v, w \in V$ and any scalar $k \in K$, we have

$$F(v+w) = 0 = 0 + 0 = F(v) + F(w) \quad \text{and} \quad F(kv) = 0 = k0 = kF(v)$$

Thus, F is linear. We call F the *zero mapping*, and we usually denote it by 0.

(b) Consider the identity mapping $I: V \to V$, which maps each $v \in V$ into itself. Then, for any vectors $v, w \in V$ and any scalars $a, b \in K$, we have

$$I(av + bw) = av + bw = aI(v) + bI(w)$$

Thus, I is linear.

Our next theorem (proved in Problem 5.13) gives us an abundance of examples of linear mappings. In particular, it tells us that a linear mapping is completely determined by its values on the elements of a basis.

THEOREM 5.2: Let V and U be vector spaces over a field K. Let $\{v_1, v_2, \ldots, v_n\}$ be a basis of V and let $u_1, u_2, \ldots, u_n$ be any vectors in U. Then there exists a unique linear mapping $F: V \to U$ such that $F(v_1) = u_1, F(v_2) = u_2, \ldots, F(v_n) = u_n$.

We emphasize that the vectors $u_1, u_2, \ldots, u_n$ in Theorem 5.2 are completely arbitrary; they may be linearly dependent or they may even be equal to each other.

Matrices as Linear Mappings

Let A be any real $m \times n$ matrix. Recall that A determines a mapping $F_A: K^n \to K^m$ by $F_A(u) = Au$ (where the vectors in K^n and K^m are written as columns). We show F_A is linear. By matrix multiplication,

$$F_A(v+w) = A(v+w) = Av + Aw = F_A(v) + F_A(w)$$
$$F_A(kv) = A(kv) = k(Av) = kF_A(v)$$

In other words, using A to represent the mapping, we have

$$A(v+w) = Av + Aw \quad \text{and} \quad A(kv) = k(Av)$$

Thus, the matrix mapping A is linear.

Vector Space Isomorphism

The notion of two vector spaces being isomorphic was defined in Chapter 4 when we investigated the coordinates of a vector relative to a basis. We now redefine this concept.

DEFINITION: Two vector spaces V and U over K are *isomorphic*, written $V \cong U$, if there exists a bijective (one-to-one and onto) linear mapping $F: V \to U$. The mapping F is then called an *isomorphism* between V and U.

Consider any vector space V of dimension n and let S be any basis of V. Then the mapping

$$v \mapsto [v]_S$$

which maps each vector $v \in V$ into its coordinate vector $[v]_S$, is an isomorphism between V and K^n.

5.4 Kernel and Image of a Linear Mapping

We begin by defining two concepts.

DEFINITION: Let $F: V \to U$ be a linear mapping. The *kernel* of F, written $\operatorname{Ker} F$, is the set of elements in V that map into the zero vector 0 in U; that is,

$$\operatorname{Ker} F = \{v \in V : F(v) = 0\}$$

The *image* (or *range*) of F, written $\operatorname{Im} F$, is the set of image points in U; that is,

$$\operatorname{Im} F = \{u \in U : \text{there exists } v \in V \text{ for which } F(v) = u\}$$

The following theorem is easily proved (Problem 5.22).

THEOREM 5.3: Let $F: V \to U$ be a linear mapping. Then the kernel of F is a subspace of V and the image of F is a subspace of U.

Now suppose that $v_1, v_2, \ldots, v_m$ span a vector space V and that $F: V \to U$ is linear. We show that $F(v_1), F(v_2), \ldots, F(v_m)$ span $\operatorname{Im} F$. Let $u \in \operatorname{Im} F$. Then there exists $v \in V$ such that $F(v) = u$. Because the v_i's span V and $v \in V$, there exist scalars $a_1, a_2, \ldots, a_m$ for which

$$v = a_1 v_1 + a_2 v_2 + \cdots + a_m v_m$$

Therefore,

$$u = F(v) = F(a_1 v_1 + a_2 v_2 + \cdots + a_m v_m) = a_1 F(v_1) + a_2 F(v_2) + \cdots + a_m F(v_m)$$

Thus, the vectors $F(v_1), F(v_2), \ldots, F(v_m)$ span $\operatorname{Im} F$.

We formally state the above result.

PROPOSITION 5.4: Suppose $v_1, v_2, \ldots, v_m$ span a vector space V, and suppose $F: V \to U$ is linear. Then $F(v_1), F(v_2), \ldots, F(v_m)$ span $\operatorname{Im} F$.

EXAMPLE 5.7

(a) Let $F: \mathbf{R}^3 \to \mathbf{R}^3$ be the projection of a vector v into the xy-plane [as pictured in Fig. 5-2(a)]; that is,

$$F(x, y, z) = (x, y, 0)$$

Clearly the image of F is the entire xy-plane—that is, points of the form $(x, y, 0)$. Moreover, the kernel of F is the z-axis—that is, points of the form $(0, 0, c)$. That is,

$$\operatorname{Im} F = \{(a, b, c) : c = 0\} = xy\text{-plane} \quad \text{and} \quad \operatorname{Ker} F = \{(a, b, c) : a = 0, b = 0\} = z\text{-axis}$$

(b) Let $G: \mathbf{R}^3 \to \mathbf{R}^3$ be the linear mapping that rotates a vector v about the z-axis through an angle θ [as pictured in Fig. 5-2(b)]; that is,

$$G(x, y, z) = (x\cos\theta - y\sin\theta,\ x\sin\theta + y\cos\theta,\ z)$$

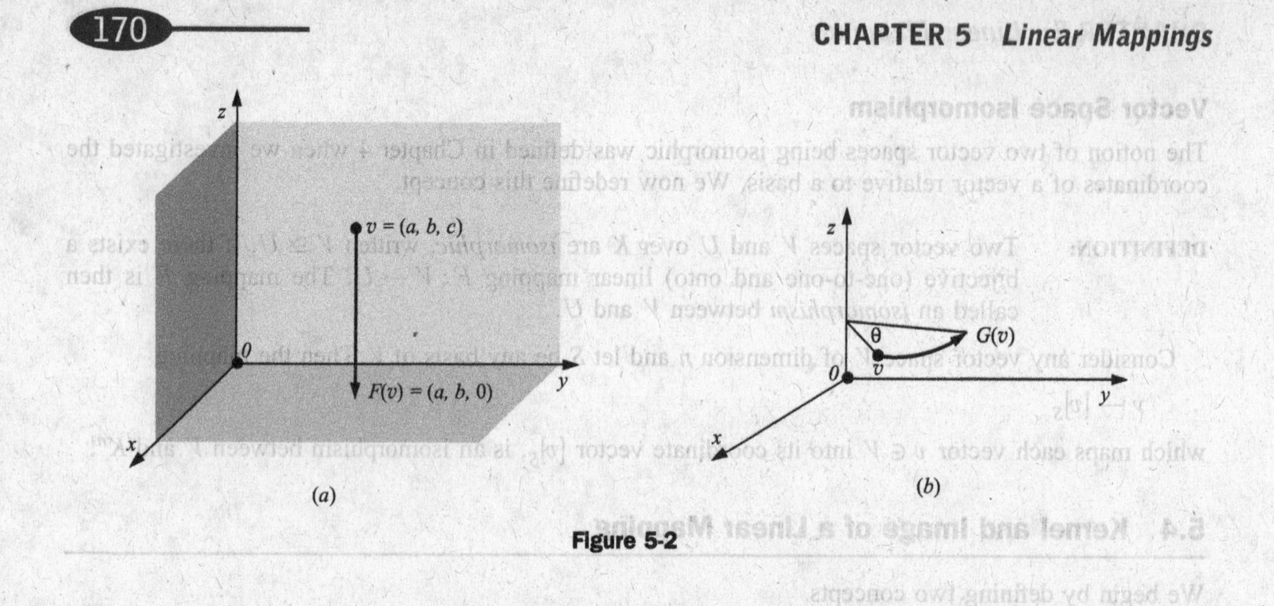

Figure 5-2

Observe that the distance of a vector v from the origin O does not change under the rotation, and so only the zero vector 0 is mapped into the zero vector 0. Thus, $\text{Ker } G = \{0\}$. On the other hand, every vector u in $\mathbf{R}^3$ is the image of a vector v in $\mathbf{R}^3$ that can be obtained by rotating u back by an angle of θ. Thus, $\text{Im } G = \mathbf{R}^3$, the entire space.

EXAMPLE 5.8 Consider the vector space $V = \mathbf{P}(t)$ of polynomials over the real field $\mathbf{R}$, and let $H: V \to V$ be the third-derivative operator; that is, $H[f(t)] = d^3f/dt^3$. [Sometimes the notation $\mathbf{D}^3$ is used for H, where $\mathbf{D}$ is the derivative operator.] We claim that

$$\text{Ker } H = \{\text{polynomials of degree} \leq 2\} = \mathbf{P}_2(t) \quad \text{and} \quad \text{Im } H = V$$

The first comes from the fact that $H(at^2 + bt + c) = 0$ but $H(t^n) \neq 0$ for $n \geq 3$. The second comes from that fact that every polynomial $g(t)$ in V is the third derivative of some polynomial $f(t)$ (which can be obtained by taking the antiderivative of $g(t)$ three times).

Kernel and Image of Matrix Mappings

Consider, say, a 3×4 matrix A and the usual basis $\{e_1, e_2, e_3, e_4\}$ of K^4 (written as columns):

$$A = \begin{bmatrix} a_1 & a_2 & a_3 & a_4 \\ b_1 & b_2 & b_3 & b_4 \\ c_1 & c_2 & c_3 & c_4 \end{bmatrix}, \quad e_1 = \begin{bmatrix} 1 \\ 0 \\ 0 \\ 0 \end{bmatrix}, \quad e_2 = \begin{bmatrix} 0 \\ 1 \\ 0 \\ 0 \end{bmatrix}, \quad e_3 = \begin{bmatrix} 0 \\ 0 \\ 1 \\ 0 \end{bmatrix}, \quad e_4 = \begin{bmatrix} 0 \\ 0 \\ 0 \\ 1 \end{bmatrix}$$

Recall that A may be viewed as a linear mapping $A: K^4 \to K^3$, where the vectors in K^4 and K^3 are viewed as column vectors. Now the usual basis vectors span K^4, so their images Ae_1, Ae_2, Ae_3, Ae_4 span the image of A. But the vectors Ae_1, Ae_2, Ae_3, Ae_4 are precisely the columns of A:

$$Ae_1 = [a_1, b_1, c_1]^T, \qquad Ae_2 = [a_2, b_2, c_2]^T, \qquad Ae_3 = [a_3, b_3, c_3]^T, \qquad Ae_4 = [a_4, b_4, c_4]^T$$

Thus, the image of A is precisely the column space of A.

On the other hand, the kernel of A consists of all vectors v for which $Av = 0$. This means that the kernel of A is the solution space of the homogeneous system $AX = 0$, called the *null space* of A.

We state the above results formally.

PROPOSITION 5.5: Let A be any $m \times n$ matrix over a field K viewed as a linear map $A: K^n \to K^m$. Then

$$\text{Ker } A = \text{nullsp}(A) \quad \text{and} \quad \text{Im } A = \text{colsp}(A)$$

Here colsp(A) denotes the column space of A, and nullsp(A) denotes the null space of A.

Rank and Nullity of a Linear Mapping

Let $F : V \to U$ be a linear mapping. The *rank* of F is defined to be the dimension of its image, and the *nullity* of F is defined to be the dimension of its kernel; namely,

$$\text{rank}(F) = \dim(\text{Im } F) \quad \text{and} \quad \text{nullity}(F) = \dim(\text{Ker } F)$$

The following important theorem (proved in Problem 5.23) holds.

THEOREM 5.6 Let V be of finite dimension, and let $F : V \to U$ be linear. Then

$$\dim V = \dim(\text{Ker } F) + \dim(\text{Im } F) = \text{nullity}(F) + \text{rank}(F)$$

Recall that the rank of a matrix A was also defined to be the dimension of its column space and row space. If we now view A as a linear mapping, then both definitions correspond, because the image of A is precisely its column space.

EXAMPLE 5.9 Let $F : \mathbf{R}^4 \to \mathbf{R}^3$ be the linear mapping defined by

$$F(x, y, z, t) = (x - y + z + t, \quad 2x - 2y + 3z + 4t, \quad 3x - 3y + 4z + 5t)$$

(a) Find a basis and the dimension of the image of F.

First find the image of the usual basis vectors of $\mathbf{R}^4$,

$$F(1, 0, 0, 0) = (1, 2, 3), \qquad F(0, 0, 1, 0) = (1, 3, 4)$$
$$F(0, 1, 0, 0) = (-1, -2, -3), \qquad F(0, 0, 0, 1) = (1, 4, 5)$$

By Proposition 5.4, the image vectors span Im F. Hence, form the matrix M whose rows are these image vectors and row reduce to echelon form:

$$M = \begin{bmatrix} 1 & 2 & 3 \\ -1 & -2 & -3 \\ 1 & 3 & 4 \\ 1 & 4 & 5 \end{bmatrix} \sim \begin{bmatrix} 1 & 2 & 3 \\ 0 & 0 & 0 \\ 0 & 1 & 1 \\ 0 & 2 & 2 \end{bmatrix} \sim \begin{bmatrix} 1 & 2 & 3 \\ 0 & 1 & 1 \\ 0 & 0 & 0 \\ 0 & 0 & 0 \end{bmatrix}$$

Thus, $(1, 2, 3)$ and $(0, 1, 1)$ form a basis of Im F. Hence, $\dim(\text{Im } F) = 2$ and $\text{rank}(F) = 2$.

(b) Find a basis and the dimension of the kernel of the map F.

Set $F(v) = 0$, where $v = (x, y, z, t)$,

$$F(x, y, z, t) = (x - y + z + t, \quad 2x - 2y + 3z + 4t, \quad 3x - 3y + 4z + 5t) = (0, 0, 0)$$

Set corresponding components equal to each other to form the following homogeneous system whose solution space is Ker F:

$$\begin{aligned} x - y + z + t &= 0 \\ 2x - 2y + 3z + 4t &= 0 \\ 3x - 3y + 4z + 5t &= 0 \end{aligned} \qquad \text{or} \qquad \begin{aligned} x - y + z + t &= 0 \\ z + 2t &= 0 \\ z + 2t &= 0 \end{aligned} \qquad \text{or} \qquad \begin{aligned} x - y + z + t &= 0 \\ z + 2t &= 0 \end{aligned}$$

The free variables are y and t. Hence, $\dim(\text{Ker } F) = 2$ or $\text{nullity}(F) = 2$.

(i) Set $y = 1$, $t = 0$ to obtain the solution $(-1, 1, 0, 0)$,

(ii) Set $y = 0$, $t = 1$ to obtain the solution $(1, 0, -2, 1)$.

Thus, $(-1, 1, 0, 0)$ and $(1, 0, -2, 1)$ form a basis for Ker F.

As expected from Theorem 5.6, $\dim(\text{Im } F) + \dim(\text{Ker } F) = 4 = \dim \mathbf{R}^4$.

Application to Systems of Linear Equations

Let $AX = B$ denote the matrix form of a system of m linear equations in n unknowns. Now the matrix A may be viewed as a linear mapping

$$A : K^n \to K^m$$

Thus, the solution of the equation $AX = B$ may be viewed as the preimage of the vector $B \in K^m$ under the linear mapping A. Furthermore, the solution of the associated homogeneous system

$$AX = 0$$

may be viewed as the kernel of the linear mapping A. Applying Theorem 5.6 to this homogeneous system yields

$$\dim(\text{Ker } A) = \dim K^n - \dim(\text{Im } A) = n - \text{rank } A$$

But n is exactly the number of unknowns in the homogeneous system $AX = 0$. Thus, we have proved the following theorem of Chapter 4.

THEOREM 4.19: The dimension of the solution space W of a homogenous system $AX = 0$ of linear equations is $s = n - r$, where n is the number of unknowns and r is the rank of the coefficient matrix A.

Observe that r is also the number of pivot variables in an echelon form of $AX = 0$, so $s = n - r$ is also the number of free variables. Furthermore, the s solution vectors of $AX = 0$ described in Theorem 3.14 are linearly independent (Problem 4.52). Accordingly, because $\dim W = s$, they form a basis for the solution space W. Thus, we have also proved Theorem 3.14.

5.5 Singular and Nonsingular Linear Mappings, Isomorphisms

Let $F: V \to U$ be a linear mapping. Recall that $F(0) = 0$. F is said to be *singular* if the image of some nonzero vector v is 0—that is, if there exists $v \neq 0$ such that $F(v) = 0$. Thus, $F: V \to U$ is *nonsingular* if the zero vector 0 is the only vector whose image under F is 0 or, in other words, if $\text{Ker } F = \{0\}$.

EXAMPLE 5.10 Consider the projection map $F: \mathbf{R}^3 \to \mathbf{R}^3$ and the rotation map $G: \mathbf{R}^3 \to \mathbf{R}^3$ appearing in Fig. 5-2. (See Example 5.7.) Because the kernel of F is the z-axis, F is singular. On the other hand, the kernel of G consists only of the zero vector 0. Thus, G is nonsingular.

Nonsingular linear mappings may also be characterized as those mappings that carry independent sets into independent sets. Specifically, we prove (Problem 5.28) the following theorem.

THEOREM 5.7: Let $F: V \to U$ be a nonsingular linear mapping. Then the image of any linearly independent set is linearly independent.

Isomorphisms

Suppose a linear mapping $F: V \to U$ is one-to-one. Then only $0 \in V$ can map into $0 \in U$, and so F is nonsingular. The converse is also true. For suppose F is nonsingular and $F(v) = F(w)$, then $F(v - w) = F(v) - F(w) = 0$, and hence, $v - w = 0$ or $v = w$. Thus, $F(v) = F(w)$ implies $v = w$—that is, F is one-to-one. We have proved the following proposition.

PROPOSITION 5.8: A linear mapping $F: V \to U$ is one-to-one if and only if F is nonsingular.

Recall that a mapping $F: V \to U$ is called an *isomorphism* if F is linear and if F is bijective (i.e., if F is one-to-one and onto). Also, recall that a vector space V is said to be *isomorphic* to a vector space U, written $V \cong U$, if there is an isomorphism $F: V \to U$.

The following theorem (proved in Problem 5.29) applies.

THEOREM 5.9: Suppose V has finite dimension and $\dim V = \dim U$. Suppose $F: V \to U$ is linear. Then F is an isomorphism if and only if F is nonsingular.

5.6 Operations with Linear Mappings

We are able to combine linear mappings in various ways to obtain new linear mappings. These operations are very important and will be used throughout the text.

Let $F: V \to U$ and $G: V \to U$ be linear mappings over a field K. The sum $F+G$ and the scalar product kF, where $k \in K$, are defined to be the following mappings from V into U:

$$(F+G)(v) \equiv F(v) + G(v) \qquad \text{and} \qquad (kF)(v) \equiv kF(v)$$

We now show that if F and G are linear, then $F+G$ and kF are also linear. Specifically, for any vectors $v, w \in V$ and any scalars $a, b \in K$,

$$\begin{aligned} (F+G)(av+bw) &= F(av+bw) + G(av+bw) \\ &= aF(v) + bF(w) + aG(v) + bG(w) \\ &= a[F(v) + G(v)] + b[F(w) + G(w)] \\ &= a(F+G)(v) + b(F+G)(w) \end{aligned}$$

and

$$\begin{aligned} (kF)(av+bw) &= kF(av+bw) = k[aF(v) + bF(w)] \\ &= akF(v) + bkF(w) = a(kF)(v) + b(kF)(w) \end{aligned}$$

Thus, $F+G$ and kF are linear.

The following theorem holds.

THEOREM 5.10: Let V and U be vector spaces over a field K. Then the collection of all linear mappings from V into U with the above operations of addition and scalar multiplication forms a vector space over K.

The vector space of linear mappings in Theorem 5.10 is usually denoted by

$$\text{Hom}(V, U)$$

Here Hom comes from the word "homomorphism." We emphasize that the proof of Theorem 5.10 reduces to showing that $\text{Hom}(V, U)$ does satisfy the eight axioms of a vector space. The zero element of $\text{Hom}(V, U)$ is the *zero mapping* from V into U, denoted by $\mathbf{0}$ and defined by

$$\mathbf{0}(v) = 0$$

for every vector $v \in V$.

Suppose V and U are of finite dimension. Then we have the following theorem.

THEOREM 5.11: Suppose $\dim V = m$ and $\dim U = n$. Then $\dim[\text{Hom}(V, U)] = mn$.

Composition of Linear Mappings

Now suppose V, U, and W are vector spaces over the same field K, and suppose $F: V \to U$ and $G: U \to W$ are linear mappings. We picture these mappings as follows:

$$V \xrightarrow{F} U \xrightarrow{G} W$$

Recall that the composition function $G \circ F$ is the mapping from V into W defined by $(G \circ F)(v) = G(F(v))$. We show that $G \circ F$ is linear whenever F and G are linear. Specifically, for any vectors $v, w \in V$ and any scalars $a, b \in K$, we have

$$\begin{aligned} (G \circ F)(av+bw) &= G(F(av+bw)) = G(aF(v) + bF(w)) \\ &= aG(F(v)) + bG(F(w)) = a(G \circ F)(v) + b(G \circ F)(w) \end{aligned}$$

Thus, $G \circ F$ is linear.

The composition of linear mappings and the operations of addition and scalar multiplication are related as follows.

THEOREM 5.12: Let V, U, W be vector spaces over K. Suppose the following mappings are linear:

$$F: V \to U, \qquad F': V \to U \qquad \text{and} \qquad G: U \to W, \qquad G': U \to W$$

Then, for any scalar $k \in K$:

(i) $G \circ (F + F') = G \circ F + G \circ F'$.
(ii) $(G + G') \circ F = G \circ F + G' \circ F$.
(iii) $k(G \circ F) = (kG) \circ F = G \circ (kF)$.

5.7 Algebra $A(V)$ of Linear Operators

Let V be a vector space over a field K. This section considers the special case of linear mappings from the vector space V into itself—that is, linear mappings of the form $F: V \to V$. They are also called *linear operators* or *linear transformations* on V. We will write $A(V)$, instead of $\text{Hom}(V, V)$, for the space of all such mappings.

Now $A(V)$ is a vector space over K (Theorem 5.8), and, if $\dim V = n$, then $\dim A(V) = n^2$. Moreover, for any mappings $F, G \in A(V)$, the composition $G \circ F$ exists and also belongs to $A(V)$. Thus, we have a "multiplication" defined in $A(V)$. [We sometimes write FG instead of $G \circ F$ in the space $A(V)$.]

Remark: An *algebra* A over a field K is a vector space over K in which an operation of multiplication is defined satisfying, for every $F, G, H \in A$ and every $k \in K$:

(i) $F(G + H) = FG + FH$,
(ii) $(G + H)F = GF + HF$,
(iii) $k(GF) = (kG)F = G(kF)$.

The algebra is said to be *associative* if, in addition, $(FG)H = F(GH)$.

The above definition of an algebra and previous theorems give us the following result.

THEOREM 5.13: Let V be a vector space over K. Then $A(V)$ is an associative algebra over K with respect to composition of mappings. If $\dim V = n$, then $\dim A(V) = n^2$.

This is why $A(V)$ is called the *algebra of linear operators* on V.

Polynomials and Linear Operators

Observe that the identity mapping $I: V \to V$ belongs to $A(V)$. Also, for any linear operator F in $A(V)$, we have $FI = IF = F$. We can also form "powers" of F. Namely, we define

$$F^0 = I, \qquad F^2 = F \circ F, \qquad F^3 = F^2 \circ F = F \circ F \circ F, \qquad F^4 = F^3 \circ F, \quad \ldots$$

Furthermore, for any polynomial $p(t)$ over K, say,

$$p(t) = a_0 + a_1 t + a_2 t^2 + \cdots + a_s t^2$$

we can form the linear operator $p(F)$ defined by

$$p(F) = a_0 I + a_1 F + a_2 F^2 + \cdots + a_s F^s$$

(For any scalar k, the operator kI is sometimes denoted simply by k.) In particular, we say F is a *zero* of the polynomial $p(t)$ if $p(F) = 0$.

EXAMPLE 5.11 Let $F: K^3 \to K^3$ be defined by $F(x, y, z) = (0, x, y)$. For any $(a, b, c) \in K^3$,

$$(F + I)(a, b, c) = (0, a, b) + (a, b, c) = (a, \ a + b, \ b + c)$$
$$F^3(a, b, c) = F^2(0, a, b) = F(0, 0, a) = (0, 0, 0)$$

Thus, $F^3 = 0$, the zero mapping in $A(V)$. This means F is a zero of the polynomial $p(t) = t^3$.

Square Matrices as Linear Operators

Let $\mathbf{M} = \mathbf{M}_{n,n}$ be the vector space of all square $n \times n$ matrices over K. Then any matrix A in M defines a linear mapping $F_A : K^n \to K^n$ by $F_A(u) = Au$ (where the vectors in K^n are written as columns). Because the mapping is from K^n into itself, the square matrix A is a linear operator, not simply a linear mapping.

Suppose A and B are matrices in M. Then the matrix product AB is defined. Furthermore, for any (column) vector u in K^n,

$$F_{AB}(u) = (AB)u = A(Bu) = A(F_B(U)) = F_A(F_B(u)) = (F_A \circ F_B)(u)$$

In other words, the matrix product AB corresponds to the composition of A and B as linear mappings. Similarly, the matrix sum $A + B$ corresponds to the sum of A and B as linear mappings, and the scalar product kA corresponds to the scalar product of A as a linear mapping.

Invertible Operators in $A(V)$

Let $F : V \to V$ be a linear operator. F is said to be *invertible* if it has an inverse—that is, if there exists F^{-1} in $A(V)$ such that $FF^{-1} = F^{-1}F = I$. On the other hand, F is invertible as a mapping if F is both one-to-one and onto. In such a case, F^{-1} is also linear and F^{-1} is the inverse of F as a linear operator (proved in Problem 5.15).

Suppose F is invertible. Then only $0 \in V$ can map into itself, and so F is nonsingular. The converse is not true, as seen by the following example.

EXAMPLE 5.12 Let $V = \mathbf{P}(t)$, the vector space of polynomials over K. Let F be the mapping on V that increases by 1 the exponent of t in each term of a polynomial; that is,

$$F(a_0 + a_1 t + a_2 t^2 + \cdots + a_s t^s) = a_0 t + a_1 t^2 + a_2 t^3 + \cdots + a_s t^{s+1}$$

Then F is a linear mapping and F is nonsingular. However, F is not onto, and so F is not invertible.

The vector space $V = \mathbf{P}(t)$ in the above example has infinite dimension. The situation changes significantly when V has finite dimension. Namely, the following theorem applies.

THEOREM 5.14: Let F be a linear operator on a finite-dimensional vector space V. Then the following four conditions are equivalent.

(i) F is nonsingular: $\text{Ker } F = \{0\}$.
(ii) F is one-to-one.
(iii) F is an onto mapping.
(iv) F is invertible.

The proof of the above theorem mainly follows from Theorem 5.6, which tells us that

$$\dim V = \dim(\text{Ker } F) + \dim(\text{Im } F)$$

By Proposition 5.8, (i) and (ii) are equivalent. Note that (iv) is equivalent to (ii) and (iii). Thus, to prove the theorem, we need only show that (i) and (iii) are equivalent. This we do below.

(a) Suppose (i) holds. Then $\dim(\text{Ker } F) = 0$, and so the above equation tells us that $\dim V = \dim(\text{Im } F)$. This means $V = \text{Im } F$ or, in other words, F is an onto mapping. Thus, (i) implies (iii).

(b) Suppose (iii) holds. Then $V = \text{Im } F$, and so $\dim V = \dim(\text{Im } F)$. Therefore, the above equation tells us that $\dim(\text{Ker } F) = 0$, and so F is nonsingular. Therefore, (iii) implies (i).

Accordingly, all four conditions are equivalent.

Remark: Suppose A is a square $n \times n$ matrix over K. Then A may be viewed as a linear operator on K^n. Because K^n has finite dimension, Theorem 5.14 holds for the square matrix A. This is why the terms "nonsingular" and "invertible" are used interchangeably when applied to square matrices.

EXAMPLE 5.13 Let F be the linear operator on $\mathbf{R}^2$ defined by $F(x, y) = (2x + y,\ 3x + 2y)$.

(a) To show that F is invertible, we need only show that F is nonsingular. Set $F(x, y) = (0, 0)$ to obtain the homogeneous system

$$2x + y = 0 \quad \text{and} \quad 3x + 2y = 0$$

Solve for x and y to get $x = 0$, $y = 0$. Hence, F is nonsingular and so invertible.

(b) To find a formula for F^{-1}, we set $F(x,y) = (s,t)$ and so $F^{-1}(s,t) = (x,y)$. We have

$$(2x+y,\ 3x+2y) = (s,t) \quad \text{or} \quad \begin{aligned} 2x + \ \ y &= s \\ 3x + 2y &= t \end{aligned}$$

Solve for x and y in terms of s and t to obtain $x = 2s - t$, $y = -3s + 2t$. Thus,

$$F^{-1}(s,t) = (2s - t,\ -3s + 2t) \quad \text{or} \quad F^{-1}(x,y) = (2x - y,\ -3x + 2y)$$

where we rewrite the formula for F^{-1} using x and y instead of s and t.

SOLVED PROBLEMS

Mappings

5.1. State whether each diagram in Fig. 5-3 defines a mapping from $A = \{a, b, c\}$ into $B = \{x, y, z\}$.

(a) No. There is nothing assigned to the element $b \in A$.

(b) No. Two elements, x and z, are assigned to $c \in A$.

(c) Yes.

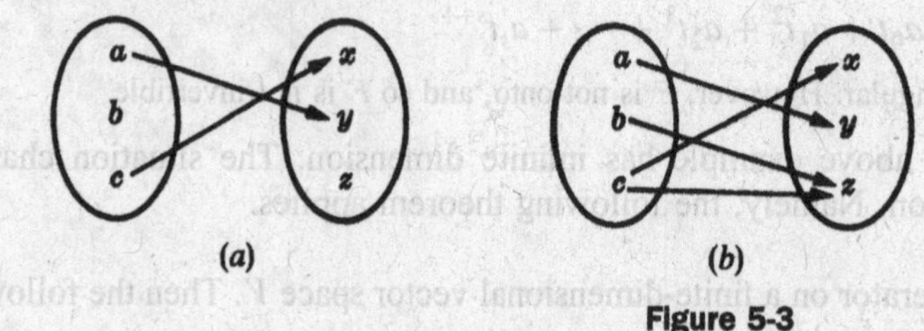

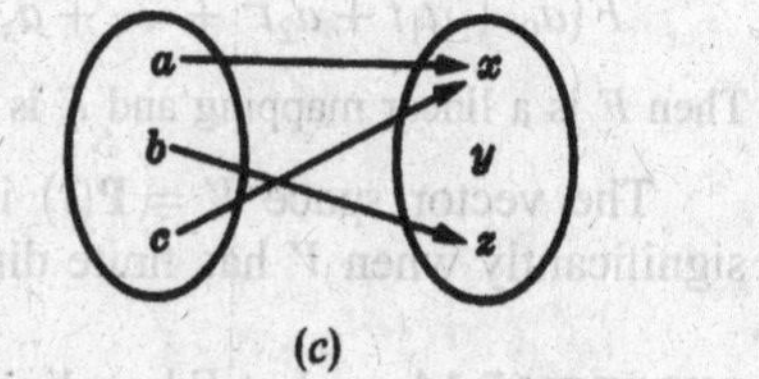

Figure 5-3

5.2. Let $f : A \to B$ and $g : B \to C$ be defined by Fig. 5-4.

(a) Find the composition mapping $(g \circ f) : A \to C$.

(b) Find the images of the mappings f, g, $g \circ f$.

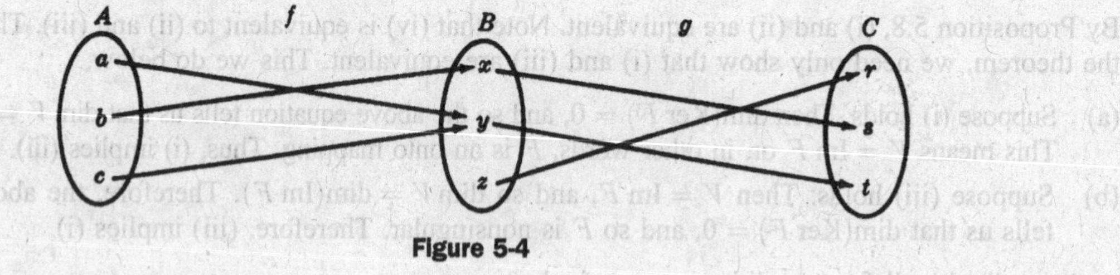

Figure 5-4

(a) Use the definition of the composition mapping to compute

$$(g \circ f)\ (a) = g(f(a)) = g(y) = t, \qquad (g \circ f)\ (b) = g(f(b)) = g(x) = s$$
$$(g \circ f)\ (c) = g(f(c)) = g(y) = t$$

Observe that we arrive at the same answer if we "follow the arrows" in Fig. 5-4:

$$a \to y \to t, \qquad b \to x \to s, \qquad c \to y \to t$$

(b) By Fig. 5-4, the image values under the mapping f are x and y, and the image values under g are r, s, t.

Hence,

$$\operatorname{Im} f = \{x, y\} \quad \text{and} \quad \operatorname{Im} g = \{r, s, t\}$$

Also, by part (a), the image values under the composition mapping $g \circ f$ are t and s; accordingly, $\operatorname{Im} g \circ f = \{s, t\}$. Note that the images of g and $g \circ f$ are different.

5.3. Consider the mapping $F: \mathbf{R}^3 \to \mathbf{R}^2$ defined by $F(x, y, z) = (yz, x^2)$. Find (a) $F(2, 3, 4)$; (b) $F(5, -2, 7)$; (c) $F^{-1}(0, 0)$, that is, all $v \in \mathbf{R}^3$ such that $F(v) = 0$.

(a) Substitute in the formula for F to get $F(2, 3, 4) = (3 \cdot 4, 2^2) = (12, 4)$.

(b) $F(5, -2, 7) = (-2 \cdot 7, 5^2) = (-14, 25)$.

(c) Set $F(v) = 0$, where $v = (x, y, z)$, and then solve for x, y, z:

$$F(x, y, z) = (yz, x^2) = (0, 0) \quad \text{or} \quad yz = 0, x^2 = 0$$

Thus, $x = 0$ and either $y = 0$ or $z = 0$. In other words, $x = 0$, $y = 0$ or $x = 0$, $z = 0$—that is, the z-axis and the y-axis.

5.4. Consider the mapping $F: \mathbf{R}^2 \to \mathbf{R}^2$ defined by $F(x, y) = (3y, 2x)$. Let S be the unit circle in $\mathbf{R}^2$, that is, the solution set of $x^2 + y^2 = 1$. (a) Describe $F(S)$. (b) Find $F^{-1}(S)$.

(a) Let (a, b) be an element of $F(S)$. Then there exists $(x, y) \in S$ such that $F(x, y) = (a, b)$. Hence,

$$(3y, 2x) = (a, b) \quad \text{or} \quad 3y = a, 2x = b \quad \text{or} \quad y = \frac{a}{3}, x = \frac{b}{2}$$

Because $(x, y) \in S$—that is, $x^2 + y^2 = 1$—we have

$$\left(\frac{b}{2}\right)^2 + \left(\frac{a}{3}\right)^2 = 1 \quad \text{or} \quad \frac{a^2}{9} + \frac{b^2}{4} = 1$$

Thus, $F(S)$ is an ellipse.

(b) Let $F(x, y) = (a, b)$, where $(a, b) \in S$. Then $(3y, 2x) = (a, b)$ or $3y = a$, $2x = b$. Because $(a, b) \in S$, we have $a^2 + b^2 = 1$. Thus, $(3y)^2 + (2x)^2 = 1$. Accordingly, $F^{-1}(S)$ is the ellipse $4x^2 + 9y^2 = 1$.

5.5. Let the mappings $f: A \to B$, $g: B \to C$, $h: C \to D$ be defined by Fig. 5-5. Determine whether or not each function is (a) one-to-one; (b) onto; (c) invertible (i.e., has an inverse).

(a) The mapping $f: A \to B$ is one-to-one, as each element of A has a different image. The mapping $g: B \to C$ is not one-to one, because x and z both have the same image 4. The mapping $h: C \to D$ is one-to-one.

(b) The mapping $f: A \to B$ is not onto, because $z \in B$ is not the image of any element of A. The mapping $g: B \to C$ is onto, as each element of C is the image of some element of B. The mapping $h: C \to D$ is also onto.

(c) A mapping has an inverse if and only if it is one-to-one and onto. Hence, only h has an inverse.

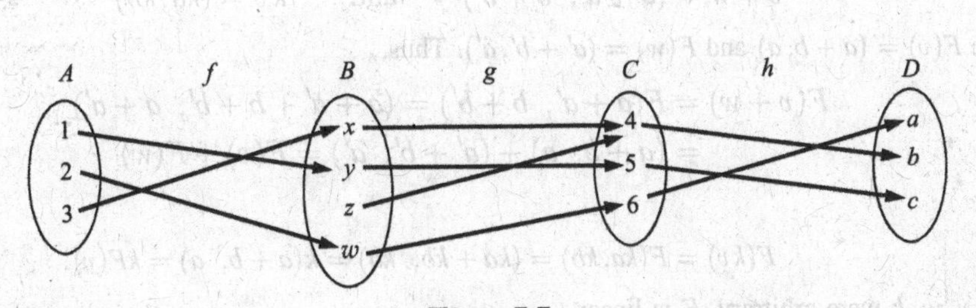

Figure 5-5

5.6. Suppose $f: A \to B$ and $g: B \to C$. Hence, $(g \circ f): A \to C$ exists. Prove

(a) If f and g are one-to-one, then $g \circ f$ is one-to-one.
(b) If f and g are onto mappings, then $g \circ f$ is an onto mapping.
(c) If $g \circ f$ is one-to-one, then f is one-to-one.
(d) If $g \circ f$ is an onto mapping, then g is an onto mapping.

(a) Suppose $(g \circ f)(x) = (g \circ f)(y)$. Then $g(f(x)) = g(f(y))$. Because g is one-to-one, $f(x) = f(y)$. Because f is one-to-one, $x = y$. We have proven that $(g \circ f)(x) = (g \circ f)(y)$ implies $x = y$; hence $g \circ f$ is one-to-one.

(b) Suppose $c \in C$. Because g is onto, there exists $b \in B$ for which $g(b) = c$. Because f is onto, there exists $a \in A$ for which $f(a) = b$. Thus, $(g \circ f)(a) = g(f(a)) = g(b) = c$. Hence, $g \circ f$ is onto.

(c) Suppose f is not one-to-one. Then there exist distinct elements $x, y \in A$ for which $f(x) = f(y)$. Thus, $(g \circ f)(x) = g(f(x)) = g(f(y)) = (g \circ f)(y)$. Hence, $g \circ f$ is not one-to-one. Therefore, if $g \circ f$ is one-to-one, then f must be one-to-one.

(d) If $a \in A$, then $(g \circ f)(a) = g(f(a)) \in g(B)$. Hence, $(g \circ f)(A) \subseteq g(B)$. Suppose g is not onto. Then $g(B)$ is properly contained in C and so $(g \circ f)(A)$ is properly contained in C; thus, $g \circ f$ is not onto. Accordingly, if $g \circ f$ is onto, then g must be onto.

5.7. Prove that $f: A \to B$ has an inverse if and only if f is one-to-one and onto.

Suppose f has an inverse—that is, there exists a function $f^{-1}: B \to A$ for which $f^{-1} \circ f = \mathbf{1}_A$ and $f \circ f^{-1} = \mathbf{1}_B$. Because $\mathbf{1}_A$ is one-to-one, f is one-to-one by Problem 5.6(c), and because $\mathbf{1}_B$ is onto, f is onto by Problem 5.6(*d*); that is, f is both one-to-one and onto.

Now suppose f is both one-to-one and onto. Then each $b \in B$ is the image of a unique element in A, say b^*. Thus, if $f(a) = b$, then $a = b^*$; hence, $f(b^*) = b$. Now let g denote the mapping from B to A defined by $b \mapsto b^*$. We have

(i) $(g \circ f)(a) = g(f(a)) = g(b) = b^* = a$ for every $a \in A$; hence, $g \circ f = \mathbf{1}_A$.

(ii) $(f \circ g)(b) = f(g(b)) = f(b^*) = b$ for every $b \in B$; hence, $f \circ g = \mathbf{1}_B$.

Accordingly, f has an inverse. Its inverse is the mapping g.

5.8. Let $f: \mathbf{R} \to \mathbf{R}$ be defined by $f(x) = 2x - 3$. Now f is one-to-one and onto; hence, f has an inverse mapping f^{-1}. Find a formula for f^{-1}.

Let y be the image of x under the mapping f; that is, $y = f(x) = 2x - 3$. Hence, x will be the image of y under the inverse mapping f^{-1}. Thus, solve for x in terms of y in the above equation to obtain $x = \frac{1}{2}(y+3)$. Then the formula defining the inverse function is $f^{-1}(y) = \frac{1}{2}(y+3)$, or, using x instead of y, $f^{-1}(x) = \frac{1}{2}(x+3)$.

Linear Mappings

5.9. Suppose the mapping $F: \mathbf{R}^2 \to \mathbf{R}^2$ is defined by $F(x, y) = (x + y,\ x)$. Show that F is linear.

We need to show that $F(v + w) = F(v) + F(w)$ and $F(kv) = kF(v)$, where u and v are any elements of $\mathbf{R}^2$ and k is any scalar. Let $v = (a, b)$ and $w = (a', b')$. Then

$$v + w = (a + a',\ b + b') \qquad \text{and} \qquad kv = (ka, kb)$$

We have $F(v) = (a + b, a)$ and $F(w) = (a' + b', a')$. Thus,

$$\begin{aligned} F(v + w) &= F(a + a',\ b + b') = (a + a' + b + b',\ a + a') \\ &= (a + b,\ a) + (a' + b',\ a') = F(v) + F(w) \end{aligned}$$

and

$$F(kv) = F(ka, kb) = (ka + kb,\ ka) = k(a + b,\ a) = kF(v)$$

Because v, w, k were arbitrary, F is linear.

5.10. Suppose $F: \mathbf{R}^3 \to \mathbf{R}^2$ is defined by $F(x,y,z) = (x+y+z,\ 2x-3y+4z)$. Show that F is linear.

We argue via matrices. Writing vectors as columns, the mapping F may be written in the form $F(v) = Av$, where $v = [x,y,z]^T$ and

$$A = \begin{bmatrix} 1 & 1 & 1 \\ 2 & -3 & 4 \end{bmatrix}$$

Then, using properties of matrices, we have

$$F(v+w) = A(v+w) = Av + Aw = F(v) + F(w)$$

and

$$F(kv) = A(kv) = k(Av) = kF(v)$$

Thus, F is linear.

5.11. Show that the following mappings are not linear:

(a) $F: \mathbf{R}^2 \to \mathbf{R}^2$ defined by $F(x,y) = (xy, x)$

(b) $F: \mathbf{R}^2 \to \mathbf{R}^3$ defined by $F(x,y) = (x+3,\ 2y,\ x+y)$

(c) $F: \mathbf{R}^3 \to \mathbf{R}^2$ defined by $F(x,y,z) = (|x|,\ y+z)$

(a) Let $v = (1,2)$ and $w = (3,4)$; then $v + w = (4,6)$. Also,

$$F(v) = (1(2), 1) = (2, 1) \quad \text{and} \quad F(w) = (3(4), 3) = (12, 3)$$

Hence,

$$F(v+w) = (4(6), 4) = (24, 6) \neq F(v) + F(w)$$

(b) Because $F(0,0) = (3,0,0) \neq (0,0,0)$, F cannot be linear.

(c) Let $v = (1,2,3)$ and $k = -3$. Then $kv = (-3,-6,-9)$. We have

$$F(v) = (1,5) \quad \text{and} \quad kF(v) = -3(1,5) = (-3,-15).$$

Thus,

$$F(kv) = F(-3,-6,-9) = (3,-15) \neq kF(v)$$

Accordingly, F is not linear.

5.12. Let V be the vector space of n-square real matrices. Let M be an arbitrary but fixed matrix in V. Let $F: V \to V$ be defined by $F(A) = AM + MA$, where A is any matrix in V. Show that F is linear.

For any matrices A and B in V and any scalar k, we have

$$\begin{aligned} F(A+B) &= (A+B)M + M(A+B) = AM + BM + MA + MB \\ &= (AM + MA) = (BM + MB) = F(A) + F(B) \end{aligned}$$

and

$$F(kA) = (kA)M + M(kA) = k(AM) + k(MA) = k(AM + MA) = kF(A)$$

Thus, F is linear.

5.13. Prove Theorem 5.2: Let V and U be vector spaces over a field K. Let $\{v_1, v_2, \ldots, v_n\}$ be a basis of V and let $u_1, u_2, \ldots, u_n$ be any vectors in U. Then there exists a unique linear mapping $F: V \to U$ such that $F(v_1) = u_1, F(v_2) = u_2, \ldots, F(v_n) = u_n$.

There are three steps to the proof of the theorem: (1) Define the mapping $F: V \to U$ such that $F(v_i) = u_i, i = 1, \ldots, n$. (2) Show that F is linear. (3) Show that F is unique.

Step 1. Let $v \in V$. Because $\{v_1, \ldots, v_n\}$ is a basis of V, there exist unique scalars $a_1, \ldots, a_n \in K$ for which $v = a_1v_1 + a_2v_2 + \cdots + a_nv_n$. We define $F: V \to U$ by

$$F(v) = a_1u_1 + a_2u_2 + \cdots + a_nu_n$$

(Because the a_i are unique, the mapping F is well defined.) Now, for $i = 1, \dots, n$,

$$v_i = 0v_1 + \cdots + 1v_i + \cdots + 0v_n$$

Hence,

$$F(v_i) = 0u_1 + \cdots + 1u_i + \cdots + 0u_n = u_i$$

Thus, the first step of the proof is complete.

Step 2. Suppose $v = a_1v_1 + a_2v_2 + \cdots + a_nv_n$ and $w = b_1v_1 + b_2v_2 + \cdots + b_nv_n$. Then

$$v + w = (a_1 + b_1)v_1 + (a_2 + b_2)v_2 + \cdots + (a_n + b_n)v_n$$

and, for any $k \in K$, $kv = ka_1v_1 + ka_2v_2 + \cdots + ka_nv_n$. By definition of the mapping F,

$$F(v) = a_1u_1 + a_2u_2 + \cdots + a_nv_n \qquad \text{and} \qquad F(w) = b_1u_1 + b_2u_2 + \cdots + b_nu_n$$

Hence,

$$\begin{aligned} F(v + w) &= (a_1 + b_1)u_1 + (a_2 + b_2)u_2 + \cdots + (a_n + b_n)u_n \\ &= (a_1u_1 + a_2u_2 + \cdots + a_nu_n) + (b_1u_1 + b_2u_2 + \cdots + b_nu_n) \\ &= F(v) + F(w) \end{aligned}$$

and

$$F(kv) = k(a_1u_1 + a_2u_2 + \cdots + a_nu_n) = kF(v)$$

Thus, F is linear.

Step 3. Suppose $G: V \to U$ is linear and $G(v_i) = u_i, i = 1, \dots, n$. Let

$$v = a_1v_1 + a_2v_2 + \cdots + a_nv_n$$

Then

$$\begin{aligned} G(v) &= G(a_1v_1 + a_2v_2 + \cdots + a_nv_n) = a_1G(v_1) + a_2G(v_2) + \cdots + a_nG(v_n) \\ &= a_1u_1 + a_2u_2 + \cdots + a_nu_n = F(v) \end{aligned}$$

Because $G(v) = F(v)$ for every $v \in V, G = F$. Thus, F is unique and the theorem is proved.

5.14. Let $F: \mathbf{R}^2 \to \mathbf{R}^2$ be the linear mapping for which $F(1,2) = (2,3)$ and $F(0,1) = (1,4)$. [Note that $\{(1,2), (0,1)\}$ is a basis of $\mathbf{R}^2$, so such a linear map F exists and is unique by Theorem 5.2.] Find a formula for F; that is, find $F(a,b)$.

Write (a,b) as a linear combination of $(1,2)$ and $(0,1)$ using unknowns x and y,

$$(a,b) = x(1,2) + y(0,1) = (x, \ 2x + y), \qquad \text{so} \qquad a = x, \ b = 2x + y$$

Solve for x and y in terms of a and b to get $x = a, \quad y = -2a + b$. Then

$$F(a,b) = xF(1,2) + yF(0,1) = a(2,3) + (-2a + b)(1,4) = (b, \ -5a + 4b)$$

5.15. Suppose a linear mapping $F: V \to U$ is one-to-one and onto. Show that the inverse mapping $F^{-1}: U \to V$ is also linear.

Suppose $u, u' \in U$. Because F is one-to-one and onto, there exist unique vectors $v, v' \in V$ for which $F(v) = u$ and $F(v') = u'$. Because F is linear, we also have

$$F(v + v') = F(v) + F(v') = u + u' \qquad \text{and} \qquad F(kv) = kF(v) = ku$$

By definition of the inverse mapping,

$$F^{-1}(u) = v, \ F^{-1}(u') = v', \ F^{-1}(u + u') = v + v', \ F^{-1}(ku) = kv.$$

Then

$$F^{-1}(u + u') = v + v' = F^{-1}(u) + F^{-1}(u') \qquad \text{and} \qquad F^{-1}(ku) = kv = kF^{-1}(u)$$

Thus, F^{-1} is linear.

Kernel and Image of Linear Mappings

5.16. Let $F: \mathbf{R}^4 \to \mathbf{R}^3$ be the linear mapping defined by

$$F(x,y,z,t) = (x - y + z + t, \quad x + 2z - t, \quad x + y + 3z - 3t)$$

Find a basis and the dimension of (a) the image of F, (b) the kernel of F.

(a) Find the images of the usual basis of $\mathbf{R}^4$:

$$F(1,0,0,0) = (1,1,1), \qquad F(0,0,1,0) = (1,2,3)$$
$$F(0,1,0,0) = (-1,0,1), \qquad F(0,0,0,1) = (1,-1,-3)$$

By Proposition 5.4, the image vectors span Im F. Hence, form the matrix whose rows are these image vectors, and row reduce to echelon form:

$$\begin{bmatrix} 1 & 1 & 1 \\ -1 & 0 & 1 \\ 1 & 2 & 3 \\ 1 & -1 & -3 \end{bmatrix} \sim \begin{bmatrix} 1 & 1 & 1 \\ 0 & 1 & 2 \\ 0 & 1 & 2 \\ 0 & -2 & -4 \end{bmatrix} \sim \begin{bmatrix} 1 & 1 & 1 \\ 0 & 1 & 2 \\ 0 & 0 & 0 \\ 0 & 0 & 0 \end{bmatrix}$$

Thus, $(1,1,1)$ and $(0,1,2)$ form a basis for Im F; hence, $\dim(\text{Im } F) = 2$.

(b) Set $F(v) = 0$, where $v = (x,y,z,t)$; that is, set

$$F(x,y,z,t) = (x - y + z + t, \; x + 2z - t, \; x + y + 3z - 3t) = (0,0,0)$$

Set corresponding entries equal to each other to form the following homogeneous system whose solution space is Ker F:

$$\begin{aligned} x - y + z + t &= 0 \\ x \quad + 2z - t &= 0 \\ x + y + 3z - 3t &= 0 \end{aligned} \quad \text{or} \quad \begin{aligned} x - y + z + t &= 0 \\ y + z - 2t &= 0 \\ 2y + 2z - 4t &= 0 \end{aligned} \quad \text{or} \quad \begin{aligned} x - y + z + t &= 0 \\ y + z - 2t &= 0 \end{aligned}$$

The free variables are z and t. Hence, $\dim(\text{Ker } F) = 2$.

(i) Set $z = -1$, $t = 0$ to obtain the solution $(2,1,-1,0)$.

(ii) Set $z = 0$, $t = 1$ to obtain the solution $(1,2,0,1)$.

Thus, $(2,1,-1,0)$ and $(1,2,0,1)$ form a basis of Ker F.
[As expected, $\dim(\text{Im } F) + \dim(\text{Ker } F) = 2 + 2 = 4 = \dim \mathbf{R}^4$, the domain of F.]

5.17. Let $G: \mathbf{R}^3 \to \mathbf{R}^3$ be the linear mapping defined by

$$G(x,y,z) = (x + 2y - z, \quad y + z, \quad x + y - 2z)$$

Find a basis and the dimension of (a) the image of G, (b) the kernel of G.

(a) Find the images of the usual basis of $\mathbf{R}^3$:

$$G(1,0,0) = (1,0,1), \qquad G(0,1,0) = (2,1,1), \qquad G(0,0,1) = (-1,1,-2)$$

By Proposition 5.4, the image vectors span Im G. Hence, form the matrix M whose rows are these image vectors, and row reduce to echelon form:

$$M = \begin{bmatrix} 1 & 0 & 1 \\ 2 & 1 & 1 \\ -1 & 1 & -2 \end{bmatrix} \sim \begin{bmatrix} 1 & 0 & 1 \\ 0 & 1 & -1 \\ 0 & 1 & -1 \end{bmatrix} \sim \begin{bmatrix} 1 & 0 & 1 \\ 0 & 1 & -1 \\ 0 & 0 & 0 \end{bmatrix}$$

Thus, $(1,0,1)$ and $(0,1,-1)$ form a basis for Im G; hence, $\dim(\text{Im } G) = 2$.

(b) Set $G(v) = 0$, where $v = (x,y,z)$; that is,

$$G(x,y,z) = (x + 2y - z, \quad y + z, \quad x + y - 2z) = (0,0,0)$$

Set corresponding entries equal to each other to form the following homogeneous system whose solution space is Ker G:

$$\begin{aligned} x+2y-\ z&=0 \\ y+\ z&=0 \\ x+\ y-2z&=0 \end{aligned} \quad \text{or} \quad \begin{aligned} x+2y-z&=0 \\ y+z&=0 \\ -y-z&=0 \end{aligned} \quad \text{or} \quad \begin{aligned} x+2y-z&=0 \\ y+z&=0 \end{aligned}$$

The only free variable is z; hence, $\dim(\text{Ker}\ G)=1$. Set $z=1$; then $y=-1$ and $x=3$. Thus, $(3,-1,1)$ forms a basis of Ker G. [As expected, $\dim(\text{Im}\ G)+\dim(\text{Ker}\ G)=2+1=3=\dim \mathbf{R}^3$, the domain of G.]

5.18. Consider the matrix mapping $A: \mathbf{R}^4 \to \mathbf{R}^3$, where $A=\begin{bmatrix} 1 & 2 & 3 & 1 \\ 1 & 3 & 5 & -2 \\ 3 & 8 & 13 & -3 \end{bmatrix}$. Find a basis and the dimension of (a) the image of A, (b) the kernel of A.

(a) The column space of A is equal to Im A. Now reduce A^T to echelon form:

$$A^T=\begin{bmatrix} 1 & 1 & 3 \\ 2 & 3 & 8 \\ 3 & 5 & 13 \\ 1 & -2 & -3 \end{bmatrix} \sim \begin{bmatrix} 1 & 1 & 3 \\ 0 & 1 & 2 \\ 0 & 2 & 4 \\ 0 & -3 & -6 \end{bmatrix} \sim \begin{bmatrix} 1 & 1 & 3 \\ 0 & 1 & 2 \\ 0 & 0 & 0 \\ 0 & 0 & 0 \end{bmatrix}$$

Thus, $\{(1,1,3),(0,1,2)\}$ is a basis of Im A, and $\dim(\text{Im}\ A)=2$.

(b) Here Ker A is the solution space of the homogeneous system $AX=0$, where $X=\{x,y,z,t\}^T$. Thus, reduce the matrix A of coefficients to echelon form:

$$\begin{bmatrix} 1 & 2 & 3 & 1 \\ 0 & 1 & 2 & -3 \\ 0 & 2 & 4 & -6 \end{bmatrix} \sim \begin{bmatrix} 1 & 2 & 3 & 1 \\ 0 & 1 & 2 & -3 \\ 0 & 0 & 0 & 0 \end{bmatrix} \quad \text{or} \quad \begin{aligned} x+2y+3z+\ t&=0 \\ y+2z-3t&=0 \end{aligned}$$

The free variables are z and t. Thus, $\dim(\text{Ker}\ A)=2$.

(i) Set $z=1$, $t=0$ to get the solution $(1,-2,1,0)$.

(ii) Set $z=0$, $t=1$ to get the solution $(-7,3,0,1)$.

Thus, $(1,-2,1,0)$ and $(-7,3,0,1)$ form a basis for Ker A.

5.19. Find a linear map $F: \mathbf{R}^3 \to \mathbf{R}^4$ whose image is spanned by $(1,2,0,-4)$ and $(2,0,-1,-3)$.

Form a 4×3 matrix whose columns consist only of the given vectors, say

$$A=\begin{bmatrix} 1 & 2 & 2 \\ 2 & 0 & 0 \\ 0 & -1 & -1 \\ -4 & -3 & -3 \end{bmatrix}$$

Recall that A determines a linear map $A: \mathbf{R}^3 \to \mathbf{R}^4$ whose image is spanned by the columns of A. Thus, A satisfies the required condition.

5.20. Suppose $f: V \to U$ is linear with kernel W, and that $f(v)=u$. Show that the "coset" $v+W=\{v+w: w\in W\}$ is the preimage of u; that is, $f^{-1}(u)=v+W$.

We must prove that (i) $f^{-1}(u)\subseteq v+W$ and (ii) $v+W\subseteq f^{-1}(u)$.

We first prove (i). Suppose $v'\in f^{-1}(u)$. Then $f(v')=u$, and so

$$f(v'-v)=f(v')-f(v)=u-u=0$$

that is, $v'-v\in W$. Thus, $v'=v+(v'-v)\in v+W$, and hence $f^{-1}(u)\subseteq v+W$.

Now we prove (ii). Suppose $v' \in v + W$. Then $v' = v + w$, where $w \in W$. Because W is the kernel of f, we have $f(w) = 0$. Accordingly,

$$f(v') = f(v + w) + f(v) + f(w) = f(v) + 0 = f(v) = u$$

Thus, $v' \in f^{-1}(u)$, and so $v + W \subseteq f^{-1}(u)$.

Both inclusions imply $f^{-1}(u) = v + W$.

5.21. Suppose $F: V \to U$ and $G: U \to W$ are linear. Prove

(a) $\text{rank}(G \circ F) \leq \text{rank}(G)$, (b) $\text{rank}(G \circ F) \leq \text{rank}(F)$.

(a) Because $F(V) \subseteq U$, we also have $G(F(V)) \subseteq G(U)$, and so $\dim[G(F(V))] \leq \dim[G(U)]$. Then $\text{rank}(G \circ F) = \dim[(G \circ F)(V)] = \dim[G(F(V))] \leq \dim[G(U)] = \text{rank}(G)$.

(b) We have $\dim[G(F(V))] \leq \dim[F(V)]$. Hence,

$$\text{rank}(G \circ F) = \dim[(G \circ F)(V)] = \dim[G(F(V))] \leq \dim[F(V)] = \text{rank}(F)$$

5.22. Prove Theorem 5.3: Let $F: V \to U$ be linear. Then,

(a) Im F is a subspace of U, (b) Ker F is a subspace of V.

(a) Because $F(0) = 0$, we have $0 \in \text{Im } F$. Now suppose $u, u' \in \text{Im } F$ and $a, b \in K$. Because u and u' belong to the image of F, there exist vectors $v, v' \in V$ such that $F(v) = u$ and $F(v') = u'$. Then

$$F(av + bv') = aF(v) + bF(v') = au + bu' \in \text{Im } F$$

Thus, the image of F is a subspace of U.

(b) Because $F(0) = 0$, we have $0 \in \text{Ker } F$. Now suppose $v, w \in \text{Ker } F$ and $a, b \in K$. Because v and w belong to the kernel of F, $F(v) = 0$ and $F(w) = 0$. Thus,

$$F(av + bw) = aF(v) + bF(w) = a0 + b0 = 0 + 0 = 0, \qquad \text{and so} \qquad av + bw \in \text{Ker } F$$

Thus, the kernel of F is a subspace of V.

5.23. Prove Theorem 5.6: Suppose V has finite dimension and $F: V \to U$ is linear. Then

$$\dim V = \dim(\text{Ker } F) + \dim(\text{Im } F) = \text{nullity}(F) + \text{rank}(F)$$

Suppose $\dim(\text{Ker } F) = r$ and $\{w_1, \ldots, w_r\}$ is a basis of Ker F, and suppose $\dim(\text{Im } F) = s$ and $\{u_1, \ldots, u_s\}$ is a basis of Im F. (By Proposition 5.4, Im F has finite dimension.) Because every $u_j \in \text{Im } F$, there exist vectors $v_1, \ldots, v_s$ in V such that $F(v_1) = u_1, \ldots, F(v_s) = u_s$. We claim that the set

$$B = \{w_1, \ldots, w_r, v_1, \ldots, v_s\}$$

is a basis of V; that is, (i) B spans V, and (ii) B is linearly independent. Once we prove (i) and (ii), then $\dim V = r + s = \dim(\text{Ker } F) + \dim(\text{Im } F)$.

(i) *B spans V.* Let $v \in V$. Then $F(v) \in \text{Im } F$. Because the u_j span Im F, there exist scalars $a_1, \ldots, a_s$ such that $F(v) = a_1u_1 + \cdots + a_su_s$. Set $\hat{v} = a_1v_1 + \cdots + a_sv_s - v$. Then

$$\begin{aligned} F(\hat{v}) &= F(a_1v_1 + \cdots + a_sv_s - v) = a_1F(v_1) + \cdots + a_sF(v_s) - F(v) \\ &= a_1u_1 + \cdots + a_su_s - F(v) = 0 \end{aligned}$$

Thus, $\hat{v} \in \text{Ker } F$. Because the w_i span Ker F, there exist scalars $b_1, \ldots, b_r$, such that

$$\hat{v} = b_1w_1 + \cdots + b_rw_r = a_1v_1 + \cdots + a_sv_s - v$$

Accordingly,

$$v = a_1v_1 + \cdots + a_sv_s - b_1w_1 - \cdots - b_rw_r$$

Thus, B spans V.

(ii) *B is linearly independent.* Suppose

$$x_1w_1 + \cdots + x_rw_r + y_1v_1 + \cdots + y_sv_s = 0 \tag{1}$$

where $x_i, y_j \in K$. Then

$$\begin{aligned} 0 = F(0) &= F(x_1w_1 + \cdots + x_rw_r + y_1v_1 + \cdots + y_sv_s) \\ &= x_1F(w_1) + \cdots + x_rF(w_r) + y_1F(v_1) + \cdots + y_sF(v_s) \end{aligned} \tag{2}$$

But $F(w_i) = 0$, since $w_i \in \operatorname{Ker} F$, and $F(v_j) = u_j$. Substituting into (2), we will obtain $y_1u_1 + \cdots + y_su_s = 0$. Since the u_j are linearly independent, each $y_j = 0$. Substitution into (1) gives $x_1w_1 + \cdots + x_rw_r = 0$. Since the w_i are linearly independent, each $x_i = 0$. Thus B is linearly independent.

Singular and Nonsingular Linear Maps, Isomorphisms

5.24. Determine whether or not each of the following linear maps is nonsingular. If not, find a nonzero vector v whose image is 0.

(a) $F: \mathbf{R}^2 \to \mathbf{R}^2$ defined by $F(x,y) = (x - y,\ x - 2y)$.

(b) $G: \mathbf{R}^2 \to \mathbf{R}^2$ defined by $G(x,y) = (2x - 4y,\ 3x - 6y)$.

(a) Find Ker F by setting $F(v) = 0$, where $v = (x,y)$,

$$(x - y,\ x - 2y) = (0,0) \quad \text{or} \quad \begin{aligned} x - y &= 0 \\ x - 2y &= 0 \end{aligned} \quad \text{or} \quad \begin{aligned} x - y &= 0 \\ -y &= 0 \end{aligned}$$

The only solution is $x = 0, y = 0$. Hence, F is nonsingular.

(b) Set $G(x,y) = (0,0)$ to find Ker G:

$$(2x - 4y,\ 3x - 6y) = (0,0) \quad \text{or} \quad \begin{aligned} 2x - 4y &= 0 \\ 3x - 6y &= 0 \end{aligned} \quad \text{or} \quad x - 2y = 0$$

The system has nonzero solutions, because y is a free variable. Hence, G is singular. Let $y = 1$ to obtain the solution $v = (2,1)$, which is a nonzero vector, such that $G(v) = 0$.

5.25. The linear map $F: \mathbf{R}^2 \to \mathbf{R}^2$ defined by $F(x,y) = (x - y,\ x - 2y)$ is nonsingular by the previous Problem 5.24. Find a formula for F^{-1}.

Set $F(x,y) = (a,b)$, so that $F^{-1}(a,b) = (x,y)$. We have

$$(x - y,\ x - 2y) = (a,b) \quad \text{or} \quad \begin{aligned} x - y &= a \\ x - 2y &= b \end{aligned} \quad \text{or} \quad \begin{aligned} x - y &= a \\ y &= a - b \end{aligned}$$

Solve for x and y in terms of a and b to get $x = 2a - b,\ y = a - b$. Thus,

$$F^{-1}(a,b) = (2a - b,\ a - b) \quad \text{or} \quad F^{-1}(x,y) = (2x - y,\ x - y)$$

(The second equation is obtained by replacing a and b by x and y, respectively.)

5.26. Let $G: \mathbf{R}^2 \to \mathbf{R}^3$ be defined by $G(x,y) = (x + y,\ x - 2y,\ 3x + y)$.

(a) Show that G is nonsingular. (b) Find a formula for G^{-1}.

(a) Set $G(x,y) = (0,0,0)$ to find Ker G. We have

$$(x + y,\ x - 2y,\ 3x + y) = (0,0,0) \quad \text{or} \quad x + y = 0,\ x - 2y = 0,\ 3x + y = 0$$

The only solution is $x = 0, y = 0$; hence, G is nonsingular.

(b) Although G is nonsingular, it is not invertible, because $\mathbf{R}^2$ and $\mathbf{R}^3$ have different dimensions. (Thus, Theorem 5.9 does not apply.) Accordingly, G^{-1} does not exist.

5.27. Suppose that $F: V \to U$ is linear and that V is of finite dimension. Show that V and the image of F have the same dimension if and only if F is nonsingular. Determine all nonsingular linear mappings $T: \mathbf{R}^4 \to \mathbf{R}^3$.

By Theorem 5.6, $\dim V = \dim(\operatorname{Im} F) + \dim(\operatorname{Ker} F)$. Hence, V and $\operatorname{Im} F$ have the same dimension if and only if $\dim(\operatorname{Ker} F) = 0$ or $\operatorname{Ker} F = \{0\}$ (i.e., if and only if F is nonsingular).

Because $\dim \mathbf{R}^3$ is less than $\dim \mathbf{R}^4$, we have that $\dim(\operatorname{Im} T)$ is less than the dimension of the domain $\mathbf{R}^4$ of T. Accordingly no linear mapping $T: \mathbf{R}^4 \to \mathbf{R}^3$ can be nonsingular.

5.28. Prove Theorem 5.7: Let $F: V \to U$ be a nonsingular linear mapping. Then the image of any linearly independent set is linearly independent.

Suppose $v_1, v_2, \ldots, v_n$ are linearly independent vectors in V. We claim that $F(v_1), F(v_2), \ldots, F(v_n)$ are also linearly independent. Suppose $a_1F(v_1) + a_2F(v_2) + \cdots + a_nF(v_n) = 0$, where $a_i \in K$. Because F is linear, $F(a_1v_1 + a_2v_2 + \cdots + a_nv_n) = 0$. Hence,

$$a_1v_1 + a_2v_2 + \cdots + a_nv_n \in \operatorname{Ker} F$$

But F is nonsingular—that is, $\operatorname{Ker} F = \{0\}$. Hence, $a_1v_1 + a_2v_2 + \cdots + a_nv_n = 0$. Because the v_i are linearly independent, all the a_i are 0. Accordingly, the $F(v_i)$ are linearly independent. Thus, the theorem is proved.

5.29. Prove Theorem 5.9: Suppose V has finite dimension and $\dim V = \dim U$. Suppose $F: V \to U$ is linear. Then F is an isomorphism if and only if F is nonsingular.

If F is an isomorphism, then only 0 maps to 0; hence, F is nonsingular. Conversely, suppose F is nonsingular. Then $\dim(\operatorname{Ker} F) = 0$. By Theorem 5.6, $\dim V = \dim(\operatorname{Ker} F) + \dim(\operatorname{Im} F)$. Thus,

$$\dim U = \dim V = \dim(\operatorname{Im} F)$$

Because U has finite dimension, $\operatorname{Im} F = U$. This means F maps V onto U. Thus, F is one-to-one and onto; that is, F is an isomorphism.

Operations with Linear Maps

5.30. Define $F: \mathbf{R}^3 \to \mathbf{R}^2$ and $G: \mathbf{R}^3 \to \mathbf{R}^2$ by $F(x,y,z) = (2x,\ y+z)$ and $G(x,y,z) = (x-z,\ y)$. Find formulas defining the maps: (a) $F+G$, (b) $3F$, (c) $2F-5G$.

(a) $(F+G)(x,y,z) = F(x,y,z) + G(x,y,z) = (2x,\ y+z) + (x-z,\ y) = (3x-z,\ 2y+z)$

(b) $(3F)(x,y,z) = 3F(x,y,z) = 3(2x,\ y+z) = (6x,\ 3y+3z)$

(c) $(2F-5G)(x,y,z) = 2F(x,y,z) - 5G(x,y,z) = 2(2x,\ y+z) - 5(x-z,\ y)$
$= (4x,\ 2y+2z) + (-5x+5z,\ -5y) = (-x+5z,\ -3y+2z)$

5.31. Let $F: \mathbf{R}^3 \to \mathbf{R}^2$ and $G: \mathbf{R}^2 \to \mathbf{R}^2$ be defined by $F(x,y,z) = (2x,\ y+z)$ and $G(x,y) = (y,x)$. Derive formulas defining the mappings: (a) $G \circ F$, (b) $F \circ G$.

(a) $(G \circ F)(x,y,z) = G(F(x,y,z)) = G(2x,\ y+z) = (y+z,\ 2x)$

(b) The mapping $F \circ G$ is not defined, because the image of G is not contained in the domain of F.

5.32. Prove: (a) The zero mapping $\mathbf{0}$, defined by $\mathbf{0}(v) = 0 \in U$ for every $v \in V$, is the zero element of $\operatorname{Hom}(V,U)$. (b) The negative of $F \in \operatorname{Hom}(V,U)$ is the mapping $(-1)F$, that is, $-F = (-1)F$.

Let $F \in \operatorname{Hom}(V,U)$. Then, for every $v \in V$:

(a) $$(F + \mathbf{0})(v) = F(v) + \mathbf{0}(v) = F(v) + 0 = F(v)$$

Because $(F+\mathbf{0})(v) = F(v)$ for every $v \in V$, we have $F + \mathbf{0} = F$. Similarly, $\mathbf{0} + F = F$.

(b) $$(F + (-1)F)(v) = F(v) + (-1)F(v) = F(v) - F(v) = 0 = \mathbf{0}(v)$$

Thus, $F + (-1)F = \mathbf{0}$. Similarly $(-1)F + F = \mathbf{0}$. Hence, $-F = (-1)F$.

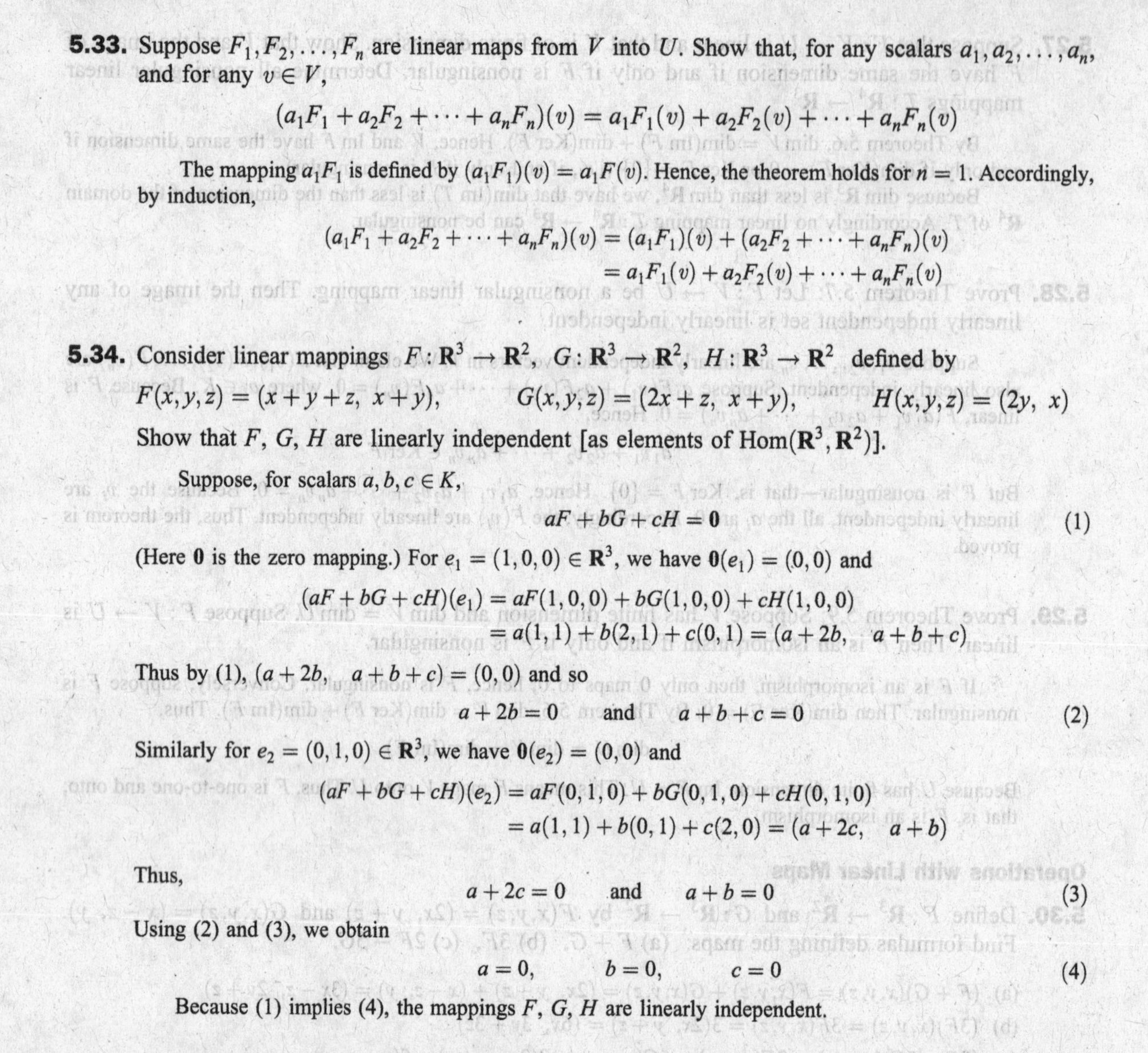

5.33. Suppose $F_1, F_2, \ldots, F_n$ are linear maps from V into U. Show that, for any scalars $a_1, a_2, \ldots, a_n$, and for any $v \in V$,

$$(a_1F_1 + a_2F_2 + \cdots + a_nF_n)(v) = a_1F_1(v) + a_2F_2(v) + \cdots + a_nF_n(v)$$

The mapping a_1F_1 is defined by $(a_1F_1)(v) = a_1F(v)$. Hence, the theorem holds for $n = 1$. Accordingly, by induction,

$$\begin{aligned}(a_1F_1 + a_2F_2 + \cdots + a_nF_n)(v) &= (a_1F_1)(v) + (a_2F_2 + \cdots + a_nF_n)(v) \\ &= a_1F_1(v) + a_2F_2(v) + \cdots + a_nF_n(v)\end{aligned}$$

5.34. Consider linear mappings $F: \mathbf{R}^3 \to \mathbf{R}^2$, $G: \mathbf{R}^3 \to \mathbf{R}^2$, $H: \mathbf{R}^3 \to \mathbf{R}^2$ defined by

$$F(x,y,z) = (x+y+z,\ x+y), \qquad G(x,y,z) = (2x+z,\ x+y), \qquad H(x,y,z) = (2y,\ x)$$

Show that F, G, H are linearly independent [as elements of $\text{Hom}(\mathbf{R}^3, \mathbf{R}^2)$].

Suppose, for scalars $a, b, c \in K$,

$$aF + bG + cH = \mathbf{0} \tag{1}$$

(Here $\mathbf{0}$ is the zero mapping.) For $e_1 = (1,0,0) \in \mathbf{R}^3$, we have $\mathbf{0}(e_1) = (0,0)$ and

$$\begin{aligned}(aF + bG + cH)(e_1) &= aF(1,0,0) + bG(1,0,0) + cH(1,0,0) \\ &= a(1,1) + b(2,1) + c(0,1) = (a+2b,\quad a+b+c)\end{aligned}$$

Thus by (1), $(a+2b, \quad a+b+c) = (0,0)$ and so

$$a + 2b = 0 \qquad \text{and} \qquad a+b+c = 0 \tag{2}$$

Similarly for $e_2 = (0,1,0) \in \mathbf{R}^3$, we have $\mathbf{0}(e_2) = (0,0)$ and

$$\begin{aligned}(aF + bG + cH)(e_2) &= aF(0,1,0) + bG(0,1,0) + cH(0,1,0) \\ &= a(1,1) + b(0,1) + c(2,0) = (a+2c,\quad a+b)\end{aligned}$$

Thus,

$$a + 2c = 0 \qquad \text{and} \qquad a + b = 0 \tag{3}$$

Using (2) and (3), we obtain

$$a = 0, \qquad b = 0, \qquad c = 0 \tag{4}$$

Because (1) implies (4), the mappings F, G, H are linearly independent.

5.35. Let k be a nonzero scalar. Show that a linear map T is singular if and only if kT is singular. Hence, T is singular if and only if $-T$ is singular.

Suppose T is singular. Then $T(v) = 0$ for some vector $v \neq 0$. Hence,

$$(kT)(v) = kT(v) = k0 = 0$$

and so kT is singular.

Now suppose kT is singular. Then $(kT)(w) = 0$ for some vector $w \neq 0$. Hence,

$$T(kw) = kT(w) = (kT)(w) = 0$$

But $k \neq 0$ and $w \neq 0$ implies $kw \neq 0$. Thus, T is also singular.

5.36. Find the dimension d of:

(a) $\text{Hom}(\mathbf{R}^3, \mathbf{R}^4)$, (b) $\text{Hom}(\mathbf{R}^5, \mathbf{R}^3)$, (c) $\text{Hom}(\mathbf{P}_3(t), \mathbf{R}^2)$, (d) $\text{Hom}(\mathbf{M}_{2,3}, \mathbf{R}^4)$.

Use $\dim[\text{Hom}(V, U)] = mn$, where $\dim V = m$ and $\dim U = n$.

(a) $d = 3(4) = 12$. (c) Because $\dim \mathbf{P}_3(t) = 4$, $d = 4(2) = 8$.

(b) $d = 5(3) = 15$. (d) Because $\dim \mathbf{M}_{2,3} = 6$, $d = 6(4) = 24$.

5.37. Prove Theorem 5.11. Suppose $\dim V = m$ and $\dim U = n$. Then $\dim[\text{Hom}(V, U)] = mn$.

Suppose $\{v_1, \ldots, v_m\}$ is a basis of V and $\{u_1, \ldots, u_n\}$ is a basis of U. By Theorem 5.2, a linear mapping in $\text{Hom}(V, U)$ is uniquely determined by arbitrarily assigning elements of U to the basis elements v_i of V. We define

$$F_{ij} \in \text{Hom}(V, U), \qquad i = 1, \ldots, m, \quad j = 1, \ldots, n$$

to be the linear mapping for which $F_{ij}(v_i) = u_j$, and $F_{ij}(v_k) = 0$ for $k \neq i$. That is, F_{ij} maps v_i into u_j and the other v's into 0. Observe that $\{F_{ij}\}$ contains exactly mn elements; hence, the theorem is proved if we show that it is a basis of $\text{Hom}(V, U)$.

Proof that $\{F_{ij}\}$ *generates* $\text{Hom}(V, U)$. Consider an arbitrary function $F \in \text{Hom}(V, U)$. Suppose $F(v_1) = w_1, F(v_2) = w_2, \ldots, F(v_m) = w_m$. Because $w_k \in U$, it is a linear combination of the u's; say,

$$w_k = a_{k1}u_1 + a_{k2}u_2 + \cdots + a_{kn}u_n, \qquad k = 1, \ldots, m, \quad a_{ij} \in K \tag{1}$$

Consider the linear mapping $G = \sum_{i=1}^{m} \sum_{j=1}^{n} a_{ij}F_{ij}$. Because G is a linear combination of the F_{ij}, the proof that $\{F_{ij}\}$ generates $\text{Hom}(V, U)$ is complete if we show that $F = G$.

We now compute $G(v_k), k = 1, \ldots, m$. Because $F_{ij}(v_k) = 0$ for $k \neq i$ and $F_{ki}(v_k) = u_i$,

$$\begin{aligned} G(v_k) &= \sum_{i=1}^{m}\sum_{j=1}^{n} a_{ij}F_{ij}(v_k) = \sum_{j=1}^{n} a_{kj}F_{kj}(v_k) = \sum_{j=1}^{n} a_{kj}u_j \\ &= a_{k1}u_1 + a_{k2}u_2 + \cdots + a_{kn}u_n \end{aligned}$$

Thus, by (1), $G(v_k) = w_k$ for each k. But $F(v_k) = w_k$ for each k. Accordingly, by Theorem 5.2, $F = G$; hence, $\{F_{ij}\}$ generates $\text{Hom}(V, U)$.

Proof that $\{F_{ij}\}$ *is linearly independent.* Suppose, for scalars $c_{ij} \in K$,

$$\sum_{i=1}^{m}\sum_{j=1}^{n} c_{ij}F_{ij} = \mathbf{0}$$

For $v_k, k = 1, \ldots, m$,

$$\begin{aligned} 0 = \mathbf{0}(v_k) &= \sum_{i=1}^{m}\sum_{j=1}^{n} c_{ij}F_{ij}(v_k) = \sum_{j=1}^{n} c_{kj}F_{kj}(v_k) = \sum_{j=1}^{n} c_{kj}u_j \\ &= c_{k1}u_1 + c_{k2}u_2 + \cdots + c_{kn}u_n \end{aligned}$$

But the u_i are linearly independent; hence, for $k = 1, \ldots, m$, we have $c_{k1} = 0, c_{k2} = 0, \ldots, c_{kn} = 0$. In other words, all the $c_{ij} = 0$, and so $\{F_{ij}\}$ is linearly independent.

5.38. Prove Theorem 5.12: (i) $G \circ (F + F') = G \circ F + G \circ F'$. (ii) $(G + G') \circ F = G \circ F + G' \circ F$. (iii) $k(G \circ F) = (kG) \circ F = G \circ (kF)$.

(i) For every $v \in V$,

$$\begin{aligned} (G \circ (F + F'))(v) &= G((F + F')(v)) = G(F(v) + F'(v)) \\ &= G(F(v)) + G(F'(v)) = (G \circ F)(v) + (G \circ F')(v) = (G \circ F + G \circ F')(v) \end{aligned}$$

Thus, $G \circ (F + F') = G \circ F + G \circ F'$.

(ii) For every $v \in V$,

$$\begin{aligned} ((G + G') \circ F)(v) &= (G + G')(F(v)) = G(F(v)) + G'(F(v)) \\ &= (G \circ F)(v) + (G' \circ F)(v) = (G \circ F + G' \circ F)(v) \end{aligned}$$

Thus, $(G + G') \circ F = G \circ F + G' \circ F$.

(iii) For every $v \in V$,

$$(k(G \circ F))(v) = k(G \circ F)(v) = k(G(F(v))) = (kG)(F(v)) = (kG \circ F)(v)$$

and

$$(k(G \circ F))(v) = k(G \circ F)(v) = k(G(F(v))) = G(kF(v)) = G((kF)(v)) = (G \circ kF)(v)$$

Accordingly, $k(G \circ F) = (kG) \circ F = G \circ (kF)$. (We emphasize that two mappings are shown to be equal by showing that each of them assigns the same image to each point in the domain.)

Algebra of Linear Maps

5.39. Let F and G be the linear operators on $\mathbf{R}^2$ defined by $F(x, y) = (y, x)$ and $G(x, y) = (0, x)$. Find formulas defining the following operators:
(a) $F + G$, (b) $2F - 3G$, (c) FG, (d) GF, (e) F^2, (f) G^2.

(a) $(F + G)(x, y) = F(x, y) + G(x, y) = (y, x) + (0, x) = (y, 2x)$.
(b) $(2F - 3G)(x, y) = 2F(x, y) - 3G(x, y) = 2(y, x) - 3(0, x) = (2y, -x)$.
(c) $(FG)(x, y) = F(G(x, y)) = F(0, x) = (x, 0)$.
(d) $(GF)(x, y) = G(F(x, y)) = G(y, x) = (0, y)$.
(e) $F^2(x, y) = F(F(x, y)) = F(y, x) = (x, y)$. (Note that $F^2 = I$, the identity mapping.)
(f) $G^2(x, y) = G(G(x, y)) = G(0, x) = (0, 0)$. (Note that $G^2 = \mathbf{0}$, the zero mapping.)

5.40. Consider the linear operator T on $\mathbf{R}^3$ defined by $T(x, y, z) = (2x,\ 4x - y,\ 2x + 3y - z)$.
(a) Show that T is invertible. Find formulas for (b) T^{-1}, (c) T^2, (d) T^{-2}.

(a) Let $W = \text{Ker}\, T$. We need only show that T is nonsingular (i.e., that $W = \{0\}$). Set $T(x, y, z) = (0, 0, 0)$, which yields

$$T(x, y, z) = (2x,\ 4x - y,\ 2x + 3y - z) = (0, 0, 0)$$

Thus, W is the solution space of the homogeneous system

$$2x = 0, \qquad 4x - y = 0, \qquad 2x + 3y - z = 0$$

which has only the trivial solution $(0, 0, 0)$. Thus, $W = \{0\}$. Hence, T is nonsingular, and so T is invertible.

(b) Set $T(x, y, z) = (r, s, t)$ [and so $T^{-1}(r, s, t) = (x, y, z)$]. We have

$$(2x,\ 4x - y,\ 2x + 3y - z) = (r, s, t) \quad \text{or} \quad 2x = r, \quad 4x - y = s, \quad 2x + 3y - z = t$$

Solve for x, y, z in terms of r, s, t to get $x = \frac{1}{2}r$, $y = 2r - s$, $z = 7r - 3s - t$. Thus,

$$T^{-1}(r, s, t) = (\tfrac{1}{2}r,\ 2r - s,\ 7r - 3s - t) \quad \text{or} \quad T^{-1}(x, y, z) = (\tfrac{1}{2}x,\ 2x - y,\ 7x - 3y - z)$$

(c) Apply T twice to get

$$\begin{aligned} T^2(x, y, z) &= T(2x,\ 4x - y,\ 2x + 3y - z) \\ &= [4x,\ \ 4(2x) - (4x - y),\ \ 2(2x) + 3(4x - y) - (2x + 3y - z)] \\ &= (4x,\ 4x + y,\ 14x - 6y + z) \end{aligned}$$

(d) Apply T^{-1} twice to get

$$\begin{aligned} T^{-1}(x, y, z) &= T^{-1}(\tfrac{1}{2}x,\ 2x - y,\ 7x - 3y - z) \\ &= [\tfrac{1}{4}x,\ \ 2(\tfrac{1}{2}x) - (2x - y),\ \ 7(\tfrac{1}{2}x) - 3(2x - y) - (7x - 3y - z)] \\ &= (\tfrac{1}{4}x,\ \ -x + y,\ \ -\tfrac{19}{2}x + 6y + z) \end{aligned}$$

5.41. Let V be of finite dimension and let T be a linear operator on V for which $TR = I$, for some operator R on V. (We call R a *right inverse* of T.)

(a) Show that T is invertible. (b) Show that $R = T^{-1}$.

(c) Give an example showing that the above need not hold if V is of infinite dimension.

(a) Let $\dim V = n$. By Theorem 5.14, T is invertible if and only if T is onto; hence, T is invertible if and only if $\text{rank}(T) = n$. We have $n = \text{rank}(I) = \text{rank}(TR) \leq \text{rank}(T) \leq n$. Hence, $\text{rank}(T) = n$ and T is invertible.

(b) $TT^{-1} = T^{-1}T = I$. Then $R = IR = (T^{-1}T)R = T^{-1}(TR) = T^{-1}I = T^{-1}$.

(c) Let V be the space of polynomials in t over K; say, $p(t) = a_0 + a_1t + a_2t^2 + \cdots + a_st^s$. Let T and R be the operators on V defined by

$$T(p(t)) = 0 + a_1 + a_2t + \cdots + a_st^{s-1} \quad \text{and} \quad R(p(t)) = a_0t + a_1t^2 + \cdots + a_st^{s+1}$$

We have

$$(TR)(p(t)) = T(R(p(t))) = T(a_0t + a_1t^2 + \cdots + a_st^{s+1}) = a_0 + a_1t + \cdots + a_st^s = p(t)$$

and so $TR = I$, the identity mapping. On the other hand, if $k \in K$ and $k \neq 0$, then

$$(RT)(k) = R(T(k)) = R(0) = 0 \neq k$$

Accordingly, $RT \neq I$.

5.42. Let F and G be linear operators on $\mathbf{R}^2$ defined by $F(x, y) = (0, x)$ and $G(x, y) = (x, 0)$. Show that (a) $GF = \mathbf{0}$, the zero mapping, but $FG \neq \mathbf{0}$. (b) $G^2 = G$.

(a) $(GF)(x, y) = G(F(x, y)) = G(0, x) = (0, 0)$. Because GF assigns $0 = (0, 0)$ to every vector (x, y) in $\mathbf{R}^2$, it is the zero mapping; that is, $GF = \mathbf{0}$.

On the other hand, $(FG)(x, y) = F(G(x, y)) = F(x, 0) = (0, x)$. For example, $(FG)(2, 3) = (0, 2)$. Thus, $FG \neq \mathbf{0}$, as it does not assign $0 = (0, 0)$ to every vector in $\mathbf{R}^2$.

(b) For any vector (x, y) in $\mathbf{R}^2$, we have $G^2(x, y) = G(G(x, y)) = G(x, 0) = (x, 0) = G(x, y)$. Hence, $G^2 = G$.

5.43. Find the dimension of (a) $A(\mathbf{R}^4)$, (b) $A(\mathbf{P}_2(t))$, (c) $A(\mathbf{M}_{2,3})$.

Use $\dim[A(V)] = n^2$ where $\dim V = n$. Hence, (a) $\dim[A(\mathbf{R}^4)] = 4^2 = 16$, (b) $\dim[A(\mathbf{P}_2(t))] = 3^2 = 9$, (c) $\dim[A(\mathbf{M}_{2,3})] = 6^2 = 36$.

5.44. Let E be a linear operator on V for which $E^2 = E$. (Such an operator is called a *projection*.) Let U be the image of E, and let W be the kernel. Prove

(a) If $u \in U$, then $E(u) = u$ (i.e., E is the identity mapping on U).

(b) If $E \neq I$, then E is singular—that is, $E(v) = 0$ for some $v \neq 0$.

(c) $V = U \oplus W$.

(a) If $u \in U$, the image of E, then $E(v) = u$ for some $v \in V$. Hence, using $E^2 = E$, we have

$$u = E(v) = E^2(v) = E(E(v)) = E(u)$$

(b) If $E \neq I$, then for some $v \in V$, $E(v) = u$, where $v \neq u$. By (i), $E(u) = u$. Thus,

$$E(v - u) = E(v) - E(u) = u - u = 0, \quad \text{where} \quad v - u \neq 0$$

(c) We first show that $V = U + W$. Let $v \in V$. Set $u = E(v)$ and $w = v - E(v)$. Then

$$v = E(v) + v - E(v) = u + w$$

By definition, $u = E(v) \in U$, the image of E. We now show that $w \in W$, the kernel of E,

$$E(w) = E(v - E(v)) = E(v) - E^2(v) = E(v) - E(v) = 0$$

and thus $w \in W$. Hence, $V = U + W$.

We next show that $U \cap W = \{0\}$. Let $v \in U \cap W$. Because $v \in U$, $E(v) = v$ by part (a). Because $v \in W$, $E(v) = 0$. Thus, $v = E(v) = 0$ and so $U \cap W = \{0\}$.

The above two properties imply that $V = U \oplus W$.

SUPPLEMENTARY PROBLEMS

Mappings

5.45. Determine the number of different mappings from (a) $\{1,2\}$ into $\{1,2,3\}$, (b) $\{1,2,\ldots,r\}$ into $\{1,2,\ldots,s\}$.

5.46. Let $f:\mathbf{R}\to\mathbf{R}$ and $g:\mathbf{R}\to\mathbf{R}$ be defined by $f(x)=x^2+3x+1$ and $g(x)=2x-3$. Find formulas defining the composition mappings: (a) $f\circ g$; (b) $g\circ f$; (c) $g\circ g$; (d) $f\circ f$.

5.47. For each mappings $f:\mathbf{R}\to\mathbf{R}$ find a formula for its inverse: (a) $f(x)=3x-7$, (b) $f(x)=x^3+2$.

5.48. For any mapping $f:A\to B$, show that $\mathbf{1}_B\circ f=f=f\circ\mathbf{1}_A$.

Linear Mappings

5.49. Show that the following mappings are linear:

(a) $F:\mathbf{R}^3\to\mathbf{R}^2$ defined by $F(x,y,z)=(x+2y-3z,\ \ 4x-5y+6z)$.
(b) $F:\mathbf{R}^2\to\mathbf{R}^2$ defined by $F(x,y)=(ax+by,\ \ cx+dy)$, where a, b, c, d belong to $\mathbf{R}$.

5.50. Show that the following mappings are not linear:

(a) $F:\mathbf{R}^2\to\mathbf{R}^2$ defined by $F(x,y)=(x^2,y^2)$.
(b) $F:\mathbf{R}^3\to\mathbf{R}^2$ defined by $F(x,y,z)=(x+1,\ \ y+z)$.
(c) $F:\mathbf{R}^2\to\mathbf{R}^2$ defined by $F(x,y)=(xy,y)$.
(d) $F:\mathbf{R}^3\to\mathbf{R}^2$ defined by $F(x,y,z)=(|x|,\ \ y+z)$.

5.51. Find $F(a,b)$, where the linear map $F:\mathbf{R}^2\to\mathbf{R}^2$ is defined by $F(1,2)=(3,-1)$ and $F(0,1)=(2,1)$.

5.52. Find a 2×2 matrix A that maps

(a) $(1,3)^T$ and $(1,4)^T$ into $(-2,5)^T$ and $(3,-1)^T$, respectively.
(b) $(2,-4)^T$ and $(-1,2)^T$ into $(1,1)^T$ and $(1,3)^T$, respectively.

5.53. Find a 2×2 singular matrix B that maps $(1,1)^T$ into $(1,3)^T$.

5.54. Let V be the vector space of real n-square matrices, and let M be a fixed nonzero matrix in V. Show that the first two of the following mappings $T:V\to V$ are linear, but the third is not:
(a) $T(A)=MA$, (b) $T(A)=AM+MA$, (c) $T(A)=M+A$.

5.55. Give an example of a nonlinear map $F:\mathbf{R}^2\to\mathbf{R}^2$ such that $F^{-1}(0)=\{0\}$ but F is not one-to-one.

5.56. Let $F:\mathbf{R}^2\to\mathbf{R}^2$ be defined by $F(x,y)=(3x+5y,\ \ 2x+3y)$, and let S be the unit circle in $\mathbf{R}^2$. (S consists of all points satisfying $x^2+y^2=1$.) Find (a) the image $F(S)$, (b) the preimage $F^{-1}(S)$.

5.57. Consider the linear map $G:\mathbf{R}^3\to\mathbf{R}^3$ defined by $G(x,y,z)=(x+y+z,\ \ y-2z,\ \ y-3z)$ and the unit sphere S_2 in $\mathbf{R}^3$, which consists of the points satisfying $x^2+y^2+z^2=1$. Find (a) $G(S_2)$, (b) $G^{-1}(S_2)$.

5.58. Let H be the plane $x+2y-3z=4$ in $\mathbf{R}^3$ and let G be the linear map in Problem 5.57. Find (a) $G(H)$, (b) $G^{-1}(H)$.

5.59. Let W be a subspace of V. The *inclusion* map, denoted by $i:W\hookrightarrow V$, is defined by $i(w)=w$ for every $w\in W$. Show that the inclusion map is linear.

5.60. Suppose $F:V\to U$ is linear. Show that $F(-v)=-F(v)$.

Kernel and Image of Linear Mappings

5.61. For each linear map F find a basis and the dimension of the kernel and the image of F:

(a) $F:\mathbf{R}^3\to\mathbf{R}^3$ defined by $F(x,y,z)=(x+2y-3z,\ \ 2x+5y-4z,\ \ x+4y+z)$,
(b) $F:\mathbf{R}^4\to\mathbf{R}^3$ defined by $F(x,y,z,t)=(x+2y+3z+2t,\ \ 2x+4y+7z+5t,\ \ x+2y+6z+5t)$.

5.62. For each linear map G, find a basis and the dimension of the kernel and the image of G:

(a) $G: \mathbf{R}^3 \to \mathbf{R}^2$ defined by $G(x, y, z) = (x + y + z,\ 2x + 2y + 2z)$,
(b) $G: \mathbf{R}^3 \to \mathbf{R}^2$ defined by $G(x, y, z) = (x + y,\ y + z)$,
(c) $G: \mathbf{R}^5 \to \mathbf{R}^3$ defined by

$$G(x, y, z, s, t) = (x + 2y + 2z + s + t,\quad x + 2y + 3z + 2s - t,\quad 3x + 6y + 8z + 5s - t).$$

5.63. Each of the following matrices determines a linear map from $\mathbf{R}^4$ into $\mathbf{R}^3$:

$$\text{(a)}\quad A = \begin{bmatrix} 1 & 2 & 0 & 1 \\ 2 & -1 & 2 & -1 \\ 1 & -3 & 2 & -2 \end{bmatrix}, \quad \text{(b)}\quad B = \begin{bmatrix} 1 & 0 & 2 & -1 \\ 2 & 3 & -1 & 1 \\ -2 & 0 & -5 & 3 \end{bmatrix}.$$

Find a basis as well as the dimension of the kernel and the image of each linear map.

5.64. Find a linear mapping $F: \mathbf{R}^3 \to \mathbf{R}^3$ whose image is spanned by $(1, 2, 3)$ and $(4, 5, 6)$.

5.65. Find a linear mapping $G: \mathbf{R}^4 \to \mathbf{R}^3$ whose kernel is spanned by $(1, 2, 3, 4)$ and $(0, 1, 1, 1)$.

5.66. Let $V = \mathbf{P}_{10}(t)$, the vector space of polynomials of degree ≤ 10. Consider the linear map $\mathbf{D}^4: V \to V$, where $\mathbf{D}^4$ denotes the fourth derivative $d^4(f)/dt^4$. Find a basis and the dimension of (a) the image of $\mathbf{D}^4$; (b) the kernel of $\mathbf{D}^4$.

5.67. Suppose $F: V \to U$ is linear. Show that (a) the image of any subspace of V is a subspace of U; (b) the preimage of any subspace of U is a subspace of V.

5.68. Show that if $F: V \to U$ is onto, then $\dim U \leq \dim V$. Determine all linear maps $F: \mathbf{R}^3 \to \mathbf{R}^4$ that are onto.

5.69. Consider the zero mapping $\mathbf{0}: V \to U$ defined by $\mathbf{0}(v) = 0, \forall\, v \in V$. Find the kernel and the image of $\mathbf{0}$.

Operations with linear Mappings

5.70. Let $F: \mathbf{R}^3 \to \mathbf{R}^2$ and $G: \mathbf{R}^3 \to \mathbf{R}^2$ be defined by $F(x, y, z) = (y,\ x + z)$ and $G(x, y, z) = (2z,\ x - y)$. Find formulas defining the mappings $F + G$ and $3F - 2G$.

5.71. Let $H: \mathbf{R}^2 \to \mathbf{R}^2$ be defined by $H(x, y) = (y, 2x)$. Using the maps F and G in Problem 5.70, find formulas defining the mappings: (a) $H \circ F$ and $H \circ G$, (b) $F \circ H$ and $G \circ H$, (c) $H \circ (F + G)$ and $H \circ F + H \circ G$.

5.72. Show that the following mappings F, G, H are linearly independent:

(a) $F, G, H \in \mathrm{Hom}(\mathbf{R}^2, \mathbf{R}^2)$ defined by $F(x, y) = (x, 2y)$, $\quad G(x, y) = (y,\ x + y)$, $\quad H(x, y) = (0, x)$,
(b) $F, G, H \in \mathrm{Hom}(\mathbf{R}^3, \mathbf{R})$ defined by $F(x, y, z) = x + y + z$, $\quad G(x, y, z) = y + z$, $\quad H(x, y, z) = x - z$.

5.73. For $F, G \in \mathrm{Hom}(V, U)$, show that $\mathrm{rank}(F + G) \leq \mathrm{rank}(F) + \mathrm{rank}(G)$. (Here V has finite dimension.)

5.74. Let $F: V \to U$ and $G: U \to V$ be linear. Show that if F and G are nonsingular, then $G \circ F$ is nonsingular. Give an example where $G \circ F$ is nonsingular but G is not. [Hint: Let $\dim V < \dim U$.]

5.75. Find the dimension d of (a) $\mathrm{Hom}(\mathbf{R}^2, \mathbf{R}^8)$, (b) $\mathrm{Hom}(\mathbf{P}_4(t), \mathbf{R}^3)$, (c) $\mathrm{Hom}(\mathbf{M}_{2,4}, \mathbf{P}_2(t))$.

5.76. Determine whether or not each of the following linear maps is nonsingular. If not, find a nonzero vector v whose image is 0; otherwise find a formula for the inverse map:

(a) $F: \mathbf{R}^3 \to \mathbf{R}^3$ defined by $F(x, y, z) = (x + y + z,\quad 2x + 3y + 5z,\quad x + 3y + 7z)$,
(b) $G: \mathbf{R}^3 \to \mathbf{P}_2(t)$ defined by $G(x, y, z) = (x + y)t^2 + (x + 2y + 2z)t + y + z$,
(c) $H: \mathbf{R}^2 \to \mathbf{P}_2(t)$ defined by $H(x, y) = (x + 2y)t^2 + (x - y)t + x + y$.

5.77. When can $\dim\,[\mathrm{Hom}(V, U)] = \dim V$?

Algebra of Linear Operators

5.78. Let F and G be the linear operators on $\mathbf{R}^2$ defined by $F(x,y) = (x+y,\ 0)$ and $G(x,y) = (-y,x)$. Find formulas defining the linear operators: (a) $F+G$, (b) $5F-3G$, (c) FG, (d) GF, (e) F^2, (f) G^2.

5.79. Show that each linear operator T on $\mathbf{R}^2$ is nonsingular and find a formula for T^{-1}, where (a) $T(x,y) = (x+2y,\ 2x+3y)$, (b) $T(x,y) = (2x-3y,\ 3x-4y)$.

5.80. Show that each of the following linear operators T on $\mathbf{R}^3$ is nonsingular and find a formula for T^{-1}, where (a) $T(x,y,z) = (x-3y-2z,\ y-4z,\ z)$; (b) $T(x,y,z) = (x+z,\ x-y,\ y)$.

5.81. Find the dimension of $A(V)$, where (a) $V = \mathbf{R}^7$, (b) $V = \mathbf{P}_5(t)$, (c) $V = \mathbf{M}_{3,4}$.

5.82. Which of the following integers can be the dimension of an algebra $A(V)$ of linear maps: 5, 9, 12, 25, 28, 36, 45, 64, 88, 100?

5.83. Let T be the linear operator on $\mathbf{R}^2$ defined by $T(x,y) = (x+2y,\ 3x+4y)$. Find a formula for $f(T)$, where (a) $f(t) = t^2+2t-3$, (b) $f(t) = t^2-5t-2$.

Miscellaneous Problems

5.84. Suppose $F: V \to U$ is linear and k is a nonzero scalar. Prove that the maps F and kF have the same kernel and the same image.

5.85. Suppose F and G are linear operators on V and that F is nonsingular. Assume that V has finite dimension. Show that $\operatorname{rank}(FG) = \operatorname{rank}(GF) = \operatorname{rank}(G)$.

5.86. Suppose V has finite dimension. Suppose T is a linear operator on V such that $\operatorname{rank}(T^2) = \operatorname{rank}(T)$. Show that $\operatorname{Ker} T \cap \operatorname{Im} T = \{0\}$.

5.87. Suppose $V = U \oplus W$. Let E_1 and E_2 be the linear operators on V defined by $E_1(v) = u$, $E_2(v) = w$, where $v = u+w$, $u \in U$, $w \in W$. Show that (a) $E_1^2 = E_1$ and $E_2^2 = E_2$ (i.e., that E_1 and E_2 are projections); (b) $E_1 + E_2 = I$, the identity mapping; (c) $E_1E_2 = \mathbf{0}$ and $E_2E_1 = \mathbf{0}$.

5.88. Let E_1 and E_2 be linear operators on V satisfying parts (a), (b), (c) of Problem 5.88. Prove

$$V = \operatorname{Im} E_1 \oplus \operatorname{Im} E_2$$

5.89. Let v and w be elements of a real vector space V. The *line segment* L from v to $v+w$ is defined to be the set of vectors $v+tw$ for $0 \le t \le 1$. (See Fig. 5.6.)

(a) Show that the line segment L between vectors v and u consists of the points:
(i) $(1-t)v + tu$ for $0 \le t \le 1$, (ii) $t_1v + t_2u$ for $t_1 + t_2 = 1$, $t_1 \ge 0$, $t_2 \ge 0$.

(b) Let $F: V \to U$ be linear. Show that the image $F(L)$ of a line segment L in V is a line segment in U.

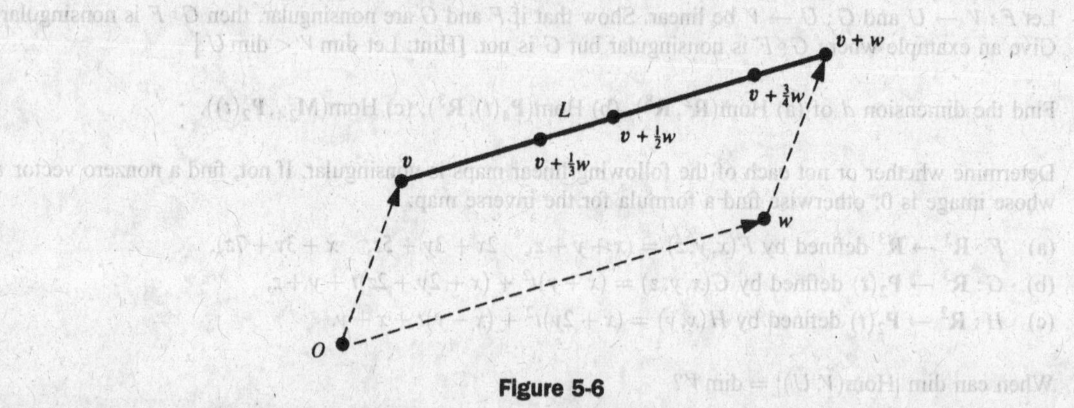

Figure 5-6

5.90. Let $F: V \to U$ be linear and let W be a subspace of V. The *restriction* of F to W is the map $F|W: W \to U$ defined by $F|W(v) = F(v)$ for every v in W. Prove the following:
(a) $F|W$ is linear; (b) $\text{Ker}(F|W) = (\text{Ker } F) \cap W$; (c) $\text{Im}(F|W) = F(W)$.

5.91. A subset X of a vector space V is said to be *convex* if the line segment L between any two points (vectors) $P, Q \in X$ is contained in X. (a) Show that the intersection of convex sets is convex; (b) suppose $F: V \to U$ is linear and X is convex. Show that $F(X)$ is convex.

ANSWERS TO SUPPLEMENTARY PROBLEMS

5.45. (a) $3^2 = 9$, (b) s^r

5.46. (a) $(f \circ g)(x) = 4x^2 + 1$, (b) $(g \circ f)(x) = 2x^2 + 6x - 1$, (c) $(g \circ g)(x) = 4x - 9$,
(d) $(f \circ f)(x) = x^4 + 6x^3 + 14x^2 + 15x + 5$

5.47. (a) $f^{-1}(x) = \frac{1}{3}(x + 7)$, (b) $f^{-1}(x) = \sqrt[3]{x - 2}$

5.49. $F(x,y,z) = A(x,y,z)^T$, where (a) $A = \begin{bmatrix} 1 & 2 & -3 \\ 4 & -5 & 6 \end{bmatrix}$, (b) $A = \begin{bmatrix} a & b \\ c & d \end{bmatrix}$

5.50. (a) $u = (2,2)$, $k = 3$; then $F(ku) = (36,36)$ but $kF(u) = (12,12)$; (b) $F(0) \neq 0$;
(c) $u = (1,2)$, $v = (3,4)$; then $F(u + v) = (24,6)$ but $F(u) + F(v) = (14,6)$;
(d) $u = (1,2,3)$, $k = -2$; then $F(ku) = (2,-10)$ but $kF(u) = (-2,-10)$.

5.51. $F(a,b) = (-a + 2b, \ -3a + b)$

5.52. (a) $A = \begin{bmatrix} -17 & 5 \\ 23 & -6 \end{bmatrix}$; (b) None. $(2,-4)$ and $(-1,2)$ are linearly dependent but not $(1,1)$ and $(1,3)$.

5.53. $B = \begin{bmatrix} 1 & 0 \\ 3 & 0 \end{bmatrix}$ [*Hint*: Send $(0,1)^T$ into $(0,0)^T$.]

5.55. $F(x,y) = (x^2, y^2)$

5.56. (a) $13x^2 - 42xy + 34y^2 = 1$, (b) $13x^2 + 42xy + 34y^2 = 1$

5.57. (a) $x^2 - 8xy + 26y^2 + 6xz - 38yz + 14z^2 = 1$, (b) $x^2 + 2xy + 3y^2 + 2xz - 8yz + 14z^2 = 1$

5.58. (a) $x - y + 2z = 4$, (b) $x + 6z = 4$

5.61. (a) $\dim(\text{Ker } F) = 1$, $\{(7,-2,1)\}$; $\dim(\text{Im } F) = 2$, $\{(1,2,1), \ (0,1,2)\}$;
(b) $\dim(\text{Ker } F) = 2$, $\{(-2,1,0,0), \ (1,0,-1,1)\}$; $\dim(\text{Im } F) = 2$, $\{(1,2,1), \ (0,1,3)\}$

5.62. (a) $\dim(\text{Ker } G) = 2$, $\{(1,0,-1), \ (1,-1,0)\}$; $\dim(\text{Im } G) = 1$, $\{(1,2)\}$;
(b) $\dim(\text{Ker } G) = 1$, $\{(1,-1,1)\}$; $\text{Im } G = \mathbf{R}^2$, $\{(1,0), \ (0,1)\}$;
(c) $\dim(\text{Ker } G) = 3$, $\{(-2,1,0,0,0), \ (1,0,-1,1,0), \ (-5,0,2,0,1)\}$; $\dim(\text{Im } G) = 2$, $\{(1,1,3), \ (0,1,2)\}$

5.63. (a) $\dim(\text{Ker } A) = 2$, $\{(4,-2,-5,0), \ (1,-3,0,5)\}$; $\dim(\text{Im } A) = 2$, $\{(1,2,1), \ (0,1,1)\}$;
(b) $\dim(\text{Ker } B) = 1$, $\{(-1, \frac{2}{3}, 1, 1)\}$; $\text{Im } B = \mathbf{R}^3$

5.64. $F(x,y,z) = (x + 4y, \ 2x + 5y, \ 3x + 6y)$

5.65. $F(x,y,z,t) = (x+y-z, \ 2x+y-t, 0)$

5.66. (a) $\{1, t, t^2, \ldots, t^6\}$, (b) $\{1, t, t^2, t^3\}$

5.68. None, because $\dim \mathbf{R}^4 > \dim \mathbf{R}^3$.

5.69. $\mathrm{Ker}\, \mathbf{0} = V$, $\mathrm{Im}\, \mathbf{0} = \{0\}$

5.70. $(F+G)(x,y,z) = (y+2z, \ 2x-y+z)$, $(3F-2G)(x,y,z) = (3y-4z, \ x+2y+3z)$

5.71. (a) $(H \circ F)(x,y,z) = (x+z, \quad 2y)$, $(H \circ G)(x,y,z) = (x-y, \quad 4z)$; (b) not defined;
(c) $(H \circ (F+G))(x,y,z) = (H \circ F + H \circ G)(x,y,z) = (2x-y+z, \quad 2y+4z)$

5.74. $F(x,y) = (x,y,y)$, $G(x,y,z) = (x,y)$

5.75. (a) 16, (b) 15, (c) 24

5.76. (a) $v = (2,-3,1)$; (b) $G^{-1}(at^2+bt+c) = (b-2c, \quad a-b+2c, \quad -a+b-c)$;
(c) H is nonsingular, but not invertible, because $\dim \mathbf{P}_2(t) > \dim \mathbf{R}^2$.

5.77. $\dim U = 1$; that is, $U = K$.

5.78. (a) $(F+G)(x,y) = (x,x)$; (b) $(5F-3G)(x,y) = (5x+8y, \quad -3x)$; (c) $(FG)(x,y) = (x-y, \ 0)$;
(d) $(GF)(x,y) = (0, \ x+y)$; (e) $F^2(x,y) = (x+y, \quad 0)$ (note that $F^2 = F$); (f) $G^2(x,y) = (-x, \quad -y)$.
[Note that $G^2 + I = 0$; hence, G is a zero of $f(t) = t^2 + 1$.]

5.79. (a) $T^{-1}(x,y) = (-3x+2y, \ 2x-y)$, (b) $T^{-1}(x,y) = (-4x+3y, \ -3x+2y)$

5.80. (a) $T^{-1}(x,y,z) = (x+3y+14z, \ y-4z, \ z)$, (b) $T^{-1}(x,y,z) = (y+z, \ y, \ x-y-z)$

5.81. (a) 49, (b) 36, (c) 144

5.82. Squares: 9, 25, 36, 64, 100

5.83. (a) $T(x,y) = (6x+14y, \quad 21x+27y)$; (b) $T(x,y) = (0,0)$—that is, $f(T) = 0$

CHAPTER 6

Linear Mappings and Matrices

6.1 Introduction

Consider a basis $S = \{u_1, u_2, \ldots, u_n\}$ of a vector space V over a field K. For any vector $v \in V$, suppose

$$v = a_1u_1 + a_2u_2 + \cdots + a_nu_n$$

Then the coordinate vector of v relative to the basis S, which we assume to be a column vector (unless otherwise stated or implied), is denoted and defined by

$$[v]_S = [a_1, a_2, \ldots, a_n]^T$$

Recall (Section 4.11) that the mapping $v \mapsto [v]_S$, determined by the basis S, is an isomorphism between V and K^n.

This chapter shows that there is also an isomorphism, determined by the basis S, between the algebra $A(V)$ of linear operators on V and the algebra M of n-square matrices over K. Thus, every linear mapping $F: V \to V$ will correspond to an n-square matrix $[F]_S$ determined by the basis S. We will also show how our matrix representation changes when we choose another basis.

6.2 Matrix Representation of a Linear Operator

Let T be a linear operator (transformation) from a vector space V into itself, and suppose $S = \{u_1, u_2, \ldots, u_n\}$ is a basis of V. Now $T(u_1), T(u_2), \ldots, T(u_n)$ are vectors in V, and so each is a linear combination of the vectors in the basis S; say,

$$\begin{aligned}
T(u_1) &= a_{11}u_1 + a_{12}u_2 + \cdots + a_{1n}u_n \\
T(u_2) &= a_{21}u_1 + a_{22}u_2 + \cdots + a_{2n}u_n \\
&\cdots\cdots\cdots\cdots\cdots\cdots\cdots\cdots \\
T(u_n) &= a_{n1}u_1 + a_{n2}u_2 + \cdots + a_{nn}u_n
\end{aligned}$$

The following definition applies.

DEFINITION: The transpose of the above matrix of coefficients, denoted by $m_S(T)$ or $[T]_S$, is called the *matrix representation* of T relative to the basis S, or simply the matrix of T in the basis S. (The subscript S may be omitted if the basis S is understood.)

Using the coordinate (column) vector notation, the matrix representation of T may be written in the form

$$m_S(T) = [T]_S = \big[[T(u_1)]_S,\ [T(u_2)]_S,\ \ldots,\ [T(u_1)]_S\big]$$

That is, the columns of $m(T)$ are the coordinate vectors of $T(u_1), T(u_2), \ldots, T(u_n)$, respectively.

EXAMPLE 6.1 Let $F: \mathbf{R}^2 \to \mathbf{R}^2$ be the linear operator defined by $F(x,y) = (2x + 3y,\ 4x - 5y)$.

(a) Find the matrix representation of F relative to the basis $S = \{u_1, u_2\} = \{(1,2),\ (2,5)\}$.

(1) First find $F(u_1)$, and then write it as a linear combination of the basis vectors u_1 and u_2. (For notational convenience, we use column vectors.) We have

$$F(u_1) = F\left(\begin{bmatrix}1\\2\end{bmatrix}\right) = \begin{bmatrix}8\\-6\end{bmatrix} = x\begin{bmatrix}1\\2\end{bmatrix} + y\begin{bmatrix}2\\5\end{bmatrix} \quad \text{and} \quad \begin{aligned} x + 2y &= 8 \\ 2x + 5y &= -6 \end{aligned}$$

Solve the system to obtain $x = 52$, $y = -22$. Hence, $F(u_1) = 52u_1 - 22u_2$.

(2) Next find $F(u_2)$, and then write it as a linear combination of u_1 and u_2:

$$F(u_2) = F\left(\begin{bmatrix}2\\5\end{bmatrix}\right) = \begin{bmatrix}19\\-17\end{bmatrix} = x\begin{bmatrix}1\\2\end{bmatrix} + y\begin{bmatrix}2\\5\end{bmatrix} \quad \text{and} \quad \begin{aligned} x + 2y &= 19 \\ 2x + 5y &= -17 \end{aligned}$$

Solve the system to get $x = 129$, $y = -55$. Thus, $F(u_2) = 129u_1 - 55u_2$.

Now write the coordinates of $F(u_1)$ and $F(u_2)$ as columns to obtain the matrix

$$[F]_S = \begin{bmatrix} 52 & 129 \\ -22 & -55 \end{bmatrix}$$

(b) Find the matrix representation of F relative to the (usual) basis $E = \{e_1, e_2\} = \{(1,0),\ (0,1)\}$.

Find $F(e_1)$ and write it as a linear combination of the usual basis vectors e_1 and e_2, and then find $F(e_2)$ and write it as a linear combination of e_1 and e_2. We have

$$\begin{aligned} F(e_1) &= F(1,0) = (2,2) \quad = 2e_1 + 4e_2 \\ F(e_2) &= F(0,1) = (3,-5) = 3e_1 - 5e_2 \end{aligned} \quad \text{and so} \quad [F]_E = \begin{bmatrix} 2 & 3 \\ 4 & -5 \end{bmatrix}$$

Note that the coordinates of $F(e_1)$ and $F(e_2)$ form the columns, not the rows, of $[F]_E$. Also, note that the arithmetic is much simpler using the usual basis of $\mathbf{R}^2$.

EXAMPLE 6.2 Let V be the vector space of functions with basis $S = \{\sin t, \cos t, e^{3t}\}$, and let $\mathbf{D}: V \to V$ be the differential operator defined by $\mathbf{D}(f(t)) = d(f(t))/dt$. We compute the matrix representing $\mathbf{D}$ in the basis S:

$$\begin{aligned} \mathbf{D}(\sin t) &= \quad \cos t = \quad 0(\sin t) + 1(\cos t) + 0(e^3 t) \\ \mathbf{D}(\cos\ t) &= -\sin t = -1(\sin t) + 0(\cos t) + 0(e^{3t}) \\ \mathbf{D}(e^{3t}) &= \quad 3e^{3t} = \quad 0(\sin t) + 0(\cos t) + 3(e^{3t}) \end{aligned}$$

and so

$$[\mathbf{D}] = \begin{bmatrix} 0 & -1 & 0 \\ 1 & 0 & 0 \\ 0 & 0 & 3 \end{bmatrix}$$

Note that the coordinates of $\mathbf{D}(\sin t)$, $\mathbf{D}(\cos t)$, $\mathbf{D}(e^{3t})$ form the columns, not the rows, of $[\mathbf{D}]$.

Matrix Mappings and Their Matrix Representation

Consider the following matrix A, which may be viewed as a linear operator on $\mathbf{R}^2$, and basis S of $\mathbf{R}^2$:

$$A = \begin{bmatrix} 3 & -2 \\ 4 & -5 \end{bmatrix} \quad \text{and} \quad S = \{u_1, u_2\} = \left\{\begin{bmatrix}1\\2\end{bmatrix}, \begin{bmatrix}2\\5\end{bmatrix}\right\}$$

(We write vectors as columns, because our map is a matrix.) We find the matrix representation of A relative to the basis S.

(1) First we write $A(u_1)$ as a linear combination of u_1 and u_2. We have

$$A(u_1) = \begin{bmatrix} 3 & -2 \\ 4 & -5 \end{bmatrix}\begin{bmatrix} 1 \\ 2 \end{bmatrix} = \begin{bmatrix} -1 \\ -6 \end{bmatrix} = x\begin{bmatrix} 1 \\ 2 \end{bmatrix} + y\begin{bmatrix} 2 \\ 5 \end{bmatrix} \quad \text{and so} \quad \begin{aligned} x + 2y &= -1 \\ 2x + 5y &= -6 \end{aligned}$$

Solving the system yields $x = 7$, $y = -4$. Thus, $A(u_1) = 7u_1 - 4u_2$.

(2) Next we write $A(u_2)$ as a linear combination of u_1 and u_2. We have

$$A(u_2) = \begin{bmatrix} 3 & -2 \\ 4 & -5 \end{bmatrix}\begin{bmatrix} 2 \\ 5 \end{bmatrix} = \begin{bmatrix} -4 \\ -17 \end{bmatrix} = x\begin{bmatrix} 1 \\ 2 \end{bmatrix} + y\begin{bmatrix} 2 \\ 5 \end{bmatrix} \quad \text{and so} \quad \begin{aligned} x + 2y &= -4 \\ 2x + 5y &= -17 \end{aligned}$$

Solving the system yields $x = 14$, $y = -9$. Thus, $A(u_2) = 14u_1 - 9u_2$. Writing the coordinates of $A(u_1)$ and $A(u_2)$ as columns gives us the following matrix representation of A:

$$[A]_S = \begin{bmatrix} 7 & 14 \\ -4 & -9 \end{bmatrix}$$

Remark: Suppose we want to find the matrix representation of A relative to the usual basis $E = \{e_1, e_2\} = \{[1, 0]^T, [0, 1]^T\}$ of $\mathbf{R}^2$. We have

$$\begin{aligned} A(e_1) &= \begin{bmatrix} 3 & -2 \\ 4 & -5 \end{bmatrix}\begin{bmatrix} 1 \\ 0 \end{bmatrix} = \begin{bmatrix} 3 \\ 4 \end{bmatrix} = 3e_1 + 4e_2 \\ A(e_2) &= \begin{bmatrix} 3 & -2 \\ 4 & -5 \end{bmatrix}\begin{bmatrix} 0 \\ 1 \end{bmatrix} = \begin{bmatrix} -2 \\ -5 \end{bmatrix} = -2e_1 - 5e_2 \end{aligned} \quad \text{and so} \quad [A]_E = \begin{bmatrix} 3 & -2 \\ 4 & -5 \end{bmatrix}$$

Note that $[A]_E$ is the original matrix A. This result is true in general:

> The matrix representation of any $n \times n$ square matrix A over a field K relative to the usual basis E of K^n is the matrix A itself; that is,
>
> $$[A]_E = A$$

Algorithm for Finding Matrix Representations

Next follows an algorithm for finding matrix representations. The first Step 0 is optional. It may be useful to use it in Step 1(b), which is repeated for each basis vector.

ALGORITHM 6.1: The input is a linear operator T on a vector space V and a basis $S = \{u_1, u_2, \ldots, u_n\}$ of V. The output is the matrix representation $[T]_S$.

Step 0. Find a formula for the coordinates of an arbitrary vector v relative to the basis S.

Step 1. Repeat for each basis vector u_k in S:

(a) Find $T(u_k)$.

(b) Write $T(u_k)$ as a linear combination of the basis vectors $u_1, u_2, \ldots, u_n$.

Step 2. Form the matrix $[T]_S$ whose columns are the coordinate vectors in Step 1(b).

EXAMPLE 6.3 Let $F: \mathbf{R}^2 \to \mathbf{R}^2$ be defined by $F(x, y) = (2x + 3y,\ 4x - 5y)$. Find the matrix representation $[F]_S$ of F relative to the basis $S = \{u_1, u_2\} = \{(1, -2),\ (2, -5)\}$.

(Step 0) First find the coordinates of $(a, b) \in \mathbf{R}^2$ relative to the basis S. We have

$$\begin{bmatrix} a \\ b \end{bmatrix} = x\begin{bmatrix} 1 \\ -2 \end{bmatrix} + y\begin{bmatrix} 2 \\ -5 \end{bmatrix} \quad \text{or} \quad \begin{aligned} x + 2y &= a \\ -2x - 5y &= b \end{aligned} \quad \text{or} \quad \begin{aligned} x + 2y &= a \\ -y &= 2a + b \end{aligned}$$

Solving for x and y in terms of a and b yields $x = 5a + 2b$, $y = -2a - b$. Thus,

$$(a, b) = (5a + 2b)u_1 + (-2a - b)u_2$$

(Step 1) Now we find $F(u_1)$ and write it as a linear combination of u_1 and u_2 using the above formula for (a, b), and then we repeat the process for $F(u_2)$. We have

$$\begin{aligned} F(u_1) &= F(1, -2) = (-4, 14) = 8u_1 - 6u_2 \\ F(u_2) &= F(2, -5) = (-11, 33) = 11u_1 - 11u_2 \end{aligned}$$

(Step 2) Finally, we write the coordinates of $F(u_1)$ and $F(u_2)$ as columns to obtain the required matrix:

$$[F]_S = \begin{bmatrix} 8 & 11 \\ -6 & -11 \end{bmatrix}$$

Properties of Matrix Representations

This subsection gives the main properties of the matrix representations of linear operators T on a vector space V. We emphasize that we are always given a particular basis S of V.

Our first theorem, proved in Problem 6.9, tells us that the "action" of a linear operator T on a vector v is preserved by its matrix representation.

THEOREM 6.1: Let $T: V \to V$ be a linear operator, and let S be a (finite) basis of V. Then, for any vector v in V, $[T]_S[v]_S = [T(v)]_S$.

EXAMPLE 6.4 Consider the linear operator F on R^2 and the basis S of Example 6.3; that is,

$$F(x, y) = (2x + 3y, \quad 4x - 5y) \qquad \text{and} \qquad S = \{u_1, u_2\} = \{(1, -2), \quad (2, -5)\}$$

Let

$$v = (5, -7), \qquad \text{and so} \qquad F(v) = (-11, 55)$$

Using the formula from Example 6.3, we get

$$[v] = [11, -3]^T \qquad \text{and} \qquad [F(v)] = [55, -33]^T$$

We verify Theorem 6.1 for this vector v (where $[F]$ is obtained from Example 6.3):

$$[F][v] = \begin{bmatrix} 8 & 11 \\ -6 & -11 \end{bmatrix} \begin{bmatrix} 11 \\ -3 \end{bmatrix} = \begin{bmatrix} 55 \\ -33 \end{bmatrix} = [F(v)]$$

Given a basis S of a vector space V, we have associated a matrix $[T]$ to each linear operator T in the algebra $A(V)$ of linear operators on V. Theorem 6.1 tells us that the "action" of an individual linear operator T is preserved by this representation. The next two theorems (proved in Problems 6.10 and 6.11) tell us that the three basic operations in $A(V)$ with these operators—namely (i) addition, (ii) scalar multiplication, and (iii) composition—are also preserved.

THEOREM 6.2: Let V be an n-dimensional vector space over K, let S be a basis of V, and let $\mathbf{M}$ be the algebra of $n \times n$ matrices over K. Then the mapping

$$m: A(V) \to \mathbf{M} \qquad \text{defined by} \qquad m(T) = [T]_S$$

is a vector space isomorphism. That is, for any $F, G \in A(V)$ and any $k \in K$,

(i) $m(F + G) = m(F) + m(G)$ or $[F + G] = [F] + [G]$

(ii) $m(kF) = km(F)$ or $[kF] = k[F]$

(iii) m is bijective (one-to-one and onto).

THEOREM 6.3: For any linear operators $F, G \in A(V)$,

$$m(G \circ F) = m(G)m(F) \quad \text{or} \quad [G \circ F] = [G][F]$$

(Here $G \circ F$ denotes the composition of the maps G and F.)

6.3 Change of Basis

Let V be an n-dimensional vector space over a field K. We have shown that once we have selected a basis S of V, every vector $v \in V$ can be represented by means of an n-tuple $[v]_S$ in K^n, and every linear operator T in $A(V)$ can be represented by an $n \times n$ matrix over K. We ask the following natural question:

How do our representations change if we select another basis?

In order to answer this question, we first need a definition.

DEFINITION: Let $S = \{u_1, u_2, \ldots, u_n\}$ be a basis of a vector space V, and let $S' = \{v_1, v_2, \ldots, v_n\}$ be another basis. (For reference, we will call S the "old" basis and S' the "new" basis.) Because S is a basis, each vector in the "new" basis S' can be written uniquely as a linear combination of the vectors in S; say,

$$\begin{aligned} v_1 &= a_{11}u_1 + a_{12}u_2 + \cdots + a_{1n}u_n \\ v_2 &= a_{21}u_1 + a_{22}u_2 + \cdots + a_{2n}u_n \\ &\cdots\cdots\cdots\cdots\cdots\cdots\cdots\cdots\cdots \\ v_n &= a_{n1}u_1 + a_{n2}u_2 + \cdots + a_{nn}u_n \end{aligned}$$

Let P be the transpose of the above matrix of coefficients; that is, let $P = [p_{ij}]$, where $p_{ij} = a_{ji}$. Then P is called the *change-of-basis matrix* (or *transition matrix*) from the "old" basis S to the "new" basis S'.

The following remarks are in order.

Remark 1: The above change-of-basis matrix P may also be viewed as the matrix whose columns are, respectively, the coordinate column vectors of the "new" basis vectors v_i relative to the "old" basis S; namely,

$$P = \big[[v_1]_S, [v_2]_S, \ldots, [v_n]_S\big]$$

Remark 2: Analogously, there is a change-of-basis matrix Q from the "new" basis S' to the "old" basis S. Similarly, Q may be viewed as the matrix whose columns are, respectively, the coordinate column vectors of the "old" basis vectors u_i relative to the "new" basis S'; namely,

$$Q = \big[[u_1]_{S'}, [u_2]_{S'}, \ldots, [u_n]_{S'}\big]$$

Remark 3: Because the vectors $v_1, v_2, \ldots, v_n$ in the new basis S' are linearly independent, the matrix P is invertible (Problem 6.18). Similarly, Q is invertible. In fact, we have the following proposition (proved in Problem 6.18).

PROPOSITION 6.4: Let P and Q be the above change-of-basis matrices. Then $Q = P^{-1}$.

Now suppose $S = \{u_1, u_2, \ldots, u_n\}$ is a basis of a vector space V, and suppose $P = [p_{ij}]$ is any nonsingular matrix. Then the n vectors

$$v_i = p_{1i}u_i + p_{2i}u_2 + \cdots + p_{ni}u_n, \qquad i = 1, 2, \ldots, n$$

corresponding to the columns of P, are linearly independent [Problem 6.21(a)]. Thus, they form another basis S' of V. Moreover, P will be the change-of-basis matrix from S to the new basis S'.

EXAMPLE 6.5 Consider the following two bases of $\mathbf{R}^2$:

$$S = \{u_1, u_2\} = \{(1,2),\ (3,5)\} \quad \text{and} \quad S' = \{v_1, v_2\} = \{(1,-1),\ (1,-2)\}$$

(a) Find the change-of-basis matrix P from S to the "new" basis S'.

Write each of the new basis vectors of S' as a linear combination of the original basis vectors u_1 and u_2 of S. We have

$$\begin{bmatrix} 1 \\ -1 \end{bmatrix} = x\begin{bmatrix} 1 \\ 2 \end{bmatrix} + y\begin{bmatrix} 3 \\ 5 \end{bmatrix} \quad \text{or} \quad \begin{matrix} x + 3y = 1 \\ 2x + 5y = -1 \end{matrix} \quad \text{yielding} \quad x = -8, \quad y = 3$$

$$\begin{bmatrix} 1 \\ -2 \end{bmatrix} = x\begin{bmatrix} 1 \\ 2 \end{bmatrix} + y\begin{bmatrix} 3 \\ 5 \end{bmatrix} \quad \text{or} \quad \begin{matrix} x + 3y = 1 \\ 2x + 5y = -2 \end{matrix} \quad \text{yielding} \quad x = -11, \quad y = 4$$

Thus,

$$\begin{aligned} v_1 &= -8u_1 + 3u_2 \\ v_2 &= -11u_1 + 4u_2 \end{aligned} \quad \text{and hence,} \quad P = \begin{bmatrix} -8 & -11 \\ 3 & 4 \end{bmatrix}.$$

Note that the coordinates of v_1 and v_2 are the columns, not rows, of the change-of-basis matrix P.

(b) Find the change-of-basis matrix Q from the "new" basis S' back to the "old" basis S.

Here we write each of the "old" basis vectors u_1 and u_2 of S' as a linear combination of the "new" basis vectors v_1 and v_2 of S'. This yields

$$\begin{aligned} u_1 &= 4v_1 - 3v_2 \\ u_2 &= 11v_1 - 8v_2 \end{aligned} \quad \text{and hence,} \quad Q = \begin{bmatrix} 4 & 11 \\ -3 & -8 \end{bmatrix}$$

As expected from Proposition 6.4, $Q = P^{-1}$. (In fact, we could have obtained Q by simply finding P^{-1}.)

EXAMPLE 6.6 Consider the following two bases of $\mathbf{R}^3$:

$$E = \{e_1, e_2, e_3\} = \{(1,0,0),\quad (0,1,0),\quad (0,0,1)\}$$

and
$$S = \{u_1, u_2, u_3\} = \{(1,0,1),\quad (2,1,2),\quad (1,2,2)\}$$

(a) Find the change-of-basis matrix P from the basis E to the basis S.

Because E is the usual basis, we can immediately write each basis element of S as a linear combination of the basis elements of E. Specifically,

$$\begin{aligned} u_1 &= (1,0,1) = e_1 + 0 + e_3 \\ u_2 &= (2,1,2) = 2e_1 + e_2 + 2e_3 \\ u_3 &= (1,2,2) = e_1 + 2e_2 + 2e_3 \end{aligned} \quad \text{and hence,} \quad P = \begin{bmatrix} 1 & 2 & 1 \\ 0 & 1 & 2 \\ 1 & 2 & 2 \end{bmatrix}$$

Again, the coordinates of u_1, u_2, u_3 appear as the columns in P. Observe that P is simply the matrix whose columns are the basis vectors of S. This is true only because the original basis was the usual basis E.

(b) Find the change-of-basis matrix Q from the basis S to the basis E.

The definition of the change-of-basis matrix Q tells us to write each of the (usual) basis vectors in E as a linear combination of the basis elements of S. This yields

$$\begin{aligned} e_1 &= (1,0,0) = -2u_1 + 2u_2 - u_3 \\ e_2 &= (0,1,0) = -2u_1 + u_2 \\ e_3 &= (0,0,1) = 3u_1 - 2u_2 + u_3 \end{aligned} \quad \text{and hence,} \quad Q = \begin{bmatrix} -2 & -2 & 3 \\ 2 & 1 & -2 \\ -1 & 0 & 1 \end{bmatrix}$$

We emphasize that to find Q, we need to solve three 3×3 systems of linear equations—one 3×3 system for each of e_1, e_2, e_3.

Alternatively, we can find $Q = P^{-1}$ by forming the matrix $M = [P, I]$ and row reducing M to row canonical form:

$$M = \begin{bmatrix} 1 & 2 & 1 & 1 & 0 & 0 \\ 0 & 1 & 2 & 0 & 1 & 0 \\ 1 & 2 & 2 & 0 & 0 & 1 \end{bmatrix} \sim \begin{bmatrix} 1 & 0 & 0 & -2 & -2 & 3 \\ 0 & 1 & 0 & 2 & 1 & -2 \\ 0 & 0 & 1 & -1 & 0 & 1 \end{bmatrix} = [I, P^{-1}]$$

thus,

$$Q = P^{-1} = \begin{bmatrix} -2 & -2 & 3 \\ 2 & 1 & -2 \\ -1 & 0 & 1 \end{bmatrix}$$

(Here we have used the fact that Q is the inverse of P.)

The result in Example 6.6(a) is true in general. We state this result formally, because it occurs often.

PROPOSITION 6.5: The change-of-basis matrix from the usual basis E of K^n to any basis S of K^n is the matrix P whose columns are, respectively, the basis vectors of S.

Applications of Change-of-Basis Matrix

First we show how a change of basis affects the coordinates of a vector in a vector space V. The following theorem is proved in Problem 6.22.

THEOREM 6.6: Let P be the change-of-basis matrix from a basis S to a basis S' in a vector space V. Then, for any vector $v \in V$, we have

$$P[v]_{S'} = [v]_S \qquad \text{and hence,} \qquad P^{-1}[v]_S = [v]_{S'}$$

Namely, if we multiply the coordinates of v in the original basis S by P^{-1}, we get the coordinates of v in the new basis S'.

Remark 1: Although P is called the change-of-basis matrix from the old basis S to the new basis S', we emphasize that P^{-1} transforms the coordinates of v in the original basis S into the coordinates of v in the new basis S'.

Remark 2: Because of the above theorem, many texts call $Q = P^{-1}$, not P, the transition matrix from the old basis S to the new basis S'. Some texts also refer to Q as the *change-of-coordinates* matrix.

We now give the proof of the above theorem for the special case that $\dim V = 3$. Suppose P is the change-of-basis matrix from the basis $S = \{u_1, u_2, u_3\}$ to the basis $S' = \{v_1, v_2, v_3\}$; say,

$$\begin{aligned} v_1 &= a_1u_1 + a_2u_2 + a_3a_3 \\ v_2 &= b_1u_1 + b_2u_2 + b_3u_3 \\ v_3 &= c_1u_1 + c_2u_2 + c_3u_3 \end{aligned} \qquad \text{and hence,} \qquad P = \begin{bmatrix} a_1 & b_1 & c_1 \\ a_2 & b_2 & c_2 \\ a_3 & b_3 & c_3 \end{bmatrix}$$

Now suppose $v \in V$ and, say, $v = k_1v_1 + k_2v_2 + k_3v_3$. Then, substituting for v_1, v_2, v_3 from above, we obtain

$$\begin{aligned} v &= k_1(a_1u_1 + a_2u_2 + a_3u_3) + k_2(b_1u_1 + b_2u_2 + b_3u_3) + k_3(c_1u_1 + c_2u_2 + c_3u_3) \\ &= (a_1k_1 + b_1k_2 + c_1k_3)u_1 + (a_2k_1 + b_2k_2 + c_2k_3)u_2 + (a_3k_1 + b_3k_2 + c_3k_3)u_3 \end{aligned}$$

Thus,

$$[v]_{S'} = \begin{bmatrix} k_1 \\ k_2 \\ k_3 \end{bmatrix} \quad \text{and} \quad [v]_S = \begin{bmatrix} a_1k_1 + b_1k_2 + c_1k_3 \\ a_2k_1 + b_2k_2 + c_2k_3 \\ a_3k_1 + b_3k_2 + c_3k_3 \end{bmatrix}$$

Accordingly,

$$P[v]_{S'} = \begin{bmatrix} a_1 & b_1 & c_1 \\ a_2 & b_2 & c_2 \\ a_3 & b_3 & c_3 \end{bmatrix} \begin{bmatrix} k_1 \\ k_2 \\ k_3 \end{bmatrix} = \begin{bmatrix} a_1k_1 + b_1k_2 + c_1k_3 \\ a_2k_1 + b_2k_2 + c_2k_3 \\ a_3k_1 + b_3k_2 + c_3k_3 \end{bmatrix} = [v]_S$$

Finally, multiplying the equation $[v]_S = P[v]_{S'}$ by P^{-1}, we get

$$P^{-1}[v]_S = P^{-1}P[v]_{S'} = I[v]_{S'} = [v]_{S'}$$

The next theorem (proved in Problem 6.26) shows how a change of basis affects the matrix representation of a linear operator.

THEOREM 6.7: Let P be the change-of-basis matrix from a basis S to a basis S' in a vector space V. Then, for any linear operator T on V,

$$[T]_{S'} = P^{-1}[T]_S P$$

That is, if A and B are the matrix representations of T relative, respectively, to S and S', then

$$B = P^{-1}AP$$

EXAMPLE 6.7 Consider the following two bases of $\mathbf{R}^3$:

$$E = \{e_1, e_2, e_3\} = \{(1,0,0), \quad (0,1,0), \quad (0,0,1)\}$$

and $$S = \{u_1, u_2, u_3\} = \{(1,0,1), \quad (2,1,2), \quad (1,2,2)\}$$

The change-of-basis matrix P from E to S and its inverse P^{-1} were obtained in Example 6.6.

(a) Write $v = (1,3,5)$ as a linear combination of u_1, u_2, u_3, or, equivalently, find $[v]_S$.

One way to do this is to directly solve the vector equation $v = xu_1 + yu_2 + zu_3$; that is,

$$\begin{bmatrix} 1 \\ 3 \\ 5 \end{bmatrix} = x\begin{bmatrix} 1 \\ 0 \\ 1 \end{bmatrix} + y\begin{bmatrix} 2 \\ 1 \\ 2 \end{bmatrix} + z\begin{bmatrix} 1 \\ 2 \\ 2 \end{bmatrix} \quad \text{or} \quad \begin{aligned} x + 2y + z &= 1 \\ y + 2z &= 3 \\ x + 2y + 2z &= 5 \end{aligned}$$

The solution is $x = 7$, $y = -5$, $z = 4$, so $v = 7u_1 - 5u_2 + 4u_3$.

On the other hand, we know that $[v]_E = [1,3,5]^T$, because E is the usual basis, and we already know P^{-1}. Therefore, by Theorem 6.6,

$$[v]_S = P^{-1}[v]_E = \begin{bmatrix} -2 & -2 & 3 \\ 2 & 1 & -2 \\ -1 & 0 & 1 \end{bmatrix} \begin{bmatrix} 1 \\ 3 \\ 5 \end{bmatrix} = \begin{bmatrix} 7 \\ -5 \\ 4 \end{bmatrix}$$

Thus, again, $v = 7u_1 - 5u_2 + 4u_3$.

(b) Let $A = \begin{bmatrix} 1 & 3 & -2 \\ 2 & -4 & 1 \\ 3 & -1 & 2 \end{bmatrix}$, which may be viewed as a linear operator on $\mathbf{R}^3$. Find the matrix B that represents A relative to the basis S.

The definition of the matrix representation of A relative to the basis S tells us to write each of $A(u_1)$, $A(u_2)$, $A(u_3)$ as a linear combination of the basis vectors u_1, u_2, u_3 of S. This yields

$$\begin{aligned} A(u_1) &= (-1,3,5) = 11u_1 - 9u_2 + 6u_3 \\ A(u_2) &= (1,2,9) \quad = 21u_1 - 14u_2 + 8u_3 \\ A(u_3) &= (3,-4,5) = 17u_1 - 8e_2 + 2u_3 \end{aligned} \quad \text{and hence,} \quad B = \begin{bmatrix} 11 & 21 & 17 \\ -9 & -14 & -8 \\ 6 & 8 & 2 \end{bmatrix}$$

We emphasize that to find B, we need to solve three 3×3 systems of linear equations—one 3×3 system for each of $A(u_1)$, $A(u_2)$, $A(u_3)$.

On the other hand, because we know P and P^{-1}, we can use Theorem 6.7. That is,

$$B = P^{-1}AP = \begin{bmatrix} -2 & -2 & 3 \\ 2 & 1 & -2 \\ -1 & 0 & 1 \end{bmatrix} \begin{bmatrix} 1 & 3 & -2 \\ 2 & -4 & 1 \\ 3 & -1 & 2 \end{bmatrix} \begin{bmatrix} 1 & 2 & 1 \\ 0 & 1 & 2 \\ 1 & 2 & 2 \end{bmatrix} = \begin{bmatrix} 11 & 21 & 17 \\ -9 & -14 & -8 \\ 6 & 8 & 2 \end{bmatrix}$$

This, as expected, gives the same result.

6.4 Similarity

Suppose A and B are square matrices for which there exists an invertible matrix P such that $B = P^{-1}AP$; then B is said to be *similar* to A, or B is said to be obtained from A by a *similarity transformation.* We show (Problem 6.29) that similarity of matrices is an equivalence relation.

By Theorem 6.7 and the above remark, we have the following basic result.

THEOREM 6.8: Two matrices represent the same linear operator if and only if the matrices are similar.

That is, all the matrix representations of a linear operator T form an equivalence class of similar matrices.

A linear operator T is said to be *diagonalizable* if there exists a basis S of V such that T is represented by a diagonal matrix; the basis S is then said to *diagonalize* T. The preceding theorem gives us the following result.

THEOREM 6.9: Let A be the matrix representation of a linear operator T. Then T is diagonalizable if and only if there exists an invertible matrix P such that $P^{-1}AP$ is a diagonal matrix.

That is, T is diagonalizable if and only if its matrix representation can be diagonalized by a similarity transformation.

We emphasize that not every operator is diagonalizable. However, we will show (Chapter 10) that every linear operator can be represented by certain "standard" matrices called its *normal* or *canonical* forms. Such a discussion will require some theory of fields, polynomials, and determinants.

Functions and Similar Matrices

Suppose f is a function on square matrices that assigns the same value to similar matrices; that is, $f(A) = f(B)$ whenever A is similar to B. Then f induces a function, also denoted by f, on linear operators T in the following natural way. We define

$$f(T) = f([T]_S)$$

where S is any basis. By Theorem 6.8, the function is well defined.

The determinant (Chapter 8) is perhaps the most important example of such a function. The trace (Section 2.7) is another important example of such a function.

EXAMPLE 6.8 Consider the following linear operator F and bases E and S of $\mathbf{R}^2$:

$$F(x,y) = (2x+3y,\ 4x-5y), \qquad E = \{(1,0),\ (0,1)\}, \qquad S = \{(1,2),\ (2,5)\}$$

By Example 6.1, the matrix representations of F relative to the bases E and S are, respectively,

$$A = \begin{bmatrix} 2 & 3 \\ 4 & -5 \end{bmatrix} \qquad \text{and} \qquad B = \begin{bmatrix} 52 & 129 \\ -22 & -55 \end{bmatrix}$$

Using matrix A, we have

(i) Determinant of $F = \det(A) = -10 - 12 = -22$; (ii) Trace of $F = \mathrm{tr}(A) = 2 - 5 = -3$.

On the other hand, using matrix B, we have

(i) Determinant of $F = \det(B) = -2860 + 2838 = -22$; (ii) Trace of $F = \mathrm{tr}(B) = 52 - 55 = -3$.

As expected, both matrices yield the same result.

6.5 Matrices and General Linear Mappings

Last, we consider the general case of linear mappings from one vector space into another. Suppose V and U are vector spaces over the same field K and, say, $\dim V = m$ and $\dim U = n$. Furthermore, suppose

$$S = \{v_1, v_2, \ldots, v_m\} \qquad \text{and} \qquad S' = \{u_1, u_2, \ldots, u_n\}$$

are arbitrary but fixed bases, respectively, of V and U.

Suppose $F: V \to U$ is a linear mapping. Then the vectors $F(v_1)$, $F(v_2)$, ..., $F(v_m)$ belong to U, and so each is a linear combination of the basis vectors in S'; say,

$$\begin{aligned}
F(v_1) &= a_{11}u_1 + a_{12}u_2 + \cdots + a_{1n}u_n \\
F(v_2) &= a_{21}u_1 + a_{22}u_2 + \cdots + a_{2n}u_n \\
&\cdots\cdots\cdots\cdots\cdots\cdots\cdots\cdots\cdots\cdots \\
F(v_m) &= a_{m1}u_1 + a_{m2}u_2 + \cdots + a_{mn}u_n
\end{aligned}$$

DEFINITION: The transpose of the above matrix of coefficients, denoted by $m_{S,S'}(F)$ or $[F]_{S,S'}$, is called the *matrix representation* of F relative to the bases S and S'. [We will use the simple notation $m(F)$ and $[F]$ when the bases are understood.]

The following theorem is analogous to Theorem 6.1 for linear operators (Problem 6.67).

THEOREM 6.10: For any vector $v \in V$, $[F]_{S,S'}[v]_S = [F(v)]_{S'}$.

That is, multiplying the coordinates of v in the basis S of V by $[F]$, we obtain the coordinates of $F(v)$ in the basis S' of U.

Recall that for any vector spaces V and U, the collection of all linear mappings from V into U is a vector space and is denoted by $\mathrm{Hom}(V,U)$. The following theorem is analogous to Theorem 6.2 for linear operators, where now we let $\mathbf{M} = \mathbf{M}_{m,n}$ denote the vector space of all $m \times n$ matrices (Problem 6.67).

THEOREM 6.11: The mapping $m: \mathrm{Hom}(V,U) \to \mathbf{M}$ defined by $m(F) = [F]$ is a vector space isomorphism. That is, for any $F, G \in \mathrm{Hom}(V,U)$ and any scalar k,

(i) $m(F+G) = m(F) + m(G)$ or $[F+G] = [F] + [G]$
(ii) $m(kF) = km(F)$ or $[kF] = k[F]$
(iii) m is bijective (one-to-one and onto).

Our next theorem is analogous to Theorem 6.3 for linear operators (Problem 6.67).

THEOREM 6.12: Let S, S', S'' be bases of vector spaces V, U, W, respectively. Let $F: V \to U$ and $G \circ U \to W$ be linear mappings. Then

$$[G \circ F]_{S,S''} = [G]_{S',S''}[F]_{S,S'}$$

That is, relative to the appropriate bases, the matrix representation of the composition of two mappings is the matrix product of the matrix representations of the individual mappings.

Next we show how the matrix representation of a linear mapping $F: V \to U$ is affected when new bases are selected (Problem 6.67).

THEOREM 6.13: Let P be the change-of-basis matrix from a basis e to a basis e' in V, and let Q be the change-of-basis matrix from a basis f to a basis f' in U. Then, for any linear map $F: V \to U$,

$$[F]_{e',f'} = Q^{-1}[F]_{e,f}P$$

In other words, if A is the matrix representation of a linear mapping F relative to the bases e and f, and B is the matrix representation of F relative to the bases e' and f', then

$$B = Q^{-1}AP$$

Our last theorem, proved in Problem 6.36, shows that any linear mapping from one vector space V into another vector space U can be represented by a very simple matrix. We note that this theorem is analogous to Theorem 3.18 for $m \times n$ matrices.

THEOREM 6.14: Let $F: V \to U$ be linear and, say, $\text{rank}(F) = r$. Then there exist bases of V and U such that the matrix representation of F has the form

$$A = \begin{bmatrix} I_r & 0 \\ 0 & 0 \end{bmatrix}$$

where I_r is the r-square identity matrix.

The above matrix A is called the *normal* or *canonical* form of the linear map F.

SOLVED PROBLEMS

Matrix Representation of Linear Operators

6.1. Consider the linear mapping $F: \mathbf{R}^2 \to \mathbf{R}^2$ defined by $F(x, y) = (3x + 4y,\ \ 2x - 5y)$ and the following bases of $\mathbf{R}^2$:

$$E = \{e_1, e_2\} = \{(1,0),\ (0,1)\} \qquad \text{and} \qquad S = \{u_1, u_2\} = \{(1,2),\ (2,3)\}$$

(a) Find the matrix A representing F relative to the basis E.

(b) Find the matrix B representing F relative to the basis S.

(a) Because E is the usual basis, the rows of A are simply the coefficients in the components of $F(x, y)$; that is, using $(a, b) = ae_1 + be_2$, we have

$$\begin{aligned} F(e_1) &= F(1,0) = (3,2) \ \ = 3e_1 + 2e_2 \\ F(e_2) &= F(0,1) = (4,-5) = 4e_1 - 5e_2 \end{aligned} \qquad \text{and so} \qquad A = \begin{bmatrix} 3 & 4 \\ 2 & -5 \end{bmatrix}$$

Note that the coefficients of the basis vectors are written as columns in the matrix representation.

(b) First find $F(u_1)$ and write it as a linear combination of the basis vectors u_1 and u_2. We have

$$F(u_1) = F(1,2) = (11,-8) = x(1,2) + y(2,3), \quad \text{and so} \quad \begin{matrix} x + 2y = 11 \\ 2x + 3y = -8 \end{matrix}$$

Solve the system to obtain $x = -49$, $y = 30$. Therefore,

$$F(u_1) = -49u_1 + 30u_2$$

Next find $F(u_2)$ and write it as a linear combination of the basis vectors u_1 and u_2. We have

$$F(u_2) = F(2,3) = (18,-11) = x(1,2) + y(2,3), \quad \text{and so} \quad \begin{matrix} x + 2y = 18 \\ 2x + 3y = -11 \end{matrix}$$

Solve for x and y to obtain $x = -76$, $y = 47$. Hence,

$$F(u_2) = -76u_1 + 47u_2$$

Write the coefficients of u_1 and u_2 as columns to obtain $B = \begin{bmatrix} -49 & -76 \\ 30 & 47 \end{bmatrix}$

(b′) Alternatively, one can first find the coordinates of an arbitrary vector (a,b) in $\mathbf{R}^2$ relative to the basis S. We have

$$(a,b) = x(1,2) + y(2,3) = (x + 2y,\ 2x + 3y), \quad \text{and so} \quad \begin{matrix} x + 2y = a \\ 2x + 3y = b \end{matrix}$$

Solve for x and y in terms of a and b to get $x = -3a + 2b$, $y = 2a - b$. Thus,

$$(a,b) = (-3a + 2b)u_1 + (2a - b)u_2$$

Then use the formula for (a,b) to find the coordinates of $F(u_1)$ and $F(u_2)$ relative to S:

$$\begin{matrix} F(u_1) = F(1,2) = (11,-8) = -49u_1 + 30u_2 \\ F(u_2) = F(2,3) = (18,-11) = -76u_1 + 47u_2 \end{matrix} \quad \text{and so} \quad B = \begin{bmatrix} -49 & -76 \\ 30 & 47 \end{bmatrix}$$

6.2. Consider the following linear operator G on $\mathbf{R}^2$ and basis S:

$$G(x,y) = (2x - 7y,\ 4x + 3y) \quad \text{and} \quad S = \{u_1, u_2\} = \{(1,3),\ (2,5)\}$$

(a) Find the matrix representation $[G]_S$ of G relative to S.

(b) Verify $[G]_S[v]_S = [G(v)]_S$ for the vector $v = (4,-3)$ in $\mathbf{R}^2$.

First find the coordinates of an arbitrary vector $v = (a,b)$ in $\mathbf{R}^2$ relative to the basis S. We have

$$\begin{bmatrix} a \\ b \end{bmatrix} = x\begin{bmatrix} 1 \\ 3 \end{bmatrix} + y\begin{bmatrix} 2 \\ 5 \end{bmatrix}, \quad \text{and so} \quad \begin{matrix} x + 2y = a \\ 3x + 5y = b \end{matrix}$$

Solve for x and y in terms of a and b to get $x = -5a + 2b$, $y = 3a - b$. Thus,

$$(a,b) = (-5a + 2b)u_1 + (3a - b)u_2, \quad \text{and so} \quad [v] = [-5a + 2b,\ 3a - b]^T$$

(a) Using the formula for (a,b) and $G(x,y) = (2x - 7y,\ 4x + 3y)$, we have

$$\begin{matrix} G(u_1) = G(1,3) = (-19,13) = 121u_1 - 70u_2 \\ G(u_2) = G(2,5) = (-31,23) = 201u_1 - 116u_2 \end{matrix} \quad \text{and so} \quad [G]_S = \begin{bmatrix} 121 & 201 \\ -70 & -116 \end{bmatrix}$$

(We emphasize that the coefficients of u_1 and u_2 are written as columns, not rows, in the matrix representation.)

(b) Use the formula $(a,b) = (-5a + 2b)u_1 + (3a - b)u_2$ to get

$$v = (4,-3) = -26u_1 + 15u_2$$
$$G(v) = G(4,-3) = (20,7) = -131u_1 + 80u_2$$

Then $\quad [v]_S = [-26, 15]^T \quad \text{and} \quad [G(v)]_S = [-131, 80]^T$

Accordingly,

$$[G]_S[v]_S = \begin{bmatrix} 121 & 201 \\ -70 & -116 \end{bmatrix}\begin{bmatrix} -26 \\ 15 \end{bmatrix} = \begin{bmatrix} -131 \\ 80 \end{bmatrix} = [G(v)]_S$$

(This is expected from Theorem 6.1.)

6.3. Consider the following 2×2 matrix A and basis S of $\mathbf{R}^2$:

$$A = \begin{bmatrix} 2 & 4 \\ 5 & 6 \end{bmatrix} \quad \text{and} \quad S = \{u_1, u_2\} = \left\{ \begin{bmatrix} 1 \\ -2 \end{bmatrix}, \begin{bmatrix} 3 \\ -7 \end{bmatrix} \right\}$$

The matrix A defines a linear operator on $\mathbf{R}^2$. Find the matrix B that represents the mapping A relative to the basis S.

First find the coordinates of an arbitrary vector $(a, b)^T$ with respect to the basis S. We have

$$\begin{bmatrix} a \\ b \end{bmatrix} = x\begin{bmatrix} 1 \\ -2 \end{bmatrix} + y\begin{bmatrix} 3 \\ -7 \end{bmatrix} \quad \text{or} \quad \begin{aligned} x + 3y &= a \\ -2x - 7y &= b \end{aligned}$$

Solve for x and y in terms of a and b to obtain $x = 7a + 3b$, $y = -2a - b$. Thus,

$$(a, b)^T = (7a + 3b)u_1 + (-2a - b)u_2$$

Then use the formula for $(a, b)^T$ to find the coordinates of Au_1 and Au_2 relative to the basis S:

$$Au_1 = \begin{bmatrix} 2 & 4 \\ 5 & 6 \end{bmatrix}\begin{bmatrix} 1 \\ -2 \end{bmatrix} = \begin{bmatrix} -6 \\ -7 \end{bmatrix} = -63u_1 + 19u_2$$

$$Au_2 = \begin{bmatrix} 2 & 4 \\ 5 & 6 \end{bmatrix}\begin{bmatrix} 3 \\ -7 \end{bmatrix} = \begin{bmatrix} -22 \\ -27 \end{bmatrix} = -235u_1 + 71u_2$$

Writing the coordinates as columns yields

$$B = \begin{bmatrix} -63 & -235 \\ 19 & 71 \end{bmatrix}$$

6.4. Find the matrix representation of each of the following linear operators F on $\mathbf{R}^3$ relative to the usual basis $E = \{e_1, e_2, e_3\}$ of $\mathbf{R}^3$; that is, find $[F] = [F]_E$:

(a) F defined by $F(x, y, z) = (x + 2y - 3z,\ 4x - 5y - 6z,\ 7x + 8y + 9z)$.

(b) F defined by the 3×3 matrix $A = \begin{bmatrix} 1 & 1 & 1 \\ 2 & 3 & 4 \\ 5 & 5 & 5 \end{bmatrix}$.

(c) F defined by $F(e_1) = (1, 3, 5), F(e_2) = (2, 4, 6), F(e_3) = (7, 7, 7)$. (Theorem 5.2 states that a linear map is completely defined by its action on the vectors in a basis.)

(a) Because E is the usual basis, simply write the coefficients of the components of $F(x, y, z)$ as rows:

$$[F] = \begin{bmatrix} 1 & 2 & -3 \\ 4 & -5 & -6 \\ 7 & 8 & 9 \end{bmatrix}$$

(b) Because E is the usual basis, $[F] = A$, the matrix A itself.

(c) Here

$$\begin{aligned} F(e_1) &= (1, 3, 5) = e_1 + 3e_2 + 5e_3 \\ F(e_2) &= (2, 4, 6) = 2e_1 + 4e_2 + 6e_3 \\ F(e_3) &= (7, 7, 7) = 7e_1 + 7e_2 + 7e_3 \end{aligned} \quad \text{and so} \quad [F] = \begin{bmatrix} 1 & 2 & 7 \\ 3 & 4 & 7 \\ 5 & 6 & 7 \end{bmatrix}$$

That is, the columns of $[F]$ are the images of the usual basis vectors.

6.5. Let G be the linear operator on $\mathbf{R}^3$ defined by $G(x, y, z) = (2y + z,\ x - 4y,\ 3x)$.

(a) Find the matrix representation of G relative to the basis

$$S = \{w_1, w_2, w_3\} = \{(1, 1, 1),\quad (1, 1, 0),\quad (1, 0, 0)\}$$

(b) Verify that $[G][v] = [G(v)]$ for any vector v in $\mathbf{R}^3$.

First find the coordinates of an arbitrary vector $(a, b, c) \in \mathbf{R}^3$ with respect to the basis S. Write (a, b, c) as a linear combination of w_1, w_2, w_3 using unknown scalars x, y, and z:

$$(a, b, c) = x(1, 1, 1) + y(1, 1, 0) + z(1, 0, 0) = (x + y + z,\ x + y,\ x)$$

Set corresponding components equal to each other to obtain the system of equations

$$x + y + z = a, \qquad x + y = b, \qquad x = c$$

Solve the system for x, y, z in terms of a, b, c to find $x = c,\ y = b - c,\ z = a - b$. Thus,

$$(a, b, c) = cw_1 + (b - c)w_2 + (a - b)w_3, \qquad \text{or equivalently,} \qquad [(a, b, c)] = [c,\ b - c,\ a - b]^T$$

(a) Because $G(x, y, z) = (2y + z,\ x - 4y,\ 3x)$,

$$\begin{aligned} G(w_1) &= G(1, 1, 1) = (3, -3, 3) = 3w_1 - 6x_2 + 6x_3 \\ G(w_2) &= G(1, 1, 0) = (2, -3, 3) = 3w_1 - 6w_2 + 5w_3 \\ G(w_3) &= G(1, 0, 0) = (0, 1, 3) = 3w_1 - 2w_2 - w_3 \end{aligned}$$

Write the coordinates $G(w_1), G(w_2), G(w_3)$ as columns to get

$$[G] = \begin{bmatrix} 3 & 3 & 3 \\ -6 & -6 & -2 \\ 6 & 5 & -1 \end{bmatrix}$$

(b) Write $G(v)$ as a linear combination of w_1, w_2, w_3, where $v = (a, b, c)$ is an arbitrary vector in $\mathbf{R}^3$,

$$G(v) = G(a, b, c) = (2b + c,\ a - 4b,\ 3a) = 3aw_1 + (-2a - 4b)w_2 + (-a + 6b + c)w_3$$

or equivalently,

$$[G(v)] = [3a,\ -2a - 4b,\ -a + 6b + c]^T$$

Accordingly,

$$[G][v] = \begin{bmatrix} 3 & 3 & 3 \\ -6 & -6 & -2 \\ 6 & 5 & -1 \end{bmatrix} \begin{bmatrix} c \\ b - c \\ a - b \end{bmatrix} = \begin{bmatrix} 3a \\ -2a - 4b \\ -a + 6b + c \end{bmatrix} = [G(v)]$$

6.6. Consider the following 3×3 matrix A and basis S of $\mathbf{R}^3$:

$$A = \begin{bmatrix} 1 & -2 & 1 \\ 3 & -1 & 0 \\ 1 & 4 & -2 \end{bmatrix} \qquad \text{and} \qquad S = \{u_1, u_2, u_3\} = \left\{ \begin{bmatrix} 1 \\ 1 \\ 1 \end{bmatrix}, \begin{bmatrix} 0 \\ 1 \\ 1 \end{bmatrix}, \begin{bmatrix} 1 \\ 2 \\ 3 \end{bmatrix} \right\}$$

The matrix A defines a linear operator on $\mathbf{R}^3$. Find the matrix B that represents the mapping A relative to the basis S. (Recall that A represents itself relative to the usual basis of $\mathbf{R}^3$.)

First find the coordinates of an arbitrary vector (a, b, c) in $\mathbf{R}^3$ with respect to the basis S. We have

$$\begin{bmatrix} a \\ b \\ c \end{bmatrix} = x \begin{bmatrix} 1 \\ 1 \\ 1 \end{bmatrix} + y \begin{bmatrix} 0 \\ 1 \\ 1 \end{bmatrix} + z \begin{bmatrix} 1 \\ 2 \\ 3 \end{bmatrix} \qquad \text{or} \qquad \begin{aligned} x + \quad\ \ z &= a \\ x + y + 2z &= b \\ x + y + 3z &= c \end{aligned}$$

Solve for x, y, z in terms of a, b, c to get

$$x = a + b - c, \quad y = -a + 2b - c, \quad z = c - b$$

thus, $\quad (a, b, c)^T = (a + b - c)u_1 + (-a + 2b - c)u_2 + (c - b)u_3$

Then use the formula for $(a,b,c)^T$ to find the coordinates of Au_1, Au_2, Au_3 relative to the basis S:

$$\begin{array}{lll} A(u_1) = A(1,1,1)^T = (0,2,3)^T & = -u_1 + u_2 + u_3 \\ A(u_2) = A(1,1,0)^T = (-1,-1,2)^T & = -4u_1 - 3u_2 + 3u_3 \\ A(u_3) = A(1,2,3)^T = (0,1,3)^T & = -2u_1 - u_2 + 2u_3 \end{array} \quad \text{so} \quad B = \begin{bmatrix} -1 & -4 & -2 \\ 1 & -3 & -1 \\ 1 & 3 & 2 \end{bmatrix}$$

6.7. For each of the following linear transformations (operators) L on $\mathbf{R}^2$, find the matrix A that represents L (relative to the usual basis of $\mathbf{R}^2$):

(a) L is defined by $L(1,0) = (2,4)$ and $L(0,1) = (5,8)$.

(b) L is the rotation in $\mathbf{R}^2$ counterclockwise by 90°.

(c) L is the reflection in $\mathbf{R}^2$ about the line $y = -x$.

(a) Because $\{(1,0),\ (0,1)\}$ is the usual basis of $\mathbf{R}^2$, write their images under L as columns to get

$$A = \begin{bmatrix} 2 & 5 \\ 4 & 8 \end{bmatrix}$$

(b) Under the rotation L, we have $L(1,0) = (0,1)$ and $L(0,1) = (-1,0)$. Thus,

$$A = \begin{bmatrix} 0 & -1 \\ 1 & 0 \end{bmatrix}$$

(c) Under the reflection L, we have $L(1,0) = (0,-1)$ and $L(0,1) = (-1,0)$. Thus,

$$A = \begin{bmatrix} 0 & -1 \\ -1 & 0 \end{bmatrix}$$

6.8. The set $S = \{e^{3t},\ te^{3t},\ t^2e^{3t}\}$ is a basis of a vector space V of functions $f\colon \mathbf{R} \to \mathbf{R}$. Let $\mathbf{D}$ be the differential operator on V; that is, $\mathbf{D}(f) = df/dt$. Find the matrix representation of $\mathbf{D}$ relative to the basis S.

Find the image of each basis function:

$$\begin{array}{lll} \mathbf{D}(e^{3t}) & = 3e^{3t} & = 3(e^{3t}) + 0(te^{3t}) + 0(t^2e^{3t}) \\ \mathbf{D}(te^{3t}) & = e^{3t} + 3te^{3t} & = 1(e^{3t}) + 3(te^{3t}) + 0(t^2e^{3t}) \\ \mathbf{D}(t^2e^{3t}) & = 2te^{3t} + 3t^2e^{3t} & = 0(e^{3t}) + 2(te^{3t}) + 3(t^2e^{3t}) \end{array} \quad \text{and thus,} \quad [\mathbf{D}] = \begin{bmatrix} 3 & 1 & 0 \\ 0 & 3 & 2 \\ 0 & 0 & 3 \end{bmatrix}$$

6.9. Prove Theorem 6.1: Let $T\colon V \to V$ be a linear operator, and let S be a (finite) basis of V. Then, for any vector v in V, $[T]_S[v]_S = [T(v)]_S$.

Suppose $S = \{u_1, u_2, \ldots, u_n\}$, and suppose, for $i = 1, \ldots, n$,

$$T(u_i) = a_{i1}u_1 + a_{i2}u_2 + \cdots + a_{in}u_n = \sum_{j=1}^{n} a_{ij}u_j$$

Then $[T]_S$ is the n-square matrix whose jth row is

$$(a_{1j}, a_{2j}, \ldots, a_{nj}) \tag{1}$$

Now suppose

$$v = k_1u_1 + k_2u_2 + \cdots + k_nu_n = \sum_{i=1}^{n} k_iu_i$$

Writing a column vector as the transpose of a row vector, we have

$$[v]_S = [k_1, k_2, \ldots, k_n]^T \tag{2}$$

Furthermore, using the linearity of T,

$$T(v) = T\left(\sum_{i=1}^{n} k_i u_i\right) = \sum_{i=1}^{n} k_i T(u_i) = \sum_{i=1}^{n} k_i \left(\sum_{j=1}^{n} a_{ij} u_j\right)$$
$$= \sum_{j=1}^{n}\left(\sum_{i=1}^{n} a_{ij} k_i\right) u_j = \sum_{j=1}^{n} (a_{1j}k_1 + a_{2j}k_2 + \cdots + a_{nj}k_n) u_j$$

Thus, $[T(v)]_S$ is the column vector whose jth entry is

$$a_{1j}k_1 + a_{2j}k_2 + \cdots + a_{nj}k_n \qquad (3)$$

On the other hand, the jth entry of $[T]_S[v]_S$ is obtained by multiplying the jth row of $[T]_S$ by $[v]_S$—that is (1) by (2). But the product of (1) and (2) is (3). Hence, $[T]_S[v]_S$ and $[T(v)]_S$ have the same entries. Thus, $[T]_S[v]_S = [T(v)]_S$.

6.10. Prove Theorem 6.2: Let $S = \{u_1, u_2, \ldots, u_n\}$ be a basis for V over K, and let $\mathbf{M}$ be the algebra of n-square matrices over K. Then the mapping $m: A(V) \to \mathbf{M}$ defined by $m(T) = [T]_S$ is a vector space isomorphism. That is, for any $F, G \in A(V)$ and any $k \in K$, we have

(i) $[F + G] = [F] + [G]$, (ii) $[kF] = k[F]$, (iii) m is one-to-one and onto.

(i) Suppose, for $i = 1, \ldots, n$,

$$F(u_i) = \sum_{j=1}^{n} a_{ij} u_j \quad \text{and} \quad G(u_i) = \sum_{j=1}^{n} b_{ij} u_j$$

Consider the matrices $A = [a_{ij}]$ and $B = [b_{ij}]$. Then $[F] = A^T$ and $[G] = B^T$. We have, for $i = 1, \ldots, n$,

$$(F + G)(u_i) = F(u_i) + G(u_i) = \sum_{j=1}^{n} (a_{ij} + b_{ij}) u_j$$

Because $A + B$ is the matrix $(a_{ij} + b_{ij})$, we have

$$[F + G] = (A + B)^T = A^T + B^T = [F] + [G]$$

(ii) Also, for $i = 1, \ldots, n$,

$$(kF)(u_i) = kF(u_i) = k\sum_{j=1}^{n} a_{ij} u_j = \sum_{j=1}^{n} (ka_{ij}) u_j$$

Because kA is the matrix (ka_{ij}), we have

$$[kF] = (kA)^T = kA^T = k[F]$$

(iii) Finally, m is one-to-one, because a linear mapping is completely determined by its values on a basis. Also, m is onto, because matrix $A = [a_{ij}]$ in $\mathbf{M}$ is the image of the linear operator,

$$F(u_i) = \sum_{j=1}^{n} a_{ij} u_j, \quad i = 1, \ldots, n$$

Thus, the theorem is proved.

6.11. Prove Theorem 6.3: For any linear operators $G, F \in A(V)$, $[G \circ F] = [G][F]$.

Using the notation in Problem 6.10, we have

$$(G \circ F)(u_i) = G(F(u_i)) = G\left(\sum_{j=1}^{n} a_{ij} u_j\right) = \sum_{j=1}^{n} a_{ij} G(u_j)$$
$$= \sum_{j=1}^{n} a_{ij}\left(\sum_{k=1}^{n} b_{jk} u_k\right) = \sum_{k=1}^{n}\left(\sum_{j=1}^{n} a_{ij} b_{jk}\right) u_k$$

Recall that AB is the matrix $AB = [c_{ik}]$, where $c_{ik} = \sum_{j=1}^{n} a_{ij} b_{jk}$. Accordingly,

$$[G \circ F] = (AB)^T = B^T A^T = [G][F]$$

The theorem is proved.

6.12. Let A be the matrix representation of a linear operator T. Prove that, for any polynomial $f(t)$, we have that $f(A)$ is the matrix representation of $f(T)$. [Thus, $f(T) = 0$ if and only if $f(A) = 0$.]

Let ϕ be the mapping that sends an operator T into its matrix representation A. We need to prove that $\phi(f(T)) = f(A)$. Suppose $f(t) = a_n t^n + \cdots + a_1 t + a_0$. The proof is by induction on n, the degree of $f(t)$.

Suppose $n = 0$. Recall that $\phi(I') = I$, where I' is the identity mapping and I is the identity matrix. Thus,

$$\phi(f(T)) = \phi(a_0 I') = a_0 \phi(I') = a_0 I = f(A)$$

and so the theorem holds for $n = 0$.

Now assume the theorem holds for polynomials of degree less than n. Then, because ϕ is an algebra isomorphism,

$$\begin{aligned}
\phi(f(T)) &= \phi(a_n T^n + a_{n-1} T^{n-1} + \cdots + a_1 T + a_0 I') \\
&= a_n \phi(T)\phi(T^{n-1}) + \phi(a_{n-1} T^{n-1} + \cdots + a_1 T + a_0 I') \\
&= a_n A A^{n-1} + (a_{n-1} A^{n-1} + \cdots + a_1 A + a_0 I) = f(A)
\end{aligned}$$

and the theorem is proved.

Change of Basis

The coordinate vector $[v]_S$ in this section will always denote a column vector; that is,

$$[v]_S = [a_1, a_2, \ldots, a_n]^T$$

6.13. Consider the following bases of $\mathbf{R}^2$:

$$E = \{e_1, e_2\} = \{(1,0),\ (0,1)\} \qquad \text{and} \qquad S = \{u_1, u_2\} = \{(1,3),\ (1,4)\}$$

(a) Find the change-of-basis matrix P from the usual basis E to S.

(b) Find the change-of-basis matrix Q from S back to E.

(c) Find the coordinate vector $[v]$ of $v = (5, -3)$ relative to S.

(a) Because E is the usual basis, simply write the basis vectors in S as columns: $P = \begin{bmatrix} 1 & 1 \\ 3 & 4 \end{bmatrix}$

(b) **Method 1.** Use the definition of the change-of-basis matrix. That is, express each vector in E as a linear combination of the vectors in S. We do this by first finding the coordinates of an arbitrary vector $v = (a, b)$ relative to S. We have

$$(a, b) = x(1,3) + y(1,4) = (x + y, 3x + 4y) \quad \text{or} \quad \begin{aligned} x + y &= a \\ 3x + 4y &= b \end{aligned}$$

Solve for x and y to obtain $x = 4a - b$, $y = -3a + b$. Thus,

$$v = (4a - b)u_1 + (-3a + b)u_2 \qquad \text{and} \qquad [v]_S = [(a, b)]_S = [4a - b,\ -3a + b]^T$$

Using the above formula for $[v]_S$ and writing the coordinates of the e_i as columns yields

$$\begin{aligned} e_1 &= (1, 0) = 4u_1 - 3u_2 \\ e_2 &= (0, 1) = -u_1 + u_2 \end{aligned} \qquad \text{and} \qquad Q = \begin{bmatrix} 4 & -1 \\ -3 & 1 \end{bmatrix}$$

Method 2. Because $Q = P^{-1}$, find P^{-1}, say by using the formula for the inverse of a 2×2 matrix. Thus,

$$P^{-1} = \begin{bmatrix} 4 & -1 \\ -3 & 1 \end{bmatrix}$$

(c) **Method 1.** Write v as a linear combination of the vectors in S, say by using the above formula for $v = (a, b)$. We have $v = (5, -3) = 23u_1 - 18u_2$, and so $[v]_S = [23, -18]^T$.

Method 2. Use, from Theorem 6.6, the fact that $[v]_S = P^{-1}[v]_E$ and the fact that $[v]_E = [5, -3]^T$:

$$[v]_S = P^{-1}[v]_E = \begin{bmatrix} 4 & -1 \\ -3 & 1 \end{bmatrix}\begin{bmatrix} 5 \\ -3 \end{bmatrix} = \begin{bmatrix} 23 \\ -18 \end{bmatrix}$$

6.14. The vectors $u_1 = (1,2,0)$, $u_2 = (1,3,2)$, $u_3 = (0,1,3)$ form a basis S of $\mathbf{R}^3$. Find

(a) The change-of-basis matrix P from the usual basis $E = \{e_1, e_2, e_3\}$ to S.

(b) The change-of-basis matrix Q from S back to E.

(a) Because E is the usual basis, simply write the basis vectors of S as columns: $P = \begin{bmatrix} 1 & 1 & 0 \\ 2 & 3 & 1 \\ 0 & 2 & 3 \end{bmatrix}$

(b) **Method 1.** Express each basis vector of E as a linear combination of the basis vectors of S by first finding the coordinates of an arbitrary vector $v = (a,b,c)$ relative to the basis S. We have

$$\begin{bmatrix} a \\ b \\ c \end{bmatrix} = x\begin{bmatrix} 1 \\ 2 \\ 0 \end{bmatrix} + y\begin{bmatrix} 1 \\ 3 \\ 2 \end{bmatrix} + z\begin{bmatrix} 0 \\ 1 \\ 3 \end{bmatrix} \quad \text{or} \quad \begin{aligned} x + y \qquad &= a \\ 2x + 3y + z &= b \\ 2y + 3z &= c \end{aligned}$$

Solve for x, y, z to get $x = 7a - 3b + c$, $y = -6a + 3b - c$, $z = 4a - 2b + c$. Thus,

$$v = (a,b,c) = (7a - 3b + c)u_1 + (-6a + 3b - c)u_2 + (4a - 2b + c)u_3$$

or
$$[v]_S = [(a,b,c)]_S = [7a - 3b + c, \quad -6a + 3b - c, \quad 4a - 2b + c]^T$$

Using the above formula for $[v]_S$ and then writing the coordinates of the e_i as columns yields

$$\begin{aligned} e_1 &= (1,0,0) = 7u_1 - 6u_2 + 4u_3 \\ e_2 &= (0,1,0) = -3u_1 + 3u_2 - 2u_3 \\ e_3 &= (0,0,1) = u_1 - u_2 + u_3 \end{aligned} \quad \text{and} \quad Q = \begin{bmatrix} 7 & -3 & 1 \\ -6 & 3 & -1 \\ 4 & -2 & 1 \end{bmatrix}$$

Method 2. Find P^{-1} by row reducing $M = [P, I]$ to the form $[I, P^{-1}]$:

$$M = \left[\begin{array}{ccc|ccc} 1 & 1 & 0 & 1 & 0 & 0 \\ 2 & 3 & 1 & 0 & 1 & 0 \\ 0 & 2 & 3 & 0 & 0 & 1 \end{array}\right] \sim \left[\begin{array}{ccc|ccc} 1 & 1 & 0 & 1 & 0 & 0 \\ 0 & 1 & 1 & -2 & 1 & 0 \\ 0 & 2 & 3 & 0 & 0 & 1 \end{array}\right]$$

$$\sim \left[\begin{array}{ccc|ccc} 1 & 1 & 0 & 1 & 0 & 0 \\ 0 & 1 & 1 & -2 & 1 & 0 \\ 0 & 0 & 1 & 4 & -2 & 1 \end{array}\right] \sim \left[\begin{array}{ccc|ccc} 1 & 0 & 0 & 7 & -3 & 1 \\ 0 & 1 & 0 & -6 & 3 & -1 \\ 0 & 0 & 1 & 4 & -2 & 1 \end{array}\right] = [I, P^{-1}]$$

Thus, $Q = P^{-1} = \begin{bmatrix} 7 & -3 & 1 \\ -6 & 3 & -1 \\ 4 & -2 & 1 \end{bmatrix}$.

6.15. Suppose the x-axis and y-axis in the plane $\mathbf{R}^2$ are rotated counterclockwise 45° so that the new x'-axis and y'-axis are along the line $y = x$ and the line $y = -x$, respectively.

(a) Find the change-of-basis matrix P.

(b) Find the coordinates of the point $A(5,6)$ under the given rotation.

(a) The unit vectors in the direction of the new x'- and y'-axes are

$$u_1 = (\tfrac{1}{2}\sqrt{2}, \tfrac{1}{2}\sqrt{2}) \quad \text{and} \quad u_2 = (-\tfrac{1}{2}\sqrt{2}, \tfrac{1}{2}\sqrt{2})$$

(The unit vectors in the direction of the original x and y axes are the usual basis of $\mathbf{R}^2$.) Thus, write the coordinates of u_1 and u_2 as columns to obtain

$$P = \begin{bmatrix} \frac{1}{2}\sqrt{2} & -\frac{1}{2}\sqrt{2} \\ \frac{1}{2}\sqrt{2} & \frac{1}{2}\sqrt{2} \end{bmatrix}$$

(b) Multiply the coordinates of the point by P^{-1}:

$$\begin{bmatrix} \frac{1}{2}\sqrt{2} & \frac{1}{2}\sqrt{2} \\ -\frac{1}{2}\sqrt{2} & \frac{1}{2}\sqrt{2} \end{bmatrix}\begin{bmatrix} 5 \\ 6 \end{bmatrix} = \begin{bmatrix} \frac{11}{2}\sqrt{2} \\ \frac{1}{2}\sqrt{2} \end{bmatrix}$$

(Because P is orthogonal, P^{-1} is simply the transpose of P.)

6.16. The vectors $u_1 = (1,1,0)$, $u_2 = (0,1,1)$, $u_3 = (1,2,2)$ form a basis S of $\mathbf{R}^3$. Find the coordinates of an arbitrary vector $v = (a,b,c)$ relative to the basis S.

Method 1. Express v as a linear combination of u_1, u_2, u_3 using unknowns x, y, z. We have

$$(a,b,c) = x(1,1,0) + y(0,1,1) + z(1,2,2) = (x+z,\ x+y+2z,\ y+2z)$$

this yields the system

$$\begin{aligned} x + \quad\ z &= a \\ x+y+2z &= b \\ y+2z &= c \end{aligned} \quad \text{or} \quad \begin{aligned} x + \quad\ z &= a \\ y+\ z &= -a+b \\ y+2z &= c \end{aligned} \quad \text{or} \quad \begin{aligned} x + \quad z &= a \\ y+z &= -a+b \\ z &= a-b+c \end{aligned}$$

Solving by back-substitution yields $x = b-c$, $y = -2a+2b-c$, $z = a-b+c$. Thus,

$$[v]_S = [b-c,\ -2a+2b-c,\ a-b+c]^T$$

Method 2. Find P^{-1} by row reducing $M = [P, I]$ to the form $[I, P^{-1}]$, where P is the change-of-basis matrix from the usual basis E to S or, in other words, the matrix whose columns are the basis vectors of S.

We have

$$M = \left[\begin{array}{ccc|ccc} 1 & 0 & 1 & 1 & 0 & 0 \\ 1 & 1 & 2 & 0 & 1 & 0 \\ 0 & 1 & 2 & 0 & 0 & 1 \end{array}\right] \sim \left[\begin{array}{ccc|ccc} 1 & 0 & 1 & 1 & 0 & 0 \\ 0 & 1 & 1 & -1 & 1 & 0 \\ 0 & 1 & 2 & 0 & 0 & 1 \end{array}\right]$$

$$\sim \left[\begin{array}{ccc|ccc} 1 & 0 & 1 & 1 & 0 & 0 \\ 0 & 1 & 1 & -1 & 1 & 0 \\ 0 & 0 & 1 & 1 & -1 & 1 \end{array}\right] \sim \left[\begin{array}{ccc|ccc} 1 & 0 & 0 & 0 & 1 & -1 \\ 0 & 1 & 0 & -2 & 2 & -1 \\ 0 & 0 & 1 & 1 & -1 & 1 \end{array}\right] = [I, P^{-1}]$$

Thus, $P^{-1} = \begin{bmatrix} 0 & 1 & -1 \\ -2 & 2 & -1 \\ 1 & -1 & 1 \end{bmatrix}$ and $[v]_S = P^{-1}[v]_E = \begin{bmatrix} 0 & 1 & -1 \\ -2 & 2 & -1 \\ 1 & -1 & 1 \end{bmatrix}\begin{bmatrix} a \\ b \\ c \end{bmatrix} = \begin{bmatrix} b-c \\ -2a+2b-c \\ a-b+c \end{bmatrix}$

6.17. Consider the following bases of $\mathbf{R}^2$:

$$S = \{u_1, u_2\} = \{(1,-2),\ (3,-4)\} \qquad \text{and} \qquad S' = \{v_1, v_2\} = \{(1,3),\ (3,8)\}$$

(a) Find the coordinates of $v = (a,b)$ relative to the basis S.
(b) Find the change-of-basis matrix P from S to S'.
(c) Find the coordinates of $v = (a,b)$ relative to the basis S'.
(d) Find the change-of-basis matrix Q from S' back to S.
(e) Verify $Q = P^{-1}$.
(f) Show that, for any vector $v = (a,b)$ in $\mathbf{R}^2$, $P^{-1}[v]_S = [v]_{S'}$. (See Theorem 6.6.)

(a) Let $v = xu_1 + yu_2$ for unknowns x and y; that is,

$$\begin{bmatrix} a \\ b \end{bmatrix} = x\begin{bmatrix} 1 \\ -2 \end{bmatrix} + y\begin{bmatrix} 3 \\ -4 \end{bmatrix} \quad \text{or} \quad \begin{aligned} x+3y &= a \\ -2x-4y &= b \end{aligned} \quad \text{or} \quad \begin{aligned} x+3y &= a \\ 2y &= 2a+b \end{aligned}$$

Solve for x and y in terms of a and b to get $x = -2a - \frac{3}{2}b$ and $y = a + \frac{1}{2}b$. Thus,

$$(a,b) = (-2a - \tfrac{3}{2})u_1 + (a + \tfrac{1}{2}b)u_2 \qquad \text{or} \qquad [(a,b)]_S = [-2a - \tfrac{3}{2}b,\ a + \tfrac{1}{2}b]^T$$

(b) Use part (a) to write each of the basis vectors v_1 and v_2 of S' as a linear combination of the basis vectors u_1 and u_2 of S; that is,

$$\begin{aligned} v_1 &= (1,3) = (-2 - \tfrac{9}{2})u_1 + (1 + \tfrac{3}{2})u_2 = -\tfrac{13}{2}u_1 + \tfrac{5}{2}u_2 \\ v_2 &= (3,8) = (-6-12)u_1 + (3+4)u_2 = -18u_1 + 7u_2 \end{aligned}$$

Then P is the matrix whose columns are the coordinates of v_1 and v_2 relative to the basis S; that is,

$$P = \begin{bmatrix} -\frac{13}{2} & -18 \\ \frac{5}{2} & 7 \end{bmatrix}$$

(c) Let $v = xv_1 + yv_2$ for unknown scalars x and y:

$$\begin{bmatrix} a \\ b \end{bmatrix} = x\begin{bmatrix} 1 \\ 3 \end{bmatrix} + y\begin{bmatrix} 3 \\ 8 \end{bmatrix} \quad \text{or} \quad \begin{matrix} x + 3y = a \\ 3x + 8y = b \end{matrix} \quad \text{or} \quad \begin{matrix} x + 3y = a \\ -y = b - 3a \end{matrix}$$

Solve for x and y to get $x = -8a + 3b$ and $y = 3a - b$. Thus,

$$(a, b) = (-8a + 3b)v_1 + (3a - b)v_2 \quad \text{or} \quad [(a, b)]_{S'} = [-8a + 3b, \ \ 3a - b]^T$$

(d) Use part (c) to express each of the basis vectors u_1 and u_2 of S as a linear combination of the basis vectors v_1 and v_2 of S':

$$u_1 = (1, -2) = (-8 - 6)v_1 + (3 + 2)v_2 = -14v_1 + 5v_2$$
$$u_2 = (3, -4) = (-24 - 12)v_1 + (9 + 4)v_2 = -36v_1 + 13v_2$$

Write the coordinates of u_1 and u_2 relative to S' as columns to obtain $Q = \begin{bmatrix} -14 & -36 \\ 5 & 13 \end{bmatrix}$.

(e) $QP = \begin{bmatrix} -14 & -36 \\ 5 & 13 \end{bmatrix}\begin{bmatrix} -\frac{13}{2} & -18 \\ \frac{5}{2} & 7 \end{bmatrix} = \begin{bmatrix} 1 & 0 \\ 0 & 1 \end{bmatrix} = I$

(f) Use parts (a), (c), and (d) to obtain

$$P^{-1}[v]_S = Q[v]_S = \begin{bmatrix} -14 & -36 \\ 5 & 13 \end{bmatrix}\begin{bmatrix} -2a - \frac{3}{2}b \\ a + \frac{1}{2}b \end{bmatrix} = \begin{bmatrix} -8a + 3b \\ 3a - b \end{bmatrix} = [v]_{S'}$$

6.18. Suppose P is the change-of-basis matrix from a basis $\{u_i\}$ to a basis $\{w_i\}$, and suppose Q is the change-of-basis matrix from the basis $\{w_i\}$ back to $\{u_i\}$. Prove that P is invertible and that $Q = P^{-1}$.

Suppose, for $i = 1, 2, \ldots, n$, that

$$w_i = a_{i1}u_1 + a_{i2}u_2 + \ldots + a_{in}u_n = \sum_{j=1}^{n} a_{ij}u_j \tag{1}$$

and, for $j = 1, 2, \ldots, n$,

$$u_j = b_{j1}w_1 + b_{j2}w_2 + \cdots + b_{jn}w_n = \sum_{k=1}^{n} b_{jk}w_k \tag{2}$$

Let $A = [a_{ij}]$ and $B = [b_{jk}]$. Then $P = A^T$ and $Q = B^T$. Substituting (2) into (1) yields

$$w_i = \sum_{j=1}^{n} a_{ij}\left(\sum_{k=1}^{n} b_{jk}w_k\right) = \sum_{k=1}^{n}\left(\sum_{j=1}^{n} a_{ij}b_{jk}\right)w_k$$

Because $\{w_i\}$ is a basis, $\sum a_{ij}b_{jk} = \delta_{ik}$, where δ_{ik} is the Kronecker delta; that is, $\delta_{ik} = 1$ if $i = k$ but $\delta_{ik} = 0$ if $i \neq k$. Suppose $AB = [c_{ik}]$. Then $c_{ik} = \delta_{ik}$. Accordingly, $AB = I$, and so

$$QP = B^TA^T = (AB)^T = I^T = I$$

Thus, $Q = P^{-1}$.

6.19. Consider a finite sequence of vectors $S = \{u_1, u_2, \ldots, u_n\}$. Let S' be the sequence of vectors obtained from S by one of the following "elementary operations":

(1) Interchange two vectors.

(2) Multiply a vector by a nonzero scalar.

(3) Add a multiple of one vector to another vector.

Show that S and S' span the same subspace W. Also, show that S' is linearly independent if and only if S is linearly independent.

Observe that, for each operation, the vectors S' are linear combinations of vectors in S. Also, because each operation has an inverse of the same type, each vector in S is a linear combination of vectors in S'. Thus, S and S' span the same subspace W. Moreover, S' is linearly independent if and only if $\dim W = n$, and this is true if and only if S is linearly independent.

6.20. Let $A = [a_{ij}]$ and $B = [b_{ij}]$ be row equivalent $m \times n$ matrices over a field K, and let $v_1, v_2, \ldots, v_n$ be any vectors in a vector space V over K. For $i = 1, 2, \ldots, m$, let u_i and w_i be defined by

$$u_i = a_{i1}v_1 + a_{i2}v_2 + \cdots + a_{in}v_n \quad \text{and} \quad w_i = b_{i1}v_1 + b_{i2}v_2 + \cdots + b_{in}v_n$$

Show that $\{u_i\}$ and $\{w_i\}$ span the same subspace of V.

Applying an "elementary operation" of Problem 6.19 to $\{u_i\}$ is equivalent to applying an elementary row operation to the matrix A. Because A and B are row equivalent, B can be obtained from A by a sequence of elementary row operations. Hence, $\{w_i\}$ can be obtained from $\{u_i\}$ by the corresponding sequence of operations. Accordingly, $\{u_i\}$ and $\{w_i\}$ span the same space.

6.21. Suppose $u_1, u_2, \ldots, u_n$ belong to a vector space V over a field K, and suppose $P = [a_{ij}]$ is an n-square matrix over K. For $i = 1, 2, \ldots, n$, let $v_i = a_{i1}u_1 + a_{i2}u_2 + \cdots + a_{in}u_n$.

(a) Suppose P is invertible. Show that $\{u_i\}$ and $\{v_i\}$ span the same subspace of V. Hence, $\{u_i\}$ is linearly independent if and only if $\{v_i\}$ is linearly independent.

(b) Suppose P is singular (not invertible). Show that $\{v_i\}$ is linearly dependent.

(c) Suppose $\{v_i\}$ is linearly independent. Show that P is invertible.

(a) Because P is invertible, it is row equivalent to the identity matrix I. Hence, by Problem 6.19, $\{v_i\}$ and $\{u_i\}$ span the same subspace of V. Thus, one is linearly independent if and only if the other is linearly independent.

(b) Because P is not invertible, it is row equivalent to a matrix with a zero row. This means $\{v_i\}$ spans a subspace that has a spanning set with less than n elements. Thus, $\{v_i\}$ is linearly dependent.

(c) This is the contrapositive of the statement of part (b), and so it follows from part (b).

6.22. Prove Theorem 6.6: Let P be the change-of-basis matrix from a basis S to a basis S' in a vector space V. Then, for any vector $v \in V$, we have $P[v]_{S'} = [v]_S$, and hence, $P^{-1}[v]_S = [v]_{S'}$.

Suppose $S = \{u_1, \ldots, u_n\}$ and $S' = \{w_1, \ldots, w_n\}$, and suppose, for $i = 1, \ldots, n$,

$$w_i = a_{i1}u_1 + a_{i2}u_2 + \cdots + a_{in}u_n = \sum_{j=1}^{n} a_{ij}u_j$$

Then P is the n-square matrix whose jth row is

$$(a_{1j}, a_{2j}, \ldots, a_{nj}) \tag{1}$$

Also suppose $v = k_1w_1 + k_2w_2 + \cdots + k_nw_n = \sum_{i=1}^{n} k_iw_i$. Then

$$[v]_{S'} = [k_1, k_2, \ldots, k_n]^T \tag{2}$$

Substituting for w_i in the equation for v, we obtain

$$v = \sum_{i=1}^{n} k_iw_i = \sum_{i=1}^{n} k_i\left(\sum_{j=1}^{n} a_{ij}u_j\right) = \sum_{j=1}^{n}\left(\sum_{i=1}^{n} a_{ij}k_i\right)u_j$$

$$= \sum_{j=1}^{n}(a_{1j}k_1 + a_{2j}k_2 + \cdots + a_{nj}k_n)u_j$$

Accordingly, $[v]_S$ is the column vector whose jth entry is

$$a_{1j}k_1 + a_{2j}k_2 + \cdots + a_{nj}k_n \tag{3}$$

On the other hand, the jth entry of $P[v]_{S'}$ is obtained by multiplying the jth row of P by $[v]_{S'}$—that is, (1) by (2). However, the product of (1) and (2) is (3). Hence, $P[v]_{S'}$ and $[v]_S$ have the same entries. Thus, $P[v]_{S'} = [v]_S$, as claimed.

Furthermore, multiplying the above by P^{-1} gives $P^{-1}[v]_S = P^{-1}P[v]_{S'} = [v]_{S'}$.

Linear Operators and Change of Basis

6.23. Consider the linear transformation F on $\mathbf{R}^2$ defined by $F(x,y) = (5x - y,\ 2x + y)$ and the following bases of $\mathbf{R}^2$:

$$E = \{e_1, e_2\} = \{(1,0),\ (0,1)\} \qquad \text{and} \qquad S = \{u_1, u_2\} = \{(1,4),\ (2,7)\}$$

(a) Find the change-of-basis matrix P from E to S and the change-of-basis matrix Q from S back to E.

(b) Find the matrix A that represents F in the basis E.

(c) Find the matrix B that represents F in the basis S.

(a) Because E is the usual basis, simply write the vectors in S as columns to obtain the change-of-basis matrix P. Recall, also, that $Q = P^{-1}$. Thus,

$$P = \begin{bmatrix} 1 & 2 \\ 4 & 7 \end{bmatrix} \quad \text{and} \quad Q = P^{-1} = \begin{bmatrix} -7 & 2 \\ 4 & -1 \end{bmatrix}$$

(b) Write the coefficients of x and y in $F(x,y) = (5x - y,\ 2x + y)$ as rows to get

$$A = \begin{bmatrix} 5 & -1 \\ 2 & 1 \end{bmatrix}$$

(c) **Method 1.** Find the coordinates of $F(u_1)$ and $F(u_2)$ relative to the basis S. This may be done by first finding the coordinates of an arbitrary vector (a,b) in $\mathbf{R}^2$ relative to the basis S. We have

$$(a,b) = x(1,4) + y(2,7) = (x + 2y,\ 4x + 7y), \quad \text{and so} \quad \begin{aligned} x + 2y &= a \\ 4x + 7y &= b \end{aligned}$$

Solve for x and y in terms of a and b to get $x = -7a + 2b,\ y = 4a - b$. Then

$$(a,b) = (-7a + 2b)u_1 + (4a - b)u_2$$

Now use the formula for (a,b) to obtain

$$\begin{aligned} F(u_1) &= F(1,4) = (1,6) = 5u_1 - 2u_2 \\ F(u_2) &= F(2,7) = (3,11) = u_1 + u_2 \end{aligned} \quad \text{and so} \quad B = \begin{bmatrix} 5 & 1 \\ -2 & 1 \end{bmatrix}$$

Method 2. By Theorem 6.7, $B = P^{-1}AP$. Thus,

$$B = P^{-1}AP = \begin{bmatrix} -7 & 2 \\ 4 & -1 \end{bmatrix} \begin{bmatrix} 5 & -1 \\ 2 & 1 \end{bmatrix} \begin{bmatrix} 1 & 2 \\ 4 & 7 \end{bmatrix} = \begin{bmatrix} 5 & 1 \\ -2 & 1 \end{bmatrix}$$

6.24. Let $A = \begin{bmatrix} 2 & 3 \\ 4 & -1 \end{bmatrix}$. Find the matrix B that represents the linear operator A relative to the basis $S = \{u_1, u_2\} = \{[1,3]^T,\ [2,5]^T\}$. [Recall A defines a linear operator $A: \mathbf{R}^2 \to \mathbf{R}^2$ relative to the usual basis E of $\mathbf{R}^2$].

Method 1. Find the coordinates of $A(u_1)$ and $A(u_2)$ relative to the basis S by first finding the coordinates of an arbitrary vector $[a,b]^T$ in $\mathbf{R}^2$ relative to the basis S. By Problem 6.2,

$$[a,b]^T = (-5a + 2b)u_1 + (3a - b)u_2$$

Using the formula for $[a,b]^T$, we obtain

$$A(u_1) = \begin{bmatrix} 2 & 3 \\ 4 & -1 \end{bmatrix} \begin{bmatrix} 1 \\ 3 \end{bmatrix} = \begin{bmatrix} 11 \\ 1 \end{bmatrix} = -53u_1 + 32u_2$$

and

$$A(u_2) = \begin{bmatrix} 2 & 3 \\ 4 & -1 \end{bmatrix} \begin{bmatrix} 2 \\ 5 \end{bmatrix} = \begin{bmatrix} 19 \\ 3 \end{bmatrix} = -89u_1 + 54u_2$$

Thus,

$$B = \begin{bmatrix} -53 & -89 \\ 32 & 54 \end{bmatrix}$$

Method 2. Use $B = P^{-1}AP$, where P is the change-of-basis matrix from the usual basis E to S. Thus, simply write the vectors in S (as columns) to obtain the change-of-basis matrix P and then use the formula for P^{-1}. This gives

$$P = \begin{bmatrix} 1 & 2 \\ 3 & 5 \end{bmatrix} \quad \text{and} \quad P^{-1} = \begin{bmatrix} -5 & 2 \\ 3 & -1 \end{bmatrix}$$

Then
$$B = P^{-1}AP = \begin{bmatrix} 1 & 2 \\ 3 & 5 \end{bmatrix}\begin{bmatrix} 2 & 3 \\ 4 & -1 \end{bmatrix}\begin{bmatrix} -5 & 2 \\ 3 & -1 \end{bmatrix} = \begin{bmatrix} -53 & -89 \\ 32 & 54 \end{bmatrix}$$

6.25. Let $A = \begin{bmatrix} 1 & 3 & 1 \\ 2 & 5 & -4 \\ 1 & -2 & 2 \end{bmatrix}$. Find the matrix B that represents the linear operator A relative to the basis

$$S = \{u_1, u_2, u_3\} = \{[1,1,0]^T, \quad [0,1,1]^T, \quad [1,2,2]^T\}$$

[Recall A that defines a linear operator $A: \mathbf{R}^3 \to \mathbf{R}^3$ relative to the usual basis E of $\mathbf{R}^3$.]

Method 1. Find the coordinates of $A(u_1)$, $A(u_2)$, $A(u_3)$ relative to the basis S by first finding the coordinates of an arbitrary vector $v = (a, b, c)$ in $\mathbf{R}^3$ relative to the basis S. By Problem 6.16,

$$[v]_S = (b - c)u_1 + (-2a + 2b - c)u_2 + (a - b + c)u_3$$

Using this formula for $[a, b, c]^T$, we obtain

$$A(u_1) = [4, 7, -1]^T = 8u_1 + 7u_2 - 4u_3, \qquad A(u_2) = [4, 1, 0]^T = u_1 - 6u_2 + 3u_3$$
$$A(u_3) = [9, 4, 1]^T = 3u_1 - 11u_2 + 6u_3$$

Writing the coefficients of u_1, u_2, u_3 as columns yields

$$B = \begin{bmatrix} 8 & 1 & 3 \\ 7 & -6 & -11 \\ -4 & 3 & 6 \end{bmatrix}$$

Method 2. Use $B = P^{-1}AP$, where P is the change-of-basis matrix from the usual basis E to S. The matrix P (whose columns are simply the vectors in S) and P^{-1} appear in Problem 6.16. Thus,

$$B = P^{-1}AP = \begin{bmatrix} 0 & 1 & -1 \\ -2 & 2 & -1 \\ 1 & -1 & 1 \end{bmatrix}\begin{bmatrix} 1 & 3 & 1 \\ 2 & 5 & -4 \\ 1 & -2 & 2 \end{bmatrix}\begin{bmatrix} 1 & 0 & 1 \\ 1 & 1 & 2 \\ 0 & 1 & 2 \end{bmatrix} = \begin{bmatrix} 8 & 1 & 3 \\ 7 & -6 & -11 \\ -4 & 3 & 6 \end{bmatrix}$$

6.26. Prove Theorem 6.7: Let P be the change-of-basis matrix from a basis S to a basis S' in a vector space V. Then, for any linear operator T on V, $[T]_{S'} = P^{-1}[T]_S P$.

Let v be a vector in V. Then, by Theorem 6.6, $P[v]_{S'} = [v]_S$. Therefore,

$$P^{-1}[T]_S P[v]_{S'} = P^{-1}[T]_S[v]_S = P^{-1}[T(v)]_S = [T(v)]_{S'}$$

But $[T]_{S'}[v]_{S'} = [T(v)]_{S'}$. Hence,

$$P^{-1}[T]_S P[v]_{S'} = [T]_{S'}[v]_{S'}$$

Because the mapping $v \mapsto [v]_{S'}$ is onto K^n, we have $P^{-1}[T]_S PX = [T]_{S'}X$ for every $X \in K^n$. Thus, $P^{-1}[T]_S P = [T]_{S'}$, as claimed.

Similarity of Matrices

6.27. Let $A = \begin{bmatrix} 4 & -2 \\ 3 & 6 \end{bmatrix}$ and $P = \begin{bmatrix} 1 & 2 \\ 3 & 4 \end{bmatrix}$.

(a) Find $B = P^{-1}AP$. (b) Verify $\operatorname{tr}(B) = \operatorname{tr}(A)$. (c) Verify $\det(B) = \det(A)$.

(a) First find P^{-1} using the formula for the inverse of a 2×2 matrix. We have

$$P^{-1} = \begin{bmatrix} -2 & 1 \\ \frac{3}{2} & -\frac{1}{2} \end{bmatrix}$$

Then

$$B = P^{-1}AP = \begin{bmatrix} -2 & 1 \\ \frac{3}{2} & -\frac{1}{2} \end{bmatrix} \begin{bmatrix} 4 & -2 \\ 3 & 6 \end{bmatrix} \begin{bmatrix} 1 & 2 \\ 3 & 4 \end{bmatrix} = \begin{bmatrix} 25 & 30 \\ -\frac{27}{2} & -15 \end{bmatrix}$$

(b) $\text{tr}(A) = 4 + 6 = 10$ and $\text{tr}(B) = 25 - 15 = 10$. Hence, $\text{tr}(B) = \text{tr}(A)$.

(c) $\det(A) = 24 + 6 = 30$ and $\det(B) = -375 + 405 = 30$. Hence, $\det(B) = \det(A)$.

6.28. Find the trace of each of the linear transformations F on $\mathbf{R}^3$ in Problem 6.4.

Find the trace (sum of the diagonal elements) of any matrix representation of F such as the matrix representation $[F] = [F]_E$ of F relative to the usual basis E given in Problem 6.4.

(a) $\text{tr}(F) = \text{tr}([F]) = 1 - 5 + 9 = 5$.

(b) $\text{tr}(F) = \text{tr}([F]) = 1 + 3 + 5 = 9$.

(c) $\text{tr}(F) = \text{tr}([F]) = 1 + 4 + 7 = 12$.

6.29. Write $A \approx B$ if A is similar to B—that is, if there exists an invertible matrix P such that $A = P^{-1}BP$. Prove that $\approx$ is an equivalence relation (on square matrices); that is,

(a) $A \approx A$, for every A. (b) If $A \approx B$, then $B \approx A$.

(c) If $A \approx B$ and $B \approx C$, then $A \approx C$.

(a) The identity matrix I is invertible, and $I^{-1} = I$. Because $A = I^{-1}AI$, we have $A \approx A$.

(b) Because $A \approx B$, there exists an invertible matrix P such that $A = P^{-1}BP$. Hence, $B = PAP^{-1} = (P^{-1})^{-1}AP^{-1}$ and P^{-1} is also invertible. Thus, $B \approx A$.

(c) Because $A \approx B$, there exists an invertible matrix P such that $A = P^{-1}BP$, and as $B \approx C$, there exists an invertible matrix Q such that $B = Q^{-1}CQ$. Thus,

$$A = P^{-1}BP = P^{-1}(Q^{-1}CQ)P = (P^{-1}Q^{-1})C(QP) = (QP)^{-1}C(QP)$$

and QP is also invertible. Thus, $A \approx C$.

6.30. Suppose B is similar to A, say $B = P^{-1}AP$. Prove

(a) $B^n = P^{-1}A^nP$, and so B^n is similar to A^n.

(b) $f(B) = P^{-1}f(A)P$, for any polynomial $f(x)$, and so $f(B)$ is similar to $f(A)$.

(c) B is a root of a polynomial $g(x)$ if and only if A is a root of $g(x)$.

(a) The proof is by induction on n. The result holds for $n = 1$ by hypothesis. Suppose $n > 1$ and the result holds for $n - 1$. Then

$$B^n = BB^{n-1} = (P^{-1}AP)(P^{-1}A^{n-1}P) = P^{-1}A^nP$$

(b) Suppose $f(x) = a_nx^n + \cdots + a_1x + a_0$. Using the left and right distributive laws and part (a), we have

$$\begin{aligned} P^{-1}f(A)P &= P^{-1}(a_nA^n + \cdots + a_1A + a_0I)P \\ &= P^{-1}(a_nA^n)P + \cdots + P^{-1}(a_1A)P + P^{-1}(a_0I)P \\ &= a_n(P^{-1}A^nP) + \cdots + a_1(P^{-1}AP) + a_0(P^{-1}IP) \\ &= a_nB^n + \cdots + a_1B + a_0I = f(B) \end{aligned}$$

(c) By part (b), $g(B) = 0$ if and only if $P^{-1}g(A)P = 0$ if and only if $g(A) = P0P^{-1} = 0$.

Matrix Representations of General Linear Mappings

6.31. Let $F: \mathbf{R}^3 \to \mathbf{R}^2$ be the linear map defined by $F(x, y, z) = (3x + 2y - 4z, \ x - 5y + 3z)$.

(a) Find the matrix of F in the following bases of $\mathbf{R}^3$ and $\mathbf{R}^2$:

$$S = \{w_1, w_2, w_3\} = \{(1, 1, 1), \ (1, 1, 0), \ (1, 0, 0)\} \quad \text{and} \quad S' = \{u_1, u_2\} = \{(1, 3), \ (2, 5)\}$$

(b) Verify Theorem 6.10: The action of F is preserved by its matrix representation; that is, for any v in $\mathbf{R}^3$, we have $[F]_{S,S'}[v]_S = [F(v)]_{S'}$.

(a) From Problem 6.2, $(a, b) = (-5a + 2b)u_1 + (3a - b)u_2$. Thus,

$$\begin{aligned} F(w_1) &= F(1,1,1) = (1,-1) = -7u_1 + 4u_2 \\ F(w_2) &= F(1,1,0) = (5,-4) = -33u_1 + 19u_2 \\ F(w_3) &= F(1,0,0) = (3,1) = -13u_1 + 8u_2 \end{aligned}$$

Write the coordinates of $F(w_1)$, $F(w_2)$, $F(w_3)$ as columns to get

$$[F]_{S,S'} = \begin{bmatrix} -7 & -33 & 13 \\ 4 & 19 & 8 \end{bmatrix}$$

(b) If $v = (x, y, z)$, then, by Problem 6.5, $v = zw_1 + (y - z)w_2 + (x - y)w_3$. Also,

$$F(v) = (3x + 2y - 4z,\ x - 5y + 3z) = (-13x - 20y + 26z)u_1 + (8x + 11y - 15z)u_2$$

Hence, $\quad [v]_S = (z,\ y - z,\ x - y)^T \quad$ and $\quad [F(v)]_{S'} = \begin{bmatrix} -13x - 20y + 26z \\ 8x + 11y - 15z \end{bmatrix}$

Thus, $\quad [F]_{S,S'}[v]_S = \begin{bmatrix} -7 & -33 & -13 \\ 4 & 19 & 8 \end{bmatrix} \begin{bmatrix} z \\ y - x \\ x - y \end{bmatrix} = \begin{bmatrix} -13x - 20y + 26z \\ 8x + 11y - 15z \end{bmatrix} = [F(v)]_{S'}$

6.32. Let $F: \mathbf{R}^n \to \mathbf{R}^m$ be the linear mapping defined as follows:

$$F(x_1, x_2, \ldots, x_n) = (a_{11}x_1 + \cdots + a_{1n}x_n,\ a_{21}x_1 + \cdots + a_{2n}x_n,\ \ldots,\ a_{m1}x_1 + \cdots + a_{mn}x_n)$$

(a) Show that the rows of the matrix $[F]$ representing F relative to the usual bases of $\mathbf{R}^n$ and $\mathbf{R}^m$ are the coefficients of the x_i in the components of $F(x_1, \ldots, x_n)$.

(b) Find the matrix representation of each of the following linear mappings relative to the usual basis of $\mathbf{R}^n$:

(i) $F: \mathbf{R}^2 \to \mathbf{R}^3$ defined by $F(x, y) = (3x - y,\ \ 2x + 4y,\ \ 5x - 6y)$.

(ii) $F: \mathbf{R}^4 \to \mathbf{R}^2$ defined by $F(x, y, s, t) = (3x - 4y + 2s - 5t,\ \ 5x + 7y - s - 2t)$.

(iii) $F: \mathbf{R}^3 \to \mathbf{R}^4$ defined by $F(x, y, z) = (2x + 3y - 8z,\ \ x + y + z,\ \ 4x - 5z,\ \ 6y)$.

(a) We have

$$\begin{aligned} F(1, 0, \ldots, 0) &= (a_{11}, a_{21}, \ldots, a_{m1}) \\ F(0, 1, \ldots, 0) &= (a_{12}, a_{22}, \ldots, a_{m2}) \\ &\cdots\cdots\cdots\cdots\cdots\cdots \\ F(0, 0, \ldots, 1) &= (a_{1n}, a_{2n}, \ldots, a_{mn}) \end{aligned} \quad \text{and thus,} \quad [F] = \begin{bmatrix} a_{11} & a_{12} & \cdots & a_{1n} \\ a_{21} & a_{22} & \cdots & a_{2n} \\ \cdots & \cdots & \cdots & \cdots \\ a_{m1} & a_{m2} & \cdots & a_{mn} \end{bmatrix}$$

(b) By part (a), we need only look at the coefficients of the unknown $x, y, \ldots$ in $F(x, y, \ldots)$. Thus,

$$\text{(i)}\ \ [F] = \begin{bmatrix} 3 & -1 \\ 2 & 4 \\ 5 & -6 \end{bmatrix}, \qquad \text{(ii)}\ \ [F] = \begin{bmatrix} 3 & -4 & 2 & -5 \\ 5 & 7 & -1 & -2 \end{bmatrix}, \qquad \text{(iii)}\ \ [F] = \begin{bmatrix} 2 & 3 & -8 \\ 1 & 1 & 1 \\ 4 & 0 & -5 \\ 0 & 6 & 0 \end{bmatrix}$$

6.33. Let $A = \begin{bmatrix} 2 & 5 & -3 \\ 1 & -4 & 7 \end{bmatrix}$. Recall that A determines a mapping $F: \mathbf{R}^3 \to \mathbf{R}^2$ defined by $F(v) = Av$, where vectors are written as columns. Find the matrix $[F]$ that represents the mapping relative to the following bases of $\mathbf{R}^3$ and $\mathbf{R}^2$:

(a) The usual bases of $\mathbf{R}^3$ and of $\mathbf{R}^2$.

(b) $S = \{w_1, w_2, w_3\} = \{(1,1,1),\ (1,1,0),\ (1,0,0)\}$ and $S' = \{u_1, u_2\} = \{(1,3),\ (2,5)\}$.

(a) Relative to the usual bases, $[F]$ is the matrix A.

(b) From Problem 6.2, $(a,b) = (-5a+2b)u_1 + (3a-b)u_2$. Thus,

$$F(w_1) = \begin{bmatrix} 2 & 5 & -3 \\ 1 & -4 & 7 \end{bmatrix} \begin{bmatrix} 1 \\ 1 \\ 1 \end{bmatrix} = \begin{bmatrix} 4 \\ 4 \end{bmatrix} = -12u_1 + 8u_2$$

$$F(w_2) = \begin{bmatrix} 2 & 5 & -3 \\ 1 & -4 & 7 \end{bmatrix} \begin{bmatrix} 1 \\ 1 \\ 0 \end{bmatrix} = \begin{bmatrix} 7 \\ -3 \end{bmatrix} = -41u_1 + 24u_2$$

$$F(w_3) = \begin{bmatrix} 2 & 5 & -3 \\ 1 & -4 & 7 \end{bmatrix} \begin{bmatrix} 1 \\ 0 \\ 0 \end{bmatrix} = \begin{bmatrix} 2 \\ 1 \end{bmatrix} = -8u_1 + 5u_2$$

Writing the coefficients of $F(w_1)$, $F(w_2)$, $F(w_3)$ as columns yields $[F] = \begin{bmatrix} -12 & -41 & -8 \\ 8 & 24 & 5 \end{bmatrix}$.

6.34. Consider the linear transformation T on $\mathbf{R}^2$ defined by $T(x,y) = (2x-3y,\ \ x+4y)$ and the following bases of $\mathbf{R}^2$:

$$E = \{e_1, e_2\} = \{(1,0),\ (0,1)\} \quad \text{and} \quad S = \{u_1, u_2\} = \{(1,3),\ (2,5)\}$$

(a) Find the matrix A representing T relative to the bases E and S.

(b) Find the matrix B representing T relative to the bases S and E.

(We can view T as a linear mapping from one space into another, each having its own basis.)

(a) From Problem 6.2, $(a,b) = (-5a+2b)u_1 + (3a-b)u_2$. Hence,

$$\begin{aligned} T(e_1) &= T(1,0) = (2,1) &= -8u_1 + 5u_2 \\ T(e_2) &= T(0,1) = (-3,4) &= 23u_1 - 13u_2 \end{aligned} \quad \text{and so} \quad A = \begin{bmatrix} -8 & 23 \\ 5 & -13 \end{bmatrix}$$

(b) We have

$$\begin{aligned} T(u_1) &= T(1,3) = (-7,13) &= -7e_1 + 13e_2 \\ T(u_2) &= T(2,5) = (-11,22) &= -11e_1 + 22e_2 \end{aligned} \quad \text{and so} \quad B = \begin{bmatrix} -7 & -11 \\ 13 & 22 \end{bmatrix}$$

6.35. How are the matrices A and B in Problem 6.34 related?

By Theorem 6.12, the matrices A and B are equivalent to each other; that is, there exist nonsingular matrices P and Q such that $B = Q^{-1}AP$, where P is the change-of-basis matrix from S to E, and Q is the change-of-basis matrix from E to S. Thus,

$$P = \begin{bmatrix} 1 & 2 \\ 3 & 5 \end{bmatrix}, \qquad Q = \begin{bmatrix} -5 & 2 \\ 3 & -1 \end{bmatrix}, \qquad Q^{-1} = \begin{bmatrix} 1 & 2 \\ 3 & 5 \end{bmatrix}$$

and

$$Q^{-1}AP = \begin{bmatrix} 1 & 2 \\ 3 & 5 \end{bmatrix} \begin{bmatrix} -8 & -23 \\ 5 & -13 \end{bmatrix} \begin{bmatrix} 1 & 2 \\ 3 & 5 \end{bmatrix} = \begin{bmatrix} -7 & -11 \\ 13 & 22 \end{bmatrix} = B$$

6.36. Prove Theorem 6.14: Let $F: V \to U$ be linear and, say, $\text{rank}(F) = r$. Then there exist bases V and of U such that the matrix representation of F has the following form, where I_r is the r-square identity matrix:

$$A = \begin{bmatrix} I_r & 0 \\ 0 & 0 \end{bmatrix}$$

Suppose $\dim V = m$ and $\dim U = n$. Let W be the kernel of F and U' the image of F. We are given that rank $(F) = r$. Hence, the dimension of the kernel of F is $m - r$. Let $\{w_1, \ldots, w_{m-r}\}$ be a basis of the kernel of F and extend this to a basis of V:

$$\{v_1, \ldots, v_r, w_1, \ldots, w_{m-r}\}$$

Set $\qquad u_1 = F(v_1),\ u_2 = F(v_2),\ \ldots,\ u_r = F(v_r)$

Then $\{u_1, \ldots, u_r\}$ is a basis of U', the image of F. Extend this to a basis of U, say

$$\{u_1, \ldots, u_r, u_{r+1}, \ldots, u_n\}$$

Observe that

$$\begin{aligned}
F(v_1) &= u_1 = 1u_1 + 0u_2 + \cdots + 0u_r + 0u_{r+1} + \cdots + 0u_n \\
F(v_2) &= u_2 = 0u_1 + 1u_2 + \cdots + 0u_r + 0u_{r+1} + \cdots + 0u_n \\
&\cdots\cdots\cdots\cdots\cdots\cdots\cdots\cdots\cdots \\
F(v_r) &= u_r = 0u_1 + 0u_2 + \cdots + 1u_r + 0u_{r+1} + \cdots + 0u_n \\
F(w_1) &= 0 = 0u_1 + 0u_2 + \cdots + 0u_r + 0u_{r+1} + \cdots + 0u_n \\
&\cdots\cdots\cdots\cdots\cdots\cdots\cdots\cdots\cdots \\
F(w_{m-r}) &= 0 = 0u_1 + 0u_2 + \cdots + 0u_r + 0u_{r+1} + \cdots + 0u_n
\end{aligned}$$

Thus, the matrix of F in the above bases has the required form.

SUPPLEMENTARY PROBLEMS

Matrices and Linear Operators

6.37. Let $F: \mathbf{R}^2 \to \mathbf{R}^2$ be defined by $F(x,y) = (4x + 5y,\ 2x - y)$.

(a) Find the matrix A representing F in the usual basis E.
(b) Find the matrix B representing F in the basis $S = \{u_1, u_2\} = \{(1,4),\ (2,9)\}$.
(c) Find P such that $B = P^{-1}AP$.
(d) For $v = (a,b)$, find $[v]_S$ and $[F(v)]_S$. Verify that $[F]_S[v]_S = [F(v)]_S$.

6.38. Let $A: \mathbf{R}^2 \to \mathbf{R}^2$ be defined by the matrix $A = \begin{bmatrix} 5 & -1 \\ 2 & 4 \end{bmatrix}$.

(a) Find the matrix B representing A relative to the basis $S = \{u_1, u_2\} = \{(1,3),\ (2,8)\}$. (Recall that A represents the mapping A relative to the usual basis E.)
(b) For $v = (a,b)$, find $[v]_S$ and $[A(v)]_S$.

6.39. For each linear transformation L on $\mathbf{R}^2$, find the matrix A representing L (relative to the usual basis of $\mathbf{R}^2$):

(a) L is the rotation in $\mathbf{R}^2$ counterclockwise by 45°.
(b) L is the reflection in $\mathbf{R}^2$ about the line $y = x$.
(c) L is defined by $L(1,0) = (3,5)$ and $L(0,1) = (7,-2)$.
(d) L is defined by $L(1,1) = (3,7)$ and $L(1,2) = (5,-4)$.

6.40. Find the matrix representing each linear transformation T on $\mathbf{R}^3$ relative to the usual basis of $\mathbf{R}^3$:

(a) $T(x,y,z) = (x,y,0)$. (b) $T(x,y,z) = (z,\ y+z,\ x+y+z)$.
(c) $T(x,y,z) = (2x - 7y - 4z,\ 3x + y + 4z,\ 6x - 8y + z)$.

6.41. Repeat Problem 6.40 using the basis $S = \{u_1, u_2, u_3\} = \{(1,1,0),\ (1,2,3),\ (1,3,5)\}$.

6.42. Let L be the linear transformation on $\mathbf{R}^3$ defined by

$$L(1,0,0) = (1,1,1), \qquad L(0,1,0) = (1,3,5), \qquad L(0,0,1) = (2,2,2)$$

(a) Find the matrix A representing L relative to the usual basis of $\mathbf{R}^3$.
(b) Find the matrix B representing L relative to the basis S in Problem 6.41.

6.43. Let $\mathbf{D}$ denote the differential operator; that is, $\mathbf{D}(f(t)) = df/dt$. Each of the following sets is a basis of a vector space V of functions. Find the matrix representing $\mathbf{D}$ in each basis:

(a) $\{e^t, e^{2t}, te^{2t}\}$. (b) $\{1, t, \sin 3t, \cos 3t\}$. (c) $\{e^{5t}, te^{5t}, t^2e^{5t}\}$.

6.44. Let **D** denote the differential operator on the vector space V of functions with basis $S = \{\sin\theta, \cos\theta\}$.

(a) Find the matrix $A = [\mathbf{D}]_S$. (b) Use A to show that **D** is a zero of $f(t) = t^2 + 1$.

6.45. Let V be the vector space of 2×2 matrices. Consider the following matrix M and usual basis E of V:

$$M = \begin{bmatrix} a & b \\ c & d \end{bmatrix} \quad \text{and} \quad E = \left\{ \begin{bmatrix} 1 & 0 \\ 0 & 0 \end{bmatrix}, \begin{bmatrix} 0 & 1 \\ 0 & 0 \end{bmatrix}, \begin{bmatrix} 0 & 0 \\ 1 & 0 \end{bmatrix}, \begin{bmatrix} 0 & 0 \\ 0 & 1 \end{bmatrix} \right\}$$

Find the matrix representing each of the following linear operators T on V relative to E:

(a) $T(A) = MA$. (b) $T(A) = AM$. (c) $T(A) = MA - AM$.

6.46. Let $\mathbf{1}_V$ and $\mathbf{0}_V$ denote the identity and zero operators, respectively, on a vector space V. Show that, for any basis S of V, (a) $[\mathbf{1}_V]_S = I$, the identity matrix. (b) $[\mathbf{0}_V]_S = 0$, the zero matrix.

Change of Basis

6.47. Find the change-of-basis matrix P from the usual basis E of $\mathbf{R}^2$ to a basis S, the change-of-basis matrix Q from S back to E, and the coordinates of $v = (a, b)$ relative to S, for the following bases S:

(a) $S = \{(1,2),\ (3,5)\}$. (c) $S = \{(2,5),\ (3,7)\}$.
(b) $S = \{(1,-3),\ (3,-8)\}$. (d) $S = \{(2,3),\ (4,5)\}$.

6.48. Consider the bases $S = \{(1,2),\ (2,3)\}$ and $S' = \{(1,3),\ (1,4)\}$ of $\mathbf{R}^2$. Find the change-of-basis matrix:

(a) P from S to S'. (b) Q from S' back to S.

6.49. Suppose that the x-axis and y-axis in the plane $\mathbf{R}^2$ are rotated counterclockwise 30° to yield new x'-axis and y'-axis for the plane. Find

(a) The unit vectors in the direction of the new x'-axis and y'-axis.
(b) The change-of-basis matrix P for the new coordinate system.
(c) The new coordinates of the points $A(1,3)$, $B(2,-5)$, $C(a,b)$.

6.50. Find the change-of-basis matrix P from the usual basis E of $\mathbf{R}^3$ to a basis S, the change-of-basis matrix Q from S back to E, and the coordinates of $v = (a,b,c)$ relative to S, where S consists of the vectors:

(a) $u_1 = (1,1,0), u_2 = (0,1,2), u_3 = (0,1,1)$.
(b) $u_1 = (1,0,1), u_2 = (1,1,2), u_3 = (1,2,4)$.
(c) $u_1 = (1,2,1), u_2 = (1,3,4), u_3 = (2,5,6)$.

6.51. Suppose S_1, S_2, S_3 are bases of V. Let P and Q be the change-of-basis matrices, respectively, from S_1 to S_2 and from S_2 to S_3. Prove that PQ is the change-of-basis matrix from S_1 to S_3.

Linear Operators and Change of Basis

6.52. Consider the linear operator F on $\mathbf{R}^2$ defined by $F(x,y) = (5x + y,\ 3x - 2y)$ and the following bases of $\mathbf{R}^2$:

$$S = \{(1,2),\ (2,3)\} \quad \text{and} \quad S' = \{(1,3),\ (1,4)\}$$

(a) Find the matrix A representing F relative to the basis S.
(b) Find the matrix B representing F relative to the basis S'.
(c) Find the change-of-basis matrix P from S to S'.
(d) How are A and B related?

6.53. Let $A: \mathbf{R}^2 \to \mathbf{R}^2$ be defined by the matrix $A = \begin{bmatrix} 1 & -1 \\ 3 & 2 \end{bmatrix}$. Find the matrix B that represents the linear operator A relative to each of the following bases: (a) $S = \{(1,3)^T,\ (2,5)^T\}$. (b) $S = \{(1,3)^T,\ (2,4)^T\}$.

6.54. Let $F: \mathbf{R}^2 \to \mathbf{R}^2$ be defined by $F(x,y) = (x - 3y,\ 2x - 4y)$. Find the matrix A that represents F relative to each of the following bases: (a) $S = \{(2,5),\ (3,7)\}$. (b) $S = \{(2,3),\ (4,5)\}$.

6.55. Let $A: \mathbf{R}^3 \to \mathbf{R}^3$ be defined by the matrix $A = \begin{bmatrix} 1 & 3 & 1 \\ 2 & 7 & 4 \\ 1 & 4 & 3 \end{bmatrix}$. Find the matrix B that represents the linear operator A relative to the basis $S = \{(1,1,1)^T,\ (0,1,1)^T,\ (1,2,3)^T\}$.

Similarity of Matrices

6.56. Let $A = \begin{bmatrix} 1 & 1 \\ 2 & -3 \end{bmatrix}$ and $P = \begin{bmatrix} 1 & -2 \\ 3 & -5 \end{bmatrix}$.

(a) Find $B = P^{-1}AP$. (b) Verify that $\operatorname{tr}(B) = \operatorname{tr}(A)$. (c) Verify that $\det(B) = \det(A)$.

6.57. Find the trace and determinant of each of the following linear maps on $\mathbf{R}^2$:

(a) $F(x,y) = (2x - 3y,\ 5x + 4y)$. (b) $G(x,y) = (ax + by,\ cx + dy)$.

6.58. Find the trace and determinant of each of the following linear maps on $\mathbf{R}^3$:

(a) $F(x,y,z) = (x + 3y,\ 3x - 2z,\ x - 4y - 3z)$.

(b) $G(x,y,z) = (y + 3z,\ 2x - 4z,\ 5x + 7y)$.

6.59. Suppose $S = \{u_1, u_2\}$ is a basis of V, and $T: V \to V$ is defined by $T(u_1) = 3u_1 - 2u_2$ and $T(u_2) = u_1 + 4u_2$. Suppose $S' = \{w_1, w_2\}$ is a basis of V for which $w_1 = u_1 + u_2$ and $w_2 = 2u_1 + 3u_2$.

(a) Find the matrices A and B representing T relative to the bases S and S', respectively.

(b) Find the matrix P such that $B = P^{-1}AP$.

6.60. Let A be a 2×2 matrix such that only A is similar to itself. Show that A is a scalar matrix, that is, that $A = \begin{bmatrix} a & 0 \\ 0 & a \end{bmatrix}$.

6.61. Show that all matrices similar to an invertible matrix are invertible. More generally, show that similar matrices have the same rank.

Matrix Representation of General Linear Mappings

6.62. Find the matrix representation of each of the following linear maps relative to the usual basis for $\mathbf{R}^n$:

(a) $F: \mathbf{R}^3 \to \mathbf{R}^2$ defined by $F(x,y,z) = (2x - 4y + 9z,\ 5x + 3y - 2z)$.

(b) $F: \mathbf{R}^2 \to \mathbf{R}^4$ defined by $F(x,y) = (3x + 4y,\ 5x - 2y,\ x + 7y,\ 4x)$.

(c) $F: \mathbf{R}^4 \to \mathbf{R}$ defined by $F(x_1, x_2, x_3, x_4) = 2x_1 + x_2 - 7x_3 - x_4$.

6.63. Let $G: \mathbf{R}^3 \to \mathbf{R}^2$ be defined by $G(x,y,z) = (2x + 3y - z,\ 4x - y + 2z)$.

(a) Find the matrix A representing G relative to the bases

$S = \{(1,1,0),\ (1,2,3),\ (1,3,5)\}$ and $S' = \{(1,2),\ (2,3)\}$

(b) For any $v = (a,b,c)$ in $\mathbf{R}^3$, find $[v]_S$ and $[G(v)]_{S'}$. (c) Verify that $A[v]_S = [G(v)]_{S'}$.

6.64. Let $H: \mathbf{R}^2 \to \mathbf{R}^2$ be defined by $H(x,y) = (2x + 7y,\ x - 3y)$ and consider the following bases of $\mathbf{R}^2$:

$S = \{(1,1),\ (1,2)\}$ and $S' = \{(1,4),\ (1,5)\}$

(a) Find the matrix A representing H relative to the bases S and S'.

(b) Find the matrix B representing H relative to the bases S' and S.

6.65. Let $F: \mathbf{R}^3 \to \mathbf{R}^2$ be defined by $F(x,y,z) = (2x + y - z, \quad 3x - 2y + 4z)$.

(a) Find the matrix A representing F relative to the bases

$$S = \{(1,1,1), \quad (1,1,0), \quad (1,0,0)\} \quad \text{and} \quad S' = (1,3), \quad (1,4)\}$$

(b) Verify that, for any $v = (a,b,c)$ in $\mathbf{R}^3$, $A[v]_S = [F(v)]_{S'}$.

6.66. Let S and S' be bases of V, and let $\mathbf{1}_V$ be the identity mapping on V. Show that the matrix A representing $\mathbf{1}_V$ relative to the bases S and S' is the inverse of the change-of-basis matrix P from S to S'; that is, $A = P^{-1}$.

6.67. Prove (a) Theorem 6.10, (b) Theorem 6.11, (c) Theorem 6.12, (d) Theorem 6.13. [*Hint:* See the proofs of the analogous Theorems 6.1 (Problem 6.9), 6.2 (Problem 6.10), 6.3 (Problem 6.11), and 6.7 (Problem 6.26).]

Miscellaneous Problems

6.68. Suppose $F: V \to V$ is linear. A subspace W of V is said to be *invariant* under F if $F(W) \subseteq W$. Suppose W is invariant under F and dim $W = r$. Show that F has a block triangular matrix representation $M = \begin{bmatrix} A & B \\ 0 & C \end{bmatrix}$ where A is an $r \times r$ submatrix.

6.69. Suppose $V = U + W$, and suppose U and V are each invariant under a linear operator $F: V \to V$. Also, suppose dim $U = r$ and dim $W = S$. Show that F has a block diagonal matrix representation $M = \begin{bmatrix} A & 0 \\ 0 & B \end{bmatrix}$ where A and B are $r \times r$ and $s \times s$ submatrices.

6.70. Two linear operators F and G on V are said to be *similar* if there exists an invertible linear operator T on V such that $G = T^{-1} \circ F \circ T$. Prove

(a) F and G are similar if and only if, for any basis S of V, $[F]_S$ and $[G]_S$ are similar matrices.

(b) If F is diagonalizable (similar to a diagonal matrix), then any similar matrix G is also diagonalizable.

ANSWERS TO SUPPLEMENTARY PROBLEMS

Notation: $M = [R_1; \quad R_2; \quad \ldots]$ represents a matrix M with rows $R_1, R_2, \ldots$.

6.37. (a) $A = [4,5; \quad 2,-1]$; (b) $B = [220, 487; \quad -98, -217]$; (c) $P = [1,2; \quad 4,9]$;
(d) $[v]_S = [9a - 2b, \quad -4a + b]^T$ and $[F(v)]_S = [32a + 47b, \quad -14a - 21b]^T$

6.38. (a) $B = [-6, -28; \quad 4, 15]$;
(b) $[v]_S = [4a - b, -\frac{3}{2}a + \frac{1}{2}b]^T$ and $[A(v)]_S = [18a - 8b, \frac{1}{2}(-13a + 7b)]$

6.39. (a) $[\sqrt{2}, -\sqrt{2}; \quad \sqrt{2}, \sqrt{2}]$; (b) $[0,1; \quad 1,0]$; (c) $[3,7; \quad 5,-2]$;
(d) $[1,2; \quad 18,-11]$

6.40. (a) $[1,0,0; \quad 0,1,0; \quad 0,0,0]$; (b) $[0,0,1; \quad 0,1,1; \quad 1,1,1]$;
(c) $[2,-7,-4; \quad 3,1,4; \quad 6,-8,1]$

6.41. (a) $[1,3,5; \quad 0,-5,-10; \quad 0,3,6]$; (b) $[0,1,2; \quad -1,2,3; \quad 1,0,0]$;
(c) $[15,65,104; \quad -49,-219,-351; \quad 29,130,208]$

6.42. (a) $[1,1,2; \quad 1,3,2; \quad 1,5,2]$; (b) $[0; \quad 2,14,22; \quad 0,-5,-8]$

6.43. (a) $[1,0,0; \quad 0,2,1; \quad 0,0,2]$; (b) $[0,1,0,0; \quad 0; \quad 0,0,0,-3; \quad 0,0,3,0]$;
(c) $[5,1,0; \quad 0,5,2; \quad 0,0,5]$

6.44. (a) $A = [0, -1; \quad 1, 0]$; (b) $A^2 + I = 0$

6.45. (a) $[a, 0, b, 0; \quad 0, a, 0, b; \quad c, 0, d, 0; \quad 0, c, 0, d]$;
(b) $[a, c, 0, 0; \quad b, d, 0, 0; \quad 0, 0, a, c; \quad 0, 0, b, d]$;
(c) $[0, -c, b, 0; \quad -b, a-d, 0, b; \quad c, 0, d-a, -c; \quad 0, c, -b, 0]$

6.47. (a) $[1, 3; \quad 2, 5]$, $[-5, 3; \quad 2, -1]$, $[v] = [-5a + 3b, \quad 2a - b]^T$;
(b) $[1, 3; \quad -3, -8]$, $[-8, -3; \quad 3, 1]$, $[v] = [-8a - 3b, \quad 3a + b]^T$;
(c) $[2, 3; \quad 5, 7]$, $[-7, 3; \quad 5, -2]$, $[v] = [-7a + 3b, \quad 5a - 2b]^T$;
(d) $[2, 4; \quad 3, 5]$, $[-\frac{5}{2}, 2; \quad \frac{3}{2}, -1]$, $[v] = [-\frac{5}{2}a + 2b, \quad \frac{3}{2}a - b]^T$

6.48. (a) $P = [3, 5; \quad -1, -2]$; (b) $Q = [2, 5; \quad -1, -3]$

6.49. Here $K = \sqrt{3}$.
(a) $\frac{1}{2}(K, 1)$, $\frac{1}{2}(-1, K)$;
(b) $P = \frac{1}{2}[K, -1; \quad 1, K]$;
(c) $\frac{1}{2}[K + 3, 3K - 1]^T$, $\frac{1}{2}[2K - 5, -5K - 2]^T$, $\frac{1}{2}[aK + b, bK - a]^T$

6.50. P is the matrix whose columns are u_1, u_2, u_3, $Q = P^{-1}$, $[v] = Q[a, b, c]^T$.
(a) $Q = [1, 0, 0; \quad 1, -1, 1; \quad -2, 2, -1]$, $[v] = [a, \quad a - b + c, \quad -2a + 2b - c]^T$;
(b) $Q = [0, -2, 1; \quad 2, 3, -2; \quad -1, -1, 1]$, $[v] = [-2b + c, \quad 2a + 3b - 2c, \quad -a - b + c]^T$;
(c) $Q = [-2, 2, -1; \quad -7, 4, -1; \quad 5, -3, 1]$, $[v] = [-2a + 2b - c, -7a + 4b - c, \quad 5a - 3b + c]^T$

6.52. (a) $[-23, -39; \quad 15, 26]$; (b) $[35, 41; \quad -27, -32]$; (c) $[3, 5; \quad -1, -2]$; (d) $B = P^{-1}AP$

6.53. (a) $[28, 47; \quad -15, -25]$; (b) $[13, 18; \quad -\frac{15}{2}, -10]$

6.54. (a) $[43, 60; \quad -33, -46]$; (b) $\frac{1}{2}[3, 7; \quad -5, -9]$

6.55. $[10, 8, 20; \quad 13, 11, 28; \quad -5, -4, -10]$

6.56. (a) $[-34, 57; \quad -19, 32]$; (b) $\mathrm{tr}(B) = \mathrm{tr}(A) = -2$; (c) $\det(B) = \det(A) = -5$

6.57. (a) $\mathrm{tr}(F) = 6, \det(F) = 23$; (b) $\mathrm{tr}(G) = a + d, \det(G) = ad - bc$

6.58. (a) $\mathrm{tr}(F) = -2$, $\det(F) = 13$; (b) $\mathrm{tr}(G) = 0$, $\det(G) = 22$

6.59. (a) $A = [3, 1; \quad -2, 4]$, $B = [8, 11; \quad -2, -1]$; (b) $P = [1, 2; \quad 1, 3]$

6.62. (a) $[2, -4, 9; \quad 5, 3, -2]$; (b) $[3, 5, 1, 4; \quad 4, -2, 7, 0]$; (c) $[2, 1, -7, -1]$

6.63. (a) $[-9, 1, 4; \quad 7, 2, 1]$; (b) $[v]_S = [-a + 2b - c, \quad 5a - 5b + 2c, \quad -3a + 3b - c]^T$, and $[G(v)]_{S'} = [2a - 11b + 7c, \quad 7b - 4c]^T$

6.64. (a) $A = [47, 85; \quad -38, -69]$; (b) $B = [71, 88; \quad -41, -51]$

6.65. $A = [3, 11, 5; \quad -1, -8, -3]$

CHAPTER 7

Inner Product Spaces, Orthogonality

7.1 Introduction

The definition of a vector space V involves an arbitrary field K. Here we first restrict K to be the real field $\mathbf{R}$, in which case V is called a *real vector space*; in the last sections of this chapter, we extend our results to the case where K is the complex field $\mathbf{C}$, in which case V is called a *complex vector space*. Also, we adopt the previous notation that

u, v, w	are vectors in V
a, b, c, k	are scalars in K

Furthermore, the vector spaces V in this chapter have finite dimension unless otherwise stated or implied.

Recall that the concepts of "length" and "orthogonality" did not appear in the investigation of arbitrary vector spaces V (although they did appear in Section 1.4 on the spaces $\mathbf{R}^n$ and $\mathbf{C}^n$). Here we place an additional structure on a vector space V to obtain an inner product space, and in this context these concepts are defined.

7.2 Inner Product Spaces

We begin with a definition.

DEFINITION: Let V be a real vector space. Suppose to each pair of vectors $u, v \in V$ there is assigned a real number, denoted by $\langle u, v\rangle$. This function is called a (*real*) *inner product* on V if it satisfies the following axioms:

[I_1] ***(Linear Property)***: $\langle au_1 + bu_2,\ v\rangle = a\langle u_1, v\rangle + b\langle u_2, v\rangle$.

[I_2] ***(Symmetric Property)***: $\langle u, v\rangle = \langle v, u\rangle$.

[I_3] ***(Positive Definite Property)***: $\langle u, u\rangle \geq 0$.; and $\langle u, u\rangle = 0$ if and only if $u = 0$.

The vector space V with an inner product is called a (*real*) *inner product space*.

Axiom [I_1] states that an inner product function is linear in the first position. Using [I_1] and the symmetry axiom [I_2], we obtain

$$\langle u,\ cv_1 + dv_2\rangle = \langle cv_1 + dv_2,\ u\rangle = c\langle v_1, u\rangle + d\langle v_2, u\rangle = c\langle u, v_1\rangle + d\langle u, v_2\rangle$$

That is, the inner product function is also linear in its second position. Combining these two properties and using induction yields the following general formula:

$$\left\langle \sum_i a_i u_i,\ \sum_j b_j v_j \right\rangle = \sum_i \sum_j a_i b_j \langle u_i, v_j\rangle$$

That is, an inner product of linear combinations of vectors is equal to a linear combination of the inner products of the vectors.

EXAMPLE 7.1 Let V be a real inner product space. Then, by linearity,

$$\begin{aligned}\langle 3u_1 - 4u_2,\ 2v_1 - 5v_2 + 6v_3\rangle &= 6\langle u_1, v_1\rangle - 15\langle u_1, v_2\rangle + 18\langle u_1, v_3\rangle \\ &\quad - 8\langle u_2, v_1\rangle + 20\langle u_2, v_2\rangle - 24\langle u_2, v_3\rangle\end{aligned}$$

$$\begin{aligned}\langle 2u - 5v,\ 4u + 6v\rangle &= 8\langle u, u\rangle + 12\langle u, v\rangle - 20\langle v, u\rangle - 30\langle v, v\rangle \\ &= 8\langle u, u\rangle - 8\langle v, u\rangle - 30\langle v, v\rangle\end{aligned}$$

Observe that in the last equation we have used the symmetry property that $\langle u, v\rangle = \langle v, u\rangle$.

Remark: Axiom $[I_1]$ by itself implies $\langle 0, 0\rangle = \langle 0v, 0\rangle = 0\langle v, 0\rangle = 0$. Thus, $[I_1]$, $[I_2]$, $[I_3]$ are equivalent to $[I_1]$, $[I_2]$, and the following axiom:

$[I_3']$ If $u \neq 0$, then $\langle u, u\rangle$ is positive.

That is, a function satisfying $[I_1]$, $[I_2]$, $[I_3']$ is an inner product.

Norm of a Vector

By the third axiom $[I_3]$ of an inner product, $\langle u, u\rangle$ is nonnegative for any vector u. Thus, its positive square root exists. We use the notation

$$\|u\| = \sqrt{\langle u, u\rangle}$$

This nonnegative number is called the *norm* or *length* of u. The relation $\|u\|^2 = \langle u, u\rangle$ will be used frequently.

Remark: If $\|u\| = 1$ or, equivalently, if $\langle u, u\rangle = 1$, then u is called a *unit vector* and it is said to be *normalized*. Every nonzero vector v in V can be multiplied by the reciprocal of its length to obtain the unit vector

$$\hat{v} = \frac{1}{\|v\|} v$$

which is a positive multiple of v. This process is called *normalizing* v.

7.3 Examples of Inner Product Spaces

This section lists the main examples of inner product spaces used in this text.

Euclidean *n*-Space $\mathbf{R}^n$

Consider the vector space $\mathbf{R}^n$. The *dot product* or *scalar product* in $\mathbf{R}^n$ is defined by

$$u \cdot v = a_1 b_1 + a_2 b_2 + \cdots + a_n b_n$$

where $u = (a_i)$ and $v = (b_i)$. This function defines an inner product on $\mathbf{R}^n$. The norm $\|u\|$ of the vector $u = (a_i)$ in this space is as follows:

$$\|u\| = \sqrt{u \cdot u} = \sqrt{a_1^2 + a_2^2 + \cdots + a_n^2}$$

On the other hand, by the Pythagorean theorem, the distance from the origin O in $\mathbf{R}^3$ to a point $P(a, b, c)$ is given by $\sqrt{a^2 + b^2 + c^2}$. This is precisely the same as the above-defined norm of the vector $v = (a, b, c)$ in $\mathbf{R}^3$. Because the Pythagorean theorem is a consequence of the axioms of

Euclidean geometry, the vector space $\mathbf{R}^n$ with the above inner product and norm is called *Euclidean n-space*. Although there are many ways to define an inner product on $\mathbf{R}^n$, we shall assume this inner product unless otherwise stated or implied. It is called the *usual* (or *standard*) *inner product* on $\mathbf{R}^n$.

Remark: Frequently the vectors in $\mathbf{R}^n$ will be represented by column vectors—that is, by $n \times 1$ column matrices. In such a case, the formula

$$\langle u, v \rangle = u^T v$$

defines the usual inner product on $\mathbf{R}^n$.

EXAMPLE 7.2 Let $u = (1, 3, -4, 2)$, $v = (4, -2, 2, 1)$, $w = (5, -1, -2, 6)$ in $\mathbf{R}^4$.

(a) Show $\langle 3u - 2v, w \rangle = 3\langle u, w \rangle - 2\langle v, w \rangle$.

By definition,

$$\langle u, w \rangle = 5 - 3 + 8 + 12 = 22 \qquad \text{and} \qquad \langle v, w \rangle = 20 + 2 - 4 + 6 = 24$$

Note that $3u - 2v = (-5, 13, -16, 4)$. Thus,

$$\langle 3u - 2v, \ w \rangle = -25 - 13 + 32 + 24 = 18$$

As expected, $3\langle u, w \rangle - 2\langle v, w \rangle = 3(22) - 2(24) = 18 = \langle 3u - 2v, \ w \rangle$.

(b) Normalize u and v.

By definition,

$$\|u\| = \sqrt{1 + 9 + 16 + 4} = \sqrt{30} \qquad \text{and} \qquad \|v\| = \sqrt{16 + 4 + 4 + 1} = 5$$

We normalize u and v to obtain the following unit vectors in the directions of u and v, respectively:

$$\hat{u} = \frac{1}{\|u\|} u = \left(\frac{1}{\sqrt{30}}, \frac{3}{\sqrt{30}}, \frac{-4}{\sqrt{30}}, \frac{2}{\sqrt{30}} \right) \qquad \text{and} \qquad \hat{v} = \frac{1}{\|v\|} v = \left(\frac{4}{5}, \frac{-2}{5}, \frac{2}{5}, \frac{1}{5} \right)$$

Function Space $C[a, b]$ and Polynomial Space $\mathbf{P}(t)$

The notation $C[a, b]$ is used to denote the vector space of all continuous functions on the closed interval $[a, b]$—that is, where $a \leq t \leq b$. The following defines an inner product on $C[a, b]$, where $f(t)$ and $g(t)$ are functions in $C[a, b]$:

$$\langle f, g \rangle = \int_a^b f(t) g(t) \, dt$$

It is called the *usual inner product* on $C[a, b]$.

The vector space $\mathbf{P}(t)$ of all polynomials is a subspace of $C[a, b]$ for any interval $[a, b]$, and hence, the above is also an inner product on $\mathbf{P}(t)$.

EXAMPLE 7.3

Consider $f(t) = 3t - 5$ and $g(t) = t^2$ in the polynomial space $\mathbf{P}(t)$ with inner product

$$\langle f, g \rangle = \int_0^1 f(t) g(t) \, dt.$$

(a) Find $\langle f, g \rangle$.

We have $f(t)g(t) = 3t^3 - 5t^2$. Hence,

$$\langle f, g \rangle = \int_0^1 (3t^3 - 5t^2) \, dt = \tfrac{3}{4} t^4 - \tfrac{5}{3} t^3 \Big|_0^1 = \tfrac{3}{4} - \tfrac{5}{3} = -\tfrac{11}{12}$$

(b) Find $\|f\|$ and $\|g\|$.

We have $[f(t)]^2 = f(t)f(t) = 9t^2 - 30t + 25$ and $[g(t)]^2 = t^4$. Then

$$\|f\|^2 = \langle f, f \rangle = \int_0^1 (9t^2 - 30t + 25)\, dt = 3t^3 - 15t^2 + 25t \Big|_0^1 = 13$$

$$\|g\|^2 = \langle g, g \rangle = \int_0^1 t^4\, dt = \tfrac{1}{5}t^5 \Big|_0^1 = \tfrac{1}{5}$$

Therefore, $\|f\| = \sqrt{13}$ and $\|g\| = \sqrt{\frac{1}{5}} = \frac{1}{5}\sqrt{5}$.

Matrix Space $\mathbf{M} = \mathbf{M}_{m,n}$

Let $\mathbf{M} = \mathbf{M}_{m,n}$, the vector space of all real $m \times n$ matrices. An inner product is defined on $\mathbf{M}$ by

$$\langle A, B \rangle = \operatorname{tr}(B^T A)$$

where, as usual, tr( ) is the trace—the sum of the diagonal elements. If $A = [a_{ij}]$ and $B = [b_{ij}]$, then

$$\langle A, B \rangle = \operatorname{tr}(B^T A) = \sum_{i=1}^{m}\sum_{j=1}^{n} a_{ij}b_{ij} \quad \text{and} \quad \|A\|^2 = \langle A, A \rangle = \sum_{i=1}^{m}\sum_{j=1}^{n} a_{ij}^2$$

That is, $\langle A, B \rangle$ is the sum of the products of the corresponding entries in A and B and, in particular, $\langle A, A \rangle$ is the sum of the squares of the entries of A.

Hilbert Space

Let V be the vector space of all infinite sequences of real numbers $(a_1, a_2, a_3, \ldots)$ satisfying

$$\sum_{i=1}^{\infty} a_i^2 = a_1^2 + a_2^2 + \cdots < \infty$$

that is, the sum converges. Addition and scalar multiplication are defined in V componentwise; that is, if

$$u = (a_1, a_2, \ldots) \qquad \text{and} \qquad v = (b_1, b_2, \ldots)$$

then

$$u + v = (a_1 + b_1,\ a_2 + b_2,\ \ldots) \qquad \text{and} \qquad ku = (ka_1, ka_2, \ldots)$$

An inner product is defined in v by

$$\langle u, v \rangle = a_1b_1 + a_2b_2 + \cdots$$

The above sum converges absolutely for any pair of points in V. Hence, the inner product is well defined. This inner product space is called l_2-*space* or *Hilbert space*.

7.4 Cauchy–Schwarz Inequality, Applications

The following formula (proved in Problem 7.8) is called the Cauchy–Schwarz inequality or Schwarz inequality. It is used in many branches of mathematics.

THEOREM 7.1: (Cauchy–Schwarz) For any vectors u and v in an inner product space V,

$$\langle u, v \rangle^2 \le \langle u, u \rangle \langle v, v \rangle \qquad \text{or} \qquad |\langle u, v \rangle| \le \|u\|\|v\|$$

Next we examine this inequality in specific cases.

EXAMPLE 7.4

(a) Consider any real numbers $a_1, \ldots, a_n$, $b_1, \ldots, b_n$. Then, by the Cauchy–Schwarz inequality,

$$(a_1b_1 + a_2b_2 + \cdots + a_nb_n)^2 \le (a_1^2 + \cdots + a_n^2)(b_1^2 + \cdots + b_n^2)$$

That is, $(u \cdot v)^2 \le \|u\|^2\|v\|^2$, where $u = (a_i)$ and $v = (b_i)$.

(b) Let f and g be continuous functions on the unit interval $[0, 1]$. Then, by the Cauchy–Schwarz inequality,

$$\left[\int_0^1 f(t)g(t)\,dt\right]^2 \le \int_0^1 f^2(t)\,dt \int_0^1 g^2(t)\,dt$$

That is, $(\langle f, g\rangle)^2 \le \|f\|^2\|v\|^2$. Here V is the inner product space $C[0, 1]$.

The next theorem (proved in Problem 7.9) gives the basic properties of a norm. The proof of the third property requires the Cauchy–Schwarz inequality.

THEOREM 7.2: Let V be an inner product space. Then the norm in V satisfies the following properties:

$[\mathrm{N}_1]$ $\|v\| \ge 0$; and $\|v\| = 0$ if and only if $v = 0$.

$[\mathrm{N}_2]$ $\|kv\| = |k|\|v\|$.

$[\mathrm{N}_3]$ $\|u + v\| \le \|u\| + \|v\|$.

The property $[\mathrm{N}_3]$ is called the *triangle inequality*, because if we view $u + v$ as the side of the triangle formed with sides u and v (as shown in Fig. 7-1), then $[\mathrm{N}_3]$ states that the length of one side of a triangle cannot be greater than the sum of the lengths of the other two sides.

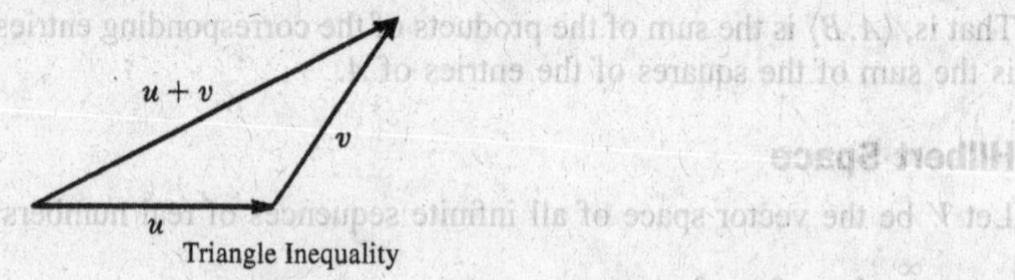

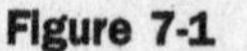

Figure 7-1

Angle Between Vectors

For any nonzero vectors u and v in an inner product space V, the *angle between u and v* is defined to be the angle θ such that $0 \le \theta \le \pi$ and

$$\cos\theta = \frac{\langle u, v\rangle}{\|u\|\|v\|}$$

By the Cauchy–Schwartz inequality, $-1 \le \cos\theta \le 1$, and so the angle exists and is unique.

EXAMPLE 7.5

(a) Consider vectors $u = (2, 3, 5)$ and $v = (1, -4, 3)$ in $\mathbf{R}^3$. Then

$$\langle u, v\rangle = 2 - 12 + 15 = 5, \qquad \|u\| = \sqrt{4 + 9 + 25} = \sqrt{38}, \qquad \|v\| = \sqrt{1 + 16 + 9} = \sqrt{26}$$

Then the angle θ between u and v is given by

$$\cos\theta = \frac{5}{\sqrt{38}\sqrt{26}}$$

Note that θ is an acute angle, because $\cos\theta$ is positive.

(b) Let $f(t) = 3t - 5$ and $g(t) = t^2$ in the polynomial space $\mathbf{P}(t)$ with inner product $\langle f, g\rangle = \int_0^1 f(t)g(t)\,dt$. By Example 7.3,

$$\langle f, g\rangle = -\tfrac{11}{12}, \qquad \|f\| = \sqrt{13}, \qquad \|g\| = \tfrac{1}{5}\sqrt{5}$$

Then the "angle" θ between f and g is given by

$$\cos\theta = \frac{-\frac{11}{12}}{(\sqrt{13})(\frac{1}{5}\sqrt{5})} = -\frac{55}{12\sqrt{13}\sqrt{5}}$$

Note that θ is an obtuse angle, because $\cos\theta$ is negative.

7.5 Orthogonality

Let V be an inner product space. The vectors $u, v \in V$ are said to be *orthogonal* and u is said to be *orthogonal* to v if

$$\langle u, v \rangle = 0$$

The relation is clearly symmetric—if u is orthogonal to v, then $\langle v, u \rangle = 0$, and so v is orthogonal to u. We note that $0 \in V$ is orthogonal to every $v \in V$, because

$$\langle 0, v \rangle = \langle 0v, v \rangle = 0\langle v, v \rangle = 0$$

Conversely, if u is orthogonal to every $v \in V$, then $\langle u, u \rangle = 0$ and hence $u = 0$ by $[I_3]$. Observe that u and v are orthogonal if and only if $\cos\theta = 0$, where θ is the angle between u and v. Also, this is true if and only if u and v are "perpendicular"—that is, $\theta = \pi/2$ (or $\theta = 90°$).

EXAMPLE 7.6

(a) Consider the vectors $u = (1, 1, 1)$, $v = (1, 2, -3)$, $w = (1, -4, 3)$ in $\mathbf{R}^3$. Then

$$\langle u, v \rangle = 1 + 2 - 3 = 0, \qquad \langle u, w \rangle = 1 - 4 + 3 = 0, \qquad \langle v, w \rangle = 1 - 8 - 9 = -16$$

Thus, u is orthogonal to v and w, but v and w are not orthogonal.

(b) Consider the functions $\sin t$ and $\cos t$ in the vector space $C[-\pi, \pi]$ of continuous functions on the closed interval $[-\pi, \pi]$. Then

$$\langle \sin t, \cos t \rangle = \int_{-\pi}^{\pi} \sin t \cos t \, dt = \tfrac{1}{2} \sin^2 t \big|_{-\pi}^{\pi} = 0 - 0 = 0$$

Thus, $\sin t$ and $\cos t$ are orthogonal functions in the vector space $C[-\pi, \pi]$.

Remark: A vector $w = (x_1, x_2, \ldots, x_n)$ is orthogonal to $u = (a_1, a_2, \ldots, a_n)$ in R^n if

$$\langle u, w \rangle = a_1x_1 + a_2x_2 + \cdots + a_nx_n = 0$$

That is, w is orthogonal to u if w satisfies a homogeneous equation whose coefficients are the elements of u.

EXAMPLE 7.7 Find a nonzero vector w that is orthogonal to $u_1 = (1, 2, 1)$ and $u_2 = (2, 5, 4)$ in R^3.

Let $w = (x, y, z)$. Then we want $\langle u_1, w \rangle = 0$ and $\langle u_2, w \rangle = 0$. This yields the homogeneous system

$$\begin{aligned} x + 2y + z &= 0 \\ 2x + 5y + 4z &= 0 \end{aligned} \qquad \text{or} \qquad \begin{aligned} x + 2y + z &= 0 \\ y + 2z &= 0 \end{aligned}$$

Here z is the only free variable in the echelon system. Set $z = 1$ to obtain $y = -2$ and $x = 3$. Thus, $w = (3, -2, 1)$ is a desired nonzero vector orthogonal to u_1 and u_2.

Any multiple of w will also be orthogonal to u_1 and u_2. Normalizing w, we obtain the following unit vector orthogonal to u_1 and u_2:

$$\hat{w} = \frac{w}{\|w\|} = \left(\frac{3}{\sqrt{14}}, -\frac{2}{\sqrt{14}}, \frac{1}{\sqrt{14}} \right)$$

Orthogonal Complements

Let S be a subset of an inner product space V. The orthogonal complement of S, denoted by $S^\perp$ (read "S perp") consists of those vectors in V that are orthogonal to every vector $u \in S$; that is,

$$S^\perp = \{v \in V : \langle v, u \rangle = 0 \text{ for every } u \in S\}$$

In particular, for a given vector u in V, we have

$$u^{\perp} = \{v \in V : \langle v, u \rangle = 0\}$$

that is, $u^{\perp}$ consists of all vectors in V that are orthogonal to the given vector u.

We show that $S^{\perp}$ is a subspace of V. Clearly $0 \in S^{\perp}$, because 0 is orthogonal to every vector in V. Now suppose $v, w \in S^{\perp}$. Then, for any scalars a and b and any vector $u \in S$, we have

$$\langle av + bw, \ u \rangle = a\langle v, u \rangle + b\langle w, u \rangle = a \cdot 0 + b \cdot 0 = 0$$

Thus, $av + bw \in S^{\perp}$, and therefore $S^{\perp}$ is a subspace of V.

We state this result formally.

PROPOSITION 7.3: Let S be a subset of a vector space V. Then $S^{\perp}$ is a subspace of V.

Remark 1: Suppose u is a nonzero vector in $\mathbf{R}^3$. Then there is a geometrical description of $u^{\perp}$. Specifically, $u^{\perp}$ is the plane in $\mathbf{R}^3$ through the origin O and perpendicular to the vector u. This is shown in Fig. 7-2.

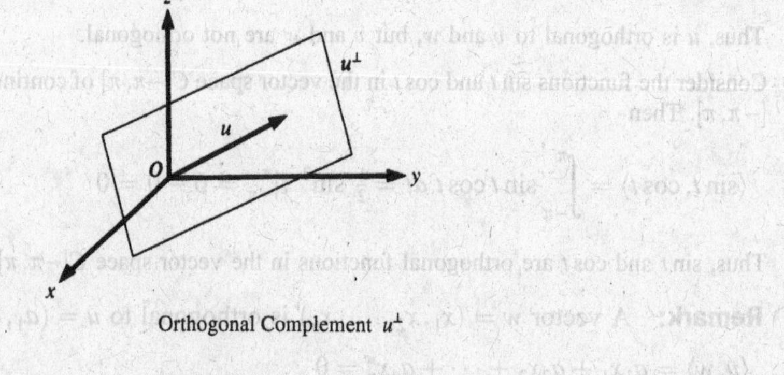

Orthogonal Complement $u^{\perp}$

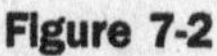

Figure 7-2

Remark 2: Let W be the solution space of an $m \times n$ homogeneous system $AX = 0$, where $A = [a_{ij}]$ and $X = [x_i]$. Recall that W may be viewed as the kernel of the linear mapping $A: \mathbf{R}^n \to \mathbf{R}^m$. Now we can give another interpretation of W using the notion of orthogonality. Specifically, each solution vector $w = (x_1, x_2, \ldots, x_n)$ is orthogonal to each row of A; hence, W is the orthogonal complement of the row space of A.

EXAMPLE 7.8 Find a basis for the subspace $u^{\perp}$ of $\mathbf{R}^3$, where $u = (1, 3, -4)$.

Note that $u^{\perp}$ consists of all vectors $w = (x, y, z)$ such that $\langle u, w \rangle = 0$, or $x + 3y - 4z = 0$. The free variables are y and z.

(1) Set $y = 1$, $z = 0$ to obtain the solution $w_1 = (-3, 1, 0)$.

(2) Set $y = 0$, $z = 1$ to obtain the solution $w_2 = (4, 0, 1)$.

The vectors w_1 and w_2 form a basis for the solution space of the equation, and hence a basis for $u^{\perp}$.

Suppose W is a subspace of V. Then both W and $W^{\perp}$ are subspaces of V. The next theorem, whose proof (Problem 7.28) requires results of later sections, is a basic result in linear algebra.

THEOREM 7.4: Let W be a subspace of V. Then V is the direct sum of W and $W^{\perp}$; that is, $V = W \oplus W^{\perp}$.

7.6 Orthogonal Sets and Bases

Consider a set $S = \{u_1, u_2, \ldots, u_r\}$ of nonzero vectors in an inner product space V. S is called *orthogonal* if each pair of vectors in S are orthogonal, and S is called *orthonormal* if S is orthogonal and each vector in S has unit length. That is,

(i) **Orthogonal:** $\langle u_i, u_j \rangle = 0$ for $i \neq j$

(ii) **Orthonormal:** $\langle u_i, u_j \rangle = \begin{cases} 0 & \text{for } i \neq j \\ 1 & \text{for } i = j \end{cases}$

Normalizing an orthogonal set S refers to the process of multiplying each vector in S by the reciprocal of its length in order to transform S into an orthonormal set of vectors.

The following theorems apply.

THEOREM 7.5: Suppose S is an orthogonal set of nonzero vectors. Then S is linearly independent.

THEOREM 7.6: (Pythagoras) Suppose $\{u_1, u_2, \ldots, u_r\}$ is an orthogonal set of vectors. Then

$$\|u_1 + u_2 + \cdots + u_r\|^2 = \|u_1\|^2 + \|u_2\|^2 + \cdots + \|u_r\|^2$$

These theorems are proved in Problems 7.15 and 7.16, respectively. Here we prove the Pythagorean theorem in the special and familiar case for two vectors. Specifically, suppose $\langle u, v \rangle = 0$. Then

$$\|u + v\|^2 = \langle u + v,\ u + v \rangle = \langle u, u \rangle + 2\langle u, v \rangle + \langle v, v \rangle = \langle u, u \rangle + \langle v, v \rangle = \|u\|^2 + \|v\|^2$$

which gives our result.

EXAMPLE 7.9

(a) Let $E = \{e_1, e_2, e_3\} = \{(1,0,0),\ (0,1,0),\ (0,0,1)\}$ be the usual basis of Euclidean space $\mathbf{R}^3$. It is clear that

$$\langle e_1, e_2 \rangle = \langle e_1, e_3 \rangle = \langle e_2, e_3 \rangle = 0 \qquad \text{and} \qquad \langle e_1, e_1 \rangle = \langle e_2, e_2 \rangle = \langle e_3, e_3 \rangle = 1$$

Namely, E is an orthonormal basis of $\mathbf{R}^3$. More generally, the usual basis of $\mathbf{R}^n$ is orthonormal for every n.

(b) Let $V = C[-\pi, \pi]$ be the vector space of continuous functions on the interval $-\pi \leq t \leq \pi$ with inner product defined by $\langle f, g \rangle = \int_{-\pi}^{\pi} f(t)g(t)\,dt$. Then the following is a classical example of an orthogonal set in V:

$$\{1, \cos t, \cos 2t, \cos 3t, \ldots, \sin t, \sin 2t, \sin 3t, \ldots\}$$

This orthogonal set plays a fundamental role in the theory of Fourier series.

Orthogonal Basis and Linear Combinations, Fourier Coefficients

Let S consist of the following three vectors in $\mathbf{R}^3$:

$$u_1 = (1,2,1), \qquad u_2 = (2,1,-4), \qquad u_3 = (3,-2,1)$$

The reader can verify that the vectors are orthogonal; hence, they are linearly independent. Thus, S is an orthogonal basis of $\mathbf{R}^3$.

Suppose we want to write $v = (7,1,9)$ as a linear combination of u_1, u_2, u_3. First we set v as a linear combination of u_1, u_2, u_3 using unknowns x_1, x_2, x_3 as follows:

$$v = x_1u_1 + x_2u_2 + x_3u_3 \qquad \text{or} \qquad (7,1,9) = x_1(1,2,1) + x_2(2,1,-4) + x_3(3,-2,1) \qquad (*)$$

We can proceed in two ways.

METHOD 1: Expand (*) (as in Chapter 3) to obtain the system

$$x_1 + 2x_2 + 3x_3 = 7, \qquad 2x_1 + x_2 - 2x_3 = 1, \qquad x_1 - 4x_2 + x_3 = 9$$

Solve the system by Gaussian elimination to obtain $x_1 = 3$, $x_2 = -1$, $x_3 = 2$. Thus, $v = 3u_1 - u_2 + 2u_3$.

METHOD 2: (This method uses the fact that the basis vectors are orthogonal, and the arithmetic is much simpler.) If we take the inner product of each side of (*) with respect to u_i, we get

$$\langle v, u_i \rangle = \langle x_1 u_2 + x_2 u_2 + x_3 u_3,\ u_i \rangle \quad \text{or} \quad \langle v, u_i \rangle = x_i \langle u_i, u_i \rangle \quad \text{or} \quad x_i = \frac{\langle v, u_i \rangle}{\langle u_i, u_i \rangle}$$

Here two terms drop out, because u_1, u_2, u_3 are orthogonal. Accordingly,

$$x_1 = \frac{\langle v, u_1 \rangle}{\langle u_1, u_1 \rangle} = \frac{7+2+9}{1+4+1} = \frac{18}{6} = 3, \qquad x_2 = \frac{\langle v, u_2 \rangle}{\langle u_2, u_2 \rangle} = \frac{14+1-36}{4+1+16} = \frac{-21}{21} = -1$$

$$x_3 = \frac{\langle v, u_3 \rangle}{\langle u_3, u_3 \rangle} = \frac{21-2+9}{9+4+1} = \frac{28}{14} = 2$$

Thus, again, we get $v = 3u_1 - u_2 + 2u_3$.

The procedure in Method 2 is true in general. Namely, we have the following theorem (proved in Problem 7.17).

THEOREM 7.7: Let $\{u_1, u_2, \ldots, u_n\}$ be an orthogonal basis of V. Then, for any $v \in V$,

$$v = \frac{\langle v, u_1 \rangle}{\langle u_1, u_1 \rangle} u_1 + \frac{\langle v, u_2 \rangle}{\langle u_2, u_2 \rangle} u_2 + \cdots + \frac{\langle v, u_n \rangle}{\langle u_n, u_n \rangle} u_n$$

Remark: The scalar $k_i \equiv \dfrac{\langle v, u_i \rangle}{\langle u_i, u_i \rangle}$ is called the *Fourier coefficient* of v with respect to u_i, because it is analogous to a coefficient in the Fourier series of a function. This scalar also has a geometric interpretation, which is discussed below.

Projections

Let V be an inner product space. Suppose w is a given nonzero vector in V, and suppose v is another vector. We seek the "projection of v along w," which, as indicated in Fig. 7-3(a), will be the multiple cw of w such that $v' = v - cw$ is orthogonal to w. This means

$$\langle v - cw,\ w \rangle = 0 \quad \text{or} \quad \langle v, w \rangle - c\langle w, w \rangle = 0 \quad \text{or} \quad c = \frac{\langle v, w \rangle}{\langle w, w \rangle}$$

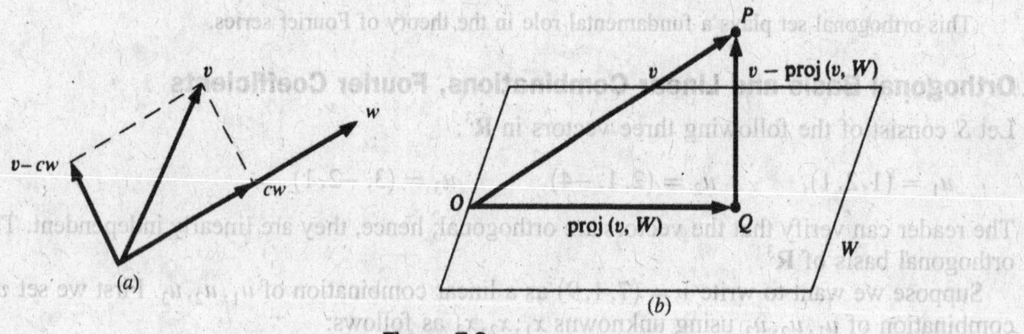

Figure 7-3

Accordingly, the *projection of v along w* is denoted and defined by

$$\text{proj}(v, w) = cw = \frac{\langle v, w \rangle}{\langle w, w \rangle} w$$

Such a scalar c is unique, and it is called the *Fourier coefficient* of v with respect to w or the *component* of v along w.

The above notion is generalized as follows (see Problem 7.25).

THEOREM 7.8: Suppose $w_1, w_2, \ldots, w_r$ form an orthogonal set of nonzero vectors in V. Let v be any vector in V. Define

$$v' = v - (c_1 w_1 + c_2 w_2 + \cdots + c_r w_r)$$

where

$$c_1 = \frac{\langle v, w_1 \rangle}{\langle w_1, w_1 \rangle}, \quad c_2 = \frac{\langle v, w_2 \rangle}{\langle w_2, w_2 \rangle}, \quad \ldots, \quad c_r = \frac{\langle v, w_r \rangle}{\langle w_r, w_r \rangle}$$

Then v' is orthogonal to $w_1, w_2, \ldots, w_r$.

Note that each c_i in the above theorem is the component (Fourier coefficient) of v along the given w_i.

Remark: The notion of the projection of a vector $v \in V$ along a subspace W of V is defined as follows. By Theorem 7.4, $V = W \oplus W^{\perp}$. Hence, v may be expressed uniquely in the form

$$v = w + w', \quad \text{where} \quad w \in W \quad \text{and} \quad w' \in W^{\perp}$$

We define w to be the *projection of v along W*, and denote it by $\text{proj}(v, W)$, as pictured in Fig. 7-3(b). In particular, if $W = \text{span}(w_1, w_2, \ldots, w_r)$, where the w_i form an orthogonal set, then

$$\text{proj}(v, W) = c_1 w_1 + c_2 w_2 + \cdots + c_r w_r$$

Here c_i is the component of v along w_i, as above.

7.7 Gram–Schmidt Orthogonalization Process

Suppose $\{v_1, v_2, \ldots, v_n\}$ is a basis of an inner product space V. One can use this basis to construct an orthogonal basis $\{w_1, w_2, \ldots, w_n\}$ of V as follows. Set

$$w_1 = v_1$$

$$w_2 = v_2 - \frac{\langle v_2, w_1 \rangle}{\langle w_1, w_1 \rangle} w_1$$

$$w_3 = v_3 - \frac{\langle v_3, w_1 \rangle}{\langle w_1, w_1 \rangle} w_1 - \frac{\langle v_3, w_2 \rangle}{\langle w_2, w_2 \rangle} w_2$$

$$\cdots\cdots\cdots\cdots\cdots\cdots\cdots\cdots\cdots\cdots\cdots\cdots$$

$$w_n = v_n - \frac{\langle v_n, w_1 \rangle}{\langle w_1, w_1 \rangle} w_1 - \frac{\langle v_n, w_2 \rangle}{\langle w_2, w_2 \rangle} w_2 - \cdots - \frac{\langle v_n, w_{n-1} \rangle}{\langle w_{n-1}, w_{n-1} \rangle} w_{n-1}$$

In other words, for $k = 2, 3, \ldots, n$, we define

$$w_k = v_k - c_{k1} w_1 - c_{k2} w_2 - \cdots - c_{k,k-1} w_{k-1}$$

where $c_{ki} = \langle v_k, w_i \rangle / \langle w_i, w_i \rangle$ is the component of v_k along w_i. By Theorem 7.8, each w_k is orthogonal to the preceding w's. Thus, $w_1, w_2, \ldots, w_n$ form an orthogonal basis for V as claimed. Normalizing each w_i will then yield an orthonormal basis for V.

The above construction is known as the *Gram–Schmidt orthogonalization process*. The following remarks are in order.

Remark 1: Each vector w_k is a linear combination of v_k and the preceding w's. Hence, one can easily show, by induction, that each w_k is a linear combination of $v_1, v_2, \ldots, v_n$.

Remark 2: Because taking multiples of vectors does not affect orthogonality, it may be simpler in hand calculations to clear fractions in any new w_k, by multiplying w_k by an appropriate scalar, before obtaining the next w_{k+1}.

Remark 3: Suppose $u_1, u_2, \ldots, u_r$ are linearly independent, and so they form a basis for $U = \text{span}(u_i)$. Applying the Gram–Schmidt orthogonalization process to the u's yields an orthogonal basis for U.

The following theorems (proved in Problems 7.26 and 7.27) use the above algorithm and remarks.

THEOREM 7.9: Let $\{v_1, v_2, \ldots, v_n\}$ be any basis of an inner product space V. Then there exists an orthonormal basis $\{u_1, u_2, \ldots, u_n\}$ of V such that the change-of-basis matrix from $\{v_i\}$ to $\{u_i\}$ is triangular; that is, for $k = 1, \ldots, n$,

$$u_k = a_{k1}v_1 + a_{k2}v_2 + \cdots + a_{kk}v_k$$

THEOREM 7.10: Suppose $S = \{w_1, w_2, \ldots, w_r\}$ is an orthogonal basis for a subspace W of a vector space V. Then one may extend S to an orthogonal basis for V; that is, one may find vectors $w_{r+1}, \ldots, w_n$ such that $\{w_1, w_2, \ldots, w_n\}$ is an orthogonal basis for V.

EXAMPLE 7.10 Apply the Gram–Schmidt orthogonalization process to find an orthogonal basis and then an orthonormal basis for the subspace U of $\mathbf{R}^4$ spanned by

$$v_1 = (1,1,1,1), \qquad v_2 = (1,2,4,5), \qquad v_3 = (1,-3,-4,-2)$$

(1) First set $w_1 = v_1 = (1,1,1,1)$.

(2) Compute

$$v_2 - \frac{\langle v_2, w_1\rangle}{\langle w_1, w_1\rangle}w_1 = v_2 - \frac{12}{4}w_1 = (-2,-1,1,2)$$

Set $w_2 = (-2,-1,1,2)$.

(3) Compute

$$v_3 - \frac{\langle v_3, w_1\rangle}{\langle w_1, w_1\rangle}w_1 - \frac{\langle v_3, w_2\rangle}{\langle w_2, w_2\rangle}w_2 = v_3 - \frac{(-8)}{4}w_1 - \frac{(-7)}{10}w_2 = \left(\tfrac{8}{5}, -\tfrac{17}{10}, -\tfrac{13}{10}, \tfrac{7}{5}\right)$$

Clear fractions to obtain $w_3 = (-6,-17,-13,14)$.

Thus, w_1, w_2, w_3 form an orthogonal basis for U. Normalize these vectors to obtain an orthonormal basis $\{u_1, u_2, u_3\}$ of U. We have $\|w_1\|^2 = 4$, $\|w_2\|^2 = 10$, $\|w_3\|^2 = 910$, so

$$u_1 = \frac{1}{2}(1,1,1,1), \qquad u_2 = \frac{1}{\sqrt{10}}(-2,-1,1,2), \qquad u_3 = \frac{1}{\sqrt{910}}(16,-17,-13,14)$$

EXAMPLE 7.11 Let V be the vector space of polynomials $f(t)$ with inner product $\langle f, g\rangle = \int_{-1}^{1} f(t)g(t)\,dt$. Apply the Gram–Schmidt orthogonalization process to $\{1, t, t^2, t^3\}$ to find an orthogonal basis $\{f_0, f_1, f_2, f_3\}$ with integer coefficients for $\mathbf{P}_3(t)$.

Here we use the fact that, for $r + s = n$,

$$\langle t^r, t^s\rangle = \int_{-1}^{1} t^n\,dt = \left.\frac{t^{n+1}}{n+1}\right|_{-1}^{1} = \begin{cases} 2/(n+1) & \text{when } n \text{ is even} \\ 0 & \text{when } n \text{ is odd} \end{cases}$$

(1) First set $f_0 = 1$.

(2) Compute $t - \dfrac{\langle t, 1\rangle}{\langle 1, 1\rangle}(1) = t - 0 = t$. Set $f_1 = t$.

(3) Compute

$$t^2 - \frac{\langle t^2, 1\rangle}{\langle 1, 1\rangle}(1) - \frac{\langle t^2, t\rangle}{\langle t, t\rangle}(t) = t^2 - \frac{\frac{2}{3}}{2}(1) + \frac{0}{\frac{2}{3}}(t) = t^2 - \tfrac{1}{3}$$

Multiply by 3 to obtain $f_2 = 3t^2 = 1$.

(4) Compute

$$t^3 - \frac{\langle t^3, 1\rangle}{\langle 1, 1\rangle}(1) - \frac{\langle t^3, t\rangle}{\langle t, t\rangle}(t) - \frac{\langle t^3, 3t^2 - 1\rangle}{\langle 3t^2 - 1,\ 3t^2 - 1\rangle}(3t^2 - 1)$$

$$= t^3 - 0(1) - \frac{\frac{2}{5}}{\frac{2}{3}}(t) - 0(3t^2 - 1) = t^3 - \tfrac{3}{5}t$$

Multiply by 5 to obtain $f_3 = 5t^3 - 3t$.

Thus, $\{1,\ t,\ 3t^2 - 1,\ 5t^3 - 3t\}$ is the required orthogonal basis.

Remark: Normalizing the polynomials in Example 7.11 so that $p(1) = 1$ yields the polynomials

$$1,\ t,\ \tfrac{1}{2}(3t^2 - 1),\ \tfrac{1}{2}(5t^3 - 3t)$$

These are the first four *Legendre polynomials*, which appear in the study of differential equations.

7.8 Orthogonal and Positive Definite Matrices

This section discusses two types of matrices that are closely related to real inner product spaces V. Here vectors in $\mathbf{R}^n$ will be represented by column vectors. Thus, $\langle u, v\rangle = u^T v$ denotes the inner product in Euclidean space $\mathbf{R}^n$.

Orthogonal Matrices

A real matrix P is *orthogonal* if P is nonsingular and $P^{-1} = P^T$, or, in other words, if $PP^T = P^T P = I$. First we recall (Theorem 2.6) an important characterization of such matrices.

THEOREM 7.11: Let P be a real matrix. Then the following are equivalent: (a) P is orthogonal; (b) the rows of P form an orthonormal set; (c) the columns of P form an orthonormal set.

(This theorem is true only using the usual inner product on $\mathbf{R}^n$. It is not true if $\mathbf{R}^n$ is given any other inner product.)

EXAMPLE 7.12

(a) Let $P = \begin{bmatrix} 1/\sqrt{3} & 1/\sqrt{3} & 1/\sqrt{3} \\ 0 & 1/\sqrt{2} & -1/\sqrt{2} \\ 2/\sqrt{6} & -1/\sqrt{6} & -1/\sqrt{6} \end{bmatrix}$. The rows of P are orthogonal to each other and are unit vectors. Thus P is an orthogonal matrix.

(b) Let P be a 2×2 orthogonal matrix. Then, for some real number θ, we have

$$P = \begin{bmatrix} \cos\theta & \sin\theta \\ -\sin\theta & \cos\theta \end{bmatrix} \quad \text{or} \quad P = \begin{bmatrix} \cos\theta & \sin\theta \\ \sin\theta & -\cos\theta \end{bmatrix}$$

The following two theorems (proved in Problems 7.37 and 7.38) show important relationships between orthogonal matrices and orthonormal bases of a real inner product space V.

THEOREM 7.12: Suppose $E = \{e_i\}$ and $E' = \{e_i'\}$ are orthonormal bases of V. Let P be the change-of-basis matrix from the basis E to the basis E'. Then P is orthogonal.

THEOREM 7.13: Let $\{e_1, \ldots, e_n\}$ be an orthonormal basis of an inner product space V. Let $P = [a_{ij}]$ be an orthogonal matrix. Then the following n vectors form an orthonormal basis for V:

$$e_i' = a_{1i}e_1 + a_{2i}e_2 + \cdots + a_{ni}e_n, \qquad i = 1, 2, \ldots, n$$

Positive Definite Matrices

Let A be a real symmetric matrix; that is, $A^T = A$. Then A is said to be *positive definite* if, for every nonzero vector u in $\mathbf{R}^n$,

$$\langle u, Au \rangle = u^T A u > 0$$

Algorithms to decide whether or not a matrix A is positive definite will be given in Chapter 13. However, for 2×2 matrices, we have simple criteria that we state formally in the following theorem (proved in Problem 7.43).

THEOREM 7.14: A 2×2 real symmetric matrix $A = \begin{bmatrix} a & b \\ c & d \end{bmatrix} = \begin{bmatrix} a & b \\ b & d \end{bmatrix}$ is positive definite if and only if the diagonal entries a and d are positive and the determinant $|A| = ad - bc = ad - b^2$ is positive.

EXAMPLE 7.13 Consider the following symmetric matrices:

$$A = \begin{bmatrix} 1 & 3 \\ 3 & 4 \end{bmatrix}, \qquad B = \begin{bmatrix} 1 & -2 \\ -2 & -3 \end{bmatrix}, \qquad C = \begin{bmatrix} 1 & -2 \\ -2 & 5 \end{bmatrix}$$

A is not positive definite, because $|A| = 4 - 9 = -5$ is negative. B is not positive definite, because the diagonal entry -3 is negative. However, C is positive definite, because the diagonal entries 1 and 5 are positive, and the determinant $|C| = 5 - 4 = 1$ is also positive.

The following theorem (proved in Problem 7.44) holds.

THEOREM 7.15: Let A be a real positive definite matrix. Then the function $\langle u, v \rangle = u^T A v$ is an inner product on $\mathbf{R}^n$.

Matrix Representation of an Inner Product (Optional)

Theorem 7.15 says that every positive definite matrix A determines an inner product on $\mathbf{R}^n$. This subsection may be viewed as giving the converse of this result.

Let V be a real inner product space with basis $S = \{u_1, u_2, \ldots, u_n\}$. The matrix

$$A = [a_{ij}], \qquad \text{where} \qquad a_{ij} = \langle u_i, u_j \rangle$$

is called the *matrix representation of the inner product on V relative to the basis S*.

Observe that A is symmetric, because the inner product is symmetric; that is, $\langle u_i, u_j \rangle = \langle u_j, u_i \rangle$. Also, A depends on both the inner product on V and the basis S for V. Moreover, if S is an orthogonal basis, then A is diagonal, and if S is an orthonormal basis, then A is the identity matrix.

EXAMPLE 7.14 The vectors $u_1 = (1, 1, 0)$, $u_2 = (1, 2, 3)$, $u_3 = (1, 3, 5)$ form a basis S for Euclidean space $\mathbf{R}^3$. Find the matrix A that represents the inner product in $\mathbf{R}^3$ relative to this basis S.

First compute each $\langle u_i, u_j \rangle$ to obtain

$$\langle u_1, u_1 \rangle = 1 + 1 + 0 = 2, \qquad \langle u_1, u_2 \rangle = 1 + 2 + 0 = 3, \qquad \langle u_1, u_3 \rangle = 1 + 3 + 0 = 4$$
$$\langle u_2, u_2 \rangle = 1 + 4 + 9 = 14, \qquad \langle u_2, u_3 \rangle = 1 + 6 + 15 = 22, \qquad \langle u_3, u_3 \rangle = 1 + 9 + 25 = 35$$

Then $A = \begin{bmatrix} 2 & 3 & 4 \\ 3 & 14 & 22 \\ 4 & 22 & 35 \end{bmatrix}$. As expected, A is symmetric.

The following theorems (proved in Problems 7.45 and 7.46, respectively) hold.

THEOREM 7.16: Let A be the matrix representation of an inner product relative to basis S for V. Then, for any vectors $u, v \in V$, we have

$$\langle u, v \rangle = [u]^T A [v]$$

where $[u]$ and $[v]$ denote the (column) coordinate vectors relative to the basis S.

THEOREM 7.17: Let A be the matrix representation of any inner product on V. Then A is a positive definite matrix.

7.9 Complex Inner Product Spaces

This section considers vector spaces over the complex field **C**. First we recall some properties of the complex numbers (Section 1.7), especially the relations between a complex number $z = a + bi$, where $a, b \in \mathbf{R}$, and its complex conjugate $\bar{z} = a - bi$:

$$z\bar{z} = a^2 + b^2, \qquad |z| = \sqrt{a^2 + b^2}, \qquad \overline{z_1 + z_2} = \bar{z}_1 + \bar{z}_2 \qquad \overline{z_1 z_2} = \bar{z}_1 \bar{z}_2, \qquad \bar{\bar{z}} = z$$

Also, z is real if and only if $\bar{z} = z$.

The following definition applies.

DEFINITION: Let V be a vector space over **C**. Suppose to each pair of vectors, $u, v \in V$ there is assigned a complex number, denoted by $\langle u, v \rangle$. This function is called a (*complex*) *inner product* on V if it satisfies the following axioms:

$[I_1^*]$ (***Linear Property***) $\langle au_1 + bu_2,\ v \rangle = a\langle u_1, v \rangle + b\langle u_2, v \rangle$

$[I_2^*]$ (***Conjugate Symmetric Property***) $\langle u, v \rangle = \overline{\langle v, u \rangle}$

$[I_3^*]$ (***Positive Definite Property***) $\langle u, u \rangle \geq 0$; and $\langle u, u \rangle = 0$ if and only if $u = 0$.

The vector space V over C with an inner product is called a (*complex*) *inner product space*. Observe that a complex inner product differs from the real case only in the second axiom $[I_2^*]$. Axiom $[I_1^*]$ (Linear Property) is equivalent to the two conditions:

$$(a) \quad \langle u_1 + u_2,\ v \rangle = \langle u_1, v \rangle + \langle u_2, v \rangle, \qquad (b) \quad \langle ku, v \rangle = k\langle u, v \rangle$$

On the other hand, applying $[I_1^*]$ and $[I_2^*]$, we obtain

$$\langle u, kv \rangle = \overline{\langle kv, u \rangle} = \overline{k\langle v, u \rangle} = \bar{k}\overline{\langle v, u \rangle} = \bar{k}\langle u, v \rangle$$

That is, we must take the conjugate of a complex number when it is taken out of the second position of a complex inner product. In fact (Problem 7.47), the inner product is *conjugate linear* in the second position; that is,

$$\langle u,\ av_1 + bv_2 \rangle = \bar{a}\langle u, v_1 \rangle + \bar{b}\langle u, v_2 \rangle$$

Combining linear in the first position and conjugate linear in the second position, we obtain, by induction,

$$\left\langle \sum_i a_i u_i,\ \sum_j b_j v_j \right\rangle = \sum_{i,j} a_i \bar{b}_j \langle u_i, v_j \rangle$$

The following remarks are in order.

Remark 1: Axiom $[I_1^*]$ by itself implies that $\langle 0, 0 \rangle = \langle 0v, 0 \rangle = 0\langle v, 0 \rangle = 0$. Accordingly, $[I_1^*]$, $[I_2^*]$, and $[I_3^*]$ are equivalent to $[I_1^*]$, $[I_2^*]$, and the following axiom:

$[I_3^{*\prime}]$ If $u \neq 0$, then $\langle u, u \rangle > 0$.

That is, a function satisfying $[I_1^*]$, $[I_2^*]$, and $[I_3^{*\prime}]$ is a (complex) inner product on V.

Remark 2: By $[I_2^*]$, $\langle u, u \rangle = \overline{\langle u, u \rangle}$. Thus, $\langle u, u \rangle$ must be real. By $[I_3^*]$, $\langle u, u \rangle$ must be nonnegative, and hence, its positive real square root exists. As with real inner product spaces, we define $\|u\| = \sqrt{\langle u, u \rangle}$ to be the norm or length of u.

Remark 3: In addition to the norm, we define the notions of orthogonality, orthogonal complement, and orthogonal and orthonormal sets as before. In fact, the definitions of distance and Fourier coefficient and projections are the same as in the real case.

EXAMPLE 7.15 (Complex Euclidean Space $\mathbf{C}^n$). Let $V = \mathbf{C}^n$, and let $u = (z_i)$ and $v = (w_i)$ be vectors in $\mathbf{C}^n$. Then

$$\langle u, v \rangle = \sum_k z_k \overline{w_k} = z_1\overline{w_1} + z_2\overline{w_2} + \cdots + z_n\overline{w_n}$$

is an inner product on V, called the *usual* or *standard inner product* on $\mathbf{C}^n$. V with this inner product is called Complex Euclidean Space. We assume this inner product on $\mathbf{C}^n$ unless otherwise stated or implied. Assuming u and v are column vectors, the above inner product may be defined by

$$\langle u, v \rangle = u^T \bar{v}$$

where, as with matrices, $\bar{v}$ means the conjugate of each element of v. If u and v are real, we have $\overline{w_i} = w_i$. In this case, the inner product reduced to the analogous one on $\mathbf{R}^n$.

EXAMPLE 7.16

(a) Let V be the vector space of complex continuous functions on the (real) interval $a \le t \le b$. Then the following is the *usual inner product* on V:

$$\langle f, g \rangle = \int_a^b f(t)\overline{g(t)}\, dt$$

(b) Let U be the vector space of $m \times n$ matrices over $\mathbf{C}$. Suppose $A = (z_{ij})$ and $B = (w_{ij})$ are elements of U. Then the following is the usual inner product on U:

$$\langle A, B \rangle = \operatorname{tr}(B^H A) = \sum_{i=1}^{m}\sum_{j=1}^{n} \bar{w}_{ij} z_{ij}$$

As usual, $B^H = \bar{B}^T$; that is, B^H is the conjugate transpose of B.

The following is a list of theorems for complex inner product spaces that are analogous to those for the real case. Here a Hermitian matrix A (i.e., one where $A^H = \bar{A}^T = A$) plays the same role that a symmetric matrix A (i.e., one where $A^T = A$) plays in the real case. (Theorem 7.18 is proved in Problem 7.50.)

THEOREM 7.18: (Cauchy–Schwarz) Let V be a complex inner product space. Then

$$|\langle u, v \rangle| \le \|u\|\|v\|$$

THEOREM 7.19: Let W be a subspace of a complex inner product space V. Then $V = W \oplus W^{\perp}$.

THEOREM 7.20: Suppose $\{u_1, u_2, \ldots, u_n\}$ is a basis for a complex inner product space V. Then, for any $v \in V$,

$$v = \frac{\langle v, u_1 \rangle}{\langle u_1, u_1 \rangle} u_1 + \frac{\langle v, u_2 \rangle}{\langle u_2, u_2 \rangle} u_2 + \cdots + \frac{\langle v, u_n \rangle}{\langle u_n, u_n \rangle} u_n$$

THEOREM 7.21: Suppose $\{u_1, u_2, \ldots, u_n\}$ is a basis for a complex inner product space V. Let $A = [a_{ij}]$ be the complex matrix defined by $a_{ij} = \langle u_i, u_j \rangle$. Then, for any $u, v \in V$,

$$\langle u, v \rangle = [u]^T A \overline{[v]}$$

where $[u]$ and $[v]$ are the coordinate column vectors in the given basis $\{u_i\}$. (*Remark*: This matrix A is said to represent the inner product on V.)

THEOREM 7.22: Let A be a Hermitian matrix (i.e., $A^H = \bar{A}^T = A$) such that $X^T A \bar{X}$ is real and positive for every nonzero vector $X \in \mathbf{C}^n$. Then $\langle u, v \rangle = u^T A \bar{v}$ is an inner product on $\mathbf{C}^n$.

THEOREM 7.23: Let A be the matrix that represents an inner product on V. Then A is Hermitian, and $X^T A X$ is real and positive for any nonzero vector in $\mathbf{C}^n$.

7.10 Normed Vector Spaces (Optional)

We begin with a definition.

DEFINITION: Let V be a real or complex vector space. Suppose to each $v \in V$ there is assigned a real number, denoted by $\|v\|$. This function $\|\cdot\|$ is called a *norm* on V if it satisfies the following axioms:

[N$_1$] $\|v\| \geq 0$; and $\|v\| = 0$ if and only if $v = 0$.

[N$_2$] $\|kv\| = |k|\|v\|$.

[N$_3$] $\|u + v\| \leq \|u\| + \|v\|$.

A vector space V with a norm is called a *normed vector space*.

Suppose V is a normed vector space. The *distance* between two vectors u and v in V is denoted and defined by

$$d(u, v) = \|u - v\|$$

The following theorem (proved in Problem 7.56) is the main reason why $d(u, v)$ is called the distance between u and v.

THEOREM 7.24: Let V be a normed vector space. Then the function $d(u, v) = \|u - v\|$ satisfies the following three axioms of a metric space:

[M$_1$] $d(u, v) \geq 0$; and $d(u, v) = 0$ if and only if $u = v$.

[M$_2$] $d(u, v) = d(v, u)$.

[M$_3$] $d(u, v) \leq d(u, w) + d(w, v)$.

Normed Vector Spaces and Inner Product Spaces

Suppose V is an inner product space. Recall that the norm of a vector v in V is defined by

$$\|v\| = \sqrt{\langle v, v\rangle}$$

One can prove (Theorem 7.2) that this norm satisfies [N$_1$], [N$_2$], and [N$_3$]. Thus, every inner product space V is a normed vector space. On the other hand, there may be norms on a vector space V that do not come from an inner product on V, as shown below.

Norms on $\mathbf{R}^n$ and $\mathbf{C}^n$

The following define three important norms on $\mathbf{R}^n$ and $\mathbf{C}^n$:

$$\|(a_1, \ldots, a_n)\|_\infty = \max(|a_i|)$$
$$\|(a_1, \ldots, a_n)\|_1 = |a_1| + |a_2| + \cdots + |a_n|$$
$$\|(a_1, \ldots, a_n)\|_2 = \sqrt{|a_1|^2 + |a_2|^2 + \cdots + |a_n|^2}$$

(Note that subscripts are used to distinguish between the three norms.) The norms $\|\cdot\|_\infty$, $\|\cdot\|_1$, and $\|\cdot\|_2$ are called the *infinity-norm*, *one-norm*, and *two-norm*, respectively. Observe that $\|\cdot\|_2$ is the norm on $\mathbf{R}^n$ (respectively, $\mathbf{C}^n$) induced by the usual inner product on $\mathbf{R}^n$ (respectively, $\mathbf{C}^n$). We will let d_∞, d_1, d_2 denote the corresponding distance functions.

EXAMPLE 7.17 Consider vectors $u = (1, -5, 3)$ and $v = (4, 2, -3)$ in $\mathbf{R}^3$.

(a) The infinity norm chooses the maximum of the absolute values of the components. Hence,

$$\|u\|_\infty = 5 \quad \text{and} \quad \|v\|_\infty = 4$$

(b) The one-norm adds the absolute values of the components. Thus,

$$\|u\|_1 = 1 + 5 + 3 = 9 \qquad \text{and} \qquad \|v\|_1 = 4 + 2 + 3 = 9$$

(c) The two-norm is equal to the square root of the sum of the squares of the components (i.e., the norm induced by the usual inner product on $\mathbf{R}^3$). Thus,

$$\|u\|_2 = \sqrt{1 + 25 + 9} = \sqrt{35} \qquad \text{and} \qquad \|v\|_2 = \sqrt{16 + 4 + 9} = \sqrt{29}$$

(d) Because $u - v = (1 - 4, \ -5 - 2, \ 3 + 3) = (-3, -7, 6)$, we have

$$d_\infty(u, v) = 7, \qquad d_1(u, v) = 3 + 7 + 6 = 16, \qquad d_2(u, v) = \sqrt{9 + 49 + 36} = \sqrt{94}$$

EXAMPLE 7.18 Consider the Cartesian plane $\mathbf{R}^2$ shown in Fig. 7-4.

(a) Let D_1 be the set of points $u = (x, y)$ in $\mathbf{R}^2$ such that $\|u\|_2 = 1$. Then D_1 consists of the points (x, y) such that $\|u\|_2^2 = x^2 + y^2 = 1$. Thus, D_1 is the unit circle, as shown in Fig. 7-4.

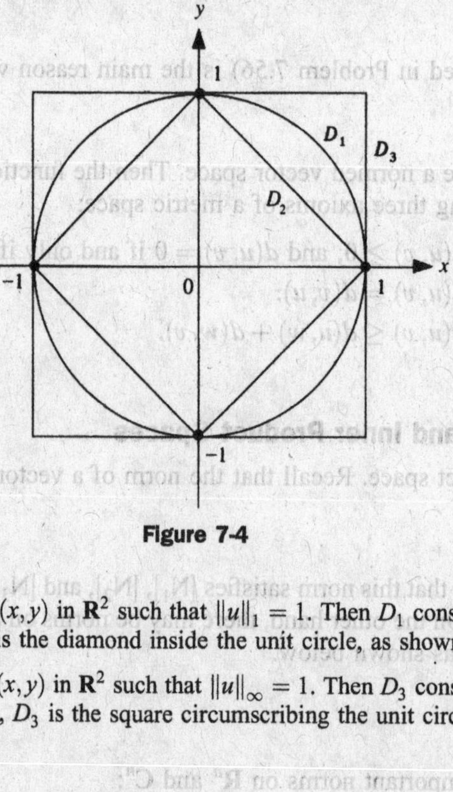

Figure 7-4

(b) Let D_2 be the set of points $u = (x, y)$ in $\mathbf{R}^2$ such that $\|u\|_1 = 1$. Then D_1 consists of the points (x, y) such that $\|u\|_1 = |x| + |y| = 1$. Thus, D_2 is the diamond inside the unit circle, as shown in Fig. 7-4.

(c) Let D_3 be the set of points $u = (x, y)$ in $\mathbf{R}^2$ such that $\|u\|_\infty = 1$. Then D_3 consists of the points (x, y) such that $\|u\|_\infty = \max(|x|, |y|) = 1$. Thus, D_3 is the square circumscribing the unit circle, as shown in Fig. 7-4.

Norms on $\mathbf{C}[a, b]$

Consider the vector space $V = C[a, b]$ of real continuous functions on the interval $a \le t \le b$. Recall that the following defines an inner product on V:

$$\langle f, g \rangle = \int_a^b f(t) g(t) \, dt$$

Accordingly, the above inner product defines the following norm on $V = C[a, b]$ (which is analogous to the $\|\cdot\|_2$ norm on $\mathbf{R}^n$):

$$\|f\|_2 = \sqrt{\int_a^b [f(t)]^2 \, dt}$$

The following define the other norms on $V = C[a, b]$:

$$\|f\|_1 = \int_a^b |f(t)|\, dt \quad \text{and} \quad \|f\|_\infty = \max(|f(t)|)$$

There are geometrical descriptions of these two norms and their corresponding distance functions, which are described below.

The first norm is pictured in Fig. 7-5. Here

$$\|f\|_1 = \text{area between the function } |f| \text{ and the } t\text{-axis}$$
$$d_1(f, g) = \text{area between the functions } f \text{ and } g$$

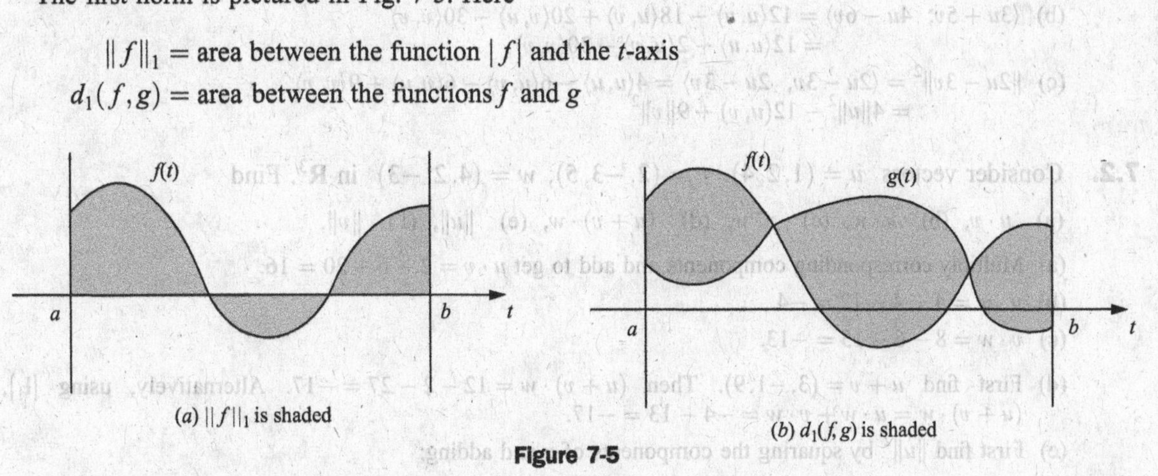

(a) $\|f\|_1$ is shaded

(b) $d_1(f, g)$ is shaded

Figure 7-5

This norm is analogous to the norm $\|\cdot\|_1$ on $\mathbf{R}^n$.

The second norm is pictured in Fig. 7-6. Here

$$\|f\|_\infty = \text{maximum distance between } f \text{ and the } t\text{-axis}$$
$$d_\infty(f, g) = \text{maximum distance between } f \text{ and } g$$

This norm is analogous to the norms $\|\cdot\|_\infty$ on $\mathbf{R}^n$.

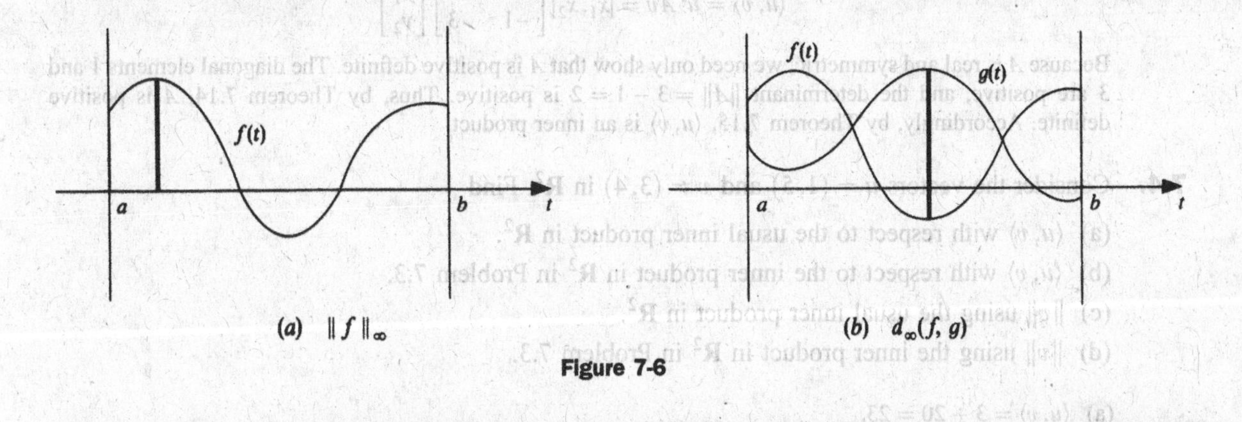

(a) $\|f\|_\infty$

(b) $d_\infty(f, g)$

Figure 7-6

SOLVED PROBLEMS

Inner Products

7.1. Expand:

(a) $\langle 5u_1 + 8u_2,\ 6v_1 - 7v_2\rangle$,

(b) $\langle 3u + 5v,\ 4u - 6v\rangle$,

(c) $\|2u - 3v\|^2$

Use linearity in both positions and, when possible, symmetry, $\langle u, v\rangle = \langle v, u\rangle$.

(a) Take the inner product of each term on the left with each term on the right:

$$\langle 5u_1 + 8u_2,\ 6v_1 - 7v_2\rangle = \langle 5u_1, 6v_1\rangle + \langle 5u_1, -7v_2\rangle + \langle 8u_2, 6v_1\rangle + \langle 8u_2, -7v_2\rangle$$
$$= 30\langle u_1, v_1\rangle - 35\langle u_1, v_2\rangle + 48\langle u_2, v_1\rangle - 56\langle u_2, v_2\rangle$$

[*Remark:* Observe the similarity between the above expansion and the expansion $(5a–8b)(6c–7d)$ in ordinary algebra.]

(b) $\langle 3u + 5v;\ 4u - 6v\rangle = 12\langle u, u\rangle - 18\langle u, v\rangle + 20\langle v, u\rangle - 30\langle v, v\rangle$
$= 12\langle u, u\rangle + 2\langle u, v\rangle - 30\langle v, v\rangle$

(c) $\|2u - 3v\|^2 = \langle 2u - 3v,\ 2u - 3v\rangle = 4\langle u, u\rangle - 6\langle u, v\rangle - 6\langle v, u\rangle + 9\langle v, v\rangle$
$= 4\|u\|^2 - 12(u, v) + 9\|v\|^2$

7.2. Consider vectors $u = (1, 2, 4)$, $v = (2, -3, 5)$, $w = (4, 2, -3)$ in $\mathbf{R}^3$. Find

(a) $u \cdot v$, (b) $u \cdot w$, (c) $v \cdot w$, (d) $(u + v) \cdot w$, (e) $\|u\|$, (f) $\|v\|$.

(a) Multiply corresponding components and add to get $u \cdot v = 2 - 6 + 20 = 16$.

(b) $u \cdot w = 4 + 4 - 12 = -4$.

(c) $v \cdot w = 8 - 6 - 15 = -13$.

(d) First find $u + v = (3, -1, 9)$. Then $(u + v) \cdot w = 12 - 2 - 27 = -17$. Alternatively, using $[I_1]$, $(u + v) \cdot w = u \cdot w + v \cdot w = -4 - 13 = -17$.

(e) First find $\|u\|^2$ by squaring the components of u and adding:

$$\|u\|^2 = 1^2 + 2^2 + 4^2 = 1 + 4 + 16 = 21, \qquad \text{and so} \qquad \|u\| = \sqrt{21}$$

(f) $\|v\|^2 = 4 + 9 + 25 = 38$, and so $\|v\| = \sqrt{38}$.

7.3. Verify that the following defines an inner product in $\mathbf{R}^2$:

$$\langle u, v\rangle = x_1y_1 - x_1y_2 - x_2y_1 + 3x_2y_2, \qquad \text{where} \qquad u = (x_1, x_2),\ v = (y_1, y_2)$$

We argue via matrices. We can write $\langle u, v\rangle$ in matrix notation as follows:

$$\langle u, v\rangle = u^TAv = [x_1, x_2]\begin{bmatrix} 1 & -1 \\ -1 & 3 \end{bmatrix}\begin{bmatrix} y_1 \\ y_2 \end{bmatrix}$$

Because A is real and symmetric, we need only show that A is positive definite. The diagonal elements 1 and 3 are positive, and the determinant $\|A\| = 3 - 1 = 2$ is positive. Thus, by Theorem 7.14, A is positive definite. Accordingly, by Theorem 7.15, $\langle u, v\rangle$ is an inner product.

7.4. Consider the vectors $u = (1, 5)$ and $v = (3, 4)$ in $\mathbf{R}^2$. Find

(a) $\langle u, v\rangle$ with respect to the usual inner product in $\mathbf{R}^2$.

(b) $\langle u, v\rangle$ with respect to the inner product in $\mathbf{R}^2$ in Problem 7.3.

(c) $\|v\|$ using the usual inner product in $\mathbf{R}^2$.

(d) $\|v\|$ using the inner product in $\mathbf{R}^2$ in Problem 7.3.

(a) $\langle u, v\rangle = 3 + 20 = 23$.

(b) $\langle u, v\rangle = 1 \cdot 3 - 1 \cdot 4 - 5 \cdot 3 + 3 \cdot 5 \cdot 4 = 3 - 4 - 15 + 60 = 44$.

(c) $\|v\|^2 = \langle v, v\rangle = \langle (3, 4), (3, 4)\rangle = 9 + 16 = 25$; hence, $|v| = 5$.

(d) $\|v\|^2 = \langle v, v\rangle = \langle (3, 4), (3, 4)\rangle = 9 - 12 - 12 + 48 = 33$; hence, $\|v\| = \sqrt{33}$.

7.5. Consider the following polynomials in $\mathbf{P}(t)$ with the inner product $\langle f, g\rangle = \int_0^1 f(t)g(t)\,dt$:

$$f(t) = t + 2, \qquad g(t) = 3t - 2, \qquad h(t) = t^2 - 2t - 3$$

(a) Find $\langle f, g\rangle$ and $\langle f, h\rangle$.

(b) Find $\|f\|$ and $\|g\|$.

(c) Normalize f and g.

(a) Integrate as follows:

$$\langle f,g\rangle = \int_0^1 (t+2)(3t-2)\,dt = \int_0^1 (3t^2+4t-4)\,dt = \left(t^3+2t^2-4t\right)\Big|_0^1 = -1$$

$$\langle f,h\rangle = \int_0^1 (t+2)(t^2-2t-3)\,dt = \left(\frac{t^4}{4}-\frac{7t^2}{2}-6t\right)\Big|_0^1 = -\frac{37}{4}$$

(b) $\langle f,f\rangle = \int_0^1 (t+2)(t+2)\,dt = \frac{19}{3}$; hence, $\|f\| = \sqrt{\frac{19}{3}} = \frac{1}{3}\sqrt{57}$

$$\langle g,g\rangle = \int_0^1 (3t-2)(3t-2) = 1; \quad \text{hence,} \quad \|g\| = \sqrt{1} = 1$$

(c) Because $\|f\| = \frac{1}{3}\sqrt{57}$ and g is already a unit vector, we have

$$\hat{f} = \frac{1}{\|f\|}f = \frac{3}{\sqrt{57}}(t+2) \quad \text{and} \quad \hat{g} = g = 3t-2$$

7.6. Find $\cos\theta$ where θ is the angle between:

(a) $u = (1,3,-5,4)$ and $v = (2,-3,4,1)$ in $\mathbf{R}^4$,

(b) $A = \begin{bmatrix} 9 & 8 & 7 \\ 6 & 5 & 4 \end{bmatrix}$ and $B = \begin{bmatrix} 1 & 2 & 3 \\ 4 & 5 & 6 \end{bmatrix}$, where $\langle A,B\rangle = \operatorname{tr}(B^TA)$.

Use $\cos\theta = \dfrac{\langle u, v\rangle}{\|u\|\|v\|}$

(a) Compute:

$$\langle u, v\rangle = 2-9-20+4 = -23, \qquad \|u\|^2 = 1+9+25+16 = 51, \qquad \|v\|^2 = 4+9+16+1 = 30$$

Thus,
$$\cos\theta = \frac{-23}{\sqrt{51}\sqrt{30}} = \frac{-23}{3\sqrt{170}}$$

(b) Use $\langle A,B\rangle = \operatorname{tr}(B^TA) = \sum_{i=1}^m \sum_{j=1}^n a_{ij}b_{ij}$, the sum of the products of corresponding entries.

$$\langle A,B\rangle = 9+16+21+24+25+24 = 119$$

Use $\|A\|^2 = \langle A,A\rangle = \sum_{i=1}^m \sum_{j=1}^n a_{ij}^2$, the sum of the squares of all the elements of A.

$$\|A\|^2 = \langle A,A\rangle = 9^2+8^2+7^2+6^2+5^2+4^2 = 271, \quad \text{and so} \quad \|A\| = \sqrt{271}$$

$$\|B\|^2 = \langle B,B\rangle = 1^2+2^2+3^2+4^2+5^2+6^2 = 91, \quad \text{and so} \quad \|B\| = \sqrt{91}$$

Thus,
$$\cos\theta = \frac{119}{\sqrt{271}\sqrt{91}}$$

7.7. Verify each of the following:

(a) Parallelogram Law (Fig. 7-7): $\|u+v\|^2 + \|u-v\|^2 = 2\|u\|^2 + 2\|v\|^2$.

(b) Polar form for $\langle u, v\rangle$ (which shows the inner product can be obtained from the norm function):

$$\langle u, v\rangle = \tfrac{1}{4}(\|u+v\|^2 - \|u-v\|^2).$$

Expand as follows to obtain

$$\|u+v\|^2 = \langle u+v, u+v\rangle = \|u\|^2 + 2\langle u, v\rangle + \|v\|^2 \tag{1}$$

$$\|u-v\|^2 = \langle u-v, u-v\rangle = \|u\|^2 - 2\langle u, v\rangle + \|v\|^2 \tag{2}$$

Add (1) and (2) to get the Parallelogram Law (a). Subtract (2) from (1) to obtain

$$\|u+v\|^2 - \|u-v\|^2 = 4\langle u, v\rangle$$

Divide by 4 to obtain the (real) polar form (b).

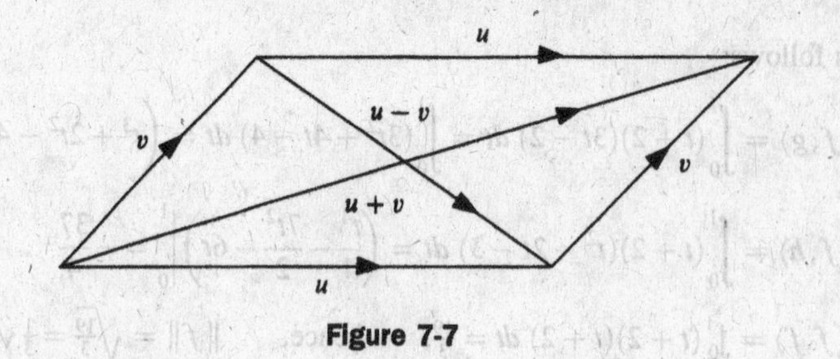

Figure 7-7

7.8. Prove Theorem 7.1 (Cauchy–Schwarz): For u and v in a real inner product space V, $\langle u, u\rangle^2 \leq \langle u, u\rangle\langle v, v\rangle$ or $|\langle u, v\rangle| \leq \|u\|\|v\|$.

For any real number t,

$$\langle tu + v,\ tu + v\rangle = t^2\langle u, u\rangle + 2t\langle u, v\rangle + \langle v, v\rangle = t^2\|u\|^2 + 2t\langle u, v\rangle + \|v\|^2$$

Let $a = \|u\|^2$, $b = 2\langle u, v\rangle$, $c = \|v\|^2$. Because $\|tu + v\|^2 \geq 0$, we have

$$at^2 + bt + c \geq 0$$

for every value of t. This means that the quadratic polynomial cannot have two real roots, which implies that $b^2 - 4ac \leq 0$ or $b^2 \leq 4ac$. Thus,

$$4\langle u, v\rangle^2 \leq 4\|u\|^2\|v\|^2$$

Dividing by 4 gives our result.

7.9. Prove Theorem 7.2: The norm in an inner product space V satisfies

(a) $[\mathrm{N}_1]$ $\|v\| \geq 0$; and $\|v\| = 0$ if and only if $v = 0$.

(b) $[\mathrm{N}_2]$ $\|kv\| = |k|\|v\|$.

(c) $[\mathrm{N}_3]$ $\|u + v\| \leq \|u\| + \|v\|$.

(a) If $v \neq 0$, then $\langle v, v\rangle > 0$, and hence, $\|v\| = \sqrt{\langle v, v\rangle} > 0$. If $v = 0$, then $\langle 0, 0\rangle = 0$. Consequently, $\|0\| = \sqrt{0} = 0$. Thus, $[\mathrm{N}_1]$ is true.

(b) We have $\|kv\|^2 = \langle kv, kv\rangle = k^2\langle v, v\rangle = k^2\|v\|^2$. Taking the square root of both sides gives $[\mathrm{N}_2]$.

(c) Using the Cauchy–Schwarz inequality, we obtain

$$\|u + v\|^2 = \langle u + v,\ u + v\rangle = \langle u, u\rangle + \langle u, v\rangle + \langle u, v\rangle + \langle v, v\rangle$$
$$\leq \|u\|^2 + 2\|u\|\|v\| + \|v\|^2 = (\|u\| + \|v\|)^2$$

Taking the square root of both sides yields $[\mathrm{N}_3]$.

Orthogonality, Orthonormal Complements, Orthogonal Sets

7.10. Find k so that $u = (1, 2, k, 3)$ and $v = (3, k, 7, -5)$ in $\mathbf{R}^4$ are orthogonal.

First find

$$\langle u, v\rangle = (1, 2, k, 3) \cdot (3, k, 7, -5) = 3 + 2k + 7k - 15 = 9k - 12$$

Then set $\langle u, v\rangle = 9k - 12 = 0$ to obtain $k = \frac{4}{3}$.

7.11. Let W be the subspace of $\mathbf{R}^5$ spanned by $u = (1, 2, 3, -1, 2)$ and $v = (2, 4, 7, 2, -1)$. Find a basis of the orthogonal complement $W^{\perp}$ of W.

We seek all vectors $w = (x, y, z, s, t)$ such that

$$\langle w, u\rangle = x + 2y + 3z - s + 2t = 0$$
$$\langle w, v\rangle = 2x + 4y + 7z + 2s - t = 0$$

Eliminating x from the second equation, we find the equivalent system

$$x + 2y + 3z - s + 2t = 0$$
$$z + 4s - 5t = 0$$

The free variables are y, s, and t. Therefore,

(1) Set $y = -1$, $s = 0$, $t = 0$ to obtain the solution $w_1 = (2, -1, 0, 0, 0)$.
(2) Set $y = 0$, $s = 1$, $t = 0$ to find the solution $w_2 = (13, 0, -4, 1, 0)$.
(3) Set $y = 0$, $s = 0$, $t = 1$ to obtain the solution $w_3 = (-17, 0, 5, 0, 1)$.

The set $\{w_1, w_2, w_3\}$ is a basis of $W^\perp$.

7.12. Let $w = (1, 2, 3, 1)$ be a vector in $\mathbf{R}^4$. Find an orthogonal basis for $w^\perp$.

Find a nonzero solution of $x + 2y + 3z + t = 0$, say $v_1 = (0, 0, 1, -3)$. Now find a nonzero solution of the system

$$x + 2y + 3z + t = 0, \qquad z - 3t = 0$$

say $v_2 = (0, -5, 3, 1)$. Last, find a nonzero solution of the system

$$x + 2y + 3z + t = 0, \qquad -5y + 3z + t = 0, \qquad z - 3t = 0$$

say $v_3 = (-14, 2, 3, 1)$. Thus, v_1, v_2, v_3 form an orthogonal basis for $w^\perp$.

7.13. Let S consist of the following vectors in $\mathbf{R}^4$:

$$u_1 = (1, 1, 0, -1), \quad u_2 = (1, 2, 1, 3), \quad u_3 = (1, 1, -9, 2), \quad u_4 = (16, -13, 1, 3)$$

(a) Show that S is orthogonal and a basis of $\mathbf{R}^4$.
(b) Find the coordinates of an arbitrary vector $v = (a, b, c, d)$ in $\mathbf{R}^4$ relative to the basis S.

(a) Compute

$$u_1 \cdot u_2 = 1 + 2 + 0 - 3 = 0, \qquad u_1 \cdot u_3 = 1 + 1 + 0 - 2 = 0, \qquad u_1 \cdot u_4 = 16 - 13 + 0 - 3 = 0$$
$$u_2 \cdot u_3 = 1 + 2 - 9 + 6 = 0, \qquad u_2 \cdot u_4 = 16 - 26 + 1 + 9 = 0, \qquad u_3 \cdot u_4 = 16 - 13 - 9 + 6 = 0$$

Thus, S is orthogonal, and S is linearly independent. Accordingly, S is a basis for $\mathbf{R}^4$ because any four linearly independent vectors form a basis of $\mathbf{R}^4$.

(b) Because S is orthogonal, we need only find the Fourier coefficients of v with respect to the basis vectors, as in Theorem 7.7. Thus,

$$k_1 = \frac{\langle v, u_1 \rangle}{\langle u_1, u_1 \rangle} = \frac{a + b - d}{3}, \qquad k_3 = \frac{\langle v, u_3 \rangle}{\langle u_3, u_3 \rangle} = \frac{a + b - 9c + 2d}{87}$$

$$k_2 = \frac{\langle v, u_2 \rangle}{\langle u_2, u_2 \rangle} = \frac{a + 2b + c + 3d}{15}, \qquad k_4 = \frac{\langle v, u_4 \rangle}{\langle u_4, u_4 \rangle} = \frac{16a - 13b + c + 3d}{435}$$

are the coordinates of v with respect to the basis S.

7.14. Suppose S, S_1, S_2 are the subsets of V. Prove the following (where $S^{\perp\perp}$ means $(S^\perp)^\perp$):

(a) $S \subseteq S^{\perp\perp}$.
(b) If $S_1 \subseteq S_2$, then $S_2^\perp \subseteq S_1^\perp$.
(c) $S^\perp = \text{span}\,(S)^\perp$.

(a) Let $w \in S$. Then $\langle w, v \rangle = 0$ for every $v \in S^\perp$; hence, $w \in S^{\perp\perp}$. Accordingly, $S \subseteq S^{\perp\perp}$.

(b) Let $w \in S_2^\perp$. Then $\langle w, v \rangle = 0$ for every $v \in S_2$. Because $S_1 \subseteq S_2$, $\langle w, v \rangle = 0$ for every $v = S_1$. Thus, $w \in S_1^\perp$, and hence, $S_2^\perp \subseteq S_1^\perp$.

(c) Because $S \subseteq \text{span}(S)$, part (b) gives us $\text{span}(S)^\perp \subseteq S^\perp$. Suppose $u \in S^\perp$ and $v \in \text{span}(S)$. Then there exist $w_1, w_2, \ldots, w_k$ in S such that $v = a_1 w_1 + a_2 w_2 + \cdots + a_k w_k$. Then, using $u \in S^\perp$, we have

$$\begin{aligned} \langle u, v \rangle &= \langle u, \ a_1 w_1 + a_2 w_2 + \cdots + a_k w_k \rangle = a_1 \langle u, w_1 \rangle + a_2 \langle u, w_2 \rangle + \cdots + a_k \langle u, w_k \rangle \\ &= a_1(0) + a_2(0) + \cdots + a_k(0) = 0 \end{aligned}$$

Thus, $u \in \text{span}(S)^\perp$. Accordingly, $S^\perp \subseteq \text{span}(S)^\perp$. Both inclusions give $S^\perp = \text{span}(S)^\perp$.

7.15. Prove Theorem 7.5: Suppose S is an orthogonal set of nonzero vectors. Then S is linearly independent.

Suppose $S = \{u_1, u_2, \ldots, u_r\}$ and suppose

$$a_1u_1 + a_2u_2 + \cdots + a_ru_r = 0 \tag{1}$$

Taking the inner product of (1) with u_1, we get

$$\begin{aligned} 0 = \langle 0, u_1 \rangle &= \langle a_1u_1 + a_2u_2 + \cdots + a_ru_r,\ u_1 \rangle \\ &= a_1\langle u_1, u_1 \rangle + a_2\langle u_2, u_1 \rangle + \cdots + a_r\langle u_r, u_1 \rangle \\ &= a_1\langle u_1, u_1 \rangle + a_2 \cdot 0 + \cdots + a_r \cdot 0 = a_1\langle u_1, u_1 \rangle \end{aligned}$$

Because $u_1 \neq 0$, we have $\langle u_1, u_1 \rangle \neq 0$. Thus, $a_1 = 0$. Similarly, for $i = 2, \ldots, r$, taking the inner product of (1) with u_i,

$$\begin{aligned} 0 = \langle 0, u_i \rangle &= \langle a_1u_1 + \cdots + a_ru_r,\ u_i \rangle \\ &= a_1\langle u_1, u_i \rangle + \cdots + a_i\langle u_i, u_i \rangle + \cdots + a_r\langle u_r, u_i \rangle = a_i\langle u_i, u_i \rangle \end{aligned}$$

But $\langle u_i, u_i \rangle \neq 0$, and hence, every $a_i = 0$. Thus, S is linearly independent.

7.16. Prove Theorem 7.6 (Pythagoras): Suppose $\{u_1, u_2, \ldots, u_r\}$ is an orthogonal set of vectors. Then

$$\|u_1 + u_2 + \cdots + u_r\|^2 = \|u_1\|^2 + \|u_2\|^2 + \cdots + \|u_r\|^2$$

Expanding the inner product, we have

$$\begin{aligned} \|u_1 + u_2 + \cdots + u_r\|^2 &= \langle u_1 + u_2 + \cdots + u_r,\ u_1 + u_2 + \cdots + u_r \rangle \\ &= \langle u_1, u_1 \rangle + \langle u_2, u_2 \rangle + \cdots + \langle u_r, u_r \rangle + \sum_{i \neq j} \langle u_i, u_j \rangle \end{aligned}$$

The theorem follows from the fact that $\langle u_i, u_i \rangle = \|u_i\|^2$ and $\langle u_i, u_j \rangle = 0$ for $i \neq j$.

7.17. Prove Theorem 7.7: Let $\{u_1, u_2, \ldots, u_n\}$ be an orthogonal basis of V. Then for any $v \in V$,

$$v = \frac{\langle v, u_1 \rangle}{\langle u_1, u_1 \rangle}u_1 + \frac{\langle v, u_2 \rangle}{\langle u_2, u_2 \rangle}u_2 + \cdots + \frac{\langle v, u_n \rangle}{\langle u_n, u_n \rangle}u_n$$

Suppose $v = k_1u_1 + k_2u_2 + \cdots + k_nu_n$. Taking the inner product of both sides with u_1 yields

$$\begin{aligned} \langle v, u_1 \rangle &= \langle k_1u_2 + k_2u_2 + \cdots + k_nu_n,\ u_1 \rangle \\ &= k_1\langle u_1, u_1 \rangle + k_2\langle u_2, u_1 \rangle + \cdots + k_n\langle u_n, u_1 \rangle \\ &= k_1\langle u_1, u_1 \rangle + k_2 \cdot 0 + \cdots + k_n \cdot 0 = k_1\langle u_1, u_1 \rangle \end{aligned}$$

Thus, $k_1 = \dfrac{\langle v, u_1 \rangle}{\langle u_1, u_1 \rangle}$. Similarly, for $i = 2, \ldots, n$,

$$\begin{aligned} \langle v, u_i \rangle &= \langle k_1u_i + k_2u_2 + \cdots + k_nu_n,\ u_i \rangle \\ &= k_1\langle u_1, u_i \rangle + k_2\langle u_2, u_i \rangle + \cdots + k_n\langle u_n, u_i \rangle \\ &= k_1 \cdot 0 + \cdots + k_i\langle u_i, u_i \rangle + \cdots + k_n \cdot 0 = k_i\langle u_i, u_i \rangle \end{aligned}$$

Thus, $k_i = \dfrac{\langle v, u_i \rangle}{\langle u_i, u_i \rangle}$. Substituting for k_i in the equation $v = k_1u_1 + \cdots + k_nu_n$, we obtain the desired result.

7.18. Suppose $E = \{e_1, e_2, \ldots, e_n\}$ is an orthonormal basis of V. Prove

(a) For any $u \in V$, we have $u = \langle u, e_1 \rangle e_1 + \langle u, e_2 \rangle e_2 + \cdots + \langle u, e_n \rangle e_n$.
(b) $\langle a_1e_1 + \cdots + a_ne_n,\ b_1e_1 + \cdots + b_ne_n \rangle = a_1b_1 + a_2b_2 + \cdots + a_nb_n$.
(c) For any $u, v \in V$, we have $\langle u, v \rangle = \langle u, e_1 \rangle \langle v, e_1 \rangle + \cdots + \langle u, e_n \rangle \langle v, e_n \rangle$.

(a) Suppose $u = k_1e_1 + k_2e_2 + \cdots + k_ne_n$. Taking the inner product of u with e_1,

$$\begin{aligned} \langle u, e_1 \rangle &= \langle k_1e_1 + k_2e_2 + \cdots + k_ne_n,\ e_1 \rangle \\ &= k_1\langle e_1, e_1 \rangle + k_2\langle e_2, e_1 \rangle + \cdots + k_n\langle e_n, e_1 \rangle \\ &= k_1(1) + k_2(0) + \cdots + k_n(0) = k_1 \end{aligned}$$

Similarly, for $i = 2, \ldots, n$,

$$\begin{aligned}\langle u, e_i\rangle &= \langle k_1e_1 + \cdots + k_ie_i + \cdots + k_ne_n,\ e_i\rangle \\ &= k_1\langle e_1, e_i\rangle + \cdots + k_i\langle e_i, e_i\rangle + \cdots + k_n\langle e_n, e_i\rangle \\ &= k_1(0) + \cdots + k_i(1) + \cdots + k_n(0) = k_i\end{aligned}$$

Substituting $\langle u, e_i\rangle$ for k_i in the equation $u = k_1e_1 + \cdots + k_ne_n$, we obtain the desired result.

(b) We have

$$\left\langle \sum_{i=1}^n a_ie_i,\ \sum_{j=1}^n b_je_j \right\rangle = \sum_{i,j=1}^n a_ib_j\langle e_i, e_j\rangle = \sum_{i=1}^n a_ib_i\langle e_i, e_i\rangle + \sum_{i\neq j} a_ib_j\langle e_i, e_j\rangle$$

But $\langle e_i, e_j\rangle = 0$ for $i \neq j$, and $\langle e_i, e_j\rangle = 1$ for $i = j$. Hence, as required,

$$\left\langle \sum_{i=1}^n a_ie_i,\ \sum_{j=1}^n b_je_j \right\rangle = \sum_{i=1}^n a_ib_i = a_1b_1 + a_2b_2 + \cdots + a_nb_n$$

(c) By part (a), we have

$$u = \langle u, e_1\rangle e_1 + \cdots + \langle u, e_n\rangle e_n \qquad \text{and} \qquad v = \langle v, e_1\rangle e_1 + \cdots + \langle v, e_n\rangle e_n$$

Thus, by part (b),

$$\langle u, v\rangle = \langle u, e_1\rangle\langle v, e_1\rangle + \langle u, e_2\rangle\langle v, e_2\rangle + \cdots + \langle u, e_n\rangle\langle v, e_n\rangle$$

Projections, Gram–Schmidt Algorithm, Applications

7.19. Suppose $w \neq 0$. Let v be any vector in V. Show that

$$c = \frac{\langle v, w\rangle}{\langle w, w\rangle} = \frac{\langle v, w\rangle}{\|w\|^2}$$

is the unique scalar such that $v' = v - cw$ is orthogonal to w.

In order for v' to be orthogonal to w we must have

$$\langle v - cw,\ w\rangle = 0 \quad \text{or} \quad \langle v, w\rangle - c\langle w, w\rangle = 0 \quad \text{or} \quad \langle v, w\rangle = c\langle w, w\rangle$$

Thus, $c\dfrac{\langle v, w\rangle}{\langle w, w\rangle}$. Conversely, suppose $c = \dfrac{\langle v, w\rangle}{\langle w, w\rangle}$. Then

$$\langle v - cw,\ w\rangle = \langle v, w\rangle - c\langle w, w\rangle = \langle v, w\rangle - \frac{\langle v, w\rangle}{\langle w, w\rangle}\langle w, w\rangle = 0$$

7.20. Find the Fourier coefficient c and the projection of $v = (1, -2, 3, -4)$ along $w = (1, 2, 1, 2)$ in $\mathbf{R}^4$.

Compute $\langle v, w\rangle = 1 - 4 + 3 - 8 = -8$ and $\|w\|^2 = 1 + 4 + 1 + 4 = 10$. Then

$$c = -\tfrac{8}{10} = -\tfrac{4}{5} \qquad \text{and} \qquad \text{proj}(v, w) = cw = (-\tfrac{4}{5}, -\tfrac{8}{5}, -\tfrac{4}{5}, -\tfrac{8}{5})$$

7.21. Consider the subspace U of $\mathbf{R}^4$ spanned by the vectors:

$$v_1 = (1, 1, 1, 1), \qquad v_2 = (1, 1, 2, 4), \qquad v_3 = (1, 2, -4, -3)$$

Find (a) an orthogonal basis of U; (b) an orthonormal basis of U.

(a) Use the Gram–Schmidt algorithm. Begin by setting $w_1 = u = (1, 1, 1, 1)$. Next find

$$v_2 - \frac{\langle v_2, w_1\rangle}{\langle w_1, w_1\rangle}w_1 = (1, 1, 2, 4) - \frac{8}{4}(1, 1, 1, 1) = (-1, -1, 0, 2)$$

Set $w_2 = (-1, -1, 0, 2)$. Then find

$$\begin{aligned}v_3 - \frac{\langle v_3, w_1\rangle}{\langle w_1, w_1\rangle}w_1 - \frac{\langle v_3, w_2\rangle}{\langle w_2, w_2\rangle}w_2 &= (1, 2, -4, -3) - \frac{(-4)}{4}(1, 1, 1, 1) - \frac{(-9)}{6}(-1, -1, 0, 2) \\ &= (\tfrac{1}{2}, \tfrac{3}{2}, -3, 1)\end{aligned}$$

Clear fractions to obtain $w_3 = (1, 3, -6, 2)$. Then w_1, w_2, w_3 form an orthogonal basis of U.

(b) Normalize the orthogonal basis consisting of w_1, w_2, w_3. Because $\|w_1\|^2 = 4$, $\|w_2\|^2 = 6$, and $\|w_3\|^2 = 50$, the following vectors form an orthonormal basis of U:

$$u_1 = \frac{1}{2}(1,1,1,1), \qquad u_2 = \frac{1}{\sqrt{6}}(-1,-1,0,2), \qquad u_3 = \frac{1}{5\sqrt{2}}(1,3,-6,2)$$

7.22. Consider the vector space $\mathbf{P}(t)$ with inner product $\langle f, g\rangle = \int_0^1 f(t)g(t)\,dt$. Apply the Gram–Schmidt algorithm to the set $\{1, t, t^2\}$ to obtain an orthogonal set $\{f_0, f_1, f_2\}$ with integer coefficients.

First set $f_0 = 1$. Then find

$$t - \frac{\langle t, 1\rangle}{\langle 1, 1\rangle}\cdot 1 = t - \frac{\frac{1}{2}}{1}\cdot 1 = t - \frac{1}{2}$$

Clear fractions to obtain $f_1 = 2t - 1$. Then find

$$t^2 - \frac{\langle t^2, 1\rangle}{\langle 1, 1\rangle}(1) - \frac{\langle t^2,\ 2t-1\rangle}{\langle 2t-1,\ 2t-1\rangle}(2t-1) = t^2 - \frac{\frac{1}{3}}{1}(1) - \frac{\frac{1}{6}}{\frac{1}{3}}(2t-1) = t^2 - t + \frac{1}{6}$$

Clear fractions to obtain $f_2 = 6t^2 - 6t + 1$. Thus, $\{1,\ 2t-1,\ 6t^2 - 6t + 1\}$ is the required orthogonal set.

7.23. Suppose $v = (1,3,5,7)$. Find the projection of v onto W or, in other words, find $w \in W$ that minimizes $\|v - w\|$, where W is the subspace of $\mathbf{R}^4$ spanned by

(a) $u_1 = (1,1,1,1)$ and $u_2 = (1,-3,4,-2)$,

(b) $v_1 = (1,1,1,1)$ and $v_2 = (1,2,3,2)$.

(a) Because u_1 and u_2 are orthogonal, we need only compute the Fourier coefficients:

$$c_1 = \frac{\langle v, u_1\rangle}{\langle u_1, u_1\rangle} = \frac{1+3+5+7}{1+1+1+1} = \frac{16}{4} = 4$$

$$c_2 = \frac{\langle v, u_2\rangle}{\langle u_2, u_2\rangle} = \frac{1-9+20-14}{1+9+16+4} = \frac{-2}{30} = -\frac{1}{15}$$

Then $w = \text{proj}(v, W) = c_1u_1 + c_2u_2 = 4(1,1,1,1) - \frac{1}{15}(1,-3,4,-2) = (\frac{59}{15}, \frac{63}{5}, \frac{56}{15}, \frac{62}{15})$.

(b) Because v_1 and v_2 are not orthogonal, first apply the Gram–Schmidt algorithm to find an orthogonal basis for W. Set $w_1 = v_1 = (1,1,1,1)$. Then find

$$v_2 - \frac{\langle v_2, w_1\rangle}{\langle w_1, w_1\rangle}w_1 = (1,2,3,2) - \frac{8}{4}(1,1,1,1) = (-1,0,1,0)$$

Set $w_2 = (-1,0,1,0)$. Now compute

$$c_1 = \frac{\langle v, w_1\rangle}{\langle w_1, w_1\rangle} = \frac{1+3+5+7}{1+1+1+1} = \frac{16}{4} = 4$$

$$c_2 = \frac{\langle v, w_2\rangle}{\langle w_2, w_2\rangle} = \frac{-1+0+5+0}{1+0+1+0} = \frac{4}{2} = 2$$

Then $w = \text{proj}(v, W) = c_1w_1 + c_2w_2 = 4(1,1,1,1) + 2(-1,0,1,0) = (2,4,6,4)$.

7.24. Suppose w_1 and w_2 are nonzero orthogonal vectors. Let v be any vector in V. Find c_1 and c_2 so that v' is orthogonal to w_1 and w_2, where $v' = v - c_1w_1 - c_2w_2$.

If v' is orthogonal to w_1, then

$$\begin{aligned} 0 &= \langle v - c_1w_1 - c_2w_2,\ w_1\rangle = \langle v, w_1\rangle - c_1\langle w_1, w_1\rangle - c_2\langle w_2, w_1\rangle \\ &= \langle v, w_1\rangle - c_1\langle w_1, w_1\rangle - c_2 0 = \langle v, w_1\rangle - c_1\langle w_1, w_1\rangle \end{aligned}$$

Thus, $c_1 = \langle v, w_1\rangle / \langle w_1, w_1\rangle$. (That is, c_1 is the component of v along w_1.) Similarly, if v' is orthogonal to w_2, then

$$0 = \langle v - c_1w_1 - c_2w_2,\ w_2\rangle = \langle v, w_2\rangle - c_2\langle w_2, w_2\rangle$$

Thus, $c_2 = \langle v, w_2\rangle / \langle w_2, w_2\rangle$. (That is, c_2 is the component of v along w_2.)

7.25. Prove Theorem 7.8: Suppose $w_1, w_2, \ldots, w_r$ form an orthogonal set of nonzero vectors in V. Let $v \in V$. Define

$$v' = v - (c_1 w_1 + c_2 w_2 + \cdots + c_r w_r), \quad \text{where} \quad c_i = \frac{\langle v, w_i \rangle}{\langle w_i, w_i \rangle}$$

Then v' is orthogonal to $w_1, w_2, \ldots, w_r$.

For $i = 1, 2, \ldots, r$ and using $\langle w_i, w_j \rangle = 0$ for $i \neq j$, we have

$$\begin{aligned}\langle v - c_1 w_1 - c_2 w_2 - \cdots - c_r w_r, \ w_i \rangle &= \langle v, w_i \rangle - c_1 \langle w_1, w_i \rangle - \cdots - c_i \langle w_i, w_i \rangle - \cdots - c_r \langle w_r, w_i \rangle \\ &= \langle v, w_i \rangle - c_1 \cdot 0 - \cdots - c_i \langle w_i, w_i \rangle - \cdots - c_r \cdot 0 \\ &= \langle v, w_i \rangle - c_i \langle w_i, w_i \rangle = \langle v, w_i \rangle - \frac{\langle v, w_i \rangle}{\langle w_i, w_i \rangle} \langle w_i, w_i \rangle = 0\end{aligned}$$

The theorem is proved.

7.26. Prove Theorem 7.9: Let $\{v_1, v_2, \ldots, v_n\}$ be any basis of an inner product space V. Then there exists an orthonormal basis $\{u_1, u_2, \ldots, u_n\}$ of V such that the change-of-basis matrix from $\{v_i\}$ to $\{u_i\}$ is triangular; that is, for $k = 1, 2, \ldots, n$,

$$u_k = a_{k1} v_1 + a_{k2} v_2 + \cdots + a_{kk} v_k$$

The proof uses the Gram–Schmidt algorithm and Remarks 1 and 3 of Section 7.7. That is, apply the algorithm to $\{v_i\}$ to obtain an orthogonal basis $\{w_i, \ldots, w_n\}$, and then normalize $\{w_i\}$ to obtain an orthonormal basis $\{u_i\}$ of V. The specific algorithm guarantees that each w_k is a linear combination of $v_1, \ldots, v_k$, and hence, each u_k is a linear combination of $v_1, \ldots, v_k$.

7.27. Prove Theorem 7.10: Suppose $S = \{w_1, w_2, \ldots, w_r\}$, is an orthogonal basis for a subspace W of V. Then one may extend S to an orthogonal basis for V; that is, one may find vectors $w_{r+1}, \ldots, w_n$ such that $\{w_1, w_2, \ldots, w_n\}$ is an orthogonal basis for V.

Extend S to a basis $S' = \{w_1, \ldots, w_r, v_{r+1}, \ldots, v_n\}$ for V. Applying the Gram–Schmidt algorithm to S', we first obtain $w_1, w_2, \ldots, w_r$ because S is orthogonal, and then we obtain vectors $w_{r+1}, \ldots, w_n$, where $\{w_1, w_2, \ldots, w_n\}$ is an orthogonal basis for V. Thus, the theorem is proved.

7.28. Prove Theorem 7.4: Let W be a subspace of V. Then $V = W \oplus W^{\perp}$.

By Theorem 7.9, there exists an orthogonal basis $\{u_1, \ldots, u_r\}$ of W, and by Theorem 7.10 we can extend it to an orthogonal basis $\{u_1, u_2, \ldots, u_n\}$ of V. Hence, $u_{r+1}, \ldots, u_n \in W^{\perp}$. If $v \in V$, then

$$v = a_1 u_1 + \cdots + a_n u_n, \quad \text{where } a_1 u_1 + \cdots + a_r u_r \in W \text{ and } a_{r+1} u_{r+1} + \cdots + a_n u_n \in W^{\perp}$$

Accordingly, $V = W + W^{\perp}$.

On the other hand, if $w \in W \cap W^{\perp}$, then $\langle w, w \rangle = 0$. This yields $w = 0$. Hence, $W \cap W^{\perp} = \{0\}$. The two conditions $V = W + W^{\perp}$ and $W \cap W^{\perp} = \{0\}$ give the desired result $V = W \oplus W^{\perp}$.

Remark: Note that we have proved the theorem for the case that V has finite dimension. We remark that the theorem also holds for spaces of arbitrary dimension.

7.29. Suppose W is a subspace of a finite-dimensional space V. Prove that $W = W^{\perp\perp}$.

By Theorem 7.4, $V = W \oplus W^{\perp}$, and also $V = W^{\perp} \oplus W^{\perp\perp}$. Hence,

$$\dim W = \dim V - \dim W^{\perp} \quad \text{and} \quad \dim W^{\perp\perp} = \dim V - \dim W^{\perp}$$

This yields $\dim W = \dim W^{\perp\perp}$. But $W \subseteq W^{\perp\perp}$ (see Problem 7.14). Hence, $W = W^{\perp\perp}$, as required.

7.30. Prove the following: Suppose $w_1, w_2, \ldots, w_r$ form an orthogonal set of nonzero vectors in V. Let v be any vector in V and let c_i be the component of v along w_i. Then, for any scalars $a_1, \ldots, a_r$, we have

$$\left\| v - \sum_{k=1}^{r} c_k w_k \right\| \leq \left\| v - \sum_{k=1}^{r} a_k w_k \right\|$$

That is, $\sum c_i w_i$ is the closest approximation to v as a linear combination of $w_1, \ldots, w_r$.

By Theorem 7.8, $v - \sum c_k w_k$ is orthogonal to every w_i and hence orthogonal to any linear combination of $w_1, w_2, \ldots, w_r$. Therefore, using the Pythagorean theorem and summing from $k = 1$ to r,

$$\begin{aligned}\|v - \sum a_k w_k\|^2 &= \|v - \sum c_k w_k + \sum (c_k - a_k) w_k\|^2 = \|v - \sum c_k w_k\|^2 + \|\sum (c_k - a_k) w_k\|^2 \\ &\geq \|v - \sum c_k w_k\|^2\end{aligned}$$

The square root of both sides gives our theorem.

7.31. Suppose $\{e_1, e_2, \ldots, e_r\}$ is an orthonormal set of vectors in V. Let v be any vector in V and let c_i be the Fourier coefficient of v with respect to e_i. Prove Bessel's inequality:

$$\sum_{k=1}^{r} c_k^2 \leq \|v\|^2$$

Note that $c_i = \langle v, e_i \rangle$, because $\|e_i\| = 1$. Then, using $\langle e_i, e_j \rangle = 0$ for $i \neq j$ and summing from $k = 1$ to r, we get

$$\begin{aligned}0 \leq \langle v - \sum c_k e_k,\ v - \sum c_k e_k \rangle &= \langle v, v \rangle - 2\langle v, \sum c_k e_k \rangle + \sum c_k^2 = \langle v, v \rangle - \sum 2c_k \langle v, e_k \rangle + \sum c_k^2 \\ &= \langle v, v \rangle - \sum 2c_k^2 + \sum c_k^2 = \langle v, v \rangle - \sum c_k^2\end{aligned}$$

This gives us our inequality.

Orthogonal Matrices

7.32. Find an orthogonal matrix P whose first row is $u_1 = (\frac{1}{3}, \frac{2}{3}, \frac{2}{3})$.

First find a nonzero vector $w_2 = (x, y, z)$ that is orthogonal to u_1—that is, for which

$$0 = \langle u_1, w_2 \rangle = \frac{x}{3} + \frac{2y}{3} + \frac{2z}{3} = 0 \qquad \text{or} \qquad x + 2y + 2z = 0$$

One such solution is $w_2 = (0, 1, -1)$. Normalize w_2 to obtain the second row of P:

$$u_2 = (0, 1/\sqrt{2}, -1/\sqrt{2})$$

Next find a nonzero vector $w_3 = (x, y, z)$ that is orthogonal to both u_1 and u_2—that is, for which

$$0 = \langle u_1, w_3 \rangle = \frac{x}{3} + \frac{2y}{3} + \frac{2z}{3} = 0 \qquad \text{or} \qquad x + 2y + 2z = 0$$

$$0 = \langle u_2, w_3 \rangle = \frac{y}{\sqrt{2}} - \frac{z}{\sqrt{2}} = 0 \qquad \text{or} \qquad y - z = 0$$

Set $z = -1$ and find the solution $w_3 = (4, -1, -1)$. Normalize w_3 and obtain the third row of P; that is,

$$u_3 = (4/\sqrt{18}, -1/\sqrt{18}, -1/\sqrt{18}).$$

Thus,
$$P = \begin{bmatrix} \frac{1}{3} & \frac{2}{3} & \frac{2}{3} \\ 0 & 1/\sqrt{2} & -1/\sqrt{2} \\ 4/3\sqrt{2} & -1/3\sqrt{2} & -1/3\sqrt{2} \end{bmatrix}$$

We emphasize that the above matrix P is not unique.

7.33. Let $A = \begin{bmatrix} 1 & 1 & -1 \\ 1 & 3 & 4 \\ 7 & -5 & 2 \end{bmatrix}$. Determine whether or not: (a) the rows of A are orthogonal;

(b) A is an orthogonal matrix; (c) the columns of A are orthogonal.

(a) Yes, because $(1, 1, -1) \cdot (1, 3, 4) = 1 + 3 - 4 = 0$, $(1, 1 - 1) \cdot (7, -5, 2) = 7 - 5 - 2 = 0$, and $(1, 3, 4) \cdot (7, -5, 2) = 7 - 15 + 8 = 0$.

(b) No, because the rows of A are not unit vectors, for example, $(1, 1, -1)^2 = 1 + 1 + 1 = 3$.

(c) No; for example, $(1, 1, 7) \cdot (1, 3, -5) = 1 + 3 - 35 = -31 \neq 0$.

7.34. Let B be the matrix obtained by normalizing each row of A in Problem 7.33.

(a) Find B.

(b) Is B an orthogonal matrix?

(c) Are the columns of B orthogonal?

(a) We have

$$\|(1,1,-1)\|^2 = 1+1+1 = 3, \qquad \|(1,3,4)\|^2 = 1+9+16 = 26$$
$$\|(7,-5,2)\|^2 = 49+25+4 = 78$$

Thus,
$$B = \begin{bmatrix} 1/\sqrt{3} & 1/\sqrt{3} & -1/\sqrt{3} \\ 1/\sqrt{26} & 3/\sqrt{26} & 4/\sqrt{26} \\ 7/\sqrt{78} & -5/\sqrt{78} & 2/\sqrt{78} \end{bmatrix}$$

(b) Yes, because the rows of B are still orthogonal and are now unit vectors.

(c) Yes, because the rows of B form an orthonormal set of vectors. Then, by Theorem 7.11, the columns of B must automatically form an orthonormal set.

7.35. Prove each of the following:

(a) P is orthogonal if and only if P^T is orthogonal.

(b) If P is orthogonal, then P^{-1} is orthogonal.

(c) If P and Q are orthogonal, then PQ is orthogonal.

(a) We have $(P^T)^T = P$. Thus, P is orthogonal if and only if $PP^T = I$ if and only if $P^{TT}P^T = I$ if and only if P^T is orthogonal.

(b) We have $P^T = P^{-1}$, because P is orthogonal. Thus, by part (a), P^{-1} is orthogonal.

(c) We have $P^T = P^{-1}$ and $Q^T = Q^{-1}$. Thus, $(PQ)(PQ)^T = PQQ^TP^T = PQQ^{-1}P^{-1} = I$. Therefore, $(PQ)^T = (PQ)^{-1}$, and so PQ is orthogonal.

7.36. Suppose P is an orthogonal matrix. Show that

(a) $\langle Pu, Pv\rangle = \langle u, v\rangle$ for any $u, v \in V$;

(b) $\|Pu\| = \|u\|$ for every $u \in V$.

Use $P^TP = I$ and $\langle u, v\rangle = u^Tv$.

(a) $\langle Pu, Pv\rangle = (Pu)^T(Pv) = u^TP^TPv = u^Tv = \langle u, v\rangle$.

(b) We have

$$\|Pu\|^2 = \langle Pu, Pu\rangle = u^TP^TPu = u^Tu = \langle u, u\rangle = \|u\|^2$$

Taking the square root of both sides gives our result.

7.37. Prove Theorem 7.12: Suppose $E = \{e_i\}$ and $E' = \{e'_i\}$ are orthonormal bases of V. Let P be the change-of-basis matrix from E to E'. Then P is orthogonal.

Suppose

$$e'_i = b_{i1}e_1 + b_{i2}e_2 + \cdots + b_{in}e_n, \qquad i = 1, \ldots, n \tag{1}$$

Using Problem 7.18(b) and the fact that E' is orthonormal, we get

$$\delta_{ij} = \langle e'_i, e'_j\rangle = b_{i1}b_{j1} + b_{i2}b_{j2} + \cdots + b_{in}b_{jn} \tag{2}$$

Let $B = [b_{ij}]$ be the matrix of the coefficients in (1). (Then $P = B^T$.) Suppose $BB^T = [c_{ij}]$. Then

$$c_{ij} = b_{i1}b_{j1} + b_{i2}b_{j2} + \cdots + b_{in}b_{jn} \tag{3}$$

By (2) and (3), we have $c_{ij} = \delta_{ij}$. Thus, $BB^T = I$. Accordingly, B is orthogonal, and hence, $P = B^T$ is orthogonal.

7.38. Prove Theorem 7.13: Let $\{e_1, \ldots, e_n\}$ be an orthonormal basis of an inner product space V. Let $P = [a_{ij}]$ be an orthogonal matrix. Then the following n vectors form an orthonormal basis for V:

$$e'_i = a_{1i}e_1 + a_{2i}e_2 + \cdots + a_{ni}e_n, \qquad i = 1, 2, \ldots, n$$

Because $\{e_i\}$ is orthonormal, we get, by Problem 7.18(b),

$$\langle e_i', e_j' \rangle = a_{1i}a_{1j} + a_{2i}a_{2j} + \cdots + a_{ni}a_{nj} = \langle C_i, C_j \rangle$$

where C_i denotes the ith column of the orthogonal matrix $P = [a_{ij}]$. Because P is orthogonal, its columns form an orthonormal set. This implies $\langle e_i', e_j' \rangle = \langle C_i, C_j \rangle = \delta_{ij}$. Thus, $\{e_i'\}$ is an orthonormal basis.

Inner Products And Positive Definite Matrices

7.39. Which of the following symmetric matrices are positive definite?

(a) $A = \begin{bmatrix} 3 & 4 \\ 4 & 5 \end{bmatrix}$, (b) $B = \begin{bmatrix} 8 & -3 \\ -3 & 2 \end{bmatrix}$, (c) $C = \begin{bmatrix} 2 & 1 \\ 1 & -3 \end{bmatrix}$, (d) $D = \begin{bmatrix} 3 & 5 \\ 5 & 9 \end{bmatrix}$

Use Theorem 7.14 that a 2×2 real symmetric matrix is positive definite if and only if its diagonal entries are positive and if its determinant is positive.

(a) No, because $|A| = 15 - 16 = -1$ is negative.

(b) Yes.

(c) No, because the diagonal entry -3 is negative.

(d) Yes.

7.40. Find the values of k that make each of the following matrices positive definite:

(a) $A = \begin{bmatrix} 2 & -4 \\ -4 & k \end{bmatrix}$, (b) $B = \begin{bmatrix} 4 & k \\ k & 9 \end{bmatrix}$, (c) $C = \begin{bmatrix} k & 5 \\ 5 & -2 \end{bmatrix}$

(a) First, k must be positive. Also, $|A| = 2k - 16$ must be positive; that is, $2k - 16 > 0$. Hence, $k > 8$.

(b) We need $|B| = 36 - k^2$ positive; that is, $36 - k^2 > 0$. Hence, $k^2 < 36$ or $-6 < k < 6$.

(c) C can never be positive definite, because C has a negative diagonal entry -2.

7.41. Find the matrix A that represents the usual inner product on $\mathbf{R}^2$ relative to each of the following bases of $\mathbf{R}^2$: (a) $\{v_1 = (1,3),\ v_2 = (2,5)\}$; (b) $\{w_1 = (1,2),\ w_2 = (4,-2)\}$.

(a) Compute $\langle v_1, v_1 \rangle = 1 + 9 = 10$, $\langle v_1, v_2 \rangle = 2 + 15 = 17$, $\langle v_2, v_2 \rangle = 4 + 25 = 29$. Thus,

$$A = \begin{bmatrix} 10 & 17 \\ 17 & 29 \end{bmatrix}.$$

(b) Compute $\langle w_1, w_1 \rangle = 1 + 4 = 5$, $\langle w_1, w_2 \rangle = 4 - 4 = 0$, $\langle w_2, w_2 \rangle = 16 + 4 = 20$. Thus, $A = \begin{bmatrix} 5 & 0 \\ 0 & 20 \end{bmatrix}$. (Because the basis vectors are orthogonal, the matrix A is diagonal.)

7.42. Consider the vector space $\mathbf{P}_2(t)$ with inner product $\langle f, g \rangle = \int_{-1}^{1} f(t)g(t)\, dt$.

(a) Find $\langle f, g \rangle$, where $f(t) = t + 2$ and $g(t) = t^2 - 3t + 4$.

(b) Find the matrix A of the inner product with respect to the basis $\{1, t, t^2\}$ of V.

(c) Verify Theorem 7.16 by showing that $\langle f, g \rangle = [f]^T A [g]$ with respect to the basis $\{1, t, t^2\}$.

(a) $\langle f, g \rangle = \int_{-1}^{1} (t+2)(t^2 - 3t + 4)\, dt = \int_{-1}^{1} (t^3 - t^2 - 2t + 8)\, dt = \left(\frac{t^4}{4} - \frac{t^3}{3} - t^2 + 8t \right) \Big|_{-1}^{1} = \frac{46}{3}$

(b) Here we use the fact that if $r + s = n$,

$$\langle t^r, t^s \rangle = \int_{-1}^{1} t^n\, dt = \frac{t^{n+1}}{n+1} \Big|_{-1}^{1} = \begin{cases} 2/(n+1) & \text{if } n \text{ is even,} \\ 0 & \text{if } n \text{ is odd.} \end{cases}$$

Then $\langle 1, 1 \rangle = 2$, $\langle 1, t \rangle = 0$, $\langle 1, t^2 \rangle = \frac{2}{3}$, $\langle t, t \rangle = \frac{2}{3}$, $\langle t, t^2 \rangle = 0$, $\langle t^2, t^2 \rangle = \frac{2}{5}$. Thus,

$$A = \begin{bmatrix} 2 & 0 & \frac{2}{3} \\ 0 & \frac{2}{3} & 0 \\ \frac{2}{3} & 0 & \frac{2}{5} \end{bmatrix}$$

(c) We have $[f]^T = (2, 1, 0)$ and $[g]^T = (4, -3, 1)$ relative to the given basis. Then

$$[f]^T A[g] = (2, 1, 0)\begin{bmatrix} 2 & 0 & \frac{2}{3} \\ 0 & \frac{2}{3} & 0 \\ \frac{2}{3} & 0 & \frac{2}{5} \end{bmatrix}\begin{bmatrix} 4 \\ -3 \\ 1 \end{bmatrix} = (4, \tfrac{2}{3}, \tfrac{4}{3})\begin{bmatrix} 4 \\ -3 \\ 1 \end{bmatrix} = \tfrac{46}{3} = \langle f, g\rangle$$

7.43. Prove Theorem 7.14: $A = \begin{bmatrix} a & b \\ b & d \end{bmatrix}$ is positive definite if and only if a and d are positive and $|A| = ad - b^2$ is positive.

Let $u = [x, y]^T$. Then

$$f(u) = u^T A u = [x, y]\begin{bmatrix} a & b \\ b & d \end{bmatrix}\begin{bmatrix} x \\ y \end{bmatrix} = ax^2 + 2bxy + dy^2$$

Suppose $f(u) > 0$ for every $u \neq 0$. Then $f(1, 0) = a > 0$ and $f(0, 1) = d > 0$. Also, we have $f(b, -a) = a(ad - b^2) > 0$. Because $a > 0$, we get $ad - b^2 > 0$.

Conversely, suppose $a > 0$, $d > 0$, $ad - b^2 > 0$. Completing the square gives us

$$f(u) = a\left(x^2 + \frac{2b}{a}xy + \frac{b^2}{a_2}y^2\right) + dy^2 - \frac{b^2}{a}y^2 = a\left(x + \frac{by}{a}\right)^2 + \frac{ad - b^2}{a}y^2$$

Accordingly, $f(u) > 0$ for every $u \neq 0$.

7.44. Prove Theorem 7.15: Let A be a real positive definite matrix. Then the function $\langle u, v\rangle = u^T A v$ is an inner product on $\mathbf{R}^n$.

For any vectors u_1, u_2, and v,

$$\langle u_1 + u_2, v\rangle = (u_1 + u_2)^T A v = (u_1^T + u_2^T)Av = u_1^T A v + u_2^T A v = \langle u_1, v\rangle + \langle u_2, v\rangle$$

and, for any scalar k and vectors u, v,

$$\langle ku, v\rangle = (ku)^T A v = k u^T A v = k\langle u, v\rangle$$

Thus $[I_1]$ is satisfied.

Because $u^T A v$ is a scalar, $(u^T A v)^T = u^T A v$. Also, $A^T = A$ because A is symmetric. Therefore,

$$\langle u, v\rangle = u^T A v = (u^T A v)^T = v^T A^T u^{TT} = v^T A u = \langle v, u\rangle$$

Thus, $[I_2]$ is satisfied.

Last, because A is positive definite, $X^T A X > 0$ for any nonzero $X \in \mathbf{R}^n$. Thus, for any nonzero vector v, $\langle v, v\rangle = v^T A v > 0$. Also, $\langle 0, 0\rangle = 0^T A 0 = 0$. Thus, $[I_3]$ is satisfied. Accordingly, the function $\langle u, v\rangle = Av$ is an inner product.

7.45. Prove Theorem 7.16: Let A be the matrix representation of an inner product relative to a basis S of V. Then, for any vectors $u, v \in V$, we have

$$\langle u, v\rangle = [u]^T A[v]$$

Suppose $S = \{w_1, w_2, \ldots, w_n\}$ and $A = [k_{ij}]$. Hence, $k_{ij} = \langle w_i, w_j\rangle$. Suppose

$$u = a_1 w_1 + a_2 w_2 + \cdots + a_n w_n \quad \text{and} \quad v = b_1 w_1 + b_2 w_2 + \cdots + b_n w_n$$

Then

$$\langle u, v\rangle = \sum_{i=1}^{n}\sum_{j=1}^{n} a_i b_j \langle w_i, w_j\rangle \tag{1}$$

On the other hand,

$$[u]^T A[v] = (a_1, a_2, \ldots, a_n)\begin{bmatrix} k_{11} & k_{12} & \cdots & k_{1n} \\ k_{21} & k_{22} & \cdots & k_{2n} \\ \cdots & \cdots & \cdots & \cdots \\ k_{n1} & k_{n2} & \cdots & k_{nn} \end{bmatrix}\begin{bmatrix} b_1 \\ b_2 \\ \vdots \\ b_n \end{bmatrix}$$

$$= \left(\sum_{i=1}^{n} a_i k_{i1}, \sum_{i=1}^{n} a_i k_{i2}, \ldots, \sum_{i=1}^{n} a_i k_{in}\right)\begin{bmatrix} b_1 \\ b_2 \\ \vdots \\ b_n \end{bmatrix} = \sum_{j=1}^{n}\sum_{i=1}^{n} a_i b_j k_{ij} \tag{2}$$

Equations (1) and (2) give us our result.

7.46. Prove Theorem 7.17: Let A be the matrix representation of any inner product on V. Then A is a positive definite matrix.

Because $\langle w_i, w_j\rangle = \langle w_j, w_i\rangle$ for any basis vectors w_i and w_j, the matrix A is symmetric. Let X be any nonzero vector in $\mathbf{R}^n$. Then $[u] = X$ for some nonzero vector $u \in V$. Theorem 7.16 tells us that $X^TAX = [u]^TA[u] = \langle u, u\rangle > 0$. Thus, A is positive definite.

Complex Inner Product Spaces

7.47. Let V be a complex inner product space. Verify the relation

$$\langle u,\ av_1 + bv_2\rangle = \bar{a}\langle u, v_1\rangle + \bar{b}\langle u, v_2\rangle$$

Using $[I_2^*]$, $[I_1^*]$, and then $[I_2^*]$, we find

$$\langle u, av_1 + bv_2\rangle = \overline{\langle av_1 + bv_2, u\rangle} = \overline{a\langle v_1, u\rangle + b\langle v_2, u\rangle} = \bar{a}\overline{\langle v_1, u\rangle} + \bar{b}\overline{\langle v_2, u\rangle} = \bar{a}\langle u, v_1\rangle + \bar{b}\langle u, v_2\rangle$$

7.48. Suppose $\langle u, v\rangle = 3 + 2i$ in a complex inner product space V. Find

(a) $\langle (2-4i)u, v\rangle$; (b) $\langle u,\ (4+3i)v\rangle$; (c) $\langle (3-6i)u,\ (5-2i)v\rangle$.

(a) $\langle (2-4i)u,\ v\rangle = (2-4i)\langle u, v\rangle = (2-4i)(3+2i) = 14 - 8i$

(b) $\langle u,\ (4+3i)v\rangle = \overline{(4+3i)}\langle u, v\rangle = (4-3i)(3+2i) = 18 - i$

(c) $\langle (3-6i)u,\ (5-2i)v\rangle = (3-6i)\overline{(5-2i)}\langle u, v\rangle = (3-6i)(5+2i)(3+2i) = 129 - 18i$

7.49. Find the Fourier coefficient (component) c and the projection cw of $v = (3+4i,\ 2-3i)$ along $w = (5+i,\ 2i)$ in $\mathbf{C}^2$.

Recall that $c = \langle v, w\rangle / \langle w, w\rangle$. Compute

$$\begin{aligned}\langle v, w\rangle &= (3+4i)\overline{(5+i)} + (2-3i)\overline{(2i)} = (3+4i)(5-i) + (2-3i)(-2i)\\ &= 19 + 17i - 6 - 4i = 13 + 13i\\ \langle w, w\rangle &= 25 + 1 + 4 = 30\end{aligned}$$

Thus, $c = (13+13i)/30 = \frac{13}{30} + \frac{13}{30}i$. Accordingly, $\text{proj}(v, w) = cw = \left(\frac{26}{15} + \frac{39}{15}i,\ -\frac{13}{15} + \frac{1}{15}i\right)$

7.50. Prove Theorem 7.18 (Cauchy–Schwarz): Let V be a complex inner product space. Then $|\langle u, v\rangle| \le \|u\|\|v\|$.

If $v = 0$, the inequality reduces to $0 \le 0$ and hence is valid. Now suppose $v \ne 0$. Using $z\bar{z} = |z|^2$ (for any complex number z) and $\langle v, u\rangle = \overline{\langle u, v\rangle}$, we expand $\|u - \langle u, v\rangle tv\|^2 \ge 0$, where t is any real value:

$$\begin{aligned}0 \le \|u - \langle u, v\rangle tv\|^2 &= \langle u - \langle u, v\rangle tv,\ u - \langle u, v\rangle tv\rangle\\ &= \langle u, u\rangle - \overline{\langle u, v\rangle}t\langle u, v\rangle - \langle u, v\rangle t\langle v, u\rangle + \langle u, v\rangle\overline{\langle u, v\rangle}t^2\langle v, v\rangle\\ &= \|u\|^2 - 2t|\langle u, v\rangle|^2 + |\langle u, v\rangle|^2t^2\|v\|^2\end{aligned}$$

Set $t = 1/\|v\|^2$ to find $0 \le \|u\|^2 - \dfrac{|\langle u, v\rangle|^2}{\|v\|^2}$, from which $|\langle u, v\rangle|^2 \le \|u\|^2\|v\|^2$. Taking the square root of both sides, we obtain the required inequality.

7.51. Find an orthogonal basis for $u^\perp$ in C^3 where $u = (1,\ i,\ 1+i)$.

Here $u^\perp$ consists of all vectors $s = (x, y, z)$ such that

$$\langle w, u\rangle = x - iy + (1-i)z = 0$$

Find one solution, say $w_1 = (0,\ 1-i,\ i)$. Then find a solution of the system

$$x - iy + (1-i)z = 0, \qquad (1+i)y - iz = 0$$

Here z is a free variable. Set $z = 1$ to obtain $y = i/(1+i) = (1+i)/2$ and $x = (3i-3)2$. Multiplying by 2 yields the solution $w_2 = (3i-3,\ 1+i,\ 2)$. The vectors w_1 and w_2 form an orthogonal basis for $u^\perp$.

7.52. Find an orthonormal basis of the subspace W of $\mathbf{C}^3$ spanned by

$$v_1 = (1, i, 0) \quad \text{and} \quad v_2 = (1,\ 2,\ 1-i).$$

Apply the Gram–Schmidt algorithm. Set $w_1 = v_1 = (1, i, 0)$. Compute

$$v_2 - \frac{\langle v_2, w_1\rangle}{\langle w_1, w_1\rangle} w_1 = (1,\ 2,\ 1-i) - \frac{1-2i}{2}(1, i, 0) = \left(\tfrac{1}{2}+i,\ 1-\tfrac{1}{2}i,\ 1-i\right)$$

Multiply by 2 to clear fractions, obtaining $w_2 = (1+2i,\ 2-i,\ 2-2i)$. Next find $\|w_1\| = \sqrt{2}$ and then $\|w_2\| = \sqrt{18}$. Normalizing $\{w_1, w_2\}$, we obtain the following orthonormal basis of W:

$$\left\{ u_1 = \left(\frac{1}{\sqrt{2}},\ \frac{i}{\sqrt{2}},\ 0\right),\ u_2 = \left(\frac{1+2i}{\sqrt{18}},\ \frac{2-i}{\sqrt{18}},\ \frac{2-2i}{\sqrt{18}}\right) \right\}$$

7.53. Find the matrix P that represents the usual inner product on $\mathbf{C}^3$ relative to the basis $\{1,\ i,\ 1-i\}$.

Compute the following six inner products:

$$\langle 1, 1\rangle = 1, \qquad \langle 1, i\rangle = \bar{i} = -i, \qquad \langle 1, 1-i\rangle = \overline{1-i} = 1+i$$
$$\langle i, i\rangle = i\bar{i} = 1, \qquad \langle i, 1-i\rangle = i(\overline{1-i}) = -1+i, \qquad \langle 1-i, 1-i\rangle = 2$$

Then, using $(u, v) = \overline{(v, u)}$, we obtain

$$P = \begin{bmatrix} 1 & -i & 1+i \\ i & 1 & -1+i \\ 1-i & -1-i & 2 \end{bmatrix}$$

(As expected, P is Hermitian; that is, $P^H = P$.)

Normed Vector Spaces

7.54. Consider vectors $u = (1, 3, -6, 4)$ and $v = (3, -5, 1, -2)$ in $\mathbf{R}^4$. Find

(a) $\|u\|_\infty$ and $\|v\|_\infty$, (b) $\|u\|_1$ and $\|v\|_1$, (c) $\|u\|_2$ and $\|v\|_2$,
(d) $d_\infty(u, v), d_1(u, v), d_2(u, v)$.

(a) The infinity norm chooses the maximum of the absolute values of the components. Hence,

$$\|u\|_\infty = 6 \quad \text{and} \quad \|v\|_\infty = 5$$

(b) The one-norm adds the absolute values of the components. Thus,

$$\|u\|_1 = 1+3+6+4 = 14 \quad \text{and} \quad \|v\|_1 = 3+5+1+2 = 11$$

(c) The two-norm is equal to the square root of the sum of the squares of the components (i.e., the norm induced by the usual inner product on $\mathbf{R}^3$). Thus,

$$\|u\|_2 = \sqrt{1+9+36+16} = \sqrt{62} \quad \text{and} \quad \|v\|_2 = \sqrt{9+25+1+4} = \sqrt{39}$$

(d) First find $u - v = (-2, 8, -7, 6)$. Then

$$d_\infty(u, v) = \|u - v\|_\infty = 8$$
$$d_1(u, v) = \|u - v\|_1 = 2+8+7+6 = 23$$
$$d_2(u, v) = \|u - v\|_2 = \sqrt{4+64+49+36} = \sqrt{153}$$

7.55. Consider the function $f(t) = t^2 - 4t$ in $C[0, 3]$.

(a) Find $\|f\|_\infty$, (b) Plot $f(t)$ in the plane $\mathbf{R}^2$, (c) Find $\|f\|_1$, (d) Find $\|f\|_2$.

(a) We seek $\|f\|_\infty = \max(|f(t)|)$. Because $f(t)$ is differentiable on $[0, 3]$, $|f(t)|$ has a maximum at a critical point of $f(t)$ (i.e., when the derivative $f'(t) = 0$), or at an endpoint of $[0, 3]$. Because $f'(t) = 2t - 4$, we set $2t - 4 = 0$ and obtain $t = 2$ as a critical point. Compute

$$f(2) = 4 - 8 = -4, \qquad f(0) = 0 - 0 = 0, \qquad f(3) = 9 - 12 = -3$$

Thus, $\|f\|_\infty = |f(2)| = |-4| = 4$.

(b) Compute $f(t)$ for various values of t in $[0, 3]$, for example,

t	0	1	2	3
$f(t)$	0	-3	-4	-3

Plot the points in $\mathbf{R}^2$ and then draw a continuous curve through the points, as shown in Fig. 7-8.

(c) We seek $\|f\|_1 = \int_0^3 |f(t)|\,dt$. As indicated in Fig. 7-3, $f(t)$ is negative in $[0, 3]$; hence, $|f(t)| = -(t^2 - 4t) = 4t - t^2$

$$\text{Thus,}\quad \|f\|_1 = \int_0^3 (4t - t^2)\,dt = \left(2t^2 - \frac{t^3}{3}\right)\Bigg|_0^3 = 18 - 9 = 9$$

(d) $$\|f\|_2^2 = \int_0^3 f(t)^2\,dt = \int_0^3 (t^4 - 8t^3 + 16t^2)\,dt = \left(\frac{t^5}{5} - 2t^4 + \frac{16t^3}{3}\right)\Bigg|_0^3 = \frac{153}{5}.$$

Thus, $\|f\|_2 = \sqrt{\dfrac{153}{5}}$.

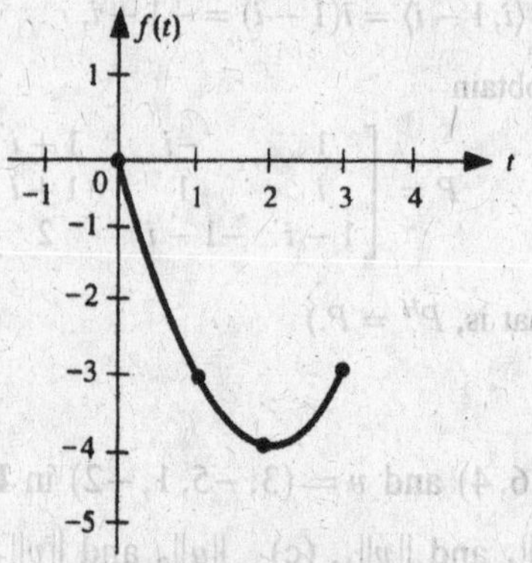

Figure 7-8

7.56. Prove Theorem 7.24: Let V be a normed vector space. Then the function $d(u, v) = \|u - v\|$ satisfies the following three axioms of a metric space:

$[M_1]$ $d(u, v) \ge 0$; and $d(u, v) = 0$ iff $u = v$.

$[M_2]$ $d(u, v) = d(v, u)$.

$[M_3]$ $d(u, v) \le d(u, w) + d(w, v)$.

If $u \ne v$, then $u - v \ne 0$, and hence, $d(u, v) = \|u - v\| > 0$. Also, $d(u, u) = \|u - u\| = \|0\| = 0$. Thus, $[M_1]$ is satisfied. We also have

$$d(u, v) = \|u - v\| = \|-1(v - u)\| = |-1|\|v - u\| = \|v - u\| = d(v, u)$$

and $$d(u, v) = \|u - v\| = \|(u - w) + (w - v)\| \le \|u - w\| + \|w - v\| = d(u, w) + d(w, v)$$

Thus, $[M_2]$ and $[M_3]$ are satisfied.

SUPPLEMENTARY PROBLEMS

Inner Products

7.57. Verify that the following is an inner product on $\mathbf{R}^2$, where $u = (x_1, x_2)$ and $v = (y_1, y_2)$:

$$f(u, v) = x_1y_1 - 2x_1y_2 - 2x_2y_1 + 5x_2y_2$$

7.58. Find the values of k so that the following is an inner product on $\mathbf{R}^2$, where $u = (x_1, x_2)$ and $v = (y_1, y_2)$:

$$f(u, v) = x_1y_1 - 3x_1y_2 - 3x_2y_1 + kx_2y_2$$

7.59. Consider the vectors $u = (1, -3)$ and $v = (2, 5)$ in $\mathbf{R}^2$. Find

(a) $\langle u, v \rangle$ with respect to the usual inner product in $\mathbf{R}^2$.

(b) $\langle u, v \rangle$ with respect to the inner product in $\mathbf{R}^2$ in Problem 7.57.

(c) $\|v\|$ using the usual inner product in $\mathbf{R}^2$.

(d) $\|v\|$ using the inner product in $\mathbf{R}^2$ in Problem 7.57.

7.60. Show that each of the following is not an inner product on $\mathbf{R}^3$, where $u = (x_1, x_2, x_3)$ and $v = (y_1, y_2, y_3)$:

(a) $\langle u, v \rangle = x_1y_1 + x_2y_2$, (b) $\langle u, v \rangle = x_1y_2x_3 + y_1x_2y_3$.

7.61. Let V be the vector space of $m \times n$ matrices over $\mathbf{R}$. Show that $\langle A, B \rangle = \mathrm{tr}(B^TA)$ defines an inner product in V.

7.62. Suppose $|\langle u, v \rangle| = \|u\|\|v\|$. (That is, the Cauchy–Schwarz inequality reduces to an equality.) Show that u and v are linearly dependent.

7.63. Suppose $f(u, v)$ and $g(u, v)$ are inner products on a vector space V over $\mathbf{R}$. Prove

(a) The sum $f + g$ is an inner product on V, where $(f + g)(u, v) = f(u, v) + g(u, v)$.

(b) The scalar product kf, for $k > 0$, is an inner product on V, where $(kf)(u, v) = kf(u, v)$.

Orthogonality, Orthogonal Complements, Orthogonal Sets

7.64. Let V be the vector space of polynomials over $\mathbf{R}$ of degree ≤ 2 with inner product defined by $\langle f, g \rangle = \int_0^1 f(t)g(t)\,dt$. Find a basis of the subspace W orthogonal to $h(t) = 2t + 1$.

7.65. Find a basis of the subspace W of $\mathbf{R}^4$ orthogonal to $u_1 = (1, -2, 3, 4)$ and $u_2 = (3, -5, 7, 8)$.

7.66. Find a basis for the subspace W of $\mathbf{R}^5$ orthogonal to the vectors $u_1 = (1, 1, 3, 4, 1)$ and $u_2 = (1, 2, 1, 2, 1)$.

7.67. Let $w = (1, -2, -1, 3)$ be a vector in $\mathbf{R}^4$. Find

(a) an orthogonal basis for $w^{\perp}$, (b) an orthonormal basis for $w^{\perp}$.

7.68. Let W be the subspace of $\mathbf{R}^4$ orthogonal to $u_1 = (1, 1, 2, 2)$ and $u_2 = (0, 1, 2, -1)$. Find

(a) an orthogonal basis for W, (b) an orthonormal basis for W. (Compare with Problem 7.65.)

7.69. Let S consist of the following vectors in $\mathbf{R}^4$:

$$u_1 = (1, 1, 1, 1), \qquad u_2 = (1, 1, -1, -1), \qquad u_3 = (1, -1, 1, -1), \qquad u_4 = (1, -1, -1, 1)$$

(a) Show that S is orthogonal and a basis of $\mathbf{R}^4$.

(b) Write $v = (1, 3, -5, 6)$ as a linear combination of u_1, u_2, u_3, u_4.

(c) Find the coordinates of an arbitrary vector $v = (a, b, c, d)$ in $\mathbf{R}^4$ relative to the basis S.

(d) Normalize S to obtain an orthonormal basis of $\mathbf{R}^4$.

7.70. Let $\mathbf{M} = \mathbf{M}_{2,2}$ with inner product $\langle A, B \rangle = \mathrm{tr}(B^TA)$. Show that the following is an orthonormal basis for $\mathbf{M}$:

$$\left\{ \begin{bmatrix} 1 & 0 \\ 0 & 0 \end{bmatrix}, \begin{bmatrix} 0 & 1 \\ 0 & 0 \end{bmatrix}, \begin{bmatrix} 0 & 0 \\ 1 & 0 \end{bmatrix}, \begin{bmatrix} 0 & 0 \\ 0 & 1 \end{bmatrix} \right\}$$

7.71. Let $\mathbf{M} = \mathbf{M}_{2,2}$ with inner product $\langle A, B \rangle = \mathrm{tr}(B^TA)$. Find an orthogonal basis for the orthogonal complement of (a) diagonal matrices, (b) symmetric matrices.

7.72. Suppose $\{u_1, u_2, \ldots, u_r\}$ is an orthogonal set of vectors. Show that $\{k_1u_1, k_2u_2, \ldots, k_ru_r\}$ is an orthogonal set for any scalars $k_1, k_2, \ldots, k_r$.

7.73. Let U and W be subspaces of a finite-dimensional inner product space V. Show that

(a) $(U + W)^{\perp} = U^{\perp} \cap W^{\perp}$, (b) $(U \cap W)^{\perp} = U^{\perp} + W^{\perp}$.

Projections, Gram–Schmidt Algorithm, Applications

7.74. Find the Fourier coefficient c and projection cw of v along w, where

(a) $v = (2, 3, -5)$ and $w = (1, -5, 2)$ in $\mathbf{R}^3$.

(b) $v = (1, 3, 1, 2)$ and $w = (1, -2, 7, 4)$ in $\mathbf{R}^4$.

(c) $v = t^2$ and $w = t + 3$ in $\mathbf{P}(t)$, with inner product $\langle f, g \rangle = \int_0^1 f(t)g(t)\,dt$

(d) $v = \begin{bmatrix} 1 & 2 \\ 3 & 4 \end{bmatrix}$ and $w = \begin{bmatrix} 1 & 1 \\ 5 & 5 \end{bmatrix}$ in $\mathbf{M} = \mathbf{M}_{2,2}$, with inner product $\langle A, B \rangle = \mathrm{tr}(B^TA)$.

7.75. Let U be the subspace of $\mathbf{R}^4$ spanned by

$$v_1 = (1, 1, 1, 1), \qquad v_2 = (1, -1, 2, 2), \qquad v_3 = (1, 2, -3, -4)$$

(a) Apply the Gram–Schmidt algorithm to find an orthogonal and an orthonormal basis for U.

(b) Find the projection of $v = (1, 2, -3, 4)$ onto U.

7.76. Suppose $v = (1, 2, 3, 4, 6)$. Find the projection of v onto W, or, in other words, find $w \in W$ that minimizes $\|v - w\|$, where W is the subspace of $\mathbf{R}^5$ spanned by

(a) $u_1 = (1, 2, 1, 2, 1)$ and $u_2 = (1, -1, 2, -1, 1)$, (b) $v_1 = (1, 2, 1, 2, 1)$ and $v_2 = (1, 0, 1, 5, -1)$.

7.77. Consider the subspace $W = \mathbf{P}_2(t)$ of $\mathbf{P}(t)$ with inner product $\langle f, g \rangle = \int_0^1 f(t)g(t)\,dt$. Find the projection of $f(t) = t^3$ onto W. (*Hint:* Use the orthogonal polynomials $1, 2t - 1, 6t^2 - 6t + 1$ obtained in Problem 7.22.)

7.78. Consider $\mathbf{P}(t)$ with inner product $\langle f, g \rangle = \int_{-1}^1 f(t)g(t)\,dt$ and the subspace $W = P_3(t)$.

(a) Find an orthogonal basis for W by applying the Gram–Schmidt algorithm to $\{1, t, t^2, t^3\}$.

(b) Find the projection of $f(t) = t^5$ onto W.

Orthogonal Matrices

7.79. Find the number and exhibit all 2×2 orthogonal matrices of the form $\begin{bmatrix} \frac{1}{3} & x \\ y & z \end{bmatrix}$.

7.80. Find a 3×3 orthogonal matrix P whose first two rows are multiples of $u = (1, 1, 1)$ and $v = (1, -3, 2)$, respectively.

7.81. Find a symmetric orthogonal matrix P whose first row is $(\frac{1}{3}, \frac{2}{3}, \frac{2}{3})$. (Compare with Problem 7.32.)

7.82. Real matrices A and B are said to be *orthogonally equivalent* if there exists an orthogonal matrix P such that $B = P^TAP$. Show that this relation is an equivalence relation.

Positive Definite Matrices and Inner Products

7.83. Find the matrix A that represents the usual inner product on $\mathbf{R}^2$ relative to each of the following bases:

(a) $\{v_1 = (1, 4),\ v_2 = (2, -3)\}$, (b) $\{w_1 = (1, -3),\ w_2 = (6, 2)\}$.

7.84. Consider the following inner product on $\mathbf{R}^2$:

$$f(u, v) = x_1y_1 - 2x_1y_2 - 2x_2y_1 + 5x_2y_2, \qquad \text{where} \qquad u = (x_1, x_2) \qquad v = (y_1, y_2)$$

Find the matrix B that represents this inner product on $\mathbf{R}^2$ relative to each basis in Problem 7.83.

7.85. Find the matrix C that represents the usual basis on $\mathbf{R}^3$ relative to the basis S of $\mathbf{R}^3$ consisting of the vectors $u_1 = (1,1,1)$, $u_2 = (1,2,1)$, $u_3 = (1,-1,3)$.

7.86. Let $V = \mathbf{P}_2(t)$ with inner product $\langle f,g\rangle = \int_0^1 f(t)g(t)\,dt$.

(a) Find $\langle f,g\rangle$, where $f(t) = t+2$ and $g(t) = t^2 - 3t + 4$.

(b) Find the matrix A of the inner product with respect to the basis $\{1, t, t^2\}$ of V.

(c) Verify Theorem 7.16 that $\langle f,g\rangle = [f]^T A[g]$ with respect to the basis $\{1, t, t^2\}$.

7.87. Determine which of the following matrices are positive definite:

(a) $\begin{bmatrix} 1 & 3 \\ 3 & 5 \end{bmatrix}$, (b) $\begin{bmatrix} 3 & 4 \\ 4 & 7 \end{bmatrix}$, (c) $\begin{bmatrix} 4 & 2 \\ 2 & 1 \end{bmatrix}$, (d) $\begin{bmatrix} 6 & -7 \\ -7 & 9 \end{bmatrix}$.

7.88. Suppose A and B are positive definite matrices. Show that:
(a) $A+B$ is positive definite and (b) kA is positive definite for $k > 0$.

7.89. Suppose B is a real nonsingular matrix. Show that: (a) $B^T B$ is symmetric and (b) $B^T B$ is positive definite.

Complex Inner Product Spaces

7.90. Verify that

$$\langle a_1u_1 + a_2u_2 \quad b_1v_1 + b_2v_2\rangle = a_1\bar{b}_1\langle u_1, v_1\rangle + a_1\bar{b}_2\langle u_1, v_2\rangle + a_2\bar{b}_1\langle u_2, v_1\rangle + a_2\bar{b}_2\langle u_2, v_2\rangle$$

More generally, prove that $\langle \sum_{i=1}^m a_iu_i,\ \sum_{j=1}^n b_jv_j\rangle = \sum_{i,j} a_i\bar{b}_j\langle u_i, v_i\rangle$.

7.91. Consider $u = (1+i,\ 3,\ 4-i)$ and $v = (3-4i,\ 1+i,\ 2i)$ in $\mathbf{C}^3$. Find

(a) $\langle u, v\rangle$, (b) $\langle v, u\rangle$, (c) $\|u\|$, (d) $\|v\|$, (e) $d(u, v)$.

7.92. Find the Fourier coefficient c and the projection cw of

(a) $u = (3+i,\ 5-2i)$ along $w = (5+i,\ 1+i)$ in $\mathbf{C}^2$,

(b) $u = (1-i,\ 3i,\ 1+i)$ along $w = (1,\ 2-i,\ 3+2i)$ in $\mathbf{C}^3$.

7.93. Let $u = (z_1, z_2)$ and $v = (w_1, w_2)$ belong to $\mathbf{C}^2$. Verify that the following is an inner product of $\mathbf{C}^2$:

$$f(u, v) = z_1\bar{w}_1 + (1+i)z_1\bar{w}_2 + (1-i)z_2\bar{w}_1 + 3z_2\bar{w}_2$$

7.94. Find an orthogonal basis and an orthonormal basis for the subspace W of $\mathbf{C}^3$ spanned by $u_1 = (1, i, 1)$ and $u_2 = (1+i,\ 0,\ 2)$.

7.95. Let $u = (z_1, z_2)$ and $v = (w_1, w_2)$ belong to $\mathbf{C}^2$. For what values of $a, b, c, d \in \mathbf{C}$ is the following an inner product on $\mathbf{C}^2$?

$$f(u, v) = az_1\bar{w}_1 + bz_1\bar{w}_2 + cz_2\bar{w}_1 + dz_2\bar{w}_2$$

7.96. Prove the following form for an inner product in a complex space V:

$$\langle u, v\rangle = \tfrac{1}{4}\|u+v\|^2 - \tfrac{1}{4}\|u-v\|^2 + \tfrac{i}{4}\|u+iv\|^2 - \tfrac{i}{4}\|u-iv\|^2$$

[Compare with Problem 7.7(b).]

7.97. Let V be a real inner product space. Show that

(i) $\|u\| = \|v\|$ if and only if $\langle u+v,\ u-v\rangle = 0$;

(ii) $\|u+v\|^2 = \|u\|^2 + \|v\|^2$ if and only if $\langle u, v\rangle = 0$.

Show by counterexamples that the above statements are not true for, say, $\mathbf{C}^2$.

7.98. Find the matrix P that represents the usual inner product on $\mathbf{C}^3$ relative to the basis $\{1,\ 1+i,\ 1-2i\}$.

7.99. A complex matrix A is *unitary* if it is invertible and $A^{-1} = A^H$. Alternatively, A is unitary if its rows (columns) form an orthonormal set of vectors (relative to the usual inner product of $\mathbf{C}^n$). Find a unitary matrix whose first row is: (a) a multiple of $(1,\ 1-i)$; (b) a multiple of $(\frac{1}{2},\ \frac{1}{2}i,\ \frac{1}{2}-\frac{1}{2}i)$.

Normed Vector Spaces

7.100. Consider vectors $u = (1, -3, 4, 1, -2)$ and $v = (3, 1, -2, -3, 1)$ in $\mathbf{R}^5$. Find

(a) $\|u\|_\infty$ and $\|v\|_\infty$, (b) $\|u\|_1$ and $\|v\|_1$, (c) $\|u\|_2$ and $\|v\|_2$, (d) $d_\infty(u, v), d_1(u, v), d_2(u, v)$

7.101. Repeat Problem 7.100 for $u = (1+i,\ 2-4i)$ and $v = (1-i,\ 2+3i)$ in $\mathbf{C}^2$.

7.102. Consider the functions $f(t) = 5t - t^2$ and $g(t) = 3t - t^2$ in $C[0, 4]$. Find

(a) $d_\infty(f, g)$, (b) $d_1(f, g)$, (c) $d_2(f, g)$

7.103. Prove (a) $\|\cdot\|_1$ is a norm on $\mathbf{R}^n$. (b) $\|\cdot\|_\infty$ is a norm on $\mathbf{R}^n$.

7.104. Prove (a) $\|\cdot\|_1$ is a norm on $C[a, b]$. (b) $\|\cdot\|_\infty$ is a norm on $C[a, b]$.

ANSWERS TO SUPPLEMENTARY PROBLEMS

Notation: $M = [R_1;\ R_2;\ \ldots]$ denotes a matrix M with rows $R_1, R_2, \ldots$. Also, basis need not be unique.

7.58. $k > 9$

7.59. (a) -13, (b) -71, (c) $\sqrt{29}$, (d) $\sqrt{89}$

7.60. Let $u = (0, 0, 1)$; then $\langle u, u \rangle = 0$ in both cases

7.64. $\{7t^2 - 5t,\ 12t^2 - 5\}$

7.65. $\{(1, 2, 1, 0),\ (4, 4, 0, 1)\}$

7.66. $(-1, 0, 0, 0, 1), (-6, 2, 0, 1, 0), (-5, 2, 1, 0, 0)$

7.67. (a) $u_1 = (0, 0, 3, 1), u_2 = (0, 5, -1, 3), u_3 = (-14, -2, -1, 3)$;
(b) $u_1/\sqrt{10}, u_2/\sqrt{35}, u_3/\sqrt{210}$

7.68. (a) $(0, 2, -1, 0), (-15, 1, 2, 5)$, (b) $(0, 2, -1, 0)/\sqrt{5}, (-15, 1, 2, 5)/\sqrt{255}$

7.69. (b) $v = \frac{1}{4}(5u_1 + 3u_2 - 13u_3 + 9u_4)$,
(c) $[v] = \frac{1}{4}[a+b+c+d,\ a+b-c-d,\ a-b+c-d,\ a-b-c+d]$

7.71. (a) $[0, 1;\ 0, 0]$, $[0, 0;\ 1, 0]$, (b) $[0, -1;\ 1, 0]$

7.74. (a) $c = -\frac{23}{30}$, (b) $c = \frac{1}{7}$, (c) $c = \frac{15}{148}$, (d) $c = \frac{19}{26}$

7.75. (a) $w_1 = (1, 1, 1, 1), w_2 = (0, -2, 1, 1), w_3 = (12, -4, -1, -7)$,
(b) $\text{proj}(v, U) = \frac{1}{5}(-1, 12, 3, 6)$

7.76. (a) $\text{proj}(v, W) = \frac{1}{8}(23, 25, 30, 25, 23)$, (b) First find an orthogonal basis for W; say, $w_1 = (1, 2, 1, 2, 1)$ and $w_2 = (0, 2, 0, -3, 2)$. Then $\text{proj}(v, W) = \frac{1}{17}(34, 76, 34, 56, 42)$

7.77. $\text{proj}(f, W) = \frac{3}{2}t^2 - \frac{3}{5}t + \frac{1}{20}$

7.78. (a) $\{1,\ t,\ 3t^2-1,\ 5t^3-3t\}$, $\quad \mathrm{proj}(f, W) = \frac{10}{9}t^3 - \frac{5}{21}t$

7.79. Four: $[a, b;\quad b, -a]$, $[a, b;\quad -b, -a]$, $[a, -b;\quad b, a]$, $[a, -b;\quad -b, -a]$, where $a = \frac{1}{3}$ and $b = \frac{1}{3}\sqrt{8}$

7.80. $P = [1/a, 1/a, 1/a;\quad 1/b, -3/b, 2/b;\quad 5/c, -1/c, -4/c]$, where $a = \sqrt{3}, b = \sqrt{14}, c = \sqrt{42}$

7.81. $\frac{1}{3}[1, 2, 2;\quad 2, -2, 1;\quad 2, 1, -2]$

7.83. (a) $[17, -10;\quad -10, 13]$, (b) $[10, 0;\quad 0, 40]$

7.84. (a) $[65, -68;\quad -68, 73]$, (b) $[58, 8;\quad 8, 8]$

7.85. $[3, 4, 3;\quad 4, 6, 2;\quad 3, 2, 11]$

7.86. (a) $\frac{83}{12}$, (b) $[1, a, b;\quad a, b, c;\ b, c, d]$, where $a = \frac{1}{2}$, $b = \frac{1}{3}$, $c = \frac{1}{4}$, $d = \frac{1}{5}$

7.87. (a) No, (b) Yes, (c) No, (d) Yes

7.91. (a) $-4i$, (b) $4i$, (c) $\sqrt{28}$, (d) $\sqrt{31}$, (e) $\sqrt{59}$

7.92. (a) $c = \frac{1}{28}(19 - 5i)$, (b) $c = \frac{1}{19}(3 + 6i)$

7.94. $\{v_1 = (1, i, 1)/\sqrt{3},\ v_2 = (2i,\ 1-3i,\ 3-i)/\sqrt{24}\}$

7.95. a and d real and positive, $c = \bar{b}$ and $ad - bc$ positive.

7.97. $u = (1, 2)$, $v = (i, 2i)$

7.98. $P = [1,\ 1-i,\ 1+2i;\quad 1+i,\ 2,\ -1+3i;\quad 1-2i,\ -1-3i,\ 5]$

7.99. (a) $(1/\sqrt{3})[1,\ 1-i;\quad 1+i,\ -1]$,
(b) $[a,\ ai,\ a-ai;\quad bi,\ b,\ 0;\ a,\ ai,\ -a-ai]$, where $a = \frac{1}{2}$ and $b = 1/\sqrt{2}$.

7.100. (a) 4 and 3, (b) 11 and 10, (c) $\sqrt{31}$ and $\sqrt{24}$, (d) 6, 19, 9

7.101. (a) $\sqrt{20}$ and $\sqrt{13}$, (b) $\sqrt{2} + \sqrt{20}$ and $\sqrt{2} + \sqrt{13}$, (c) $\sqrt{22}$ and $\sqrt{15}$, (d) $7, 9, \sqrt{53}$

7.102. (a) 8, (b) 16, (c) $16/\sqrt{3}$

Determinants

8.1 Introduction

Each n-square matrix $A = [a_{ij}]$ is assigned a special scalar called the *determinant* of A, denoted by $\det(A)$ or $|A|$ or

$$\begin{vmatrix} a_{11} & a_{12} & \cdots & a_{1n} \\ a_{21} & a_{22} & \cdots & a_{2n} \\ \cdots & \cdots & \cdots & \cdots \\ a_{n1} & a_{n2} & \cdots & a_{nn} \end{vmatrix}$$

We emphasize that an $n \times n$ array of scalars enclosed by straight lines, called a *determinant of order n*, is not a matrix but denotes the determinant of the enclosed array of scalars (i.e., the enclosed matrix).

The determinant function was first discovered during the investigation of systems of linear equations. We shall see that the determinant is an indispensable tool in investigating and obtaining properties of square matrices.

The definition of the determinant and most of its properties also apply in the case where the entries of a matrix come from a commutative ring.

We begin with a special case of determinants of orders 1, 2, and 3. Then we define a determinant of arbitrary order. This general definition is preceded by a discussion of permutations, which is necessary for our general definition of the determinant.

8.2 Determinants of Orders 1 and 2

Determinants of orders 1 and 2 are defined as follows:

$$|a_{11}| = a_{11} \qquad \text{and} \qquad \begin{vmatrix} a_{11} & a_{12} \\ a_{21} & a_{22} \end{vmatrix} = a_{11}a_{22} - a_{12}a_{21}$$

Thus, the determinant of a 1×1 matrix $A = [a_{11}]$ is the scalar a_{11}; that is, $\det(A) = |a_{11}| = a_{11}$. The determinant of order two may easily be remembered by using the following diagram:

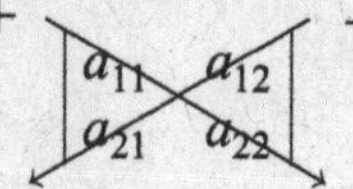

That, is, the determinant is equal to the product of the elements along the plus-labeled arrow minus the product of the elements along the minus-labeled arrow. (There is an analogous diagram for determinants of order 3, but not for higher-order determinants.)

EXAMPLE 8.1

(a) Because the determinant of order 1 is the scalar itself, we have:

$$\det(27) = 27, \qquad \det(-7) = -7, \qquad \det(t-3) = t-3$$

(b) $\begin{vmatrix} 5 & 3 \\ 4 & 6 \end{vmatrix} = 5(6) - 3(4) = 30 - 12 = 18, \qquad \begin{vmatrix} 3 & 2 \\ -5 & 7 \end{vmatrix} = 21 + 10 = 31$

Application to Linear Equations

Consider two linear equations in two unknowns, say

$$a_1x + b_1y = c_1$$
$$a_2x + b_2y = c_2$$

Let $D = a_1b_2 - a_2b_1$, the determinant of the matrix of coefficients. Then the system has a unique solution if and only if $D \neq 0$. In such a case, the unique solution may be expressed completely in terms of determinants as follows:

$$x = \frac{N_x}{D} = \frac{b_2c_1 - b_1c_2}{a_1b_2 - a_2b_1} = \frac{\begin{vmatrix} c_1 & b_1 \\ c_2 & b_2 \end{vmatrix}}{\begin{vmatrix} a_1 & b_1 \\ a_2 & b_2 \end{vmatrix}}, \qquad y = \frac{N_y}{D} = \frac{a_1c_2 - a_2c_1}{a_1b_2 - a_2b_1} = \frac{\begin{vmatrix} a_1 & c_1 \\ a_2 & c_2 \end{vmatrix}}{\begin{vmatrix} a_1 & b_1 \\ a_2 & b_2 \end{vmatrix}}$$

Here D appears in the denominator of both quotients. The numerators N_x and N_y of the quotients for x and y, respectively, can be obtained by substituting the column of constant terms in place of the column of coefficients of the given unknown in the matrix of coefficients. On the other hand, if $D = 0$, then the system may have no solution or more than one solution.

EXAMPLE 8.2 Solve by determinants the system $\begin{cases} 4x - 3y = 15 \\ 2x + 5y = 1 \end{cases}$

First find the determinant D of the matrix of coefficients:

$$D = \begin{vmatrix} 4 & -3 \\ 2 & 5 \end{vmatrix} = 4(5) - (-3)(2) = 20 + 6 = 26$$

Because $D \neq 0$, the system has a unique solution. To obtain the numerators N_x and N_y, simply replace, in the matrix of coefficients, the coefficients of x and y, respectively, by the constant terms, and then take their determinants:

$$N_x = \begin{vmatrix} 15 & -3 \\ 1 & 5 \end{vmatrix} = 75 + 3 = 78 \qquad N_y = \begin{vmatrix} 4 & 15 \\ 2 & 1 \end{vmatrix} = 4 - 30 = -26$$

Then the unique solution of the system is

$$x = \frac{N_x}{D} = \frac{78}{26} = 3, \qquad y = \frac{N_y}{D} = \frac{-26}{26} = -1$$

8.3 Determinants of Order 3

Consider an arbitrary 3×3 matrix $A = [a_{ij}]$. The determinant of A is defined as follows:

$$\det(A) = \begin{vmatrix} a_{11} & a_{12} & a_{13} \\ a_{21} & a_{22} & a_{23} \\ a_{31} & a_{32} & a_{33} \end{vmatrix} = a_{11}a_{22}a_{33} + a_{12}a_{23}a_{31} + a_{13}a_{21}a_{32} - a_{13}a_{22}a_{31} - a_{12}a_{21}a_{33} - a_{11}a_{23}a_{32}$$

Observe that there are six products, each product consisting of three elements of the original matrix. Three of the products are plus-labeled (keep their sign) and three of the products are minus-labeled (change their sign).

The diagrams in Fig. 8-1 may help us to remember the above six products in $\det(A)$. That is, the determinant is equal to the sum of the products of the elements along the three plus-labeled arrows in

Fig. 8-1 plus the sum of the negatives of the products of the elements along the three minus-labeled arrows. We emphasize that there are no such diagrammatic devices with which to remember determinants of higher order.

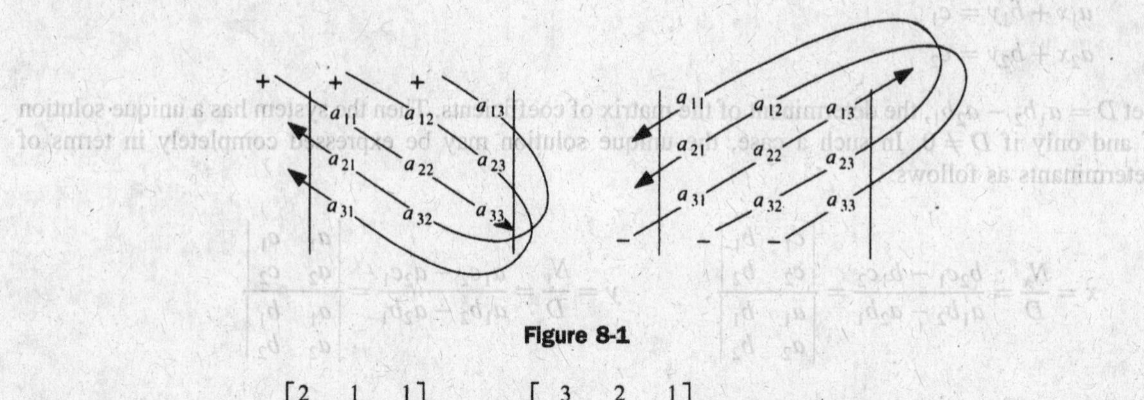

Figure 8-1

EXAMPLE 8.3 Let $A = \begin{bmatrix} 2 & 1 & 1 \\ 0 & 5 & -2 \\ 1 & -3 & 4 \end{bmatrix}$ and $B = \begin{bmatrix} 3 & 2 & 1 \\ -4 & 5 & -1 \\ 2 & -3 & 4 \end{bmatrix}$. Find $\det(A)$ and $\det(B)$.

Use the diagrams in Fig. 8-1:

$$\begin{aligned} \det(A) &= 2(5)(4) + 1(-2)(1) + 1(-3)(0) - 1(5)(1) - (-3)(-2)(2) - 4(1)(0) \\ &= 40 - 2 + 0 - 5 - 12 - 0 = 21 \\ \det(B) &= 60 - 4 + 12 - 10 - 9 + 32 = 81 \end{aligned}$$

Alternative Form for a Determinant of Order 3

The determinant of the 3×3 matrix $A = [a_{ij}]$ may be rewritten as follows:

$$\begin{aligned} \det(A) &= a_{11}(a_{22}a_{23} - a_{23}a_{32}) - a_{12}(a_{21}a_{33} - a_{23}a_{31}) + a_{13}(a_{21}a_{32} - a_{22}a_{31}) \\ &= a_{11}\begin{vmatrix} a_{22} & a_{23} \\ a_{32} & a_{33} \end{vmatrix} - a_{12}\begin{vmatrix} a_{21} & a_{23} \\ a_{31} & a_{33} \end{vmatrix} + a_{13}\begin{vmatrix} a_{21} & a_{22} \\ a_{31} & a_{32} \end{vmatrix} \end{aligned}$$

which is a linear combination of three determinants of order 2 whose coefficients (with alternating signs) form the first row of the given matrix. This linear combination may be indicated in the form

$$a_{11}\begin{vmatrix} a_{11} & a_{12} & a_{13} \\ a_{21} & a_{22} & a_{23} \\ a_{31} & a_{32} & a_{33} \end{vmatrix} - a_{12}\begin{vmatrix} a_{11} & a_{12} & a_{13} \\ a_{21} & a_{22} & a_{23} \\ a_{31} & a_{32} & a_{33} \end{vmatrix} + a_{13}\begin{vmatrix} a_{11} & a_{12} & a_{13} \\ a_{21} & a_{22} & a_{23} \\ a_{31} & a_{32} & a_{33} \end{vmatrix}$$

Note that each 2×2 matrix can be obtained by deleting, in the original matrix, the row and column containing its coefficient.

EXAMPLE 8.4

$$\begin{aligned} \begin{vmatrix} 1 & 2 & 3 \\ 4 & -2 & 3 \\ 0 & 5 & -1 \end{vmatrix} &= 1\begin{vmatrix} 1 & 2 & 3 \\ 4 & -2 & 3 \\ 0 & 5 & -1 \end{vmatrix} - 2\begin{vmatrix} 1 & 2 & 3 \\ 4 & -2 & 3 \\ 0 & 5 & -1 \end{vmatrix} + 3\begin{vmatrix} 1 & 2 & 3 \\ 4 & -2 & 3 \\ 0 & 5 & -1 \end{vmatrix} \\ &= 1\begin{vmatrix} -2 & 3 \\ 5 & -1 \end{vmatrix} - 2\begin{vmatrix} 4 & 3 \\ 0 & -1 \end{vmatrix} + 3\begin{vmatrix} 4 & -2 \\ 0 & 5 \end{vmatrix} \\ &= 1(2 - 15) - 2(-4 + 0) + 3(20 + 0) = -13 + 8 + 60 = 55 \end{aligned}$$

8.4 Permutations

A permutation σ of the set $\{1, 2, \ldots, n\}$ is a one-to-one mapping of the set onto itself or, equivalently, a rearrangement of the numbers $1, 2, \ldots, n$. Such a permutation σ is denoted by

$$\sigma = \begin{pmatrix} 1 & 2 & \cdots & n \\ j_1 & j_2 & \cdots & j_n \end{pmatrix} \quad \text{or} \quad \sigma = j_1 j_2 \cdots j_n, \quad \text{where } j_i = \sigma(i)$$

The set of all such permutations is denoted by S_n, and the number of such permutations is $n!$. If $\sigma \in S_n$, then the inverse mapping $\sigma^{-1} \in S_n$; and if $\sigma, \tau \in S_n$, then the composition mapping $\sigma \circ \tau \in S_n$. Also, the identity mapping $\varepsilon = \sigma \circ \sigma^{-1} \in S_n$. (In fact, $\varepsilon = 123 \ldots n$.)

EXAMPLE 8.5

(a) There are $2! = 2 \cdot 1 = 2$ permutations in S_2; they are 12 and 21.

(b) There are $3! = 3 \cdot 2 \cdot 1 = 6$ permutations in S_3; they are 123, 132, 213, 231, 312, 321.

Sign (Parity) of a Permutation

Consider an arbitrary permutation σ in S_n, say $\sigma = j_1 j_2 \cdots j_n$. We say σ is an even or odd permutation according to whether there is an even or odd number of inversions in σ. By an *inversion* in σ we mean a pair of integers (i, k) such that $i > k$, but i precedes k in σ. We then define the sign or parity of σ, written sgn σ, by

$$\operatorname{sgn} \sigma = \begin{cases} 1 & \text{if } \sigma \text{ is even} \\ -1 & \text{if } \sigma \text{ is odd} \end{cases}$$

EXAMPLE 8.6

(a) Find the sign of $\sigma = 35142$ in S_5.

For each element k, we count the number of elements i such that $i > k$ and i precedes k in σ. There are

2 numbers (3 and 5) greater than and preceding 1,
3 numbers (3, 5, and 4) greater than and preceding 2,
1 number (5) greater than and preceding 4.

(There are no numbers greater than and preceding either 3 or 5.) Because there are, in all, six inversions, σ is even and sgn $\sigma = 1$.

(b) The identity permutation $\varepsilon = 123 \ldots n$ is even because there are no inversions in ε.

(c) In S_2, the permutation 12 is even and 21 is odd. In S_3, the permutations 123, 231, 312 are even and the permutations 132, 213, 321 are odd.

(d) Let τ be the permutation that interchanges two numbers i and j and leaves the other numbers fixed. That is,

$$\tau(i) = j, \qquad \tau(j) = i, \qquad \tau(k) = k, \quad \text{where} \quad k \neq i, j$$

We call τ a *transposition*. If $i < j$, then there are $2(j - i) - 1$ inversions in τ, and hence, the transposition τ is odd.

Remark: One can show that, for any n, half of the permutations in S_n are even and half of them are odd. For example, 3 of the 6 permutations in S_3 are even, and 3 are odd.

8.5. Determinants of Arbitrary Order

Let $A = [a_{ij}]$ be a square matrix of order n over a field K.

Consider a product of n elements of A such that one and only one element comes from each row and one and only one element comes from each column. Such a product can be written in the form

$$a_{1j_1} a_{2j_2} \cdots a_{nj_n}$$

that is, where the factors come from successive rows, and so the first subscripts are in the natural order $1, 2, \ldots, n$. Now because the factors come from different columns, the sequence of second subscripts forms a permutation $\sigma = j_1 j_2 \cdots j_n$ in S_n. Conversely, each permutation in S_n determines a product of the above form. Thus, the matrix A contains $n!$ such products.

DEFINITION: The determinant of $A = [a_{ij}]$, denoted by $\det(A)$ or $|A|$, is the sum of all the above $n!$ products, where each such product is multiplied by $\operatorname{sgn} \sigma$. That is,

$$|A| = \sum_{\sigma} (\operatorname{sgn} \sigma) a_{1j_1} a_{2j_2} \cdots a_{nj_n}$$

or

$$|A| = \sum_{\sigma \in S_n} (\operatorname{sgn} \sigma) a_{1\sigma(1)} a_{2\sigma(2)} \cdots a_{n\sigma(n)}$$

The determinant of the n-square matrix A is said to be of order n.

The next example shows that the above definition agrees with the previous definition of determinants of orders 1, 2, and 3.

EXAMPLE 8.7

(a) Let $A = [a_{11}]$ be a 1×1 matrix. Because S_1 has only one permutation, which is even, $\det(A) = a_{11}$, the number itself.

(b) Let $A = [a_{ij}]$ be a 2×2 matrix. In S_2, the permutation 12 is even and the permutation 21 is odd. Hence,

$$\det(A) = \begin{vmatrix} a_{11} & a_{12} \\ a_{21} & a_{22} \end{vmatrix} = a_{11}a_{22} - a_{12}a_{21}$$

(c) Let $A = [a_{ij}]$ be a 3×3 matrix. In S_3, the permutations 123, 231, 312 are even, and the permutations 321, 213, 132 are odd. Hence,

$$\det(A) = \begin{vmatrix} a_{11} & a_{12} & a_{13} \\ a_{21} & a_{22} & a_{23} \\ a_{31} & a_{32} & a_{33} \end{vmatrix} = a_{11}a_{22}a_{33} + a_{12}a_{23}a_{31} + a_{13}a_{21}a_{32} - a_{13}a_{22}a_{31} - a_{12}a_{21}a_{33} - a_{11}a_{23}a_{32}$$

Remark: As n increases, the number of terms in the determinant becomes astronomical. Accordingly, we use indirect methods to evaluate determinants rather than the definition of the determinant. In fact, we prove a number of properties about determinants that will permit us to shorten the computation considerably. In particular, we show that a determinant of order n is equal to a linear combination of determinants of order $n - 1$, as in the case $n = 3$ above.

8.6 Properties of Determinants

We now list basic properties of the determinant.

THEOREM 8.1: The determinant of a matrix A and its transpose A^T are equal; that is, $|A| = |A^T|$.

By this theorem (proved in Problem 8.22), any theorem about the determinant of a matrix A that concerns the rows of A will have an analogous theorem concerning the columns of A.

The next theorem (proved in Problem 8.24) gives certain cases for which the determinant can be obtained immediately.

THEOREM 8.2: Let A be a square matrix.

(i) If A has a row (column) of zeros, then $|A| = 0$.

(ii) If A has two identical rows (columns), then $|A| = 0$.

(iii) If A is triangular (i.e., A has zeros above or below the diagonal), then $|A|$ = product of diagonal elements. Thus, in particular, $|I| = 1$, where I is the identity matrix.

The next theorem (proved in Problems 8.23 and 8.25) shows how the determinant of a matrix is affected by the elementary row and column operations.

THEOREM 8.3: Suppose B is obtained from A by an elementary row (column) operation.

(i) If two rows (columns) of A were interchanged, then $|B| = -|A|$.

(ii) If a row (column) of A were multiplied by a scalar k, then $|B| = k|A|$.

(iii) If a multiple of a row (column) of A were added to another row (column) of A, then $|B| = |A|$.

Major Properties of Determinants

We now state two of the most important and useful theorems on determinants.

THEOREM 8.4: The determinant of a product of two matrices A and B is the product of their determinants; that is,

$$\det(AB) = \det(A)\det(B)$$

The above theorem says that the determinant is a multiplicative function.

THEOREM 8.5: Let A be a square matrix. Then the following are equivalent:

(i) A is invertible; that is, A has an inverse A^{-1}.

(ii) $AX = 0$ has only the zero solution.

(iii) The determinant of A is not zero; that is, $\det(A) \neq 0$.

Remark: Depending on the author and the text, a nonsingular matrix A is defined to be an invertible matrix A, or a matrix A for which $|A| \neq 0$, or a matrix A for which $AX = 0$ has only the zero solution. The above theorem shows that all such definitions are equivalent.

We will prove Theorems 8.4 and 8.5 (in Problems 8.29 and 8.28, respectively) using the theory of elementary matrices and the following lemma (proved in Problem 8.26), which is a special case of Theorem 8.4.

LEMMA 8.6: Let E be an elementary matrix. Then, for any matrix A, $|EA| = |E||A|$.

Recall that matrices A and B are similar if there exists a nonsingular matrix P such that $B = P^{-1}AP$. Using the multiplicative property of the determinant (Theorem 8.4), one can easily prove (Problem 8.31) the following theorem.

THEOREM 8.7: Suppose A and B are similar matrices. Then $|A| = |B|$.

8.7 Minors and Cofactors

Consider an n-square matrix $A = [a_{ij}]$. Let M_{ij} denote the $(n-1)$-square submatrix of A obtained by deleting its ith row and jth column. The determinant $|M_{ij}|$ is called the *minor* of the element a_{ij} of A, and we define the *cofactor* of a_{ij}, denoted by A_{ij}, to be the "signed" minor:

$$A_{ij} = (-1)^{i+j}|M_{ij}|$$

Note that the "signs" $(-1)^{i+j}$ accompanying the minors form a chessboard pattern with +'s on the main diagonal:

$$\begin{bmatrix} + & - & + & - & \cdots \\ - & + & - & + & \cdots \\ + & - & + & - & \cdots \\ \cdots & \cdots & \cdots & \cdots & \cdots \end{bmatrix}$$

We emphasize that M_{ij} denotes a matrix, whereas A_{ij} denotes a scalar.

Remark: The sign $(-1)^{i+j}$ of the cofactor A_{ij} is frequently obtained using the checkerboard pattern. Specifically, beginning with + and alternating signs:

$$+, -, +, -, \ldots,$$

count from the main diagonal to the appropriate square.

EXAMPLE 8.8 Let $A = \begin{bmatrix} 1 & 2 & 3 \\ 4 & 5 & 6 \\ 7 & 8 & 9 \end{bmatrix}$. Find the following minors and cofactors: (a) $|M_{23}|$ and A_{23}, (b) $|M_{31}|$ and A_{31}.

(a) $|M_{23}| = \begin{vmatrix} 1 & 2 & 3 \\ 4 & 5 & 6 \\ 7 & 8 & 9 \end{vmatrix} = \begin{vmatrix} 1 & 2 \\ 7 & 8 \end{vmatrix} = 8 - 14 = -6$, and so $A_{23} = (-1)^{2+3}|M_{23}| = -(-6) = 6$

(b) $|M_{31}| = \begin{vmatrix} 1 & 2 & 3 \\ 4 & 5 & 6 \\ 7 & 8 & 9 \end{vmatrix} = \begin{vmatrix} 2 & 3 \\ 5 & 6 \end{vmatrix} = 12 - 15 = -3$, and so $A_{31} = (-1)^{1+3}|M_{31}| = +(-3) = -3$

Laplace Expansion

The following theorem (proved in Problem 8.32) holds.

THEOREM 8.8: (Laplace) The determinant of a square matrix $A = [a_{ij}]$ is equal to the sum of the products obtained by multiplying the elements of any row (column) by their respective cofactors:

$$|A| = a_{i1}A_{i1} + a_{i2}A_{i2} + \cdots + a_{in}A_{in} = \sum_{j=1}^{n} a_{ij}A_{ij}$$

$$|A| = a_{1j}A_{1j} + a_{2j}A_{2j} + \cdots + a_{nj}A_{nj} = \sum_{i=1}^{n} a_{ij}A_{ij}$$

The above formulas for $|A|$ are called the *Laplace expansions* of the determinant of A by the ith row and the jth column. Together with the elementary row (column) operations, they offer a method of simplifying the computation of $|A|$, as described below.

8.8 Evaluation of Determinants

The following algorithm reduces the evaluation of a determinant of order n to the evaluation of a determinant of order $n - 1$.

ALGORITHM 8.1: (Reduction of the order of a determinant) The input is a nonzero n-square matrix $A = [a_{ij}]$ with $n > 1$.

Step 1. Choose an element $a_{ij} = 1$ or, if lacking, $a_{ij} \neq 0$.

Step 2. Using a_{ij} as a pivot, apply elementary row (column) operations to put 0's in all the other positions in the column (row) containing a_{ij}.

Step 3. Expand the determinant by the column (row) containing a_{ij}.

The following remarks are in order.

Remark 1: Algorithm 8.1 is usually used for determinants of order 4 or more. With determinants of order less than 4, one uses the specific formulas for the determinant.

Remark 2: Gaussian elimination or, equivalently, repeated use of Algorithm 8.1 together with row interchanges can be used to transform a matrix A into an upper triangular matrix whose determinant is the product of its diagonal entries. However, one must keep track of the number of row interchanges, because each row interchange changes the sign of the determinant.

EXAMPLE 8.9 Use Algorithm 8.1 to find the determinant of $A = \begin{bmatrix} 5 & 4 & 2 & 1 \\ 2 & 3 & 1 & -2 \\ -5 & -7 & -3 & 9 \\ 1 & -2 & -1 & 4 \end{bmatrix}$.

Use $a_{23} = 1$ as a pivot to put 0's in the other positions of the third column; that is, apply the row operations "Replace R_1 by $-2R_2 + R_1$," "Replace R_3 by $3R_2 + R_3$," and "Replace R_4 by $R_2 + R_4$." By Theorem 8.3(iii), the value of the determinant does not change under these operations. Thus,

$$|A| = \begin{vmatrix} 5 & 4 & 2 & 1 \\ 2 & 3 & 1 & -2 \\ -5 & -7 & -3 & 9 \\ 1 & -2 & -1 & 4 \end{vmatrix} = \begin{vmatrix} 1 & -2 & 0 & 5 \\ 2 & 3 & 1 & -2 \\ 1 & 2 & 0 & 3 \\ 3 & 1 & 0 & 2 \end{vmatrix}$$

Now expand by the third column. Specifically, neglect all terms that contain 0 and use the fact that the sign of the minor M_{23} is $(-1)^{2+3} = -1$. Thus,

$$|A| = -\begin{vmatrix} 1 & 2 & 0 & 5 \\ 2 & 3 & 1 & -2 \\ 1 & 2 & 0 & 3 \\ 3 & 1 & 0 & 2 \end{vmatrix} = -\begin{vmatrix} 1 & -2 & 5 \\ 1 & 2 & 3 \\ 3 & 1 & 2 \end{vmatrix} = -(4 - 18 + 5 - 30 - 3 + 4) = -(-38) = 38$$

8.9 Classical Adjoint

Let $A = [a_{ij}]$ be an $n \times n$ matrix over a field K and let A_{ij} denote the cofactor of a_{ij}. The *classical adjoint* of A, denoted by adj A, is the transpose of the matrix of cofactors of A. Namely,

$$\text{adj } A = [A_{ij}]^T$$

We say "classical adjoint" instead of simply "adjoint" because the term "adjoint" is currently used for an entirely different concept.

EXAMPLE 8.10 Let $A = \begin{bmatrix} 2 & 3 & -4 \\ 0 & -4 & 2 \\ 1 & -1 & 5 \end{bmatrix}$. The cofactors of the nine elements of A follow:

$$A_{11} = +\begin{vmatrix} -4 & 2 \\ -1 & 5 \end{vmatrix} = -18, \qquad A_{12} = -\begin{vmatrix} 0 & 2 \\ 1 & 5 \end{vmatrix} = 2, \qquad A_{13} = +\begin{vmatrix} 0 & -4 \\ 1 & -1 \end{vmatrix} = 4$$

$$A_{21} = -\begin{vmatrix} 3 & -4 \\ -1 & 5 \end{vmatrix} = -11, \qquad A_{22} = +\begin{vmatrix} 2 & -4 \\ 1 & 5 \end{vmatrix} = 14, \qquad A_{23} = -\begin{vmatrix} 2 & 3 \\ 1 & -1 \end{vmatrix} = 5$$

$$A_{31} = +\begin{vmatrix} 3 & -4 \\ -4 & 2 \end{vmatrix} = -10, \qquad A_{32} = -\begin{vmatrix} 2 & -4 \\ 0 & 2 \end{vmatrix} = -4, \qquad A_{33} = +\begin{vmatrix} 2 & 3 \\ 0 & -4 \end{vmatrix} = -8$$

The transpose of the above matrix of cofactors yields the classical adjoint of A; that is,

$$\text{adj } A = \begin{bmatrix} -18 & -11 & -10 \\ 2 & 14 & -4 \\ 4 & 5 & -8 \end{bmatrix}$$

The following theorem (proved in Problem 8.34) holds.

THEOREM 8.9: Let A be any square matrix. Then

$$A(\text{adj } A) = (\text{adj } A)A = |A|I$$

where I is the identity matrix. Thus, if $|A| \neq 0$,

$$A^{-1} = \frac{1}{|A|}(\text{adj } A)$$

EXAMPLE 8.11 Let A be the matrix in Example 8.10. We have

$$\det(A) = -40 + 6 + 0 - 16 + 4 + 0 = -46$$

Thus, A does have an inverse, and, by Theorem 8.9,

$$A^{-1} = \frac{1}{|A|}(\text{adj } A) = -\frac{1}{46}\begin{bmatrix} -18 & -11 & -10 \\ 2 & 14 & -4 \\ 4 & 5 & -8 \end{bmatrix} = \begin{bmatrix} \frac{9}{23} & \frac{11}{46} & \frac{5}{23} \\ -\frac{1}{23} & -\frac{7}{23} & \frac{2}{23} \\ -\frac{2}{23} & -\frac{5}{46} & \frac{4}{23} \end{bmatrix}$$

8.10 Applications to Linear Equations, Cramer's Rule

Consider a system $AX = B$ of n linear equations in n unknowns. Here $A = [a_{ij}]$ is the (square) matrix of coefficients and $B = [b_i]$ is the column vector of constants. Let A_i be the matrix obtained from A by replacing the ith column of A by the column vector B. Furthermore, let

$$D = \det(A), \qquad N_1 = \det(A_1), \qquad N_2 = \det(A_2), \qquad \ldots, \qquad N_n = \det(A_n)$$

The fundamental relationship between determinants and the solution of the system $AX = B$ follows.

THEOREM 8.10: The (square) system $AX = B$ has a solution if and only if $D \neq 0$. In this case, the unique solution is given by

$$x_1 = \frac{N_1}{D}, \qquad x_2 = \frac{N_2}{D}, \qquad \ldots, \qquad x_n = \frac{N_n}{D}$$

The above theorem (proved in Problem 8.10) is known as *Cramer's rule* for solving systems of linear equations. We emphasize that the theorem only refers to a system with the same number of equations as unknowns, and that it only gives the solution when $D \neq 0$. In fact, if $D = 0$, the theorem does not tell us whether or not the system has a solution. However, in the case of a homogeneous system, we have the following useful result (to be proved in Problem 8.54).

THEOREM 8.11: A square homogeneous system $AX = 0$ has a nonzero solution if and only if $D = |A| = 0$.

EXAMPLE 8.12 Solve the system using determinants $\begin{cases} x+\ y+\ z= \ 5 \\ x-2y-3z=-1 \\ 2x+\ y-\ z= \ 3 \end{cases}$

First compute the determinant D of the matrix of coefficients:

$$D = \begin{vmatrix} 1 & 1 & 1 \\ 1 & -2 & -3 \\ 2 & 1 & -1 \end{vmatrix} = 2 - 6 + 1 + 4 + 3 + 1 = 5$$

Because $D \neq 0$, the system has a unique solution. To compute N_x, N_y, N_z, we replace, respectively, the coefficients of x, y, z in the matrix of coefficients by the constant terms. This yields

$$N_x = \begin{vmatrix} 5 & 1 & 1 \\ -1 & -2 & -3 \\ 3 & 1 & -1 \end{vmatrix} = 20, \qquad N_y = \begin{vmatrix} 1 & 5 & 1 \\ 1 & -1 & -3 \\ 2 & 3 & -1 \end{vmatrix} = -10, \qquad N_z = \begin{vmatrix} 1 & 1 & 5 \\ 1 & -2 & -1 \\ 2 & 1 & 3 \end{vmatrix} = 15$$

Thus, the unique solution of the system is $x = N_x/D = 4$, $y = N_y/D = -2$, $z = N_z/D = 3$; that is, the vector $u = (4, -2, 3)$.

8.11 Submatrices, Minors, Principal Minors

Let $A = [a_{ij}]$ be a square matrix of order n. Consider any r rows and r columns of A. That is, consider any set $I = (i_1, i_2, \ldots, i_r)$ of r row indices and any set $J = (j_1, j_2, \ldots, j_r)$ of r column indices. Then I and J define an $r \times r$ submatrix of A, denoted by $A(I; J)$, obtained by deleting the rows and columns of A whose subscripts do not belong to I or J, respectively. That is,

$$A(I; J) = [a_{st} : s \in I, t \in J]$$

The determinant $|A(I; J)|$ is called a *minor* of A of order r and

$$(-1)^{i_1+i_2+\cdots+i_r+j_1+j_2+\cdots+j_r}|A(I; J)|$$

is the corresponding signed minor. (Note that a minor of order $n-1$ is a minor in the sense of Section 8.7, and the corresponding signed minor is a cofactor.) Furthermore, if I' and J' denote, respectively, the remaining row and column indices, then

$$|A(I'; J')|$$

denotes the *complementary minor*, and its sign (Problem 8.74) is the same sign as the minor.

EXAMPLE 8.13 Let $A = [a_{ij}]$ be a 5-square matrix, and let $I = \{1, 2, 4\}$ and $J = \{2, 3, 5\}$. Then $I' = \{3, 5\}$ and $J' = \{1, 4\}$, and the corresponding minor $|M|$ and complementary minor $|M'|$ are as follows:

$$|M| = |A(I; J)| = \begin{vmatrix} a_{12} & a_{13} & a_{15} \\ a_{22} & a_{23} & a_{25} \\ a_{42} & a_{43} & a_{45} \end{vmatrix} \qquad \text{and} \qquad |M'| = |A(I'; J')| = \begin{vmatrix} a_{31} & a_{34} \\ a_{51} & a_{54} \end{vmatrix}$$

Because $1 + 2 + 4 + 2 + 3 + 5 = 17$ is odd, $-|M|$ is the signed minor, and $-|M'|$ is the signed complementary minor.

Principal Minors

A minor is *principal* if the row and column indices are the same, or equivalently, if the diagonal elements of the minor come from the diagonal of the matrix. We note that the sign of a principal minor is always $+1$, because the sum of the row and identical column subscripts must always be even.

EXAMPLE 8.14 Let $A = \begin{bmatrix} 1 & 2 & -1 \\ 3 & 5 & 4 \\ -3 & 1 & -2 \end{bmatrix}$. Find the sums C_1, C_2, and C_3 of the principal minors of A of orders 1, 2, and 3, respectively.

(a) There are three principal minors of order 1. These are

$$|1| = 1, \quad |5| = 5, \quad |-2| = -2, \quad \text{and so} \quad C_1 = 1 + 5 - 2 = 4$$

Note that C_1 is simply the trace of A. Namely, $C_1 = \operatorname{tr}(A)$.

(b) There are three ways to choose two of the three diagonal elements, and each choice gives a minor of order 2. These are

$$\begin{vmatrix} 1 & 2 \\ 3 & 5 \end{vmatrix} = -1, \quad \begin{vmatrix} 1 & -1 \\ -3 & -2 \end{vmatrix} = 1, \quad \begin{vmatrix} 5 & 4 \\ 1 & -2 \end{vmatrix} = -14$$

(Note that these minors of order 2 are the cofactors A_{33}, A_{22}, and A_{11} of A, respectively.) Thus,

$$C_2 = -1 + 1 - 14 = -14$$

(c) There is only one way to choose three of the three diagonal elements. Thus, the only minor of order 3 is the determinant of A itself. Thus,

$$C_3 = |A| = -10 - 24 - 3 - 15 - 4 + 12 = -44$$

8.12 Block Matrices and Determinants

The following theorem (proved in Problem 8.36) is the main result of this section.

THEOREM 8.12: Suppose M is an upper (lower) triangular block matrix with the diagonal blocks $A_1, A_2, \ldots, A_n$. Then

$$\det(M) = \det(A_1)\det(A_2)\ldots\det(A_n)$$

EXAMPLE 8.15 Find $|M|$ where $M = \left[\begin{array}{cc|ccc} 2 & 3 & 4 & 7 & 8 \\ -1 & 5 & 3 & 2 & 1 \\ \hline 0 & 0 & 2 & 1 & 5 \\ 0 & 0 & 3 & -1 & 4 \\ 0 & 0 & 5 & 2 & 6 \end{array}\right]$

Note that M is an upper triangular block matrix. Evaluate the determinant of each diagonal block:

$$\begin{vmatrix} 2 & 3 \\ -1 & 5 \end{vmatrix} = 10 + 3 = 13, \quad \begin{vmatrix} 2 & 1 & 5 \\ 3 & -1 & 4 \\ 5 & 2 & 6 \end{vmatrix} = -12 + 20 + 30 + 25 - 16 - 18 = 29$$

Then $|M| = 13(29) = 377$.

Remark: Suppose $M = \begin{bmatrix} A & B \\ C & D \end{bmatrix}$, where A, B, C, D are square matrices. Then it is not generally true that $|M| = |A||D| - |B||C|$. (See Problem 8.68.)

8.13 Determinants and Volume

Determinants are related to the notions of area and volume as follows. Let $u_1, u_2, \ldots, u_n$ be vectors in $\mathbf{R}^n$. Let S be the (solid) parallelopiped determined by the vectors; that is,

$$S = \{a_1u_1 + a_2u_2 + \cdots + a_nu_n : 0 \le a_i \le 1 \text{ for } i = 1, \ldots, n\}$$

(When $n = 2$, S is a parallelogram.) Let $V(S)$ denote the volume of S (or area of S when $n = 2$). Then

$$V(S) = \text{absolute value of } \det(A)$$

where A is the matrix with rows $u_1, u_2, \ldots, u_n$. In general, $V(S) = 0$ if and only if the vectors $u_1, \ldots, u_n$ do not form a coordinate system for $\mathbf{R}^n$ (i.e., if and only if the vectors are linearly dependent).

EXAMPLE 8.16 Let $u_1 = (1,1,0)$, $u_2 = (1,1,1)$, $u_3 = (0,2,3)$. Find the volume $V(S)$ of the parallelepiped S in $\mathbf{R}^3$ (Fig. 8-2) determined by the three vectors.

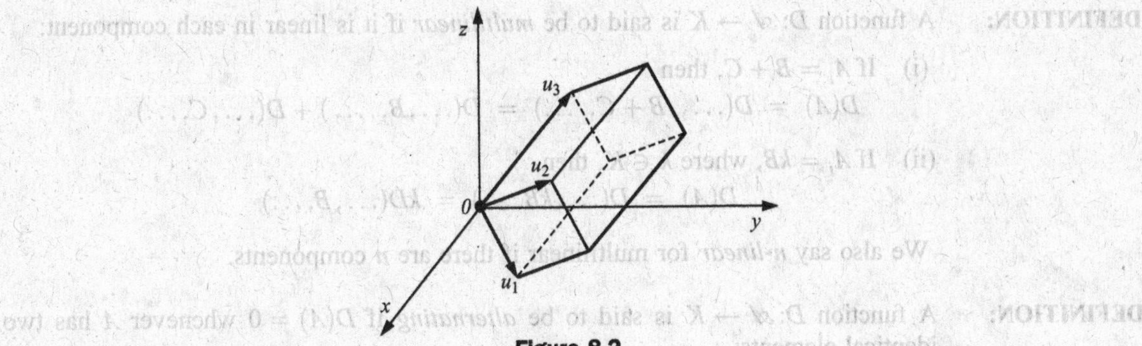

Figure 8-2

Evaluate the determinant of the matrix whose rows are u_1, u_2, u_3:

$$\begin{vmatrix} 1 & 1 & 0 \\ 1 & 1 & 1 \\ 0 & 2 & 3 \end{vmatrix} = 3 + 0 + 0 - 0 - 2 - 3 = -2$$

Hence, $V(S) = |-2| = 2$.

8.14 Determinant of a Linear Operator

Let F be a linear operator on a vector space V with finite dimension. Let A be the matrix representation of F relative to some basis S of V. Then we define the determinant of F, written $\det(F)$, by

$$\det(F) = |A|$$

If B were another matrix representation of F relative to another basis S' of V, then A and B are similar matrices (Theorem 6.7) and $|B| = |A|$ (Theorem 8.7). In other words, the above definition $\det(F)$ is independent of the particular basis S of V. (We say that the definition is *well defined.*)

The next theorem (to be proved in Problem 8.62) follows from analogous theorems on matrices.

THEOREM 8.13: Let F and G be linear operators on a vector space V. Then

(i) $\det(F \circ G) = \det(F)\det(G)$.

(ii) F is invertible if and only if $\det(F) \neq 0$.

EXAMPLE 8.17 Let F be the following linear operator on $\mathbf{R}^3$ and let A be the matrix that represents F relative to the usual basis of $\mathbf{R}^3$:

$$F(x,y,z) = (2x - 4y + z,\ x - 2y + 3z,\ 5x + y - z) \quad \text{and} \quad A = \begin{bmatrix} 2 & -4 & 1 \\ 1 & -2 & 3 \\ 5 & 1 & -1 \end{bmatrix}$$

Then

$$\det(F) = |A| = 4 - 60 + 1 + 10 - 6 - 4 = -55$$

8.15 Multilinearity and Determinants

Let V be a vector space over a field K. Let $\mathscr{A} = V^n$; that is, $\mathscr{A}$ consists of all the n-tuples

$$A = (A_1, A_2, \ldots, A_n)$$

where the A_i are vectors in V. The following definitions apply.

DEFINITION: A function $D: \mathscr{A} \to K$ is said to be *multilinear* if it is linear in each component:

(i) If $A_i = B + C$, then

$$D(A) = D(\ldots, B + C, \ldots) = D(\ldots, B, \ldots) + D(\ldots, C, \ldots)$$

(ii) If $A_i = kB$, where $k \in K$, then

$$D(A) = D(\ldots, kB, \ldots) = kD(\ldots, B, \ldots)$$

We also say *n-linear* for multilinear if there are n components.

DEFINITION: A function $D: \mathscr{A} \to K$ is said to be *alternating* if $D(A) = 0$ whenever A has two identical elements:

$$D(A_1, A_2, \ldots, A_n) = 0 \quad \text{whenever} \quad A_i = A_j, \quad i \neq j$$

Now let **M** denote the set of all n-square matrices A over a field K. We may view A as an n-tuple consisting of its row vectors $A_1, A_2, \ldots, A_n$; that is, we may view A in the form $A = (A_1, A_2, \ldots, A_n)$.

The following theorem (proved in Problem 8.37) characterizes the determinant function.

THEOREM 8.14: There exists a unique function $D: M \to K$ such that

(i) D is multilinear, (ii) D is alternating, (iii) $D(I) = 1$.

This function D is the determinant function; that is, $D(A) = |A|$, for any matrix $A \in M$.

SOLVED PROBLEMS

Computation of Determinants

8.1. Evaluate the determinant of each of the following matrices:

(a) $A = \begin{bmatrix} 6 & 5 \\ 2 & 3 \end{bmatrix}$, (b) $B = \begin{bmatrix} 2 & -3 \\ 4 & 7 \end{bmatrix}$, (c) $C = \begin{bmatrix} 4 & -5 \\ -1 & -2 \end{bmatrix}$, (d) $D = \begin{bmatrix} t-5 & 6 \\ 3 & t+2 \end{bmatrix}$

Use the formula $\begin{vmatrix} a & b \\ c & d \end{vmatrix} = ad - bc$:

(a) $|A| = 6(3) - 5(2) = 18 - 10 = 8$
(b) $|B| = 14 + 12 = 26$
(c) $|C| = -8 - 5 = -13$
(d) $|D| = (t-5)(t+2) - 18 = t^2 - 3t - 10 - 18 = t^2 - 10t - 28$

8.2. Evaluate the determinant of each of the following matrices:

(a) $A = \begin{bmatrix} 2 & 3 & 4 \\ 5 & 4 & 3 \\ 1 & 2 & 1 \end{bmatrix}$, (b) $B = \begin{bmatrix} 1 & -2 & 3 \\ 2 & 4 & -1 \\ 1 & 5 & -2 \end{bmatrix}$, (c) $C = \begin{bmatrix} 1 & 3 & -5 \\ 3 & -1 & 2 \\ 1 & -2 & 1 \end{bmatrix}$

Use the diagram in Fig. 8-1 to obtain the six products:

(a) $|A| = 2(4)(1) + 3(3)(1) + 4(2)(5) - 1(4)(4) - 2(3)(2) - 1(3)(5) = 8 + 9 + 40 - 16 - 12 - 15 = 14$

(b) $|B| = -8 + 2 + 30 - 12 + 5 - 8 = 9$

(c) $|C| = -1 + 6 + 30 - 5 + 4 - 9 = 25$

8.3. Compute the determinant of each of the following matrices:

$$\text{(a)}\quad A = \begin{bmatrix} 2 & 3 & 4 \\ 5 & 6 & 7 \\ 8 & 9 & 1 \end{bmatrix}, \quad \text{(b)}\quad B = \begin{bmatrix} 4 & -6 & 8 & 9 \\ 0 & -2 & 7 & -3 \\ 0 & 0 & 5 & 6 \\ 0 & 0 & 0 & 3 \end{bmatrix}, \quad \text{(c)}\quad C = \begin{bmatrix} \frac{1}{2} & -1 & -\frac{1}{3} \\ \frac{3}{4} & \frac{1}{2} & -1 \\ 1 & -4 & 1 \end{bmatrix}.$$

(a) One can simplify the entries by first subtracting twice the first row from the second row—that is, by applying the row operation "Replace R_2 by $-2_1 + R_2$." Then

$$|A| = \begin{vmatrix} 2 & 3 & 4 \\ 5 & 6 & 7 \\ 8 & 9 & 1 \end{vmatrix} = \begin{vmatrix} 2 & 3 & 4 \\ 1 & 0 & -1 \\ 8 & 9 & 1 \end{vmatrix} = 0 - 24 + 36 - 0 + 18 - 3 = 27$$

(b) B is triangular, so $|B| =$ product of the diagonal entries $= -120$.

(c) The arithmetic is simpler if fractions are first eliminated. Hence, multiply the first row R_1 by 6 and the second row R_2 by 4. Then

$$|24C| = \begin{vmatrix} 3 & -6 & -2 \\ 3 & 2 & -4 \\ 1 & -4 & 1 \end{vmatrix} = 6 + 24 + 24 + 4 - 48 + 18 = 28, \qquad \text{so } |C| = \frac{28}{24} = \frac{7}{6}$$

8.4. Compute the determinant of each of the following matrices:

$$\text{(a)}\quad A = \begin{bmatrix} 2 & 5 & -3 & -2 \\ -2 & -3 & 2 & -5 \\ 1 & 3 & -2 & 2 \\ -1 & -6 & 4 & 3 \end{bmatrix}, \quad \text{(b)}\quad B = \begin{bmatrix} 6 & 2 & 1 & 0 & 5 \\ 2 & 1 & 1 & -2 & 1 \\ 1 & 1 & 2 & -2 & 3 \\ 3 & 0 & 2 & 3 & -1 \\ -1 & -1 & -3 & 4 & 2 \end{bmatrix}$$

(a) Use $a_{31} = 1$ as a pivot to put 0's in the first column, by applying the row operations "Replace R_1 by $-2R_3 + R_1$," "Replace R_2 by $2R_3 + R_2$," and "Replace R_4 by $R_3 + R_4$." Then

$$|A| = \begin{vmatrix} 2 & 5 & -3 & -2 \\ -2 & -3 & 2 & -5 \\ 1 & 3 & -2 & 2 \\ -1 & -6 & 4 & 3 \end{vmatrix} = \begin{vmatrix} 0 & -1 & 1 & -6 \\ 0 & 3 & -2 & -1 \\ 1 & 3 & -2 & 2 \\ 0 & -3 & 2 & 5 \end{vmatrix} = \begin{vmatrix} -1 & 1 & -6 \\ 3 & -2 & -1 \\ -3 & 2 & 5 \end{vmatrix}$$

$$= 10 + 3 - 36 + 36 - 2 - 15 = -4$$

(b) First reduce $|B|$ to a determinant of order 4, and then to a determinant of order 3, for which we can use Fig. 8-1. First use $c_{22} = 1$ as a pivot to put 0's in the second column, by applying the row operations "Replace R_1 by $-2R_2 + R_1$," "Replace R_3 by $-R_2 + R_3$," and "Replace R_5 by $R_2 + R_5$." Then

$$|B| = \begin{vmatrix} 2 & 0 & -1 & 4 & 3 \\ 2 & 1 & 1 & -2 & 1 \\ -1 & 0 & 1 & 0 & 2 \\ 3 & 0 & 2 & 3 & -1 \\ 1 & 0 & -2 & 2 & 3 \end{vmatrix} = \begin{vmatrix} 2 & -1 & 4 & 3 \\ -1 & 1 & 0 & 2 \\ 3 & 2 & 3 & -1 \\ 1 & -2 & 2 & 3 \end{vmatrix} = \begin{vmatrix} 1 & 1 & 4 & 5 \\ 0 & 1 & 0 & 0 \\ 5 & 2 & 3 & -5 \\ -1 & -2 & 2 & 7 \end{vmatrix}$$

$$= \begin{vmatrix} 1 & 4 & 5 \\ 5 & 3 & -5 \\ -1 & 2 & 7 \end{vmatrix} = 21 + 20 + 50 + 15 + 10 - 140 = -34$$

Cofactors, Classical Adjoints, Minors, Principal Minors

8.5. Let $A = \begin{bmatrix} 2 & 1 & -3 & 4 \\ 5 & -4 & 7 & -2 \\ 4 & 0 & 6 & -3 \\ 3 & -2 & 5 & 2 \end{bmatrix}$.

(a) Find A_{23}, the cofactor (signed minor) of 7 in A.

(b) Find the minor and the signed minor of the submatrix $M = A(2,4;\ 2,3)$.

(c) Find the principal minor determined by the first and third diagonal entries—that is, by $M = A(1,3;\ \ 1,3)$.

(a) Take the determinant of the submatrix of A obtained by deleting row 2 and column 3 (those which contain the 7), and multiply the determinant by $(-1)^{2+3}$:

$$A_{23} = -\begin{vmatrix} 2 & 1 & 4 \\ 4 & 0 & -3 \\ 3 & -2 & 2 \end{vmatrix} = -(-61) = 61$$

The exponent $2+3$ comes from the subscripts of A_{23}—that is, from the fact that 7 appears in row 2 and column 3.

(b) The row subscripts are 2 and 4 and the column subscripts are 2 and 3. Hence, the minor is the determinant

$$|M| = \begin{vmatrix} a_{22} & a_{23} \\ a_{42} & a_{43} \end{vmatrix} = \begin{vmatrix} -4 & 7 \\ -2 & 5 \end{vmatrix} = -20 + 14 = -6$$

and the signed minor is $(-1)^{2+4+2+3}|M| = -|M| = -(-6) = 6$.

(c) The principal minor is the determinant

$$|M| = \begin{vmatrix} a_{11} & a_{13} \\ a_{31} & a_{33} \end{vmatrix} = \begin{vmatrix} 2 & -3 \\ 4 & 6 \end{vmatrix} = 12 + 12 = 24$$

Note that now the diagonal entries of the submatrix are diagonal entries of the original matrix. Also, the sign of the principal minor is positive.

8.6. Let $B = \begin{bmatrix} 1 & 1 & 1 \\ 2 & 3 & 4 \\ 5 & 8 & 9 \end{bmatrix}$. Find: (a) $|B|$, (b) adj B, (c) B^{-1} using adj B.

(a) $|B| = 27 + 20 + 16 - 15 - 32 - 18 = -2$

(b) Take the transpose of the matrix of cofactors:

$$\text{adj } B = \begin{bmatrix} \begin{vmatrix} 3 & 4 \\ 8 & 9 \end{vmatrix} & -\begin{vmatrix} 2 & 4 \\ 5 & 9 \end{vmatrix} & \begin{vmatrix} 2 & 3 \\ 5 & 8 \end{vmatrix} \\ -\begin{vmatrix} 1 & 1 \\ 8 & 9 \end{vmatrix} & \begin{vmatrix} 1 & 1 \\ 5 & 9 \end{vmatrix} & -\begin{vmatrix} 1 & 1 \\ 5 & 8 \end{vmatrix} \\ \begin{vmatrix} 1 & 1 \\ 3 & 4 \end{vmatrix} & -\begin{vmatrix} 1 & 1 \\ 2 & 4 \end{vmatrix} & \begin{vmatrix} 1 & 1 \\ 2 & 3 \end{vmatrix} \end{bmatrix}^T = \begin{bmatrix} -5 & 2 & 1 \\ -1 & 4 & -3 \\ 1 & -2 & 1 \end{bmatrix}^T = \begin{bmatrix} -5 & -1 & 1 \\ 2 & 4 & -2 \\ 1 & -3 & 1 \end{bmatrix}$$

(c) Because $|B| \neq 0$, $B^{-1} = \frac{1}{|B|}(\text{adj } B) = \frac{1}{-2}\begin{bmatrix} -5 & -1 & 1 \\ 2 & 4 & -2 \\ 1 & -3 & 1 \end{bmatrix} = \begin{bmatrix} \frac{5}{2} & \frac{1}{2} & -\frac{1}{2} \\ -1 & -2 & 1 \\ -\frac{1}{2} & \frac{3}{2} & -\frac{1}{2} \end{bmatrix}$

8.7. Let $A = \begin{bmatrix} 1 & 2 & 3 \\ 4 & 5 & 6 \\ 0 & 7 & 8 \end{bmatrix}$, and let S_k denote the sum of its principal minors of order k. Find S_k for

(a) $k = 1$, (b) $k = 2$, (c) $k = 3$.

(a) The principal minors of order 1 are the diagonal elements. Thus, S_1 is the trace of A; that is,

$$S_1 = \operatorname{tr}(A) = 1 + 5 + 8 = 14$$

(b) The principal minors of order 2 are the cofactors of the diagonal elements. Thus,

$$S_2 = A_{11} + A_{22} + A_{33} = \begin{vmatrix} 5 & 6 \\ 7 & 8 \end{vmatrix} + \begin{vmatrix} 1 & 3 \\ 0 & 8 \end{vmatrix} + \begin{vmatrix} 1 & 2 \\ 4 & 5 \end{vmatrix} = -2 + 8 - 3 = 3$$

(c) There is only one principal minor of order 3, the determinant of A. Then

$$S_3 = |A| = 40 + 0 + 84 - 0 - 42 - 64 = 18$$

8.8. Let $A = \begin{bmatrix} 1 & 3 & 0 & -1 \\ -4 & 2 & 5 & 1 \\ 1 & 0 & 3 & -2 \\ 3 & -2 & 1 & 4 \end{bmatrix}$. Find the number N_k and sum S_k of principal minors of order:

(a) $k = 1$, (b) $k = 2$, (c) $k = 3$, (d) $k = 4$.

Each (nonempty) subset of the diagonal (or equivalently, each nonempty subset of $\{1, 2, 3, 4\}$) determines a principal minor of A, and $N_k = \binom{n}{k} = \dfrac{n!}{k!(n-k)!}$ of them are of order k.

Thus, $N_1 = \binom{4}{1} = 4$, $N_2 = \binom{4}{2} = 6$, $N_3 = \binom{4}{3} = 4$, $N_4 = \binom{4}{4} = 1$

(a) $S_1 = |1| + |2| + |3| + |4| = 1 + 2 + 3 + 4 = 10$

(b) $S_2 = \begin{vmatrix} 1 & 3 \\ -4 & 2 \end{vmatrix} + \begin{vmatrix} 1 & 0 \\ 1 & 3 \end{vmatrix} + \begin{vmatrix} 1 & -1 \\ 3 & 4 \end{vmatrix} + \begin{vmatrix} 2 & 5 \\ 0 & 3 \end{vmatrix} + \begin{vmatrix} 2 & 1 \\ -2 & 4 \end{vmatrix} + \begin{vmatrix} 3 & -2 \\ 1 & 4 \end{vmatrix}$

$= 14 + 3 + 7 + 6 + 10 + 14 = 54$

(c) $S_3 = \begin{vmatrix} 1 & 3 & 0 \\ -4 & 2 & 5 \\ 1 & 0 & 3 \end{vmatrix} + \begin{vmatrix} 1 & 3 & -1 \\ -4 & 2 & 1 \\ 3 & -2 & 4 \end{vmatrix} + \begin{vmatrix} 1 & 0 & -1 \\ 1 & 3 & -2 \\ 3 & 1 & 4 \end{vmatrix} + \begin{vmatrix} 2 & 5 & 1 \\ 0 & 3 & -2 \\ -2 & 1 & 4 \end{vmatrix}$

$= 57 + 65 + 22 + 54 = 198$

(d) $S_4 = \det(A) = 378$

Determinants and Systems of Linear Equations

8.9. Use determinants to solve the system $\begin{cases} 3y + 2x = z + 1 \\ 3x + 2z = 8 - 5y \\ 3z - 1 = x - 2y \end{cases}$.

First arrange the equation in standard form, then compute the determinant D of the matrix of coefficients:

$$\begin{aligned} 2x + 3y - z &= 1 \\ 3x + 5y + 2z &= 8 \\ x - 2y - 3z &= -1 \end{aligned} \quad \text{and} \quad D = \begin{vmatrix} 2 & 3 & -1 \\ 3 & 5 & 2 \\ 1 & -2 & -3 \end{vmatrix} = -30 + 6 + 6 + 5 + 8 + 27 = 22$$

Because $D \neq 0$, the system has a unique solution. To compute N_x, N_y, N_z, we replace, respectively, the coefficients of x, y, z in the matrix of coefficients by the constant terms. Then

$$N_x = \begin{vmatrix} 1 & 3 & -1 \\ 8 & 5 & 2 \\ -1 & -2 & -1 \end{vmatrix} = 66, \qquad N_y = \begin{vmatrix} 2 & 1 & -1 \\ 3 & 8 & 2 \\ 1 & -1 & -3 \end{vmatrix} = -22, \qquad N_z = \begin{vmatrix} 2 & 3 & 1 \\ 3 & 5 & 8 \\ 1 & -2 & -1 \end{vmatrix} = 44$$

Thus,

$$x = \frac{N_x}{D} = \frac{66}{22} = 3, \qquad y = \frac{N_y}{D} = \frac{-22}{22} = -1, \qquad z = \frac{N_z}{D} = \frac{44}{22} = 2$$

8.10. Consider the system $\begin{cases} kx + y + z = 1 \\ x + ky + z = 1 \\ x + y + kz = 1 \end{cases}$

Use determinants to find those values of k for which the system has

(a) a unique solution, (b) more than one solution, (c) no solution.

(a) The system has a unique solution when $D \neq 0$, where D is the determinant of the matrix of coefficients. Compute

$$D = \begin{vmatrix} k & 1 & 1 \\ 1 & k & 1 \\ 1 & 1 & k \end{vmatrix} = k^3 + 1 + 1 - k - k - k = k^3 - 3k + 2 = (k-1)^2(k+2)$$

Thus, the system has a unique solution when

$$(k-1)^2(k+2) \neq 0, \quad \text{when } k \neq 1 \text{ and } k \neq 2$$

(b and c) Gaussian elimination shows that the system has more than one solution when $k = 1$, and the system has no solution when $k = -2$.

Miscellaneous Problems

8.11. Find the volume $V(S)$ of the parallelepiped S in $\mathbf{R}^3$ determined by the vectors:

(a) $u_1 = (1,1,1), u_2 = (1,3,-4), u_3 = (1,2,-5)$.
(b) $u_1 = (1,2,4), u_2 = (2,1,-3), u_3 = (5,7,9)$.

$V(S)$ is the absolute value of the determinant of the matrix M whose rows are the given vectors. Thus,

(a) $|M| = \begin{vmatrix} 1 & 1 & 1 \\ 1 & 3 & -4 \\ 1 & 2 & -5 \end{vmatrix} = -15 - 4 + 2 - 3 + 8 + 5 = -7$. Hence, $V(S) = |-7| = 7$.

(b) $|M| = \begin{vmatrix} 1 & 2 & 4 \\ 2 & 1 & -3 \\ 5 & 7 & 9 \end{vmatrix} = 9 - 30 + 56 - 20 + 21 - 36 = 0$. Thus, $V(S) = 0$, or, in other words, u_1, u_2, u_3 lie in a plane and are linearly dependent.

8.12. Find $\det(M)$ where $M = \begin{bmatrix} 3 & 4 & 0 & 0 & 0 \\ 2 & 5 & 0 & 0 & 0 \\ 0 & 9 & 2 & 0 & 0 \\ 0 & 5 & 0 & 6 & 7 \\ 0 & 0 & 4 & 3 & 4 \end{bmatrix} = \left[\begin{array}{cc|c|cc} 3 & 4 & 0 & 0 & 0 \\ 2 & 5 & 0 & 0 & 0 \\ \hline 0 & 9 & 2 & 0 & 0 \\ \hline 0 & 5 & 0 & 6 & 7 \\ 0 & 0 & 4 & 3 & 4 \end{array}\right]$

M is a (lower) triangular block matrix; hence, evaluate the determinant of each diagonal block:

$$\begin{vmatrix} 3 & 4 \\ 2 & 5 \end{vmatrix} = 15 - 8 = 7, \qquad |2| = 2, \qquad \begin{vmatrix} 6 & 7 \\ 3 & 4 \end{vmatrix} = 24 - 21 = 3$$

Thus, $|M| = 7(2)(3) = 42$.

8.13. Find the determinant of $F: \mathbf{R}^3 \to \mathbf{R}^3$ defined by

$$F(x,y,z) = (x + 3y - 4z, \; 2y + 7z, \; x + 5y - 3z)$$

The determinant of a linear operator F is equal to the determinant of any matrix that represents F. Thus first find the matrix A representing F in the usual basis (whose rows, respectively, consist of the coefficients of x, y, z). Then

$$A = \begin{bmatrix} 1 & 3 & -4 \\ 0 & 2 & 7 \\ 1 & 5 & -3 \end{bmatrix}, \quad \text{and so} \quad \det(F) = |A| = -6 + 21 + 0 + 8 - 35 - 0 = -8$$

8.14. Write out $g = g(x_1, x_2, x_3, x_4)$ explicitly where

$$g(x_1, x_2, \ldots, x_n) = \prod_{i<j}(x_i - x_j).$$

The symbol $\prod$ is used for a product of terms in the same way that the symbol $\sum$ is used for a sum of terms. That is, $\prod_{i<j}(x_i - x_j)$ means the product of all terms $(x_i - x_j)$ for which $i < j$. Hence,

$$g = g(x_1, \ldots, x_4) = (x_1 - x_2)(x_1 - x_3)(x_1 - x_4)(x_2 - x_3)(x_2 - x_4)(x_3 - x_4)$$

8.15. Let D be a 2-linear, alternating function. Show that $D(A, B) = -D(B, A)$.

Because D is alternating, $D(A, A) = 0$, $D(B, B) = 0$. Hence,

$$D(A+B, A+B) = D(A,A) + D(A,B) + D(B,A) + D(B,B) = D(A,B) + D(B,A)$$

However, $D(A+B,\ A+B) = 0$. Hence, $D(A,B) = -D(B,A)$, as required.

Permutations

8.16. Determine the parity (sign) of the permutation $\sigma = 364152$.

Count the number of inversions. That is, for each element k, count the number of elements i in σ such that $i > k$ and i precedes k in σ. Namely,

$$\begin{array}{llll} k=1: & 3 \text{ numbers } (3,6,4) & k=4: & 1 \text{ number } (6) \\ k=2: & 4 \text{ numbers } (3,6,4,5) & k=5: & 1 \text{ number } (6) \\ k=3: & 0 \text{ numbers} & k=6: & 0 \text{ numbers} \end{array}$$

Because $3+4+0+1+1+0 = 9$ is odd, σ is an odd permutation, and $\text{sgn}\,\sigma = -1$.

8.17. Let $\sigma = 24513$ and $\tau = 41352$ be permutations in S_5. Find (a) $\tau \circ \sigma$, (b) σ^{-1}.

Recall that $\sigma = 24513$ and $\tau = 41352$ are short ways of writing

$$\sigma = \begin{pmatrix} 1 & 2 & 3 & 4 & 5 \\ 2 & 4 & 5 & 1 & 3 \end{pmatrix} \quad \text{or} \quad \sigma(1) = 2, \quad \sigma(2) = 4, \quad \sigma(3) = 5, \quad \sigma(4) = 1, \quad \sigma(5) = 3$$

$$\tau = \begin{pmatrix} 1 & 2 & 3 & 4 & 5 \\ 4 & 1 & 3 & 5 & 2 \end{pmatrix} \quad \text{or} \quad \tau(1) = 4, \quad \tau(2) = 1, \quad \tau(3) = 3, \quad \tau(4) = 5, \quad \tau(5) = 2$$

(a) The effects of σ and then τ on $1,2,3,4,5$ are as follows:

$$1 \to 2 \to 1, \qquad 2 \to 4 \to 5, \qquad 3 \to 5 \to 2, \qquad 4 \to 1 \to 4, \qquad 5 \to 3 \to 3$$

[That is, for example, $(\tau \circ \sigma)(1) = \tau(\sigma(1)) = \tau(2) = 1$.] Thus, $\tau \circ \sigma = 15243$.

(b) By definition, $\sigma^{-1}(j) = k$ if and only if $\sigma(k) = j$. Hence,

$$\sigma^{-1} = \begin{pmatrix} 2 & 4 & 5 & 1 & 3 \\ 1 & 2 & 3 & 4 & 5 \end{pmatrix} = \begin{pmatrix} 1 & 2 & 3 & 4 & 5 \\ 4 & 1 & 5 & 2 & 3 \end{pmatrix} \quad \text{or} \quad \sigma^{-1} = 41523$$

8.18. Let $\sigma = j_1 j_2 \ldots j_n$ be any permutation in S_n. Show that, for each inversion (i, k) where $i > k$ but i precedes k in σ, there is a pair (i^*, j^*) such that

$$i^* < k^* \quad \text{and} \quad \sigma(i^*) > \sigma(j^*) \tag{1}$$

and vice versa. Thus, σ is even or odd according to whether there is an even or an odd number of pairs satisfying (1).

Choose i^* and k^* so that $\sigma(i^*) = i$ and $\sigma(k^*) = k$. Then $i > k$ if and only if $\sigma(i^*) > \sigma(k^*)$, and i precedes k in σ if and only if $i^* < k^*$.

8.19. Consider the polynomials $g = g(x_1, \ldots, x_n)$ and $\sigma(g)$, defined by

$$g = g(x_1, \ldots, x_n) = \prod_{i<j}(x_i - x_j) \qquad \text{and} \qquad \sigma(g) = \prod_{i<j}(x_{\sigma(i)} - x_{\sigma(j)})$$

(See Problem 8.14.) Show that $\sigma(g) = g$ when σ is an even permutation, and $\sigma(g) = -g$ when σ is an odd permutation. That is, $\sigma(g) = (\text{sgn } \sigma)g$.

Because σ is one-to-one and onto,

$$\sigma(g) = \prod_{i<j}(x_{\sigma(i)} - x_{\sigma(j)}) = \prod_{i<j \text{ or } i>j}(x_i - x_j)$$

Thus, $\sigma(g)$ or $\sigma(g) = -g$ according to whether there is an even or an odd number of terms of the form $x_i - x_j$, where $i > j$. Note that for each pair (i, j) for which

$$i < j \qquad \text{and} \qquad \sigma(i) > \sigma(j)$$

there is a term $(x_{\sigma(i)} - x_{\sigma(j)})$ in $\sigma(g)$ for which $\sigma(i) > \sigma(j)$. Because σ is even if and only if there is an even number of pairs satisfying (1), we have $\sigma(g) = g$ if and only if σ is even. Hence, $\sigma(g) = -g$ if and only if σ is odd.

8.20. Let $\sigma, \tau \in S_n$. Show that $\text{sgn}(\tau \circ \sigma) = (\text{sgn } \tau)(\text{sgn } \sigma)$. Thus, the product of two even or two odd permutations is even, and the product of an odd and an even permutation is odd.

Using Problem 8.19, we have

$$\text{sgn}(\tau \circ \sigma)\, g = (\tau \circ \sigma)(g) = \tau(\sigma(g)) = \tau((\text{sgn } \sigma)g) = (\text{sgn } \tau)(\text{sgn } \sigma)g$$

Accordingly, $\text{sgn}\,(\tau \circ \sigma) = (\text{sgn } \tau)(\text{sgn } \sigma)$.

8.21. Consider the permutation $\sigma = j_1 j_2 \cdots j_n$. Show that $\text{sgn } \sigma^{-1} = \text{sgn } \sigma$ and, for scalars a_{ij}, show that

$$a_{j_1 1} a_{j_2 2} \cdots a_{j_n n} = a_{1k_1} a_{2k_2} \cdots a_{nk_n}$$

where $\sigma^{-1} = k_1 k_2 \cdots k_n$.

We have $\sigma^{-1} \circ \sigma = \varepsilon$, the identity permutation. Because ε is even, σ^{-1} and σ are both even or both odd. Hence $\text{sgn } \sigma^{-1} = \text{sgn } \sigma$.

Because $\sigma = j_1 j_2 \cdots j_n$ is a permutation, $a_{j_1 1} a_{j_2 2} \cdots a_{j_n n} = a_{1k_1} a_{2k_2} \cdots a_{nk_n}$. Then $k_1, k_2, \ldots, k_n$ have the property that

$$\sigma(k_1) = 1, \qquad \sigma(k_2) = 2, \qquad \ldots, \qquad \sigma(k_n) = n$$

Let $\tau = k_1 k_2 \cdots k_n$. Then, for $i = 1, \ldots, n$,

$$(\sigma \circ \tau)(i) = \sigma(\tau(i)) = \sigma(k_i) = i$$

Thus, $\sigma \circ \tau = \varepsilon$, the identity permutation. Hence, $\tau = \sigma^{-1}$.

Proofs of Theorems

8.22. Prove Theorem 8.1: $|A^T| = |A|$.

If $A = [a_{ij}]$, then $A^T = [b_{ij}]$, with $b_{ij} = a_{ji}$. Hence,

$$|A^T| = \sum_{\sigma \in S_n} (\text{sgn } \sigma) b_{1\sigma(1)} b_{2\sigma(2)} \cdots b_{n\sigma(n)} = \sum_{\sigma \in S_n} (\text{sgn } \sigma) a_{\sigma(1),1} a_{\sigma(2),2} \cdots a_{\sigma(n),n}$$

Let $\tau = \sigma^{-1}$. By Problem 8.21 $\text{sgn } \tau = \text{sgn } \sigma$, and $a_{\sigma(1),1} a_{\sigma(2),2} \cdots a_{\sigma(n),n} = a_{1\tau(1)} a_{2\tau(2)} \cdots a_{n\tau(n)}$. Hence,

$$|A^T| = \sum_{\sigma \in S_n} (\text{sgn } \tau) a_{1\tau(1)} a_{2\tau(2)} \cdots a_{n\tau(n)}$$

However, as σ runs through all the elements of S_n, $\tau = \sigma^{-1}$ also runs through all the elements of S_n. Thus, $|A^T| = |A|$.

8.23. Prove Theorem 8.3(i): If two rows (columns) of A are interchanged, then $|B| = -|A|$.

We prove the theorem for the case that two columns are interchanged. Let τ be the transposition that interchanges the two numbers corresponding to the two columns of A that are interchanged. If $A = [a_{ij}]$ and $B = [b_{ij}]$, then $b_{ij} = a_{i\tau(j)}$. Hence, for any permutation σ,

$$b_{1\sigma(1)}b_{2\sigma(2)}\cdots b_{n\sigma(n)} = a_{1(\tau\circ\sigma)(1)}a_{2(\tau\circ\sigma)(2)}\cdots a_{n(\tau\circ\sigma)(n)}$$

Thus,

$$|B| = \sum_{\sigma\in S_n} (\text{sgn } \sigma) b_{1\sigma(1)}b_{2\sigma(2)}\cdots b_{n\sigma(n)} = \sum_{\sigma\in S_n} (\text{sgn } \sigma) a_{1(\tau\circ\sigma)(1)}a_{2(\tau\circ\sigma)(2)}\cdots a_{n(\tau\circ\sigma)(n)}$$

Because the transposition τ is an odd permutation, $\text{sgn}(\tau\circ\sigma) = (\text{sgn } \tau)(\text{sgn } \sigma) = -\text{sgn } \sigma$. Accordingly, $\text{sgn } \sigma = -\text{sgn } (\tau\circ\sigma)$, and so

$$|B| = -\sum_{\sigma\in S_n} [\text{sgn}(\tau\circ\sigma)] a_{1(\tau\circ\sigma)(1)}a_{2(\tau\circ\sigma)(2)}\cdots a_{n(\tau\circ\sigma)(n)}$$

But as σ runs through all the elements of S_n, $\tau\circ\sigma$ also runs through all the elements of S_n. Hence, $|B| = -|A|$.

8.24. Prove Theorem 8.2.

(i) If A has a row (column) of zeros, then $|A| = 0$.

(ii) If A has two identical rows (columns), then $|A| = 0$.

(iii) If A is triangular, then $|A|$ = product of diagonal elements. Thus, $|I| = 1$.

(i) Each term in $|A|$ contains a factor from every row, and so from the row of zeros. Thus, each term of $|A|$ is zero, and so $|A| = 0$.

(ii) Suppose $1 + 1 \neq 0$ in K. If we interchange the two identical rows of A, we still obtain the matrix A. Hence, by Problem 8.23, $|A| = -|A|$, and so $|A| = 0$.

Now suppose $1 + 1 = 0$ in K. Then $\text{sgn } \sigma = 1$ for every $\sigma \in S_n$. Because A has two identical rows, we can arrange the terms of A into pairs of equal terms. Because each pair is 0, the determinant of A is zero.

(iii) Suppose $A = [a_{ij}]$ is lower triangular; that is, the entries above the diagonal are all zero: $a_{ij} = 0$ whenever $i < j$. Consider a term t of the determinant of A:

$$t = (\text{sgn } \sigma) a_{1i_1}a_{2i_2}\cdots a_{ni_n}, \qquad \text{where} \qquad \sigma = i_1 i_2 \cdots i_n$$

Suppose $i_1 \neq 1$. Then $1 < i_1$ and so $a_{1i_1} = 0$; hence, $t = 0$. That is, each term for which $i_1 \neq 1$ is zero.

Now suppose $i_1 = 1$ but $i_2 \neq 2$. Then $2 < i_2$, and so $a_{2i_2} = 0$; hence, $t = 0$. Thus, each term for which $i_1 \neq 1$ or $i_2 \neq 2$ is zero.

Similarly, we obtain that each term for which $i_1 \neq 1$ or $i_2 \neq 2$ or ... or $i_n \neq n$ is zero. Accordingly, $|A| = a_{11}a_{22}\cdots a_{nn}$ = product of diagonal elements.

8.25. Prove Theorem 8.3: B is obtained from A by an elementary operation.

(i) If two rows (columns) of A were interchanged, then $|B| = -|A|$.

(ii) If a row (column) of A were multiplied by a scalar k, then $|B| = k|A|$.

(iii) If a multiple of a row (column) of A were added to another row (column) of A, then $|B| = |A|$.

(i) This result was proved in Problem 8.23.

(ii) If the jth row of A is multiplied by k, then every term in $|A|$ is multiplied by k, and so $|B| = k|A|$. That is,

$$|B| = \sum_{\sigma} (\text{sgn } \sigma) a_{1i_1}a_{2i_2}\cdots (ka_{ji_j})\cdots a_{ni_n} = k\sum_{\sigma} (\text{sgn } \sigma) a_{1i_1}a_{2i_2}\cdots a_{ni_n} = k|A|$$

(iii) Suppose c times the kth row is added to the jth row of A. Using the symbol ^ to denote the jth position in a determinant term, we have

$$|B| = \sum_\sigma (\operatorname{sgn} \sigma) a_{1i_1} a_{2i_2} \cdots (\widehat{c a_{ki_k} + a_{ji_j}}) \ldots a_{ni_n}$$
$$= c \sum_\sigma (\operatorname{sgn} \sigma) a_{1i_1} a_{2i_2} \cdots \widehat{a_{ki_k}} \cdots a_{ni_n} + \sum_\sigma (\operatorname{sgn} \sigma) a_{1i_1} a_{2i_2} \cdots a_{ji_j} \cdots a_{ni_n}$$

The first sum is the determinant of a matrix whose kth and jth rows are identical. Accordingly, by Theorem 8.2(ii), the sum is zero. The second sum is the determinant of A. Thus, $|B| = c \cdot 0 + |A| = |A|$.

8.26. Prove Lemma 8.6: Let E be an elementary matrix. Then $|EA| = |E||A|$.

Consider the elementary row operations: (i) Multiply a row by a constant $k \neq 0$,

(ii) Interchange two rows, (iii) Add a multiple of one row to another.

Let E_1, E_2, E_3 be the corresponding elementary matrices That is, E_1, E_2, E_3 are obtained by applying the above operations to the identity matrix I. By Problem 8.25,

$$|E_1| = k|I| = k, \qquad |E_2| = -|I| = -1, \qquad |E_3| = |I| = 1$$

Recall (Theorem 3.11) that E_iA is identical to the matrix obtained by applying the corresponding operation to A. Thus, by Theorem 8.3, we obtain the following which proves our lemma:

$$|E_1A| = k|A| = |E_1||A|, \qquad |E_2A| = -|A| = |E_2||A|, \qquad |E_3A| = |A| = 1|A| = |E_3||A|$$

8.27. Suppose B is row equivalent to a square matrix A. Prove that $|B| = 0$ if and only if $|A| = 0$.

By Theorem 8.3, the effect of an elementary row operation is to change the sign of the determinant or to multiply the determinant by a nonzero scalar. Hence, $|B| = 0$ if and only if $|A| = 0$.

8.28. Prove Theorem 8.5: Let A be an n-square matrix. Then the following are equivalent:

(i) A is invertible, (ii) $AX = 0$ has only the zero solution, (iii) $\det(A) \neq 0$.

The proof is by the Gaussian algorithm. If A is invertible, it is row equivalent to I. But $|I| \neq 0$. Hence, by Problem 8.27, $|A| \neq 0$. If A is not invertible, it is row equivalent to a matrix with a zero row. Hence, $\det(A) = 0$. Thus, (i) and (iii) are equivalent.

If $AX = 0$ has only the solution $X = 0$, then A is row equivalent to I and A is invertible. Conversely, if A is invertible with inverse A^{-1}, then

$$X = IX = (A^{-1}A)X = A^{-1}(AX) = A^{-1}0 = 0$$

is the only solution of $AX = 0$. Thus, (i) and (ii) are equivalent.

8.29. Prove Theorem 8.4: $|AB| = |A||B|$.

If A is singular, then AB is also singular, and so $|AB| = 0 = |A||B|$. On the other hand, if A is nonsingular, then $A = E_n \cdots E_2E_1$, a product of elementary matrices. Then, Lemma 8.6 and induction yields

$$|AB| = |E_n \cdots E_2E_1B| = |E_n| \cdots |E_2||E_1||B| = |A||B|$$

8.30. Suppose P is invertible. Prove that $|P^{-1}| = |P|^{-1}$.

$P^{-1}P = I$. Hence, $1 = |I| = |P^{-1}P| = |P^{-1}||P|$, and so $|P^{-1}| = |P|^{-1}$.

8.31. Prove Theorem 8.7: Suppose A and B are similar matrices. Then $|A| = |B|$.

Because A and B are similar, there exists an invertible matrix P such that $B = P^{-1}AP$. Therefore, using Problem 8.30, we get $|B| = |P^{-1}AP| = |P^{-1}||A||P| = |A||P^{-1}||P = |A|$.

We remark that although the matrices P^{-1} and A may not commute, their determinants $|P^{-1}|$ and $|A|$ do commute, because they are scalars in the field K.

8.32. Prove Theorem 8.8 (Laplace): Let $A = [a_{ij}]$, and let A_{ij} denote the cofactor of a_{ij}. Then, for any i or j

$$|A| = a_{i1}A_{i1} + \cdots + a_{in}A_{in} \qquad \text{and} \qquad |A| = a_{1j}A_{1j} + \cdots + a_{nj}A_{nj}$$

Because $|A| = |A^T|$, we need only prove one of the expansions, say, the first one in terms of rows of A. Each term in $|A|$ contains one and only one entry of the ith row $(a_{i1}, a_{i2}, \ldots, a_{in})$ of A. Hence, we can write $|A|$ in the form

$$|A| = a_{i1}A_{i1}^* + a_{i2}A_{i2}^* + \cdots + a_{in}A_{in}^*$$

(Note that A_{ij}^* is a sum of terms involving no entry of the ith row of A.) Thus, the theorem is proved if we can show that

$$A_{ij}^* = A_{ij} = (-1)^{i+j}|M_{ij}|$$

where M_{ij} is the matrix obtained by deleting the row and column containing the entry a_{ij}. (Historically, the expression A_{ij}^* was defined as the cofactor of a_{ij}, and so the theorem reduces to showing that the two definitions of the cofactor are equivalent.)

First we consider the case that $i = n, j = n$. Then the sum of terms in $|A|$ containing a_{nn} is

$$a_{nn}A_{nn}^* = a_{nn}\sum_{\sigma}(\operatorname{sgn}\sigma)a_{1\sigma(1)}a_{2\sigma(2)}\cdots a_{n-1,\sigma(n-1)}$$

where we sum over all permutations $\sigma \in S_n$ for which $\sigma(n) = n$. However, this is equivalent (Prove!) to summing over all permutations of $\{1, \ldots, n-1\}$. Thus, $A_{nn}^* = |M_{nn}| = (-1)^{n+n}|M_{nn}|$.

Now we consider any i and j. We interchange the ith row with each succeeding row until it is last, and we interchange the jth column with each succeeding column until it is last. Note that the determinant $|M_{ij}|$ is not affected, because the relative positions of the other rows and columns are not affected by these interchanges. However, the "sign" of $|A|$ and of A_{ij}^* is changed $n-1$ and then $n-j$ times. Accordingly,

$$A_{ij}^* = (-1)^{n-i+n-j}|M_{ij}| = (-1)^{i+j}|M_{ij}|$$

8.33. Let $A = [a_{ij}]$ and let B be the matrix obtained from A by replacing the ith row of A by the row vector $(b_{i1}, \ldots, b_{in})$. Show that

$$|B| = b_{i1}A_{i1} + b_{i2}A_{i2} + \cdots + b_{in}A_{in}$$

Furthermore, show that, for $j \neq i$,

$$a_{j1}A_{i1} + a_{j2}A_{i2} + \cdots + a_{jn}A_{in} = 0 \qquad \text{and} \qquad a_{1j}A_{1i} + a_{2j}A_{2i} + \cdots + a_{nj}A_{ni} = 0$$

Let $B = [b_{ij}]$. By Theorem 8.8,

$$|B| = b_{i1}B_{i1} + b_{i2}B_{i2} + \cdots + b_{in}B_{in}$$

Because B_{ij} does not depend on the ith row of B, we get $B_{ij} = A_{ij}$ for $j = 1, \ldots, n$. Hence,

$$|B| = b_{i1}A_{i1} + b_{i2}A_{i2} + \cdots + b_{in}A_{in}$$

Now let A' be obtained from A by replacing the ith row of A by the jth row of A. Because A' has two identical rows, $|A'| = 0$. Thus, by the above result,

$$|A'| = a_{j1}A_{i1} + a_{j2}A_{i2} + \cdots + a_{jn}A_{in} = 0$$

Using $|A^T| = |A|$, we also obtain that $a_{1j}A_{1i} + a_{2j}A_{2i} + \cdots + a_{nj}A_{ni} = 0$.

8.34. Prove Theorem 8.9: $A(\operatorname{adj} A) = (\operatorname{adj} A)A = |A|I$.

Let $A = [a_{ij}]$ and let $A(\operatorname{adj} A) = [b_{ij}]$. The ith row of A is

$$(a_{i1}, a_{i2}, \ldots, a_{in}) \tag{1}$$

Because adj A is the transpose of the matrix of cofactors, the jth column of adj A is the tranpose of the cofactors of the jth row of A:

$$(A_j, A_{j2}, \ldots, A_{jn})^T \tag{2}$$

Now b_{ij}, the ij entry in $A(\operatorname{adj} A)$, is obtained by multiplying expressions (1) and (2):

$$b_{ij} = a_{i1}A_{j1} + a_{i2}A_{j2} + \cdots + a_{in}A_{jn}$$

By Theorem 8.8 and Problem 8.33,

$$b_{ij} = \begin{cases} |A| & \text{if } i = j \\ 0 & \text{if } i \neq j \end{cases}$$

Accordingly, $A(\text{adj } A)$ is the diagonal matrix with each diagonal element $|A|$. In other words, $A(\text{adj } A) = |A|I$. Similarly, $(\text{adj } A)A = |A|I$.

8.35. Prove Theorem 8.10 (Cramer's rule): The (square) system $AX = B$ has a unique solution if and only if $D \neq 0$. In this case, $x_i = N_i/D$ for each i.

By previous results, $AX = B$ has a unique solution if and only if A is invertible, and A is invertible if and only if $D = |A| \neq 0$.

Now suppose $D \neq 0$. By Theorem 8.9, $A^{-1} = (1/D)(\text{adj } A)$. Multiplying $AX = B$ by A^{-1}, we obtain

$$X = A^{-1}AX = (1/D)(\text{adj } A)B \tag{1}$$

Note that the ith row of $(1/D)(\text{adj } A)$ is $(1/D)(A_{1i}, A_{2i}, \ldots, A_{ni})$. If $B = (b_1, b_2, \ldots, b_n)^T$, then, by (1),

$$x_i = (1/D)(b_1A_{1i} + b_2A_{2i} + \cdots + b_nA_{ni})$$

However, as in Problem 8.33, $b_1A_{1i} + b_2A_{2i} + \cdots + b_nA_{ni} = N_i$, the determinant of the matrix obtained by replacing the ith column of A by the column vector B. Thus, $x_i = (1/D)N_i$, as required.

8.36. Prove Theorem 8.12: Suppose M is an upper (lower) triangular block matrix with diagonal blocks $A_1, A_2, \ldots, A_n$. Then

$$\det(M) = \det(A_1)\det(A_2)\cdots\det(A_n)$$

We need only prove the theorem for $n = 2$—that is, when M is a square matrix of the form $M = \begin{bmatrix} A & C \\ 0 & B \end{bmatrix}$. The proof of the general theorem follows easily by induction.

Suppose $A = [a_{ij}]$ is r-square, $B = [b_{ij}]$ is s-square, and $M = [m_{ij}]$ is n-square, where $n = r + s$. By definition,

$$\det(M) = \sum_{\sigma \in S_n} (\text{sgn } \sigma) m_{1\sigma(1)} m_{2\sigma(2)} \cdots m_{n\sigma(n)}$$

If $i > r$ and $j \leq r$, then $m_{ij} = 0$. Thus, we need only consider those permutations σ such that

$$\sigma\{r+1, r+2, \ldots, r+s\} = \{r+1, r+2, \ldots, r+s\} \quad \text{and} \quad \sigma\{1, 2, \ldots, r\} = \{1, 2, \ldots, r\}$$

Let $\sigma_1(k) = \sigma(k)$ for $k \leq r$, and let $\sigma_2(k) = \sigma(r+k) - r$ for $k \leq s$. Then

$$(\text{sgn } \sigma) m_{1\sigma(1)} m_{2\sigma(2)} \cdots m_{n\sigma(n)} = (\text{sgn } \sigma_1) a_{1\sigma_1(1)} a_{2\sigma_1(2)} \cdots a_{r\sigma_1(r)} (\text{sgn } \sigma_2) b_{1\sigma_2(1)} b_{2\sigma_2(2)} \cdots b_{s\sigma_2(s)}$$

which implies $\det(M) = \det(A)\det(B)$.

8.37. Prove Theorem 8.14: There exists a unique function $D : \mathbf{M} \to K$ such that

(i) D is multilinear, (ii) D is alternating, (iii) $D(I) = 1$.

This function D is the determinant function; that is, $D(A) = |A|$.

Let D be the determinant function, $D(A) = |A|$. We must show that D satisfies (i), (ii), and (iii), and that D is the only function satisfying (i), (ii), and (iii).

By Theorem 8.2, D satisfies (ii) and (iii). Hence, we show that it is multilinear. Suppose the ith row of $A = [a_{ij}]$ has the form $(b_{i1} + c_{i1},\ b_{i2} + c_{i2},\ \ldots,\ b_{in} + c_{in})$. Then

$$\begin{aligned} D(A) &= D(A_1, \ldots, B_i + C_i, \ldots, A_n) \\ &= \sum_{S_n} (\text{sgn } \sigma) a_{1\sigma(1)} \cdots a_{i-1,\sigma(i-1)} (b_{i\sigma(i)} + c_{i\sigma(i)}) \cdots a_{n\sigma(n)} \\ &= \sum_{S_n} (\text{sgn } \sigma) a_{1\sigma(1)} \cdots b_{i\sigma(i)} \cdots a_{n\sigma(n)} + \sum_{S_n} (\text{sgn } \sigma) a_{1\sigma(1)} \cdots c_{i\sigma(i)} \cdots a_{n\sigma(n)} \\ &= D(A_1, \ldots, B_i, \ldots, A_n) + D(A_1, \ldots, C_i, \ldots, A_n) \end{aligned}$$

Also, by Theorem 8.3(ii),

$$D(A_1, \ldots, kA_i, \ldots, A_n) = kD(A_1, \ldots, A_i, \ldots, A_n)$$

Thus, D is multilinear—D satisfies (i).

We next must prove the uniqueness of D. Suppose D satisfies (i), (ii), and (iii). If $\{e_1, \ldots, e_n\}$ is the usual basis of K^n, then, by (iii), $D(e_1, e_2, \ldots, e_n) = D(I) = 1$. Using (ii), we also have that

$$D(e_{i_1}, e_{i_2}, \ldots, e_{i_n}) = \operatorname{sgn} \sigma, \qquad \text{where} \qquad \sigma = i_1 i_2 \cdots i_n \tag{1}$$

Now suppose $A = [a_{ij}]$. Observe that the kth row A_k of A is

$$A_k = (a_{k1}, a_{k2}, \ldots, a_{kn}) = a_{k1}e_1 + a_{k2}e_2 + \cdots + a_{kn}e_n$$

Thus,

$$D(A) = D(a_{11}e_1 + \cdots + a_{1n}e_n, \; a_{21}e_1 + \cdots + a_{2n}e_n, \; \ldots, \; a_{n1}e_1 + \cdots + a_{nn}e_n)$$

Using the multilinearity of D, we can write $D(A)$ as a sum of terms of the form

$$\begin{aligned} D(A) &= \sum D(a_{1i_1}e_{i_1}, a_{2i_2}e_{i_2}, \ldots, a_{ni_n}e_{i_n}) \\ &= \sum (a_{1i_1}a_{2i_2} \cdots a_{ni_n}) D(e_{i_1}, e_{i_2}, \ldots, e_{i_n}) \end{aligned} \tag{2}$$

where the sum is summed over all sequences $i_1 i_2 \ldots i_n$, where $i_k \in \{1, \ldots, n\}$. If two of the indices are equal, say $i_j = i_k$ but $j \neq k$, then, by (ii),

$$D(e_{i_1}, e_{i_2}, \ldots, e_{i_n}) = 0$$

Accordingly, the sum in (2) need only be summed over all permutations $\sigma = i_1 i_2 \cdots i_n$. Using (1), we finally have that

$$\begin{aligned} D(A) &= \sum_{\sigma} (a_{1i_1}a_{2i_2} \cdots a_{ni_n}) D(e_{i_1}, e_{i_2}, \ldots, e_{i_n}) \\ &= \sum_{\sigma} (\operatorname{sgn} \sigma) a_{1i_1}a_{2i_2} \cdots a_{ni_n}, \qquad \text{where} \qquad \sigma = i_1 i_2 \cdots i_n \end{aligned}$$

Hence, D is the determinant function, and so the theorem is proved.

SUPPLEMENTARY PROBLEMS

Computation of Determinants

8.38. Evaluate:

(a) $\begin{vmatrix} 2 & 6 \\ 4 & 1 \end{vmatrix}$, (b) $\begin{vmatrix} 5 & 1 \\ 3 & -2 \end{vmatrix}$, (c) $\begin{vmatrix} -2 & 8 \\ -5 & -3 \end{vmatrix}$, (d) $\begin{vmatrix} 4 & 9 \\ 1 & -3 \end{vmatrix}$, (e) $\begin{vmatrix} a+b & a \\ b & a+b \end{vmatrix}$

8.39. Find all t such that (a) $\begin{vmatrix} t-4 & 3 \\ 2 & t-9 \end{vmatrix} = 0$, (b) $\begin{vmatrix} t-1 & 4 \\ 3 & t-2 \end{vmatrix} = 0$

8.40. Compute the determinant of each of the following matrices:

(a) $\begin{bmatrix} 2 & 1 & 1 \\ 0 & 5 & -2 \\ 1 & -3 & 4 \end{bmatrix}$, (b) $\begin{bmatrix} 3 & -2 & -4 \\ 2 & 5 & -1 \\ 0 & 6 & 1 \end{bmatrix}$, (c) $\begin{bmatrix} -2 & -1 & 4 \\ 6 & -3 & -2 \\ 4 & 1 & 2 \end{bmatrix}$, (d) $\begin{bmatrix} 7 & 6 & 5 \\ 1 & 2 & 1 \\ 3 & -2 & 1 \end{bmatrix}$

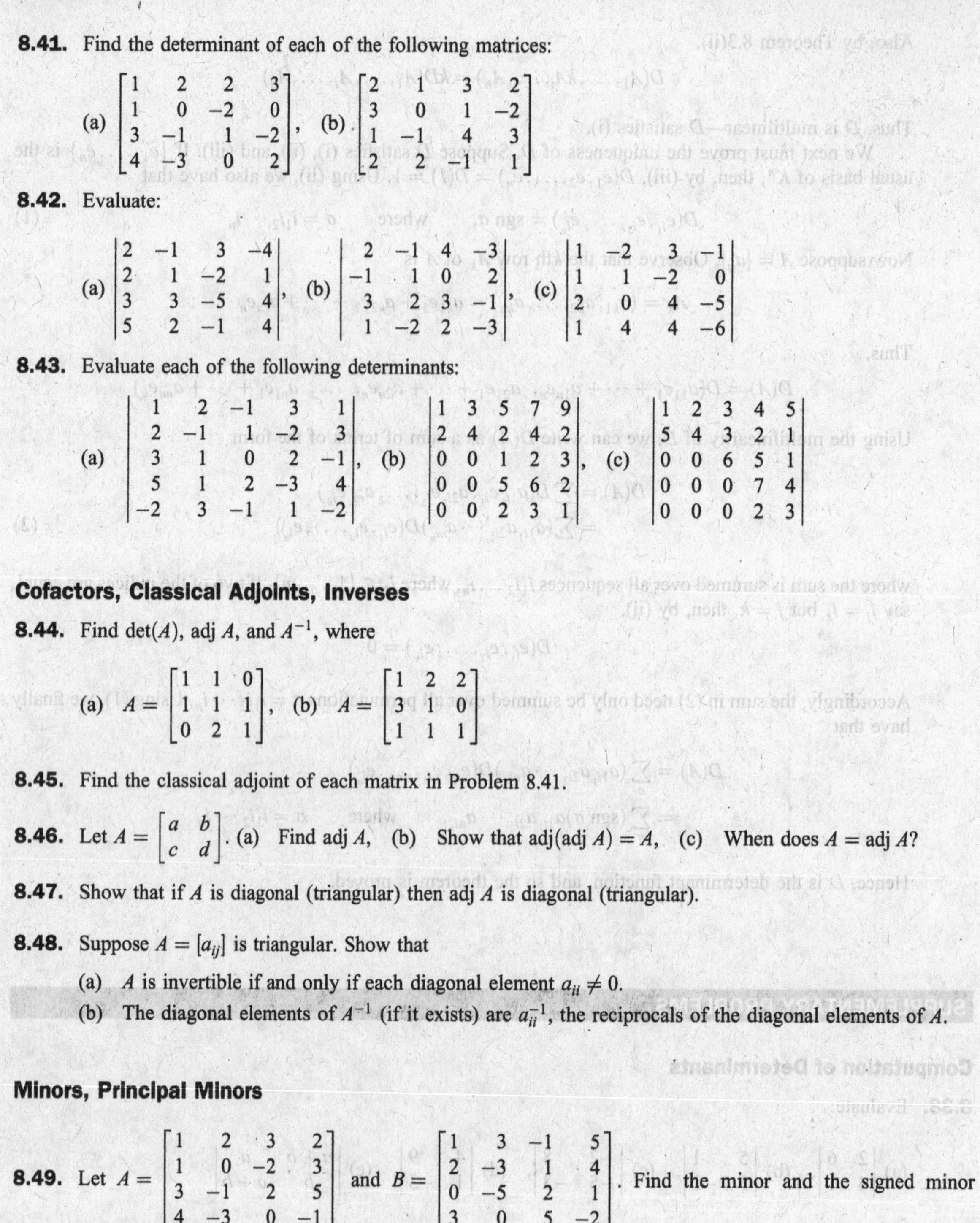

8.41. Find the determinant of each of the following matrices:

$$\text{(a)}\ \begin{bmatrix} 1 & 2 & 2 & 3 \\ 1 & 0 & -2 & 0 \\ 3 & -1 & 1 & -2 \\ 4 & -3 & 0 & 2 \end{bmatrix}, \quad \text{(b)}\ \begin{bmatrix} 2 & 1 & 3 & 2 \\ 3 & 0 & 1 & -2 \\ 1 & -1 & 4 & 3 \\ 2 & 2 & -1 & 1 \end{bmatrix}$$

8.42. Evaluate:

$$\text{(a)}\ \begin{vmatrix} 2 & -1 & 3 & -4 \\ 2 & 1 & -2 & 1 \\ 3 & 3 & -5 & 4 \\ 5 & 2 & -1 & 4 \end{vmatrix}, \quad \text{(b)}\ \begin{vmatrix} 2 & -1 & 4 & -3 \\ -1 & 1 & 0 & 2 \\ 3 & 2 & 3 & -1 \\ 1 & -2 & 2 & -3 \end{vmatrix}, \quad \text{(c)}\ \begin{vmatrix} 1 & -2 & 3 & -1 \\ 1 & 1 & -2 & 0 \\ 2 & 0 & 4 & -5 \\ 1 & 4 & 4 & -6 \end{vmatrix}$$

8.43. Evaluate each of the following determinants:

$$\text{(a)}\ \begin{vmatrix} 1 & 2 & -1 & 3 & 1 \\ 2 & -1 & 1 & -2 & 3 \\ 3 & 1 & 0 & 2 & -1 \\ 5 & 1 & 2 & -3 & 4 \\ -2 & 3 & -1 & 1 & -2 \end{vmatrix}, \quad \text{(b)}\ \begin{vmatrix} 1 & 3 & 5 & 7 & 9 \\ 2 & 4 & 2 & 4 & 2 \\ 0 & 0 & 1 & 2 & 3 \\ 0 & 0 & 5 & 6 & 2 \\ 0 & 0 & 2 & 3 & 1 \end{vmatrix}, \quad \text{(c)}\ \begin{vmatrix} 1 & 2 & 3 & 4 & 5 \\ 5 & 4 & 3 & 2 & 1 \\ 0 & 0 & 6 & 5 & 1 \\ 0 & 0 & 0 & 7 & 4 \\ 0 & 0 & 0 & 2 & 3 \end{vmatrix}$$

Cofactors, Classical Adjoints, Inverses

8.44. Find $\det(A)$, adj A, and A^{-1}, where

$$\text{(a)}\ A = \begin{bmatrix} 1 & 1 & 0 \\ 1 & 1 & 1 \\ 0 & 2 & 1 \end{bmatrix}, \quad \text{(b)}\ A = \begin{bmatrix} 1 & 2 & 2 \\ 3 & 1 & 0 \\ 1 & 1 & 1 \end{bmatrix}$$

8.45. Find the classical adjoint of each matrix in Problem 8.41.

8.46. Let $A = \begin{bmatrix} a & b \\ c & d \end{bmatrix}$. (a) Find adj A, (b) Show that $\text{adj}(\text{adj } A) = A$, (c) When does $A = \text{adj } A$?

8.47. Show that if A is diagonal (triangular) then adj A is diagonal (triangular).

8.48. Suppose $A = [a_{ij}]$ is triangular. Show that

(a) A is invertible if and only if each diagonal element $a_{ii} \neq 0$.

(b) The diagonal elements of A^{-1} (if it exists) are a_{ii}^{-1}, the reciprocals of the diagonal elements of A.

Minors, Principal Minors

8.49. Let $A = \begin{bmatrix} 1 & 2 & 3 & 2 \\ 1 & 0 & -2 & 3 \\ 3 & -1 & 2 & 5 \\ 4 & -3 & 0 & -1 \end{bmatrix}$ and $B = \begin{bmatrix} 1 & 3 & -1 & 5 \\ 2 & -3 & 1 & 4 \\ 0 & -5 & 2 & 1 \\ 3 & 0 & 5 & -2 \end{bmatrix}$. Find the minor and the signed minor corresponding to the following submatrices:

(a) $A(1,4;\ 3,4)$, (b) $B(1,4;\ 3,4)$, (c) $A(2,3;\ 2,4)$, (d) $B(2,3;\ 2,4)$.

8.50. For $k = 1,2,3$, find the sum S_k of all principal minors of order k for

$$\text{(a)}\ A = \begin{bmatrix} 1 & 3 & 2 \\ 2 & -4 & 3 \\ 5 & -2 & 1 \end{bmatrix}, \quad \text{(b)}\ B = \begin{bmatrix} 1 & 5 & -4 \\ 2 & 6 & 1 \\ 3 & -2 & 0 \end{bmatrix}, \quad \text{(c)}\ C = \begin{bmatrix} 1 & -4 & 3 \\ 2 & 1 & 5 \\ 4 & -7 & 11 \end{bmatrix}$$

8.51. For $k = 1, 2, 3, 4$, find the sum S_k of all principal minors of order k for

(a) $A = \begin{bmatrix} 1 & 2 & 3 & -1 \\ 1 & -2 & 0 & 5 \\ 0 & 1 & -2 & 2 \\ 4 & 0 & -1 & -3 \end{bmatrix}$, (b) $B = \begin{bmatrix} 1 & 2 & 1 & 2 \\ 0 & 1 & 2 & 3 \\ 1 & 3 & 0 & 4 \\ 2 & 7 & 4 & 5 \end{bmatrix}$

Determinants and Linear Equations

8.52. Solve the following systems by determinants:

(a) $\begin{cases} 3x + 5y = 8 \\ 4x - 2y = 1 \end{cases}$, (b) $\begin{cases} 2x - 3y = -1 \\ 4x + 7y = -1 \end{cases}$, (c) $\begin{cases} ax - 2by = c \\ 3ax - 5by = 2c \end{cases}$ $(ab \neq 0)$

8.53. Solve the following systems by determinants:

(a) $\begin{cases} 2x - 5y + 2z = 2 \\ x + 2y - 4z = 5 \\ 3x - 4y - 6z = 1 \end{cases}$, (b) $\begin{cases} 2z + 3 = y + 3x \\ x - 3z = 2y + 1 \\ 3y + z = 2 - 2x \end{cases}$

8.54. Prove Theorem 8.11: The system $AX = 0$ has a nonzero solution if and only if $D = |A| = 0$.

Permutations

8.55. Find the parity of the permutations $\sigma = 32154$, $\tau = 13524$, $\pi = 42531$ in S_5.

8.56. For the permutations in Problem 8.55, find

(a) $\tau \circ \sigma$, (b) $\pi \circ \sigma$, (c) σ^{-1}, (d) τ^{-1}.

8.57. Let $\tau \in S_n$. Show that $\tau \circ \sigma$ runs through S_n as σ runs through S_n, that is, $S_n = \{\tau \circ \sigma : \sigma \in S_n\}$.

8.58. Let $\sigma \in S_n$ have the property that $\sigma(n) = n$. Let $\sigma^* \in S_{n-1}$ be defined by $\sigma^*(x) = \sigma(x)$.

(a) Show that $\operatorname{sgn} \sigma^* = \operatorname{sgn} \sigma$,

(b) Show that as σ runs through S_n, where $\sigma(n) = n$, σ^* runs through S_{n-1}; that is,

$$S_{n-1} = \{\sigma^* : \sigma \in S_n, \sigma(n) = n\}.$$

8.59. Consider a permutation $\sigma = j_1 j_2 \ldots j_n$. Let $\{e_i\}$ be the usual basis of K^n, and let A be the matrix whose ith row is e_{j_i} [i.e., $A = (e_{j_1}, e_{j_2}, \ldots, e_{j_n})$]. Show that $|A| = \operatorname{sgn} \sigma$.

Determinant of Linear Operators

8.60. Find the determinant of each of the following linear transformations:

(a) $T:\mathbf{R}^2 \to \mathbf{R}^2$ defined by $T(x, y) = (2x - 9y,\ 3x - 5y)$,

(b) $T:\mathbf{R}^3 \to \mathbf{R}^3$ defined by $T(x, y, z) = (3x - 2z,\ 5y + 7z,\ x + y + z)$,

(c) $T:\mathbf{R}^3 \to \mathbf{R}^2$ defined by $T(x, y, z) = (2x + 7y - 4z,\ 4x - 6y + 2z)$.

8.61. Let $\mathbf{D}:V \to V$ be the differential operator; that is, $\mathbf{D}(f(t)) = df/dt$. Find $\det(\mathbf{D})$ if V is the vector space of functions with the following bases: (a) $\{1, t, \ldots, t^5\}$, (b) $\{e^t, e^{2t}, e^{3t}\}$, (c) $\{\sin t, \cos t\}$.

8.62. Prove Theorem 8.13: Let F and G be linear operators on a vector space V. Then

(i) $\det(F \circ G) = \det(F)\det(G)$, (ii) F is invertible if and only if $\det(F) \neq 0$.

8.63. Prove (a) $\det(\mathbf{1}_V) = 1$, where $\mathbf{1}_V$ is the identity operator, (b) $\det(T^{-1}) = \det(T)^{-1}$ when T is invertible.

Miscellaneous Problems

8.64. Find the volume $V(S)$ of the parallelopiped S in $\mathbf{R}^3$ determined by the following vectors:

(a) $u_1 = (1,2,-3)$, $u_2 = (3,4,-1)$, $u_3 = (2,-1,5)$,
(b) $u_1 = (1,1,3)$, $u_2 = (1,-2,-4)$, $u_3 = (4,1,5)$.

8.65. Find the volume $V(S)$ of the parallelepiped S in $\mathbf{R}^4$ determined by the following vectors:

$$u_1 = (1,-2,5,-1),\ u_2 = (2,1,-2,1),\ u_3 = (3,0,1-2),\ u_4 = (1,-1,4,-1)$$

8.66. Let V be the space of 2×2 matrices $M = \begin{bmatrix} a & b \\ c & d \end{bmatrix}$ over $\mathbf{R}$. Determine whether $D: V \to \mathbf{R}$ is 2-linear (with respect to the rows), where

(a) $D(M) = a + d$, (c) $D(M) = ac - bd$, (e) $D(M) = 0$
(b) $D(M) = ad$, (d) $D(M) = ab - cd$, (f) $D(M) = 1$

8.67. Let A be an n-square matrix. Prove $|kA| = k^n|A|$.

8.68. Let A, B, C, D be commuting n-square matrices. Consider the $2n$-square block matrix $M = \begin{bmatrix} A & B \\ C & D \end{bmatrix}$. Prove that $|M| = |A||D| - |B||C|$. Show that the result may not be true if the matrices do not commute.

8.69. Suppose A is orthogonal; that is, $A^T A = I$. Show that $\det(A) = \pm 1$.

8.70. Let V be the space of m-square matrices viewed as m-tuples of row vectors. Suppose $D: V \to K$ is m-linear and alternating. Show that

(a) $D(\ldots, A, \ldots, B, \ldots) = -D(\ldots, B, \ldots, A, \ldots)$; sign changed when two rows are interchanged.
(b) If $A_1, A_2, \ldots, A_m$ are linearly dependent, then $D(A_1, A_2, \ldots, A_m) = 0$.

8.71. Let V be the space of m-square matrices (as above), and suppose $D: V \to K$. Show that the following weaker statement is equivalent to D being alternating:

$$D(A_1, A_2, \ldots, A_n) = 0 \quad \text{whenever} \quad A_i = A_{i+1} \text{ for some } i$$

Let V be the space of n-square matrices over K. Suppose $B \in V$ is invertible and so $\det(B) \neq 0$. Define $D: V \to K$ by $D(A) = \det(AB)/\det(B)$, where $A \in V$. Hence,

$$D(A_1, A_2, \ldots, A_n) = \det(A_1 B, A_2 B, \ldots, A_n B)/\det(B)$$

where A_i is the ith row of A, and so $A_i B$ is the ith row of AB. Show that D is multilinear and alternating, and that $D(I) = 1$. (This method is used by some texts to prove that $|AB| = |A||B|$.)

8.72. Show that $g = g(x_1, \ldots, x_n) = (-1)^n V_{n-1}(x)$ where $g = g(x_i)$ is the difference product in Problem 8.19, $x = x_n$, and V_{n-1} is the *Vandermonde determinant* defined by

$$V_{n-1}(x) \equiv \begin{vmatrix} 1 & 1 & \cdots & 1 & 1 \\ x_1 & x_2 & \cdots & x_{n-1} & x \\ x_1^2 & x_2^2 & \cdots & x_{n-1}^2 & x^2 \\ \cdots & \cdots & \cdots & \cdots & \cdots \\ x_1^{n-1} & x_2^{n-1} & \cdots & x_{n-1}^{n-1} & x^{n-1} \end{vmatrix}$$

8.73. Let A be any matrix. Show that the signs of a minor $A[I, J]$ and its complementary minor $A[I', J']$ are equal.

8.74. Let A be an n-square matrix. The *determinantal rank* of A is the order of the largest square submatrix of A (obtained by deleting rows and columns of A) whose determinant is not zero. Show that the determinantal rank of A is equal to its rank—the maximum number of linearly independent rows (or columns).

ANSWERS TO SUPPLEMENTARY PROBLEMS

Notation: $M = [R_1; \quad R_2; \quad \ldots]$ denotes a matrix with rows $R_1, R_2, \ldots$.

8.38. (a) -22, (b) -13, (c) 46, (d) -21, (e) $a^2 + ab + b^2$

8.39. (a) $3, 10$; (b) $5, -2$

8.40. (a) 21, (b) -11, (c) 100, (d) 0

8.41. (a) -131, (b) -55

8.42. (a) 33, (b) 0, (c) 45

8.43. (a) -32, (b) -14, (c) -468

8.44. (a) $|A| = -2$, $\quad \text{adj}\, A = [-1, -1, 1; \quad -1, 1, -1; \quad 2, -2, 0]$,
(b) $|A| = -1$, $\quad \text{adj}\, A = [1, 0, -2; \quad -3, -1, 6; \quad 2, 1, -5]$. Also, $A^{-1} = (\text{adj}\, A)/|A|$

8.45. (a) $[-16, -29, -26, -2; \quad -30, -38, -16, 29; \quad -8, 51, -13, -1; \quad -13, 1, 28, -18]$,
(b) $[21, -14, -17, -19; \quad -44, 11, 33, 11; \quad -29, 1, 13, 21; \quad 17, 7, -19, -18]$

8.46. (a) $\text{adj}\, A = [d, -b; \quad -c, a]$, (c) $A = kI$

8.49. (a) $-3, -3$, (b) $-23, -23$, (c) $3, -3$, (d) $17, -17$

8.50. (a) $-2, -17, 73$, (b) $7, 10, 105$, (c) $13, 54, 0$

8.51. (a) $-6, 13, 62, -219$; (b) $7, -37, 30, 20$

8.52. (a) $x = \frac{21}{26}, y = \frac{29}{26}$; (b) $x = -\frac{5}{13}, y = \frac{1}{13}$; (c) $x = -\frac{c}{a}, y = -\frac{c}{b}$

8.53. (a) $x = 5, y = 2, z = 1$, (b) Because $D = 0$, the system cannot be solved by determinants.

8.55. (a) $\text{sgn}\, \sigma = 1, \text{sgn}\, \tau = -1, \text{sgn}\, \pi = -1$

8.56. (a) $\tau \circ \sigma = 53142$, (b) $\pi \circ \sigma = 52413$, (c) $\sigma^{-1} = 32154$, (d) $\tau^{-1} = 14253$

8.60. (a) $\det(T) = 17$, (b) $\det(T) = 4$, (c) not defined

8.61. (a) 0, (b) 6, (c) 1

8.64. (a) 18, (b) 0

8.65. 17

8.66. (a) no, (b) yes, (c) yes, (d) no, (e) yes, (f) no

CHAPTER 9

Diagonalization: Eigenvalues and Eigenvectors

9.1 Introduction

The ideas in this chapter can be discussed from two points of view.

Matrix Point of View

Suppose an n-square matrix A is given. The matrix A is said to be *diagonalizable* if there exists a nonsingular matrix P such that

$$B = P^{-1}AP$$

is diagonal. This chapter discusses the diagonalization of a matrix A. In particular, an algorithm is given to find the matrix P when it exists.

Linear Operator Point of View

Suppose a linear operator $T: V \to V$ is given. The linear operator T is said to be *diagonalizable* if there exists a basis S of V such that the matrix representation of T relative to the basis S is a diagonal matrix D. This chapter discusses conditions under which the linear operator T is diagonalizable.

Equivalence of the Two Points of View

The above two concepts are essentially the same. Specifically, a square matrix A may be viewed as a linear operator F defined by

$$F(X) = AX$$

where X is a column vector, and $B = P^{-1}AP$ represents F relative to a new coordinate system (basis) S whose elements are the columns of P. On the other hand, any linear operator T can be represented by a matrix A relative to one basis and, when a second basis is chosen, T is represented by the matrix

$$B = P^{-1}AP$$

where P is the change-of-basis matrix.

Most theorems will be stated in two ways: one in terms of matrices A and again in terms of linear mappings T.

Role of Underlying Field K

The underlying number field K did not play any special role in our previous discussions on vector spaces and linear mappings. However, the diagonalization of a matrix A or a linear operator T will depend on the

roots of a polynomial $\Delta(t)$ over K, and these roots do depend on K. For example, suppose $\Delta(t) = t^2 + 1$. Then $\Delta(t)$ has no roots if $K = \mathbf{R}$, the real field; but $\Delta(t)$ has roots $\pm i$ if $K = \mathbf{C}$, the complex field. Furthermore, finding the roots of a polynomial with degree greater than two is a subject unto itself (frequently discussed in numerical analysis courses). Accordingly, our examples will usually lead to those polynomials $\Delta(t)$ whose roots can be easily determined.

9.2 Polynomials of Matrices

Consider a polynomial $f(t) = a_n t^n + \cdots + a_1 t + a_0$ over a field K. Recall (Section 2.8) that if A is any square matrix, then we define

$$f(A) = a_n A^n + \cdots + a_1 A + a_0 I$$

where I is the identity matrix. In particular, we say that A is a *root* of $f(t)$ if $f(A) = 0$, the zero matrix.

EXAMPLE 9.1 Let $A = \begin{bmatrix} 1 & 2 \\ 3 & 4 \end{bmatrix}$. Then $A^2 = \begin{bmatrix} 7 & 10 \\ 15 & 22 \end{bmatrix}$. Let

$$f(t) = 2t^2 - 3t + 5 \quad \text{and} \quad g(t) = t^2 - 5t - 2$$

Then

$$f(A) = 2A^2 - 3A + 5I = \begin{bmatrix} 14 & 20 \\ 30 & 44 \end{bmatrix} + \begin{bmatrix} -3 & -6 \\ -9 & -12 \end{bmatrix} + \begin{bmatrix} 5 & 0 \\ 0 & 5 \end{bmatrix} = \begin{bmatrix} 16 & 14 \\ 21 & 37 \end{bmatrix}$$

and

$$g(A) = A^2 - 5A - 2I = \begin{bmatrix} 7 & 10 \\ 15 & 22 \end{bmatrix} + \begin{bmatrix} -5 & -10 \\ -15 & -20 \end{bmatrix} + \begin{bmatrix} -2 & 0 \\ 0 & -2 \end{bmatrix} = \begin{bmatrix} 0 & 0 \\ 0 & 0 \end{bmatrix}$$

Thus, A is a zero of $g(t)$.

The following theorem (proved in Problem 9.7) applies.

THEOREM 9.1: Let f and g be polynomials. For any square matrix A and scalar k,

(i) $(f + g)(A) = f(A) + g(A)$ (iii) $(kf)(A) = kf(A)$

(ii) $(fg)(A) = f(A)g(A)$ (iv) $f(A)g(A) = g(A)f(A)$.

Observe that (iv) tells us that any two polynomials in A commute.

Matrices and Linear Operators

Now suppose that $T: V \to V$ is a linear operator on a vector space V. Powers of T are defined by the composition operation:

$$T^2 = T \circ T, \qquad T^3 = T^2 \circ T, \qquad \ldots$$

Also, for any polynomial $f(t) = a_n t^n + \cdots + a_1 t + a_0$, we define $f(T)$ in the same way as we did for matrices:

$$f(T) = a_n T^n + \cdots + a_1 T + a_0 I$$

where I is now the identity mapping. We also say that T is a *zero* or *root* of $f(t)$ if $f(T) = 0$, the zero mapping. We note that the relations in Theorem 9.1 hold for linear operators as they do for matrices.

Remark: Suppose A is a matrix representation of a linear operator T. Then $f(A)$ is the matrix representation of $f(T)$, and, in particular, $f(T) = 0$ if and only if $f(A) = 0$.

9.3 Characteristic Polynomial, Cayley–Hamilton Theorem

Let $A = [a_{ij}]$ be an n-square matrix. The matrix $M = A - tI_n$, where I_n is the n-square identity matrix and t is an indeterminate, may be obtained by subtracting t down the diagonal of A. The negative of M is the matrix $tI_n - A$, and its determinant

$$\Delta(t) = \det(tI_n - A) = (-1)^n \det(A - tI_n)$$

which is a polynomial in t of degree n and is called the *characteristic polynomial* of A.

We state an important theorem in linear algebra (proved in Problem 9.8).

THEOREM 9.2: (Cayley–Hamilton) Every matrix A is a root of its characteristic polynomial.

Remark: Suppose $A = [a_{ij}]$ is a triangular matrix. Then $tI - A$ is a triangular matrix with diagonal entries $t - a_{ii}$; hence,

$$\Delta(t) = \det(tI - A) = (t - a_{11})(t - a_{22}) \cdots (t - a_{nn})$$

Observe that the roots of $\Delta(t)$ are the diagonal elements of A.

EXAMPLE 9.2 Let $A = \begin{bmatrix} 1 & 3 \\ 4 & 5 \end{bmatrix}$. Its characteristic polynomial is

$$\Delta(t) = |tI - A| = \begin{vmatrix} t-1 & -3 \\ -4 & t-5 \end{vmatrix} = (t-1)(t-5) - 12 = t^2 - 6t - 7$$

As expected from the Cayley–Hamilton theorem, A is a root of $\Delta(t)$; that is,

$$\Delta(A) = A^2 - 6A - 7I = \begin{bmatrix} 13 & 18 \\ 24 & 37 \end{bmatrix} + \begin{bmatrix} -6 & -18 \\ -24 & -30 \end{bmatrix} + \begin{bmatrix} -7 & 0 \\ 0 & -7 \end{bmatrix} = \begin{bmatrix} 0 & 0 \\ 0 & 0 \end{bmatrix}$$

Now suppose A and B are similar matrices, say $B = P^{-1}AP$, where P is invertible. We show that A and B have the same characteristic polynomial. Using $tI = P^{-1}tIP$, we have

$$\begin{aligned} \Delta_B(t) &= \det(tI - B) = \det(tI - P^{-1}AP) = \det(P^{-1}tIP - P^{-1}AP) \\ &= \det[P^{-1}(tI - A)P] = \det(P^{-1}) \det(tI - A) \det(P) \end{aligned}$$

Using the fact that determinants are scalars and commute and that $\det(P^{-1}) \det(P) = 1$, we finally obtain

$$\Delta_B(t) = \det(tI - A) = \Delta_A(t)$$

Thus, we have proved the following theorem.

THEOREM 9.3: Similar matrices have the same characteristic polynomial.

Characteristic Polynomials of Degrees 2 and 3

There are simple formulas for the characteristic polynomials of matrices of orders 2 and 3.

(a) Suppose $A = \begin{bmatrix} a_{11} & a_{12} \\ a_{21} & a_{22} \end{bmatrix}$. Then

$$\Delta(t) = t^2 - (a_{11} + a_{22})t + \det(A) = t^2 - \text{tr}(A)\, t + \det(A)$$

Here $\text{tr}(A)$ denotes the trace of A—that is, the sum of the diagonal elements of A.

(b) Suppose $A = \begin{bmatrix} a_{11} & a_{12} & a_{13} \\ a_{21} & a_{22} & a_{23} \\ a_{31} & a_{32} & a_{33} \end{bmatrix}$. Then

$$\Delta(t) = t^3 - \text{tr}(A)\, t^2 + (A_{11} + A_{22} + A_{33})t - \det(A)$$

(Here A_{11}, A_{22}, A_{33} denote, respectively, the cofactors of a_{11}, a_{22}, a_{33}.)

EXAMPLE 9.3 Find the characteristic polynomial of each of the following matrices:

(a) $A = \begin{bmatrix} 5 & 3 \\ 2 & 10 \end{bmatrix}$, (b) $B = \begin{bmatrix} 7 & -1 \\ 6 & 2 \end{bmatrix}$, (c) $C = \begin{bmatrix} 5 & -2 \\ 4 & -4 \end{bmatrix}$.

(a) We have $\text{tr}(A) = 5 + 10 = 15$ and $|A| = 50 - 6 = 44$; hence, $\Delta(t) + t^2 - 15t + 44$.

(b) We have $\text{tr}(B) = 7 + 2 = 9$ and $|B| = 14 + 6 = 20$; hence, $\Delta(t) = t^2 - 9t + 20$.

(c) We have $\text{tr}(C) = 5 - 4 = 1$ and $|C| = -20 + 8 = -12$; hence, $\Delta(t) = t^2 - t - 12$.

EXAMPLE 9.4 Find the characteristic polynomial of $A = \begin{bmatrix} 1 & 1 & 2 \\ 0 & 3 & 2 \\ 1 & 3 & 9 \end{bmatrix}$.

We have $\text{tr}(A) = 1 + 3 + 9 = 13$. The cofactors of the diagonal elements are as follows:

$$A_{11} = \begin{vmatrix} 3 & 2 \\ 3 & 9 \end{vmatrix} = 21, \qquad A_{22} = \begin{vmatrix} 1 & 2 \\ 1 & 9 \end{vmatrix} = 7, \qquad A_{33} = \begin{vmatrix} 1 & 1 \\ 0 & 3 \end{vmatrix} = 3$$

Thus, $A_{11} + A_{22} + A_{33} = 31$. Also, $|A| = 27 + 2 + 0 - 6 - 6 - 0 = 17$. Accordingly,

$$\Delta(t) = t^3 - 13t^2 + 31t - 17$$

Remark: The coefficients of the characteristic polynomial $\Delta(t)$ of the 3-square matrix A are, with alternating signs, as follows:

$$S_1 = \text{tr}(A), \qquad S_2 = A_{11} + A_{22} + A_{33}, \qquad S_3 = \det(A)$$

We note that each S_k is the sum of all principal minors of A of order k.

The next theorem, whose proof lies beyond the scope of this text, tells us that this result is true in general.

THEOREM 9.4: Let A be an n-square matrix. Then its characteristic polynomial is

$$\Delta(t) = t^n - S_1 t^{n-1} + S_2 t^{n-2} + \cdots + (-1)^n S_n$$

where S_k is the sum of the principal minors of order k.

Characteristic Polynomial of a Linear Operator

Now suppose $T: V \to V$ is a linear operator on a vector space V of finite dimension. We define the *characteristic polynomial* $\Delta(t)$ of T to be the characteristic polynomial of any matrix representation of T. Recall that if A and B are matrix representations of T, then $B = P^{-1}AP$, where P is a change-of-basis matrix. Thus, A and B are similar, and by Theorem 9.3, A and B have the same characteristic polynomial. Accordingly, the characteristic polynomial of T is independent of the particular basis in which the matrix representation of T is computed.

Because $f(T) = 0$ if and only if $f(A) = 0$, where $f(t)$ is any polynomial and A is any matrix representation of T, we have the following analogous theorem for linear operators.

THEOREM 9.2′: (Cayley–Hamilton) A linear operator T is a zero of its characteristic polynomial.

9.4 Diagonalization, Eigenvalues and Eigenvectors

Let A be any n-square matrix. Then A can be represented by (or is similar to) a diagonal matrix $D = \text{diag}(k_1, k_2, \ldots, k_n)$ if and only if there exists a basis S consisting of (column) vectors $u_1, u_2, \ldots, u_n$ such that

$$\begin{aligned} Au_1 &= k_1u_1 \\ Au_2 &= \quad k_2u_2 \\ &\cdots\cdots\cdots\cdots \\ Au_n &= \qquad k_nu_n \end{aligned}$$

In such a case, A is said to be *diagonizable*. Furthermore, $D = P^{-1}AP$, where P is the nonsingular matrix whose columns are, respectively, the basis vectors $u_1, u_2, \ldots, u_n$.

The above observation leads us to the following definition.

DEFINITION: Let A be any square matrix. A scalar λ is called an *eigenvalue* of A if there exists a nonzero (column) vector v such that

$$Av = \lambda v$$

Any vector satisfying this relation is called an *eigenvector* of A *belonging* to the eigenvalue λ.

We note that each scalar multiple kv of an eigenvector v belonging to λ is also such an eigenvector, because

$$A(kv) = k(Av) = k(\lambda v) = \lambda(kv)$$

The set E_λ of all such eigenvectors is a subspace of V (Problem 9.19), called the *eigenspace* of λ. (If $\dim E_\lambda = 1$, then E_λ is called an *eigenline* and λ is called a *scaling factor*.)

The terms *characteristic value* and *characteristic vector* (or *proper value* and *proper vector*) are sometimes used instead of eigenvalue and eigenvector.

The above observation and definitions give us the following theorem.

THEOREM 9.5: An n-square matrix A is similar to a diagonal matrix D if and only if A has n linearly independent eigenvectors. In this case, the diagonal elements of D are the corresponding eigenvalues and $D = P^{-1}AP$, where P is the matrix whose columns are the eigenvectors.

Suppose a matrix A can be diagonalized as above, say $P^{-1}AP = D$, where D is diagonal. Then A has the extremely useful *diagonal factorization*:

$$A = PDP^{-1}$$

Using this factorization, the algebra of A reduces to the algebra of the diagonal matrix D, which can be easily calculated. Specifically, suppose $D = \text{diag}(k_1, k_2, \ldots, k_n)$. Then

$$A^m = (PDP^{-1})^m = PD^mP^{-1} = P\,\text{diag}(k_1^m, \ldots, k_n^m)P^{-1}$$

More generally, for any polynomial $f(t)$,

$$f(A) = f(PDP^{-1}) = Pf(D)P^{-1} = P\,\text{diag}(f(k_1), f(k_2), \ldots, f(k_n))P^{-1}$$

Furthermore, if the diagonal entries of D are nonnegative, let

$$B = P\,\text{diag}(\sqrt{k_1}, \sqrt{k_2}, \ldots, \sqrt{k_n})\,P^{-1}$$

Then B is a *nonnegative square root* of A; that is, $B^2 = A$ and the eigenvalues of B are nonnegative.

EXAMPLE 9.5 Let $A = \begin{bmatrix} 3 & 1 \\ 2 & 2 \end{bmatrix}$ and let $v_1 = \begin{bmatrix} 1 \\ -2 \end{bmatrix}$ and $v_2 = \begin{bmatrix} 1 \\ 1 \end{bmatrix}$. Then

$$Av_1 = \begin{bmatrix} 3 & 1 \\ 2 & 2 \end{bmatrix}\begin{bmatrix} 1 \\ -2 \end{bmatrix} = \begin{bmatrix} 1 \\ -2 \end{bmatrix} = v_1 \quad \text{and} \quad Av_2 = \begin{bmatrix} 3 & 1 \\ 2 & 2 \end{bmatrix}\begin{bmatrix} 1 \\ 1 \end{bmatrix} = \begin{bmatrix} 4 \\ 4 \end{bmatrix} = 4v_2$$

Thus, v_1 and v_2 are eigenvectors of A belonging, respectively, to the eigenvalues $\lambda_1 = 1$ and $\lambda_2 = 4$. Observe that v_1 and v_2 are linearly independent and hence form a basis of $\mathbf{R}^2$. Accordingly, A is diagonalizable. Furthermore, let P be the matrix whose columns are the eigenvectors v_1 and v_2. That is, let

$$P = \begin{bmatrix} 1 & 1 \\ -2 & 1 \end{bmatrix}, \quad \text{and so} \quad P^{-1} = \begin{bmatrix} \frac{1}{3} & -\frac{1}{3} \\ \frac{2}{3} & \frac{1}{3} \end{bmatrix}$$

Then A is similar to the diagonal matrix

$$D = P^{-1}AP = \begin{bmatrix} \frac{1}{3} & -\frac{1}{3} \\ \frac{2}{3} & \frac{1}{3} \end{bmatrix}\begin{bmatrix} 3 & 1 \\ 2 & 2 \end{bmatrix}\begin{bmatrix} 1 & 1 \\ -2 & 1 \end{bmatrix} = \begin{bmatrix} 1 & 0 \\ 0 & 4 \end{bmatrix}$$

As expected, the diagonal elements 1 and 4 in D are the eigenvalues corresponding, respectively, to the eigenvectors v_1 and v_2, which are the columns of P. In particular, A has the factorization

$$A = PDP^{-1} = \begin{bmatrix} 1 & 1 \\ -2 & 1 \end{bmatrix}\begin{bmatrix} 1 & 0 \\ 0 & 4 \end{bmatrix}\begin{bmatrix} \frac{1}{3} & -\frac{1}{3} \\ \frac{2}{3} & \frac{1}{3} \end{bmatrix}$$

Accordingly,

$$A^4 = \begin{bmatrix} 1 & 1 \\ -2 & 1 \end{bmatrix}\begin{bmatrix} 1 & 0 \\ 0 & 256 \end{bmatrix}\begin{bmatrix} \frac{1}{3} & -\frac{1}{3} \\ \frac{2}{3} & \frac{1}{3} \end{bmatrix} = \begin{bmatrix} 171 & 85 \\ 170 & 86 \end{bmatrix}$$

Moreover, suppose $f(t) = t^3 - 5t^2 + 3t + 6$; hence, $f(1) = 5$ and $f(4) = 2$. Then

$$f(A) = Pf(D)P^{-1} = \begin{bmatrix} 1 & 1 \\ -2 & 1 \end{bmatrix}\begin{bmatrix} 5 & 0 \\ 0 & 2 \end{bmatrix}\begin{bmatrix} \frac{1}{3} & -\frac{1}{3} \\ \frac{2}{3} & \frac{1}{3} \end{bmatrix} = \begin{bmatrix} 3 & -1 \\ -2 & 4 \end{bmatrix}$$

Last, we obtain a "positive square root" of A. Specifically, using $\sqrt{1} = 1$ and $\sqrt{4} = 2$, we obtain the matrix

$$B = P\sqrt{D}P^{-1} = \begin{bmatrix} 1 & 1 \\ -2 & 1 \end{bmatrix}\begin{bmatrix} 1 & 0 \\ 0 & 2 \end{bmatrix}\begin{bmatrix} \frac{1}{3} & -\frac{1}{3} \\ \frac{2}{3} & \frac{1}{3} \end{bmatrix} = \begin{bmatrix} \frac{5}{3} & \frac{1}{3} \\ \frac{2}{3} & \frac{4}{3} \end{bmatrix}$$

where $B^2 = A$ and where B has positive eigenvalues 1 and 2.

Remark: Throughout this chapter, we use the following fact:

$$\text{If } P = \begin{bmatrix} a & b \\ c & d \end{bmatrix}, \text{ then } P^{-1} = \begin{bmatrix} d/|P| & -b/|P| \\ -c/|P| & a/|P| \end{bmatrix}.$$

That is, P^{-1} is obtained by interchanging the diagonal elements a and d of P, taking the negatives of the nondiagonal elements b and c, and dividing each element by the determinant $|P|$.

Properties of Eigenvalues and Eigenvectors

Example 9.5 indicates the advantages of a diagonal representation (factorization) of a square matrix. In the following theorem (proved in Problem 9.20), we list properties that help us to find such a representation.

THEOREM 9.6: Let A be a square matrix. Then the following are equivalent.

(i) A scalar λ is an eigenvalue of A.

(ii) The matrix $M = A - \lambda I$ is singular.

(iii) The scalar λ is a root of the characteristic polynomial $\Delta(t)$ of A.

The eigenspace E_λ of an eigenvalue λ is the solution space of the homogeneous system $MX = 0$, where $M = A - \lambda I$; that is, M is obtained by subtracting λ down the diagonal of A.

Some matrices have no eigenvalues and hence no eigenvectors. However, using Theorem 9.6 and the Fundamental Theorem of Algebra (every polynomial over the complex field **C** has a root), we obtain the following result.

THEOREM 9.7: Let A be a square matrix over the complex field **C**. Then A has at least one eigenvalue.

The following theorems will be used subsequently. (The theorem equivalent to Theorem 9.8 for linear operators is proved in Problem 9.21, and Theorem 9.9 is proved in Problem 9.22.)

THEOREM 9.8: Suppose $v_1, v_2, \ldots, v_n$ are nonzero eigenvectors of a matrix A belonging to distinct eigenvalues $\lambda_1, \lambda_2, \ldots, \lambda_n$. Then $v_1, v_2, \ldots, v_n$ are linearly independent.

THEOREM 9.9: Suppose the characteristic polynomial $\Delta(t)$ of an n-square matrix A is a product of n distinct factors, say, $\Delta(t) = (t - a_1)(t - a_2) \cdots (t - a_n)$. Then A is similar to the diagonal matrix $D = \text{diag}(a_1, a_2, \ldots, a_n)$.

If λ is an eigenvalue of a matrix A, then the *algebraic multiplicity* of λ is defined to be the multiplicity of λ as a root of the characteristic polynomial of A, and the *geometric multiplicity* of λ is defined to be the dimension of its eigenspace, $\dim E_\lambda$. The following theorem (whose equivalent for linear operators is proved in Problem 9.23) holds.

THEOREM 9.10: The geometric multiplicity of an eigenvalue λ of a matrix A does not exceed its algebraic multiplicity.

Diagonalization of Linear Operators

Consider a linear operator $T: V \to V$. Then T is said to be *diagonalizable* if it can be represented by a diagonal matrix D. Thus, T is diagonalizable if and only if there exists a basis $S = \{u_1, u_2, \ldots, u_n\}$ of V for which

$$\begin{aligned} T(u_1) &= k_1 u_1 \\ T(u_2) &= \qquad k_2 u_2 \\ &\cdots\cdots\cdots\cdots \\ T(u_n) &= \qquad\qquad k_n u_n \end{aligned}$$

In such a case, T is represented by the diagonal matrix

$$D = \text{diag}(k_1, k_2, \ldots, k_n)$$

relative to the basis S.

The above observation leads us to the following definitions and theorems, which are analogous to the definitions and theorems for matrices discussed above.

DEFINITION: Let T be a linear operator. A scalar λ is called an *eigenvalue* of T if there exists a nonzero vector v such that $T(v) = \lambda v$.

Every vector satisfying this relation is called an *eigenvector* of T *belonging* to the eigenvalue λ.

The set E_λ of all eigenvectors belonging to an eigenvalue λ is a subspace of V, called the *eigenspace* of λ. (Alternatively, λ is an eigenvalue of T if $\lambda I - T$ is singular, and, in this case, E_λ is the kernel of $\lambda I - T$.) The *algebraic* and *geometric multiplicities* of an eigenvalue λ of a linear operator T are defined in the same way as those of an eigenvalue of a matrix A.

The following theorems apply to a linear operator T on a vector space V of finite dimension.

THEOREM 9.5′: T can be represented by a diagonal matrix D if and only if there exists a basis S of V consisting of eigenvectors of T. In this case, the diagonal elements of D are the corresponding eigenvalues.

THEOREM 9.6′: Let T be a linear operator. Then the following are equivalent:

(i) A scalar λ is an eigenvalue of T.

(ii) The linear operator $\lambda I - T$ is singular.

(iii) The scalar λ is a root of the characteristic polynomial $\Delta(t)$ of T.

THEOREM 9.7′: Suppose V is a complex vector space. Then T has at least one eigenvalue.

THEOREM 9.8′: Suppose $v_1, v_2, \ldots, v_n$ are nonzero eigenvectors of a linear operator T belonging to distinct eigenvalues $\lambda_1, \lambda_2, \ldots, \lambda_n$. Then $v_1, v_2, \ldots, v_n$ are linearly independent.

THEOREM 9.9′: Suppose the characteristic polynomial $\Delta(t)$ of T is a product of n distinct factors, say, $\Delta(t) = (t - a_1)(t - a_2) \cdots (t - a_n)$. Then T can be represented by the diagonal matrix $D = \operatorname{diag}(a_1, a_2, \ldots, a_n)$.

THEOREM 9.10′: The geometric multiplicity of an eigenvalue λ of T does not exceed its algebraic multiplicity.

Remark: The following theorem reduces the investigation of the diagonalization of a linear operator T to the diagonalization of a matrix A.

THEOREM 9.11: Suppose A is a matrix representation of T. Then T is diagonalizable if and only if A is diagonalizable.

9.5 Computing Eigenvalues and Eigenvectors, Diagonalizing Matrices

This section gives an algorithm for computing eigenvalues and eigenvectors for a given square matrix A and for determining whether or not a nonsingular matrix P exists such that $P^{-1}AP$ is diagonal.

ALGORITHM 9.1: (Diagonalization Algorithm) The input is an n-square matrix A.

Step 1. Find the characteristic polynomial $\Delta(t)$ of A.

Step 2. Find the roots of $\Delta(t)$ to obtain the eigenvalues of A.

Step 3. Repeat (a) and (b) for each eigenvalue λ of A.

(a) Form the matrix $M = A - \lambda I$ by subtracting λ down the diagonal of A.

(b) Find a basis for the solution space of the homogeneous system $MX = 0$. (These basis vectors are linearly independent eigenvectors of A belonging to λ.)

Step 4. Consider the collection $S = \{v_1, v_2, \ldots, v_m\}$ of all eigenvectors obtained in Step 3.

(a) If $m \neq n$, then A is not diagonalizable.

(b) If $m = n$, then A is diagonalizable. Specifically, let P be the matrix whose columns are the eigenvectors $v_1, v_2, \ldots, v_n$. Then

$$D = P^{-1}AP = \text{diag}(\lambda_1, \lambda_2, \ldots, \lambda_n)$$

where λ_i is the eigenvalue corresponding to the eigenvector v_i.

EXAMPLE 9.6 The diagonalization algorithm is applied to $A = \begin{bmatrix} 4 & 2 \\ 3 & -1 \end{bmatrix}$.

(1) The characteristic polynomial $\Delta(t)$ of A is computed:

$$\text{tr}(A) = 4 - 1 = 3, \quad |A| = -4 - 6 = -10, \quad \text{so} \quad \Delta(t) = t^2 - 3t - 10 = (t-5)(t+2)$$

(2) Set $\Delta(t) = t^2 - 3t - 10 = (t-5)(t+2) = 0$. The roots $\lambda_1 = 5$ and $\lambda_2 = 2$ are the eigenvalues of A.

(3) (i) We find an eigenvector v_1 of A belonging to the eigenvalue $\lambda_1 = 5$. Subtract $\lambda_1 = 5$ down the diagonal of A to obtain the matrix

$$M = \begin{bmatrix} -1 & 2 \\ 3 & -6 \end{bmatrix} \quad \text{and the homogeneous system} \quad \begin{aligned} -x + 2y &= 0 \\ 3x - 6y &= 0 \end{aligned} \quad \text{or} \quad -x + 2y = 0$$

The system has only one independent solution, for example, $v_1 = (2, 1)$.

(ii) We find an eigenvector v_2 of A belonging to the eigenvalue $\lambda_2 = -2$. Subtract -2 (or add 2) down the diagonal of A to obtain the matrix

$$M = \begin{bmatrix} 6 & 2 \\ 3 & 1 \end{bmatrix} \quad \text{and the homogeneous system} \quad \begin{aligned} 6x + 2y &= 0 \\ 3x + y &= 0 \end{aligned} \quad \text{or} \quad 3x + y = 0$$

The system has only one independent solution, for example, $v_2 = (-1, 3)$.

(4) Let P be the matrix whose columns are the eigenvectors v_1 and v_2. Then $P = \begin{bmatrix} 2 & -1 \\ 1 & 3 \end{bmatrix}$. Thus $D = P^{-1}AP$ is the diagonal matrix whose diagonal entries are the corresponding eigenvalues:

$$D = P^{-1}AP = \begin{bmatrix} 3/7 & 1/7 \\ -1/7 & 2/7 \end{bmatrix} \begin{bmatrix} 4 & 2 \\ 3 & -1 \end{bmatrix} \begin{bmatrix} 2 & -1 \\ 1 & 3 \end{bmatrix} = \begin{bmatrix} 5 & 0 \\ 0 & -2 \end{bmatrix}$$

Remark: Find a 2×2 matrix A with eigenvalues $\lambda_1 = 2$ and $\lambda_2 = 3$ and corresponding eigenvectors $v_1 = (1, 3)$ and $v_2 = (1, 4)$. We know that $P^{-1}AP = D$ where $P = \begin{bmatrix} 1 & 1 \\ 3 & 4 \end{bmatrix}$ and $D = \begin{bmatrix} 2 & 0 \\ 0 & 3 \end{bmatrix}$. [Here $P^{-1} = \begin{bmatrix} 4 & -1 \\ -3 & 1 \end{bmatrix}$.] Thus $A = PDP^{-1} = \begin{bmatrix} 1 & 1 \\ 3 & 4 \end{bmatrix} \begin{bmatrix} 2 & 0 \\ 0 & 3 \end{bmatrix} \begin{bmatrix} 4 & -1 \\ -3 & 1 \end{bmatrix} = \begin{bmatrix} -1 & 1 \\ -12 & 6 \end{bmatrix}$.

EXAMPLE 9.7 Consider the matrix $B = \begin{bmatrix} 5 & -1 \\ 1 & 3 \end{bmatrix}$. We have

$$\text{tr}(B) = 5 + 3 = 8, \quad |B| = 15 + 1 = 16; \quad \text{so} \quad \Delta(t) = t^2 - 8t + 16 = (t-4)^2$$

Accordingly, $\lambda = 4$ is the only eigenvalue of B.

Subtract $\lambda = 4$ down the diagonal of B to obtain the matrix

$$M = \begin{bmatrix} 1 & -1 \\ 1 & -1 \end{bmatrix} \quad \text{and the homogeneous system} \quad \begin{matrix} x - y = 0 \\ x - y = 0 \end{matrix} \quad \text{or} \quad x - y = 0$$

The system has only one independent solution; for example, $x = 1, y = 1$. Thus, $v = (1, 1)$ and its multiples are the only eigenvectors of B. Accordingly, B is not diagonalizable, because there does not exist a basis consisting of eigenvectors of B.

EXAMPLE 9.8 Consider the matrix $A = \begin{bmatrix} 3 & -5 \\ 2 & -3 \end{bmatrix}$. Here $\text{tr}(A) = 3 - 3 = 0$ and $|A| = -9 + 10 = 1$. Thus, $\Delta(t) = t^2 + 1$ is the characteristic polynomial of A. We consider two cases:

(a) A is a matrix over the real field **R**. Then $\Delta(t)$ has no (real) roots. Thus, A has no eigenvalues and no eigenvectors, and so A is not diagonalizable.

(b) A is a matrix over the complex field **C**. Then $\Delta(t) = (t - i)(t + i)$ has two roots, i and $-i$. Thus, A has two distinct eigenvalues i and $-i$, and hence, A has two independent eigenvectors. Accordingly there exists a nonsingular matrix P over the complex field **C** for which

$$P^{-1}AP = \begin{bmatrix} i & 0 \\ 0 & -i \end{bmatrix}$$

Therefore, A is diagonalizable (over **C**).

9.6 Diagonalizing Real Symmetric Matrices and Quadratic Forms

There are many real matrices A that are not diagonalizable. In fact, some real matrices may not have any (real) eigenvalues. However, if A is a real symmetric matrix, then these problems do not exist. Namely, we have the following theorems.

THEOREM 9.12: Let A be a real symmetric matrix. Then each root λ of its characteristic polynomial is real.

THEOREM 9.13: Let A be a real symmetric matrix. Suppose u and v are eigenvectors of A belonging to distinct eigenvalues λ_1 and λ_2. Then u and v are orthogonal, that; is, $\langle u, v \rangle = 0$.

The above two theorems give us the following fundamental result.

THEOREM 9.14: Let A be a real symmetric matrix. Then there exists an orthogonal matrix P such that $D = P^{-1}AP$ is diagonal.

The orthogonal matrix P is obtained by normalizing a basis of orthogonal eigenvectors of A as illustrated below. In such a case, we say that A is "orthogonally diagonalizable."

EXAMPLE 9.9 Let $A = \begin{bmatrix} 2 & -2 \\ -2 & 5 \end{bmatrix}$, a real symmetric matrix. Find an orthogonal matrix P such that $P^{-1}AP$ is diagonal.

First we find the characteristic polynomial $\Delta(t)$ of A. We have

$$\text{tr}(A) = 2 + 5 = 7, \qquad |A| = 10 - 4 = 6; \qquad \text{so} \qquad \Delta(t) = t^2 - 7t + 6 = (t - 6)(t - 1)$$

Accordingly, $\lambda_1 = 6$ and $\lambda_2 = 1$ are the eigenvalues of A.

(a) Subtracting $\lambda_1 = 6$ down the diagonal of A yields the matrix

$$M = \begin{bmatrix} -4 & -2 \\ -2 & -1 \end{bmatrix} \quad \text{and the homogeneous system} \quad \begin{matrix} -4x - 2y = 0 \\ -2x - \ \ y = 0 \end{matrix} \quad \text{or} \quad 2x + y = 0$$

A nonzero solution is $u_1 = (1, -2)$.

(b) Subtracting $\lambda_2 = 1$ down the diagonal of A yields the matrix

$$M = \begin{bmatrix} 1 & -2 \\ -2 & 4 \end{bmatrix} \quad \text{and the homogeneous system} \quad x - 2y = 0$$

(The second equation drops out, because it is a multiple of the first equation.) A nonzero solution is $u_2 = (2, 1)$.

As expected from Theorem 9.13, u_1 and u_2 are orthogonal. Normalizing u_1 and u_2 yields the orthonormal vectors

$$\hat{u}_1 = (1/\sqrt{5}, -2/\sqrt{5}) \quad \text{and} \quad \hat{u}_2 = (2/\sqrt{5}, 1/\sqrt{5})$$

Finally, let P be the matrix whose columns are $\hat{u}_1$ and $\hat{u}_2$, respectively. Then

$$P = \begin{bmatrix} 1/\sqrt{5} & 2/\sqrt{5} \\ -2/\sqrt{5} & 1/\sqrt{5} \end{bmatrix} \quad \text{and} \quad P^{-1}AP = \begin{bmatrix} 6 & 0 \\ 0 & 1 \end{bmatrix}$$

As expected, the diagonal entries of $P^{-1}AP$ are the eigenvalues corresponding to the columns of P.

The procedure in the above Example 9.9 is formalized in the following algorithm, which finds an orthogonal matrix P such that $P^{-1}AP$ is diagonal.

ALGORITHM 9.2: (Orthogonal Diagonalization Algorithm) The input is a real symmetric matrix A.

Step 1. Find the characteristic polynomial $\Delta(t)$ of A.

Step 2. Find the eigenvalues of A, which are the roots of $\Delta(t)$.

Step 3. For each eigenvalue λ of A in Step 2, find an orthogonal basis of its eigenspace.

Step 4. Normalize all eigenvectors in Step 3, which then forms an orthonormal basis of $\mathbf{R}^n$.

Step 5. Let P be the matrix whose columns are the normalized eigenvectors in Step 4.

Application to Quadratic Forms

Let q be a real polynomial in variables $x_1, x_2, \ldots, x_n$ such that every term in q has degree two; that is,

$$q(x_1, x_2, \ldots, x_n) = \sum_i c_i x_i^2 + \sum_{i<j} d_{ij} x_i x_j, \qquad \text{where} \qquad c_i, d_{ij} \in \mathbf{R}$$

Then q is called a *quadratic form*. If there are no cross-product terms $x_i x_j$ (i.e., all $d_{ij} = 0$), then q is said to be *diagonal*.

The above quadratic form q determines a real symmetric matrix $A = [a_{ij}]$, where $a_{ii} = c_i$ and $a_{ij} = a_{ji} = \frac{1}{2} d_{ij}$. Namely, q can be written in the matrix form

$$q(X) = X^T AX$$

where $X = [x_1, x_2, \ldots, x_n]^T$ is the column vector of the variables. Furthermore, suppose $X = PY$ is a linear substitution of the variables. Then substitution in the quadratic form yields

$$q(Y) = (PY)^T A(PY) = Y^T (P^T AP)Y$$

Thus, $P^T AP$ is the matrix representation of q in the new variables.

We seek an orthogonal matrix P such that the *orthogonal substitution* $X = PY$ yields a diagonal quadratic form for which $P^T AP$ is diagonal. Because P is orthogonal, $P^T = P^{-1}$, and hence, $P^T AP = P^{-1}AP$. The above theory yields such an orthogonal matrix P.

EXAMPLE 9.10 Consider the quadratic form

$$q(x,y) = 2x^2 - 4xy + 5y^2 = X^TAX, \quad \text{where} \quad A = \begin{bmatrix} 2 & -2 \\ -2 & 5 \end{bmatrix} \quad \text{and} \quad X = \begin{bmatrix} x \\ y \end{bmatrix}$$

By Example 9.9,

$$P^{-1}AP = \begin{bmatrix} 6 & 0 \\ 0 & 1 \end{bmatrix} = P^TAP, \quad \text{where} \quad P = \begin{bmatrix} 1/\sqrt{5} & 2/\sqrt{5} \\ -2/\sqrt{5} & 1/\sqrt{5} \end{bmatrix}$$

Let $Y = [s, t]^T$. Then matrix P corresponds to the following linear orthogonal substitution $x = PY$ of the variables x and y in terms of the variables s and t:

$$x = \frac{1}{\sqrt{5}}s + \frac{2}{\sqrt{5}}t, \qquad y = -\frac{2}{\sqrt{5}}s + \frac{1}{\sqrt{5}}t$$

This substitution in $q(x,y)$ yields the diagonal quadratic form $q(s,t) = 6s^2 + t^2$.

9.7 Minimal Polynomial

Let A be any square matrix. Let $J(A)$ denote the collection of all polynomials $f(t)$ for which A is a root—that is, for which $f(A) = 0$. The set $J(A)$ is not empty, because the Cayley–Hamilton Theorem 9.1 tells us that the characteristic polynomial $\Delta_A(t)$ of A belongs to $J(A)$. Let $m(t)$ denote the monic polynomial of lowest degree in $J(A)$. (Such a polynomial $m(t)$ exists and is unique.) We call $m(t)$ the *minimal polynomial* of the matrix A.

Remark: A polynomial $f(t) \neq 0$ is *monic* if its leading coefficient equals one.

The following theorem (proved in Problem 9.33) holds.

THEOREM 9.15: The minimal polynomial $m(t)$ of a matrix (linear operator) A divides every polynomial that has A as a zero. In particular, $m(t)$ divides the characteristic polynomial $\Delta(t)$ of A.

There is an even stronger relationship between $m(t)$ and $\Delta(t)$.

THEOREM 9.16: The characteristic polynomial $\Delta(t)$ and the minimal polynomial $m(t)$ of a matrix A have the same irreducible factors.

This theorem (proved in Problem 9.35) does not say that $m(t) = \Delta(t)$, only that any irreducible factor of one must divide the other. In particular, because a linear factor is irreducible, $m(t)$ and $\Delta(t)$ have the same linear factors. Hence, they have the same roots. Thus, we have the following theorem.

THEOREM 9.17: A scalar λ is an eigenvalue of the matrix A if and only if λ is a root of the minimal polynomial of A.

EXAMPLE 9.11 Find the minimal polynomial $m(t)$ of $A = \begin{bmatrix} 2 & 2 & -5 \\ 3 & 7 & -15 \\ 1 & 2 & -4 \end{bmatrix}$.

First find the characteristic polynomial $\Delta(t)$ of A. We have

$$\text{tr}(A) = 5, \quad A_{11} + A_{22} + A_{33} = 2 - 3 + 8 = 7, \quad \text{and} \quad |A| = 3$$

Hence,

$$\Delta(t) = t^3 - 5t^2 + 7t - 3 = (t-1)^2(t-3)$$

The minimal polynomial $m(t)$ must divide $\Delta(t)$. Also, each irreducible factor of $\Delta(t)$ (i.e., $t-1$ and $t-3$) must also be a factor of $m(t)$. Thus, $m(t)$ is exactly one of the following:

$$f(t) = (t-3)(t-1) \quad \text{or} \quad g(t) = (t-3)(t-1)^2$$

We know, by the Cayley–Hamilton theorem, that $g(A) = \Delta(A) = 0$. Hence, we need only test $f(t)$. We have

$$f(A) = (A-I)(A-3I) = \begin{bmatrix} 1 & 2 & -5 \\ 3 & 6 & -15 \\ 1 & 2 & -5 \end{bmatrix} \begin{bmatrix} -1 & 2 & -5 \\ 3 & 4 & -15 \\ 1 & 2 & -7 \end{bmatrix} = \begin{bmatrix} 0 & 0 & 0 \\ 0 & 0 & 0 \\ 0 & 0 & 0 \end{bmatrix}$$

Thus, $f(t) = m(t) = (t-1)(t-3) = t^2 - 4t + 3$ is the minimal polynomial of A.

EXAMPLE 9.12

(a) Consider the following two r-square matrices, where $a \neq 0$:

$$J(\lambda, r) = \begin{bmatrix} \lambda & 1 & 0 & \dots & 0 & 0 \\ 0 & \lambda & 1 & \dots & 0 & 0 \\ \dots & \dots & \dots & \dots & \dots & \dots \\ 0 & 0 & 0 & \dots & \lambda & 1 \\ 0 & 0 & 0 & \dots & 0 & \lambda \end{bmatrix} \quad \text{and} \quad A = \begin{bmatrix} \lambda & a & 0 & \dots & 0 & 0 \\ 0 & \lambda & a & \dots & 0 & 0 \\ \dots & \dots & \dots & \dots & \dots & \dots \\ 0 & 0 & 0 & \dots & \lambda & a \\ 0 & 0 & 0 & \dots & 0 & \lambda \end{bmatrix}$$

The first matrix, called a Jordan Block, has λ's on the diagonal, 1's on the *superdiagonal* (consisting of the entries above the diagonal entries), and 0's elsewhere. The second matrix A has λ's on the diagonal, a's on the superdiagonal, and 0's elsewhere. [Thus, A is a generalization of $J(\lambda, r)$.] One can show that

$$f(t) = (t-\lambda)^r$$

is both the characteristic and minimal polynomial of both $J(\lambda, r)$ and A.

(b) Consider an arbitrary monic polynomial:

$$f(t) = t^n + a_{n-1}t^{n-1} + \cdots + a_1 t + a_0$$

Let $C(f)$ be the n-square matrix with 1's on the *subdiagonal* (consisting of the entries below the diagonal entries), the negatives of the coefficients in the last column, and 0's elsewhere as follows:

$$C(f) = \begin{bmatrix} 0 & 0 & \dots & 0 & -a_0 \\ 1 & 0 & \dots & 0 & -a_1 \\ 0 & 1 & \dots & 0 & -a_2 \\ \dots & \dots & \dots & \dots & \dots \\ 0 & 0 & \dots & 1 & -a_{n-1} \end{bmatrix}$$

Then $C(f)$ is called the *companion matrix* of the polynomial $f(t)$. Moreover, the minimal polynomial $m(t)$ and the characteristic polynomial $\Delta(t)$ of the companion matrix $C(f)$ are both equal to the original polynomial $f(t)$.

Minimal Polynomial of a Linear Operator

The *minimal polynomial* $m(t)$ of a linear operator T is defined to be the monic polynomial of lowest degree for which T is a root. However, for any polynomial $f(t)$, we have

$$f(T) = 0 \quad \text{if and only if} \quad f(A) = 0$$

where A is any matrix representation of T. Accordingly, T and A have the same minimal polynomials. Thus, the above theorems on the minimal polynomial of a matrix also hold for the minimal polynomial of a linear operator. That is, we have the following theorems.

THEOREM 9.15′: The minimal polynomial $m(t)$ of a linear operator T divides every polynomial that has T as a root. In particular, $m(t)$ divides the characteristic polynomial $\Delta(t)$ of T.

THEOREM 9.16′: The characteristic and minimal polynomials of a linear operator T have the same irreducible factors.

THEOREM 9.17′: A scalar λ is an eigenvalue of a linear operator T if and only if λ is a root of the minimal polynomial $m(t)$ of T.

9.8 Characteristic and Minimal Polynomials of Block Matrices

This section discusses the relationship of the characteristic polynomial and the minimal polynomial to certain (square) block matrices.

Characteristic Polynomial and Block Triangular Matrices

Suppose M is a block triangular matrix, say $M = \begin{bmatrix} A_1 & B \\ 0 & A_2 \end{bmatrix}$, where A_1 and A_2 are square matrices. Then $tI - M$ is also a block triangular matrix, with diagonal blocks $tI - A_1$ and $tI - A_2$. Thus,

$$|tI - M| = \begin{vmatrix} tI - A_1 & -B \\ 0 & tI - A_2 \end{vmatrix} = |tI - A_1||tI - A_2|$$

That is, the characteristic polynomial of M is the product of the characteristic polynomials of the diagonal blocks A_1 and A_2.

By induction, we obtain the following useful result.

THEOREM 9.18: Suppose M is a block triangular matrix with diagonal blocks $A_1, A_2, \ldots, A_r$. Then the characteristic polynomial of M is the product of the characteristic polynomials of the diagonal blocks A_i; that is,

$$\Delta_M(t) = \Delta_{A_1}(t)\Delta_{A_2}(t)\ldots\Delta_{A_r}(t)$$

EXAMPLE 9.13 Consider the matrix $M = \left[\begin{array}{cc|cc} 9 & -1 & 5 & 7 \\ 8 & 3 & 2 & -4 \\ \hline 0 & 0 & 3 & 6 \\ 0 & 0 & -1 & 8 \end{array}\right]$.

Then M is a block triangular matrix with diagonal blocks $A = \begin{bmatrix} 9 & -1 \\ 8 & 3 \end{bmatrix}$ and $B = \begin{bmatrix} 3 & 6 \\ -1 & 8 \end{bmatrix}$. Here

$$\begin{aligned} &\mathrm{tr}(A) = 9 + 3 = 12, \quad \det(A) = 27 + 8 = 35, \quad \text{and so} \quad \Delta_A(t) = t^2 - 12t + 35 = (t-5)(t-7) \\ &\mathrm{tr}(B) = 3 + 8 = 11, \quad \det(B) = 24 + 6 = 30, \quad \text{and so} \quad \Delta_B(t) = t^2 - 11t + 30 = (t-5)(t-6) \end{aligned}$$

Accordingly, the characteristic polynomial of M is the product

$$\Delta_M(t) = \Delta_A(t)\Delta_B(t) = (t-5)^2(t-6)(t-7)$$

Minimal Polynomial and Block Diagonal Matrices

The following theorem (proved in Problem 9.36) holds.

THEOREM 9.19: Suppose M is a block diagonal matrix with diagonal blocks $A_1, A_2, \ldots, A_r$. Then the minimal polynomial of M is equal to the least common multiple (LCM) of the minimal polynomials of the diagonal blocks A_i.

Remark: We emphasize that this theorem applies to block diagonal matrices, whereas the analogous Theorem 9.18 on characteristic polynomials applies to block triangular matrices.

EXAMPLE 9.14 Find the characteristic polynomal $\Delta(t)$ and the minimal polynomial $m(t)$ of the block diagonal matrix:

$$M = \begin{bmatrix} 2 & 5 & 0 & 0 & 0 \\ 0 & 2 & 0 & 0 & 0 \\ 0 & 0 & 4 & 2 & 0 \\ 0 & 0 & 3 & 5 & 0 \\ 0 & 0 & 0 & 0 & 7 \end{bmatrix} = \text{diag}(A_1, A_2, A_3), \text{ where } A_1 = \begin{bmatrix} 2 & 5 \\ 0 & 2 \end{bmatrix}, A_2 = \begin{bmatrix} 4 & 2 \\ 3 & 5 \end{bmatrix}, A_3 = [7]$$

Then $\Delta(t)$ is the product of the characterization polynomials $\Delta_1(t)$, $\Delta_2(t)$, $\Delta_3(t)$ of A_1, A_2, A_3, respectively. One can show that

$$\Delta_1(t) = (t-2)^2, \qquad \Delta_2(t) = (t-2)(t-7), \qquad \Delta_3(t) = t-7$$

Thus, $\Delta(t) = (t-2)^3(t-7)^2$. [As expected, $\deg \Delta(t) = 5$.]

The minimal polynomials $m_1(t)$, $m_2(t)$, $m_3(t)$ of the diagonal blocks A_1, A_2, A_3, respectively, are equal to the characteristic polynomials; that is,

$$m_1(t) = (t-2)^2, \qquad m_2(t) = (t-2)(t-7), \qquad m_3(t) = t-7$$

But $m(t)$ is equal to the least common multiple of $m_1(t), m_2(t), m_3(t)$. Thus, $m(t) = (t-2)^2(t-7)$.

SOLVED PROBLEMS

Polynomials of Matrices, Characteristic Polynomials

9.1. Let $A = \begin{bmatrix} 1 & -2 \\ 4 & 5 \end{bmatrix}$. Find $f(A)$, where

(a) $f(t) = t^2 - 3t + 7$, (b) $f(t) = t^2 - 6t + 13$

First find $A^2 = \begin{bmatrix} 1 & -2 \\ 4 & 5 \end{bmatrix}\begin{bmatrix} 1 & -2 \\ 4 & 5 \end{bmatrix} = \begin{bmatrix} -7 & -12 \\ 24 & 17 \end{bmatrix}$. Then

(a) $f(A) = A^2 - 3A + 7I = \begin{bmatrix} -7 & -12 \\ 24 & 17 \end{bmatrix} + \begin{bmatrix} -3 & 6 \\ -12 & -15 \end{bmatrix} + \begin{bmatrix} 7 & 0 \\ 0 & 7 \end{bmatrix} = \begin{bmatrix} -3 & -6 \\ 12 & 9 \end{bmatrix}$

(b) $f(A) = A^2 - 6A + 13I = \begin{bmatrix} -7 & -12 \\ 24 & 17 \end{bmatrix} + \begin{bmatrix} -6 & 12 \\ -24 & -30 \end{bmatrix} + \begin{bmatrix} 13 & 0 \\ 0 & 13 \end{bmatrix} = \begin{bmatrix} 0 & 0 \\ 0 & 0 \end{bmatrix}$

[Thus, A is a root of $f(t)$.]

9.2. Find the characteristic polynomial $\Delta(t)$ of each of the following matrices:

(a) $A = \begin{bmatrix} 2 & 5 \\ 4 & 1 \end{bmatrix}$, (b) $B = \begin{bmatrix} 7 & -3 \\ 5 & -2 \end{bmatrix}$, (c) $C = \begin{bmatrix} 3 & -2 \\ 9 & -3 \end{bmatrix}$

Use the formula $(t) = t^2 - \text{tr}(M)\, t + |M|$ for a 2×2 matrix M:

(a) $\text{tr}(A) = 2 + 1 = 3$, $|A| = 2 - 20 = -18$, so $\Delta(t) = t^2 - 3t - 18$
(b) $\text{tr}(B) = 7 - 2 = 5$, $|B| = -14 + 15 = 1$, so $\Delta(t) = t^2 - 5t + 1$
(c) $\text{tr}(C) = 3 - 3 = 0$, $|C| = -9 + 18 = 9$, so $\Delta(t) = t^2 + 9$

9.3. Find the characteristic polynomial $\Delta(t)$ of each of the following matrices:

(a) $A = \begin{bmatrix} 1 & 2 & 3 \\ 3 & 0 & 4 \\ 6 & 4 & 5 \end{bmatrix}$, (b) $B = \begin{bmatrix} 1 & 6 & -2 \\ -3 & 2 & 0 \\ 0 & 3 & -4 \end{bmatrix}$

Use the formula $\Delta(t) = t^3 - \text{tr}(A)t^2 + (A_{11} + A_{22} + A_{33})t - |A|$, where A_{ii} is the cofactor of a_{ii} in the 3×3 matrix $A = [a_{ij}]$.

(a) $\text{tr}(A) = 1 + 0 + 5 = 6$,

$$A_{11} = \begin{vmatrix} 0 & 4 \\ 4 & 5 \end{vmatrix} = -16, \qquad A_{22} = \begin{vmatrix} 1 & 3 \\ 6 & 5 \end{vmatrix} = -13, \qquad A_{33} = \begin{vmatrix} 1 & 2 \\ 3 & 0 \end{vmatrix} = -6$$

$$A_{11} + A_{22} + A_{33} = -35, \quad \text{and} \quad |A| = 48 + 36 - 16 - 30 = 38$$

Thus,
$$\Delta(t) = t^3 - 6t^2 - 35t - 38$$

(b) $\text{tr}(B) = 1 + 2 - 4 = -1$

$$B_{11} = \begin{vmatrix} 2 & 0 \\ 3 & -4 \end{vmatrix} = -8, \qquad B_{22} = \begin{vmatrix} 1 & -2 \\ 0 & -4 \end{vmatrix} = -4, \qquad B_{33} = \begin{vmatrix} 1 & 6 \\ -3 & 2 \end{vmatrix} = 20$$

$$B_{11} + B_{22} + B_{33} = 8, \quad \text{and} \quad |B| = -8 + 18 - 72 = -62$$

Thus,
$$\Delta(t) = t^3 + t^2 - 8t + 62$$

9.4. Find the characteristic polynomial $\Delta(t)$ of each of the following matrices:

$$\text{(a) } A = \begin{bmatrix} 2 & 5 & 1 & 1 \\ 1 & 4 & 2 & 2 \\ 0 & 0 & 6 & -5 \\ 0 & 0 & 2 & 3 \end{bmatrix}, \quad \text{(b) } B = \begin{bmatrix} 1 & 1 & 2 & 2 \\ 0 & 3 & 3 & 4 \\ 0 & 0 & 5 & 5 \\ 0 & 0 & 0 & 6 \end{bmatrix}$$

(a) A is block triangular with diagonal blocks

$$A_1 = \begin{bmatrix} 2 & 5 \\ 1 & 4 \end{bmatrix} \quad \text{and} \quad A_2 = \begin{bmatrix} 6 & -5 \\ 2 & 3 \end{bmatrix}$$

Thus,
$$\Delta(t) = \Delta_{A_1}(t)\Delta_{A_2}(t) = (t^2 - 6t + 3)(t^2 - 9t + 28)$$

(b) Because B is triangular, $\Delta(t) = (t-1)(t-3)(t-5)(t-6)$.

9.5. Find the characteristic polynomial $\Delta(t)$ of each of the following linear operators:

(a) $F\colon \mathbf{R}^2 \to \mathbf{R}^2$ defined by $F(x, y) = (3x + 5y, \quad 2x - 7y)$.

(b) $\mathbf{D}\colon V \to V$ defined by $\mathbf{D}(f) = df/dt$, where V is the space of functions with basis $S = \{\sin t, \cos t\}$.

The characteristic polynomial $\Delta(t)$ of a linear operator is equal to the characteristic polynomial of any matrix A that represents the linear operator.

(a) Find the matrix A that represents T relative to the usual basis of $\mathbf{R}^2$. We have

$$A = \begin{bmatrix} 3 & 5 \\ 2 & -7 \end{bmatrix}, \quad \text{so} \quad \Delta(t) = t^2 - \text{tr}(A)\,t + |A| = t^2 + 4t - 31$$

(b) Find the matrix A representing the differential operator $\mathbf{D}$ relative to the basis S. We have

$$\begin{aligned} \mathbf{D}(\sin t) &= \cos t \;\;\, = 0(\sin t) + 1(\cos t) \\ \mathbf{D}(\cos t) &= -\sin t = -1(\sin t) + 0(\cos t) \end{aligned} \quad \text{and so} \quad A = \begin{bmatrix} 0 & -1 \\ 1 & 0 \end{bmatrix}$$

Therefore,
$$\Delta(t) = t^2 - \text{tr}(A)\,t + |A| = t^2 + 1$$

9.6. Show that a matrix A and its transpose A^T have the same characteristic polynomial.

By the transpose operation, $(tI - A)^T = tI^T - A^T = tI - A^T$. Because a matrix and its transpose have the same determinant,

$$\Delta_A(t) = |tI - A| = |(tI - A)^T| = |tI - A^T| = \Delta_{A^T}(t)$$

9.7. Prove Theorem 9.1: Let f and g be polynomials. For any square matrix A and scalar k,

(i) $(f+g)(A) = f(A) + g(A)$, (iii) $(kf)(A) = kf(A)$,
(ii) $(fg)(A) = f(A)g(A)$, (iv) $f(A)g(A) = g(A)f(A)$.

Suppose $f = a_n t^n + \cdots + a_1 t + a_0$ and $g = b_m t^m + \cdots + b_1 t + b_0$. Then, by definition,

$$f(A) = a_n A^n + \cdots + a_1 A + a_0 I \quad \text{and} \quad g(A) = b_m A^m + \cdots + b_1 A + b_0 I$$

(i) Suppose $m \le n$ and let $b_i = 0$ if $i > m$. Then

$$f + g = (a_n + b_n)t^n + \cdots + (a_1 + b_1)t + (a_0 + b_0)$$

Hence,

$$\begin{aligned}(f+g)(A) &= (a_n + b_n)A^n + \cdots + (a_1 + b_1)A + (a_0 + b_0)I \\ &= a_n A^n + b_n A^n + \cdots + a_1 A + b_1 A + a_0 I + b_0 I = f(A) + g(A)\end{aligned}$$

(ii) By definition, $fg = c_{n+m}t^{n+m} + \cdots + c_1 t + c_0 = \sum_{k=0}^{n+m} c_k t^k$, where

$$c_k = a_0 b_k + a_1 b_{k-1} + \cdots + a_k b_0 = \sum_{i=0}^{k} a_i b_{k-i}$$

Hence, $(fg)(A) = \sum_{k=0}^{n+m} c_k A^k$ and

$$f(A)g(A) = \left(\sum_{i=0}^{n} a_i A^i\right)\left(\sum_{j=0}^{m} b_j A^j\right) = \sum_{i=0}^{n}\sum_{j=0}^{m} a_i b_j A^{i+j} = \sum_{k=0}^{n+m} c_k A^k = (fg)(A)$$

(iii) By definition, $kf = ka_n t^n + \cdots + ka_1 t + ka_0$, and so

$$(kf)(A) = ka_n A^n + \cdots + ka_1 A + ka_0 I = k(a_n A^n + \cdots + a_1 A + a_0 I) = kf(A)$$

(iv) By (ii), $g(A)f(A) = (gf)(A) = (fg)(A) = f(A)g(A)$.

9.8. Prove the Cayley–Hamilton Theorem 9.2: Every matrix A is a root of its characterstic polynomial $\Delta(t)$.

Let A be an arbitrary n-square matrix and let $\Delta(t)$ be its characteristic polynomial, say,

$$\Delta(t) = |tI - A| = t^n + a_{n-1}t^{n-1} + \cdots + a_1 t + a_0$$

Now let $B(t)$ denote the classical adjoint of the matrix $tI - A$. The elements of $B(t)$ are cofactors of the matrix $tI - A$ and hence are polynomials in t of degree not exceeding $n - 1$. Thus,

$$B(t) = B_{n-1}t^{n-1} + \cdots + B_1 t + B_0$$

where the B_i are n-square matrices over K which are independent of t. By the fundamental property of the classical adjoint (Theorem 8.9), $(tI - A)B(t) = |tI - A|I$, or

$$(tI - A)(B_{n-1}t^{n-1} + \cdots + B_1 t + B_0) = (t^n + a_{n-1}t^{n-1} + \cdots + a_1 t + a_0)I$$

Removing the parentheses and equating corresponding powers of t yields

$$B_{n-1} = I, \quad B_{n-2} - AB_{n-1} = a_{n-1}I, \quad \ldots, \quad B_0 - AB_1 = a_1 I, \quad -AB_0 = a_0 I$$

Multiplying the above equations by A^n, A^{n-1}, ..., A, I, respectively, yields

$$A^n B_{n-1} = A_n I, \quad A^{n-1}B_{n-2} - A^n B_{n-1} = a_{n-1}A^{n-1}, \quad \ldots, \quad AB_0 - A^2 B_1 = a_1 A, \quad -AB_0 = a_0 I$$

Adding the above matrix equations yields 0 on the left-hand side and $\Delta(A)$ on the right-hand side; that is,

$$0 = A^n + a_{n-1}A^{n-1} + \cdots + a_1 A + a_0 I$$

Therefore, $\Delta(A) = 0$, which is the Cayley–Hamilton theorem.

Eigenvalues and Eigenvectors of 2 × 2 Matrices

9.9. Let $A = \begin{bmatrix} 3 & -4 \\ 2 & -6 \end{bmatrix}$.

(a) Find all eigenvalues and corresponding eigenvectors.

(b) Find matrices P and D such that P is nonsingular and $D = P^{-1}AP$ is diagonal.

(a) First find the characteristic polynomial $\Delta(t)$ of A:

$$\Delta(t) = t^2 - \mathrm{tr}(A)\, t + |A| = t^2 + 3t - 10 = (t-2)(t+5)$$

The roots $\lambda = 2$ and $\lambda = -5$ of $\Delta(t)$ are the eigenvalues of A. We find corresponding eigenvectors.

(i) Subtract $\lambda = 2$ down the diagonal of A to obtain the matrix $M = A - 2I$, where the corresponding homogeneous system $MX = 0$ yields the eigenvectors corresponding to $\lambda = 2$. We have

$$M = \begin{bmatrix} 1 & -4 \\ 2 & -8 \end{bmatrix}, \quad \text{corresponding to} \quad \begin{aligned} x - 4y &= 0 \\ 2x - 8y &= 0 \end{aligned} \quad \text{or} \quad x - 4y = 0$$

The system has only one free variable, and $v_1 = (4, 1)$ is a nonzero solution. Thus, $v_1 = (4, 1)$ is an eigenvector belonging to (and spanning the eigenspace of) $\lambda = 2$.

(ii) Subtract $\lambda = -5$ (or, equivalently, add 5) down the diagonal of A to obtain

$$M = \begin{bmatrix} 8 & -4 \\ 2 & -1 \end{bmatrix}, \quad \text{corresponding to} \quad \begin{aligned} 8x - 4y &= 0 \\ 2x - \ \, y &= 0 \end{aligned} \quad \text{or} \quad 2x - y = 0$$

The system has only one free variable, and $v_2 = (1, 2)$ is a nonzero solution. Thus, $v_2 = (1, 2)$ is an eigenvector belonging to $\lambda = 5$.

(b) Let P be the matrix whose columns are v_1 and v_2. Then

$$P = \begin{bmatrix} 4 & 1 \\ 1 & 2 \end{bmatrix} \quad \text{and} \quad D = P^{-1}AP = \begin{bmatrix} 2 & 0 \\ 0 & -5 \end{bmatrix}$$

Note that D is the diagonal matrix whose diagonal entries are the eigenvalues of A corresponding to the eigenvectors appearing in P.

Remark: Here P is the change-of-basis matrix from the usual basis of $\mathbf{R}^2$ to the basis $S = \{v_1, v_2\}$, and D is the matrix that represents (the matrix function) A relative to the new basis S.

9.10. Let $A = \begin{bmatrix} 2 & 2 \\ 1 & 3 \end{bmatrix}$.

(a) Find all eigenvalues and corresponding eigenvectors.

(b) Find a nonsingular matrix P such that $D = P^{-1}AP$ is diagonal, and P^{-1}.

(c) Find A^6 and $f(A)$, where $t^4 - 3t^3 - 6t^2 + 7t + 3$.

(d) Find a "real cube root" of B—that is, a matrix B such that $B^3 = A$ and B has real eigenvalues.

(a) First find the characteristic polynomial $\Delta(t)$ of A:

$$\Delta(t) = t^2 - \mathrm{tr}(A)\, t + |A| = t^2 - 5t + 4 = (t-1)(t-4)$$

The roots $\lambda = 1$ and $\lambda = 4$ of $\Delta(t)$ are the eigenvalues of A. We find corresponding eigenvectors.

(i) Subtract $\lambda = 1$ down the diagonal of A to obtain the matrix $M = A - \lambda I$, where the corresponding homogeneous system $MX = 0$ yields the eigenvectors belonging to $\lambda = 1$. We have

$$M = \begin{bmatrix} 1 & 2 \\ 1 & 2 \end{bmatrix}, \quad \text{corresponding to} \quad \begin{aligned} x + 2y &= 0 \\ x + 2y &= 0 \end{aligned} \quad \text{or} \quad x + 2y = 0$$

The system has only one independent solution; for example, $x = 2, y = -1$. Thus, $v_1 = (2, -1)$ is an eigenvector belonging to (and spanning the eigenspace of) $\lambda = 1$.

(ii) Subtract $\lambda = 4$ down the diagonal of A to obtain

$$M = \begin{bmatrix} -2 & 2 \\ 1 & -1 \end{bmatrix}, \quad \text{corresponding to} \quad \begin{matrix} -2x + 2y = 0 \\ x - y = 0 \end{matrix} \quad \text{or} \quad x - y = 0$$

The system has only one independent solution; for example, $x = 1, y = 1$. Thus, $v_2 = (1, 1)$ is an eigenvector belonging to $\lambda = 4$.

(b) Let P be the matrix whose columns are v_1 and v_2. Then

$$P = \begin{bmatrix} 2 & 1 \\ -1 & 1 \end{bmatrix} \quad \text{and} \quad D = P^{-1}AP = \begin{bmatrix} 1 & 0 \\ 0 & 4 \end{bmatrix}, \quad \text{where} \quad P^{-1} = \begin{bmatrix} \frac{1}{3} & -\frac{1}{3} \\ \frac{1}{3} & \frac{2}{3} \end{bmatrix}$$

(c) Using the diagonal factorization $A = PDP^{-1}$, and $1^6 = 1$ and $4^6 = 4096$, we get

$$A^6 = PD^6P^{-1} = \begin{bmatrix} 2 & 1 \\ -1 & 1 \end{bmatrix} \begin{bmatrix} 1 & 0 \\ 0 & 4096 \end{bmatrix} \begin{bmatrix} \frac{1}{3} & -\frac{1}{3} \\ \frac{1}{3} & \frac{2}{3} \end{bmatrix} = \begin{bmatrix} 1366 & 2230 \\ 1365 & 2731 \end{bmatrix}$$

Also, $f(1) = 2$ and $f(4) = -1$. Hence,

$$f(A) = Pf(D)P^{-1} = \begin{bmatrix} 2 & 1 \\ -1 & 1 \end{bmatrix} \begin{bmatrix} 2 & 0 \\ 0 & -1 \end{bmatrix} \begin{bmatrix} \frac{1}{3} & -\frac{1}{3} \\ \frac{1}{3} & \frac{2}{3} \end{bmatrix} = \begin{bmatrix} 1 & 2 \\ -1 & 0 \end{bmatrix}$$

(d) Here $\begin{bmatrix} 1 & 0 \\ 0 & \sqrt[3]{4} \end{bmatrix}$ is the real cube root of D. Hence the real cube root of A is

$$B = P\sqrt[3]{D}P^{-1} = \begin{bmatrix} 2 & 1 \\ -1 & 1 \end{bmatrix} \begin{bmatrix} 1 & 0 \\ 0 & \sqrt[3]{4} \end{bmatrix} \begin{bmatrix} \frac{1}{3} & -\frac{1}{3} \\ \frac{1}{3} & \frac{2}{3} \end{bmatrix} = \frac{1}{3} \begin{bmatrix} 2 + \sqrt[3]{4} & -2 + 2\sqrt[3]{4} \\ -1 + \sqrt[3]{4} & 1 + 2\sqrt[3]{4} \end{bmatrix}$$

9.11. Each of the following real matrices defines a linear transformation on $\mathbf{R}^2$:

(a) $A = \begin{bmatrix} 5 & 6 \\ 3 & -2 \end{bmatrix}$, (b) $B = \begin{bmatrix} 1 & -1 \\ 2 & -1 \end{bmatrix}$, (c) $C = \begin{bmatrix} 5 & -1 \\ 1 & 3 \end{bmatrix}$

Find, for each matrix, all eigenvalues and a maximum set S of linearly independent eigenvectors. Which of these linear operators are diagonalizable—that is, which can be represented by a diagonal matrix?

(a) First find $\Delta(t) = t^2 - 3t - 28 = (t - 7)(t + 4)$. The roots $\lambda = 7$ and $\lambda = -4$ are the eigenvalues of A. We find corresponding eigenvectors.

(i) Subtract $\lambda = 7$ down the diagonal of A to obtain

$$M = \begin{bmatrix} -2 & 6 \\ 3 & -9 \end{bmatrix}, \quad \text{corresponding to} \quad \begin{matrix} -2x + 6y = 0 \\ 3x - 9y = 0 \end{matrix} \quad \text{or} \quad x - 3y = 0$$

Here $v_1 = (3, 1)$ is a nonzero solution.

(ii) Subtract $\lambda = -4$ (or add 4) down the diagonal of A to obtain

$$M = \begin{bmatrix} 9 & 6 \\ 3 & 2 \end{bmatrix}, \quad \text{corresponding to} \quad \begin{matrix} 9x + 6y = 0 \\ 3x + 2y = 0 \end{matrix} \quad \text{or} \quad 3x + 2y = 0$$

Here $v_2 = (2, -3)$ is a nonzero solution.

Then $S = \{v_1, v_2\} = \{(3, 1), \ (2, -3)\}$ is a maximal set of linearly independent eigenvectors. Because S is a basis of $\mathbf{R}^2$, A is diagonalizable. Using the basis S, A is represented by the diagonal matrix $D = \mathrm{diag}(7, -4)$.

(b) First find the characteristic polynomial $\Delta(t) = t^2 + 1$. There are no real roots. Thus B, a real matrix representing a linear transformation on $\mathbf{R}^2$, has no eigenvalues and no eigenvectors. Hence, in particular, B is not diagonalizable.

(c) First find $\Delta(t) = t^2 - 8t + 16 = (t-4)^2$. Thus, $\lambda = 4$ is the only eigenvalue of C. Subtract $\lambda = 4$ down the diagonal of C to obtain

$$M = \begin{bmatrix} 1 & -1 \\ 1 & -1 \end{bmatrix}, \qquad \text{corresponding to} \qquad x - y = 0$$

The homogeneous system has only one independent solution; for example, $x = 1$, $y = 1$. Thus, $v = (1, 1)$ is an eigenvector of C. Furthermore, as there are no other eigenvalues, the singleton set $S = \{v\} = \{(1, 1)\}$ is a maximal set of linearly independent eigenvectors of C. Furthermore, because S is not a basis of $\mathbf{R}^2$, C is not diagonalizable.

9.12. Suppose the matrix B in Problem 9.11 represents a linear operator on complex space $\mathbf{C}^2$. Show that, in this case, B is diagonalizable by finding a basis S of $\mathbf{C}^2$ consisting of eigenvectors of B.

The characteristic polynomial of B is still $\Delta(t) = t^2 + 1$. As a polynomial over $\mathbf{C}$, $\Delta(t)$ *does* factor; specifically, $\Delta(t) = (t-i)(t+i)$. Thus, $\lambda = i$ and $\lambda = -i$ are the eigenvalues of B.

(i) Subtract $\lambda = i$ down the diagonal of B to obtain the homogeneous system

$$\begin{aligned} (1-i)x - \qquad\quad y &= 0 \\ 2x + (-1-i)y &= 0 \end{aligned} \qquad \text{or} \qquad (1-i)x - y = 0$$

The system has only one independent solution; for example, $x = 1, y = 1 - i$. Thus, $v_1 = (1,\ 1-i)$ is an eigenvector that spans the eigenspace of $\lambda = i$.

(ii) Subtract $\lambda = -i$ (or add i) down the diagonal of B to obtain the homogeneous system

$$\begin{aligned} (1+i)x - \qquad\quad y &= 0 \\ 2x + (-1+i)y &= 0 \end{aligned} \qquad \text{or} \qquad (1+i)x - y = 0$$

The system has only one independent solution; for example, $x = 1, y = 1 + i$. Thus, $v_2 = (1,\ 1+i)$ is an eigenvector that spans the eigenspace of $\lambda = -i$.

As a complex matrix, B is diagonalizable. Specifically, $S = \{v_1, v_2\} = \{(1, 1-i),\ (1, 1+i)\}$ is a basis of $\mathbf{C}^2$ consisting of eigenvectors of B. Using this basis S, B is represented by the diagonal matrix $D = \mathrm{diag}(i, -i)$.

9.13. Let L be the linear transformation on $\mathbf{R}^2$ that reflects each point P across the line $y = kx$, where $k > 0$. (See Fig. 9-1.)

(a) Show that $v_1 = (k, 1)$ and $v_2 = (1, -k)$ are eigenvectors of L.

(b) Show that L is diagonalizable, and find a diagonal representation D.

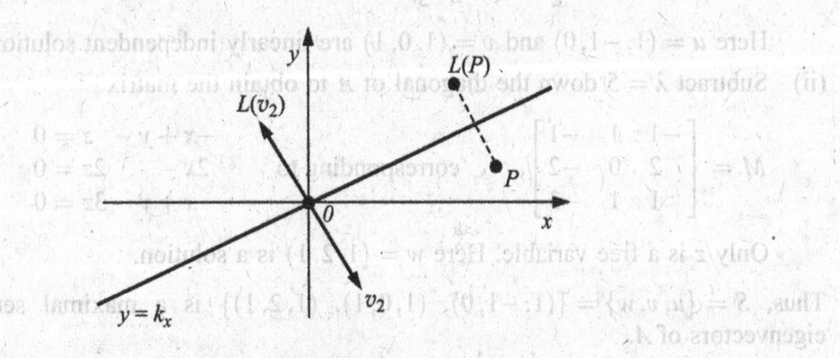

Figure 9-1

(a) The vector $v_1 = (k, 1)$ lies on the line $y = kx$, and hence is left fixed by L; that is, $L(v_1) = v_1$. Thus, v_1 is an eigenvector of L belonging to the eigenvalue $\lambda_1 = 1$.

The vector $v_2 = (1, -k)$ is perpendicular to the line $y = kx$, and hence, L reflects v_2 into its negative; that is, $L(v_2) = -v_2$. Thus, v_2 is an eigenvector of L belonging to the eigenvalue $\lambda_2 = -1$.

(b) Here $S = \{v_1, v_2\}$ is a basis of $\mathbf{R}^2$ consisting of eigenvectors of L. Thus, L is diagonalizable, with the diagonal representation $D = \begin{bmatrix} 1 & 0 \\ 0 & -1 \end{bmatrix}$ (relative to the basis S).

Eigenvalues and Eigenvectors

9.14. Let $A = \begin{bmatrix} 4 & 1 & -1 \\ 2 & 5 & -2 \\ 1 & 1 & 2 \end{bmatrix}$. (a) Find all eigenvalues of A.

(b) Find a maximum set S of linearly independent eigenvectors of A.

(c) Is A diagonalizable? If yes, find P such that $D = P^{-1}AP$ is diagonal.

(a) First find the characteristic polynomial $\Delta(t)$ of A. We have

$$\mathrm{tr}(A) = 4 + 5 + 2 = 11 \quad \text{and} \quad |A| = 40 - 2 - 2 + 5 + 8 - 4 = 45$$

Also, find each cofactor A_{ii} of a_{ii} in A:

$$A_{11} = \begin{vmatrix} 5 & -2 \\ 1 & 2 \end{vmatrix} = 12, \qquad A_{22} = \begin{vmatrix} 4 & -1 \\ 1 & 2 \end{vmatrix} = 9, \qquad A_{33} = \begin{vmatrix} 4 & 1 \\ 2 & 5 \end{vmatrix} = 18$$

Hence, $\quad \Delta(t) = t^3 - \mathrm{tr}(A)\,t^2 + (A_{11} + A_{22} + A_{33})t - |A| = t^3 - 11t^2 + 39t - 45$

Assuming Δt has a rational root, it must be among $\pm 1, \pm 3, \pm 5, \pm 9, \pm 15, \pm 45$. Testing, by synthetic division, we get

$$\begin{array}{r|l} 3 & 1 - 11 + 39 - 45 \\ & \quad\;\; 3 - 24 + 45 \\ \hline & 1 - 8 + 15 + \;\; 0 \end{array}$$

Thus, $t = 3$ is a root of $\Delta(t)$. Also, $t - 3$ is a factor and $t^2 - 8t + 15$ is a factor. Hence,

$$\Delta(t) = (t-3)(t^2 - 8t + 15) = (t-3)(t-5)(t-3) = (t-3)^2(t-5)$$

Accordingly, $\lambda = 3$ and $\lambda = 5$ are eigenvalues of A.

(b) Find linearly independent eigenvectors for each eigenvalue of A.

(i) Subtract $\lambda = 3$ down the diagonal of A to obtain the matrix

$$M = \begin{bmatrix} 1 & 1 & -1 \\ 2 & 2 & -2 \\ 1 & 1 & -1 \end{bmatrix}, \qquad \text{corresponding to} \qquad x + y - z = 0$$

Here $u = (1, -1, 0)$ and $v = (1, 0, 1)$ are linearly independent solutions.

(ii) Subtract $\lambda = 5$ down the diagonal of A to obtain the matrix

$$M = \begin{bmatrix} -1 & 1 & -1 \\ 2 & 0 & -2 \\ 1 & 1 & -3 \end{bmatrix}, \qquad \text{corresponding to} \qquad \begin{array}{r} -x + y - \;\; z = 0 \\ 2x - \quad\;\; 2z = 0 \\ x + y - 3z = 0 \end{array} \qquad \text{or} \qquad \begin{array}{r} x \qquad - \;\; z = 0 \\ y - 2z = 0 \end{array}$$

Only z is a free variable. Here $w = (1, 2, 1)$ is a solution.

Thus, $S = \{u, v, w\} = \{(1, -1, 0),\ (1, 0, 1),\ (1, 2, 1)\}$ is a maximal set of linearly independent eigenvectors of A.

Remark: The vectors u and v were chosen so that they were independent solutions of the system $x + y - z = 0$. On the other hand, w is automatically independent of u and v because w belongs to a different eigenvalue of A. Thus, the three vectors are linearly independent.

(c) A is diagonalizable, because it has three linearly independent eigenvectors. Let P be the matrix with columns u, v, w. Then

$$P = \begin{bmatrix} 1 & 1 & 1 \\ -1 & 0 & 2 \\ 0 & 1 & 1 \end{bmatrix} \quad \text{and} \quad D = P^{-1}AP = \begin{bmatrix} 3 & & \\ & 3 & \\ & & 5 \end{bmatrix}$$

9.15. Repeat Problem 9.14 for the matrix $B = \begin{bmatrix} 3 & -1 & 1 \\ 7 & -5 & 1 \\ 6 & -6 & 2 \end{bmatrix}$.

(a) First find the characteristic polynomial $\Delta(t)$ of B. We have

$$\text{tr}(B) = 0, \quad |B| = -16, \quad B_{11} = -4, \quad B_{22} = 0, \quad B_{33} = -8, \quad \text{so} \quad \sum_i B_{ii} = -12$$

Therefore, $\Delta(t) = t^3 - 12t + 16 = (t-2)^2(t+4)$. Thus, $\lambda_1 = 2$ and $\lambda_2 = -4$ are the eigenvalues of B.

(b) Find a basis for the eigenspace of each eigenvalue of B.

(i) Subtract $\lambda_1 = 2$ down the diagonal of B to obtain

$$M = \begin{bmatrix} 1 & -1 & 1 \\ 7 & -7 & 1 \\ 6 & -6 & 0 \end{bmatrix}, \quad \text{corresponding to} \quad \begin{aligned} x - y + z &= 0 \\ 7x - 7y + z &= 0 \\ 6x - 6y &= 0 \end{aligned} \quad \text{or} \quad \begin{aligned} x - y + z &= 0 \\ z &= 0 \end{aligned}$$

The system has only one independent solution; for example, $x = 1, y = 1, z = 0$. Thus, $u = (1, 1, 0)$ forms a basis for the eigenspace of $\lambda_1 = 2$.

(ii) Subtract $\lambda_2 = -4$ (or add 4) down the diagonal of B to obtain

$$M = \begin{bmatrix} 7 & -1 & 1 \\ 7 & -1 & 1 \\ 6 & -6 & 6 \end{bmatrix}, \quad \text{corresponding to} \quad \begin{aligned} 7x - y + z &= 0 \\ 7x - y + z &= 0 \\ 6x - 6y + 6z &= 0 \end{aligned} \quad \text{or} \quad \begin{aligned} x - y + z &= 0 \\ 6y - 6z &= 0 \end{aligned}$$

The system has only one independent solution; for example, $x = 0, y = 1, z = 1$. Thus, $v = (0, 1, 1)$ forms a basis for the eigenspace of $\lambda_2 = -4$.

Thus $S = \{u, v\}$ is a maximal set of linearly independent eigenvectors of B.

(c) Because B has at most two linearly independent eigenvectors, B is not similar to a diagonal matrix; that is, B is not diagonalizable.

9.16. Find the algebraic and geometric multiplicities of the eigenvalue $\lambda_1 = 2$ of the matrix B in Problem 9.15.

The algebraic multiplicity of $\lambda_1 = 2$ is 2, because $t - 2$ appears with exponent 2 in $\Delta(t)$. However, the geometric multiplicity of $\lambda_1 = 2$ is 1, because $\dim E_{\lambda_1} = 1$ (where E_{λ_1} is the eigenspace of λ_1).

9.17. Let $T: \mathbf{R}^3 \to \mathbf{R}^3$ be defined by $T(x, y, z) = (2x + y - 2z, \quad 2x + 3y - 4z, \quad x + y - z)$. Find all eigenvalues of T, and find a basis of each eigenspace. Is T diagonalizable? If so, find the basis S of $\mathbf{R}^3$ that diagonalizes T, and find its diagonal representation D.

First find the matrix A that represents T relative to the usual basis of $\mathbf{R}^3$ by writing down the coefficients of x, y, z as rows, and then find the characteristic polynomial of A (and T). We have

$$A = [T] = \begin{bmatrix} 2 & 1 & -2 \\ 2 & 3 & -4 \\ 1 & 1 & -1 \end{bmatrix} \quad \text{and} \quad \begin{aligned} &\text{tr}(A) = 4, \quad |A| = 2 \\ &A_{11} = 1, \quad A_{22} = 0, \quad A_{33} = 4 \\ &\sum_i A_{ii} = 5 \end{aligned}$$

Therefore, $\Delta(t) = t^3 - 4t^2 + 5t - 2 = (t-1)^2(t-2)$, and so $\lambda = 1$ and $\lambda = 2$ are the eigenvalues of A (and T). We next find linearly independent eigenvectors for each eigenvalue of A.

(i) Subtract $\lambda = 1$ down the diagonal of A to obtain the matrix

$$M = \begin{bmatrix} 1 & 1 & -2 \\ 2 & 2 & -4 \\ 1 & 1 & -2 \end{bmatrix}, \quad \text{corresponding to} \quad x + y - 2z = 0$$

Here y and z are free variables, and so there are two linearly independent eigenvectors belonging to $\lambda = 1$. For example, $u = (1, -1, 0)$ and $v = (2, 0, 1)$ are two such eigenvectors.

(ii) Subtract $\lambda = 2$ down the diagonal of A to obtain

$$M = \begin{bmatrix} 0 & 1 & -2 \\ 2 & 1 & -4 \\ 1 & 1 & -3 \end{bmatrix}, \quad \text{corresponding to} \quad \begin{array}{r} y - 2z = 0 \\ 2x + y - 4z = 0 \\ x + y - 3z = 0 \end{array} \quad \text{or} \quad \begin{array}{r} x + y - 3z = 0 \\ y - 2z = 0 \end{array}$$

Only z is a free variable. Here $w = (1, 2, 1)$ is a solution.

Thus, T is diagonalizable, because it has three independent eigenvectors. Specifically, choosing

$$S = \{u, v, w\} = \{(1, -1, 0), \quad (2, 0, 1), \quad (1, 2, 1)\}$$

as a basis, T is represented by the diagonal matrix $D = \text{diag}(1, 1, 2)$.

9.18. Prove the following for a linear operator (matrix) T:

(a) The scalar 0 is an eigenvalue of T if and only if T is singular.

(b) If λ is an eigenvalue of T, where T is invertible, then λ^{-1} is an eigenvalue of T^{-1}.

(a) We have that 0 is an eigenvalue of T if and only if there is a vector $v \neq 0$ such that $T(v) = 0v$—that is, if and only if T is singular.

(b) Because T is invertible, it is nonsingular; hence, by (a), $\lambda \neq 0$. By definition of an eigenvalue, there exists $v \neq 0$ such that $T(v) = \lambda v$. Applying T^{-1} to both sides, we obtain

$$v = T^{-1}(\lambda v) = \lambda T^{-1}(v), \quad \text{and so} \quad T^{-1}(v) = \lambda^{-1} v$$

Therefore, λ^{-1} is an eigenvalue of T^{-1}.

9.19. Let λ be an eigenvalue of a linear operator $T: V \to V$, and let E_λ consists of all the eigenvectors belonging to λ (called the *eigenspace* of λ). Prove that E_λ is a subspace of V. That is, prove

(a) If $u \in E_\lambda$, then $ku \in E_\lambda$ for any scalar k. (b) If $u, v, \in E_\lambda$, then $u + v \in E_\lambda$.

(a) Because $u \in E_\lambda$, we have $T(u) = \lambda u$. Then $T(ku) = kT(u) = k(\lambda u) = \lambda(ku)$, and so $ku \in E_\lambda$. (We view the zero vector $0 \in V$ as an "eigenvector" of λ in order for E_λ to be a subspace of V.)

(b) As $u, v \in E_\lambda$, we have $T(u) = \lambda u$ and $T(v) = \lambda v$. Then $T(u + v) = T(u) + T(v) = \lambda u + \lambda v = \lambda(u + v)$, and so $u + v \in E_\lambda$

9.20. Prove Theorem 9.6: The following are equivalent: (i) The scalar λ is an eigenvalue of A.

(ii) The matrix $\lambda I - A$ is singular.

(iii) The scalar λ is a root of the characteristic polynomial $\Delta(t)$ of A.

The scalar λ is an eigenvalue of A if and only if there exists a nonzero vector v such that

$$Av = \lambda v \quad \text{or} \quad (\lambda I)v - Av = 0 \quad \text{or} \quad (\lambda I - A)v = 0$$

or $\lambda I - A$ is singular. In such a case, λ is a root of $\Delta(t) = |tI - A|$. Also, v is in the eigenspace E_λ of λ if and only if the above relations hold. Hence, v is a solution of $(\lambda I - A)X = 0$.

9.21. Prove Theorem 9.8′: Suppose $v_1, v_2, \ldots, v_n$ are nonzero eigenvectors of T belonging to distinct eigenvalues $\lambda_1, \lambda_2, \ldots, \lambda_n$. Then $v_1, v_2, \ldots, v_n$ are linearly independent.

Suppose the theorem is not true. Let $v_1, v_2, \ldots, v_s$ be a minimal set of vectors for which the theorem is not true. We have $s > 1$, because $v_1 \neq 0$. Also, by the minimality condition, $v_2, \ldots, v_s$ are linearly independent. Thus, v_1 is a linear combination of $v_2, \ldots, v_s$, say,

$$v_1 = a_2 v_2 + a_3 v_3 + \cdots + a_s v_s \tag{1}$$

(where some $a_k \neq 0$). Applying T to (1) and using the linearity of T yields

$$T(v_1) = T(a_2 v_2 + a_3 v_3 + \cdots + a_s v_s) = a_2 T(v_2) + a_3 T(v_3) + \cdots + a_s T(v_s) \tag{2}$$

Because v_j is an eigenvector of T belonging to λ_j, we have $T(v_j) = \lambda_j v_j$. Substituting in (2) yields

$$\lambda_1 v_1 = a_2 \lambda_2 v_2 + a_3 \lambda_3 v_3 + \cdots + a_s \lambda_s v_s \tag{3}$$

Multiplying (1) by λ_1 yields

$$\lambda_1 v_1 = a_2 \lambda_1 v_2 + a_3 \lambda_1 v_3 + \cdots + a_s \lambda_1 v_s \tag{4}$$

Setting the right-hand sides of (3) and (4) equal to each other, or subtracting (3) from (4) yields

$$a_2(\lambda_1 - \lambda_2)v_2 + a_3(\lambda_1 - \lambda_3)v_3 + \cdots + a_s(\lambda_1 - \lambda_s)v_s = 0 \tag{5}$$

Because $v_2, v_3, \ldots, v_s$ are linearly independent, the coefficients in (5) must all be zero. That is,

$$a_2(\lambda_1 - \lambda_2) = 0, \qquad a_3(\lambda_1 - \lambda_3) = 0, \qquad \ldots, \qquad a_s(\lambda_1 - \lambda_s) = 0$$

However, the λ_i are distinct. Hence $\lambda_1 - \lambda_j \neq 0$ for $j > 1$. Hence, $a_2 = 0$, $a_3 = 0, \ldots, a_s = 0$. This contradicts the fact that some $a_k \neq 0$. The theorem is proved.

9.22. Prove Theorem 9.9. Suppose $\Delta(t) = (t - a_1)(t - a_2) \ldots (t - a_n)$ is the characteristic polynomial of an n-square matrix A, and suppose the n roots a_i are distinct. Then A is similar to the diagonal matrix $D = \text{diag}(a_1, a_2, \ldots, a_n)$.

Let $v_1, v_2, \ldots, v_n$ be (nonzero) eigenvectors corresponding to the eigenvalues a_i. Then the n eigenvectors v_i are linearly independent (Theorem 9.8), and hence form a basis of K^n. Accordingly, A is diagonalizable (i.e., A is similar to a diagonal matrix D), and the diagonal elements of D are the eigenvalues a_i.

9.23. Prove Theorem 9.10′: The geometric multiplicity of an eigenvalue λ of T does not exceed its algebraic multiplicity.

Suppose the geometric multiplicity of λ is r. Then its eigenspace E_λ contains r linearly independent eigenvectors $v_1, \ldots, v_r$. Extend the set $\{v_i\}$ to a basis of V, say, $\{v_i, \ldots, v_r, w_1, \ldots, w_s\}$. We have

$$T(v_1) = \lambda v_1, \quad T(v_2) = \lambda v_2, \quad \ldots, \quad T(v_r) = \lambda v_r,$$

$$\begin{aligned} T(w_1) &= a_{11} v_1 + \cdots + a_{1r} v_r + b_{11} w_1 + \cdots + b_{1s} w_s \\ T(w_2) &= a_{21} v_1 + \cdots + a_{2r} v_r + b_{21} w_1 + \cdots + b_{2s} w_s \\ &\cdots\cdots\cdots\cdots\cdots\cdots\cdots\cdots\cdots\cdots\cdots\cdots \\ T(w_s) &= a_{s1} v_1 + \cdots + a_{sr} v_r + b_{s1} w_1 + \cdots + b_{ss} w_s \end{aligned}$$

Then $M = \begin{bmatrix} \lambda I_r & A \\ 0 & B \end{bmatrix}$ is the matrix of T in the above basis, where $A = [a_{ij}]^T$ and $B = [b_{ij}]^T$.

Because M is block diagonal, the characteristic polynomial $(t - \lambda)^r$ of the block λI_r must divide the characteristic polynomial of M and hence of T. Thus, the algebraic multiplicity of λ for T is at least r, as required.

Diagonalizing Real Symmetric Matrices and Quadratic Forms

9.24. Let $A = \begin{bmatrix} 7 & 3 \\ 3 & -1 \end{bmatrix}$. Find an orthogonal matrix P such that $D = P^{-1}AP$ is diagonal.

First find the characteristic polynomial $\Delta(t)$ of A. We have

$$\Delta(t) = t^2 - \text{tr}(A)\, t + |A| = t^2 - 6t - 16 = (t-8)(t+2)$$

Thus, the eigenvalues of A are $\lambda = 8$ and $\lambda = -2$. We next find corresponding eigenvectors.
Subtract $\lambda = 8$ down the diagonal of A to obtain the matrix

$$M = \begin{bmatrix} -1 & 3 \\ 3 & -9 \end{bmatrix}, \quad \text{corresponding to} \quad \begin{matrix} -x + 3y = 0 \\ 3x - 9y = 0 \end{matrix} \quad \text{or} \quad x - 3y = 0$$

A nonzero solution is $u_1 = (3, 1)$.

Subtract $\lambda = -2$ (or add 2) down the diagonal of A to obtain the matrix

$$M = \begin{bmatrix} 9 & 3 \\ 3 & 1 \end{bmatrix}, \quad \text{corresponding to} \quad \begin{matrix} 9x + 3y = 0 \\ 3x + \ \ y = 0 \end{matrix} \quad \text{or} \quad 3x + y = 0$$

A nonzero solution is $u_2 = (1, -3)$.
As expected, because A is symmetric, the eigenvectors u_1 and u_2 are orthogonal. Normalize u_1 and u_2 to obtain, respectively, the unit vectors

$$\hat{u}_1 = (3/\sqrt{10}, 1/\sqrt{10}) \quad \text{and} \quad \hat{u}_2 = (1/\sqrt{10}, -3/\sqrt{10}).$$

Finally, let P be the matrix whose columns are the unit vectors $\hat{u}_1$ and $\hat{u}_2$, respectively. Then

$$P = \begin{bmatrix} 3/\sqrt{10} & 1/\sqrt{10} \\ 1/\sqrt{10} & -3/\sqrt{10} \end{bmatrix} \quad \text{and} \quad D = P^{-1}AP = \begin{bmatrix} 8 & 0 \\ 0 & -2 \end{bmatrix}$$

As expected, the diagonal entries in D are the eigenvalues of A.

9.25. Let $B = \begin{bmatrix} 11 & -8 & 4 \\ -8 & -1 & -2 \\ 4 & -2 & -4 \end{bmatrix}$. (a) Find all eigenvalues of B.

(b) Find a maximal set S of nonzero orthogonal eigenvectors of B.

(c) Find an orthogonal matrix P such that $D = P^{-1}BP$ is diagonal.

(a) First find the characteristic polynomial of B. We have

$$\text{tr}(B) = 6, \quad |B| = 400, \quad B_{11} = 0, \quad B_{22} = -60, \quad B_{33} = -75, \quad \text{so} \quad \sum_i B_{ii} = -135$$

Hence, $\Delta(t) = t^3 - 6t^2 - 135t - 400$. If $\Delta(t)$ has an integer root it must divide 400. Testing $t = -5$, by synthetic division, yields

$$\begin{array}{r|l} -5 & 1 - \ \ 6 - 135 - 400 \\ & \ \ \ - \ \ 5 + \ \ 55 + 400 \\ \hline & 1 - 11 - \ \ 80 + \ \ \ 0 \end{array}$$

Thus, $t + 5$ is a factor of $\Delta(t)$, and $t^2 - 11t - 80$ is a factor. Thus,

$$\Delta(t) = (t+5)(t^2 - 11t - 80) = (t+5)^2(t-16)$$

The eigenvalues of B are $\lambda = -5$ (multiplicity 2), and $\lambda = 16$ (multiplicity 1).

(b) Find an orthogonal basis for each eigenspace. Subtract $\lambda = -5$ (or, add 5) down the diagonal of B to obtain the homogeneous system

$$16x - 8y + 4z = 0, \qquad -8x + 4y - 2z = 0, \qquad 4x - 2y + z = 0$$

That is, $4x - 2y + z = 0$. The system has two independent solutions. One solution is $v_1 = (0, 1, 2)$. We seek a second solution $v_2 = (a, b, c)$, which is orthogonal to v_1, such that

$$4a - 2b + c = 0, \qquad \text{and also} \qquad b - 2c = 0$$

One such solution is $v_2 = (-5, -8, 4)$.

Subtract $\lambda = 16$ down the diagonal of B to obtain the homogeneous system

$$-5x - 8y + 4z = 0, \qquad -8x - 17y - 2z = 0, \qquad 4x - 2y - 20z = 0$$

This system yields a nonzero solution $v_3 = (4, -2, 1)$. (As expected from Theorem 9.13, the eigenvector v_3 is orthogonal to v_1 and v_2.)

Then v_1, v_2, v_3 form a maximal set of nonzero orthogonal eigenvectors of B.

(c) Normalize v_1, v_2, v_3 to obtain the orthonormal basis:

$$\hat{v}_1 = v_1/\sqrt{5}, \qquad \hat{v}_2 = v_2/\sqrt{105}, \qquad \hat{v}_3 = v_3/\sqrt{21}$$

Then P is the matrix whose columns are $\hat{v}_1, \hat{v}_2, \hat{v}_3$. Thus,

$$P = \begin{bmatrix} 0 & -5/\sqrt{105} & 4/\sqrt{21} \\ 1/\sqrt{5} & -8/\sqrt{105} & -2/\sqrt{21} \\ 2/\sqrt{5} & 4/\sqrt{105} & 1/\sqrt{21} \end{bmatrix} \quad \text{and} \quad D = P^{-1}BP = \begin{bmatrix} -5 & & \\ & -5 & \\ & & 16 \end{bmatrix}$$

9.26. Let $q(x, y) = x^2 + 6xy - 7y^2$. Find an orthogonal substitution that diagonalizes q.

Find the symmetric matrix A that represents q and its characteristic polynomial $\Delta(t)$. We have

$$A = \begin{bmatrix} 1 & 3 \\ 3 & -7 \end{bmatrix} \quad \text{and} \quad \Delta(t) = t^2 + 6t - 16 = (t-2)(t+8)$$

The eigenvalues of A are $\lambda = 2$ and $\lambda = -8$. Thus, using s and t as new variables, a diagonal form of q is

$$q(s, t) = 2s^2 - 8t^2$$

The corresponding orthogonal substitution is obtained by finding an orthogonal set of eigenvectors of A.

(i) Subtract $\lambda = 2$ down the diagonal of A to obtain the matrix

$$M = \begin{bmatrix} -1 & 3 \\ 3 & -9 \end{bmatrix}, \quad \text{corresponding to} \quad \begin{matrix} -x + 3y = 0 \\ 3x - 9y = 0 \end{matrix} \quad \text{or} \quad -x + 3y = 0$$

A nonzero solution is $u_1 = (3, 1)$.

(ii) Subtract $\lambda = -8$ (or add 8) down the diagonal of A to obtain the matrix

$$M = \begin{bmatrix} 9 & 3 \\ 3 & 1 \end{bmatrix}, \quad \text{corresponding to} \quad \begin{matrix} 9x + 3y = 0 \\ 3x + y = 0 \end{matrix} \quad \text{or} \quad 3x + y = 0$$

A nonzero solution is $u_2 = (-1, 3)$.

As expected, because A is symmetric, the eigenvectors u_1 and u_2 are orthogonal.

Now normalize u_1 and u_2 to obtain, respectively, the unit vectors

$$\hat{u}_1 = (3/\sqrt{10},\ 1/\sqrt{10}) \quad \text{and} \quad \hat{u}_2 = (-1/\sqrt{10},\ 3/\sqrt{10}).$$

Finally, let P be the matrix whose columns are the unit vectors $\hat{u}_1$ and $\hat{u}_2$, respectively, and then $[x, y]^T = P[s, t]^T$ is the required orthogonal change of coordinates. That is,

$$P = \begin{bmatrix} 3/\sqrt{10} & -1/\sqrt{10} \\ 1/\sqrt{10} & 3/\sqrt{10} \end{bmatrix} \quad \text{and} \quad x = \frac{3s - t}{\sqrt{10}}, \qquad y = \frac{s + 3t}{\sqrt{10}}$$

One can also express s and t in terms of x and y by using $P^{-1} = P^T$. That is,

$$s = \frac{3x + y}{\sqrt{10}}, \qquad t = \frac{-x + 3t}{\sqrt{10}}$$

Minimal Polynomial

9.27. Let $A = \begin{bmatrix} 4 & -2 & 2 \\ 6 & -3 & 4 \\ 3 & -2 & 3 \end{bmatrix}$ and $B = \begin{bmatrix} 3 & -2 & 2 \\ 4 & -4 & 6 \\ 2 & -3 & 5 \end{bmatrix}$. The characteristic polynomial of both matrices is $\Delta(t) = (t-2)(t-1)^2$. Find the minimal polynomial $m(t)$ of each matrix.

The minimal polynomial $m(t)$ must divide $\Delta(t)$. Also, each factor of $\Delta(t)$ (i.e., $t-2$ and $t-1$) must also be a factor of $m(t)$. Thus, $m(t)$ must be exactly one of the following:

$$f(t) = (t-2)(t-1) \quad \text{or} \quad g(t) = (t-2)(t-1)^2$$

(a) By the Cayley–Hamilton theorem, $g(A) = \Delta(A) = 0$, so we need only test $f(t)$. We have

$$f(A) = (A-2I)(A-I) = \begin{bmatrix} 2 & -2 & 2 \\ 6 & -5 & 4 \\ 3 & -2 & 1 \end{bmatrix}\begin{bmatrix} 3 & -2 & 2 \\ 6 & -4 & 4 \\ 3 & -2 & 2 \end{bmatrix} = \begin{bmatrix} 0 & 0 & 0 \\ 0 & 0 & 0 \\ 0 & 0 & 0 \end{bmatrix}$$

Thus, $m(t) = f(t) = (t-2)(t-1) = t^2 - 3t + 2$ is the minimal polynomial of A.

(b) Again $g(B) = \Delta(B) = 0$, so we need only test $f(t)$. We get

$$f(B) = (B-2I)(B-I) = \begin{bmatrix} 1 & -2 & 2 \\ 4 & -6 & 6 \\ 2 & -3 & 3 \end{bmatrix}\begin{bmatrix} 2 & -2 & 2 \\ 4 & -5 & 6 \\ 2 & -3 & 4 \end{bmatrix} = \begin{bmatrix} -2 & 2 & -2 \\ -4 & 4 & -4 \\ -2 & 2 & -2 \end{bmatrix} \neq 0$$

Thus, $m(t) \neq f(t)$. Accordingly, $m(t) = g(t) = (t-2)(t-1)^2$ is the minimal polynomial of B. [We emphasize that we do not need to compute $g(B)$; we know $g(B) = 0$ from the Cayley–Hamilton theorem.]

9.28. Find the minimal polynomial $m(t)$ of each of the following matrices:

(a) $A = \begin{bmatrix} 5 & 1 \\ 3 & 7 \end{bmatrix}$, (b) $B = \begin{bmatrix} 1 & 2 & 3 \\ 0 & 2 & 3 \\ 0 & 0 & 3 \end{bmatrix}$, (c) $C = \begin{bmatrix} 4 & -1 \\ 1 & 2 \end{bmatrix}$

(a) The characteristic polynomial of A is $\Delta(t) = t^2 - 12t + 32 = (t-4)(t-8)$. Because $\Delta(t)$ has distinct factors, the minimal polynomial $m(t) = \Delta(t) = t^2 - 12t + 32$.

(b) Because B is triangular, its eigenvalues are the diagonal elements $1, 2, 3$; and so its characteristic polynomial is $\Delta(t) = (t-1)(t-2)(t-3)$. Because $\Delta(t)$ has distinct factors, $m(t) = \Delta(t)$.

(c) The characteristic polynomial of C is $\Delta(t) = t^2 - 6t + 9 = (t-3)^2$. Hence the minimal polynomial of C is $f(t) = t - 3$ or $g(t) = (t-3)^2$. However, $f(C) \neq 0$; that is, $C - 3I \neq 0$. Hence,

$$m(t) = g(t) = \Delta(t) = (t-3)^2.$$

9.29. Suppose $S = \{u_1, u_2, \ldots, u_n\}$ is a basis of V, and suppose F and G are linear operators on V such that $[F]$ has 0's on and below the diagonal, and $[G]$ has $a \neq 0$ on the superdiagonal and 0's elsewhere. That is,

$$[F] = \begin{bmatrix} 0 & a_{21} & a_{31} & \cdots & a_{n1} \\ 0 & 0 & a_{32} & \cdots & a_{n2} \\ \cdots & \cdots & \cdots & \cdots & \cdots \\ 0 & 0 & 0 & \cdots & a_{n,n-1} \\ 0 & 0 & 0 & \cdots & 0 \end{bmatrix}, \qquad [G] = \begin{bmatrix} 0 & a & 0 & \cdots & 0 \\ 0 & 0 & a & \cdots & 0 \\ \cdots & \cdots & \cdots & \cdots & \cdots \\ 0 & 0 & 0 & \cdots & a \\ 0 & 0 & 0 & \cdots & 0 \end{bmatrix}$$

Show that (a) $F^n = 0$, (b) $G^{n-1} \neq 0$, but $G^n = 0$. (These conditions also hold for $[F]$ and $[G]$.)

(a) We have $F(u_1) = 0$ and, for $r > 1$, $F(u_r)$ is a linear combination of vectors preceding u_r in S. That is,

$$F(u_r) = a_{r1}u_1 + a_{r2}u_2 + \cdots + a_{r,r-1}u_{r-1}$$

Hence, $F^2(u_r) = F(F(u_r))$ is a linear combination of vectors preceding u_{r-1}, and so on. Hence, $F^r(u_r) = 0$ for each r. Thus, for each r, $F^n(u_r) = F^{n-r}(0) = 0$, and so $F^n = 0$, as claimed.

(b) We have $G(u_1) = 0$ and, for each $k > 1$, $G(u_k) = au_{k-1}$. Hence, $G^r(u_k) = a^r u_{k-r}$ for $r < k$. Because $a \neq 0$, $a^{n-1} \neq 0$. Therefore, $G^{n-1}(u_n) = a^{n-1}u_1 \neq 0$, and so $G^{n-1} \neq 0$. On the other hand, by (a), $G^n = 0$.

9.30. Let B be the matrix in Example 9.12(a) that has 1's on the diagonal, a's on the superdiagonal, where $a \neq 0$, and 0's elsewhere. Show that $f(t) = (t - \lambda)^n$ is both the characteristic polynomial $\Delta(t)$ and the minimum polynomial $m(t)$ of A.

Because A is triangular with λ's on the diagonal, $\Delta(t) = f(t) = (t - \lambda)^n$ is its characteristic polynomial. Thus, $m(t)$ is a power of $t - \lambda$. By Problem 9.29, $(A - \lambda I)^{r-1} \neq 0$. Hence, $m(t) = \Delta(t) = (t - \lambda)^n$.

9.31. Find the characteristic polynomial $\Delta(t)$ and minimal polynomial $m(t)$ of each matrix:

$$\text{(a)}\ M = \begin{bmatrix} 4 & 1 & 0 & 0 & 0 \\ 0 & 4 & 1 & 0 & 0 \\ 0 & 0 & 4 & 0 & 0 \\ 0 & 0 & 0 & 4 & 1 \\ 0 & 0 & 0 & 0 & 4 \end{bmatrix}, \quad \text{(b)}\ M' = \begin{bmatrix} 2 & 7 & 0 & 0 \\ 0 & 2 & 0 & 0 \\ 0 & 0 & 1 & 1 \\ 0 & 0 & -2 & 4 \end{bmatrix}$$

(a) M is block diagonal with diagonal blocks

$$A = \begin{bmatrix} 4 & 1 & 0 \\ 0 & 4 & 1 \\ 0 & 0 & 4 \end{bmatrix} \quad \text{and} \quad B = \begin{bmatrix} 4 & 1 \\ 0 & 4 \end{bmatrix}$$

The characteristic and minimal polynomial of A is $f(t) = (t - 4)^3$ and the characteristic and minimal polynomial of B is $g(t) = (t - 4)^2$. Then

$$\Delta(t) = f(t)g(t) = (t - 4)^5 \quad \text{but} \quad m(t) = \text{LCM}[f(t), g(t)] = (t - 4)^3$$

(where LCM means least common multiple). We emphasize that the exponent in $m(t)$ is the size of the largest block.

(b) Here M' is block diagonal with diagonal blocks $A' = \begin{bmatrix} 2 & 7 \\ 0 & 2 \end{bmatrix}$ and $B' = \begin{bmatrix} 1 & 1 \\ -2 & 4 \end{bmatrix}$. The characteristic and minimal polynomial of A' is $f(t) = (t - 2)^2$. The characteristic polynomial of B' is $g(t) = t^2 - 5t + 6 = (t - 2)(t - 3)$, which has distinct factors. Hence, $g(t)$ is also the minimal polynomial of B. Accordingly,

$$\Delta(t) = f(t)g(t) = (t - 2)^3(t - 3) \quad \text{but} \quad m(t) = \text{LCM}[f(t), g(t)] = (t - 2)^2(t - 3)$$

9.32. Find a matrix A whose minimal polynomial is $f(t) = t^3 - 8t^2 + 5t + 7$.

Simply let $A = \begin{bmatrix} 0 & 0 & -7 \\ 1 & 0 & -5 \\ 0 & 1 & 8 \end{bmatrix}$, the companion matrix of $f(t)$ [defined in Example 9.12(b)].

9.33. Prove Theorem 9.15: The minimal polynomial $m(t)$ of a matrix (linear operator) A divides every polynomial that has A as a zero. In particular (by the Cayley–Hamilton theorem), $m(t)$ divides the characteristic polynomial $\Delta(t)$ of A.

Suppose $f(t)$ is a polynomial for which $f(A) = 0$. By the division algorithm, there exist polynomials $q(t)$ and $r(t)$ for which $f(t) = m(t)q(t) + r(t)$ and $r(t) = 0$ or $\deg r(t) < \deg m(t)$. Substituting $t = A$ in this equation, and using that $f(A) = 0$ and $m(A) = 0$, we obtain $r(A) = 0$. If $r(t) \neq 0$, then $r(t)$ is a polynomial of degree less than $m(t)$ that has A as a zero. This contradicts the definition of the minimal polynomial. Thus, $r(t) = 0$, and so $f(t) = m(t)q(t)$; that is, $m(t)$ divides $f(t)$.

9.34. Let $m(t)$ be the minimal polynomial of an n-square matrix A. Prove that the characteristic polynomial $\Delta(t)$ of A divides $[m(t)]^n$.

Suppose $m(t) = t^r + c_1 t^{r-1} + \cdots + c_{r-1}t + c_r$. Define matrices B_j as follows:

$$\begin{array}{lll} B_0 = I & \text{so} & I = B_0 \\ B_1 = A + c_1 I & \text{so} & c_1 I = B_1 - A = B_1 - AB_0 \\ B_2 = A^2 + c_1 A + c_2 I & \text{so} & c_2 I = B_2 - A(A + c_1 I) = B_2 - AB_1 \\ \ldots\ldots\ldots & & \ldots\ldots\ldots \\ B_{r-1} = A^{r-1} + c_1 A^{r-2} + \cdots + c_{r-1} I & \text{so} & c_{r-1} I = B_{r-1} - AB_{r-2} \end{array}$$

Then

$$-AB_{r-1} = c_r I - (A^r + c_1 A^{r-1} + \cdots + c_{r-1}A + c_r I) = c_r I - m(A) = c_r I$$

Set

$$B(t) = t^{r-1}B_0 + t^{r-2}B_1 + \cdots + tB_{r-2} + B_{r-1}$$

Then

$$\begin{aligned} (tI - A)B(t) &= (t^r B_0 + t^{r-1}B_1 + \cdots + tB_{r-1}) - (t^{r-1}AB_0 + t^{r-2}AB_1 + \cdots + AB_{r-1}) \\ &= t^r B_0 + t^{r-1}(B_1 - AB_0) + t^{r-2}(B_2 - AB_1) + \cdots + t(B_{r-1} - AB_{r-2}) - AB_{r-1} \\ &= t^r I + c_1 t^{r-1} I + c_2 t^{r-2} I + \cdots + c_{r-1} tI + c_r I = m(t)I \end{aligned}$$

Taking the determinant of both sides gives $|tI - A||B(t)| = |m(t)I| = [m(t)]^n$. Because $|B(t)|$ is a polynomial, $|tI - A|$ divides $[m(t)]^n$; that is, the characteristic polynomial of A divides $[m(t)]^n$.

9.35. Prove Theorem 9.16: The characteristic polynomial $\Delta(t)$ and the minimal polynomial $m(t)$ of A have the same irreducible factors.

Suppose $f(t)$ is an irreducible polynomial. If $f(t)$ divides $m(t)$, then $f(t)$ also divides $\Delta(t)$ [because $m(t)$ divides $\Delta(t)$]. On the other hand, if $f(t)$ divides $\Delta(t)$, then by Problem 9.34, $f(t)$ also divides $[m(t)]^n$. But $f(t)$ is irreducible; hence, $f(t)$ also divides $m(t)$. Thus, $m(t)$ and $\Delta(t)$ have the same irreducible factors.

9.36. Prove Theorem 9.19: The minimal polynomial $m(t)$ of a block diagonal matrix M with diagonal blocks A_i is equal to the least common multiple (LCM) of the minimal polynomials of the diagonal blocks A_i.

We prove the theorem for the case $r = 2$. The general theorem follows easily by induction. Suppose $M = \begin{bmatrix} A & 0 \\ 0 & B \end{bmatrix}$, where A and B are square matrices. We need to show that the minimal polynomial $m(t)$ of M is the LCM of the minimal polynomials $g(t)$ and $h(t)$ of A and B, respectively.

Because $m(t)$ is the minimal polynomial of $M, m(M) = \begin{bmatrix} m(A) & 0 \\ 0 & m(B) \end{bmatrix} = 0$, and $m(A) = 0$ and $m(B) = 0$. Because $g(t)$ is the minimal polynomial of A, $g(t)$ divides $m(t)$. Similarly, $h(t)$ divides $m(t)$. Thus $m(t)$ is a multiple of $g(t)$ and $h(t)$.

Now let $f(t)$ be another multiple of $g(t)$ and $h(t)$. Then $f(M) = \begin{bmatrix} f(A) & 0 \\ 0 & f(B) \end{bmatrix} = \begin{bmatrix} 0 & 0 \\ 0 & 0 \end{bmatrix} = 0$. But $m(t)$ is the minimal polynomial of M; hence, $m(t)$ divides $f(t)$. Thus, $m(t)$ is the LCM of $g(t)$ and $h(t)$.

9.37. Suppose $m(t) = t^r + a_{r-1}t^{r-1} + \cdots + a_1 t + a_0$ is the minimal polynomial of an n-square matrix A. Prove the following:

(a) A is nonsingular if and only if the constant term $a_0 \neq 0$.

(b) If A is nonsingular, then A^{-1} is a polynomial in A of degree $r - 1 < n$.

(a) The following are equivalent: (i) A is nonsingular, (ii) 0 is not a root of $m(t)$, (iii) $a_0 \neq 0$. Thus, the statement is true.

(b) Because A is nonsingular, $a_0 \neq 0$ by (a). We have

$$m(A) = A^r + a_{r-1}A^{r-1} + \cdots + a_1 A + a_0 I = 0$$

Thus, $$-\frac{1}{a_0}(A^{r-1} + a_{r-1}A^{r-2} + \cdots + a_1 I)A = I$$

Accordingly, $$A^{-1} = -\frac{1}{a_0}(A^{r-1} + a_{r-1}A^{r-2} + \cdots + a_1 I)$$

SUPPLEMENTARY PROBLEMS

Polynomials of Matrices

9.38. Let $A = \begin{bmatrix} 2 & -3 \\ 5 & 1 \end{bmatrix}$ and $B = \begin{bmatrix} 1 & 2 \\ 0 & 3 \end{bmatrix}$. Find $f(A)$, $g(A)$, $f(B)$, $g(B)$, where $f(t) = 2t^2 - 5t + 6$ and $g(t) = t^3 - 2t^2 + t + 3$.

9.39. Let $A = \begin{bmatrix} 1 & 2 \\ 0 & 1 \end{bmatrix}$. Find A^2, A^3, A^n, where $n > 3$, and A^{-1}.

9.40. Let $B = \begin{bmatrix} 8 & 12 & 0 \\ 0 & 8 & 12 \\ 0 & 0 & 8 \end{bmatrix}$. Find a real matrix A such that $B = A^3$.

9.41. For each matrix, find a polynomial having the following matrix as a root:

(a) $A = \begin{bmatrix} 2 & 5 \\ 1 & -3 \end{bmatrix}$, (b) $B = \begin{bmatrix} 2 & -3 \\ 7 & -4 \end{bmatrix}$, (c) $C = \begin{bmatrix} 1 & 1 & 2 \\ 1 & 2 & 3 \\ 2 & 1 & 4 \end{bmatrix}$

9.42. Let A be any square matrix and let $f(t)$ be any polynomial. Prove (a) $(P^{-1}AP)^n = P^{-1}A^nP$. (b) $f(P^{-1}AP) = P^{-1}f(A)P$. (c) $f(A^T) = [f(A)]^T$. (d) If A is symmetric, then $f(A)$ is symmetric.

9.43. Let $M = \text{diag}[A_1, \ldots, A_r]$ be a block diagonal matrix, and let $f(t)$ be any polynomial. Show that $f(M)$ is block diagonal and $f(M) = \text{diag}[f(A_1), \ldots, f(A_r)]$.

9.44. Let M be a block triangular matrix with diagonal blocks $A_1, \ldots, A_r$, and let $f(t)$ be any polynomial. Show that $f(M)$ is also a block triangular matrix, with diagonal blocks $f(A_1), \ldots, f(A_r)$.

Eigenvalues and Eigenvectors

9.45. For each of the following matrices, find all eigenvalues and corresponding linearly independent eigenvectors:

(a) $A = \begin{bmatrix} 2 & -3 \\ 2 & -5 \end{bmatrix}$, (b) $B = \begin{bmatrix} 2 & 4 \\ -1 & 6 \end{bmatrix}$, (c) $C = \begin{bmatrix} 1 & -4 \\ 3 & -7 \end{bmatrix}$

When possible, find the nonsingular matrix P that diagonalizes the matrix.

9.46. Let $A = \begin{bmatrix} 2 & -1 \\ -2 & 3 \end{bmatrix}$.

(a) Find eigenvalues and corresponding eigenvectors.

(b) Find a nonsingular matrix P such that $D = P^{-1}AP$ is diagonal.

(c) Find A^8 and $f(A)$ where $f(t) = t^4 - 5t^3 + 7t^2 - 2t + 5$.

(d) Find a matrix B such that $B^2 = A$.

9.47. Repeat Problem 9.46 for $A = \begin{bmatrix} 5 & 6 \\ -2 & -2 \end{bmatrix}$.

9.48. For each of the following matrices, find all eigenvalues and a maximum set S of linearly independent eigenvectors:

(a) $A = \begin{bmatrix} 1 & -3 & 3 \\ 3 & -5 & 3 \\ 6 & -6 & 4 \end{bmatrix}$, (b) $B = \begin{bmatrix} 3 & -1 & 1 \\ 7 & -5 & 1 \\ 6 & -6 & 2 \end{bmatrix}$, (c) $C = \begin{bmatrix} 1 & 2 & 2 \\ 1 & 2 & -1 \\ -1 & 1 & 4 \end{bmatrix}$

Which matrices can be diagonalized, and why?

9.49. Find: (a) 2×2 matrix A with eigenvalues $\lambda_1 = 1$ and $\lambda_2 = -2$ and corresponding eigenvectors $v_1 = (1, 2)$ and $v_2 = (3, 7)$.

(b) 3×3 matrix A with eigenvalues $\lambda_1 = 1, \lambda_2 = 2, \lambda_2 = 3$ and corresponding eigenvectors $v_1 = (1, 0, 1)$, $v_2 = (1, 1, 2)$, $v_3 = (1, 2, 4)$. [Hint: See Problem 2.19.]

9.50. Let $A = \begin{bmatrix} a & b \\ c & d \end{bmatrix}$ be a real matrix. Find necessary and sufficient conditions on a, b, c, d so that A is diagonalizable—that is, so that A has two (real) linearly independent eigenvectors.

9.51. Show that matrices A and A^T have the same eigenvalues. Give an example of a 2×2 matrix A where A and A^T have different eigenvectors.

9.52. Suppose v is an eigenvector of linear operators F and G. Show that v is also an eigenvector of the linear operator $kF + k'G$, where k and k' are scalars.

9.53. Suppose v is an eigenvector of a linear operator T belonging to the eigenvalue λ. Prove

(a) For $n > 0$, v is an eigenvector of T^n belonging to λ^n.

(b) $f(\lambda)$ is an eigenvalue of $f(T)$ for any polynomial $f(t)$.

9.54. Suppose $\lambda \neq 0$ is an eigenvalue of the composition $F \circ G$ of linear operators F and G. Show that λ is also an eigenvalue of the composition $G \circ F$. [*Hint:* Show that $G(v)$ is an eigenvector of $G \circ F$.]

9.55. Let $E: V \to V$ be a projection mapping; that is, $E^2 = E$. Show that E is diagonalizable and, in fact, can be represented by the diagonal matrix $M = \begin{bmatrix} I_r & 0 \\ 0 & 0 \end{bmatrix}$, where r is the rank of E.

Diagonalizing Real Symmetric Matrices and Quadratic Forms

9.56. For each of the following symmetric matrices A, find an orthogonal matrix P and a diagonal matrix D such that $D = P^{-1}AP$:

(a) $A = \begin{bmatrix} 5 & 4 \\ 4 & -1 \end{bmatrix}$, (b) $A = \begin{bmatrix} 4 & -1 \\ -1 & 4 \end{bmatrix}$, (c) $A = \begin{bmatrix} 7 & 3 \\ 3 & -1 \end{bmatrix}$

9.57. For each of the following symmetric matrices B, find its eigenvalues, a maximal orthogonal set S of eigenvectors, and an orthogonal matrix P such that $D = P^{-1}BP$ is diagonal:

(a) $B = \begin{bmatrix} 0 & 1 & 1 \\ 1 & 0 & 1 \\ 1 & 1 & 0 \end{bmatrix}$, (b) $B = \begin{bmatrix} 2 & 2 & 4 \\ 2 & 5 & 8 \\ 4 & 8 & 17 \end{bmatrix}$

9.58. Using variables s and t, find an orthogonal substitution that diagonalizes each of the following quadratic forms:

(a) $q(x, y) = 4x^2 + 8xy - 11y^2$, (b) $q(x, y) = 2x^2 - 6xy + 10y^2$

9.59. For each of the following quadratic forms $q(x, y, z)$, find an orthogonal substitution expressing x, y, z in terms of variables r, s, t, and find $q(r, s, t)$:

(a) $q(x, y, z) = 5x^2 + 3y^2 + 12xz$, (b) $q(x, y, z) = 3x^2 - 4xy + 6y^2 + 2xz - 4yz + 3z^2$

9.60. Find a real 2×2 symmetric matrix A with eigenvalues:

(a) $\lambda = 1$ and $\lambda = 4$ and eigenvector $u = (1, 1)$ belonging to $\lambda = 1$;

(b) $\lambda = 2$ and $\lambda = 3$ and eigenvector $u = (1, 2)$ belonging to $\lambda = 2$.

In each case, find a matrix B for which $B^2 = A$.

Characteristic and Minimal Polynomials

9.61. Find the characteristic and minimal polynomials of each of the following matrices:

(a) $A = \begin{bmatrix} 3 & 1 & -1 \\ 2 & 4 & -2 \\ -1 & -1 & 3 \end{bmatrix}$, (b) $B = \begin{bmatrix} 3 & 2 & -1 \\ 3 & 8 & -3 \\ 3 & 6 & -1 \end{bmatrix}$

9.62. Find the characteristic and minimal polynomials of each of the following matrices:

(a) $A = \begin{bmatrix} 2 & 5 & 0 & 0 & 0 \\ 0 & 2 & 0 & 0 & 0 \\ 0 & 0 & 4 & 2 & 0 \\ 0 & 0 & 3 & 5 & 0 \\ 0 & 0 & 0 & 0 & 7 \end{bmatrix}$, (b) $B = \begin{bmatrix} 4 & -1 & 0 & 0 & 0 \\ 1 & 2 & 0 & 0 & 0 \\ 0 & 0 & 3 & 1 & 0 \\ 0 & 0 & 0 & 3 & 1 \\ 0 & 0 & 0 & 0 & 3 \end{bmatrix}$, (c) $C = \begin{bmatrix} 3 & 2 & 0 & 0 & 0 \\ 1 & 4 & 0 & 0 & 0 \\ 0 & 0 & 3 & 1 & 0 \\ 0 & 0 & 1 & 3 & 0 \\ 0 & 0 & 0 & 0 & 4 \end{bmatrix}$

9.63. Let $A = \begin{bmatrix} 1 & 1 & 0 \\ 0 & 2 & 0 \\ 0 & 0 & 1 \end{bmatrix}$ and $B = \begin{bmatrix} 2 & 0 & 0 \\ 0 & 2 & 2 \\ 0 & 0 & 1 \end{bmatrix}$. Show that A and B have different characteristic polynomials (and so are not similar) but have the same minimal polynomial. Thus, nonsimilar matrices may have the same minimal polynomial.

9.64. Let A be an n-square matrix for which $A^k = 0$ for some $k > n$. Show that $A^n = 0$.

9.65. Show that a matrix A and its transpose A^T have the same minimal polynomial.

9.66. Suppose $f(t)$ is an irreducible monic polynomial for which $f(A) = 0$ for a matrix A. Show that $f(t)$ is the minimal polynomial of A.

9.67. Show that A is a scalar matrix kI if and only if the minimal polynomial of A is $m(t) = t - k$.

9.68. Find a matrix A whose minimal polynomial is (a) $t^3 - 5t^2 + 6t + 8$, (b) $t^4 - 5t^3 - 2t + 7t + 4$.

9.69. Let $f(t)$ and $g(t)$ be monic polynomials (leading coefficient one) of minimal degree for which A is a root. Show $f(t) = g(t)$. [Thus, the minimal polynomial of A is unique.]

ANSWERS TO SUPPLEMENTARY PROBLEMS

Notation: $M = [R_1; \quad R_2; \quad \ldots]$ denotes a matrix M with rows $R_1, R_2, \ldots$.

9.38. $f(A) = [-26, -3; \quad 5, -27]$, $\quad g(A) = [-40, 39; \quad -65, -27]$,
$f(B) = [3, 6; \quad 0, 9]$, $\quad g(B) = [3, 12; \quad 0, 15]$

9.39. $A^2 = [1, 4; \quad 0, 1]$, $\quad A^3 = [1, 6; \quad 0, 1]$, $\quad A^n = [1, 2n; \quad 0, 1]$, $\quad A^{-1} = [1, -2; \quad 0, 1]$

9.40. Let $A = [2, a, b; \quad 0, 2, c; \quad 0, 0, 2]$. Set $B = A^3$ and then $a = 1$, $b = -\frac{1}{2}$, $c = 1$

9.41. Find $\Delta(t)$: (a) $t^2 + t - 11$, (b) $t^2 + 2t + 13$, (c) $t^3 - 7t^2 + 6t - 1$

9.45. (a) $\lambda = 1, u = (3, 1)$; $\lambda = -4, v = (1, 2)$, (b) $\lambda = 4, u = (2, 1)$,
(c) $\lambda = -1, u = (2, 1)$; $\lambda = -5, v = (2, 3)$. Only A and C can be diagonalized; use $P = [u, v]$.

9.46. (a) $\lambda = 1, u = (1, 1)$; $\lambda = 4, v = (1, -2)$,
(b) $P = [u, v]$,
(c) $f(A) = [3, 1;\ \ 2, 1]$, $A^8 = [21\,846, -21\,845;\ \ -43\,690, 43\,691]$,
(d) $B = [\frac{4}{3}, -\frac{1}{3};\ \ -\frac{2}{3}, \frac{5}{3}]$

9.47. (a) $\lambda = 1, u = (3, -2)$; $\lambda = 2, v = (2, -1)$, (b) $P = [u, v]$,
(c) $f(A) = [2, -6;\ \ 2, 9]$, $A^8 = [1021, 1530;\ \ -510, -764]$,
(d) $B = [-3 + 4\sqrt{2},\ \ -6 + 6\sqrt{2};\ \ 2 - 2\sqrt{2},\ \ 4 - 3\sqrt{2}]$

9.48. (a) $\lambda = -2, u = (1, 1, 0), v = (1, 0, -1); \lambda = 4, w = (1, 1, 2)$,
(b) $\lambda = 2, u = (1, 1, 0)$; $\lambda = -4, v = (0, 1, 1)$,
(c) $\lambda = 3, u = (1, 1, 0), v = (1, 0, 1)$; $\lambda = 1, w = (2, -1, 1)$. Only A and C can be diagonalized; use $P = [u, v, w]$.

9.49. (a) $[19, -9; 42, -20]$, (b) $[1, 1, 0; -2, 0, 2; -4, -2, 5]$

9.50. We need $[-\mathrm{tr}(A)]^2 - 4[\det(A)] \geq 0$ or $(a - d)^2 + 4bc \geq 0$.

9.51. $A = [1, 1;\ \ 0, 1]$

9.56. (a) $P = [2, -1;\ \ 1, 2]/\sqrt{5}$, $D = [7, 0;\ \ 0, 3]$,
(b) $P = [1, 1;\ \ 1, -1]/\sqrt{2}$, $D = [3, 0;\ \ 0, 5]$,
(c) $P = [3, -1;\ \ 1, 3]/\sqrt{10}$, $D = [8, 0;\ \ 0, 2]$

9.57. (a) $\lambda = -1$, $u = (1, -1, 0)$, $v = (1, 1, -2)$; $\lambda = 2$, $w = (1, 1, 1)$,
(b) $\lambda = 1$, $u = (2, 1, -1)$, $v = (2, -3, 1)$; $\lambda = 22$, $w = (1, 2, 4)$;
Normalize u, v, w, obtaining $\hat{u}, \hat{v}, \hat{w}$, and set $P = [\hat{u}, \hat{v}, \hat{w}]$. (*Remark*: u and v are not unique.)

9.58. (a) $x = (4s + t)/\sqrt{17}$, $y = (-s + 4t)/\sqrt{17}$, $q(s, t) = 5s^2 - 12t^2$,
(b) $x = (3s - t)/\sqrt{10}$, $y = (s + 3t)/\sqrt{10}$, $q(s, t) = s^2 + 11t^2$

9.59. (a) $x = (3s + 2t)/\sqrt{13}$, $y = r$, $z = (2s - 3t)/\sqrt{13}$, $q(r, s, t) = 3r^2 + 9s^2 - 4t^2$,
(b) $x = 5Ks + Lt$, $y = Jr + 2Ks - 2Lt$, $z = 2Jr - Ks - Lt$, where $J = 1/\sqrt{5}$, $K = 1/\sqrt{30}$, $L = 1/\sqrt{6}$; $q(r, s, t) = 2r^2 + 2s^2 + 8t^2$

9.60. (a) $A = \frac{1}{2}[5, -3;\ \ -3, 5]$, $B = \frac{1}{2}[3, -1;\ \ -1, 3]$,
(b) $A = \frac{1}{5}[14, -2;\ -2, 11]$, $B = \frac{1}{5}[\sqrt{2} + 4\sqrt{3}, 2\sqrt{2} - 2\sqrt{3};\ \ 2\sqrt{2} - 2\sqrt{3}, 4\sqrt{2} + \sqrt{3}]$

9.61. (a) $\Delta(t) = m(t) = (t - 2)^2(t - 6)$, (b) $\Delta(t) = (t - 2)^2(t - 6)$, $m(t) = (t - 2)(t - 6)$

9.62. (a) $\Delta(t) = (t - 2)^3(t - 7)^2$, $m(t) = (t - 2)^2(t - 7)$,
(b) $\Delta(t) = (t - 3)^5$, $m(t) = (t - 3)^3$,
(c) $\Delta(t) = (t - 2)^2(t - 4)^2(t - 5)$, $m(t) = (t - 2)(t - 4)(t - 5)$

9.68. Let A be the companion matrix [Example 9.12(b)] with last column: (a) $[-8, -6, 5]^T$, (b) $[-4, -7, 2, 5]^T$

9.69. *Hint*: A is a root of $h(t) = f(t) - g(t)$, where $h(t) \equiv 0$ or the degree of $h(t)$ is less than the degree of $f(t)$.

CHAPTER 10

Canonical Forms

10.1 Introduction

Let T be a linear operator on a vector space of finite dimension. As seen in Chapter 6, T may not have a diagonal matrix representation. However, it is still possible to "simplify" the matrix representation of T in a number of ways. This is the main topic of this chapter. In particular, we obtain the primary decomposition theorem, and the triangular, Jordan, and rational canonical forms.

We comment that the triangular and Jordan canonical forms exist for T if and only if the characteristic polynomial $\Delta(t)$ of T has all its roots in the base field K. This is always true if K is the complex field **C** but may not be true if K is the real field **R**.

We also introduce the idea of a *quotient space*. This is a very powerful tool, and it will be used in the proof of the existence of the triangular and rational canonical forms.

10.2 Triangular Form

Let T be a linear operator on an n-dimensional vector space V. Suppose T can be represented by the triangular matrix

$$A = \begin{bmatrix} a_{11} & a_{12} & \dots & a_{1n} \\ & a_{22} & \dots & a_{2n} \\ & & \dots & \dots \\ & & & a_{nn} \end{bmatrix}$$

Then the characteristic polynomial $\Delta(t)$ of T is a product of linear factors; that is,

$$\Delta(t) = \det(tI - A) = (t - a_{11})(t - a_{22}) \cdots (t - a_{nn})$$

The converse is also true and is an important theorem (proved in Problem 10.28).

THEOREM 10.1: Let $T:V \to V$ be a linear operator whose characteristic polynomial factors into linear polynomials. Then there exists a basis of V in which T is represented by a triangular matrix.

THEOREM 10.1: (Alternative Form) Let A be a square matrix whose characteristic polynomial factors into linear polynomials. Then A is similar to a triangular matrix—that is, there exists an invertible matrix P such that $P^{-1}AP$ is triangular.

We say that an operator T can be brought into triangular form if it can be represented by a triangular matrix. Note that in this case, the eigenvalues of T are precisely those entries appearing on the main diagonal. We give an application of this remark.

EXAMPLE 10.1 Let A be a square matrix over the complex field **C**. Suppose λ is an eigenvalue of A^2. Show that $\sqrt{\lambda}$ or $-\sqrt{\lambda}$ is an eigenvalue of A.

By Theorem 10.1, A and A^2 are similar, respectively, to triangular matrices of the form

$$B = \begin{bmatrix} \mu_1 & * & \cdots & * \\ & \mu_2 & \cdots & * \\ & & \cdots & \cdots \\ & & & \mu_n \end{bmatrix} \quad \text{and} \quad B^2 = \begin{bmatrix} \mu_1^2 & * & \cdots & * \\ & \mu_2^2 & \cdots & * \\ & & \cdots & \cdots \\ & & & \mu_n^2 \end{bmatrix}$$

Because similar matrices have the same eigenvalues, $\lambda = \mu_i^2$ for some i. Hence, $\mu_i = \sqrt{\lambda}$ or $\mu_i = -\sqrt{\lambda}$ is an eigenvalue of A.

10.3 Invariance

Let $T:V \to V$ be linear. A subspace W of V is said to be *invariant under* T or T*-invariant* if T maps W into itself—that is, if $v \in W$ implies $T(v) \in W$. In this case, T restricted to W defines a linear operator on W; that is, T induces a linear operator $\hat{T}:W \to W$ defined by $\hat{T}(w) = T(w)$ for every $w \in W$.

EXAMPLE 10.2

(a) Let $T: \mathbf{R}^3 \to \mathbf{R}^3$ be the following linear operator, which rotates each vector v about the z-axis by an angle θ (shown in Fig. 10-1):

$$T(x, y, z) = (x\cos\theta - y\sin\theta,\ \ x\sin\theta + y\cos\theta,\ \ z)$$

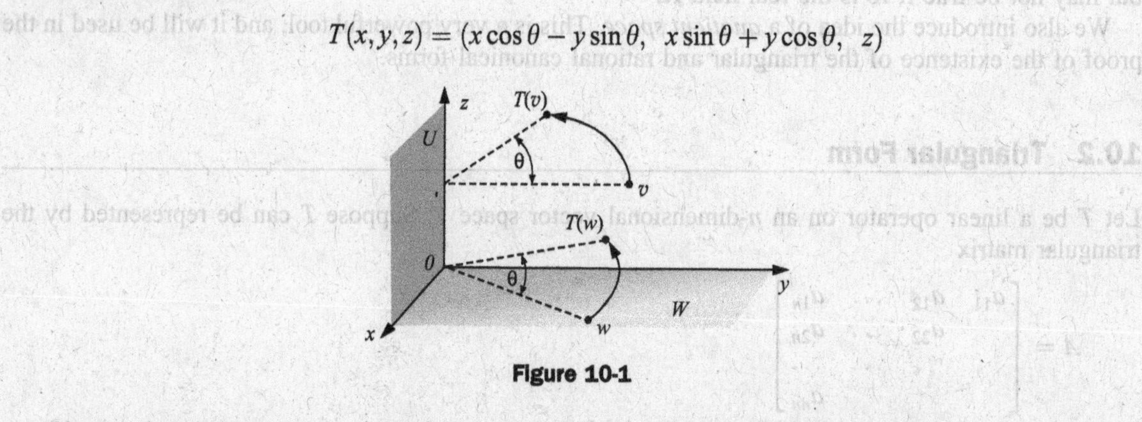

Figure 10-1

Observe that each vector $w = (a, b, 0)$ in the xy-plane W remains in W under the mapping T; hence, W is T-invariant. Observe also that the z-axis U is invariant under T. Furthermore, the restriction of T to W rotates each vector about the origin O, and the restriction of T to U is the identity mapping of U.

(b) Nonzero eigenvectors of a linear operator $T:V \to V$ may be characterized as generators of T-invariant one-dimensional subspaces. Suppose $T(v) = \lambda v,\ v \neq 0$. Then $W = \{kv,\ k \in K\}$, the one-dimensional subspace generated by v, is invariant under T because

$$T(kv) = kT(v) = k(\lambda v) = k\lambda v \in W$$

Conversely, suppose $\dim U = 1$ and $u \neq 0$ spans U, and U is invariant under T. Then $T(u) \in U$ and so $T(u)$ is a multiple of u—that is, $T(u) = \mu u$. Hence, u is an eigenvector of T.

The next theorem (proved in Problem 10.3) gives us an important class of invariant subspaces.

THEOREM 10.2: Let $T:V \to V$ be any linear operator, and let $f(t)$ be any polynomial. Then the kernel of $f(T)$ is invariant under T.

The notion of invariance is related to matrix representations (Problem 10.5) as follows.

THEOREM 10.3: Suppose W is an invariant subspace of $T:V \to V$. Then T has a block matrix representation $\begin{bmatrix} A & B \\ 0 & C \end{bmatrix}$, where A is a matrix representation of the restriction $\hat{T}$ of T to W.

10.4 Invariant Direct-Sum Decompositions

A vector space V is termed the *direct sum* of subspaces $W_1, \ldots, W_r$, written

$$V = W_1 \oplus W_2 \oplus \ldots \oplus W_r$$

if every vector $v \in V$ can be written uniquely in the form

$$v = w_1 + w_2 + \ldots + w_r, \qquad \text{with} \qquad w_i \in W_i.$$

The following theorem (proved in Problem 10.7) holds.

THEOREM 10.4: Suppose $W_1, W_2, \ldots, W_r$ are subspaces of V, and suppose

$$B_1 = \{w_{11}, w_{12}, \ldots, w_{1n_1}\}, \quad \ldots, \quad B_r = \{w_{r1}, w_{r2}, \ldots, w_{rn_r}\}$$

are bases of $W_1, W_2, \ldots, W_r$, respectively. Then V is the direct sum of the W_i if and only if the union $B = B_1 \cup \ldots \cup B_r$ is a basis of V.

Now suppose $T: V \to V$ is linear and V is the direct sum of (nonzero) T-invariant subspaces $W_1, W_2, \ldots, W_r$; that is,

$$V = W_1 \oplus \ldots \oplus W_r \qquad \text{and} \qquad T(W_i) \subseteq W_i, \qquad i = 1, \ldots, r$$

Let T_i denote the restriction of T to W_i. Then T is said to be *decomposable* into the operators T_i or T is said to be the *direct sum* of the T_i, written $T = T_1 \oplus \ldots \oplus T_r$. Also, the subspaces $W_1, \ldots, W_r$ are said to *reduce* T or to form a *T-invariant direct-sum decomposition* of V.

Consider the special case where two subspaces U and W reduce an operator $T: V \to V$; say $\dim U = 2$ and $\dim W = 3$, and suppose $\{u_1, u_2\}$ and $\{w_1, w_2, w_3\}$ are bases of U and W, respectively. If T_1 and T_2 denote the restrictions of T to U and W, respectively, then

$$\begin{aligned} T_1(u_1) &= a_{11}u_1 + a_{12}u_2 \\ T_1(u_2) &= a_{21}u_1 + a_{22}u_2 \end{aligned} \qquad \begin{aligned} T_2(w_1) &= b_{11}w_1 + b_{12}w_2 + b_{13}w_3 \\ T_2(w_2) &= b_{21}w_1 + b_{22}w_2 + b_{23}w_3 \\ T_2(w_3) &= b_{31}w_1 + b_{32}w_2 + b_{33}w_3 \end{aligned}$$

Accordingly, the following matrices A, B, M are the matrix representations of T_1, T_2, T, respectively,

$$A = \begin{bmatrix} a_{11} & a_{21} \\ a_{12} & a_{22} \end{bmatrix}, \qquad B = \begin{bmatrix} b_{11} & b_{21} & b_{31} \\ b_{12} & b_{22} & b_{32} \\ b_{13} & b_{23} & b_{33} \end{bmatrix}, \qquad M = \begin{bmatrix} A & 0 \\ 0 & B \end{bmatrix}$$

The block diagonal matrix M results from the fact that $\{u_1, u_2, w_1, w_2, w_3\}$ is a basis of V (Theorem 10.4), and that $T(u_i) = T_1(u_i)$ and $T(w_j) = T_2(w_j)$.

A generalization of the above argument gives us the following theorem.

THEOREM 10.5: Suppose $T: V \to V$ is linear and suppose V is the direct sum of T-invariant subspaces, say, $W_1, \ldots, W_r$. If A_i is a matrix representation of the restriction of T to W_i, then T can be represented by the block diagonal matrix:

$$M = \operatorname{diag}(A_1, A_2, \ldots, A_r)$$

10.5 Primary Decomposition

The following theorem shows that any operator $T: V \to V$ is decomposable into operators whose minimum polynomials are powers of irreducible polynomials. This is the first step in obtaining a canonical form for T.

THEOREM 10.6: (Primary Decomposition Theorem) Let $T:V \to V$ be a linear operator with minimal polynomial

$$m(t) = f_1(t)^{n_1} f_2(t)^{n_2} \cdots f_r(t)^{n_r}$$

where the $f_i(t)$ are distinct monic irreducible polynomials. Then V is the direct sum of T-invariant subspaces $W_1, \ldots, W_r$, where W_i is the kernel of $f_i(T)^{n_i}$. Moreover, $f_i(t)^{n_i}$ is the minimal polynomial of the restriction of T to W_i.

The above polynomials $f_i(t)^{n_i}$ are relatively prime. Therefore, the above fundamental theorem follows (Problem 10.11) from the next two theorems (proved in Problems 10.9 and 10.10, respectively).

THEOREM 10.7: Suppose $T:V \to V$ is linear, and suppose $f(t) = g(t)h(t)$ are polynomials such that $f(T) = \mathbf{0}$ and $g(t)$ and $h(t)$ are relatively prime. Then V is the direct sum of the T-invariant subspace U and W, where $U = \operatorname{Ker} g(T)$ and $W = \operatorname{Ker} h(T)$.

THEOREM 10.8: In Theorem 10.7, if $f(t)$ is the minimal polynomial of T [and $g(t)$ and $h(t)$ are monic], then $g(t)$ and $h(t)$ are the minimal polynomials of the restrictions of T to U and W, respectively.

We will also use the primary decomposition theorem to prove the following useful characterization of diagonalizable operators (see Problem 10.12 for the proof).

THEOREM 10.9: A linear operator $T:V \to V$ is diagonalizable if and only if its minimal polynomial $m(t)$ is a product of distinct linear polynomials.

THEOREM 10.9: (Alternative Form) A matrix A is similar to a diagonal matrix if and only if its minimal polynomial is a product of distinct linear polynomials.

EXAMPLE 10.3 Suppose $A \neq I$ is a square matrix for which $A^3 = I$. Determine whether or not A is similar to a diagonal matrix if A is a matrix over: (i) the real field $\mathbf{R}$, (ii) the complex field $\mathbf{C}$.

Because $A^3 = I$, A is a zero of the polynomial $f(t) = t^3 - 1 = (t-1)(t^2 + t + 1)$. The minimal polynomial $m(t)$ of A cannot be $t - 1$, because $A \neq I$. Hence,

$$m(t) = t^2 + t + 1 \qquad \text{or} \qquad m(t) = t^3 - 1$$

Because neither polynomial is a product of linear polynomials over $\mathbf{R}$, A is not diagonalizable over $\mathbf{R}$. On the other hand, each of the polynomials is a product of distinct linear polynomials over $\mathbf{C}$. Hence, A is diagonalizable over $\mathbf{C}$.

10.6 Nilpotent Operators

A linear operator $T:V \to V$ is termed *nilpotent* if $T^n = \mathbf{0}$ for some positive integer n; we call k the *index of nilpotency* of T if $T^k = \mathbf{0}$ but $T^{k-1} \neq \mathbf{0}$. Analogously, a square matrix A is termed nilpotent if $A^n = 0$ for some positive integer n, and of index k if $A^k = 0$ but $A^{k-1} \neq 0$. Clearly the minimum polynomial of a nilpotent operator (matrix) of index k is $m(t) = t^k$; hence, 0 is its only eigenvalue.

EXAMPLE 10.4 The following two r-square matrices will be used throughout the chapter:

$$N = N(r) = \begin{bmatrix} 0 & 1 & 0 & \ldots & 0 & 0 \\ 0 & 0 & 1 & \ldots & 0 & 0 \\ \ldots & \ldots & \ldots & \ldots & \ldots & \ldots \\ 0 & 0 & 0 & \ldots & 0 & 1 \\ 0 & 0 & 0 & \ldots & 0 & 0 \end{bmatrix} \qquad \text{and} \qquad J(\lambda) = \begin{bmatrix} \lambda & 1 & 0 & \ldots & 0 & 0 \\ 0 & \lambda & 1 & \ldots & 0 & 0 \\ \ldots & \ldots & \ldots & \ldots & \ldots & \ldots \\ 0 & 0 & 0 & \ldots & \lambda & 1 \\ 0 & 0 & 0 & \ldots & 0 & \lambda \end{bmatrix}$$

The first matrix N, called a *Jordan nilpotent block*, consists of 1's above the diagonal (called the *superdiagonal*), and 0's elsewhere. It is a nilpotent matrix of index r. (The matrix N of order 1 is just the 1×1 zero matrix [0].)

The second matrix $J(\lambda)$, called a *Jordan block* belonging to the eigenvalue λ, consists of λ's on the diagonal, 1's on the superdiagonal, and 0's elsewhere. Observe that

$$J(\lambda) = \lambda I + N$$

In fact, we will prove that any linear operator T can be decomposed into operators, each of which is the sum of a scalar operator and a nilpotent operator.

The following (proved in Problem 10.16) is a fundamental result on nilpotent operators.

THEOREM 10.10: Let $T{:}V \to V$ be a nilpotent operator of index k. Then T has a block diagonal matrix representation in which each diagonal entry is a Jordan nilpotent block N. There is at least one N of order k, and all other N are of orders $\leq k$. The number of N of each possible order is uniquely determined by T. The total number of N of all orders is equal to the nullity of T.

The proof of Theorem 10.10 shows that the number of N of order i is equal to $2m_i - m_{i+1} - m_{i-1}$, where m_i is the nullity of T^i.

10.7 Jordan Canonical Form

An operator T can be put into Jordan canonical form if its characteristic and minimal polynomials factor into linear polynomials. This is always true if K is the complex field **C**. In any case, we can always extend the base field K to a field in which the characteristic and minimal polynomials do factor into linear factors; thus, in a broad sense, every operator has a Jordan canonical form. Analogously, every matrix is similar to a matrix in Jordan canonical form.

The following theorem (proved in Problem 10.18) describes the *Jordan canonical form* J of a linear operator T.

THEOREM 10.11: Let $T{:}V \to V$ be a linear operator whose characteristic and minimal polynomials are, respectively,

$$\Delta(t) = (t-\lambda_1)^{n_1} \cdots (t-\lambda_r)^{n_r} \quad \text{and} \quad m(t) = (t-\lambda_1)^{m_1} \cdots (t-\lambda_r)^{m_r}$$

where the λ_i are distinct scalars. Then T has a block diagonal matrix representation J in which each diagonal entry is a Jordan block $J_{ij} = J(\lambda_i)$. For each λ_{ij}, the corresponding J_{ij} have the following properties:

(i) There is at least one J_{ij} of order m_i; all other J_{ij} are of order $\leq m_i$.

(ii) The sum of the orders of the J_{ij} is n_i.

(iii) The number of J_{ij} equals the geometric multiplicity of λ_i.

(iv) The number of J_{ij} of each possible order is uniquely determined by T.

EXAMPLE 10.5 Suppose the characteristic and minimal polynomials of an operator T are, respectively,

$$\Delta(t) = (t-2)^4(t-5)^3 \quad \text{and} \quad m(t) = (t-2)^2(t-5)^3$$

Then the Jordan canonical form of T is one of the following block diagonal matrices:

$$\operatorname{diag}\left(\begin{bmatrix}2 & 1\\ 0 & 2\end{bmatrix}, \begin{bmatrix}2 & 1\\ 0 & 2\end{bmatrix}, \begin{bmatrix}5 & 1 & 0\\ 0 & 5 & 1\\ 0 & 0 & 5\end{bmatrix}\right) \quad \text{or} \quad \operatorname{diag}\left(\begin{bmatrix}2 & 1\\ 0 & 2\end{bmatrix}, [2], [2], \begin{bmatrix}5 & 1 & 0\\ 0 & 5 & 1\\ 0 & 0 & 5\end{bmatrix}\right)$$

The first matrix occurs if T has two independent eigenvectors belonging to the eigenvalue 2; and the second matrix occurs if T has three independent eigenvectors belonging to the eigenvalue 2.

10.8 Cyclic Subspaces

Let T be a linear operator on a vector space V of finite dimension over K. Suppose $v \in V$ and $v \neq 0$. The set of all vectors of the form $f(T)(v)$, where $f(t)$ ranges over all polynomials over K, is a T-invariant subspace of V called the *T-cyclic subspace of V generated by v*; we denote it by $Z(v, T)$ and denote the restriction of T to $Z(v, T)$ by T_v. By Problem 10.56, we could equivalently define $Z(v, T)$ as the intersection of all T-invariant subspaces of V containing v.

Now consider the sequence

$$v, \quad T(v), \quad T^2(v), \quad T^3(v), \quad \ldots$$

of powers of T acting on v. Let k be the least integer such that $T^k(v)$ is a linear combination of those vectors that precede it in the sequence, say,

$$T^k(v) = -a_{k-1}T^{k-1}(v) - \cdots - a_1 T(v) - a_0 v$$

Then

$$m_v(t) = t^k + a_{k-1}t^{k-1} + \cdots + a_1 t + a_0$$

is the unique monic polynomial of lowest degree for which $m_v(T)(v) = 0$. We call $m_v(t)$ the *T-annihilator of v and $Z(v, T)$*.

The following theorem (proved in Problem 10.29) holds.

THEOREM 10.12: Let $Z(v, T)$, T_v, $m_v(t)$ be defined as above. Then

(i) The set $\{v, T(v), \ldots, T^{k-1}(v)\}$ is a basis of $Z(v, T)$; hence, $\dim Z(v, T) = k$.

(ii) The minimal polynomial of T_v is $m_v(t)$.

(iii) The matrix representation of T_v in the above basis is just the *companion matrix* $C(m_v)$ of $m_v(t)$; that is,

$$C(m_v) = \begin{bmatrix} 0 & 0 & 0 & & 0 & -a_0 \\ 1 & 0 & 0 & & 0 & -a_1 \\ 0 & 1 & 0 & & 0 & -a_2 \\ \cdots & \cdots & \cdots & \cdots & \cdots & \cdots \\ 0 & 0 & 0 & \ldots & 0 & -a_{k-2} \\ 0 & 0 & 0 & \ldots & 1 & -a_{k-1} \end{bmatrix}$$

10.9 Rational Canonical Form

In this section, we present the rational canonical form for a linear operator $T: V \to V$. We emphasize that this form exists even when the minimal polynomial cannot be factored into linear polynomials. (Recall that this is not the case for the Jordan canonical form.)

LEMMA 10.13: Let $T:V \to V$ be a linear operator whose minimal polynomial is $f(t)^n$, where $f(t)$ is a monic irreducible polynomial. Then V is the direct sum

$$V = Z(v_1, T) \oplus \cdots \oplus Z(v_r, T)$$

of T-cyclic subspaces $Z(v_i, T)$ with corresponding T-annihilators

$$f(t)^{n_1}, \ f(t)^{n_2}, \ \ldots, \ f(t)^{n_r}, \qquad n = n_1 \geq n_2 \geq \ldots \geq n_r$$

Any other decomposition of V into T-cyclic subspaces has the same number of components and the same set of T-annihilators.

We emphasize that the above lemma (proved in Problem 10.31) does not say that the vectors v_i or other T-cyclic subspaces $Z(v_i, T)$ are uniquely determined by T, but it does say that the set of T-annihilators is uniquely determined by T. Thus, T has a unique block diagonal matrix representation:

$$M = \operatorname{diag}(C_1, C_2, \ldots, C_r)$$

where the C_i are companion matrices. In fact, the C_i are the companion matrices of the polynomials $f(t)^{n_i}$.

Using the Primary Decomposition Theorem and Lemma 10.13, we obtain the following result.

THEOREM 10.14: Let $T:V \to V$ be a linear operator with minimal polynomial

$$m(t) = f_1(t)^{m_1} f_2(t)^{m_2} \cdots f_s(t)^{m_s}$$

where the $f_i(t)$ are distinct monic irreducible polynomials. Then T has a unique block diagonal matrix representation:

$$M = \operatorname{diag}(C_{11}, C_{12}, \ldots, C_{1r_1}, \ldots, C_{s1}, C_{s2}, \ldots, C_{sr_s})$$

where the C_{ij} are companion matrices. In particular, the C_{ij} are the companion matrices of the polynomials $f_i(t)^{n_{ij}}$, where

$$m_1 = n_{11} \geq n_{12} \geq \cdots \geq n_{1r_1}, \qquad \ldots, \qquad m_s = n_{s1} \geq n_{s2} \geq \cdots \geq n_{sr_s}$$

The above matrix representation of T is called its *rational canonical form*. The polynomials $f_i(t)^{n_{ij}}$ are called the *elementary divisors* of T.

EXAMPLE 10.6 Let V be a vector space of dimension 8 over the rational field $\mathbf{Q}$, and let T be a linear operator on V whose minimal polynomial is

$$m(t) = f_1(t) f_2(t)^2 = (t^4 - 4t^3 + 6t^2 - 4t - 7)(t - 3)^2$$

Thus, because $\dim V = 8$, the characteristic polynomial $\Delta(t) = f_1(t) f_2(t)^4$. Also, the rational canonical form M of T must have one block the companion matrix of $f_1(t)$ and one block the companion matrix of $f_2(t)^2$. There are two possibilities:

(a) $\operatorname{diag}[C(t^4 - 4t^3 + 6t^2 - 4t - 7), \quad C((t-3)^2), \quad C((t-3)^2)]$

(b) $\operatorname{diag}[C(t^4 - 4t^3 + 6t^2 - 4t - 7), \quad C((t-3)^2), \quad C(t-3), C(t-3)]$

That is,

$$\text{(a) diag}\left(\begin{bmatrix} 0 & 0 & 0 & 7 \\ 1 & 0 & 0 & 4 \\ 0 & 1 & 0 & -6 \\ 0 & 0 & 1 & 4 \end{bmatrix}, \begin{bmatrix} 0 & -9 \\ 1 & 6 \end{bmatrix}, \begin{bmatrix} 0 & -9 \\ 1 & 6 \end{bmatrix} \right), \quad \text{(b) diag}\left(\begin{bmatrix} 0 & 0 & 0 & 7 \\ 1 & 0 & 0 & 4 \\ 0 & 1 & 0 & -6 \\ 0 & 0 & 1 & 4 \end{bmatrix}, \begin{bmatrix} 0 & -9 \\ 1 & 6 \end{bmatrix}, [3], [3] \right)$$

10.10 Quotient Spaces

Let V be a vector space over a field K and let W be a subspace of V. If v is any vector in V, we write $v + W$ for the set of sums $v + w$ with $w \in W$; that is,

$$v + W = \{v + w : w \in W\}$$

These sets are called the *cosets* of W in V. We show (Problem 10.22) that these cosets partition V into mutually disjoint subsets.

EXAMPLE 10.7 Let W be the subspace of $\mathbf{R}^2$ defined by

$$W = \{(a, b) : a = b\},$$

that is, W is the line given by the equation $x - y = 0$. We can view $v + W$ as a translation of the line obtained by adding the vector v to each point in W. As shown in Fig. 10-2, the coset $v + W$ is also a line, and it is parallel to W. Thus, the cosets of W in $\mathbf{R}^2$ are precisely all the lines parallel to W.

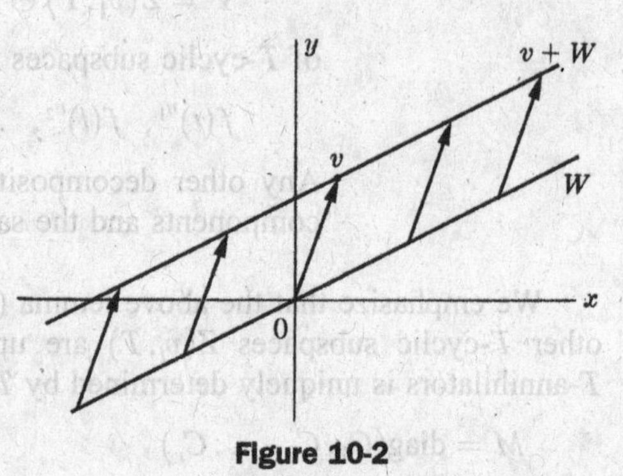

Figure 10-2

In the following theorem, we use the cosets of a subspace W of a vector space V to define a new vector space; it is called the *quotient space* of V by W and is denoted by V/W.

THEOREM 10.15: Let W be a subspace of a vector space over a field K. Then the cosets of W in V form a vector space over K with the following operations of addition and scalar multiplication:

$$\text{(i) } (u + w) + (v + W) = (u + v) + W, \quad \text{(ii) } k(u + W) = ku + W, \quad \text{where } k \in K$$

We note that, in the proof of Theorem 10.15 (Problem 10.24), it is first necessary to show that the operations are well defined; that is, whenever $u + W = u' + W$ and $v + W = v' + W$, then

$$\text{(i) } (u + v) + W = (u' + v') + W \quad \text{and} \quad \text{(ii) } ku + W = ku' + W \quad \text{for any } k \in K$$

In the case of an invariant subspace, we have the following useful result (proved in Problem 10.27).

THEOREM 10.16: Suppose W is a subspace invariant under a linear operator $T:V \to V$. Then T induces a linear operator $\bar{T}$ on V/W defined by $\bar{T}(v + W) = T(v) + W$. Moreover, if T is a zero of any polynomial, then so is $\bar{T}$. Thus, the minimal polynomial of $\bar{T}$ divides the minimal polynomial of T.

SOLVED PROBLEMS

Invariant Subspaces

10.1. Suppose $T:V \to V$ is linear. Show that each of the following is invariant under T:

(a) $\{0\}$, (b) V, (c) kernel of T, (d) image of T.

(a) We have $T(0) = 0 \in \{0\}$; hence, $\{0\}$ is invariant under T.

(b) For every $v \in V$, $T(v) \in V$; hence, V is invariant under T.

(c) Let $u \in \text{Ker } T$. Then $T(u) = 0 \in \text{Ker } T$ because the kernel of T is a subspace of V. Thus, Ker T is invariant under T.

(d) Because $T(v) \in \text{Im } T$ for every $v \in V$, it is certainly true when $v \in \text{Im } T$. Hence, the image of T is invariant under T.

10.2. Suppose $\{W_i\}$ is a collection of T-invariant subspaces of a vector space V. Show that the intersection $W = \bigcap_i W_i$ is also T-invariant.

Suppose $v \in W$; then $v \in W_i$ for every i. Because W_i is T-invariant, $T(v) \in W_i$ for every i. Thus, $T(v) \in W$ and so W is T-invariant.

10.3. Prove Theorem 10.2: Let $T:V \to V$ be linear. For any polynomial $f(t)$, the kernel of $f(T)$ is invariant under T.

Suppose $v \in \text{Ker} f(T)$—that is, $f(T)(v) = 0$. We need to show that $T(v)$ also belongs to the kernel of $f(T)$—that is, $f(T)(T(v)) = (f(T) \circ T)(v) = 0$. Because $f(t)t = tf(t)$, we have $f(T) \circ T = T \circ f(T)$. Thus, as required,

$$(f(T) \circ T)(v) = (T \circ f(T))(v) = T(f(T)(v)) = T(0) = 0$$

10.4. Find all invariant subspaces of $A = \begin{bmatrix} 2 & -5 \\ 1 & -2 \end{bmatrix}$ viewed as an operator on $\mathbf{R}^2$.

By Problem 10.1, $\mathbf{R}^2$ and $\{0\}$ are invariant under A. Now if A has any other invariant subspace, it must be one-dimensional. However, the characteristic polynomial of A is

$$\Delta(t) = t^2 - \text{tr}(A)\, t + |A| = t^2 + 1$$

Hence, A has no eigenvalues (in $\mathbf{R}$) and so A has no eigenvectors. But the one-dimensional invariant subspaces correspond to the eigenvectors; thus, $\mathbf{R}^2$ and $\{0\}$ are the only subspaces invariant under A.

10.5. Prove Theorem 10.3: Suppose W is T-invariant. Then T has a triangular block representation $\begin{bmatrix} A & B \\ 0 & C \end{bmatrix}$, where A is the matrix representation of the restriction $\hat{T}$ of T to W.

We choose a basis $\{w_1, \ldots, w_r\}$ of W and extend it to a basis $\{w_1, \ldots, w_r, v_1, \ldots, v_s\}$ of V. We have

$$\begin{aligned}
\hat{T}(w_1) &= T(w_1) = a_{11}w_1 + \cdots + a_{1r}w_r \\
\hat{T}(w_2) &= T(w_2) = a_{21}w_1 + \cdots + a_{2r}w_r \\
&\cdots\cdots\cdots\cdots\cdots\cdots\cdots\cdots\cdots\cdots \\
\hat{T}(w_r) &= T(w_r) = a_{r1}w_1 + \cdots + a_{rr}w_r \\
T(v_1) &= b_{11}w_1 + \cdots + b_{1r}w_r + c_{11}v_1 + \cdots + c_{1s}v_s \\
T(v_2) &= b_{21}w_1 + \cdots + b_{2r}w_r + c_{21}v_1 + \cdots + c_{2s}v_s \\
&\cdots\cdots\cdots\cdots\cdots\cdots\cdots\cdots\cdots\cdots\cdots\cdots\cdots\cdots \\
T(v_s) &= b_{s1}w_1 + \cdots + b_{sr}w_r + c_{s1}v_1 + \cdots + c_{ss}v_s
\end{aligned}$$

But the matrix of T in this basis is the transpose of the matrix of coefficients in the above system of equations (Section 6.2). Therefore, it has the form $\begin{bmatrix} A & B \\ 0 & C \end{bmatrix}$, where A is the transpose of the matrix of coefficients for the obvious subsystem. By the same argument, A is the matrix of $\hat{T}$ relative to the basis $\{w_i\}$ of W.

10.6. Let $\hat{T}$ denote the restriction of an operator T to an invariant subspace W. Prove

(a) For any polynomial $f(t)$, $f(\hat{T})(w) = f(T)(w)$.

(b) The minimal polynomial of $\hat{T}$ divides the minimal polynomial of T.

(a) If $f(t) = 0$ or if $f(t)$ is a constant (i.e., of degree 1), then the result clearly holds. Assume $\deg f = n > 1$ and that the result holds for polynomials of degree less than n. Suppose that

$$f(t) = a_n t^n + a_{n-1}t^{n-1} + \cdots + a_1 t + a_0$$

Then

$$\begin{aligned}
f(\hat{T})(w) &= (a_n\hat{T}^n + a_{n-1}\hat{T}^{n-1} + \cdots + a_0 I)(w) \\
&= (a_n\hat{T}^{n-1})(\hat{T}(w)) + (a_{n-1}\hat{T}^{n-1} + \cdots + a_0 I)(w) \\
&= (a_n T^{n-1})(T(w)) + (a_{n-1}T^{n-1} + \cdots + a_0 I)(w) = f(T)(w)
\end{aligned}$$

(b) Let $m(t)$ denote the minimal polynomial of T. Then by (a), $m(\hat{T})(w) = m(T)(w) = \mathbf{0}(w) = 0$ for every $w \in W$; that is, $\hat{T}$ is a zero of the polynomial $m(t)$. Hence, the minimal polynomial of $\hat{T}$ divides $m(t)$.

Invariant Direct-Sum Decompositions

10.7. Prove Theorem 10.4: Suppose $W_1, W_2, \ldots, W_r$ are subspaces of V with respective bases

$$B_1 = \{w_{11}, w_{12}, \ldots, w_{1n_1}\}, \quad \ldots, \quad B_r = \{w_{r1}, w_{r2}, \ldots, w_{rn_r}\}$$

Then V is the direct sum of the W_i if and only if the union $B = \bigcup_i B_i$ is a basis of V.

Suppose B is a basis of V. Then, for any $v \in V$,

$$v = a_{11}w_{11} + \cdots + a_{1n_1}w_{1n_1} + \cdots + a_{r1}w_{r1} + \cdots + a_{rn_r}w_{rn_r} = w_1 + w_2 + \cdots + w_r$$

where $w_i = a_{i1}w_{i1} + \cdots + a_{in_i}w_{in_i} \in W_i$. We next show that such a sum is unique. Suppose

$$v = w_1' + w_2' + \cdots + w_r', \quad \text{where} \quad w_i' \in W_i$$

Because $\{w_{i1}, \ldots, w_{in_i}\}$ is a basis of W_i, $w_i' = b_{i1}w_{i1} + \cdots + b_{in_i}w_{in_i}$, and so

$$v = b_{11}w_{11} + \cdots + b_{1n_1}w_{1n_1} + \cdots + b_{r1}w_{r1} + \cdots + b_{rn_r}w_{rn_r}$$

Because B is a basis of V, $a_{ij} = b_{ij}$, for each i and each j. Hence, $w_i = w_i'$, and so the sum for v is unique. Accordingly, V is the direct sum of the W_i.

Conversely, suppose V is the direct sum of the W_i. Then for any $v \in V$, $v = w_1 + \cdots + w_r$, where $w_i \in W_i$. Because $\{w_{ij_i}\}$ is a basis of W_i, each w_i is a linear combination of the w_{ij_i}, and so v is a linear combination of the elements of B. Thus, B spans V. We now show that B is linearly independent. Suppose

$$a_{11}w_{11} + \cdots + a_{1n_1}w_{1n_1} + \cdots + a_{r1}w_{r1} + \cdots + a_{rn_r}w_{rn_r} = 0$$

Note that $a_{i1}w_{i1} + \cdots + a_{in_i}w_{in_i} \in W_i$. We also have that $0 = 0 + 0 \cdots 0 \in W_i$. Because such a sum for 0 is unique,

$$a_{i1}w_{i1} + \cdots + a_{in_i}w_{in_i} = 0 \quad \text{for } i = 1, \ldots, r$$

The independence of the bases $\{w_{ij_i}\}$ implies that all the a's are 0. Thus, B is linearly independent and is a basis of V.

10.8. Suppose $T:V \to V$ is linear and suppose $T = T_1 \oplus T_2$ with respect to a T-invariant direct-sum decomposition $V = U \oplus W$. Show that

(a) $m(t)$ is the least common multiple of $m_1(t)$ and $m_2(t)$, where $m(t)$, $m_1(t)$, $m_2(t)$ are the minimum polynomials of T, T_1, T_2, respectively.

(b) $\Delta(t) = \Delta_1(t)\Delta_2(t)$, where $\Delta(t), \Delta_1(t), \Delta_2(t)$ are the characteristic polynomials of T, T_1, T_2, respectively.

(a) By Problem 10.6, each of $m_1(t)$ and $m_2(t)$ divides $m(t)$. Now suppose $f(t)$ is a multiple of both $m_1(t)$ and $m_2(t)$, then $f(T_1)(U) = 0$ and $f(T_2)(W) = 0$. Let $v \in V$, then $v = u + w$ with $u \in U$ and $w \in W$. Now

$$f(T)v = f(T)u + f(T)w = f(T_1)u + f(T_2)w = 0 + 0 = 0$$

That is, T is a zero of $f(t)$. Hence, $m(t)$ divides $f(t)$, and so $m(t)$ is the least common multiple of $m_1(t)$ and $m_2(t)$.

(b) By Theorem 10.5, T has a matrix representation $M = \begin{bmatrix} A & 0 \\ 0 & B \end{bmatrix}$, where A and B are matrix representations of T_1 and T_2, respectively. Then, as required,

$$\Delta(t) = |tI - M| = \begin{vmatrix} tI - A & 0 \\ 0 & tI - B \end{vmatrix} = |tI - A||tI - B| = \Delta_1(t)\Delta_2(t)$$

10.9. Prove Theorem 10.7: Suppose $T:V \to V$ is linear, and suppose $f(t) = g(t)h(t)$ are polynomials such that $f(T) = \mathbf{0}$ and $g(t)$ and $h(t)$ are relatively prime. Then V is the direct sum of the T-invariant subspaces U and W where $U = \text{Ker}\, g(T)$ and $W = \text{Ker}\, h(T)$.

Note first that U and W are T-invariant by Theorem 10.2. Now, because $g(t)$ and $h(t)$ are relatively prime, there exist polynomials $r(t)$ and $s(t)$ such that

$$r(t)g(t) + s(t)h(t) = 1$$

Hence, for the operator T, $$r(T)g(T) + s(T)h(T) = I \qquad (*)$$

Let $v \in V$; then, by (*), $$v = r(T)g(T)v + s(T)h(T)v$$

But the first term in this sum belongs to $W = \text{Ker } h(T)$, because

$$h(T)r(T)g(T)v = r(T)g(T)h(T)v = r(T)f(T)v = r(T)\mathbf{0}v = 0$$

Similarly, the second term belongs to U. Hence, V is the sum of U and W.

To prove that $V = U \oplus W$, we must show that a sum $v = u + w$ with $u \in U$, $w \in W$, is uniquely determined by v. Applying the operator $r(T)g(T)$ to $v = u + w$ and using $g(T)u = 0$, we obtain

$$r(T)g(T)v = r(T)g(T)u + r(T)g(T)w = r(T)g(T)w$$

Also, applying (*) to w alone and using $h(T)w = 0$, we obtain

$$w = r(T)g(T)w + s(T)h(T)w = r(T)g(T)w$$

Both of the above formulas give us $w = r(T)g(T)v$, and so w is uniquely determined by v. Similarly u is uniquely determined by v. Hence, $V = U \oplus W$, as required.

10.10. Prove Theorem 10.8: In Theorem 10.7 (Problem 10.9), if $f(t)$ is the minimal polynomial of T (and $g(t)$ and $h(t)$ are monic), then $g(t)$ is the minimal polynomial of the restriction T_1 of T to U and $h(t)$ is the minimal polynomial of the restriction T_2 of T to W.

Let $m_1(t)$ and $m_2(t)$ be the minimal polynomials of T_1 and T_2, respectively. Note that $g(T_1) = 0$ and $h(T_2) = 0$ because $U = \text{Ker } g(T)$ and $W = \text{Ker } h(T)$. Thus,

$$m_1(t) \text{ divides } g(t) \qquad \text{and} \qquad m_2(t) \text{ divides } h(t) \qquad (1)$$

By Problem 10.9, $f(t)$ is the least common multiple of $m_1(t)$ and $m_2(t)$. But $m_1(t)$ and $m_2(t)$ are relatively prime because $g(t)$ and $h(t)$ are relatively prime. Accordingly, $f(t) = m_1(t)m_2(t)$. We also have that $f(t) = g(t)h(t)$. These two equations together with (1) and the fact that all the polynomials are monic imply that $g(t) = m_1(t)$ and $h(t) = m_2(t)$, as required.

10.11. Prove the Primary Decomposition Theorem 10.6: Let $T:V \to V$ be a linear operator with minimal polynomial

$$m(t) = f_1(t)^{n_1} f_2(t)^{n_2} \ldots f_r(t)^{n_r}$$

where the $f_i(t)$ are distinct monic irreducible polynomials. Then V is the direct sum of T-invariant subspaces $W_1, \ldots, W_r$ where W_i is the kernel of $f_i(T)^{n_i}$. Moreover, $f_i(t)^{n_i}$ is the minimal polynomial of the restriction of T to W_i.

The proof is by induction on r. The case $r = 1$ is trivial. Suppose that the theorem has been proved for $r - 1$. By Theorem 10.7, we can write V as the direct sum of T-invariant subspaces W_1 and V_1, where W_1 is the kernel of $f_1(T)^{n_1}$ and where V_1 is the kernel of $f_2(T)^{n_2} \cdots f_r(T)^{n_r}$. By Theorem 10.8, the minimal polynomials of the restrictions of T to W_1 and V_1 are $f_1(t)^{n_1}$ and $f_2(t)^{n_2} \cdots f_r(t)^{n_r}$, respectively.

Denote the restriction of T to V_1 by $\hat{T}_1$. By the inductive hypothesis, V_1 is the direct sum of subspaces $W_2, \ldots, W_r$ such that W_i is the kernel of $f_i(T_1)^{n_i}$ and such that $f_i(t)^{n_i}$ is the minimal polynomial for the restriction of $\hat{T}_1$ to W_i. But the kernel of $f_i(T)^{n_i}$, for $i = 2, \ldots, r$ is necessarily contained in V_1, because $f_i(t)^{n_i}$ divides $f_2(t)^{n_2} \cdots f_r(t)^{n_r}$. Thus, the kernel of $f_i(T)^{n_i}$ is the same as the kernel of $f_i(T_1)^{n_i}$, which is W_i. Also, the restriction of T to W_i is the same as the restriction of $\hat{T}_1$ to W_i (for $i = 2, \ldots, r$); hence, $f_i(t)^{n_i}$ is also the minimal polynomial for the restriction of T to W_i. Thus, $V = W_1 \oplus W_2 \oplus \cdots \oplus W_r$ is the desired decomposition of T.

10.12. Prove Theorem 10.9: A linear operator $T:V \to V$ has a diagonal matrix representation if and only if its minimal polynomal $m(t)$ is a product of distinct linear polynomials.

Suppose $m(t)$ is a product of distinct linear polynomials, say,

$$m(t) = (t - \lambda_1)(t - \lambda_2) \cdots (t - \lambda_r)$$

where the λ_i are distinct scalars. By the Primary Decomposition Theorem, V is the direct sum of subspaces $W_1, \ldots, W_r$, where $W_i = \text{Ker}(T - \lambda_i I)$. Thus, if $v \in W_i$, then $(T - \lambda_i I)(v) = 0$ or $T(v) = \lambda_i v$. In other words, every vector in W_i is an eigenvector belonging to the eigenvalue λ_i. By Theorem 10.4, the union of bases for $W_1, \ldots, W_r$ is a basis of V. This basis consists of eigenvectors, and so T is diagonalizable.

Conversely, suppose T is diagonalizable (i.e., V has a basis consisting of eigenvectors of T). Let $\lambda_1, \ldots, \lambda_s$ be the distinct eigenvalues of T. Then the operator

$$f(T) = (T - \lambda_1 I)(T - \lambda_2 I) \cdots (T - \lambda_s I)$$

maps each basis vector into 0. Thus, $f(T) = 0$, and hence, the minimal polynomial $m(t)$ of T divides the polynomial

$$f(t) = (t - \lambda_1)(t - \lambda_2) \cdots (t - \lambda_s I)$$

Accordingly, $m(t)$ is a product of distinct linear polynomials.

Nilpotent Operators, Jordan Canonical Form

10.13. Let $T:V$ be linear. Suppose, for $v \in V$, $T^k(v) = 0$ but $T^{k-1}(v) \neq 0$. Prove

(a) The set $S = \{v, T(v), \ldots, T^{k-1}(v)\}$ is linearly independent.
(b) The subspace W generated by S is T-invariant.
(c) The restriction $\hat{T}$ of T to W is nilpotent of index k.
(d) Relative to the basis $\{T^{k-1}(v), \ldots, T(v), v\}$ of W, the matrix of T is the k-square Jordan nilpotent block N_k of index k (see Example 10.5).

(a) Suppose

$$av + a_1 T(v) + a_2 T^2(v) + \cdots + a_{k-1} T^{k-1}(v) = 0 \qquad (*)$$

Applying T^{k-1} to (*) and using $T^k(v) = 0$, we obtain $aT^{k-1}(v) = 0$; because $T^{k-1}(v) \neq 0$, $a = 0$. Now applying T^{k-2} to (*) and using $T^k(v) = 0$ and $a = 0$, we fiind $a_1 T^{k-1}(v) = 0$; hence, $a_1 = 0$. Next applying T^{k-3} to (*) and using $T^k(v) = 0$ and $a = a_1 = 0$, we obtain $a_2 T^{k-1}(v) = 0$; hence, $a_2 = 0$. Continuing this process, we find that all the a's are 0; hence, S is independent.

(b) Let $v \in W$. Then

$$v = bv + b_1 T(v) + b_2 T^2(v) + \cdots + b_{k-1} T^{k-1}(v)$$

Using $T^k(v) = 0$, we have

$$T(v) = bT(v) + b_1 T^2(v) + \cdots + b_{k-2} T^{k-1}(v) \in W$$

Thus, W is T-invariant.

(c) By hypothesis, $T^k(v) = 0$. Hence, for $i = 0, \ldots, k-1$,

$$\hat{T}^k(T^i(v)) = T^{k+i}(v) = 0$$

That is, applying $\hat{T}^k$ to each generator of W, we obtain 0; hence, $\hat{T}^k = \mathbf{0}$ and so $\hat{T}$ is nilpotent of index at most k. On the other hand, $\hat{T}^{k-1}(v) = T^{k-1}(v) \neq 0$; hence, T is nilpotent of index exactly k.

(d) For the basis $\{T^{k-1}(v), T^{k-2}(v), \ldots, T(v), v\}$ of W,

$$\begin{aligned}
\hat{T}(T^{k-1}(v)) &= T^k(v) = 0 \\
\hat{T}(T^{k-2}(v)) &= T^{k-1}(v) \\
\hat{T}(T^{k-3}(v)) &= T^{k-2}(v) \\
&\cdots\cdots\cdots \\
\hat{T}(T(v)) &= T^2(v) \\
\hat{T}(v) &= T(v)
\end{aligned}$$

Hence, as required, the matrix of T in this basis is the k-square Jordan nilpotent block N_k.

10.14. Let $T:V \to V$ be linear. Let $U = \text{Ker } T^i$ and $W = \text{Ker } T^{i+1}$. Show that

(a) $U \subseteq W$, (b) $T(W) \subseteq U$.

(a) Suppose $u \in U = \text{Ker } T^i$. Then $T^i(u) = 0$ and so $T^{i+1}(u) = T(T^i(u)) = T(0) = 0$. Thus, $u \in \text{Ker } T^{i+1} = W$. But this is true for every $u \in U$; hence, $U \subseteq W$.

(b) Similarly, if $w \in W = \text{Ker } T^{i+1}$, then $T^{i+1}(w) = 0$. Thus, $T^{i+1}(w) = T^i(T(w)) = T^i(0) = 0$ and so $T(W) \subseteq U$.

10.15. Let $T:V$ be linear. Let $X = \text{Ker } T^{i-2}$, $Y = \text{Ker } T^{i-1}$, $Z = \text{Ker } T^i$. Therefore (Problem 10.14), $X \subseteq Y \subseteq Z$. Suppose

$$\{u_1, \dots, u_r\}, \qquad \{u_1, \dots, u_r, v_1, \dots, v_s\}, \qquad \{u_1, \dots, u_r, v_1, \dots, v_s, w_1, \dots, w_t\}$$

are bases of X, Y, Z, respectively. Show that

$$S = \{u_1, \dots, u_r, T(w_1), \dots, T(w_t)\}$$

is contained in Y and is linearly independent.

By Problem 10.14, $T(Z) \subseteq Y$, and hence $S \subseteq Y$. Now suppose S is linearly dependent. Then there exists a relation

$$a_1u_1 + \cdots + a_ru_r + b_1T(w_1) + \cdots + b_tT(w_t) = 0$$

where at least one coefficient is not zero. Furthermore, because $\{u_i\}$ is independent, at least one of the b_k must be nonzero. Transposing, we find

$$b_1T(w_1) + \cdots + b_tT(w_t) = -a_1u_1 - \cdots - a_ru_r \in X = \text{Ker } T^{i-2}$$

Hence, $$T^{i-2}(b_1T(w_1) + \cdots + b_tT(w_t)) = 0$$

Thus, $$T^{i-1}(b_1w_1 + \cdots + b_tw_t) = 0, \quad \text{and so} \quad b_1w_1 + \cdots + b_tw_t \in Y = \text{Ker } T^{i-1}$$

Because $\{u_i, v_j\}$ generates Y, we obtain a relation among the u_i, v_j, w_k where one of the coefficients (i.e., one of the b_k) is not zero. This contradicts the fact that $\{u_i, v_j, w_k\}$ is independent. Hence, S must also be independent.

10.16. Prove Theorem 10.10: Let $T:V \to V$ be a nilpotent operator of index k. Then T has a unique block diagonal matrix representation consisting of Jordan nilpotent blocks N. There is at least one N of order k, and all other N are of orders $\leq k$. The total number of N of all orders is equal to the nullity of T.

Suppose $\dim V = n$. Let $W_1 = \text{Ker } T$, $W_2 = \text{Ker } T^2, \dots, W_k = \text{Ker } T^k$. Let us set $m_i = \dim W_i$, for $i = 1, \dots, k$. Because T is of index k, $W_k = V$ and $W_{k-1} \neq V$ and so $m_{k-1} < m_k = n$. By Problem 10.14,

$$W_1 \subseteq W_2 \subseteq \cdots \subseteq W_k = V$$

Thus, by induction, we can choose a basis $\{u_1, \dots, u_n\}$ of V such that $\{u_1, \dots, u_{m_i}\}$ is a basis of W_i.

We now choose a new basis for V with respect to which T has the desired form. It will be convenient to label the members of this new basis by pairs of indices. We begin by setting

$$v(1,k) = u_{m_{k-1}+1}, \quad v(2,k) = u_{m_{k-1}+2}, \quad \dots, \quad v(m_k - m_{k-1}, k) = u_{m_k}$$

and setting

$$v(1,k-1) = Tv(1,k), \quad v(2,k-1) = Tv(2,k), \quad \dots, \quad v(m_k - m_{k-1}, k-1) = Tv(m_k - m_{k-1}, k)$$

By the preceding problem,

$$S_1 = \{u_1 \dots, u_{m_{k-2}}, v(1,k-1), \dots, v(m_k - m_{k-1}, k-1)\}$$

is a linearly independent subset of W_{k-1}. We extend S_1 to a basis of W_{k-1} by adjoining new elements (if necessary), which we denote by

$$v(m_k - m_{k-1} + 1,\ k-1), \quad v(m_k - m_{k-1} + 2,\ k-1), \quad \dots, \quad v(m_{k-1} - m_{k-2},\ k-1)$$

Next we set

$$v(1,k-2) = Tv(1,k-1), \quad v(2,k-2) = Tv(2,k-1), \quad \dots,$$
$$v(m_{k-1} - m_{k-2}, k-2) = Tv(m_{k-1} - m_{k-2}, k-1)$$

Again by the preceding problem,

$$S_2 = \{u_1, \ldots, u_{m_{k-2}}, v(1, k-2), \ldots, v(m_{k-1} - m_{k-2}, k-2)\}$$

is a linearly independent subset of W_{k-2}, which we can extend to a basis of W_{k-2} by adjoining elements

$$v(m_{k-1} - m_{k-2} + 1, k-2), \quad v(m_{k-1} - m_{k-2} + 2, k-2), \quad \ldots, \quad v(m_{k-2} - m_{k-3}, k-2)$$

Continuing in this manner, we get a new basis for V, which for convenient reference we arrange as follows:

$$\begin{array}{llll}
v(1,k) & \ldots, v(m_k - m_{k-1}, k) & & \\
v(1,k-1), & \ldots, v(m_k - m_{k-1}, k-1) & \ldots, v(m_{k-1} - m_{k-2}, k-1) & \\
\cdots & \cdots & \cdots & \\
v(1,2), & \ldots, v(m_k - m_{k-1}, 2), & \ldots, v(m_{k-1} - m_{k-2}, 2), & \ldots, v(m_2 - m_1, 2) \\
v(1,1), & \ldots, v(m_k - m_{k-1}, 1), & \ldots, v(m_{k-1} - m_{k-2}, 1), & \ldots, v(m_2 - m_1, 1), \quad \ldots, v(m_1, 1)
\end{array}$$

The bottom row forms a basis of W_1, the bottom two rows form a basis of W_2, and so forth. But what is important for us is that T maps each vector into the vector immediately below it in the table or into 0 if the vector is in the bottom row. That is,

$$Tv(i,j) = \begin{cases} v(i, j-1) & \text{for } j > 1 \\ 0 & \text{for } j = 1 \end{cases}$$

Now it is clear [see Problem 10.13(d)] that T will have the desired form if the $v(i,j)$ are ordered lexicographically: beginning with $v(1,1)$ and moving up the first column to $v(1,k)$, then jumping to $v(2,1)$ and moving up the second column as far as possible.

Moreover, there will be exactly $m_k - m_{k-1}$ diagonal entries of order k. Also, there will be

$$\begin{array}{lll}
(m_{k-1} - m_{k-2}) - (m_k - m_{k-1}) = & 2m_{k-1} - m_k - m_{k-2} & \text{diagonal entries of order } k-1 \\
\cdots & \cdots & \cdots \\
 & 2m_2 - m_1 - m_3 & \text{diagonal entries of order 2} \\
 & 2m_1 - m_2 & \text{diagonal entries of order 1}
\end{array}$$

as can be read off directly from the table. In particular, because the numbers $m_1, \ldots, m_k$ are uniquely determined by T, the number of diagonal entries of each order is uniquely determined by T. Finally, the identity

$$m_1 = (m_k - m_{k-1}) + (2m_{k-1} - m_k - m_{k-2}) + \cdots + (2m_2 - m_1 - m_3) + (2m_1 - m_2)$$

shows that the nullity m_1 of T is the total number of diagonal entries of T.

10.17. Let $A = \begin{bmatrix} 0 & 1 & 1 & 0 & 1 \\ 0 & 0 & 1 & 1 & 1 \\ 0 & 0 & 0 & 0 & 0 \\ 0 & 0 & 0 & 0 & 0 \\ 0 & 0 & 0 & 0 & 0 \end{bmatrix}$ and $B = \begin{bmatrix} 0 & 1 & 1 & 0 & 0 \\ 0 & 0 & 1 & 1 & 1 \\ 0 & 0 & 0 & 1 & 1 \\ 0 & 0 & 0 & 0 & 0 \\ 0 & 0 & 0 & 0 & 0 \end{bmatrix}$. The reader can verify that A and B are both nilpotent of index 3; that is, $A^3 = 0$ but $A^2 \neq 0$, and $B^3 = 0$ but $B^2 \neq 0$. Find the nilpotent matrices M_A and M_B in canonical form that are similar to A and B, respectively.

Because A and B are nilpotent of index 3, M_A and M_B must each contain a Jordan nilpotent block of order 3, and none greater then 3. Note that $\operatorname{rank}(A) = 2$ and $\operatorname{rank}(B) = 3$, so $\operatorname{nullity}(A) = 5 - 2 = 3$ and $\operatorname{nullity}(B) = 5 - 3 = 2$. Thus, M_A must contain three diagonal blocks, which must be one of order 3 and two of order 1; and M_B must contain two diagonal blocks, which must be one of order 3 and one of order 2. Namely,

$$M_A = \begin{bmatrix} 0 & 1 & 0 & 0 & 0 \\ 0 & 0 & 1 & 0 & 0 \\ 0 & 0 & 0 & 0 & 0 \\ 0 & 0 & 0 & 0 & 0 \\ 0 & 0 & 0 & 0 & 0 \end{bmatrix} \quad \text{and} \quad M_B = \begin{bmatrix} 0 & 1 & 0 & 0 & 0 \\ 0 & 0 & 1 & 0 & 0 \\ 0 & 0 & 0 & 0 & 0 \\ 0 & 0 & 0 & 0 & 1 \\ 0 & 0 & 0 & 0 & 0 \end{bmatrix}$$

10.18. Prove Theorem 10.11 on the Jordan canonical form for an operator T.

By the primary decomposition theorem, T is decomposable into operators $T_1, \ldots, T_r$; that is, $T = T_1 \oplus \cdots \oplus T_r$, where $(t - \lambda_i)^{m_i}$ is the minimal polynomial of T_i. Thus, in particular,

$$(T_1 - \lambda_1 I)^{m_1} = \mathbf{0}, \quad \ldots, \quad (T_r - \lambda_r I)^{m_r} = \mathbf{0}$$

Set $N_i = T_i - \lambda_i I$. Then, for $i = 1, \ldots, r$,

$$T_i = N_i + \lambda_i I, \qquad \text{where} \qquad N_i^{m^i} = \mathbf{0}$$

That is, T_i is the sum of the scalar operator $\lambda_i I$ and a nilpotent operator N_i, which is of index m_i because $(t - \lambda_i)_i^m$ is the minimal polynomial of T_i.

Now, by Theorem 10.10 on nilpotent operators, we can choose a basis so that N_i is in canonical form. In this basis, $T_i = N_i + \lambda_i I$ is represented by a block diagonal matrix M_i whose diagonal entries are the matrices J_{ij}. The direct sum J of the matrices M_i is in Jordan canonical form and, by Theorem 10.5, is a matrix representation of T.

Last, we must show that the blocks J_{ij} satisfy the required properties. Property (i) follows from the fact that N_i is of index m_i. Property (ii) is true because T and J have the same characteristic polynomial. Property (iii) is true because the nullity of $N_i = T_i - \lambda_i I$ is equal to the geometric multiplicity of the eigenvalue λ_i. Property (iv) follows from the fact that the T_i and hence the N_i are uniquely determined by T.

10.19. Determine all possible Jordan canonical forms J for a linear operator $T: V \to V$ whose characteristic polynomial $\Delta(t) = (t-2)^5$ and whose minimal polynomial $m(t) = (t-2)^2$.

J must be a 5×5 matrix, because $\Delta(t)$ has degree 5, and all diagonal elements must be 2, because 2 is the only eigenvalue. Moreover, because the exponent of $t - 2$ in $m(t)$ is 2, J must have one Jordan block of order 2, and the others must be of order 2 or 1. Thus, there are only two possibilities:

$$J = \operatorname{diag}\left(\begin{bmatrix} 2 & 1 \\ & 2 \end{bmatrix}, \begin{bmatrix} 2 & 1 \\ & 2 \end{bmatrix}, [2]\right) \qquad \text{or} \qquad J = \operatorname{diag}\left(\begin{bmatrix} 2 & 1 \\ & 2 \end{bmatrix}, [2], [2], [2]\right)$$

10.20. Determine all possible Jordan canonical forms for a linear operator $T: V \to V$ whose characteristic polynomial $\Delta(t) = (t-2)^3(t-5)^2$. In each case, find the minimal polynomial $m(t)$.

Because $t - 2$ has exponent 3 in $\Delta(t)$, 2 must appear three times on the diagonal. Similarly, 5 must appear twice. Thus, there are six possibilities:

$$\text{(a)}\ \operatorname{diag}\left(\begin{bmatrix} 2 & 1 & \\ & 2 & 1 \\ & & 2 \end{bmatrix}, \begin{bmatrix} 5 & 1 \\ & 5 \end{bmatrix}\right), \quad \text{(b)}\ \operatorname{diag}\left(\begin{bmatrix} 2 & 1 & \\ & 2 & 1 \\ & & 2 \end{bmatrix}, [5], [5]\right),$$

$$\text{(c)}\ \operatorname{diag}\left(\begin{bmatrix} 2 & 1 \\ & 2 \end{bmatrix}, [2], \begin{bmatrix} 5 & 1 \\ & 5 \end{bmatrix}\right), \quad \text{(d)}\ \operatorname{diag}\left(\begin{bmatrix} 2 & 1 \\ & 2 \end{bmatrix}, [2], [5], [5]\right),$$

$$\text{(e)}\ \operatorname{diag}\left([2], [2], [2], \begin{bmatrix} 5 & 1 \\ & 5 \end{bmatrix}\right), \quad \text{(f)}\ \operatorname{diag}([2], [2], [2], [5], [5])$$

The exponent in the minimal polynomial $m(t)$ is equal to the size of the largest block. Thus,

$$\text{(a)}\ m(t) = (t-2)^3(t-5)^2, \quad \text{(b)}\ m(t) = (t-2)^3(t-5), \quad \text{(c)}\ m(t) = (t-2)^2(t-5)^2,$$
$$\text{(d)}\ m(t) = (t-2)^2(t-5), \quad \text{(e)}\ m(t) = (t-2)(t-5)^2, \quad \text{(f)}\ m(t) = (t-2)(t-5)$$

Quotient Space and Triangular Form

10.21. Let W be a subspace of a vector space V. Show that the following are equivalent:

(i) $u \in v + W$, (ii) $u - v \in W$, (iii) $v \in u + W$.

Suppose $u \in v + W$. Then there exists $w_0 \in W$ such that $u = v + w_0$. Hence, $u - v = w_0 \in W$. Conversely, suppose $u - v \in W$. Then $u - v = w_0$ where $w_0 \in W$. Hence, $u = v + w_0 \in v + W$. Thus, (i) and (ii) are equivalent.

We also have $u - v \in W$ iff $-(u - v) = v - u \in W$ iff $v \in u + W$. Thus, (ii) and (iii) are also equivalent.

10.22. Prove the following: The cosets of W in V partition V into mutually disjoint sets. That is,

(a) Any two cosets $u + W$ and $v + W$ are either identical or disjoint.
(b) Each $v \in V$ belongs to a coset; in fact, $v \in v + W$.

Furthermore, $u + W = v + W$ if and only if $u - v \in W$, and so $(v + w) + W = v + W$ for any $w \in W$.

Let $v \in V$. Because $0 \in W$, we have $v = v + 0 \in v + W$, which proves (b).

Now suppose the cosets $u + W$ and $v + W$ are not disjoint; say, the vector x belongs to both $u + W$ and $v + W$. Then $u - x \in W$ and $x - v \in W$. The proof of (a) is complete if we show that $u + W = v + W$. Let $u + w_0$ be any element in the coset $u + W$. Because $u - x$, $x - v$, w_0 belongs to W,

$$(u + w_0) - v = (u - x) + (x - v) + w_0 \in W$$

Thus, $u + w_0 \in v + W$, and hence the cost $u + W$ is contained in the coset $v + W$. Similarly, $v + W$ is contained in $u + W$, and so $u + W = v + W$.

The last statement follows from the fact that $u + W = v + W$ if and only if $u \in v + W$, and, by Problem 10.21, this is equivalent to $u - v \in W$.

10.23. Let W be the solution space of the homogeneous equation $2x + 3y + 4z = 0$. Describe the cosets of W in $\mathbf{R}^3$.

W is a plane through the origin $O = (0, 0, 0)$, and the cosets of W are the planes parallel to W. Equivalently, the cosets of W are the solution sets of the family of equations

$$2x + 3y + 4z = k, \qquad k \in \mathbf{R}$$

In fact, the coset $v + W$, where $v = (a, b, c)$, is the solution set of the linear equation

$$2x + 3y + 4z = 2a + 3b + 4c \qquad \text{or} \qquad 2(x - a) + 3(y - b) + 4(z - c) = 0$$

10.24. Suppose W is a subspace of a vector space V. Show that the operations in Theorem 10.15 are well defined; namely, show that if $u + W = u' + W$ and $v + W = v' + W$, then

(a) $(u + v) + W = (u' + v') + W$ and (b) $ku + W = ku' + W$ for any $k \in K$

(a) Because $u + W = u' + W$ and $v + W = v' + W$, both $u - u'$ and $v - v'$ belong to W. But then $(u + v) - (u' + v') = (u - u') + (v - v') \in W$. Hence, $(u + v) + W = (u' + v') + W$.
(b) Also, because $u - u' \in W$ implies $k(u - u') \in W$, then $ku - ku' = k(u - u') \in W$; accordingly, $ku + W = ku' + W$.

10.25. Let V be a vector space and W a subspace of V. Show that the natural map $\eta: V \to V/W$, defined by $\eta(v) = v + W$, is linear.

For any $u, v \in V$ and any $k \in K$, we have

$$n(u + v) = u + v + W = u + W + v + W = \eta(u) + \eta(v)$$

and

$$\eta(kv) = kv + W = k(v + W) = k\eta(v)$$

Accordingly, η is linear.

10.26. Let W be a subspace of a vector space V. Suppose $\{w_1, \ldots, w_r\}$ is a basis of W and the set of cosets $\{\bar{v}_1, \ldots, \bar{v}_s\}$, where $\bar{v}_j = v_j + W$, is a basis of the quotient space. Show that the set of vectors $B = \{v_1, \ldots, v_s, w_1, \ldots, w_r\}$ is a basis of V. Thus, $\dim V = \dim W + \dim(V/W)$.

Suppose $u \in V$. Because $\{\bar{v}_j\}$ is a basis of V/W,

$$\bar{u} = u + W = a_1\bar{v}_1 + a_2\bar{v}_2 + \cdots + a_s\bar{v}_s$$

Hence, $u = a_1 v_1 + \cdots + a_s v_s + w$, where $w \in W$. Since $\{w_i\}$ is a basis of W,

$$u = a_1 v_1 + \cdots + a_s v_s + b_1 w_1 + \cdots + b_r w_r$$

Accordingly, B spans V.

We now show that B is linearly independent. Suppose

$$c_1 v_1 + \cdots + c_s v_s + d_1 w_1 + \cdots + d_r w_r = 0 \tag{1}$$

Then

$$c_1 \bar{v}_1 + \cdots + c_s \bar{v}_s = \bar{0} = W$$

Because $\{\bar{v}_j\}$ is independent, the c's are all 0. Substituting into (1), we find $d_1 w_1 + \cdots + d_r w_r = 0$. Because $\{w_i\}$ is independent, the d's are all 0. Thus, B is linearly independent and therefore a basis of V.

10.27. Prove Theorem 10.16: Suppose W is a subspace invariant under a linear operator $T:V \to V$. Then T induces a linear operator $\bar{T}$ on V/W defined by $\bar{T}(v + W) = T(v) + W$. Moreover, if T is a zero of any polynomial, then so is $\bar{T}$. Thus, the minimal polynomial of $\bar{T}$ divides the minimal polynomial of T.

We first show that $\bar{T}$ is well defined; that is, if $u + W = v + W$, then $\bar{T}(u + W) = \bar{T}(v + W)$. If $u + W = v + W$, then $u - v \in W$, and, as W is T-invariant, $T(u - v) = T(u) - T(v) \in W$. Accordingly,

$$\bar{T}(u + W) = T(u) + W = T(v) + W = \bar{T}(v + W)$$

as required.

We next show that $\bar{T}$ is linear. We have

$$\begin{aligned}\bar{T}((u + W) + (v + W)) &= \bar{T}(u + v + W) = T(u + v) + W = T(u) + T(v) + W \\ &= T(u) + W + T(v) + W = \bar{T}(u + W) + \bar{T}(v + W)\end{aligned}$$

Furthermore,

$$\bar{T}(k(u + W)) = \bar{T}(ku + W) = T(ku) + W = kT(u) + W = k(T(u) + W) = k\hat{T}(u + W)$$

Thus, $\bar{T}$ is linear.

Now, for any coset $u + W$ in V/W,

$$\overline{T^2}(u + W) = T^2(u) + W = T(T(u)) + W = \bar{T}(T(u) + W) = \bar{T}(\bar{T}(u + W)) = \bar{T}^2(u + W)$$

Hence, $\overline{T^2} = \bar{T}^2$. Similarly, $\overline{T^n} = \bar{T}^n$ for any n. Thus, for any polynomial

$$f(t) = a_n t^n + \cdots + a_0 = \sum a_i t^i$$

$$\begin{aligned}\overline{f(T)}(u + W) &= f(T)(u) + W = \sum a_i T^i(u) + W = \sum a_i (T^i(u) + W) \\ &= \sum a_i \overline{T^i}(u + W) = \sum a_i \bar{T}^i(u + W) = \left(\sum a_i \bar{T}^i\right)(u + W) = f(\bar{T})(u + W)\end{aligned}$$

and so $\overline{f(T)} = f(\bar{T})$. Accordingly, if T is a root of $f(t)$ then $\overline{f(T)} = \bar{\mathbf{0}} = W = f(\bar{T})$; that is, $\bar{T}$ is also a root of $f(t)$. The theorem is proved.

10.28. Prove Theorem 10.1: Let $T:V \to V$ be a linear operator whose characteristic polynomial factors into linear polynomials. Then V has a basis in which T is represented by a triangular matrix.

The proof is by induction on the dimension of V. If $\dim V = 1$, then every matrix representation of T is a 1×1 matrix, which is triangular.

Now suppose $\dim V = n > 1$ and that the theorem holds for spaces of dimension less than n. Because the characteristic polynomial of T factors into linear polynomials, T has at least one eigenvalue and so at least one nonzero eigenvector v, say $T(v) = a_{11} v$. Let W be the one-dimensional subspace spanned by v. Set $\bar{V} = V/W$. Then (Problem 10.26) $\dim \bar{V} = \dim V - \dim W = n - 1$. Note also that W is invariant under T. By Theorem 10.16, T induces a linear operator $\bar{T}$ on $\bar{V}$ whose minimal polynomial divides the minimal polynomial of T. Because the characteristic polynomial of T is a product of linear polynomials, so is its minimal polynomial, and hence, so are the minimal and characteristic polynomials of $\bar{T}$. Thus, $\bar{V}$ and $\bar{T}$ satisfy the hypothesis of the theorem. Hence, by induction, there exists a basis $\{\bar{v}_2, \ldots, \bar{v}_n\}$ of $\bar{V}$ such that

$$\begin{aligned}&\bar{T}(\bar{v}_2) = a_{22} \bar{v}_2 \\ &\bar{T}(\bar{v}_3) = a_{32} \bar{v}_2 + a_{33} \bar{v}_3 \\ &\ldots\ldots\ldots\ldots\ldots\ldots\ldots\ldots\ldots\ldots \\ &\bar{T}(\bar{v}_n) = a_{n2} \bar{v}_n + a_{n3} \bar{v}_3 + \cdots + a_{nn} \bar{v}_n\end{aligned}$$

Now let $v_2, \ldots, v_n$ be elements of V that belong to the cosets $\bar{v}_2, \ldots, \bar{v}_n$, respectively. Then $\{v, v_2, \ldots, v_n\}$ is a basis of V (Problem 10.26). Because $\bar{T}(\bar{v}_2) = a_{22}\bar{v}_2$, we have

$$\bar{T}(\bar{v}_2) - a_{22}\bar{v}_{22} = 0, \quad \text{and so} \quad T(v_2) - a_{22}v_2 \in W$$

But W is spanned by v; hence, $T(v_2) - a_{22}v_2$ is a multiple of v, say,

$$T(v_2) - a_{22}v_2 = a_{21}v, \quad \text{and so} \quad T(v_2) = a_{21}v + a_{22}v_2$$

Similarly, for $i = 3, \ldots, n$

$$T(v_i) - a_{i2}v_2 - a_{i3}v_3 - \cdots - a_{ii}v_i \in W, \quad \text{and so} \quad T(v_i) = a_{i1}v + a_{i2}v_2 + \cdots + a_{ii}v_i$$

Thus,

$$\begin{aligned} T(v) &= a_{11}v \\ T(v_2) &= a_{21}v + a_{22}v_2 \\ &\cdots\cdots\cdots\cdots\cdots \\ T(v_n) &= a_{n1}v + a_{n2}v_2 + \cdots + a_{nn}v_n \end{aligned}$$

and hence the matrix of T in this basis is triangular.

Cyclic Subspaces, Rational Canonical Form

10.29. Prove Theorem 10.12: Let $Z(v, T)$ be a T-cyclic subspace, T_v the restriction of T to $Z(v, T)$, and $m_v(t) = t^k + a_{k-1}t^{k-1} + \cdots + a_0$ the T-annihilator of v. Then,

(i) The set $\{v, T(v), \ldots, T^{k-1}(v)\}$ is a basis of $Z(v, T)$; hence, $\dim Z(v, T) = k$.

(ii) The minimal polynomial of T_v is $m_v(t)$.

(iii) The matrix of T_v in the above basis is the companion matrix $C = C(m_v)$ of $m_v(t)$ [which has 1's below the diagonal, the negative of the coefficients $a_0, a_1, \ldots, a_{k-1}$ of $m_v(t)$ in the last column, and 0's elsewhere].

(i) By definition of $m_v(t)$, $T^k(v)$ is the first vector in the sequence $v, T(v), T^2(v), \ldots$ that, is a linear combination of those vectors that precede it in the sequence; hence, the set $B = \{v, T(v), \ldots, T^{k-1}(v)\}$ is linearly independent. We now only have to show that $Z(v, T) = L(B)$, the linear span of B. By the above, $T^k(v) \in L(B)$. We prove by induction that $T^n(v) \in L(B)$ for every n. Suppose $n > k$ and $T^{n-1}(v) \in L(B)$—that is, $T^{n-1}(v)$ is a linear combination of $v, \ldots, T^{k-1}(v)$. Then $T^n(v) = T(T^{n-1}(v))$ is a linear combination of $T(v), \ldots, T^k(v)$. But $T^k(v) \in L(B)$; hence, $T^n(v) \in L(B)$ for every n. Consequently, $f(T)(v) \in L(B)$ for any polynomial $f(t)$. Thus, $Z(v, T) = L(B)$, and so B is a basis, as claimed.

(ii) Suppose $m(t) = t^s + b_{s-1}t^{s-1} + \cdots + b_0$ is the minimal polynomial of T_v. Then, because $v \in Z(v, T)$,

$$0 = m(T_v)(v) = m(T)(v) = T^s(v) + b_{s-1}T^{s-1}(v) + \cdots + b_0 v$$

Thus, $T^s(v)$ is a linear combination of $v, T(v), \ldots, T^{s-1}(v)$, and therefore $k \le s$. However, $m_v(T) = \mathbf{0}$ and so $m_v(T_v) = \mathbf{0}$. Then $m(t)$ divides $m_v(t)$, and so $s \le k$. Accordingly, $k = s$ and hence $m_v(t) = m(t)$.

(iii)

$$\begin{aligned} T_v(v) &= T(v) \\ T_v(T(v)) &= T^2(v) \\ &\cdots\cdots\cdots\cdots\cdots \\ T_v(T^{k-2}(v)) &= T^{k-1}(v) \\ T_v(T^{k-1}(v)) &= T^k(v) = -a_0 v - a_1 T(v) - a_2 T^2(v) - \cdots - a_{k-1}T^{k-1}(v) \end{aligned}$$

By definition, the matrix of T_v in this basis is the tranpose of the matrix of coefficients of the above system of equations; hence, it is C, as required.

10.30. Let $T: V \to V$ be linear. Let W be a T-invariant subspace of V and $\bar{T}$ the induced operator on V/W. Prove

(a) The T-annihilator of $v \in V$ divides the minimal polynomial of T.

(b) The $\bar{T}$-annihilator of $\bar{v} \in V/W$ divides the minimal polynomial of T.

(a) The T-annihilator of $v \in V$ is the minimal polynomial of the restriction of T to $Z(v, T)$; therefore, by Problem 10.6, it divides the minimal polynomial of T.

(b) The $\bar{T}$-annihilator of $\bar{v} \in V/W$ divides the minimal polynomial of $\bar{T}$, which divides the minimal polynomial of T by Theorem 10.16.

Remark: In the case where the minimum polynomial of T is $f(t)^n$, where $f(t)$ is a monic irreducible polynomial, then the T-annihilator of $v \in V$ and the $\bar{T}$-annihilator of $\bar{v} \in V/W$ are of the form $f(t)^m$, where $m \leq n$.

10.31. Prove Lemma 10.13: Let $T:V \to V$ be a linear operator whose minimal polynomial is $f(t)^n$, where $f(t)$ is a monic irreducible polynomial. Then V is the direct sum of T-cyclic subspaces $Z_i = Z(v_i, T)$, $i = 1, \dots, r$, with corresponding T-annihilators

$$f(t)^{n_1}, f(t)^{n_2}, \dots, f(t)^{n_r}, \qquad n = n_1 \geq n_2 \geq \dots \geq n_r$$

Any other decomposition of V into the direct sum of T-cyclic subspaces has the same number of components and the same set of T-annihilators.

The proof is by induction on the dimension of V. If $\dim V = 1$, then V is T-cyclic and the lemma holds. Now suppose $\dim V > 1$ and that the lemma holds for those vector spaces of dimension less than that of V.

Because the minimal polynomial of T is $f(t)^n$, there exists $v_1 \in V$ such that $f(T)^{n-1}(v_1) \neq 0$; hence, the T-annihilator of v_1 is $f(t)^n$. Let $Z_1 = Z(v_1, T)$ and recall that Z_1 is T-invariant. Let $\bar{V} = V/Z_1$ and let $\bar{T}$ be the linear operator on $\bar{V}$ induced by T. By Theorem 10.16, the minimal polynomial of $\bar{T}$ divides $f(t)^n$; hence, the hypothesis holds for $\bar{V}$ and $\bar{T}$. Consequently, by induction, $\bar{V}$ is the direct sum of $\bar{T}$-cyclic subspaces; say,

$$\bar{V} = Z(\bar{v}_2, \bar{T}) \oplus \dots \oplus Z(\bar{v}_r, \bar{T})$$

where the corresponding $\bar{T}$-annihilators are $f(t)^{n_2}, \dots, f(t)^{n_r}$, $n \geq n_2 \geq \dots \geq n_r$.

We claim that there is a vector v_2 in the coset $\bar{v}_2$ whose T-annihilator is $f(t)^{n_2}$, the $\bar{T}$-annihilator of $\bar{v}_2$. Let w be any vector in $\bar{v}_2$. Then $f(T)^{n_2}(w) \in Z_1$. Hence, there exists a polynomial $g(t)$ for which

$$f(T)^{n_2}(w) = g(T)(v_1) \tag{1}$$

Because $f(t)^n$ is the minimal polynomial of T, we have, by (1),

$$0 = f(T)^n(w) = f(T)^{n-n_2} g(T)(v_1)$$

But $f(t)^n$ is the T-annihilator of v_1; hence, $f(t)^n$ divides $f(t)^{n-n_2} g(t)$, and so $g(t) = f(t)^{n_2} h(t)$ for some polynomial $h(t)$. We set

$$v_2 = w - h(T)(v_1)$$

Because $w - v_2 = h(T)(v_1) \in Z_1$, v_2 also belongs to the coset $\bar{v}_2$. Thus, the T-annihilator of v_2 is a multiple of the $\bar{T}$-annihilator of $\bar{v}_2$. On the other hand, by (1),

$$f(T)^{n_2}(v_2) = f(T)^{n_s}(w - h(T)(v_1)) = f(T)^{n_2}(w) - g(T)(v_1) = 0$$

Consequently, the T-annihilator of v_2 is $f(t)^{n_2}$, as claimed.

Similarly, there exist vectors $v_3, \dots, v_r \in V$ such that $v_i \in \bar{v_i}$ and that the T-annihilator of v_i is $f(t)^{n_i}$, the $\bar{T}$-annihilator of $\bar{v_i}$. We set

$$Z_2 = Z(v_2, T), \qquad \dots, \qquad Z_r = Z(v_r, T)$$

Let d denote the degree of $f(t)$, so that $f(t)^{n_i}$ has degree dn_i. Then, because $f(t)^{n_i}$ is both the T-annihilator of v_i and the $\bar{T}$-annihilator of $\bar{v_i}$, we know that

$$\{v_i, T(v_i), \dots, T^{dn_i-1}(v_i)\} \qquad \text{and} \qquad \{\bar{v_i}, \bar{T}(\bar{v_i}), \dots, \bar{T}^{dn_i-1}(\bar{v_i})\}$$

are bases for $Z(v_i, T)$ and $Z(\bar{v_i}, \bar{T})$, respectively, for $i = 2, \dots, r$. But $\bar{V} = Z(\bar{v_2}, \bar{T}) \oplus \dots \oplus Z(\bar{v_r}, \bar{T})$; hence,

$$\{\bar{v}_2, \dots, \bar{T}^{dn_2-1}(\bar{v}_2), \dots, \bar{v}_r, \dots, \bar{T}^{dn_r-1}(\bar{v}_r)\}$$

is a basis for $\bar{V}$. Therefore, by Problem 10.26 and the relation $\bar{T}^i(\bar{v}) = \overline{T^i(v)}$ (see Problem 10.27),

$$\{v_1, \ldots, T^{dn_1-1}(v_1), v_2, \ldots, T^{en_2-1}(v_2), \ldots, v_r, \ldots, T^{dn_r-1}(v_r)\}$$

is a basis for V. Thus, by Theorem 10.4, $V = Z(v_1, T) \oplus \cdots \oplus Z(v_r, T)$, as required.

It remains to show that the exponents $n_1, \ldots, n_r$ are uniquely determined by T. Because $d =$ degree of $f(t)$,

$$\dim V = d(n_1 + \cdots + n_r) \quad \text{and} \quad \dim Z_i = dn_i, \quad i = 1, \ldots, r$$

Also, if s is any positive integer, then (Problem 10.59) $f(T)^s(Z_i)$ is a cyclic subspace generated by $f(T)^s(v_i)$, and it has dimension $d(n_i - s)$ if $n_i > s$ and dimension 0 if $n_i \leq s$.

Now any vector $v \in V$ can be written uniquely in the form $v = w_1 + \cdots + w_r$, where $w_i \in Z_i$. Hence, any vector in $f(T)^s(V)$ can be written uniquely in the form

$$f(T)^s(v) = f(T)^s(w_1) + \cdots + f(T)^s(w_r)$$

where $f(T)^s(w_i) \in f(T)^s(Z_i)$. Let t be the integer, dependent on s, for which

$$n_1 > s, \quad \ldots, \quad n_t > s, \quad n_{t+1} \geq s$$

Then
$$f(T)^s(V) = f(T)^s(Z_1) \oplus \cdots \oplus f(T)^s(Z_t)$$

and so
$$\dim[f(T)^s(V)] = d[(n_1 - s) + \cdots + (n_t - s)] \tag{2}$$

The numbers on the left of (2) are uniquely determined by T. Set $s = n - 1$, and (2) determines the number of n_i equal to n. Next set $s = n - 2$, and (2) determines the number of n_i (if any) equal to $n - 1$. We repeat the process until we set $s = 0$ and determine the number of n_i equal to 1. Thus, the n_i are uniquely determined by T and V, and the lemma is proved.

10.32. Let V be a seven-dimensional vector space over $\mathbf{R}$, and let $T: V \to V$ be a linear operator with minimal polynomial $m(t) = (t^2 - 2t + 5)(t - 3)^3$. Find all possible rational canonical forms M of T.

Because $\dim V = 7$, there are only two possible characteristic polynomials, $\Delta_1(t) = (t^2 - 2t + 5)^2 (t - 3)^3$ or $\Delta_1(t) = (t^2 - 2t + 5)(t - 3)^5$. Moreover, the sum of the orders of the companion matrices must add up to 7. Also, one companion matrix must be $C(t^2 - 2t + 5)$ and one must be $C((t - 3)^3) = C(t^3 - 9t^2 + 27t - 27)$. Thus, M must be one of the following block diagonal matrices:

(a) $\operatorname{diag}\left(\begin{bmatrix} 0 & -5 \\ 1 & 2 \end{bmatrix}, \begin{bmatrix} 0 & -5 \\ 1 & 2 \end{bmatrix}, \begin{bmatrix} 0 & 0 & 27 \\ 1 & 0 & -27 \\ 0 & 1 & 9 \end{bmatrix} \right)$,

(b) $\operatorname{diag}\left(\begin{bmatrix} 0 & -5 \\ 1 & 2 \end{bmatrix}, \begin{bmatrix} 0 & 0 & 27 \\ 1 & 0 & -27 \\ 0 & 1 & 9 \end{bmatrix}, \begin{bmatrix} 0 & -9 \\ 1 & 6 \end{bmatrix} \right)$,

(c) $\operatorname{diag}\left(\begin{bmatrix} 0 & -5 \\ 1 & 2 \end{bmatrix}, \begin{bmatrix} 0 & 0 & 27 \\ 1 & 0 & -27 \\ 0 & 1 & 9 \end{bmatrix}, [3], [3] \right)$

Projections

10.33. Suppose $V = W_1 \oplus \cdots \oplus W_r$. The *projection* of V into its subspace W_k is the mapping $E: V \to V$ defined by $E(v) = w_k$, where $v = w_1 + \cdots + w_r$, $w_i \in W_i$. Show that (a) E is linear, (b) $E^2 = E$.

(a) Because the sum $v = w_1 + \cdots + w_r$, $w_i \in W$ is uniquely determined by v, the mapping E is well defined. Suppose, for $u \in V$, $u = w_1' + \cdots + w_r'$, $w_i' \in W_i$. Then

$$v + u = (w_1 + w_1') + \cdots + (w_r + w_r') \quad \text{and} \quad kv = kw_1 + \cdots + kw_r, \quad kw_i, w_i + w_i' \in W_i$$

are the unique sums corresponding to $v + u$ and kv. Hence,

$$E(v + u) = w_k + w_k' = E(v) + E(u) \quad \text{and} \quad E(kv) = kw_k + kE(v)$$

and therefore E is linear.

(b) We have that

$$w_k = 0 + \cdots + 0 + w_k + 0 + \cdots + 0$$

is the unique sum corresponding to $w_k \in W_k$; hence, $E(w_k) = w_k$. Then, for any $v \in V$,

$$E^2(v) = E(E(v)) = E(w_k) = w_k = E(v)$$

Thus, $E^2 = E$, as required.

10.34. Suppose $E:V \to V$ is linear and $E^2 = E$. Show that (a) $E(u) = u$ for any $u \in \operatorname{Im} E$ (i.e., the restriction of E to its image is the identity mapping); (b) V is the direct sum of the image and kernel of $E:V = \operatorname{Im} E \oplus \operatorname{Ker} E$; (c) E is the projection of V into $\operatorname{Im} E$, its image. Thus, by the preceding problem, a linear mapping $T:V \to V$ is a projection if and only if $T^2 = T$; this characterization of a projection is frequently used as its definition.

(a) If $u \in \operatorname{Im} E$, then there exists $v \in V$ for which $E(v) = u$; hence, as required,

$$E(u) = E(E(v)) = E^2(v) = E(v) = u$$

(b) Let $v \in V$. We can write v in the form $v = E(v) + v - E(v)$. Now $E(v) \in \operatorname{Im} E$ and, because

$$E(v - E(v)) = E(v) - E^2(v) = E(v) - E(v) = 0$$

$v - E(v) \in \operatorname{Ker} E$. Accordingly, $V = \operatorname{Im} E + \operatorname{Ker} E$.

Now suppose $w \in \operatorname{Im} E \cap \operatorname{Ker} E$. By (*i*), $E(w) = w$ because $w \in \operatorname{Im} E$. On the other hand, $E(w) = 0$ because $w \in \operatorname{Ker} E$. Thus, $w = 0$, and so $\operatorname{Im} E \cap \operatorname{Ker} E = \{0\}$. These two conditions imply that V is the direct sum of the image and kernel of E.

(c) Let $v \in V$ and suppose $v = u + w$, where $u \in \operatorname{Im} E$ and $w \in \operatorname{Ker} E$. Note that $E(u) = u$ by (i), and $E(w) = 0$ because $w \in \operatorname{Ker} E$. Hence,

$$E(v) = E(u + w) = E(u) + E(w) = u + 0 = u$$

That is, E is the projection of V into its image.

10.35. Suppose $V = U \oplus W$ and suppose $T:V \to V$ is linear. Show that U and W are both T-invariant if and only if $TE = ET$, where E is the projection of V into U.

Observe that $E(v) \in U$ for every $v \in V$, and that (i) $E(v) = v$ iff $v \in U$, (ii) $E(v) = 0$ iff $v \in W$. Suppose $ET = TE$. Let $u \in U$. Because $E(u) = u$,

$$T(u) = T(E(u)) = (TE)(u) = (ET)(u) = E(T(u)) \in U$$

Hence, U is T-invariant. Now let $w \in W$. Because $E(w) = 0$,

$$E(T(w)) = (ET)(w) = (TE)(w) = T(E(w)) = T(0) = 0, \qquad \text{and so} \qquad T(w) \in W$$

Hence, W is also T-invariant.

Conversely, suppose U and W are both T-invariant. Let $v \in V$ and suppose $v = u + w$, where $u \in T$ and $w \in W$. Then $T(u) \in U$ and $T(w) \in W$; hence, $E(T(u)) = T(u)$ and $E(T(w)) = 0$. Thus,

$$(ET)(v) = (ET)(u + w) = (ET)(u) + (ET)(w) = E(T(u)) + E(T(w)) = T(u)$$

and

$$(TE)(v) = (TE)(u + w) = T(E(u + w)) = T(u)$$

That is, $(ET)(v) = (TE)(v)$ for every $v \in V$; therefore, $ET = TE$, as required.

SUPPLEMENTARY PROBLEMS

Invariant Subspaces

10.36. Suppose W is invariant under $T:V \to V$. Show that W is invariant under $f(T)$ for any polynomial $f(t)$.

10.37. Show that every subspace of V is invariant under I and $\mathbf{0}$, the identity and zero operators.

10.38. Let W be invariant under $T_1: V \to V$ and $T_2: V \to V$. Prove W is also invariant under $T_1 + T_2$ and $T_1 T_2$.

10.39. Let $T:V \to V$ be linear. Prove that any eigenspace, E_λ is T-invariant.

10.40. Let V be a vector space of odd dimension (greater than 1) over the real field **R**. Show that any linear operator on V has an invariant subspace other than V or $\{0\}$.

10.41. Determine the invariant subspace of $A = \begin{bmatrix} 2 & -4 \\ 5 & -2 \end{bmatrix}$ viewed as a linear operator on (a) $\mathbf{R}^2$, (b) $\mathbf{C}^2$.

10.42. Suppose $\dim V = n$. Show that $T:V \to V$ has a triangular matrix representation if and only if there exist T-invariant subspaces $W_1 \subset W_2 \subset \cdots \subset W_n = V$ for which $\dim W_k = k$, $k = 1, \ldots, n$.

Invariant Direct Sums

10.43. The subspaces $W_1, \ldots, W_r$ are said to be *independent* if $w_1 + \cdots + w_r = 0$, $w_i \in W_i$, implies that each $w_i = 0$. Show that $\text{span}(W_i) = W_1 \oplus \cdots \oplus W_r$ if and only if the W_i are independent. [Here $\text{span}(W_i)$ denotes the linear span of the W_i.]

10.44. Show that $V = W_1 \oplus \cdots \oplus W_r$ if and only if (i) $V = \text{span}(W_i)$ and (ii) for $k = 1, 2, \ldots, r$, $W_k \cap \text{span}(W_1, \ldots, W_{k-1}, W_{k+1}, \ldots, W_r) = \{0\}$.

10.45. Show that $\text{span}(W_i) = W_1 \oplus \cdots \oplus W_r$ if and only if $\dim[\text{span}(W_i)] = \dim W_1 + \cdots + \dim W_r$.

10.46. Suppose the characteristic polynomial of $T:V \to V$ is $\Delta(t) = f_1(t)^{n_1} f_2(t)^{n_2} \cdots f_r(t)^{n_r}$, where the $f_i(t)$ are distinct monic irreducible polynomials. Let $V = W_1 \oplus \cdots \oplus W_r$ be the primary decomposition of V into T-invariant subspaces. Show that $f_i(t)^{n_i}$ is the characteristic polynomial of the restriction of T to W_i.

Nilpotent Operators

10.47. Suppose T_1 and T_2 are nilpotent operators that commute (i.e., $T_1 T_2 = T_2 T_1$). Show that $T_1 + T_2$ and $T_1 T_2$ are also nilpotent.

10.48. Suppose A is a supertriangular matrix (i.e., all entries on and below the main diagonal are 0). Show that A is nilpotent.

10.49. Let V be the vector space of polynomials of degree $\leq n$. Show that the derivative operator on V is nilpotent of index $n + 1$.

10.50. Show that any Jordan nilpotent block matrix N is similar to its transpose N^T (the matrix with 1's below the diagonal and 0's elsewhere).

10.51. Show that two nilpotent matrices of order 3 are similar if and only if they have the same index of nilpotency. Show by example that the statement is not true for nilpotent matrices of order 4.

Jordan Canonical Form

10.52. Find all possible Jordan canonical forms for those matrices whose characteristic polynomial $\Delta(t)$ and minimal polynomial $m(t)$ are as follows:

(a) $\Delta(t) = (t-2)^4(t-3)^2$, $m(t) = (t-2)^2(t-3)^2$,

(b) $\Delta(t) = (t-7)^5$, $m(t) = (t-7)^2$, (c) $\Delta(t) = (t-2)^7$, $m(t) = (t-2)^3$

10.53. Show that every complex matrix is similar to its transpose. (*Hint:* Use its Jordan canonical form.)

10.54. Show that all $n \times n$ complex matrices A for which $A^n = I$ but $A_k \neq I$ for $k < n$ are similar.

10.55. Suppose A is a complex matrix with only real eigenvalues. Show that A is similar to a matrix with only real entries.

Cyclic Subspaces

10.56. Suppose $T:V \to V$ is linear. Prove that $Z(v, T)$ is the intersection of all T-invariant subspaces containing v.

10.57. Let $f(t)$ and $g(t)$ be the T-annihilators of u and v, respectively. Show that if $f(t)$ and $g(t)$ are relatively prime, then $f(t)g(t)$ is the T-annihilator of $u+v$.

10.58. Prove that $Z(u, T) = Z(v, T)$ if and only if $g(T)(u) = v$ where $g(t)$ is relatively prime to the T-annihilator of u.

10.59. Let $W = Z(v, T)$, and suppose the T-annihilator of v is $f(t)^n$, where $f(t)$ is a monic irreducible polynomial of degree d. Show that $f(T)^s(W)$ is a cyclic subspace generated by $f(T)^s(v)$ and that it has dimension $d(n-s)$ if $n > s$ and dimension 0 if $n \le s$.

Rational Canonical Form

10.60. Find all possible rational forms for a 6×6 matrix over **R** with minimal polynomial:

(a) $m(t) = (t^2 - 2t + 3)(t+1)^2$, (b) $m(t) = (t-2)^3$.

10.61. Let A be a 4×4 matrix with minimal polynomial $m(t) = (t^2+1)(t^2-3)$. Find the rational canonical form for A if A is a matrix over (a) the rational field **Q**, (b) the real field **R**, (c) the complex field **C**.

10.62. Find the rational canonical form for the four-square Jordan block with λ's on the diagonal.

10.63. Prove that the characteristic polynomial of an operator $T:V \to V$ is a product of its elementary divisors.

10.64. Prove that two 3×3 matrices with the same minimal and characteristic polynomials are similar.

10.65. Let $C(f(t))$ denote the companion matrix to an arbitrary polynomial $f(t)$. Show that $f(t)$ is the characteristic polynomial of $C(f(t))$.

Projections

10.66. Suppose $V = W_1 \oplus \cdots \oplus W_r$. Let E_i denote the projection of V into W_i. Prove (i) $E_iE_j = 0$, $i \ne j$; (ii) $I = E_1 + \cdots + E_r$.

10.67. Let $E_1, \ldots, E_r$ be linear operators on V such that

(i) $E_i^2 = E_i$ (i.e., the E_i are projections); (ii) $E_iE_j = 0$, $i \ne j$; (iii) $I = E_1 + \cdots + E_r$

Prove that $V = \text{Im}\, E_1 \oplus \cdots \oplus \text{Im}\, E_r$.

10.68. Suppose $E: V \to V$ is a projection (i.e., $E^2 = E$). Prove that E has a matrix representation of the form $\begin{bmatrix} I_r & 0 \\ 0 & 0 \end{bmatrix}$, where r is the rank of E and I_r is the r-square identity matrix.

10.69. Prove that any two projections of the same rank are similar. (*Hint:* Use the result of Problem 10.68.)

10.70. Suppose $E: V \to V$ is a projection. Prove

(i) $I - E$ is a projection and $V = \text{Im}\, E \oplus \text{Im}\,(I-E)$, (ii) $I + E$ is invertible (if $1 + 1 \ne 0$).

Quotient Spaces

10.71. Let W be a subspace of V. Suppose the set of cosets $\{v_1 + W,\ v_2 + W,\ \ldots,\ v_n + W\}$ in V/W is linearly independent. Show that the set of vectors $\{v_1, v_2, \ldots, v_n\}$ in V is also linearly independent.

10.72. Let W be a substance of V. Suppose the set of vectors $\{u_1, u_2, \ldots, u_n\}$ in V is linearly independent, and that $L(u_i) \cap W = \{0\}$. Show that the set of cosets $\{u_1 + W,\ \ldots,\ u_n + W\}$ in V/W is also linearly independent.

10.73. Suppose $V = U \oplus W$ and that $\{u_1, \ldots, u_n\}$ is a basis of U. Show that $\{u_1 + W, \ \ldots, \ u_n + W\}$ is a basis of the quotient spaces V/W. (Observe that no condition is placed on the dimensionality of V or W.)

10.74. Let W be the solution space of the linear equation

$$a_1x_1 + a_2x_2 + \cdots + a_nx_n = 0, \qquad a_i \in K$$

and let $v = (b_1, b_2, \ldots, b_n) \in K^n$. Prove that the coset $v + W$ of W in K^n is the solution set of the linear equation

$$a_1x_1 + a_2x_2 + \cdots + a_nx_n = b, \qquad \text{where} \qquad b = a_1b_1 + \cdots + a_nb_n$$

10.75. Let V be the vector space of polynomials over $\mathbf{R}$ and let W be the subspace of polynomials divisible by t^4 (i.e., of the form $a_0t^4 + a_1t^5 + \cdots + a_{n-4}t^n$). Show that the quotient space V/W has dimension 4.

10.76. Let U and W be subspaces of V such that $W \subset U \subset V$. Note that any coset $u + W$ of W in U may also be viewed as a coset of W in V, because $u \in U$ implies $u \in V$; hence, U/W is a subset of V/W. Prove that (i) U/W is a subspace of V/W, (ii) $\dim(V/W) - \dim(U/W) = \dim(V/U)$.

10.77. Let U and W be subspaces of V. Show that the cosets of $U \cap W$ in V can be obtained by intersecting each of the cosets of U in V by each of the cosets of W in V:

$$V/(U \cap W) = \{(v + U) \cap (v' + W) : v, v' \in V\}$$

10.78. Let $T{:}V \to V'$ be linear with kernel W and image U. Show that the quotient space V/W is isomorphic to U under the mapping $\theta{:}V/W \to U$ defined by $\theta(v + W) = T(v)$. Furthermore, show that $T = i \circ \theta \circ \eta$, where $\eta{:}V \to V/W$ is the natural mapping of V into V/W (i.e., $\eta(v) = v + W$), and $i{:}U \hookrightarrow V'$ is the inclusion mapping (i.e., $i(u) = u$). (See diagram.)

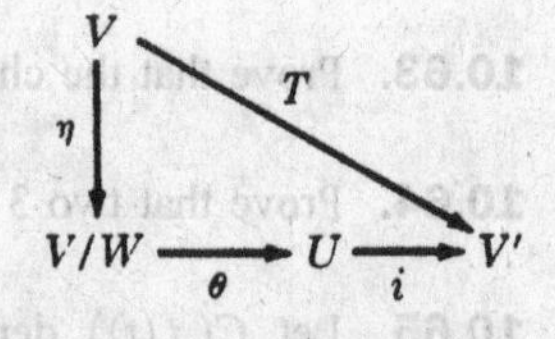

ANSWERS TO SUPPLEMENTARY PROBLEMS

10.41. (a) $\mathbf{R}^2$ and $\{0\}$, (b) $\mathbf{C}^2, \{0\}, W_1 = \text{span}(2, \ 1 - 2i), W_2 = \text{span}(2, \ 1 + 2i)$

10.52. (a) $\text{diag}\left(\begin{bmatrix} 2 & 1 \\ & 2 \end{bmatrix}, \begin{bmatrix} 2 & 1 \\ & 2 \end{bmatrix}, \begin{bmatrix} 3 & 1 \\ & 3 \end{bmatrix}\right), \quad \text{diag}\left(\begin{bmatrix} 2 & 1 \\ & 2 \end{bmatrix}, [2], [2], \begin{bmatrix} 3 & 1 \\ & 3 \end{bmatrix}\right);$

(b) $\text{diag}\left(\begin{bmatrix} 7 & 1 \\ & 7 \end{bmatrix}, \begin{bmatrix} 7 & 1 \\ & 7 \end{bmatrix}, [7]\right), \quad \text{diag}\left(\begin{bmatrix} 7 & 1 \\ & 7 \end{bmatrix}, [7], [7], [7]\right);$

(c) Let M_k denote a Jordan block with $\lambda = 2$ and order k. Then $\text{diag}(M_3, M_3, M_1)$, $\text{diag}(M_3, M_2, M_2)$, $\text{diag}(M_3, M_2, M_1, M_1)$, $\text{diag}(M_3, M_1, M_1, M_1, M_1)$

10.60. Let $A = \begin{bmatrix} 0 & -3 \\ 1 & 2 \end{bmatrix}, \quad B = \begin{bmatrix} 0 & -1 \\ 1 & -2 \end{bmatrix}, \quad C = \begin{bmatrix} 0 & 0 & 8 \\ 1 & 0 & -12 \\ 0 & 1 & 6 \end{bmatrix}, \quad D = \begin{bmatrix} 0 & -4 \\ 1 & 4 \end{bmatrix}$

(a) $\text{diag}(A, A, B), \text{diag}(A, B, B), \text{diag}(A, B, -1, -1)$; (b) $\text{diag}(C, C), \text{diag}(C, D, 2), \text{diag}(C, 2, 2, 2)$

10.61. Let $A = \begin{bmatrix} 0 & -1 \\ 1 & 0 \end{bmatrix}, \quad B = \begin{bmatrix} 0 & 3 \\ 1 & 0 \end{bmatrix}.$

(a) $\text{diag}(A, B)$, (b) $\text{diag}(A, \sqrt{3}, -\sqrt{3})$, (c) $\text{diag}(i, -i, \sqrt{3}, -\sqrt{3})$

10.62. Companion matrix with the last column $[-\lambda^4, 4\lambda^3, -6\lambda^2, 4\lambda]^T$

CHAPTER 11

Linear Functionals and the Dual Space

11.1 Introduction

In this chapter, we study linear mappings from a vector space V into its field K of scalars. (Unless otherwise stated or implied, we view K as a vector space over itself.) Naturally all the theorems and results for arbitrary mappings on V hold for this special case. However, we treat these mappings separately because of their fundamental importance and because the special relationship of V to K gives rise to new notions and results that do not apply in the general case.

11.2 Linear Functionals and the Dual Space

Let V be a vector space over a field K. A mapping $\phi: V \to K$ is termed a *linear functional* (or *linear form*) if, for every $u, v \in V$ and every $a, b, \in K$,

$$\phi(au + bv) = a\phi(u) + b\phi(v)$$

In other words, a linear functional on V is a linear mapping from V into K.

EXAMPLE 11.1

(a) Let $\pi_i: K^n \to K$ be the ith projection mapping; that is, $\pi_i(a_1, a_2, \ldots a_n) = a_i$. Then π_i is linear and so it is a linear functional on K^n.

(b) Let V be the vector space of polynomials in t over $\mathbf{R}$. Let $\mathbf{J}: V \to \mathbf{R}$ be the integral operator defined by $\mathbf{J}(p(t)) = \int_0^1 p(t)\, dt$. Recall that $\mathbf{J}$ is linear; and hence, it is a linear functional on V.

(c) Let V be the vector space of n-square matrices over K. Let $T: V \to K$ be the trace mapping

$$T(A) = a_{11} + a_{22} + \cdots + a_{nn}, \qquad \text{where} \qquad A = [a_{ij}]$$

That is, T assigns to a matrix A the sum of its diagonal elements. This map is linear (Problem 11.24), and so it is a linear functional on V.

By Theorem 5.10, the set of linear functionals on a vector space V over a field K is also a vector space over K, with addition and scalar multiplication defined by

$$(\phi + \sigma)(v) = \phi(v) + \sigma(v) \qquad \text{and} \qquad (k\phi)(v) = k\phi(v)$$

where ϕ and σ are linear functionals on V and $k \in K$. This space is called the *dual space* of V and is denoted by V^*.

EXAMPLE 11.2 Let $V = K^n$, the vector space of n-tuples, which we write as column vectors. Then the dual space V^* can be identified with the space of row vectors. In particular, any linear functional $\phi = (a_1, \ldots, a_n)$ in V^* has the representation

$$\phi(x_1, x_2, \ldots, x_n) = [a_1, a_2, \ldots, a_n][x_2, x_2, \ldots, x_n]^T = a_1x_1 + a_2x_2 + \cdots + a_nx_n$$

Historically, the formal expression on the right was termed a *linear form*.

11.3 Dual Basis

Suppose V is a vector space of dimension n over K. By Theorem 5.11, the dimension of the dual space V^* is also n (because K is of dimension 1 over itself). In fact, each basis of V determines a basis of V^* as follows (see Problem 11.3 for the proof).

THEOREM 11.1: Suppose $\{v_1, \ldots, v_n\}$ is a basis of V over K. Let $\phi_1, \ldots, \phi_n \in V^*$ be the linear functionals as defined by

$$\phi_i(v_j) = \delta_{ij} = \begin{cases} 1 & \text{if } i = j \\ 0 & \text{if } i \neq j \end{cases}$$

Then $\{\phi_1, \ldots, \phi_n\}$ is a basis of V^*.

The above basis $\{\phi_i\}$ is termed the basis *dual* to $\{v_i\}$ or the *dual basis*. The above formula, which uses the Kronecker delta δ_{ij}, is a short way of writing

$$\phi_1(v_1) = 1, \quad \phi_1(v_2) = 0, \quad \phi_1(v_3) = 0, \quad \ldots, \quad \phi_1(v_n) = 0$$
$$\phi_2(v_1) = 0, \quad \phi_2(v_2) = 1, \quad \phi_2(v_3) = 0, \quad \ldots, \quad \phi_2(v_n) = 0$$
$$\cdots\cdots\cdots\cdots\cdots\cdots\cdots\cdots\cdots\cdots\cdots\cdots\cdots\cdots\cdots\cdots\cdots\cdots$$
$$\phi_n(v_1) = 0, \quad \phi_n(v_2) = 0, \quad \ldots, \phi_n(v_{n-1}) = 0, \quad \phi_n(v_n) = 1$$

By Theorem 5.2, these linear mappings ϕ_i are unique and well defined.

EXAMPLE 11.3 Consider the basis $\{v_1 = (2, 1), \ v_2 = (3, 1)\}$ of $\mathbf{R}^2$. Find the dual basis $\{\phi_1, \phi_2\}$.

We seek linear functionals $\phi_1(x, y) = ax + by$ and $\phi_2(x, y) = cx + dy$ such that

$$\phi_1(v_1) = 1, \qquad \phi_1(v_2) = 0, \qquad \phi_2(v_2) = 0, \qquad \phi_2(v_2) = 1$$

These four conditions lead to the following two systems of linear equations:

$$\left.\begin{array}{l} \phi_1(v_1) = \phi_1(2, 1) = 2a + b = 1 \\ \phi_1(v_2) = \phi_1(3, 1) = 3a + b = 0 \end{array}\right\} \quad \text{and} \quad \left.\begin{array}{l} \phi_2(v_1) = \phi_2(2, 1) = 2c + d = 0 \\ \phi_2(v_2) = \phi_2(3, 1) = 3c + d = 1 \end{array}\right\}$$

The solutions yield $a = -1$, $b = 3$ and $c = 1$, $d = -2$. Hence, $\phi_1(x, y) = -x + 3y$ and $\phi_2(x, y) = x - 2y$ form the dual basis.

The next two theorems (proved in Problems 11.4 and 11.5, respectively) give relationships between bases and their duals.

THEOREM 11.2: Let $\{v_1, \ldots, v_n\}$ be a basis of V and let $\{\phi_1, \ldots, \phi_n\}$ be the dual basis in V^*. Then

(i) For any vector $u \in V$, $u = \phi_1(u)v_1 + \phi_2(u)v_2 + \cdots + \phi_n(u)v_n$.

(ii) For any linear functional $\sigma \in V^*$, $\sigma = \sigma(v_1)\phi_1 + \sigma(v_2)\phi_2 + \cdots + \sigma(v_n)\phi_n$.

THEOREM 11.3: Let $\{v_1, \ldots, v_n\}$ and $\{w_1, \ldots, w_n\}$ be bases of V and let $\{\phi_1, \ldots, \phi_n\}$ and $\{\sigma_1, \ldots, \sigma_n\}$ be the bases of V^* dual to $\{v_i\}$ and $\{w_i\}$, respectively. Suppose P is the change-of-basis matrix from $\{v_i\}$ to $\{w_i\}$. Then $(P^{-1})^T$ is the change-of-basis matrix from $\{\phi_i\}$ to $\{\sigma_i\}$.

11.4 Second Dual Space

We repeat: Every vector space V has a dual space V^*, which consists of all the linear functionals on V. Thus, V^* has a dual space V^{**}, called the *second dual* of V, which consists of all the linear functionals on V^*.

We now show that each $v \in V$ determines a specific element $\hat{v} \in V^{**}$. First, for any $\phi \in V^*$, we define

$$\hat{v}(\phi) = \phi(v)$$

It remains to be shown that this map $\hat{v}:V^* \to K$ is linear. For any scalars $a, b \in K$ and any linear functionals $\phi, \sigma \in V^*$, we have

$$\hat{v}(a\phi + b\sigma) = (a\phi + b\sigma)(v) = a\phi(v) + b\sigma(v) = a\hat{v}(\phi) + b\hat{v}(\sigma)$$

That is, $\hat{v}$ is linear and so $\hat{v} \in V^{**}$. The following theorem (proved in Problem 12.7) holds.

THEOREM 11.4: If V has finite dimensions, then the mapping $v \mapsto \hat{v}$ is an isomorphism of V onto V^{**}.

The above mapping $v \mapsto \hat{v}$ is called the *natural mapping* of V into V^{**}. We emphasize that this mapping is never onto V^{**} if V is not finite-dimensional. However, it is always linear, and moreover, it is always one-to-one.

Now suppose V does have finite dimension. By Theorem 11.4, the natural mapping determines an isomorphism between V and V^{**}. Unless otherwise stated, we will identify V with V^{**} by this mapping. Accordingly, we will view V as the space of linear functionals on V^* and write $V = V^{**}$. We remark that if $\{\phi_i\}$ is the basis of V^* dual to a basis $\{v_i\}$ of V, then $\{v_i\}$ is the basis of $V^{**} = V$ that is dual to $\{\phi_i\}$.

11.5 Annihilators

Let W be a subset (not necessarily a subspace) of a vector space V. A linear functional $\phi \in V^*$ is called *an annihilator* of W if $\phi(w) = 0$ for every $w \in W$—that is, if $\phi(W) = \{0\}$. We show that the set of all such mappings, denoted by W^0 and called the *annihilator* of W, is a subspace of V^*. Clearly, $0 \in W^0$. Now suppose $\phi, \sigma \in W^0$. Then, for any scalars $a, b, \in K$ and for any $w \in W$,

$$(a\phi + b\sigma)(w) = a\phi(w) + b\sigma(w) = a0 + b0 = 0$$

Thus, $a\phi + b\sigma \in W^0$, and so W^0 is a subspace of V^*.

In the case that W is a subspace of V, we have the following relationship between W and its annihilator W^0 (see Problem 11.11 for the proof).

THEOREM 11.5: Suppose V has finite dimension and W is a subspace of V. Then

(i) $\dim W + \dim W^0 = \dim V$ and (ii) $W^{00} = W$

Here $W^{00} = \{v \in V : \phi(v) = 0 \text{ for every } \phi \in W^0\}$ or, equivalently, $W^{00} = (W^0)^0$, where W^{00} is viewed as a subspace of V under the identification of V and V^{**}.

11.6 Transpose of a Linear Mapping

Let $T:V \to U$ be an arbitrary linear mapping from a vector space V into a vector space U. Now for any linear functional $\phi \in U^*$, the composition $\phi \circ T$ is a linear mapping from V into K:

$$V \xrightarrow{T} U \xrightarrow{\phi} K, \qquad \phi \circ T : V \to K$$

That is, $\phi \circ T \in V^*$. Thus, the correspondence

$$\phi \mapsto \phi \circ T$$

is a mapping from U^* into V^*; we denote it by T^t and call it the transpose of T. In other words, $T^t:U^* \to V^*$ is defined by

$$T^t(\phi) = \phi \circ T$$

Thus, $(T^t(\phi))(v) = \phi(T(v))$ for every $v \in V$.

THEOREM 11.6: The transpose mapping T^t defined above is linear.

Proof. For any scalars $a, b \in K$ and any linear functionals $\phi, \sigma \in U^*$,

$$T^t(a\phi + b\sigma) = (a\phi + b\sigma) \circ T = a(\phi \circ T) + b(\sigma \circ T) = aT^t(\phi) + bT^t(\sigma)$$

That is, T^t is linear, as claimed.

We emphasize that if T is a linear mapping from V into U, then T^t is a linear mapping from U^* into V^*. The same "transpose" for the mapping T^t no doubt derives from the following theorem (proved in Problem 11.16).

THEOREM 11.7: Let $T{:}V \to U$ be linear, and let A be the matrix representation of T relative to bases $\{v_i\}$ of V and $\{u_i\}$ of U. Then the transpose matrix A^T is the matrix representation of $T^t{:}U^* \to V^*$ relative to the bases dual to $\{u_i\}$ and $\{v_i\}$.

SOLVED PROBLEMS

Dual Spaces and Dual Bases

11.1. Find the basis $\{\phi_1, \phi_2, \phi_3\}$ that is dual to the following basis of $\mathbf{R}^3$:

$$\{v_1 = (1, -1, 3), \quad v_2 = (0, 1, -1), \quad v_3 = (0, 3, -2)\}$$

The linear functionals may be expressed in the form

$$\phi_1(x, y, z) = a_1x + a_2y + a_3z, \qquad \phi_2(x, y, z) = b_1x + b_2y + b_3z, \qquad \phi_3(x, y, z) = c_1x + c_2y + c_3z$$

By definition of the dual basis, $\phi_i(v_j) = 0$ for $i \neq j$, but $\phi_i(v_j) = 1$ for $i = j$.

We find ϕ_1 by setting $\phi_1(v_1) = 1$, $\phi_1(v_2) = 0$, $\phi_1(v_3) = 0$. This yields

$$\phi_1(1, -1, 3) = a_1 - a_2 + 3a_3 = 1, \qquad \phi_1(0, 1, -1) = a_2 - a_3 = 0, \qquad \phi_1(0, 3, -2) = 3a_2 - 2a_3 = 0$$

Solving the system of equations yields $a_1 = 1$, $a_2 = 0$, $a_3 = 0$. Thus, $\phi_1(x, y, z) = x$.

We find ϕ_2 by setting $\phi_2(v_1) = 0$, $\phi_2(v_2) = 1$, $\phi_2(v_3) = 0$. This yields

$$\phi_2(1, -1, 3) = b_1 - b_2 + 3b_3 = 0, \qquad \phi_2(0, 1, -1) = b_2 - b_3 = 1, \qquad \phi_2(0, 3, -2) = 3b_2 - 2b_3 = 0$$

Solving the system of equations yields $b_1 = 7$, $b_2 = -2$, $a_3 = -3$. Thus, $\phi_2(x, y, z) = 7x - 2y - 3z$.

We find ϕ_3 by setting $\phi_3(v_1) = 0$, $\phi_3(v_2) = 0$, $\phi_3(v_3) = 1$. This yields

$$\phi_3(1, -1, 3) = c_1 - c_2 + 3c_3 = 0, \qquad \phi_3(0, 1, -1) = c_2 - c_3 = 0, \qquad \phi_3(0, 3, -2) = 3c_2 - 2c_3 = 1$$

Solving the system of equations yields $c_1 = -2$, $c_2 = 1$, $c_3 = 1$. Thus, $\phi_3(x, y, z) = -2x + y + z$.

11.2. Let $V = \{a + bt : a, b \in \mathbf{R}\}$, the vector space of real polynomials of degree ≤ 1. Find the basis $\{v_1, v_2\}$ of V that is dual to the basis $\{\phi_1, \phi_2\}$ of V^* defined by

$$\phi_1(f(t)) = \int_0^1 f(t)\, dt \qquad \text{and} \qquad \phi_2(f(t)) = \int_0^2 f(t)\, dt$$

Let $v_1 = a + bt$ and $v_2 = c + dt$. By definition of the dual basis,

$$\phi_1(v_1) = 1, \quad \phi_1(v_2) = 0 \quad \text{and} \quad \phi_2(v_1) = 0, \quad \phi_i(v_j) = 1$$

Thus,

$$\left.\begin{aligned} \phi_1(v_1) &= \textstyle\int_0^1 (a + bt)\, dt = a + \frac{1}{2}b = 1 \\ \phi_2(v_1) &= \textstyle\int_0^2 (a + bt)\, dt = 2a + 2b = 0 \end{aligned}\right\} \quad \text{and} \quad \left.\begin{aligned} \phi_1(v_2) &= \textstyle\int_0^1 (c + dt)\, dt = c + \frac{1}{2}d = 0 \\ \phi_2(v_2) &= \textstyle\int_0^2 (c + dt)\, dt = 2c + 2d = 1 \end{aligned}\right\}$$

Solving each system yields $a = 2$, $b = -2$ and $c = -\frac{1}{2}$, $d = 1$. Thus, $\{v_1 = 2 - 2t, \ v_2 = -\frac{1}{2} + t\}$ is the basis of V that is dual to $\{\phi_1, \phi_2\}$.

11.3. Prove Theorem 11.1: Suppose $\{v_1, \ldots, v_n\}$ is a basis of V over K. Let $\phi_1, \ldots, \phi_n \in V^*$ be defined by $\phi_i(v_j) = 0$ for $i \neq j$, but $\phi_i(v_j) = 1$ for $i = j$. Then $\{\phi_1, \ldots, \phi_n\}$ is a basis of V^*.

We first show that $\{\phi_1, \ldots, \phi_n\}$ spans V^*. Let ϕ be an arbitrary element of V^*, and suppose

$$\phi(v_1) = k_1, \qquad \phi(v_2) = k_2, \qquad \ldots, \qquad \phi(v_n) = k_n$$

Set $\sigma = k_1\phi_1 + \cdots + k_n\phi_n$. Then

$$\begin{aligned}\sigma(v_1) &= (k_1\phi_1 + \cdots + k_n\phi_n)(v_1) = k_1\phi_1(v_1) + k_2\phi_2(v_1) + \cdots + k_n\phi_n(v_1)\\ &= k_1 \cdot 1 + k_2 \cdot 0 + \cdots + k_n \cdot 0 = k_1\end{aligned}$$

Similarly, for $i = 2, \ldots, n$,

$$\sigma(v_i) = (k_1\phi_1 + \cdots + k_n\phi_n)(v_i) = k_1\phi_1(v_i) + \cdots + k_i\phi_i(v_i) + \cdots + k_n\phi_n(v_i) = k_i$$

Thus, $\phi(v_i) = \sigma(v_i)$ for $i = 1, \ldots, n$. Because ϕ and σ agree on the basis vectors, $\phi = \sigma = k_1\phi_1 + \cdots + k_n\phi_n$. Accordingly, $\{\phi_1, \ldots, \phi_n\}$ spans V^*.

It remains to be shown that $\{\phi_1, \ldots, \phi_n\}$ is linearly independent. Suppose

$$a_1\phi_1 + a_2\phi_2 + \cdots + a_n\phi_n = \mathbf{0}$$

Applying both sides to v_1, we obtain

$$\begin{aligned}0 = \mathbf{0}(v_1) &= (a_1\phi_1 + \cdots + a_n\phi_n)(v_1) = a_1\phi_1(v_1) + a_2\phi_2(v_1) + \cdots + a_n\phi_n(v_1)\\ &= a_1 \cdot 1 + a_2 \cdot 0 + \cdots + a_n \cdot 0 = a_1\end{aligned}$$

Similarly, for $i = 2, \ldots, n$,

$$0 = \mathbf{0}(v_i) = (a_1\phi_1 + \cdots + a_n\phi_n)(v_i) = a_1\phi_1(v_i) + \cdots + a_i\phi_i(v_i) + \cdots + a_n\phi_n(v_i) = a_i$$

That is, $a_1 = 0, \ldots, a_n = 0$. Hence, $\{\phi_1, \ldots, \phi_n\}$ is linearly independent, and so it is a basis of V^*.

11.4. Prove Theorem 11.2: Let $\{v_1, \ldots, v_n\}$ be a basis of V and let $\{\phi_1, \ldots, \phi_n\}$ be the dual basis in V^*. For any $u \in V$ and any $\sigma \in V^*$, (i) $u = \sum_i \phi_i(u)v_i$. (ii) $\sigma = \sum_i \phi(v_i)\phi_i$.

Suppose

$$u = a_1v_1 + a_2v_2 + \cdots + a_nv_n \tag{1}$$

Then

$$\phi_1(u) = a_1\phi_1(v_1) + a_2\phi_1(v_2) + \cdots + a_n\phi_1(v_n) = a_1 \cdot 1 + a_2 \cdot 0 + \cdots + a_n \cdot 0 = a_1$$

Similarly, for $i = 2, \ldots, n$,

$$\phi_i(u) = a_1\phi_i(v_1) + \cdots + a_i\phi_i(v_i) + \cdots + a_n\phi_i(v_n) = a_i$$

That is, $\phi_1(u) = a_1, \phi_2(u) = a_2, \ldots, \phi_n(u) = a_n$. Substituting these results into (1), we obtain (i).

Next we prove (ii). Applying the linear functional σ to both sides of (i),

$$\begin{aligned}\sigma(u) &= \phi_1(u)\sigma(v_1) + \phi_2(u)\sigma(v_2) + \cdots + \phi_n(u)\sigma(v_n)\\ &= \sigma(v_1)\phi_1(u) + \sigma(v_2)\phi_2(u) + \cdots + \sigma(v_n)\phi_n(u)\\ &= (\sigma(v_1)\phi_1 + \sigma(v_2)\phi_2 + \cdots + \sigma(v_n)\phi_n)(u)\end{aligned}$$

Because the above holds for every $u \in V$, $\sigma = \sigma(v_1)\phi_2 + \sigma(v_2)\phi_2 + \cdots + \sigma(v_n)\phi_n$, as claimed.

11.5. Prove Theorem 11.3. Let $\{v_i\}$ and $\{w_i\}$ be bases of V and let $\{\phi_i\}$ and $\{\sigma_i\}$ be the respective dual bases in V^*. Let P be the change-of-basis matrix from $\{v_i\}$ to $\{w_i\}$. Then $(P^{-1})^T$ is the change-of-basis matrix from $\{\phi_i\}$ to $\{\sigma_i\}$.

Suppose, for $i = 1, \ldots, n$,

$$w_i = a_{i1}v_1 + a_{i2}v_2 + \cdots + a_{in}v_n \qquad \text{and} \qquad \sigma_i = b_{i1}\phi_1 + b_{i2}\phi_2 + \cdots + a_{in}v_n$$

Then $P = [a_{ij}]$ and $Q = [b_{ij}]$. We seek to prove that $Q = (P^{-1})^T$.

Let R_i denote the ith row of Q and let C_j denote the jth column of P^T. Then

$$R_i = (b_{i1}, b_{i2}, \ldots, b_{in}) \qquad \text{and} \qquad C_j = (a_{j1}, a_{j2}, \ldots, a_{jn})^T$$

By definition of the dual basis,

$$\sigma_i(w_j) = (b_{i1}\phi_1 + b_{i2}\phi_2 + \cdots + b_{in}\phi_n)(a_{j1}v_1 + a_{j2}v_2 + \cdots + a_{jn}v_n)$$
$$= b_{i1}a_{j1} + b_{i2}a_{j2} + \cdots + b_{in}a_{jn} = R_iC_j = \delta_{ij}$$

where δ_{ij} is the Kronecker delta. Thus,

$$QP^T = [R_iC_j] = [\delta_{ij}] = I$$

Therefore, $Q = (P^T)^{-1} = (P^{-1})^T$, as claimed.

11.6. Suppose $v \in V$, $v \neq 0$, and $\dim V = n$. Show that there exists $\phi \in V^*$ such that $\phi(v) \neq 0$.

We extend $\{v\}$ to a basis $\{v, v_2, \ldots, v_n\}$ of V. By Theorem 5.2, there exists a unique linear mapping $\phi: V \to K$ such that $\phi(v) = 1$ and $\phi(v_i) = 0$, $i = 2, \ldots, n$. Hence, ϕ has the desired property.

11.7. Prove Theorem 11.4: Suppose $\dim V = n$. Then the natural mapping $v \mapsto \hat{v}$ is an isomorphism of V onto V^{**}.

We first prove that the map $v \mapsto \hat{v}$ is linear—that is, for any vectors $v, w \in V$ and any scalars $a, b \in K$, $\widehat{av + bw} = a\hat{v} + b\hat{w}$. For any linear functional $\phi \in V^*$,

$$\widehat{av + bw}(\phi) = \phi(av + bw) = a\phi(v) + b\phi(w) = a\hat{v}(\phi) + b\hat{w}(\phi) = (a\hat{v} + b\hat{w})(\phi)$$

Because $\widehat{av + bw}(\phi) = (a\hat{v} + b\hat{w})(\phi)$ for every $\phi \in V^*$, we have $\widehat{av + bw} = a\hat{v} + b\hat{w}$. Thus, the map $v \mapsto \hat{v}$ is linear.

Now suppose $v \in V$, $v \neq 0$. Then, by Problem 11.6, there exists $\phi \in V^*$ for which $\phi(v) \neq 0$. Hence, $\hat{v}(\phi) = \phi(v) \neq 0$, and thus $\hat{v} \neq 0$. Because $v \neq 0$ implies $\hat{v} \neq 0$, the map $v \mapsto \hat{v}$ is nonsingular and hence an isomorphism (Theorem 5.64).

Now $\dim V = \dim V^* = \dim V^{**}$, because V has finite dimension. Accordingly, the mapping $v \mapsto \hat{v}$ is an isomorphism of V onto V^{**}.

Annihilators

11.8. Show that if $\phi \in V^*$ annihilates a subset S of V, then ϕ annihilates the linear span $L(S)$ of S. Hence, $S^0 = [\text{span}(S)]^0$.

Suppose $v \in \text{span}(S)$. Then there exists $w_1, \ldots, w_r \in S$ for which $v = a_1w_1 + a_2w_2 + \cdots + a_rw_r$.

$$\phi(v) = a_1\phi(w_1) + a_2\phi(w_2) + \cdots + a_r\phi(w_r) = a_10 + a_20 + \cdots + a_r0 = 0$$

Because v was an arbitrary element of $\text{span}(S)$, ϕ annihilates $\text{span}(S)$, as claimed.

11.9. Find a basis of the annihilator W^0 of the subspace W of $\mathbf{R}^4$ spanned by

$$v_1 = (1, 2, -3, 4) \quad \text{and} \quad v_2 = (0, 1, 4, -1)$$

By Problem 11.8, it suffices to find a basis of the set of linear functionals ϕ such that $\phi(v_1) = 0$ and $\phi(v_2) = 0$, where $\phi(x_1, x_2, x_3, x_4) = ax_1 + bx_2 + cx_3 + dx_4$. Thus,

$$\phi(1, 2, -3, 4) = a + 2b - 3c + 4d = 0 \qquad \text{and} \qquad \phi(0, 1, 4, -1) = b + 4c - d = 0$$

The system of two equations in the unknowns a, b, c, d is in echelon form with free variables c and d.

(1) Set $c = 1$, $d = 0$ to obtain the solution $a = 11$, $b = -4$, $c = 1$, $d = 0$.
(2) Set $c = 0$, $d = 1$ to obtain the solution $a = 6$, $b = -1$, $c = 0$, $d = 1$.

The linear functions $\phi_1(x_i) = 11x_1 - 4x_2 + x_3$ and $\phi_2(x_i) = 6x_1 - x_2 + x_4$ form a basis of W^0.

11.10. Show that (a) For any subset S of V, $S \subseteq S^{00}$. (b) If $S_1 \subseteq S_2$, then $S_2^0 \subseteq S_1^0$.

(a) Let $v \in S$. Then for every linear functional $\phi \in S^0$, $\hat{v}(\phi) = \phi(v) = 0$. Hence, $\hat{v} \in (S^0)^0$. Therefore, under the identification of V and V^{**}, $v \in S^{00}$. Accordingly, $S \subseteq S^{00}$.

(b) Let $\phi \in S_2^0$. Then $\phi(v) = 0$ for every $v \in S_2$. But $S_1 \subseteq S_2$; hence, ϕ annihilates every element of S_1 (i.e., $\phi \in S_1^0$). Therefore, $S_2^0 \subseteq S_1^0$.

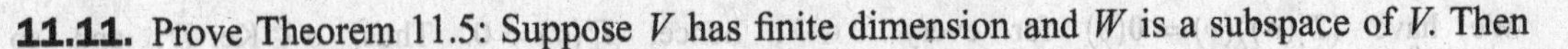

11.11. Prove Theorem 11.5: Suppose V has finite dimension and W is a subspace of V. Then

(i) $\dim W + \dim W^0 = \dim V$, (ii) $W^{00} = W$.

(i) Suppose $\dim V = n$ and $\dim W = r \leq n$. We want to show that $\dim W^0 = n - r$. We choose a basis $\{w_1, \ldots, w_r\}$ of W and extend it to a basis of V, say $\{w_1, \ldots, w_r, v_1, \ldots, v_{n-r}\}$. Consider the dual basis

$$\{\phi_1, \ldots, \phi_r, \sigma_1, \ldots, \sigma_{n-r}\}$$

By definition of the dual basis, each of the above σ's annihilates each w_i; hence, $\sigma_1, \ldots, \sigma_{n-r} \in W^0$. We claim that $\{\sigma_i\}$ is a basis of W^0. Now $\{\sigma_j\}$ is part of a basis of V^*, and so it is linearly independent.

We next show that $\{\phi_j\}$ spans W^0. Let $\sigma \in W^0$. By Theorem 11.2,

$$\begin{aligned}
\sigma &= \sigma(w_1)\phi_1 + \cdots + \sigma(w_r)\phi_r + \sigma(v_1)\sigma_1 + \cdots + \sigma(v_{n-r})\sigma_{n-r} \\
&= 0\phi_1 + \cdots + 0\phi_r + \sigma(v_1)\sigma_1 + \cdots + \sigma(v_{n-r})\sigma_{n-r} \\
&= \sigma(v_1)\sigma_1 + \cdots + \sigma(v_{n-r})\sigma_{n-r}
\end{aligned}$$

Consequently, $\{\sigma_1, \ldots, \sigma_{n-r}\}$ spans W^0 and so it is a basis of W^0. Accordingly, as required

$$\dim W^0 = n - r = \dim V - \dim W.$$

(ii) Suppose $\dim V = n$ and $\dim W = r$. Then $\dim V^* = n$ and, by (i), $\dim W^0 = n - r$. Thus, by (i), $\dim W^{00} = n - (n - r) = r$; therefore, $\dim W = \dim W^{00}$. By Problem 11.10, $W \subseteq W^{00}$. Accordingly, $W = W^{00}$.

11.12. Let U and W be subspaces of V. Prove that $(U + W)^0 = U^0 \cap W^0$.

Let $\phi \in (U + W)^0$. Then ϕ annihilates $U + W$, and so, in particular, ϕ annihilates U and W. That is, $\phi \in U^0$ and $\phi \in W^0$; hence, $\phi \in U^0 \cap W^0$. Thus, $(U + W)^0 \subseteq U^0 \cap W^0$.

On the other hand, suppose $\sigma \in U^0 \cap W^0$. Then σ annihilates U and also W. If $v \in U + W$, then $v = u + w$, where $u \in U$ and $w \in W$. Hence, $\sigma(v) = \sigma(u) + \sigma(w) = 0 + 0 = 0$. Thus, σ annihilates $U + W$; that is, $\sigma \in (U + W)^0$. Accordingly, $U^0 + W^0 \subseteq (U + W)^0$.

The two inclusion relations together give us the desired equality.

Remark: Observe that no dimension argument is employed in the proof; hence, the result holds for spaces of finite or infinite dimension.

Transpose of a Linear Mapping

11.13. Let ϕ be the linear functional on $\mathbf{R}^2$ defined by $\phi(x, y) = x - 2y$. For each of the following linear operators T on $\mathbf{R}^2$, find $(T^t(\phi))(x, y)$:

(a) $T(x, y) = (x, 0)$, (b) $T(x, y) = (y, \ x + y)$, (c) $T(x, y) = (2x - 3y, \ 5x + 2y)$

By definition, $T^t(\phi) = \phi \circ T$; that is, $(T^t(\phi))(v) = \phi(T(v))$ for every v. Hence,

(a) $(T^t(\phi))(x, y) = \phi(T(x, y)) = \phi(x, 0) = x$

(b) $(T^t(\phi))(x, y) = \phi(T(x, y)) = \phi(y, \ x + y) = y - 2(x + y) = -2x - y$

(c) $(T^t(\phi))(x, y) = \phi(T(x, y)) = \phi(2x - 3y, \ 5x + 2y) = (2x - 3y) - 2(5x + 2y) = -8x - 7y$

11.14. Let $T: V \to U$ be linear and let $T^t: U^* \to V^*$ be its transpose. Show that the kernel of T^t is the annihilator of the image of T—that is, $\text{Ker } T^t = (\text{Im } T)^0$.

Suppose $\phi \in \text{Ker } T^t$; that is, $T^t(\phi) = \phi \circ T = 0$. If $u \in \text{Im } T$, then $u = T(v)$ for some $v \in V$; hence,

$$\phi(u) = \phi(T(v)) = (\phi \circ T)(v) = \mathbf{0}(v) = 0$$

We have that $\phi(u) = 0$ for every $u \in \text{Im } T$; hence, $\phi \in (\text{Im } T)^0$. Thus, $\text{Ker } T^t \subseteq (\text{Im } T)^0$.

On the other hand, suppose $\sigma \in (\text{Im } T)^0$; that is, $\sigma(\text{Im } T) = \{0\}$. Then, for every $v \in V$,

$$(T^t(\sigma))(v) = (\sigma \circ T)(v) = \sigma(T(v)) = 0 = \mathbf{0}(v)$$

We have $(T^t(\sigma))(v) = \mathbf{0}(v)$ for every $v \in V$; hence, $T^t(\sigma) = 0$. Thus, $\sigma \in \text{Ker } T^t$, and so $(\text{Im } T)^0 \subseteq \text{Ker } T^t$.

The two inclusion relations together give us the required equality.

11.15. Suppose V and U have finite dimension and $T:V \to U$ is linear. Prove $\text{rank}(T) = \text{rank}(T^t)$.

Suppose $\dim V = n$ and $\dim U = m$, and suppose $\text{rank}(T) = r$. By Theorem 11.5,

$$\dim(\text{Im } T)^0 = \dim u - \dim(\text{Im } T) = m - \text{rank}(T) = m - r$$

By Problem 11.14, $\text{Ker } T^t = (\text{Im } T)^0$. Hence, nullity $(T^t) = m - r$. It then follows that, as claimed,

$$\text{rank}(T^t) = \dim U^* - \text{nullity}(T^t) = m - (m - r) = r = \text{rank}(T)$$

11.16. Prove Theorem 11.7: Let $T:V \to U$ be linear and let A be the matrix representation of T in the bases $\{v_j\}$ of V and $\{u_i\}$ of U. Then the transpose matrix A^T is the matrix representation of $T^t:U^* \to V^*$ in the bases dual to $\{u_i\}$ and $\{v_j\}$.

Suppose, for $j = 1, \dots, m$,

$$T(v_j) = a_{j1}u_1 + a_{j2}u_2 + \cdots + a_{jn}u_n \tag{1}$$

We want to prove that, for $i = 1, \dots, n$,

$$T^t(\sigma_i) = a_{1i}\phi_1 + a_{2i}\phi_2 + \cdots + a_{mi}\phi_m \tag{2}$$

where $\{\sigma_i\}$ and $\{\phi_j\}$ are the bases dual to $\{u_i\}$ and $\{v_j\}$, respectively.

Let $v \in V$ and suppose $v = k_1v_1 + k_2v_2 + \cdots + k_mv_m$. Then, by (1),

$$\begin{aligned} T(v) &= k_1T(v_1) + k_2T(v_2) + \cdots + k_mT(v_m) \\ &= k_1(a_{11}u_1 + \cdots + a_{1n}u_n) + k_2(a_{21}u_1 + \cdots + a_{2n}u_n) + \cdots + k_m(a_{m1}u_1 + \cdots + a_{mn}u_n) \\ &= (k_1a_{11} + k_2a_{21} + \cdots + k_ma_{m1})u_1 + \cdots + (k_1a_{1n} + k_2a_{2n} + \cdots + k_ma_{mn})u_n \\ &= \sum_{i=1}^{n}(k_1a_{1i} + k_2a_{2i} + \cdots + k_ma_{mi})u_i \end{aligned}$$

Hence, for $j = 1, \dots, n$.

$$\begin{aligned} (T^t(\sigma_j))(v) = \sigma_j(T(v)) &= \sigma_j\left(\sum_{i=1}^{n}(k_1a_{1i} + k_2a_{2i} + \cdots + k_ma_{mi})u_i\right) \\ &= k_1a_{1j} + k_2a_{2j} + \cdots + k_ma_{mj} \end{aligned} \tag{3}$$

On the other hand, for $j = 1, \dots, n$,

$$\begin{aligned} (a_{1j}\phi_1 + a_{2j}\phi_2 + \cdots + a_{mj}\phi_m)(v) &= (a_{1j}\phi_1 + a_{2j}\phi_2 + \cdots + a_{mj}\phi_m)(k_1v_1 + k_2v_2 + \cdots + k_mv_m) \\ &= k_1a_{1j} + k_2a_{2j} + \cdots + k_ma_{mj} \end{aligned} \tag{4}$$

Because $v \in V$ was arbitrary, (3) and (4) imply that

$$T^t(\sigma_j) = a_{1j}\phi_1 + a_{2j}\phi_2 + \cdots + a_{mj}\phi_m, \qquad j = 1, \dots, n$$

which is (2). Thus, the theorem is proved.

SUPPLEMENTARY PROBLEMS

Dual Spaces and Dual Bases

11.17. Find (a) $\phi + \sigma$, (b) 3ϕ, (c) $2\phi - 5\sigma$, where $\phi:\mathbf{R}^3 \to \mathbf{R}$ and $\sigma:\mathbf{R}^3 \to \mathbf{R}$ are defined by

$$\phi(x, y, z) = 2x - 3y + z \qquad \text{and} \qquad \sigma(x, y, z) = 4x - 2y + 3z$$

11.18. Find the dual basis of each of the following bases of $\mathbf{R}^3$: (a) $\{(1, 0, 0), \ (0, 1, 0), \ (0, 0, 1)\}$, (b) $\{(1, -2, 3), \ (1, -1, 1), \ (2, -4, 7)\}$.

11.19. Let V be the vector space of polynomials over $\mathbf{R}$ of degree ≤ 2. Let ϕ_1, ϕ_2, ϕ_3 be the linear functionals on V defined by

$$\phi_1(f(t)) = \int_0^1 f(t)\,dt, \qquad \phi_2(f(t)) = f'(1), \qquad \phi_3(f(t)) = f(0)$$

Here $f(t) = a + bt + ct^2 \in V$ and $f'(t)$ denotes the derivative of $f(t)$. Find the basis $\{f_1(t), f_2(t), f_3(t)\}$ of V that is dual to $\{\phi_1, \phi_2, \phi_3\}$.

11.20. Suppose $u, v \in V$ and that $\phi(u) = 0$ implies $\phi(v) = 0$ for all $\phi \in V^*$. Show that $v = ku$ for some scalar k.

11.21. Suppose $\phi, \sigma \in V^*$ and that $\phi(v) = 0$ implies $\sigma(v) = 0$ for all $v \in V$. Show that $\sigma = k\phi$ for some scalar k.

11.22. Let V be the vector space of polynomials over K. For $a \in K$, define $\phi_a : V \to K$ by $\phi_a(f(t)) = f(a)$. Show that (a) ϕ_a is linear; (b) if $a \neq b$, then $\phi_a \neq \phi_b$.

11.23. Let V be the vector space of polynomials of degree ≤ 2. Let $a, b, c \in K$ be distinct scalars. Let ϕ_a, ϕ_b, ϕ_c be the linear functionals defined by $\phi_a(f(t)) = f(a)$, $\phi_b(f(t)) = f(b)$, $\phi_c(f(t)) = f(c)$. Show that $\{\phi_a, \phi_b, \phi_c\}$ is linearly independent, and find the basis $\{f_1(t), f_2(t), f_3(t)\}$ of V that is its dual.

11.24. Let V be the vector space of square matrices of order n. Let $T : V \to K$ be the trace mapping; that is, $T(A) = a_{11} + a_{22} + \cdots + a_{nn}$, where $A = (a_{ij})$. Show that T is linear.

11.25. Let W be a subspace of V. For any linear functional ϕ on W, show that there is a linear functional σ on V such that $\sigma(w) = \phi(w)$ for any $w \in W$; that is, ϕ is the restriction of σ to W.

11.26. Let $\{e_1, \ldots, e_n\}$ be the usual basis of K^n. Show that the dual basis is $\{\pi_1, \ldots, \pi_n\}$ where π_i is the ith projection mapping; that is, $\pi_i(a_1, \ldots, a_n) = a_i$.

11.27. Let V be a vector space over $\mathbf{R}$. Let $\phi_1, \phi_2 \in V^*$ and suppose $\sigma : V \to \mathbf{R}$, defined by $\sigma(v) = \phi_1(v)\phi_2(v)$, also belongs to V^*. Show that either $\phi_1 = \mathbf{0}$ or $\phi_2 = \mathbf{0}$.

Annihilators

11.28. Let W be the subspace of $\mathbf{R}^4$ spanned by $(1, 2, -3, 4)$, $(1, 3, -2, 6)$, $(1, 4, -1, 8)$. Find a basis of the annihilator of W.

11.29. Let W be the subspace of $\mathbf{R}^3$ spanned by $(1, 1, 0)$ and $(0, 1, 1)$. Find a basis of the annihilator of W.

11.30. Show that, for any subset S of V, $\operatorname{span}(S) = S^{00}$, where $\operatorname{span}(S)$ is the linear span of S.

11.31. Let U and W be subspaces of a vector space V of finite dimension. Prove that $(U \cap W)^0 = U^0 + W^0$.

11.32. Suppose $V = U \oplus W$. Prove that $V^* = U^0 \oplus W^0$.

Transpose of a Linear Mapping

11.33. Let ϕ be the linear functional on $\mathbf{R}^2$ defined by $\phi(x, y) = 3x - 2y$. For each of the following linear mappings $T : \mathbf{R}^3 \to \mathbf{R}^2$, find $(T^t(\phi))(x, y, z)$:

(a) $T(x, y, z) = (x + y,\ y + z)$, (b) $T(x, y, z) = (x + y + z,\ 2x - y)$

11.34. Suppose $T_1 : U \to V$ and $T_2 : V \to W$ are linear. Prove that $(T_2 \circ T_1)^t = T_1^t \circ T_2^t$.

11.35. Suppose $T : V \to U$ is linear and V has finite dimension. Prove that $\operatorname{Im} T^t = (\operatorname{Ker} T)^0$.

11.36. Suppose $T:V \to U$ is linear and $u \in U$. Prove that $u \in \text{Im}\, T$ or there exists $\phi \in V^*$ such that $T^t(\phi) = 0$ and $\phi(u) = 1$.

11.37. Let V be of finite dimension. Show that the mapping $T \mapsto T^t$ is an isomorphism from $\text{Hom}(V, V)$ onto $\text{Hom}(V^*, V^*)$. (Here T is any linear operator on V.)

Miscellaneous Problems

11.38. Let V be a vector space over **R**. The *line segment* $\overline{uv}$ joining points $u, v \in V$ is defined by $\overline{uv} = \{tu + (1-t)v : 0 \le t \le 1\}$. A subset S of V is *convex* if $u, v \in S$ implies $\overline{uv} \subseteq S$. Let $\phi \in V^*$. Define

$$W^+ = \{v \in V : \phi(v) > 0\}, \qquad W = \{v \in V : \phi(v) = 0\}, \qquad W^- = \{v \in V : \phi(v) < 0\}$$

Prove that W^+, W, and W^- are convex.

11.39. Let V be a vector space of finite dimension. A *hyperplane* H of V may be defined as the kernel of a nonzero linear functional ϕ on V. Show that every subspace of V is the intersection of a finite number of hyperplanes.

ANSWERS TO SUPPLEMENTARY PROBLEMS

11.17. (a) $6x - 5y + 4z$, (b) $6x - 9y + 3z$, (c) $-16x + 4y - 13z$

11.18. (a) $\phi_1 = x$, $\phi_2 = y$, $\phi_3 = z$; (b) $\phi_1 = -3x - 5y - 2z$, $\phi_2 = 2x + y$, $\phi_3 = x + 2y + z$

11.19. $f_1(t) = 3t - \frac{3}{2}t^2$, $f_2(t) = -\frac{1}{2}t + \frac{3}{4}t^2$, $f_3(t) = 1 - 3t + \frac{3}{2}t^2$

11.22. (b) Let $f(t) = t$. Then $\phi_a(f(t)) = a \neq b = \phi_b(f(t))$; and therefore, $\phi_a \neq \phi_b$

11.23. $\left\{ f_1(t) = \dfrac{t^2 - (b+c)t + bc}{(a-b)(a-c)},\ f_2(t) = \dfrac{t^2 - (a+c)t + ac}{(b-a)(b-c)},\ f_3(t) = \dfrac{t^2 - (a+b)t + ab}{(c-a)(c-b)} \right\}$

11.28. $\{\phi_1(x,y,z,t) = 5x - y + z,\ \phi_2(x,y,z,t) = 2y - t\}$

11.29. $\{\phi(x,y,z) = x - y + z\}$

11.33. (a) $(T^t(\phi))(x,y,z) = 3x + y - 2z$, (b) $(T^t(\phi))(x,y,z) = -x + 5y + 3z$

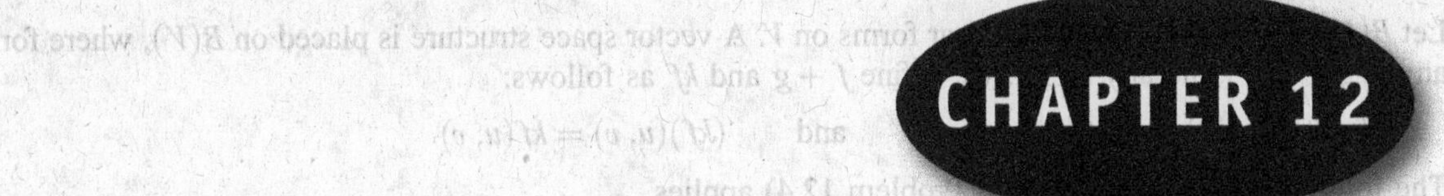

CHAPTER 12

Bilinear, Quadratic, and Hermitian Forms

12.1 Introduction

This chapter generalizes the notions of linear mappings and linear functionals. Specifically, we introduce the notion of a bilinear form. These bilinear maps also give rise to quadratic and Hermitian forms. Although quadratic forms were discussed previously, this chapter is treated independently of the previous results.

Although the field K is arbitrary, we will later specialize to the cases $K = \mathbf{R}$ and $K = \mathbf{C}$. Furthermore, we may sometimes need to divide by 2. In such cases, we must assume that $1 + 1 \neq 0$, which is true when $K = \mathbf{R}$ or $K = \mathbf{C}$.

12.2 Bilinear Forms

Let V be a vector space of finite dimension over a field K. A *bilinear form* on V is a mapping $f: V \times V \to K$ such that, for all $a, b \in K$ and all $u_i, v_i \in V$:

(i) $f(au_1 + bu_2, \ v) = af(u_1, v) + bf(u_2, v)$,

(ii) $f(u, \ av_1 + bv_2) = af(u, v_1) + bf(u, v_2)$

We express condition (i) by saying f is *linear in the first variable*, and condition (ii) by saying f is *linear in the second variable.*

EXAMPLE 12.1

(a) Let f be the dot product on $\mathbf{R}^n$; that is, for $u = (a_i)$ and $v = (b_i)$,

$$f(u, v) = u \cdot v = a_1b_1 + a_2b_2 + \cdots + a_nb_n$$

Then f is a bilinear form on $\mathbf{R}^n$. (In fact, any inner product on a real vector space V is a bilinear form on V.)

(b) Let ϕ and σ be arbitrarily linear functionals on V. Let $f: V \times V \to K$ be defined by $f(u, v) = \phi(u)\sigma(v)$. Then f is a bilinear form, because ϕ and σ are each linear.

(c) Let $A = [a_{ij}]$ be any $n \times n$ matrix over a field K. Then A may be identified with the following bilinear form F on K^n, where $X = [x_i]$ and $Y = [y_i]$ are column vectors of variables:

$$f(X, Y) = X^TAY = \sum_{i,j} a_{ij}x_iy_i = a_{11}x_1y_1 + a_{12}x_1y_2 + \cdots + a_{nn}x_ny_n$$

The above formal expression in the variables x_i, y_i is termed the *bilinear polynomial* corresponding to the matrix A. Equation (12.1) shows that, in a certain sense, every bilinear form is of this type.

Space of Bilinear Forms

Let $B(V)$ denote the set of all bilinear forms on V. A vector space structure is placed on $B(V)$, where for any $f, g \in B(V)$ and any $k \in K$, we define $f + g$ and kf as follows:

$$(f+g)(u, v) = f(u, v) + g(u, v) \qquad \text{and} \qquad (kf)(u, v) = kf(u, v)$$

The following theorem (proved in Problem 12.4) applies.

THEOREM 12.1: Let V be a vector space of dimension n over K. Let $\{\phi_1, \ldots, \phi_n\}$ be any basis of the dual space V^*. Then $\{f_{ij} : i, j = 1, \ldots, n\}$ is a basis of $B(V)$, where f_{ij} is defined by $f_{ij}(u, v) = \phi_i(u)\phi_j(v)$. Thus, in particular, $\dim B(V) = n^2$.

12.3 Bilinear Forms and Matrices

Let f be a bilinear form on V and let $S = \{u_1, \ldots, u_n\}$ be a basis of V. Suppose $u, v \in V$ and

$$u = a_1u_1 + \cdots + a_nu_n \qquad \text{and} \qquad v = b_1u_1 + \cdots + b_nu_n$$

Then

$$f(u, v) = f(a_1u_1 + \cdots + a_nu_n, \; b_1u_1 + \cdots + b_nu_n) = \sum_{i,j} a_ib_jf(u_i, u_j)$$

Thus, f is completely determined by the n^2 values $f(u_i, u_j)$.

The matrix $A = [a_{ij}]$ where $a_{ij} = f(u_i, u_j)$ is called the *matrix representation* of f relative to the basis S or, simply, the "matrix of f in S." It "represents" f in the sense that, for all $u, v \in V$,

$$f(u, v) = \sum_{i,j} a_ib_j f(u_i, u_j) = [u]_S^T A[v]_S \tag{12.1}$$

[As usual, $[u]_S$ denotes the coordinate (column) vector of u in the basis S.]

Change of Basis, Congruent Matrices

We now ask, how does a matrix representing a bilinear form transform when a new basis is selected? The answer is given in the following theorem (proved in Problem 12.5).

THEOREM 12.2: Let P be a change-of-basis matrix from one basis S to another basis S'. If A is the matrix representing a bilinear form f in the original basis S, then $B = P^TAP$ is the matrix representing f in the new basis S'.

The above theorem motivates the following definition.

DEFINITION: A matrix B is *congruent* to a matrix A, written $B \simeq A$, if there exists a nonsingular matrix P such that $B = P^TAP$.

Thus, by Theorem 12.2, matrices representing the same bilinear form are congruent. We remark that congruent matrices have the same rank, because P and P^T are nonsingular; hence, the following definition is well defined.

DEFINITION: The *rank* of a bilinear form f on V, written $\text{rank}(f)$, is the rank of any matrix representation of f. We say f is *degenerate* or *nondegenerate* according to whether $\text{rank}(f) < \dim V$ or $\text{rank}(f) = \dim V$.

12.4 Alternating Bilinear Forms

Let f be a bilinear form on V. Then f is called

(i) *alternating* if $f(v, v) = 0$ for every $v \in V$;
(ii) *skew-symmetric* if $f(u, v) = -f(v, u)$ for every $u, v \in V$.

Now suppose (i) is true. Then (ii) is true, because, for any $u, v, \in V$,

$$0 = f(u+v, \ u+v) = f(u,u) + f(u,v) + f(v,u) + f(v,v) = f(u,v) + f(v,u)$$

On the other hand, suppose (ii) is true and also $1+1 \neq 0$. Then (i) is true, because, for every $v \in V$, we have $f(v,v) = -f(v,v)$. In other words, alternating and skew-symmetric are equivalent when $1+1 \neq 0$.

The main structure theorem of alternating bilinear forms (proved in Problem 12.23) is as follows.

THEOREM 12.3: Let f be an alternating bilinear form on V. Then there exists a basis of V in which f is represented by a block diagonal matrix M of the form

$$M = \text{diag}\left(\begin{bmatrix} 0 & 1 \\ -1 & 0 \end{bmatrix}, \begin{bmatrix} 0 & 1 \\ -1 & 0 \end{bmatrix}, \ldots, \begin{bmatrix} 0 & 1 \\ -1 & 0 \end{bmatrix}, [0], [0], \ldots [0] \right)$$

Moreover, the number of nonzero blocks is uniquely determined by f [because it is equal to $\frac{1}{2}$ rank(f)].

In particular, the above theorem shows that any alternating bilinear form must have even rank.

12.5 Symmetric Bilinear Forms, Quadratic Forms

This section investigates the important notions of symmetric bilinear forms and quadratic forms and their representation by means of symmetric matrices. The only restriction on the field K is that $1+1 \neq 0$. In Section 12.6, we will restrict K to be the real field $\mathbf{R}$, which yields important special results.

Symmetric Bilinear Forms

Let f be a bilinear form on V. Then f is said to be *symmetric* if, for every $u, v \in V$,

$$f(u,v) = f(v,u)$$

One can easily show that f is symmetric if and only if any matrix representation A of f is a symmetric matrix.

The main result for symmetric bilinear forms (proved in Problem 12.10) is as follows. (We emphasize that we are assuming that $1+1 \neq 0$.)

THEOREM 12.4: Let f be a symmetric bilinear form on V. Then V has a basis $\{v_1, \ldots, v_n\}$ in which f is represented by a diagonal matrix—that is, where $f(v_i, v_j) = 0$ for $i \neq j$.

THEOREM 12.4: (Alternative Form) Let A be a symmetric matrix over K. Then A is congruent to a diagonal matrix; that is, there exists a nonsingular matrix P such that P^TAP is diagonal.

Diagonalization Algorithm

Recall that a nonsingular matrix P is a product of elementary matrices. Accordingly, one way of obtaining the diagonal form $D = P^TAP$ is by a sequence of elementary row operations and the same sequence of elementary column operations. This same sequence of elementary row operations on the identity matrix I will yield P^T. This algorithm is formalized below.

ALGORITHM 12.1: (Congruence Diagonalization of a Symmetric Matrix) The input is a symmetric matrix $A = [a_{ij}]$ of order n.

Step 1. Form the $n \times 2n$ (block) matrix $M = [A_1, I]$, where $A_1 = A$ is the left half of M and the identity matrix I is the right half of M.

Step 2. Examine the entry a_{11}. There are three cases.

Case I: $a_{11} \neq 0$. (Use a_{11} as a pivot to put 0's below a_{11} in M and to the right of a_{11} in A_1.) For $i = 2, \ldots, n$:

(a) Apply the row operation "Replace R_i by $-a_{i1}R_1 + a_{11}R_i$."

(b) Apply the corresponding column operation "Replace C_i by $-a_{i1}C_1 + a_{11}C_i$."

These operations reduce the matrix M to the form

$$M \sim \begin{bmatrix} a_{11} & 0 & * & * \\ 0 & A_1 & * & * \end{bmatrix} \qquad (*)$$

Case II: $a_{11} = 0$ but $a_{kk} \neq 0$, for some $k > 1$.

(a) Apply the row operation "Interchange R_1 and R_k."

(b) Apply the corresponding column operation "Interchange C_1 and C_k."

(These operations bring a_{kk} into the first diagonal position, which reduces the matrix to Case I.)

Case III: All diagonal entries $a_{ii} = 0$ but some $a_{ij} \neq 0$.

(a) Apply the row operation "Replace R_i by $R_j + R_i$."

(b) Apply the corresponding column operation "Replace C_i by $C_j + C_i$."

(These operations bring $2a_{ij}$ into the ith diagonal position, which reduces the matrix to Case II.)

Thus, M is finally reduced to the form (*), where A_2 is a symmetric matrix of order less than A.

Step 3. Repeat Step 2 with each new matrix A_k (by neglecting the first row and column of the preceding matrix) until A is diagonalized. Then M is transformed into the form $M' = [D, Q]$, where D is diagonal.

Step 4. Set $P = Q^T$. Then $D = P^TAP$.

Remark 1: We emphasize that in Step 2, the row operations will change both sides of M, but the column operations will only change the left half of M.

Remark 2: The condition $1 + 1 \neq 0$ is used in Case III, where we assume that $2a_{ij} \neq 0$ when $a_{ij} \neq 0$.

The justification for the above algorithm appears in Problem 12.9.

EXAMPLE 12.2 Let $A = \begin{bmatrix} 1 & 2 & -3 \\ 2 & 5 & -4 \\ -3 & -4 & 8 \end{bmatrix}$. Apply Algorithm 9.1 to find a nonsingular matrix P such that $D = P^TAP$ is diagonal.

First form the block matrix $M = [A, I]$; that is, let

$$M = [A, I] = \left[\begin{array}{ccc|ccc} 1 & 2 & -3 & 1 & 0 & 0 \\ 2 & 5 & -4 & 0 & 1 & 0 \\ -3 & -4 & 8 & 0 & 0 & 1 \end{array}\right]$$

Apply the row operations "Replace R_2 by $-2R_1 + R_2$" and "Replace R_3 by $3R_1 + R_3$" to M, and then apply the corresponding column operations "Replace C_2 by $-2C_1 + C_2$" and "Replace C_3 by $3C_1 + C_3$" to obtain

$$\left[\begin{array}{ccc|ccc} 1 & 2 & -3 & 1 & 0 & 0 \\ 0 & 1 & 2 & -2 & 1 & 0 \\ 0 & 2 & -1 & 3 & 0 & 1 \end{array}\right] \quad \text{and then} \quad \left[\begin{array}{ccc|ccc} 1 & 0 & 0 & 1 & 0 & 0 \\ 0 & 1 & 2 & -2 & 1 & 0 \\ 0 & 2 & -1 & 3 & 0 & 1 \end{array}\right]$$

Next apply the row operation "Replace R_3 by $-2R_2 + R_3$" and then the corresponding column operation "Replace C_3 by $-2C_2 + C_3$" to obtain

$$\left[\begin{array}{ccc|ccc} 1 & 0 & 0 & 1 & 0 & 0 \\ 0 & 1 & 2 & -2 & 1 & 0 \\ 0 & 0 & -5 & 7 & -2 & 1 \end{array}\right] \quad \text{and then} \quad \left[\begin{array}{ccc|ccc} 1 & 0 & 0 & 1 & 0 & 0 \\ 0 & 1 & 0 & -2 & 1 & 0 \\ 0 & 0 & -5 & 7 & -2 & 1 \end{array}\right]$$

Now A has been diagonalized. Set

$$P = \begin{bmatrix} 1 & -2 & 7 \\ 0 & 1 & -2 \\ 0 & 0 & 1 \end{bmatrix} \quad \text{and then} \quad D = P^{-1}AP = \begin{bmatrix} 1 & 0 & 0 \\ 0 & 1 & 0 \\ 0 & 0 & -5 \end{bmatrix}$$

We emphasize that P is the transpose of the right half of the final matrix.

Quadratic Forms

We begin with a definition.

DEFINITION A: A mapping $q: V \to K$ is a *quadratic form* if $q(v) = f(v, v)$ for some symmetric bilinear form f on V.

If $1 + 1 \neq 0$ in K, then the bilinear form f can be obtained from the quadratic form q by the following *polar form* of f:

$$f(u, v) = \tfrac{1}{2}[q(u + v) - q(u) - q(v)]$$

Now suppose f is represented by a symmetric matrix $A = [a_{ij}]$, and $1 + 1 \neq 0$. Letting $X = [x_i]$ denote a column vector of variables, q can be represented in the form

$$q(X) = f(X, X) = X^T AX = \sum_{i,j} a_{ij}x_i x_j = \sum_i a_{ii}x_i^2 + 2\sum_{i<j} a_{ij}x_i x_j$$

The above formal expression in the variables x_i is also called a quadratic form. Namely, we have the following second definition.

DEFINITION B: A *quadratic form* q in variables $x_1, x_2, \ldots, x_n$ is a polynomial such that every term has degree two; that is,

$$q(x_1, x_2, \ldots, x_n) = \sum_i c_i x_i^2 + \sum_{i<j} d_{ij}x_i x_j$$

Using $1 + 1 \neq 0$, the quadratic form q in Definition B determines a symmetric matrix $A = [a_{ij}]$ where $a_{ii} = c_i$ and $a_{ij} = a_{ji} = \frac{1}{2}d_{ij}$. Thus, Definitions A and B are essentially the same.

If the matrix representation A of q is diagonal, then q has the diagonal representation

$$q(X) = X^T AX = a_{11}x_1^2 + a_{22}x_2^2 + \cdots + a_{nn}x_n^2$$

That is, the quadratic polynomial representing q will contain no "cross product" terms. Moreover, by Theorem 12.4, every quadratic form has such a representation (when $1 + 1 \neq 0$).

12.6 Real Symmetric Bilinear Forms, Law of Inertia

This section treats symmetric bilinear forms and quadratic forms on vector spaces V over the real field $\mathbf{R}$. The special nature of $\mathbf{R}$ permits an independent theory. The main result (proved in Problem 12.14) is as follows.

THEOREM 12.5: Let f be a symmetric form on V over $\mathbf{R}$. Then there exists a basis of V in which f is represented by a diagonal matrix. Every other diagonal matrix representation of f has the same number $\mathbf{p}$ of positive entries and the same number $\mathbf{n}$ of negative entries.

The above result is sometimes called the *Law of Inertia* or *Sylvester's Theorem*. The *rank* and *signature* of the symmetric bilinear form f are denoted and defined by

$$\text{rank}(f) = \mathbf{p} + \mathbf{n} \quad \text{and} \quad \text{sig}(f) = \mathbf{p} - \mathbf{n}$$

These are uniquely defined by Theorem 12.5.

A real symmetric bilinear form f is said to be

(i) *positive definite* if $q(v) = f(v, v) > 0$ for every $v \neq 0$,

(ii) *nonnegative semidefinite* if $q(v) = f(v, v) \geq 0$ for every v.

EXAMPLE 12.3 Let f be the dot product on $\mathbf{R}^n$. Recall that f is a symmetric bilinear form on $\mathbf{R}^n$. We note that f is also positive definite. That is, for any $u = (a_i) \neq 0$ in $\mathbf{R}^n$,

$$f(u, u) = a_1^2 + a_2^2 + \cdots + a_n^2 > 0$$

Section 12.5 and Chapter 13 tell us how to diagonalize a real quadratic form q or, equivalently, a real symmetric matrix A by means of an orthogonal transition matrix P. If P is merely nonsingular, then q can be represented in diagonal form with only 1's and -1's as nonzero coefficients. Namely, we have the following corollary.

COROLLARY 12.6: Any real quadratic form q has a unique representation in the form

$$q(x_1, x_2, \ldots, x_n) = x_1^2 + \cdots + x_{\mathbf{p}}^2 - x_{\mathbf{p}+1}^2 - \cdots - x_r^2$$

where $r = \boldsymbol{p} + \boldsymbol{n}$ is the rank of the form.

COROLLARY 12.6: (Alternative Form) Any real symmetric matrix A is congruent to the unique diagonal matrix

$$D = \text{diag}(I_{\mathbf{p}}, -I_{\mathbf{n}}, 0)$$

where $r = \boldsymbol{p} + \boldsymbol{n}$ is the rank of A.

12.7 Hermitian Forms

Let V be a vector space of finite dimension over the complex field $\mathbf{C}$. A *Hermitian form* on V is a mapping $f: V \times V \to \mathbf{C}$ such that, for all $a, b \in C$ and all $u_i, v \in V$,

(i) $f(au_1 + bu_2, \ v) = af(u_1, v) + bf(u_2, v)$,

(ii) $f(u, v) = \overline{f(v, u)}$.

(As usual, $\bar{k}$ denotes the complex conjugate of $k \in \mathbf{C}$.)

Using (i) and (ii), we get

$$\begin{aligned} f(u, \ av_1 + bv_2) &= \overline{f(av_1 + bv_2, \ u)} = \overline{af(v_1, u) + bf(v_2, u)} \\ &= \bar{a}\overline{f(v_1, u)} + \bar{b}\overline{f(v_2, u)} = \bar{a}f(u, v_1) + \bar{b}f(u, v_2) \end{aligned}$$

That is,

(iii) $f(u, \ av_1 + bv_2) = \bar{a}f(u, v_1) + \bar{b}f(u, v_2)$.

As before, we express condition (i) by saying f is linear in the first variable. On the other hand, we express condition (iii) by saying f is "conjugate linear" in the second variable. Moreover, condition (ii) tells us that $f(v, v) = \overline{f(v, v)}$, and hence, $f(v, v)$ is real for every $v \in V$.

The results of Sections 12.5 and 12.6 for symmetric forms have their analogues for Hermitian forms. Thus, the mapping $q: V \to \mathbf{C}$, defined by $q(v) = f(v, v)$, is called the *Hermitian quadratic form* or *complex quadratic form* associated with the Hermitian form f. We can obtain f from q by the polar form

$$f(u, v) = \tfrac{1}{4}[q(u + v) - q(u - v)] + \tfrac{1}{4}[q(u + iv) - q(u - iv)]$$

Now suppose $S = \{u_1, \ldots, u_n\}$ is a basis of V. The matrix $H = [h_{ij}]$ where $h_{ij} = f(u_i, u_j)$ is called the *matrix representation* of f in the basis S. By (ii), $f(u_i, u_j) = \overline{f(u_j, u_i)}$; hence, H is Hermitian and, in particular, the diagonal entries of H are real. Thus, any diagonal representation of f contains only real entries.

The next theorem (to be proved in Problem 12.47) is the complex analog of Theorem 12.5 on real symmetric bilinear forms.

THEOREM 12.7: Let f be a Hermitian form on V over $\mathbf{C}$. Then there exists a basis of V in which f is represented by a diagonal matrix. Every other diagonal matrix representation of f has the same number $\mathbf{p}$ of positive entries and the same number $\mathbf{n}$ of negative entries.

Again the *rank* and *signature* of the Hermitian form f are denoted and defined by

$$\operatorname{rank}(f) = \mathbf{p} + \mathbf{n} \qquad \text{and} \qquad \operatorname{sig}(f) = \mathbf{p} - \mathbf{n}$$

These are uniquely defined by Theorem 12.7.

Analogously, a Hermitian form f is said to be

(i) *positive definite* if $q(v) = f(v, v) > 0$ for every $v \neq 0$,

(ii) *nonnegative semidefinite* if $q(v) = f(v, v) \geq 0$ for every v.

EXAMPLE 12.4 Let f be the dot product on $\mathbf{C}^n$; that is, for any $u = (z_i)$ and $v = (w_i)$ in $\mathbf{C}^n$,

$$f(u, v) = u \cdot v = z_1\bar{w}_1 + z_2\bar{w}_2 + \cdots + z_n\bar{w}_n$$

Then f is a Hermitian form on $\mathbf{C}^n$. Moreover, f is also positive definite, because, for any $u = (z_i) \neq 0$ in $\mathbf{C}^n$,

$$f(u, u) = z_1\bar{z}_1 + z_2\bar{z}_2 + \cdots + z_n\bar{z}_n = |z_1|^2 + |z_2|^2 + \cdots + |z_n|^2 > 0$$

SOLVED PROBLEMS

Bilinear Forms

12.1. Let $u = (x_1, x_2, x_3)$ and $v = (y_1, y_2, y_3)$. Express f in matrix notation, where

$$f(u, v) = 3x_1y_1 - 2x_1y_3 + 5x_2y_1 + 7x_2y_2 - 8x_2y_3 + 4x_3y_2 - 6x_3y_3$$

Let $A = [a_{ij}]$, where a_{ij} is the coefficient of x_iy_j. Then

$$f(u, v) = X^TAY = [x_1, x_2, x_3] \begin{bmatrix} 3 & 0 & -2 \\ 5 & 7 & -8 \\ 0 & 4 & -6 \end{bmatrix} \begin{bmatrix} y_1 \\ y_2 \\ y_3 \end{bmatrix}$$

12.2. Let A be an $n \times n$ matrix over K. Show that the mapping f defined by $f(X, Y) = X^TAY$ is a bilinear form on K^n.

For any $a, b \in K$ and any $X_i, Y_i \in K^n$,

$$\begin{aligned} f(aX_1 + bX_2,\ Y) &= (aX_1 + bX_2)^TAY = (aX_1^T + bX_2^T)AY \\ &= aX_1^TAY + bX_2^TAY = af(X_1, Y) + bf(X_2, Y) \end{aligned}$$

Hence, f is linear in the first variable. Also,

$$f(X,\ aY_1 + bY_2) = X^TA(aY_1 + bY_2) = aX^TAY_1 + bX^TAY_2 = af(X, Y_1) + bf(X, Y_2)$$

Hence, f is linear in the second variable, and so f is a bilinear form on K^n.

12.3. Let f be the bilinear form on $\mathbf{R}^2$ defined by

$$f[(x_1, x_2),\ (y_1, y_2)] = 2x_1y_1 - 3x_1y_2 + 4x_2y_2$$

(a) Find the matrix A of f in the basis $\{u_1 = (1, 0),\ u_2 = (1, 1)\}$.
(b) Find the matrix B of f in the basis $\{v_1 = (2, 1),\ v_2 = (1, -1)\}$.
(c) Find the change-of-basis matrix P from the basis $\{u_i\}$ to the basis $\{v_i\}$, and verify that $B = P^TAP$.

(a) Set $A = [a_{ij}]$, where $a_{ij} = f(u_i, u_j)$. This yields

$$a_{11} = f[(1,0),\ (1,0)] = 2 - 0 - 0 = 2, \qquad a_{21} = f[(1,1),\ (1,0)] = 2 - 0 + 0 = 2$$
$$a_{12} = f[(1,0),\ (1,1)] = 2 - 3 - 0 = -1, \qquad a_{22} = f[(1,1),\ (1,1)] = 2 - 3 + 4 = 3$$

Thus, $A = \begin{bmatrix} 2 & -1 \\ 2 & 3 \end{bmatrix}$ is the matrix of f in the basis $\{u_1, u_2\}$.

(b) Set $B = [b_{ij}]$, where $b_{ij} = f(v_i, v_j)$. This yields

$$b_{11} = f[(2,1),\ (2,1)] = 8 - 6 + 4 = 6, \qquad b_{21} = f[(1,-1),\ (2,1)] = 4 - 3 - 4 = -3$$
$$b_{12} = f[(2,1),\ (1,-1)] = 4 + 6 - 4 = 6, \qquad b_{22} = f[(1,-1),\ (1,-1)] = 2 + 3 + 4 = 9$$

Thus, $B = \begin{bmatrix} 6 & 6 \\ -3 & 9 \end{bmatrix}$ is the matrix of f in the basis $\{v_1, v_2\}$.

(c) Writing v_1 and v_2 in terms of the u_i yields $v_1 = u_1 + u_2$ and $v_2 = 2u_1 - u_2$. Then

$$P = \begin{bmatrix} 1 & 2 \\ 1 & -1 \end{bmatrix}, \qquad P^T = \begin{bmatrix} 1 & 1 \\ 2 & -1 \end{bmatrix}$$

and

$$P^TAP = \begin{bmatrix} 1 & 1 \\ 2 & -1 \end{bmatrix} \begin{bmatrix} 2 & -1 \\ 2 & 3 \end{bmatrix} \begin{bmatrix} 1 & 2 \\ 1 & -1 \end{bmatrix} = \begin{bmatrix} 6 & 6 \\ -3 & 9 \end{bmatrix} = B$$

12.4. Prove Theorem 12.1: Let V be an n-dimensional vector space over K. Let $\{\phi_1, \ldots, \phi_n\}$ be any basis of the dual space V^*. Then $\{f_{ij} : i, j = 1, \ldots, n\}$ is a basis of $B(V)$, where f_{ij} is defined by $f_{ij}(u, v) = \phi_i(u)\phi_j(v)$. Thus, $\dim B(V) = n^2$.

Let $\{u_1, \ldots, u_n\}$ be the basis of V dual to $\{\phi_i\}$. We first show that $\{f_{ij}\}$ spans $B(V)$. Let $f \in B(V)$ and suppose $f(u_i, u_j) = a_{ij}$. We claim that $f = \sum_{i,j} a_{ij} f_{ij}$. It suffices to show that

$$f(u_s, u_t) = \left(\sum a_{ij} f_{ij}\right)(u_s, u_t) \qquad \text{for} \quad s, t = 1, \ldots, n$$

We have

$$\left(\sum a_{ij} f_{ij}\right)(u_s, u_t) = \sum a_{ij} f_{ij}(u_s, u_t) = \sum a_{ij}\phi_i(u_s)\phi_j(u_t) = \sum a_{ij}\delta_{is}\delta_{jt} = a_{st} = f(u_s, u_t)$$

as required. Hence, $\{f_{ij}\}$ spans $B(V)$. Next, suppose $\sum a_{ij} f_{ij} = \mathbf{0}$. Then for $s, t = 1, \ldots, n$,

$$0 = \mathbf{0}(u_s, u_t) = \left(\sum a_{ij} f_{ij}\right)(u_s, u_t) = a_{rs}$$

The last step follows as above. Thus, $\{f_{ij}\}$ is independent, and hence is a basis of $B(V)$.

12.5. Prove Theorem 12.2. Let P be the change-of-basis matrix from a basis S to a basis S'. Let A be the matrix representing a bilinear form in the basis S. Then $B = P^TAP$ is the matrix representing f in the basis S'.

Let $u, v \in V$. Because P is the change-of-basis matrix from S to S', we have $P[u]_{S'} = [u]_S$ and also $P[v]_{S'} = [v]_S$; hence, $[u]_S^T = [u]_{S'}^T P^T$. Thus,

$$f(u, v) = [u]_S^T A [v]_S = [u]_{S'}^T P^T A P [v]_{S'}$$

Because u and v are arbitrary elements of V, P^TAP is the matrix of f in the basis S'.

Symmetric Bilinear Forms, Quadratic Forms

12.6. Find the symmetric matrix that corresponds to each of the following quadratic forms:

(a) $q(x,y,z) = 3x^2 + 4xy - y^2 + 8xz - 6yz + z^2$,

(b) $q'(x,y,z) = 3x^2 + xz - 2yz$, (c) $q''(x,y,z) = 2x^2 - 5y^2 - 7z^2$

The symmetric matrix $A = [a_{ij}]$ that represents $q(x_1, \ldots, x_n)$ has the diagonal entry a_{ii} equal to the coefficient of the square term x_i^2 and the nondiagonal entries a_{ij} and a_{ji} each equal to half of the coefficient of the cross-product term x_ix_j. Thus,

$$\text{(a)}\ A = \begin{bmatrix} 3 & 2 & 4 \\ 2 & -1 & -3 \\ 4 & -3 & 1 \end{bmatrix}, \text{(b)}\ A' = \begin{bmatrix} 3 & 0 & \frac{1}{2} \\ 0 & 0 & -1 \\ \frac{1}{2} & -1 & 0 \end{bmatrix}, \text{(c)}\ A'' = \begin{bmatrix} 2 & 0 & 0 \\ 0 & -5 & 0 \\ 0 & 0 & -7 \end{bmatrix}$$

The third matrix A'' is diagonal, because the quadratic form q'' is diagonal; that is, q'' has no cross-product terms.

12.7. Find the quadratic form $q(X)$ that corresponds to each of the following symmetric matrices:

$$\text{(a)}\ A = \begin{bmatrix} 5 & -3 \\ -3 & 8 \end{bmatrix}, \text{(b)}\ B = \begin{bmatrix} 4 & -5 & 7 \\ -5 & -6 & 8 \\ 7 & 8 & -9 \end{bmatrix}, \text{(c)}\ C = \begin{bmatrix} 2 & 4 & -1 & 5 \\ 4 & -7 & -6 & 8 \\ -1 & -6 & 3 & 9 \\ 5 & 8 & 9 & 1 \end{bmatrix}$$

The quadratic form $q(X)$ that corresponds to a symmetric matrix M is defined by $q(X) = X^TMX$, where $X = [x_i]$ is the column vector of unknowns.

(a) Compute as follows:

$$q(x,y) = X^TAX = [x,y]\begin{bmatrix} 5 & -3 \\ -3 & 8 \end{bmatrix}\begin{bmatrix} x \\ y \end{bmatrix} = [5x - 3y,\ -3x + 8y]\begin{bmatrix} x \\ y \end{bmatrix}$$
$$= 5x^2 - 3xy - 3xy + 8y^2 = 5x^2 - 6xy + 8y^2$$

As expected, the coefficient 5 of the square term x^2 and the coefficient 8 of the square term y^2 are the diagonal elements of A, and the coefficient -6 of the cross-product term xy is the sum of the nondiagonal elements -3 and -3 of A (or twice the nondiagonal element -3, because A is symmetric).

(b) Because B is a three-square matrix, there are three unknowns, say x, y, z or x_1, x_2, x_3. Then

$$q(x,y,z) = 4x^2 - 10xy - 6y^2 + 14xz + 16yz - 9z^2$$

or

$$q(x_1,x_2,x_3) = 4x_1^2 - 10x_1x_2 - 6x_2^2 + 14x_1x_3 + 16x_2x_3 - 9x_3^2$$

Here we use the fact that the coefficients of the square terms x_1^2, x_2^2, x_3^2 (or x^2, y^2, z^2) are the respective diagonal elements $4, -6, -9$ of B, and the coefficient of the cross-product term x_ix_j is the sum of the nondiagonal elements b_{ij} and b_{ji} (or twice b_{ij}, because $b_{ij} = b_{ji}$).

(c) Because C is a four-square matrix, there are four unknowns. Hence,

$$q(x_1,x_2,x_3,x_4) = 2x_1^2 - 7x_2^2 + 3x_3^2 + x_4^2 + 8x_1x_2 - 2x_1x_3 + 10x_1x_4 - 12x_2x_3 + 16x_2x_4 + 18x_3x_4$$

12.8. Let $A = \begin{bmatrix} 1 & -3 & 2 \\ -3 & 7 & -5 \\ 2 & -5 & 8 \end{bmatrix}$. Apply Algorithm 12.1 to find a nonsingular matrix P such that $D = P^TAP$ is diagonal, and find $\text{sig}(A)$, the signature of A.

First form the block matrix $M = [A, I]$:

$$M = [A, I] = \left[\begin{array}{rrr|rrr} 1 & -3 & 2 & 1 & 0 & 0 \\ -3 & 7 & -5 & 0 & 1 & 0 \\ 2 & -5 & 8 & 0 & 0 & 1 \end{array}\right]$$

Using $a_{11} = 1$ as a pivot, apply the row operations "Replace R_2 by $3R_1 + R_2$" and "Replace R_3 by $-2R_1 + R_3$" to M and then apply the corresponding column operations "Replace C_2 by $3C_1 + C_2$" and "Replace C_3 by $-2C_1 + C_3$" to A to obtain

$$\left[\begin{array}{rrr|rrr} 1 & -3 & 2 & 1 & 0 & 0 \\ 0 & -2 & 1 & 3 & 1 & 0 \\ 0 & 1 & 4 & -2 & 0 & 1 \end{array}\right] \quad \text{and then} \quad \left[\begin{array}{rrr|rrr} 1 & 0 & 0 & 1 & 0 & 0 \\ 0 & -2 & 1 & 3 & 1 & 0 \\ 0 & 1 & 4 & -2 & 0 & 1 \end{array}\right].$$

Next apply the row operation "Replace R_3 by $R_2 + 2R_3$" and then the corresponding column operation "Replace C_3 by $C_2 + 2C_3$" to obtain

$$\begin{bmatrix} 1 & 0 & 0 & 1 & 0 & 0 \\ 0 & -2 & 1 & 3 & 1 & 0 \\ 0 & 0 & 9 & -1 & 1 & 2 \end{bmatrix} \quad \text{and then} \quad \begin{bmatrix} 1 & 0 & 0 & 1 & 0 & 0 \\ 0 & -2 & 0 & 3 & 1 & 0 \\ 0 & 0 & 18 & -1 & 1 & 2 \end{bmatrix}$$

Now A has been diagonalized and the transpose of P is in the right half of M. Thus, set

$$P = \begin{bmatrix} 1 & 3 & -1 \\ 0 & 1 & 1 \\ 0 & 0 & 2 \end{bmatrix} \quad \text{and then} \quad D = P^TAP = \begin{bmatrix} 1 & 0 & 0 \\ 0 & -2 & 0 \\ 0 & 0 & 18 \end{bmatrix}$$

Note D has $\mathbf{p} = 2$ positive and $\mathbf{n} = 1$ negative diagonal elements. Thus, the signature of A is $\text{sig}(A) = \mathbf{p} - \mathbf{n} = 2 - 1 = 1$.

12.9. Justify Algorithm 12.1, which diagonalizes (under congruence) a symmetric matrix A.

Consider the block matrix $M = [A, I]$. The algorithm applies a sequence of elementary row operations and the corresponding column operations to the left side of M, which is the matrix A. This is equivalent to premultiplying A by a sequence of elementary matrices, say, $E_1, E_2, \ldots, E_r$, and postmultiplying A by the transposes of the E_i. Thus, when the algorithm ends, the diagonal matrix D on the left side of M is equal to

$$D = E_r \cdots E_2 E_1 A E_1^T E_2^T \cdots E_r^T = QAQ^T, \qquad \text{where} \qquad Q = E_r \cdots E_2 E_1$$

On the other hand, the algorithm only applies the elementary row operations to the identity matrix I on the right side of M. Thus, when the algorithm ends, the matrix on the right side of M is equal to

$$E_r \cdots E_2 E_1 I = E_r \cdots E_2 E_1 = Q$$

Setting $P = Q^T$, we get $D = P^TAP$, which is a diagonalization of A under congruence.

12.10. Prove Theorem 12.4: Let f be a symmetric bilinear form on V over K (where $1 + 1 \neq 0$). Then V has a basis in which f is represented by a diagonal matrix.

Algorithm 12.1 shows that every symmetric matrix over K is congruent to a diagonal matrix. This is equivalent to the statement that f has a diagonal representation.

12.11. Let q be the quadratic form associated with the symmetric bilinear form f. Verify the polar identity $f(u, v) = \frac{1}{2}[q(u + v) - q(u) - q(v)]$. (Assume that $1 + 1 \neq 0$.)

We have

$$\begin{aligned} q(u + v) - q(u) - q(v) &= f(u + v,\ u + v) - f(u, u) - f(v, v) \\ &= f(u, u) + f(u, v) + f(v, u) + f(v, v) - f(u, u) - f(v, v) = 2f(u, v) \end{aligned}$$

If $1 + 1 \neq 0$, we can divide by 2 to obtain the required identity.

12.12. Consider the quadratic form $q(x,y) = 3x^2 + 2xy - y^2$ and the linear substitution

$$x = s - 3t, \qquad y = 2s + t$$

(a) Rewrite $q(x,y)$ in matrix notation, and find the matrix A representing $q(x,y)$.

(b) Rewrite the linear substitution using matrix notation, and find the matrix P corresponding to the substitution.

(c) Find $q(s,t)$ using direct substitution.

(d) Find $q(s,t)$ using matrix notation.

(a) Here $q(x,y) = [x,y]\begin{bmatrix} 3 & 1 \\ 1 & -1 \end{bmatrix}\begin{bmatrix} x \\ y \end{bmatrix}$. Thus, $A = \begin{bmatrix} 3 & 1 \\ 1 & -1 \end{bmatrix}$; and $q(X) = X^TAX$, where $X = [x,y]^T$.

(b) Here $\begin{bmatrix} x \\ y \end{bmatrix} = \begin{bmatrix} 1 & -3 \\ 2 & 1 \end{bmatrix}\begin{bmatrix} s \\ t \end{bmatrix}$. Thus, $P = \begin{bmatrix} 1 & -3 \\ 2 & 1 \end{bmatrix}$; and $X = \begin{bmatrix} x \\ y \end{bmatrix}$, $Y = \begin{bmatrix} s \\ t \end{bmatrix}$ and $X = PY$.

(c) Substitute for x and y in q to obtain

$$\begin{aligned} q(s,t) &= 3(s-3t)^2 + 2(s-3t)(2s+t) - (2s+t)^2 \\ &= 3(s^2 - 6st + 9t^2) + 2(2s^2 - 5st - 3t^2) - (4s^2 + 4st + t^2) = 3s^2 - 32st + 20t^2 \end{aligned}$$

(d) Here $q(X) = X^TAX$ and $X = PY$. Thus, $X^T = Y^TP^T$. Therefore,

$$\begin{aligned} q(s,t) = q(Y) = Y^TP^TAPY &= [s,t]\begin{bmatrix} 1 & 2 \\ -3 & 1 \end{bmatrix}\begin{bmatrix} 3 & 1 \\ 1 & -1 \end{bmatrix}\begin{bmatrix} 1 & -3 \\ 2 & 1 \end{bmatrix}\begin{bmatrix} s \\ t \end{bmatrix} \\ &= [s,t]\begin{bmatrix} 3 & -16 \\ -16 & 20 \end{bmatrix}\begin{bmatrix} s \\ t \end{bmatrix} = 3s^2 - 32st + 20t^2 \end{aligned}$$

[As expected, the results in parts (c) and (d) are equal.]

12.13. Consider any diagonal matrix $A = \operatorname{diag}(a_1, \ldots, a_n)$ over K. Show that for any nonzero scalars $k_1, \ldots, k_n \in K$, A is congruent to a diagonal matrix D with diagonal entries $a_1k_1^2, \ldots, a_nk_n^2$. Furthermore, show that

(a) If $K = \mathbf{C}$, then we can choose D so that its diagonal entries are only 1's and 0's.

(b) If $K = \mathbf{R}$, then we can choose D so that its diagonal entries are only 1's, -1's, and 0's.

Let $P = \operatorname{diag}(k_1, \ldots, k_n)$. Then, as required,

$$D = P^TAP = \operatorname{diag}(k_i)\operatorname{diag}(a_i)\operatorname{diag}(k_i) = \operatorname{diag}(a_1k_1^2, \ldots, a_nk_n^2)$$

(a) Let $P = \operatorname{diag}(b_i)$, where $b_i = \begin{cases} 1/\sqrt{a_i} & \text{if } a_i \neq 0 \\ 1 & \text{if } a_i = 0 \end{cases}$

Then P^TAP has the required form.

(b) Let $P = \operatorname{diag}(b_i)$, where $b_i = \begin{cases} 1/\sqrt{|a_i|} & \text{if } a_i \neq 0 \\ 1 & \text{if } a_i = 0 \end{cases}$

Then P^TAP has the required form.

Remark: We emphasize that (b) is no longer true if "congruence" is replaced by "Hermitian congruence."

12.14. Prove Theorem 12.5: Let f be a symmetric bilinear form on V over $\mathbf{R}$. Then there exists a basis of V in which f is represented by a diagonal matrix. Every other diagonal matrix representation of f has the same number $\mathbf{p}$ of positive entries and the same number $\mathbf{n}$ of negative entries.

By Theorem 12.4, there is a basis $\{u_1, \ldots, u_n\}$ of V in which f is represented by a diagonal matrix with, say, $\mathbf{p}$ positive and $\mathbf{n}$ negative entries. Now suppose $\{w_1, \ldots, w_n\}$ is another basis of V, in which f is represented by a diagonal matrix with $\mathbf{p}'$ positive and $\mathbf{n}'$ negative entries. We can assume without loss of generality that the positive entries in each matrix appear first. Because $\operatorname{rank}(f) = \mathbf{p} + \mathbf{n} = \mathbf{p}' + \mathbf{n}'$, it suffices to prove that $\mathbf{p} = \mathbf{p}'$.

Let U be the linear span of $u_1, \ldots, u_{\mathbf{p}}$ and let W be the linear span of $w_{\mathbf{p}'+1}, \ldots, w_n$. Then $f(v, v) > 0$ for every nonzero $v \in U$, and $f(v, v) \leq 0$ for every nonzero $v \in W$. Hence, $U \cap W = \{0\}$. Note that $\dim U = \mathbf{p}$ and $\dim W = n - \mathbf{p}'$. Thus,

$$\dim(U + W) = \dim U + \dim W - \dim(U \cap W) = \mathbf{p} + (n - \mathbf{p}') - 0 = \mathbf{p} - \mathbf{p}' + n$$

But $\dim(U + W) \leq \dim V = n$; hence, $\mathbf{p} - \mathbf{p}' + n \leq n$ or $\mathbf{p} \leq \mathbf{p}'$. Similarly, $\mathbf{p}' \leq \mathbf{p}$ and therefore $\mathbf{p} = \mathbf{p}'$, as required.

Remark: The above theorem and proof depend only on the concept of positivity. Thus, the theorem is true for any subfield K of the real field $\mathbf{R}$ such as the rational field $\mathbf{Q}$.

Positive Definite Real Quadratic Forms

12.15. Prove that the following definitions of a positive definite quadratic form q are equivalent:

(a) The diagonal entries are all positive in any diagonal representation of q.

(b) $q(Y) > 0$, for any nonzero vector Y in $\mathbf{R}^n$.

Suppose $q(Y) = a_1 y_1^2 + a_2 y_2^2 + \cdots + a_n y_n^2$. If all the coefficients are positive, then clearly $q(Y) > 0$ whenever $Y \neq 0$. Thus, (a) implies (b). Conversely, suppose (a) is not true; that is, suppose some diagonal entry $a_k \leq 0$. Let $e_k = (0, \ldots, 1, \ldots 0)$ be the vector whose entries are all 0 except 1 in the kth position. Then $q(e_k) = a_k$ is not positive, and so (b) is not true. That is, (b) implies (a). Accordingly, (a) and (b) are equivalent.

12.16. Determine whether each of the following quadratic forms q is positive definite:

(a) $q(x, y, z) = x^2 + 2y^2 - 4xz - 4yz + 7z^2$

(b) $q(x, y, z) = x^2 + y^2 + 2xz + 4yz + 3z^2$

Diagonalize (under congruence) the symmetric matrix A corresponding to q.

(a) Apply the operations "Replace R_3 by $2R_1 + R_3$" and "Replace C_3 by $2C_1 + C_3$," and then "Replace R_3 by $R_2 + R_3$" and "Replace C_3 by $C_2 + C_3$." These yield

$$A = \begin{bmatrix} 1 & 0 & -2 \\ 0 & 2 & -2 \\ -2 & -2 & 7 \end{bmatrix} \simeq \begin{bmatrix} 1 & 0 & 0 \\ 0 & 2 & -2 \\ 0 & -2 & 3 \end{bmatrix} \simeq \begin{bmatrix} 1 & 0 & 0 \\ 0 & 2 & 0 \\ 0 & 0 & 1 \end{bmatrix}$$

The diagonal representation of q only contains positive entries, $1, 2, 1$, on the diagonal. Thus, q is positive definite.

(b) We have

$$A = \begin{bmatrix} 1 & 0 & 1 \\ 0 & 1 & 2 \\ 1 & 2 & 3 \end{bmatrix} \simeq \begin{bmatrix} 1 & 0 & 0 \\ 0 & 1 & 2 \\ 0 & 2 & 2 \end{bmatrix} \simeq \begin{bmatrix} 1 & 0 & 0 \\ 0 & 1 & 0 \\ 0 & 0 & -2 \end{bmatrix}$$

There is a negative entry -2 on the diagonal representation of q. Thus, q is not positive definite.

12.17. Show that $q(x, y) = ax^2 + bxy + cy^2$ is positive definite if and only if $a > 0$ and the discriminant $D = b^2 - 4ac < 0$.

Suppose $v = (x, y) \neq 0$. Then either $x \neq 0$ or $y \neq 0$; say, $y \neq 0$. Let $t = x/y$. Then

$$q(v) = y^2[a(x/y)^2 + b(x/y) + c] = y^2(at^2 + bt + c)$$

However, the following are equivalent:

(i) $s = at^2 + bt + c$ is positive for every value of t.

(ii) $s = at^2 + bt + c$ lies above the t-axis.

(iii) $a > 0$ and $D = b^2 - 4ac < 0$.

Thus, q is positive definite if and only if $a > 0$ and $D < 0$. [*Remark*: $D < 0$ is the same as $\det(A) > 0$, where A is the symmetric matrix corresponding to q.]

12.18. Determine whether or not each of the following quadratic forms q is positive definite:

(a) $q(x,y) = x^2 - 4xy + 7y^2$, (b) $q(x,y) = x^2 + 8xy + 5y^2$, (c) $q(x,y) = 3x^2 + 2xy + y^2$

Compute the discriminant $D = b^2 - 4ac$, and then use Problem 12.17.

(a) $D = 16 - 28 = -12$. Because $a = 1 > 0$ and $D < 0, q$ is positive definite.

(b) $D = 64 - 20 = 44$. Because $D > 0, q$ is not positive definite.

(c) $D = 4 - 12 = -8$. Because $a = 3 > 0$ and $D < 0, q$ is positive definite.

Hermitian Forms

12.19. Determine whether the following matrices are Hermitian:

$$\text{(a)} \begin{bmatrix} 2 & 2+3i & 4-5i \\ 2-3i & 5 & 6+2i \\ 4+5i & 6-2i & -7 \end{bmatrix}, \text{(b)} \begin{bmatrix} 3 & 2-i & 4+i \\ 2-i & 6 & i \\ 4+i & i & 7 \end{bmatrix}, \text{(c)} \begin{bmatrix} 4 & -3 & 5 \\ -3 & 2 & 1 \\ 5 & 1 & -6 \end{bmatrix}$$

A complex matrix $A = [a_{ij}]$ is Hermitian if $A^* = A$—that is, if $a_{ij} = \bar{a}_{ji}$.

(a) Yes, because it is equal to its conjugate transpose.

(b) No, even though it is symmetric.

(c) Yes. In fact, a real matrix is Hermitian if and only if it is symmetric.

12.20. Let A be a Hermitian matrix. Show that f is a Hermitian form on $\mathbf{C}^n$ where f is defined by $f(X, Y) = X^T A\bar{Y}$.

For all $a, b \in \mathbf{C}$ and all $X_1, X_2, Y \in \mathbf{C}^n$,

$$f(aX_1 + bX_2, \ Y) = (aX_1 + bX_2)^T A\bar{Y} = (aX_1^T + bX_2^T)A\bar{Y}$$
$$= aX_1^T A\bar{Y} + bX_2^T A\bar{Y} = af(X_1, Y) + bf(X_2, Y)$$

Hence, f is linear in the first variable. Also,

$$\overline{f(X, Y)} = \overline{X^T A\bar{Y}} = \overline{(X^T A\bar{Y})^T} = \overline{\bar{Y}^T A^T X} = Y^T A^* \bar{X} = Y^T A\bar{X} = f(Y, X)$$

Hence, f is a Hermitian form on $\mathbf{C}^n$.

Remark: We use the fact that $X^T A\bar{Y}$ is a scalar and so it is equal to its transpose.

12.21. Let f be a Hermitian form on V. Let H be the matrix of f in a basis $S = \{u_i\}$ of V. Prove the following:

(a) $f(u, v) = [u]_S^T H\overline{[v]}_S$ for all $u, v \in V$.

(b) If P is the change-of-basis matrix from S to a new basis S' of V, then $B = P^T H\bar{P}$ (or $B = Q^* HQ$, where $Q = \bar{P}$) is the matrix of f in the new basis S'.

Note that (b) is the complex analog of Theorem 12.2.

(a) Let $u, v \in V$ and suppose $u = a_1 u_1 + \cdots + a_n u_n$ and $v = b_1 u_1 + \cdots + b_n u_n$. Then, as required,

$$f(u, v) = f(a_1 u_1 + \cdots + a_n u_n, \ b_1 u_1 + \cdots + b_n u_n)$$
$$= \sum_{i,j} a_i \bar{b}_j f(u_i, v_j) = [a_1, \ldots, a_n] H [\bar{b}_1, \ldots, \bar{b}_n]^T = [u]_S^T H\overline{[v]}_S$$

(b) Because P is the change-of-basis matrix from S to S', we have $P[u]_{S'} = [u]_S$ and $P[v]_{S'} = [v]_S$; hence, $[u]_S^T = [u]_{S'}^T P^T$ and $\overline{[v]_S} = \bar{P}\overline{[v]_{S'}}$. Thus, by (a),

$$f(u, v) = [u]_S^T H\overline{[v]_S} = [u]_{S'}^T P^T H\bar{P}\overline{[v]_{S'}}$$

But u and v are arbitrary elements of V; hence, $P^T H\bar{P}$ is the matrix of f in the basis S'.

12.22. Let $H = \begin{bmatrix} 1 & 1+i & 2i \\ 1-i & 4 & 2-3i \\ -2i & 2+3i & 7 \end{bmatrix}$, a Hermitian matrix.

Find a nonsingular matrix P such that $D = P^T H\bar{P}$ is diagonal. Also, find the signature of H.

Use the modified Algorithm 12.1 that applies the same row operations but the corresponding conjugate column operations. Thus, first form the block matrix $M = [H, I]$:

$$M = \left[\begin{array}{ccc|ccc} 1 & 1+i & 2i & 1 & 0 & 0 \\ 1-i & 4 & 2-3i & 0 & 1 & 0 \\ -2i & 2+3i & 7 & 0 & 0 & 1 \end{array}\right]$$

Apply the row operations "Replace R_2 by $(-1+i)R_1 + R_2$" and "Replace R_3 by $2iR_1 + R_3$" and then the corresponding conjugate column operations "Replace C_2 by $(-1-i)C_1 + C_2$" and "Replace C_3 by $-2iC_1 + C_3$" to obtain

$$\begin{bmatrix} 1 & 1+i & 2i & 1 & 0 & 0 \\ 0 & 2 & -5i & -1+i & 1 & 0 \\ 0 & 5i & 3 & 2i & 0 & 1 \end{bmatrix} \quad \text{and then} \quad \begin{bmatrix} 1 & 0 & 0 & 1 & 0 & 0 \\ 0 & 2 & -5i & -1+i & 1 & 0 \\ 0 & 5i & 3 & 2i & 0 & 1 \end{bmatrix}$$

Next apply the row operation "Replace R_3 by $-5iR_2 + 2R_3$" and the corresponding conjugate column operation "Replace C_3 by $5iC_2 + 2C_3$" to obtain

$$\begin{bmatrix} 1 & 0 & 0 & 1 & 0 & 0 \\ 0 & 2 & -5i & -1+i & 1 & 0 \\ 0 & 0 & -19 & 5+9i & -5i & 2 \end{bmatrix} \quad \text{and then} \quad \begin{bmatrix} 1 & 0 & 0 & 1 & 0 & 0 \\ 0 & 2 & 0 & -1+i & 1 & 0 \\ 0 & 0 & -38 & 5+9i & -5i & 2 \end{bmatrix}$$

Now H has been diagonalized, and the transpose of the right half of M is P. Thus, set

$$P = \begin{bmatrix} 1 & -1+i & 5+9i \\ 0 & 1 & -5i \\ 0 & 0 & 2 \end{bmatrix}, \quad \text{and then} \quad D = P^T H\bar{P} = \begin{bmatrix} 1 & 0 & 0 \\ 0 & 2 & 0 \\ 0 & 0 & -38 \end{bmatrix}.$$

Note D has $\mathbf{p} = 2$ positive elements and $\mathbf{n} = 1$ negative elements. Thus, the signature of H is $\text{sig}(H) = 2 - 1 = 1$.

Miscellaneous Problems

12.23. Prove Theorem 12.3: Let f be an alternating form on V. Then there exists a basis of V in which f is represented by a block diagonal matrix M with blocks of the form $\begin{bmatrix} 0 & 1 \\ -1 & 0 \end{bmatrix}$ or 0. The number of nonzero blocks is uniquely determined by f [because it is equal to $\frac{1}{2}$ rank(f)].

If $f = 0$, then the theorem is obviously true. Also, if $\dim V = 1$, then $f(k_1 u, k_2 u) = k_1 k_2 f(u, u) = 0$ and so $f = 0$. Accordingly, we can assume that $\dim V > 1$ and $f \neq 0$.

Because $f \neq 0$, there exist (nonzero) $u_1, u_2 \in V$ such that $f(u_1, u_2) \neq 0$. In fact, multiplying u_1 by an appropriate factor, we can assume that $f(u_1, u_2) = 1$ and so $f(u_2, u_1) = -1$. Now u_1 and u_2 are linearly independent; because if, say, $u_2 = ku_1$, then $f(u_1, u_2) = f(u_1, ku_1) = kf(u_1, u_1) = 0$. Let $U = \text{span}(u_1, u_2)$; then,

(i) The matrix representation of the restriction of f to U in the basis $\{u_1, u_2\}$ is $\begin{bmatrix} 0 & 1 \\ -1 & 0 \end{bmatrix}$,

(ii) If $u \in U$, say $u = au_1 + bu_2$, then

$$f(u, u_1) = f(au_1 + bu_2,\ u_1) = -b \quad \text{and} \quad f(u, u_2) = f(au_1 + bu_2,\ u_2) = a$$

Let W consists of those vectors $w \in V$ such that $f(w, u_1) = 0$ and $f(w, u_2) = 0$. Equivalently,

$$W = \{w \in V : f(w, u) = 0 \text{ for every } u \in U\}$$

We claim that $V = U \oplus W$. It is clear that $U \cap W = \{0\}$, and so it remains to show that $V = U + W$. Let $v \in V$. Set

$$u = f(v, u_2)u_1 - f(v, u_1)u_2 \quad \text{and} \quad w = v - u \tag{1}$$

Because u is a linear combination of u_1 and u_2, $u \in U$.

We show next that $w \in W$. By (1) and (ii), $f(u, u_1) = f(v, u_1)$; hence,

$$f(w, u_1) = f(v - u,\ u_1) = f(v, u_1) - f(u, u_1) = 0$$

Similarly, $f(u, u_2) = f(v, u_2)$ and so

$$f(w, u_2) + f(v - u,\ u_2) = f(v, u_2) - f(u, u_2) = 0$$

Then $w \in W$ and so, by (1), $v = u + w$, where $u \in W$. This shows that $V = U + W$; therefore, $V = U \oplus W$.

Now the restriction of f to W is an alternating bilinear form on W. By induction, there exists a basis $u_3, \ldots, u_n$ of W in which the matrix representing f restricted to W has the desired form. Accordingly, $u_1, u_2, u_3, \ldots, u_n$ is a basis of V in which the matrix representing f has the desired form.

SUPPLEMENTARY PROBLEMS

Bilinear Forms

12.24. Let $u = (x_1, x_2)$ and $v = (y_1, y_2)$. Determine which of the following are bilinear forms on $\mathbf{R}^2$:

(a) $f(u, v) = 2x_1y_2 - 3x_2y_1$, (c) $f(u, v) = 3x_2y_2$, (e) $f(u, v) = 1$,

(b) $f(u, v) = x_1 + y_2$, (d) $f(u, v) = x_1x_2 + y_1y_2$, (f) $f(u, v) = 0$

12.25. Let f be the bilinear form on $\mathbf{R}^2$ defined by

$$f[(x_1, x_2),\ (y_1, y_2)] = 3x_1y_1 - 2x_1y_2 + 4x_2y_1 - x_2y_2$$

(a) Find the matrix A of f in the basis $\{u_1 = (1, 1),\ u_2 = (1, 2)\}$.

(b) Find the matrix B of f in the basis $\{v_1 = (1, -1),\ v_2 = (3, 1)\}$.

(c) Find the change-of-basis matrix P from $\{u_i\}$ to $\{v_i\}$, and verify that $B = P^TAP$.

12.26. Let V be the vector space of two-square matrices over $\mathbf{R}$. Let $M = \begin{bmatrix} 1 & 2 \\ 3 & 5 \end{bmatrix}$, and let $f(A, B) = \text{tr}(A^TMB)$, where $A, B \in V$ and "tr" denotes trace. (a) Show that f is a bilinear form on V. (b) Find the matrix of f in the basis

$$\left\{ \begin{bmatrix} 1 & 0 \\ 0 & 0 \end{bmatrix}, \begin{bmatrix} 0 & 1 \\ 0 & 0 \end{bmatrix}, \begin{bmatrix} 0 & 0 \\ 1 & 0 \end{bmatrix}, \begin{bmatrix} 0 & 0 \\ 0 & 1 \end{bmatrix} \right\}$$

12.27. Let $B(V)$ be the set of bilinear forms on V over K. Prove the following:

(a) If $f, g \in B(V)$, then $f + g$, $kg \in B(V)$ for any $k \in K$.

(b) If ϕ and σ are linear functions on V, then $f(u, v) = \phi(u)\sigma(v)$ belongs to $B(V)$.

12.28. Let $[f]$ denote the matrix representation of a bilinear form f on V relative to a basis $\{u_i\}$. Show that the mapping $f \mapsto [f]$ is an isomorphism of $B(V)$ onto the vector space V of n-square matrices.

12.29. Let f be a bilinear form on V. For any subset S of V, let

$$S^{\perp} = \{v \in V : f(u,v) = 0 \text{ for every } u \in S\} \text{ and } S^{\top} = \{v \in V : f(v,u) = 0 \text{ for every } u \in S\}$$

Show that: (a) $S^{\top}$ and $S^{\top}$ are subspaces of V; (b) $S_1 \subseteq S_2$ implies $S_2^{\perp} \subseteq S_1^{\perp}$ and $S_2^{\top} \subseteq S_1^{\top}$; (c) $\{0\}^{\perp} = \{0\}^{\top} = V$.

12.30. Suppose f is a bilinear form on V. Prove that: $\text{rank}(f) = \dim V - \dim V^{\perp} = \dim V - \dim V^{\top}$, and hence, $\dim V^{\perp} = \dim V^{\top}$.

12.31. Let f be a bilinear form on V. For each $u \in V$, let $\hat{u}: V \to K$ and $\tilde{u}: V \to K$ be defined by $\hat{u}(x) = f(x,u)$ and $\tilde{u}(x) = f(u,x)$. Prove the following:

(a) $\hat{u}$ and $\tilde{u}$ are each linear; i.e., $\hat{u}, \tilde{u} \in V^*$,

(b) $u \mapsto \hat{u}$ and $u \mapsto \tilde{u}$ are each linear mappings from V into V^*,

(c) $\text{rank}(f) = \text{rank}(u \mapsto \hat{u}) = \text{rank}(u \mapsto \tilde{u})$.

12.32. Show that congruence of matrices (denoted by $\simeq$) is an equivalence relation; that is, (i) $A \simeq A$; (ii) If $A \simeq B$, then $B \simeq A$; (iii) If $A \simeq B$ and $B \simeq C$, then $A \simeq C$.

Symmetric Bilinear Forms, Quadratic Forms

12.33. Find the symmetric matrix A belonging to each of the following quadratic forms:

(a) $q(x,y,z) - 2x^2 - 8xy + y^2 - 16xz + 14yz + 5z^2$, (c) $q(x,y,z) = xy + y^2 + 4xz + z^2$

(b) $q(x,y,z) = x^2 - xz + y^2$, (d) $q(x,y,z) = xy + yz$

12.34. For each of the following symmetric matrices A, find a nonsingular matrix P such that $D = P^T AP$ is diagonal:

$$\text{(a)} \quad A = \begin{bmatrix} 1 & 0 & 2 \\ 0 & 3 & 6 \\ 2 & 6 & 7 \end{bmatrix}, \text{(b)} \quad A = \begin{bmatrix} 1 & -2 & 1 \\ -2 & 5 & 3 \\ 1 & 3 & -2 \end{bmatrix}, \text{(c)} \quad A = \begin{bmatrix} 1 & -1 & 0 & 2 \\ -1 & 2 & 1 & 0 \\ 0 & 1 & 1 & 2 \\ 2 & 0 & 2 & -1 \end{bmatrix}$$

12.35. Let $q(x,y) = 2x^2 - 6xy - 3y^2$ and $x = s + 2t$, $y = 3s - t$.

(a) Rewrite $q(x,y)$ in matrix notation, and find the matrix A representing the quadratic form.

(b) Rewrite the linear substitution using matrix notation, and find the matrix P corresponding to the substitution.

(c) Find $q(s,t)$ using (i) direct substitution, (ii) matrix notation.

12.36. For each of the following quadratic forms $q(x,y,z)$, find a nonsingular linear substitution expressing the variables x, y, z in terms of variables r, s, t such that $q(r,s,t)$ is diagonal:

(a) $q(x,y,z) = x^2 + 6xy + 8y^2 - 4xz + 2yz - 9z^2$,

(b) $q(x,y,z) = 2x^2 - 3y^2 + 8xz + 12yz + 25z^2$,

(c) $q(x,y,z) = x^2 + 2xy + 3y^2 + 4xz + 8yz + 6z^2$.

In each case, find the rank and signature.

12.37. Give an example of a quadratic form $q(x,y)$ such that $q(u) = 0$ and $q(v) = 0$ but $q(u+v) \neq 0$.

12.38. Let $S(V)$ denote all symmetric bilinear forms on V. Show that

(a) $S(V)$ is a subspace of $B(V)$; (b) If $\dim V = n$, then $\dim S(V) = \frac{1}{2}n(n+1)$.

12.39. Consider a real quadratic polynomial $q(x_1, \ldots, x_n) = \sum_{i,j=1}^{n} a_{ij} x_i x_j$, where $a_{ij} = a_{ji}$.

(a) If $a_{11} \neq 0$, show that the substitution

$$x_1 = y_1 - \frac{1}{a_{11}}(a_{12}y_2 + \cdots + a_{1n}y_n), \qquad x_2 = y_2, \qquad \ldots, \qquad x_n = y_n$$

yields the equation $q(x_1, \ldots, x_n) = a_{11}y_1^2 + q'(y_2, \ldots, y_n)$, where q' is also a quadratic polynomial.

(b) If $a_{11} = 0$ but, say, $a_{12} \neq 0$, show that the substitution

$$x_1 = y_1 + y_2, \qquad x_2 = y_1 - y_2, \qquad x_3 = y_3, \qquad \ldots, \qquad x_n = y_n$$

yields the equation $q(x_1, \ldots, x_n) = \sum b_{ij}y_iy_j$, where $b_{11} \neq 0$, which reduces this case to case (a).

Remark: This method of diagonalizing q is known as *completing the square.*

Positive Definite Quadratic Forms

12.40. Determine whether or not each of the following quadratic forms is positive definite:

(a) $q(x,y) = 4x^2 + 5xy + 7y^2$, (c) $q(x,y,z) = x^2 + 4xy + 5y^2 + 6xz + 2yz + 4z^2$
(b) $q(x,y) = 2x^2 - 3xy - y^2$, (d) $q(x,y,z) = x^2 + 2xy + 2y^2 + 4xz + 6yz + 7z^2$

12.41. Find those values of k such that the given quadratic form is positive definite:

(a) $q(x,y) = 2x^2 - 5xy + ky^2$, (b) $q(x,y) = 3x^2 - kxy + 12y^2$
(c) $q(x,y,z) = x^2 + 2xy + 2y^2 + 2xz + 6yz + kz^2$

12.42. Suppose A is a real symmetric positive definite matrix. Show that $A = P^TP$ for some nonsingular matrix P.

Hermitian Forms

12.43. Modify Algorithm 12.1 so that, for a given Hermitian matrix H, it finds a nonsingular matrix P for which $D = P^TA\bar{P}$ is diagonal.

12.44. For each Hermitian matrix H, find a nonsingular matrix P such that $D = P^TH\bar{P}$ is diagonal:

(a) $H = \begin{bmatrix} 1 & i \\ -i & 2 \end{bmatrix}$, (b) $H = \begin{bmatrix} 1 & 2+3i \\ 2-3i & -1 \end{bmatrix}$, (c) $H = \begin{bmatrix} 1 & i & 2+i \\ -i & 2 & 1-i \\ 2-i & 1+i & 2 \end{bmatrix}$

Find the rank and signature in each case.

12.45. Let A be a complex nonsingular matrix. Show that $H = A^*A$ is Hermitian and positive definite.

12.46. We say that B is *Hermitian congruent* to A if there exists a nonsingular matrix P such that $B = P^TA\bar{P}$ or, equivalently, if there exists a nonsingular matrix Q such that $B = Q^*AQ$. Show that Hermitian congruence is an equivalence relation. (*Note*: If $P = \bar{Q}$, then $P^TA\bar{P} = Q^*AQ$.)

12.47. Prove Theorem 12.7: Let f be a Hermitian form on V. Then there is a basis S of V in which f is represented by a diagonal matrix, and every such diagonal representation has the same number **p** of positive entries and the same number **n** of negative entries.

Miscellaneous Problems

12.48. Let e denote an elementary row operation, and let f^* denote the corresponding conjugate column operation (where each scalar k in e is replaced by $\bar{k}$ in f^*). Show that the elementary matrix corresponding to f^* is the conjugate transpose of the elementary matrix corresponding to e.

12.49. Let V and W be vector spaces over K. A mapping $f: V \times W \to K$ is called a *bilinear form* on V and W if

(i) $f(av_1 + bv_2,\ w) = af(v_1, w) + bf(v_2, w)$,

(ii) $f(v,\ aw_1 + bw_2) = af(v, w_1) + bf(v, w_2)$

for every $a, b \in K, v_i \in V, w_j \in W$. Prove the following:

(a) The set $B(V, W)$ of bilinear forms on V and W is a subspace of the vector space of functions from $V \times W$ into K.

(b) If $\{\phi_1, \ldots, \phi_m\}$ is a basis of V^* and $\{\sigma_1, \ldots, \sigma_n\}$ is a basis of W^*, then $\{f_{ij} : i = 1, \ldots, m, j = 1, \ldots, n\}$ is a basis of $B(V, W)$, where f_{ij} is defined by $f_{ij}(v, w) = \phi_i(v)\sigma_j(w)$. Thus, $\dim B(V, W) = \dim V \dim W$.

[Note that if $V = W$, then we obtain the space $B(V)$ investigated in this chapter.]

12.50. Let V be a vector space over K. A mapping $f : \overbrace{V \times V \times \ldots \times V}^{m \text{ times}} \to K$ is called a *multilinear* (or *m-linear*) *form* on V if f is linear in each variable; that is, for $i = 1, \ldots, m$,

$$f(\ldots, \widehat{au + bv}, \ldots) = af(\ldots, \hat{u}, \ldots) + bf(\ldots, \hat{v}, \ldots)$$

where $\widehat{\ldots}$ denotes the ith element, and other elements are held fixed. An m-linear form f is said to be *alternating* if $f(v_1, \ldots v_m) = 0$ whenever $v_i = v_j$ for $i \neq j$. Prove the following:

(a) The set $B_m(V)$ of m-linear forms on V is a subspace of the vector space of functions from $V \times V \times \cdots \times V$ into K.

(b) The set $A_m(V)$ of alternating m-linear forms on V is a subspace of $B_m(V)$.

Remark 1: If $m = 2$, then we obtain the space $B(V)$ investigated in this chapter.

Remark 2: If $V = K^m$, then the determinant function is an alternating m-linear form on V.

ANSWERS TO SUPPLEMENTARY PROBLEMS

Notation: $M = [R_1;\ \ R_2;\ \ \ldots]$ denotes a matrix M with rows $R_1, R_2, \ldots$.

12.24. (a) yes, (b) no, (c) yes, (d) no, (e) no, (f) yes

12.25. (a) $A = [4, 1;\ 7, 3]$, (b) $B = [0, -4;\ 20, 32]$, (c) $P = [3, 5;\ -2, -2]$

12.26. (b) $[1, 0, 2, 0;\ 0, 1, 0, 2;\ 3, 0, 5, 0;\ 0, 3, 0, 5]$

12.33. (a) $[2, -4, -8;\ -4, 1, 7;\ -8, 7, 5]$, (b) $[1, 0, -\frac{1}{2};\ 0, 1, 0;\ -\frac{1}{2}, 0, 0]$,
(c) $[0, \frac{1}{2}, 2;\ \frac{1}{2}, 1, 0;\ 2, 0, 1]$, (d) $[0, \frac{1}{2}, 0;\ \frac{1}{2}, 0, 1;\ \frac{1}{2}, 0, \frac{1}{2};\ 0, \frac{1}{2}, 0]$

12.34. (a) $P = [1, 0, -2;\ 0, 1, -2;\ 0, 0, 1]$, $D = \text{diag}(1, 3, -9)$;
(b) $P = [1, 2, -11;\ 0, 1, -5;\ 0, 0, 1]$, $D = \text{diag}(1, 1, -28)$;
(c) $P = [1, 1, -1, -4;\ 0, 1, -1, -2;\ 0, 0, 1, 0;\ 0, 0, 0, 1]$, $D = \text{diag}(1, 1, 0, -9)$

12.35. $A = [2, -3;\ -3, -3]$, $P = [1, 2;\ 3, -1]$, $q(s, t) = -43s^2 - 4st + 17t^2$

12.36. (a) $x = r - 3s - 19t$, $y = s + 7t$, $z = t$; $q(r, s, t) = r^2 - s^2 + 36t^2$;
(b) $x = r - 2t$, $y = s + 2t$, $z = t$; $q(r, s, t) = 2r^2 - 3s^2 + 29t^2$;
(c) $x = r - s - t$, $y = s - t$, $z = t$; $q(r, s, t) = r^2 - 2s^2$

12.37. $q(x, y) = x^2 - y^2$, $u = (1, 1)$, $v = (1, -1)$

12.40. (a) yes, (b) no, (c) no, (d) yes

12.41. (a) $k > \frac{25}{8}$, (b) $-12 < k < 12$, (c) $k > 5$

12.44. (a) $P = [1, i;\ 0, 1]$, $D = I$, $s = 2$; (b) $P = [1, -2 + 3i;\ 0, 1]$, $D = \text{diag}(1, -14)$, $s = 0$;
(c) $P = [1, i, -3 + i;\ 0, 1, i;\ 0, 0, 1]$, $D = \text{diag}(1, 1, -4)$, $s = 1$

CHAPTER 13

Linear Operators on Inner Product Spaces

13.1 Introduction

This chapter investigates the space $A(V)$ of linear operators T on an inner product space V. (See Chapter 7.) Thus, the base field K is either the real numbers **R** or the complex numbers **C**. In fact, different terminologies will be used for the real case and the complex case. We also use the fact that the inner products on real Euclidean space $\mathbf{R}^n$ and complex Euclidean space $\mathbf{C}^n$ may be defined, respectively, by

$$\langle u, v \rangle = u^T v \qquad \text{and} \qquad \langle u, v \rangle = u^T \bar{v}$$

where u and v are column vectors.

The reader should review the material in Chapter 7 and be very familiar with the notions of norm (length), orthogonality, and orthonormal bases. We also note that Chapter 7 mainly dealt with real inner product spaces, whereas here we assume that V is a complex inner product space unless otherwise stated or implied.

Lastly, we note that in Chapter 2, we used A^H to denote the conjugate transpose of a complex matrix A; that is, $A^H = \overline{A^T}$. This notation is not standard. Many texts, expecially advanced texts, use A^* to denote such a matrix; we will use that notation in this chapter. That is, now $A^* = \overline{A^T}$.

13.2 Adjoint Operators

We begin with the following basic definition.

DEFINITION: A linear operator T on an inner product space V is said to have an *adjoint operator* T^* on V if $\langle T(u), v \rangle = \langle u, T^*(v) \rangle$ for every $u, v \in V$.

The following example shows that the adjoint operator has a simple description within the context of matrix mappings.

EXAMPLE 13.1

(a) Let A be a real n-square matrix viewed as a linear operator on $\mathbf{R}^n$. Then, for every $u, v \in \mathbf{R}_n$,

$$\langle Au, v \rangle = (Au)^T v = u^T A^T v = \langle u, A^T v \rangle$$

Thus, the transpose A^T of A is the adjoint of A.

(b) Let B be a complex n-square matrix viewed as a linear operator on $\mathbf{C}^n$. Then for every $u, v, \in \mathbf{C}^n$,

$$\langle Bu, v \rangle = (Bu)^T \bar{v} = u^T B^T \bar{v} = u^T \overline{B^* v} = \langle u, B^* v \rangle$$

Thus, the conjugate transpose B^* of B is the adjoint of B.

Remark: B^* may mean either the adjoint of B as a linear operator or the conjugate transpose of B as a matrix. By Example 13.1(b), the ambiguity makes no difference, because they denote the same object.

The following theorem (proved in Problem 13.4) is the main result in this section.

THEOREM 13.1: Let T be a linear operator on a finite-dimensional inner product space V over K. Then

(i) There exists a unique linear operator T^* on V such that $\langle T(u), v\rangle = \langle u, T^*(v)\rangle$ for every $u, v \in V$. (That is, T has an adjoint T^*.)

(ii) If A is the matrix representation T with respect to any orthonormal basis $S = \{u_i\}$ of V, then the matrix representation of T^* in the basis S is the conjugate transpose A^* of A (or the transpose A^T of A when K is real).

We emphasize that no such simple relationship exists between the matrices representing T and T^* if the basis is not orthonormal. Thus, we see one useful property of orthonormal bases. We also emphasize that this theorem is not valid if V has infinite dimension (Problem 13.31).

The following theorem (proved in Problem 13.5) summarizes some of the properties of the adjoint.

THEOREM 13.2: Let T, T_1, T_2 be linear operators on V and let $k \in K$. Then

(i) $(T_1 + T_2)^* = T_1^* + T_2^*$, (iii) $(T_1T_2)^* = T_2^*T_1^*$,

(ii) $(kT)^* = \bar{k}T^*$, (iv) $(T^*)^* = T$.

Observe the similarity between the above theorem and Theorem 2.3 on properties of the transpose operation on matrices.

Linear Functionals and Inner Product Spaces

Recall (Chapter 11) that a linear functional ϕ on a vector space V is a linear mapping $\phi: V \to K$. This subsection contains an important result (Theorem 13.3) that is used in the proof of the above basic Theorem 13.1.

Let V be an inner product space. Each $u \in V$ determines a mapping $\hat{u}: V \to K$ defined by

$$\hat{u}(v) = \langle v, u\rangle$$

Now, for any $a, b \in K$ and any $v_1, v_2 \in V$,

$$\hat{u}(av_1 + bv_2) = \langle av_1 + bv_2,\ u\rangle = a\langle v_1, u\rangle + b\langle v_2, u\rangle = a\hat{u}(v_1) + b\hat{u}(v_2)$$

That is, $\hat{u}$ is a linear functional on V. The converse is also true for spaces of finite dimension and it is contained in the following important theorem (proved in Problem 13.3).

THEOREM 13.3: Let ϕ be a linear functional on a finite-dimensional inner product space V. Then there exists a unique vector $u \in V$ such that $\phi(v) = \langle v, u\rangle$ for every $v \in V$.

We remark that the above theorem is not valid for spaces of infinite dimension (Problem 13.24).

13.3 Analogy Between $A(V)$ and C, Special Linear Operators

Let $A(V)$ denote the algebra of all linear operators on a finite-dimensional inner product space V. The adjoint mapping $T \mapsto T^*$ on $A(V)$ is quite analogous to the conjugation mapping $z \mapsto \bar{z}$ on the complex field **C**. To illustrate this analogy we identify in Table 13-1 certain classes of operators $T \in A(V)$ whose behavior under the adjoint map imitates the behavior under conjugation of familiar classes of complex numbers.

The analogy between these operators T and complex numbers z is reflected in the next theorem.

Table 13-1

Class of complex numbers	Behavior under conjugation	Class of operators in $A(V)$	Behavior under the adjoint map
Unit circle ($\lvert z\rvert = 1$)	$\bar{z} = 1/z$	Orthogonal operators (real case) Unitary operators (complex case)	$T^* = T^{-1}$
Real axis	$\bar{z} = z$	Self-adjoint operators Also called: symmetric (real case) Hermitian (complex case)	$T^* = T$
Imaginary axis	$\bar{z} = -z$	Skew-adjoint operators Also called: skew-symmetric (real case) skew-Hermitian (complex case)	$T^* = -T$
Positive real axis $(0, \infty)$	$z = \bar{w}w, w \neq 0$	Positive definite operators	$T = S^*S$ with S nonsingular

THEOREM 13.4: Let λ be an eigenvalue of a linear operator T on V.

(i) If $T^* = T^{-1}$ (i.e., T is orthogonal or unitary), then $|\lambda| = 1$.

(ii) If $T^* = T$ (i.e., T is self-adjoint), then λ is real.

(iii) If $T^* = -T$ (i.e., T is skew-adjoint), then λ is pure imaginary.

(iv) If $T = S^*S$ with S nonsingular (i.e., T is positive definite), then λ is real and positive.

Proof. In each case let v be a nonzero eigenvector of T belonging to λ; that is, $T(v) = \lambda v$ with $v \neq 0$. Hence, $\langle v, v\rangle$ is positive.

Proof of (i). We show that $\lambda\bar{\lambda}\langle v, v\rangle = \langle v, v\rangle$:

$$\lambda\bar{\lambda}\langle v, v\rangle = \langle \lambda v, \lambda v\rangle = \langle T(v), T(v)\rangle = \langle v, T^*T(v)\rangle = \langle v, I(v)\rangle = \langle v, v\rangle$$

But $\langle v, v\rangle \neq 0$; hence, $\lambda\bar{\lambda} = 1$ and so $|\lambda| = 1$.

Proof of (ii). We show that $\lambda\langle v, v\rangle = \bar{\lambda}\langle v, v\rangle$:

$$\lambda\langle v, v\rangle = \langle \lambda v, v\rangle = \langle T(v), v\rangle = \langle v, T^*(v)\rangle = \langle v, T(v)\rangle = \langle v, \lambda v\rangle = \bar{\lambda}\langle v, v\rangle$$

But $\langle v, v\rangle \neq 0$; hence, $\lambda = \bar{\lambda}$ and so λ is real.

Proof of (iii). We show that $\lambda\langle v, v\rangle = -\bar{\lambda}\langle v, v\rangle$:

$$\lambda\langle v, v\rangle = \langle \lambda v, v\rangle = \langle T(v), v\rangle = \langle v, T^*(v)\rangle = \langle v, -T(v)\rangle = \langle v, -\lambda v\rangle = -\bar{\lambda}\langle v, v\rangle$$

But $\langle v, v\rangle \neq 0$; hence, $\lambda = -\bar{\lambda}$ or $\bar{\lambda} = -\lambda$, and so λ is pure imaginary.

Proof of (iv). Note first that $S(v) \neq 0$ because S is nonsingular; hence, $\langle S(v), S(v)\rangle$ is positive. We show that $\lambda\langle v, v\rangle = \langle S(v), S(v)\rangle$:

$$\lambda\langle v, v\rangle = \langle \lambda v, v\rangle = \langle T(v), v\rangle = \langle S^*S(v), v\rangle = \langle S(v), S(v)\rangle$$

But $\langle v, v\rangle$ and $\langle S(v), S(v)\rangle$ are positive; hence, λ is positive.

Remark: Each of the above operators T commutes with its adjoint; that is, $TT^* = T^*T$. Such operators are called *normal* operators.

13.4 Self-Adjoint Operators

Let T be a *self-adjoint* operator on an inner product space V; that is, suppose

$$T^* = T$$

(If T is defined by a matrix A, then A is symmetric or Hermitian according as A is real or complex.) By Theorem 13.4, the eigenvalues of T are real. The following is another important property of T.

THEOREM 13.5: Let T be a self-adjoint operator on V. Suppose u and v are eigenvectors of T belonging to distinct eigenvalues. Then u and v are orthogonal; that is, $\langle u, v\rangle = 0$.

Proof. Suppose $T(u) = \lambda_1 u$ and $T(v) = \lambda_2 v$, where $\lambda_1 \neq \lambda_2$. We show that $\lambda_1\langle u, v\rangle = \lambda_2\langle u, v\rangle$:

$$\begin{aligned}\lambda_1\langle u, v\rangle &= \langle \lambda_1 u, v\rangle = \langle T(u), v\rangle = \langle u, T^*(v)\rangle = \langle u, T(v)\rangle \\ &= \langle u, \lambda_2 v\rangle = \bar{\lambda}_2\langle u, v\rangle = \lambda_2\langle u, v\rangle\end{aligned}$$

(The fourth equality uses the fact that $T^* = T$, and the last equality uses the fact that the eigenvalue λ_2 is real.) Because $\lambda_1 \neq \lambda_2$, we get $\langle u, v\rangle = 0$. Thus, the theorem is proved.

13.5 Orthogonal and Unitary Operators

Let U be a linear operator on a finite-dimensional inner product space V. Suppose

$$U^* = U^{-1} \qquad \text{or equivalently} \qquad UU^* = U^*U = I$$

Recall that U is said to be orthogonal or unitary according as the underlying field is real or complex. The next theorem (proved in Problem 13.10) gives alternative characterizations of these operators.

THEOREM 13.6: The following conditions on an operator U are equivalent:

(i) $U^* = U^{-1}$; that is, $UU^* = U^*U = I$. [U is unitary (orthogonal).]

(ii) U preserves inner products; that is, for every $v, w \in V$, $\langle U(v), U(w)\rangle = \langle v, w\rangle$.

(iii) U preserves lengths; that is, for every $v \in V$, $\|U(v)\| = \|v\|$.

EXAMPLE 13.2

(a) Let $T:\mathbf{R}^3 \to \mathbf{R}^3$ be the linear operator that rotates each vector v about the z-axis by a fixed angle θ as shown in Fig. 10-1 (Section 10.3). That is, T is defined by

$$T(x, y, z) = (x\cos\theta - y\sin\theta,\ \ x\sin\theta + y\cos\theta,\ \ z)$$

We note that lengths (distances from the origin) are preserved under T. Thus, T is an orthogonal operator.

(b) Let V be l_2-space (Hilbert space), defined in Section 7.3. Let $T:V \to V$ be the linear operator defined by

$$T(a_1, a_2, a_3, \ldots) = (0, a_1, a_2, a_3, \ldots)$$

Clearly, T preserves inner products and lengths. However, T is not surjective, because, for example, $(1, 0, 0, \ldots)$ does not belong to the image of T; hence, T is not invertible. Thus, we see that Theorem 13.6 is not valid for spaces of infinite dimension.

An isomorphism from one inner product space into another is a bijective mapping that preserves the three basic operations of an inner product space: vector addition, scalar multiplication, and inner products. Thus, the above mappings (orthogonal and unitary) may also be characterized as the isomorphisms of V into itself. Note that such a mapping U also preserves distances, because

$$\|U(v) - U(w)\| = \|U(v - w)\| = \|v - w\|$$

Hence, U is called an *isometry*.

13.6 Orthogonal and Unitary Matrices

Let U be a linear operator on an inner product space V. By Theorem 13.1, we obtain the following results.

THEOREM 13.7A: A complex matrix A represents a unitary operator U (relative to an orthonormal basis) if and only if $A^* = A^{-1}$.

THEOREM 13.7B: A real matrix A represents an orthogonal operator U (relative to an orthonormal basis) if and only if $A^T = A^{-1}$.

The above theorems motivate the following definitions (which appeared in Sections 2.10 and 2.11).

DEFINITION: A complex matrix A for which $A^* = A^{-1}$ is called a *unitary matrix*.

DEFINITION: A real matrix A for which $A^T = A^{-1}$ is called an *orthogonal matrix*.

We repeat Theorem 2.6, which characterizes the above matrices.

THEOREM 13.8: The following conditions on a matrix A are equivalent:

(i) A is unitary (orthogonal).

(ii) The rows of A form an orthonormal set.

(iii) The columns of A form an orthonormal set.

13.7 Change of Orthonormal Basis

Orthonormal bases play a special role in the theory of inner product spaces V. Thus, we are naturally interested in the properties of the change-of-basis matrix from one such basis to another. The following theorem (proved in Problem 13.12) holds.

THEOREM 13.9: Let $\{u_1, \ldots, u_n\}$ be an orthonormal basis of an inner product space V. Then the change-of-basis matrix from $\{u_i\}$ into another orthonormal basis is unitary (orthogonal). Conversely, if $P = [a_{ij}]$ is a unitary (orthogonal) matrix, then the following is an orthonormal basis:

$$\{u_i' = a_{1i}u_1 + a_{2i}u_2 + \cdots + a_{ni}u_n : i = 1, \ldots, n\}$$

Recall that matrices A and B representing the same linear operator T are similar; that is, $B = P^{-1}AP$, where P is the (nonsingular) change-of-basis matrix. On the other hand, if V is an inner product space, we are usually interested in the case when P is unitary (or orthogonal) as suggested by Theorem 13.9. (Recall that P is unitary if the conjugate tranpose $P^* = P^{-1}$, and P is orthogonal if the transpose $P^T = P^{-1}$.) This leads to the following definition.

DEFINITION: Complex matrices A and B are *unitarily equivalent* if there exists a unitary matrix P for which $B = P^*AP$. Analogously, real matrices A and B are *orthogonally equivalent* if there exists an orthogonal matrix P for which $B = P^TAP$.

Note that orthogonally equivalent matrices are necessarily congruent.

13.8 Positive Definite and Positive Operators

Let P be a linear operator on an inner product space V. Then

(i) P is said to be *positive definite* if $P = S^*S$ for some nonsingular operators S.
(ii) P is said to be *positive* (or *nonnegative* or *semidefinite*) if $P = S^*S$ for some operator S.

The following theorems give alternative characterizations of these operators.

THEOREM 13.10A: The following conditions on an operator P are equivalent:

(i) $P = T^2$ for some nonsingular self-adjoint operator T.
(ii) P is positive definite.
(iii) P is self-adjoint and $\langle P(u), u\rangle > 0$ for every $u \neq 0$ in V.

The corresponding theorem for positive operators (proved in Problem 13.21) follows.

THEOREM 13.10B: The following conditions on an operator P are equivalent:

(i) $P = T^2$ for some self-adjoint operator T.
(ii) P is positive; that is, $P = S^*S$.
(iii) P is self-adjoint and $\langle P(u), u\rangle \geq 0$ for every $u \in V$.

13.9 Diagonalization and Canonical Forms in Inner Product Spaces

Let T be a linear operator on a finite-dimensional inner product space V over K. Representing T by a diagonal matrix depends upon the eigenvectors and eigenvalues of T, and hence, upon the roots of the characteristic polynomial $\Delta(t)$ of T. Now $\Delta(t)$ always factors into linear polynomials over the complex field **C** but may not have any linear polynomials over the real field **R**. Thus, the situation for real inner product spaces (sometimes called Euclidean spaces) is inherently different than the situation for complex inner product spaces (sometimes called unitary spaces). Thus, we treat them separately.

Real Inner Product Spaces, Symmetric and Orthogonal Operators

The following theorem (proved in Problem 13.14) holds.

THEOREM 13.11: Let T be a symmetric (self-adjoint) operator on a real finite-dimensional product space V. Then there exists an orthonormal basis of V consisting of eigenvectors of T; that is, T can be represented by a diagonal matrix relative to an orthonormal basis.

We give the corresponding statement for matrices.

THEOREM 13.11: (Alternative Form) Let A be a real symmetric matrix. Then there exists an orthogonal matrix P such that $B = P^{-1}AP = P^TAP$ is diagonal.

We can choose the columns of the above matrix P to be normalized orthogonal eigenvectors of A; then the diagonal entries of B are the corresponding eigenvalues.

On the other hand, an orthogonal operator T need not be symmetric, and so it may not be represented by a diagonal matrix relative to an orthonormal matrix. However, such a matrix T does have a simple canonical representation, as described in the following theorem (proved in Problem 13.16).

THEOREM 13.12: Let T be an orthogonal operator on a real inner product space V. Then there exists an orthonormal basis of V in which T is represented by a block diagonal matrix M of the form

$$M = \operatorname{diag}\left(I_s, \ -I_t, \ \begin{bmatrix} \cos\theta_1 & -\sin\theta_1 \\ \sin\theta_1 & \cos\theta_1 \end{bmatrix}, \ \ldots, \ \begin{bmatrix} \cos\theta_r & -\sin\theta_r \\ \sin\theta_r & \cos\theta_r \end{bmatrix}\right)$$

The reader may recognize that each of the 2×2 diagonal blocks represents a rotation in the corresponding two-dimensional subspace, and each diagonal entry -1 represents a reflection in the corresponding one-dimensional subspace.

Complex Inner Product Spaces, Normal and Triangular Operators

A linear operator T is said to be *normal* if it commutes with its adjoint—that is, if $TT^* = T^*T$. We note that normal operators include both self-adjoint and unitary operators.

Analogously, a complex matrix A is said to be *normal* if it commutes with its conjugate transpose—that is, if $AA^* = A^*A$.

EXAMPLE 13.3 Let $A = \begin{bmatrix} 1 & 1 \\ i & 3+2i \end{bmatrix}$. Then $A^* = \begin{bmatrix} 1 & -i \\ 1 & 3-2i \end{bmatrix}$.

Also $AA^* = \begin{bmatrix} 2 & 3-3i \\ 3+3i & 14 \end{bmatrix} = A^*A$. Thus, A is normal.

The following theorem (proved in Problem 13.19) holds.

THEOREM 13.13: Let T be a normal operator on a complex finite-dimensional inner product space V. Then there exists an orthonormal basis of V consisting of eigenvectors of T; that is, T can be represented by a diagonal matrix relative to an orthonormal basis.

We give the corresponding statement for matrices.

THEOREM 13.13: (Alternative Form) Let A be a normal matrix. Then there exists a unitary matrix P such that $B = P^{-1}AP = P^*AP$ is diagonal.

The following theorem (proved in Problem 13.20) shows that even nonnormal operators on unitary spaces have a relatively simple form.

THEOREM 13.14: Let T be an arbitrary operator on a complex finite-dimensional inner product space V. Then T can be represented by a triangular matrix relative to an orthonormal basis of V.

THEOREM 13.14: (Alternative Form) Let A be an arbitrary complex matrix. Then there exists a unitary matrix P such that $B = P^{-1}AP = P^*AP$ is triangular.

13.10 Spectral Theorem

The Spectral Theorem is a reformulation of the diagonalization Theorems 13.11 and 13.13.

THEOREM 13.15: (Spectral Theorem) Let T be a normal (symmetric) operator on a complex (real) finite-dimensional inner product space V. Then there exists linear operators $E_1, \ldots, E_r$ on V and scalars $\lambda_1, \ldots, \lambda_r$ such that

(i) $T = \lambda_1 E_1 + \lambda_2 E_2 + \cdots + \lambda_r E_r$, (iii) $E_1^2 = E_1, E_2^2 = E_2, \ldots, E_r^2 = E_r$,

(ii) $E_1 + E_2 + \cdots + E_r = I$, (iv) $E_iE_j = 0$ for $i \neq j$.

The above linear operators $E_1, \ldots, E_r$ are *projections* in the sense that $E_i^2 = E_i$. Moreover, they are said to be *orthogonal projections* because they have the additional property that $E_iE_j = 0$ for $i \neq j$.

The following example shows the relationship between a diagonal matrix representation and the corresponding orthogonal projections.

EXAMPLE 13.4 Consider the following diagonal matrices A, E_1, E_2, E_3:

$$A = \begin{bmatrix} 2 & & & \\ & 3 & & \\ & & 3 & \\ & & & 5 \end{bmatrix}, \quad E_1 = \begin{bmatrix} 1 & & & \\ & 0 & & \\ & & 0 & \\ & & & 0 \end{bmatrix}, \quad E_2 = \begin{bmatrix} 0 & & & \\ & 1 & & \\ & & 1 & \\ & & & 0 \end{bmatrix}, \quad E_3 = \begin{bmatrix} 0 & & & \\ & 0 & & \\ & & 0 & \\ & & & 1 \end{bmatrix}$$

The reader can verify that

(i) $A = 2E_1 + 3E_2 + 5E_3$, (ii) $E_1 + E_2 + E_3 = I$, (iii) $E_i^2 = E_i$, (iv) $E_iE_j = 0$ for $i \neq j$.

SOLVED PROBLEMS

Adjoints

13.1. Find the adjoint of $F: \mathbf{R}^3 \to \mathbf{R}^3$ defined by

$$F(x,y,z) = (3x + 4y - 5z, \quad 2x - 6y + 7z, \quad 5x - 9y + z)$$

First find the matrix A that represents F in the usual basis of $\mathbf{R}^3$—that is, the matrix A whose rows are the coefficients of x, y, z—and then form the transpose A^T of A. This yields

$$A = \begin{bmatrix} 3 & 4 & -5 \\ 2 & -6 & 7 \\ 5 & -9 & 1 \end{bmatrix} \quad \text{and then} \quad A^T = \begin{bmatrix} 3 & 2 & 5 \\ 4 & -6 & -9 \\ -5 & 7 & 1 \end{bmatrix}$$

The adjoint F^* is represented by the transpose of A; hence,

$$F^*(x,y,z) = (3x + 2y + 5z, \quad 4x - 6y - 9z, \quad -5x + 7y + z)$$

13.2. Find the adjoint of $G: \mathbf{C}^3 \to \mathbf{C}^3$ defined by

$$G(x,y,z) = [2x + (1-i)y, \quad (3+2i)x - 4iz, \quad 2ix + (4-3i)y - 3z]$$

First find the matrix B that represents G in the usual basis of $\mathbf{C}^3$, and then form the conjugate transpose B^* of B. This yields

$$B = \begin{bmatrix} 2 & 1-i & 0 \\ 3+2i & 0 & -4i \\ 2i & 4-3i & -3 \end{bmatrix} \quad \text{and then} \quad B^* = \begin{bmatrix} 2 & 3-2i & -2i \\ 1+i & 0 & 4+3i \\ 0 & 4i & -3 \end{bmatrix}$$

Then $G^*(x,y,z) = [2x + (3-2i)y - 2iz, \quad (1+i)x + (4+3i)z, \quad 4iy - 3z]$.

13.3. Prove Theorem 13.3: Let ϕ be a linear functional on an n-dimensional inner product space V. Then there exists a unique vector $u \in V$ such that $\phi(v) = \langle v, u \rangle$ for every $v \in V$.

Let $\{w_1, \ldots, w_n\}$ be an orthonormal basis of V. Set

$$u = \overline{\phi(w_1)}w_1 + \overline{\phi(w_2)}w_2 + \cdots + \overline{\phi(w_n)}w_n$$

Let $\hat{u}$ be the linear functional on V defined by $\hat{u}(v) = \langle v, u \rangle$ for every $v \in V$. Then, for $i = 1, \ldots, n$,

$$\hat{u}(w_i) = \langle w_i, u \rangle = \langle w_i, \quad \overline{\phi(w_1)}w_1 + \cdots + \overline{\phi(w_n)}w_n \rangle = \phi(w_i)$$

Because $\hat{u}$ and ϕ agree on each basis vector, $\hat{u} = \phi$.

Now suppose u' is another vector in V for which $\phi(v) = \langle v, u' \rangle$ for every $v \in V$. Then $\langle v, u \rangle = \langle v, u' \rangle$ or $\langle v, \; u - u' \rangle = 0$. In particular, this is true for $v = u - u'$, and so $\langle u - u', \; u - u' \rangle = 0$. This yields $u - u' = 0$ and $u = u'$. Thus, such a vector u is unique, as claimed.

13.4. Prove Theorem 13.1: Let T be a linear operator on an n-dimensional inner product space V. Then

(a) There exists a unique linear operator T^* on V such that

$$\langle T(u), v \rangle = \langle u, T^*(v) \rangle \quad \text{for all } u, v \in V.$$

(b) Let A be the matrix that represents T relative to an orthonormal basis $S = \{u_i\}$. Then the conjugate transpose A^* of A represents T^* in the basis S.

(a) We first define the mapping T^*. Let v be an arbitrary but fixed element of V. The map $u \mapsto \langle T(u), v \rangle$ is a linear functional on V. Hence, by Theorem 13.3, there exists a unique element $v' \in V$ such that $\langle T(u), v \rangle = \langle u, v' \rangle$ for every $u \in V$. We define $T^* : V \to V$ by $T^*(v) = v'$. Then $\langle T(u), v \rangle = \langle u, T^*(v) \rangle$ for every $u, v \in V$.

We next show that T^* is linear. For any $u, v_i \in V$, and any $a, b \in K$,

$$\begin{aligned}\langle u, \; T^*(av_1 + bv_2) \rangle &= \langle T(u), \; av_1 + bv_2 \rangle = \bar{a}\langle T(u), v_1 \rangle + \bar{b}\langle T(u), v_2 \rangle \\ &= \bar{a}\langle u, T^*(v_1) \rangle + \bar{b}\langle u, T^*(v_2) \rangle = \langle u, aT^*(v_1) + bT^*(v_2) \rangle\end{aligned}$$

But this is true for every $u \in V$; hence, $T^*(av_1 + bv_2) = aT^*(v_1) + bT^*(v_2)$. Thus, T^* is linear.

(b) The matrices $A = [a_{ij}]$ and $B = [b_{ij}]$ that represent T and T^*, respectively, relative to the orthonormal basis S are given by $a_{ij} = \langle T(u_j), u_i \rangle$ and $b_{ij} = \langle T^*(u_j), u_i \rangle$ (Problem 13.67). Hence,

$$b_{ij} = \langle T^*(u_j), u_i \rangle = \overline{\langle u_i, T^*(u_j) \rangle} = \overline{\langle T(u_i), u_j \rangle} = \overline{a_{ji}}$$

Thus, $B = A^*$, as claimed.

13.5. Prove Theorem 13.2:

(i) $(T_1 + T_2)^* = T_1^* + T_2^*$, (iii) $(T_1 T_2)^* = T_2^* T_1^*$,
(ii) $(kT)^* = \bar{k}T^*$, (iv) $(T^*)^* = T$.

(i) For any $u, v \in V$,

$$\begin{aligned}\langle (T_1 + T_2)(u), v \rangle &= \langle T_1(u) + T_2(u), \; v \rangle = \langle T_1(u), v \rangle + \langle T_2(u), v \rangle \\ &= \langle u, T_1^*(v) \rangle + \langle u, T_2^*(v) \rangle = \langle u, \; T_1^*(v) + T_2^*(v) \rangle \\ &= \langle u, \; (T_1^* + T_2^*)(v) \rangle\end{aligned}$$

The uniqueness of the adjoint implies $(T_1 + T_2)^* = T_1^* + T_2^*$.

(ii) For any $u, v \in V$,

$$\langle (kT)(u), \; v \rangle = \langle kT(u), v \rangle = k\langle T(u), v \rangle = k\langle u, T^*(v) \rangle = \langle u, \bar{k}T^*(v) \rangle = \langle u, (\bar{k}T^*)(v) \rangle$$

The uniqueness of the adjoint implies $(kT)^* = \bar{k}T^*$.

(iii) For any $u, v \in V$,

$$\begin{aligned}\langle (T_1 T_2)(u), v \rangle &= \langle T_1(T_2(u)), v \rangle = \langle T_2(u), T_1^*(v) \rangle \\ &= \langle u, T_2^*(T_1^*(v)) \rangle = \langle u, (T_2^* T_1^*)(v) \rangle\end{aligned}$$

The uniqueness of the adjoint implies $(T_1 T_2)^* = T_2^* T_1^*$.

(iv) For any $u, v \in V$,

$$\langle T^*(u), v \rangle = \overline{\langle v, T^*(u) \rangle} = \overline{\langle T(v), u \rangle} = \langle u, T(v) \rangle$$

The uniqueness of the adjoint implies $(T^*)^* = T$.

13.6. Show that (a) $I^* = I$, and (b) $\mathbf{0}^* = \mathbf{0}$.

(a) For every $u, v \in V$, $\langle I(u), v\rangle = \langle u, v\rangle = \langle u, I(v)\rangle$; hence, $I^* = I$.

(b) For every $u, v \in V$, $\langle \mathbf{0}(u), v\rangle = \langle 0, v\rangle = 0 = \langle u, 0\rangle = \langle u, \mathbf{0}(v)\rangle$; hence, $\mathbf{0}^* = \mathbf{0}$.

13.7. Suppose T is invertible. Show that $(T^{-1})^* = (T^*)^{-1}$.

$$I = I^* = (TT^{-1})^* = (T^{-1})^*T^*; \text{hence, } (T^{-1})^* = (T^*)^{-1}.$$

13.8. Let T be a linear operator on V, and let W be a T-invariant subspace of V. Show that $W^\perp$ is invariant under T^*.

Let $u \in W^\perp$. If $w \in W$, then $T(w) \in W$ and so $\langle w, T^*(u)\rangle = \langle T(w), u\rangle = 0$. Thus, $T^*(u) \in W^\perp$ because it is orthogonal to every $w \in W$. Hence, $W^\perp$ is invariant under T^*.

13.9. Let T be a linear operator on V. Show that each of the following conditions implies $T = \mathbf{0}$:

(i) $\langle T(u), v\rangle = 0$ for every $u, v \in V$.

(ii) V is a complex space, and $\langle T(u), u\rangle = 0$ for every $u \in V$.

(iii) T is self-adjoint and $\langle T(u), u\rangle = 0$ for every $u \in V$.

Give an example of an operator T on a real space V for which $\langle T(u), u\rangle = 0$ for every $u \in V$ but $T \neq \mathbf{0}$. [Thus, (ii) need not hold for a real space V.]

(i) Set $v = T(u)$. Then $\langle T(u), T(u)\rangle = 0$, and hence, $T(u) = 0$, for every $u \in V$. Accordingly, $T = \mathbf{0}$.

(ii) By hypothesis, $\langle T(v+w),\ v+w\rangle = 0$ for any $v, w \in V$. Expanding and setting $\langle T(v), v\rangle = 0$ and $\langle T(w), w\rangle = 0$, we find

$$\langle T(v), w\rangle + \langle T(w), v\rangle = 0 \tag{1}$$

Note w is arbitrary in (1). Substituting iw for w, and using $\langle T(v), iw\rangle = \bar{i}\langle T(v), w\rangle = -i\langle T(v), w\rangle$ and $\langle T(iw), v\rangle = \langle iT(w), v\rangle = i\langle T(w), v\rangle$, we find

$$-i\langle T(v), w\rangle + i\langle T(w), v\rangle = 0$$

Dividing through by i and adding to (1), we obtain $\langle T(w), v\rangle = 0$ for any $v, w, \in V$. By (i), $T = \mathbf{0}$.

(iii) By (ii), the result holds for the complex case; hence we need only consider the real case. Expanding $\langle T(v+w),\ v+w\rangle = 0$, we again obtain (1). Because T is self-adjoint and as it is a real space, we have $\langle T(w), v\rangle = \langle w, T(v)\rangle = \langle T(v), w\rangle$. Substituting this into (1), we obtain $\langle T(v), w\rangle = 0$ for any $v, w \in V$. By (i), $T = \mathbf{0}$.

For an example, consider the linear operator T on $\mathbf{R}^2$ defined by $T(x, y) = (y, -x)$. Then $\langle T(u), u\rangle = 0$ for every $u \in V$, but $T \neq \mathbf{0}$.

Orthogonal and Unitary Operators and Matrices

13.10. Prove Theorem 13.6: The following conditions on an operator U are equivalent:

(i) $U^* = U^{-1}$; that is, U is unitary. (ii) $\langle U(v), U(w)\rangle = \langle u, w\rangle$. (iii) $\|U(v)\| = \|v\|$.

Suppose (i) holds. Then, for every $v, w, \in V$,

$$\langle U(v), U(w)\rangle = \langle v, U^*U(w)\rangle = \langle v, I(w)\rangle = \langle v, w\rangle$$

Thus, (i) implies (ii). Now if (ii) holds, then

$$\|U(v)\| = \sqrt{\langle U(v), U(v)\rangle} = \sqrt{\langle v, v\rangle} = \|v\|$$

Hence, (ii) implies (iii). It remains to show that (iii) implies (i).

Suppose (iii) holds. Then for every $v \in V$,

$$\langle U^*U(v)\rangle = \langle U(v), U(v)\rangle = \langle v, v\rangle = \langle I(v), v\rangle$$

Hence, $\langle (U^*U - I)(v),\ v\rangle = 0$ for every $v \in V$. But $U^*U - I$ is self-adjoint (Prove!); then, by Problem 13.9, we have $U^*U - I = 0$ and so $U^*U = I$. Thus, $U^* = U^{-1}$, as claimed.

13.11. Let U be a unitary (orthogonal) operator on V, and let W be a subspace invariant under U. Show that $W^{\perp}$ is also invariant under U.

Because U is nonsingular, $U(W) = W$; that is, for any $w \in W$, there exists $w' \in W$ such that $U(w') = w$. Now let $v \in W^{\perp}$. Then, for any $w \in W$,

$$\langle U(v), w\rangle = \langle U(v), U(w')\rangle = \langle v, w'\rangle = 0$$

Thus, $U(v)$ belongs to $W^{\perp}$. Therefore, $W^{\perp}$ is invariant under U.

13.12. Prove Theorem 13.9: The change-of-basis matrix from an orthonormal basis $\{u_1, \ldots, u_n\}$ into another orthonormal basis is unitary (orthogonal). Conversely, if $P = [a_{ij}]$ is a unitary (orthogonal) matrix, then the vectors $u_{i'} = \sum_j a_{ji} u_j$ form an orthonormal basis.

Suppose $\{v_i\}$ is another orthonormal basis and suppose

$$v_i = b_{i1}u_1 + b_{i2}u_2 + \cdots + b_{in}u_n, \quad i = 1, \ldots, n \tag{1}$$

Because $\{v_i\}$ is orthonormal,

$$\delta_{ij} = \langle v_i, v_j\rangle = b_{i1}\overline{b_{j1}} + b_{i2}\overline{b_{j2}} + \cdots + b_{in}\overline{b_{jn}} \tag{2}$$

Let $B = [b_{ij}]$ be the matrix of coefficients in (1). (Then B^T is the change-of-basis matrix from $\{u_i\}$ to $\{v_i\}$.) Then $BB^* = [c_{ij}]$, where $c_{ij} = b_{i1}\overline{b_{j1}} + b_{i2}\overline{b_{j2}} + \cdots + b_{in}\overline{b_{jn}}$. By (2), $c_{ij} = \delta_{ij}$, and therefore $BB^* = I$. Accordingly, B, and hence, B^T, is unitary.

It remains to prove that $\{u'_i\}$ is orthonormal. By Problem 13.67,

$$\langle u'_i, u'_j\rangle = a_{1i}\overline{a_{1j}} + a_{2i}\overline{a_{2j}} + \cdots + a_{ni}\overline{a_{nj}} = \langle C_i, C_j\rangle$$

where C_i denotes the ith column of the unitary (orthogonal) matrix $P = [a_{ij}]$. Because P is unitary (orthogonal), its columns are orthonormal; hence, $\langle u'_i, u'_j\rangle = \langle C_i, C_j\rangle = \delta_{ij}$. Thus, $\{u'_i\}$ is an orthonormal basis.

Symmetric Operators and Canonical Forms in Euclidean Spaces

13.13. Let T be a symmetric operator. Show that (a) The characteristic polynomial $\Delta(t)$ of T is a product of linear polynomials (over $\mathbf{R}$); (b) T has a nonzero eigenvector.

(a) Let A be a matrix representing T relative to an orthonormal basis of V; then $A = A^T$. Let $\Delta(t)$ be the characteristic polynomial of A. Viewing A as a complex self-adjoint operator, A has only real eigenvalues by Theorem 13.4. Thus,

$$\Delta(t) = (t - \lambda_1)(t - \lambda_2)\cdots(t - \lambda_n)$$

where the λ_i are all real. In other words, $\Delta(t)$ is a product of linear polynomials over $\mathbf{R}$.

(b) By (a), T has at least one (real) eigenvalue. Hence, T has a nonzero eigenvector.

13.14. Prove Theorem 13.11: Let T be a symmetric operator on a real n-dimensional inner product space V. Then there exists an orthonormal basis of V consisting of eigenvectors of T. (Hence, T can be represented by a diagonal matrix relative to an orthonormal basis.)

The proof is by induction on the dimension of V. If $\dim V = 1$, the theorem trivially holds. Now suppose $\dim V = n > 1$. By Problem 13.13, there exists a nonzero eigenvector v_1 of T. Let W be the space spanned by v_1, and let u_1 be a unit vector in W, e.g., let $u_1 = v_1/\|v_1\|$.

Because v_1 is an eigenvector of T, the subspace W of V is invariant under T. By Problem 13.8, $W^{\perp}$ is invariant under $T^* = T$. Thus, the restriction $\hat{T}$ of T to $W^{\perp}$ is a symmetric operator. By Theorem 7.4, $V = W \oplus W^{\perp}$. Hence, $\dim W^{\perp} = n - 1$, because $\dim W = 1$. By induction, there exists an orthonormal basis $\{u_2, \ldots, u_n\}$ of $W^{\perp}$ consisting of eigenvectors of $\hat{T}$ and hence of T. But $\langle u_1, u_i\rangle = 0$ for $i = 2, \ldots, n$ because $u_i \in W^{\perp}$. Accordingly $\{u_1, u_2, \ldots, u_n\}$ is an orthonormal set and consists of eigenvectors of T. Thus, the theorem is proved.

13.15. Let $q(x,y) = 3x^2 - 6xy + 11y^2$. Find an orthonormal change of coordinates (linear substitution) that diagonalizes the quadratic form q.

Find the symmetric matrix A representing q and its characteristic polynomial $\Delta(t)$. We have

$$A = \begin{bmatrix} 3 & -3 \\ -3 & 11 \end{bmatrix} \quad \text{and} \quad \Delta(t) = t^2 - \text{tr}(A)\, t + |A| = t^2 - 14t + 24 = (t-2)(t-12)$$

The eigenvalues are $\lambda = 2$ and $\lambda = 12$. Hence, a diagonal form of q is

$$q(s,t) = 2s^2 + 12t^2$$

(where we use s and t as new variables). The corresponding orthogonal change of coordinates is obtained by finding an orthogonal set of eigenvectors of A.

Subtract $\lambda = 2$ down the diagonal of A to obtain the matrix

$$M = \begin{bmatrix} 1 & -3 \\ -3 & 9 \end{bmatrix} \quad \text{corresponding to} \quad \begin{array}{r} x - 3y = 0 \\ -3x + 9y = 0 \end{array} \quad \text{or} \quad x - 3y = 0$$

A nonzero solution is $u_1 = (3, 1)$. Next subtract $\lambda = 12$ down the diagonal of A to obtain the matrix

$$M = \begin{bmatrix} -9 & -3 \\ -3 & -1 \end{bmatrix} \quad \text{corresponding to} \quad \begin{array}{r} -9x - 3y = 0 \\ -3x - y = 0 \end{array} \quad \text{or} \quad -3x - y = 0$$

A nonzero solution is $u_2 = (-1, 3)$. Normalize u_1 and u_2 to obtain the orthonormal basis

$$\hat{u}_1 = (3/\sqrt{10},\ 1/\sqrt{10}), \qquad \hat{u}_2 = (-1/\sqrt{10},\ 3/\sqrt{10})$$

Now let P be the matrix whose columns are $\hat{u}_1$ and $\hat{u}_2$. Then

$$P = \begin{bmatrix} 3/\sqrt{10} & -1/\sqrt{10} \\ 1/\sqrt{10} & 3/\sqrt{10} \end{bmatrix} \quad \text{and} \quad D = P^{-1}AP = P^TAP = \begin{bmatrix} 2 & 0 \\ 0 & 12 \end{bmatrix}$$

Thus, the required orthogonal change of coordinates is

$$\begin{bmatrix} x \\ y \end{bmatrix} = P \begin{bmatrix} s \\ t \end{bmatrix} \quad \text{or} \quad x = \frac{3s - t}{\sqrt{10}}, \quad y = \frac{s + 3t}{\sqrt{10}}$$

One can also express s and t in terms of x and y by using $P^{-1} = P^T$; that is,

$$s = \frac{3x + y}{\sqrt{10}}, \qquad t = \frac{-x + 3y}{\sqrt{10}}$$

13.16. Prove Theorem 13.12: Let T be an orthogonal operator on a real inner product space V. Then there exists an orthonormal basis of V in which T is represented by a block diagonal matrix M of the form

$$M = \text{diag}\left(1, \ldots, 1, -1, \ldots, -1, \begin{bmatrix} \cos\theta_1 & -\sin\theta_1 \\ \sin\theta_1 & \cos\theta_1 \end{bmatrix}, \ldots, \begin{bmatrix} \cos\theta_r & -\sin\theta_r \\ \sin\theta_r & \cos\theta_r \end{bmatrix}\right)$$

Let $S = T + T^{-1} = T + T^*$. Then $S^* = (T + T^*)^* = T^* + T = S$. Thus, S is a symmetric operator on V. By Theorem 13.11, there exists an orthonormal basis of V consisting of eigenvectors of S. If $\lambda_1, \ldots, \lambda_m$ denote the distinct eigenvalues of S, then V can be decomposed into the direct sum $V = V_1 \oplus V_2 \oplus \cdots \oplus V_m$ where the V_i consists of the eigenvectors of S belonging to λ_i. We claim that each V_i is invariant under T. For suppose $v \in V$; then $S(v) = \lambda_i v$ and

$$S(T(v)) = (T + T^{-1})T(v) = T(T + T^{-1})(v) = TS(v) = T(\lambda_i v) = \lambda_i T(v)$$

That is, $T(v) \in V_i$. Hence, V_i is invariant under T. Because the V_i are orthogonal to each other, we can restrict our investigation to the way that T acts on each individual V_i.

On a given V_i, we have $(T + T^{-1})v = S(v) = \lambda_i v$. Multiplying by T, we get

$$(T^2 - \lambda_i T + I)(v) = 0 \tag{1}$$

We consider the cases $\lambda_i = \pm 2$ and $\lambda_i \neq \pm 2$ separately. If $\lambda_i = \pm 2$, then $(T \pm I)^2(v) = 0$, which leads to $(T \pm I)(v) = 0$ or $T(v) = \pm v$. Thus, T restricted to this V_i is either I or $-I$.

If $\lambda_i \neq \pm 2$, then T has no eigenvectors in V_i, because, by Theorem 13.4, the only eigenvalues of T are 1 or -1. Accordingly, for $v \neq 0$, the vectors v and $T(v)$ are linearly independent. Let W be the subspace spanned by v and $T(v)$. Then W is invariant under T, because using (1) we get

$$T(T(v)) = T^2(v) = \lambda_i T(v) - v \in W$$

By Theorem 7.4, $V_i = W \oplus W^{\perp}$. Furthermore, by Problem 13.8, $W^{\perp}$ is also invariant under T. Thus, we can decompose V_i into the direct sum of two-dimensional subspaces W_j where the W_j are orthogonal to each other and each W_j is invariant under T. Thus, we can restrict our investigation to the way in which T acts on each individual W_j.

Because $T^2 - \lambda_i T + I = 0$, the characteristic polynomial $\Delta(t)$ of T acting on W_j is $\Delta(t) = t^2 - \lambda_i t + 1$. Thus, the determinant of T is 1, the constant term in $\Delta(t)$. By Theorem 2.7, the matrix A representing T acting on W_j relative to any orthogonal basis of W_j must be of the form

$$\begin{bmatrix} \cos\theta & -\sin\theta \\ \sin\theta & \cos\theta \end{bmatrix}$$

The union of the bases of the W_j gives an orthonormal basis of V_i, and the union of the bases of the V_i gives an orthonormal basis of V in which the matrix representing T is of the desired form.

Normal Operators and Canonical Forms in Unitary Spaces

13.17. Determine which of the following matrices is normal:

(a) $A = \begin{bmatrix} 1 & i \\ 0 & 1 \end{bmatrix}$, (b) $B = \begin{bmatrix} 1 & i \\ 1 & 2+i \end{bmatrix}$

(a) $AA^* = \begin{bmatrix} 1 & i \\ 0 & 1 \end{bmatrix}\begin{bmatrix} 1 & 0 \\ -i & 1 \end{bmatrix} = \begin{bmatrix} 2 & i \\ -i & 1 \end{bmatrix}, \qquad A^*A = \begin{bmatrix} 1 & 0 \\ -i & 1 \end{bmatrix}\begin{bmatrix} 1 & i \\ 0 & 1 \end{bmatrix} = \begin{bmatrix} 1 & i \\ -i & 2 \end{bmatrix}$

Because $AA^* \neq A^*A$, the matrix A is not normal.

(b) $BB^*\begin{bmatrix} 1 & i \\ 1 & 2+i \end{bmatrix}\begin{bmatrix} 1 & 1 \\ -i & 2-i \end{bmatrix} = \begin{bmatrix} 2 & 2+2i \\ 2-2i & 6 \end{bmatrix} = \begin{bmatrix} 1 & 1 \\ -i & 2-i \end{bmatrix}\begin{bmatrix} 1 & i \\ 1 & 2+i \end{bmatrix} = B^*B$

Because $BB^* = B^*B$, the matrix B is normal.

13.18. Let T be a normal operator. Prove the following:

(a) $T(v) = 0$ if and only if $T^*(v) = 0$. (b) $T - \lambda I$ is normal.

(c) If $T(v) = \lambda v$, then $T^*(v) = \bar{\lambda} v$; hence, any eigenvector of T is also an eigenvector of T^*.

(d) If $T(v) = \lambda_1 v$ and $T(w) = \lambda_2 w$ where $\lambda_1 \neq \lambda_2$, then $\langle v, w \rangle = 0$; that is, eigenvectors of T belonging to distinct eigenvalues are orthogonal.

(a) We show that $\langle T(v), T(v) \rangle = \langle T^*(v), T^*(v) \rangle$:

$$\langle T(v), T(v) \rangle = \langle v, T^*T(v) \rangle = \langle v, TT^*(v) \rangle = \langle T^*(v), T^*(v) \rangle$$

Hence, by $[I_3]$ in the definition of the inner product in Section 7.2, $T(v) = 0$ if and only if $T^*(v) = 0$.

(b) We show that $T - \lambda I$ commutes with its adjoint:

$$\begin{aligned} (T - \lambda I)(T - \lambda I)^* &= (T - \lambda I)(T^* - \bar{\lambda} I) = TT^* - \lambda T^* - \bar{\lambda} T + \lambda \bar{\lambda} I \\ &= T^*T - \bar{\lambda} T - \lambda T^* + \bar{\lambda}\lambda I = (T^* - \bar{\lambda} I)(T - \lambda I) \\ &= (T - \lambda I)^*(T - \lambda I) \end{aligned}$$

Thus, $T - \lambda I$ is normal.

(c) If $T(v) = \lambda v$, then $(T - \lambda I)(v) = 0$. Now $T - \lambda I$ is normal by (b); therefore, by (a), $(T - \lambda I)^*(v) = 0$. That is, $(T^* - \bar{\lambda} I)(v) = 0$; hence, $T^*(v) = \bar{\lambda} v$.

(d) We show that $\lambda_1 \langle v, w \rangle = \lambda_2 \langle v, w \rangle$:

$$\lambda_1 \langle v, w \rangle = \langle \lambda_1 v, w \rangle = \langle T(v), w \rangle = \langle v, T^*(w) \rangle = \langle v, \bar{\lambda}_2 w \rangle = \lambda_2 \langle v, w \rangle$$

But $\lambda_1 \neq \lambda_2$; hence, $\langle v, w \rangle = 0$.

13.19. Prove Theorem 13.13: Let T be a normal operator on a complex finite-dimensional inner product space V. Then there exists an orthonormal basis of V consisting of eigenvectors of T. (Thus, T can be represented by a diagonal matrix relative to an orthonormal basis.)

The proof is by induction on the dimension of V. If $\dim V = 1$, then the theorem trivially holds. Now suppose $\dim V = n > 1$. Because V is a complex vector space, T has at least one eigenvalue and hence a nonzero eigenvector v. Let W be the subspace of V spanned by v, and let u_1 be a unit vector in W.

Because v is an eigenvector of T, the subspace W is invariant under T. However, v is also an eigenvector of T^* by Problem 13.18; hence, W is also invariant under T^*. By Problem 13.8, $W^\perp$ is invariant under $T^{**} = T$. The remainder of the proof is identical with the latter part of the proof of Theorem 13.11 (Problem 13.14).

13.20. Prove Theorem 13.14: Let T be any operator on a complex finite-dimensional inner product space V. Then T can be represented by a triangular matrix relative to an orthonormal basis of V.

The proof is by induction on the dimension of V. If $\dim V = 1$, then the theorem trivially holds. Now suppose $\dim V = n > 1$. Because V is a complex vector space, T has at least one eigenvalue and hence at least one nonzero eigenvector v. Let W be the subspace of V spanned by v, and let u_1 be a unit vector in W. Then u_1 is an eigenvector of T and, say, $T(u_1) = a_{11} u_1$.

By Theorem 7.4, $V = W \oplus W^\perp$. Let E denote the orthogonal projection V into $W^\perp$. Clearly $W^\perp$ is invariant under the operator ET. By induction, there exists an orthonormal basis $\{u_2, \ldots, u_n\}$ of $W^\perp$ such that, for $i = 2, \ldots, n$,

$$ET(u_i) = a_{i2} u_2 +_{i3} u_3 + \cdots + a_{ii} u_i$$

(Note that $\{u_1, u_2, \ldots, u_n\}$ is an orthonormal basis of V.) But E is the orthogonal projection of V onto $W^\perp$; hence, we must have

$$T(u_i) = a_{i1} u_1 + a_{i2} u_2 + \cdots + a_{ii} u_i$$

for $i = 2, \ldots, n$. This with $T(u_1) = a_{11} u_1$ gives us the desired result.

Miscellaneous Problems

13.21. Prove Theorem 13.10B: The following are equivalent:

(i) $P = T^2$ for some self-adjoint operator T.

(ii) $P = S^*S$ for some operator S; that is, P is positive.

(iii) P is self-adjoint and $\langle P(u), u \rangle \geq 0$ for every $u \in V$.

Suppose (i) holds; that is, $P = T^2$ where $T = T^*$. Then $P = TT = T^*T$, and so (i) implies (ii). Now suppose (ii) holds. Then $P^* = (S^*S)^* = S^*S^{**} = S^*S = P$, and so P is self-adjoint. Furthermore,

$$\langle P(u), u \rangle = \langle S^*S(u), u \rangle = \langle S(u), S(u) \rangle \geq 0$$

Thus, (ii) implies (iii), and so it remains to prove that (iii) implies (i).

Now suppose (iii) holds. Because P is self-adjoint, there exists an orthonormal basis $\{u_1, \ldots, u_n\}$ of V consisting of eigenvectors of P; say, $P(u_i) = \lambda_i u_i$. By Theorem 13.4, the λ_i are real. Using (iii), we show that the λ_i are nonnegative. We have, for each i,

$$0 \leq \langle P(u_i), u_i \rangle = \langle \lambda_i u_i, u_i \rangle = \lambda_i \langle u_i, u_i \rangle$$

Thus, $\langle u_i, u_i \rangle \geq 0$ forces $\lambda_i \geq 0$, as claimed. Accordingly, $\sqrt{\lambda_i}$ is a real number. Let T be the linear operator defined by

$$T(u_i) = \sqrt{\lambda_i} u_i \quad \text{for } i = 1, \ldots, n$$

Because T is represented by a real diagonal matrix relative to the orthonormal basis $\{u_i\}$, T is self-adjoint. Moreover, for each i,

$$T^2(u_i) = T(\sqrt{\lambda_i}u_i) = \sqrt{\lambda_i}T(i_i) = \sqrt{\lambda_i}\sqrt{\lambda_i}u_i = \lambda_i u_i = P(u_i)$$

Because T^2 and P agree on a basis of $V, P = T^2$. Thus, the theorem is proved.

Remark: The above operator T is the unique positive operator such that $P = T^2$; it is called the *positive square root* of P.

13.22. Show that any operator T is the sum of a self-adjoint operator and a skew-adjoint operator.

Set $S = \frac{1}{2}(T + T^*)$ and $U = \frac{1}{2}(T - T^*)$. Then $T = S + U$, where

$$S^* = [\tfrac{1}{2}(T + T^*)]^* = \tfrac{1}{2}(T^* + T^{**}) = \tfrac{1}{2}(T^* + T) = S$$

and

$$U^* = [\tfrac{1}{2}(T - T^*)]^* = \tfrac{1}{2}(T^* - T) = -\tfrac{1}{2}(T - T^*) = -U$$

that is, S is self-adjoint and U is skew-adjoint.

13.23. Prove: Let T be an arbitrary linear operator on a finite-dimensional inner product space V. Then T is a product of a unitary (orthogonal) operator U and a unique positive operator P; that is, $T = UP$. Furthermore, if T is invertible, then U is also uniquely determined.

By Theorem 13.10, T^*T is a positive operator; hence, there exists a (unique) positive operator P such that $P^2 = T^*T$ (Problem 13.43). Observe that

$$\|P(v)\|^2 = \langle P(v), P(v)\rangle = \langle P^2(v), v\rangle = \langle T^*T(v), v\rangle = \langle T(v), T(v)\rangle = \|T(v)\|^2 \tag{1}$$

We now consider separately the cases when T is invertible and noninvertible.

If T is invertible, then we set $\hat{U} = PT^{-1}$. We show that $\hat{U}$ is unitary:

$$\hat{U}^* = (PT^{-1})^* = T^{-1*}P^* = (T^*)^{-1}P \qquad \text{and} \qquad \hat{U}*\hat{U} = (T^*)^{-1}PPT^{-1} = (T^*)^{-1}T^*TT^{-1} = I$$

Thus, $\hat{U}$ is unitary. We next set $U = \hat{U}^{-1}$. Then U is also unitary, and $T = UP$ as required.

To prove uniqueness, we assume $T = U_0P_0$, where U_0 is unitary and P_0 is positive. Then

$$T^*T = P_0^*U_0^*U_0P_0 = P_0IP_0 = P_0^2$$

But the positive square root of T^*T is unique (Problem 13.43); hence, $P_0 = P$. (Note that the invertibility of T is not used to prove the uniqueness of P.) Now if T is invertible, then P is also invertible by (1). Multiplying $U_0P = UP$ on the right by P^{-1} yields $U_0 = U$. Thus, U is also unique when T is invertible.

Now suppose T is not invertible. Let W be the image of P; that is, $W = \text{Im}\, P$. We define $U_1 : W \to V$ by

$$U_1(w) = T(v), \qquad \text{where} \qquad P(v) = w \tag{2}$$

We must show that U_1 is well defined; that is, that $P(v) = P(v')$ implies $T(v) = T(v')$. This follows from the fact that $P(v - v') = 0$ is equivalent to $\|P(v - v')\| = 0$, which forces $\|T(v - v')\| = 0$ by (1). Thus, U_1 is well defined. We next define $U_2 : W \to V$. Note that, by (1), P and T have the same kernels. Hence, the images of P and T have the same dimension; that is, $\dim(\text{Im}\, P) = \dim W = \dim(\text{Im}\, T)$. Consequently, $W^\perp$ and $(\text{Im}\, T)^\perp$ also have the same dimension. We let U_2 be any isomorphism between $W^\perp$ and $(\text{Im}\, T)^\perp$.

We next set $U = U_1 \oplus U_2$. [Here U is defined as follows: If $v \in V$ and $v = w + w'$, where $w \in W$, $w' \in W^\perp$, then $U(v) = U_1(w) + U_2(w')$.] Now U is linear (Problem 13.69), and, if $v \in V$ and $P(v) = w$, then, by (2),

$$T(v) = U_1(w) = U(w) = UP(v)$$

Thus, $T = UP$, as required.

It remains to show that U is unitary. Now every vector $x \in V$ can be written in the form $x = P(v) + w'$, where $w' \in W^\perp$. Then $U(x) = UP(v) + U_2(w') = T(v) + U_2(w')$, where $\langle T(v), U_2(w')\rangle = 0$ by definition of U_2. Also, $\langle T(v), T(v)\rangle = \langle P(v), P(v)\rangle$ by (1). Thus,

$$\begin{aligned}\langle U(x), U(x)\rangle &= \langle T(v) + U_2(w'),\ T(v) + U_2(w')\rangle = \langle T(v), T(v)\rangle + \langle U_2(w'), U_2(w')\rangle \\ &= \langle P(v), P(v)\rangle + \langle w', w'\rangle = \langle P(v) + w',\ P(v) + w'\rangle = \langle x, x\rangle\end{aligned}$$

[We also used the fact that $\langle P(v), w'\rangle = 0$.] Thus, U is unitary, and the theorem is proved.

13.24. Let V be the vector space of polynomials over $\mathbf{R}$ with inner product defined by

$$\langle f, g\rangle = \int_0^1 f(t)g(t)\,dt$$

Give an example of a linear functional ϕ on V for which Theorem 13.3 does not hold—that is, for which there is no polynomial $h(t)$ such that $\phi(f) = \langle f, h\rangle$ for every $f \in V$.

Let $\phi: V \to \mathbf{R}$ be defined by $\phi(f) = f(0)$; that is, ϕ evaluates $f(t)$ at 0, and hence maps $f(t)$ into its constant term. Suppose a polynomial $h(t)$ exists for which

$$\phi(f) = f(0) = \int_0^1 f(t)h(t)\,dt \tag{1}$$

for every polynomial $f(t)$. Observe that ϕ maps the polynomial $tf(t)$ into 0; hence, by (1),

$$\int_0^1 tf(t)h(t)\,dt = 0 \tag{2}$$

for every polynomial $f(t)$. In particular (2) must hold for $f(t) = th(t)$; that is,

$$\int_0^1 t^2h^2(t)\,dt = 0$$

This integral forces $h(t)$ to be the zero polynomial; hence, $\phi(f) = \langle f, h\rangle = \langle f, \mathbf{0}\rangle = 0$ for every polynomial $f(t)$. This contradicts the fact that ϕ is not the zero functional; hence, the polynomial $h(t)$ does not exist.

SUPPLEMENTARY PROBLEMS

Adjoint Operators

13.25. Find the adjoint of:

(a) $A = \begin{bmatrix} 5-2i & 3+7i \\ 4-6i & 8+3i \end{bmatrix}$, (b) $B = \begin{bmatrix} 3 & 5i \\ i & -2i \end{bmatrix}$, (c) $C = \begin{bmatrix} 1 & 1 \\ 2 & 3 \end{bmatrix}$

13.26. Let $T: \mathbf{R}^3 \to \mathbf{R}^3$ be defined by $T(x,y,z) = (x + 2y,\ 3x - 4z,\ y)$. Find $T^*(x,y,z)$.

13.27. Let $T: \mathbf{C}^3 \to \mathbf{C}^3$ be defined by $T(x,y,z) = [ix + (2+3i)y,\ 3x + (3-i)z,\ (2-5i)y + iz]$. Find $T^*(x,y,z)$.

13.28. For each linear function ϕ on V, find $u \in V$ such that $\phi(v) = \langle v, u\rangle$ for every $v \in V$:

(a) $\phi: \mathbf{R}^3 \to \mathbf{R}$ defined by $\phi(x,y,z) = x + 2y - 3z$.

(b) $\phi: \mathbf{C}^3 \to \mathbf{C}$ defined by $\phi(x,y,z) = ix + (2+3i)y + (1-2i)z$.

13.29. Suppose V has finite dimension. Prove that the image of T^* is the orthogonal complement of the kernel of T; that is, $\mathrm{Im}\, T^* = (\mathrm{Ker}\, T)^{\perp}$. Hence, $\mathrm{rank}(T) = \mathrm{rank}(T^*)$.

13.30. Show that $T^*T = 0$ implies $T = 0$.

13.31. Let V be the vector space of polynomials over $\mathbf{R}$ with inner product defined by $\langle f, g\rangle = \int_0^1 f(t)g(t)\,dt$. Let $\mathbf{D}$ be the derivative operator on V; that is, $\mathbf{D}(f) = df/dt$. Show that there is no operator $\mathbf{D}^*$ on V such that $\langle \mathbf{D}(f), g\rangle = \langle f, \mathbf{D}^*(g)\rangle$ for every $f, g \in V$. That is, $\mathbf{D}$ has no adjoint.

Unitary and Orthogonal Operators and Matrices

13.32. Find a unitary (orthogonal) matrix whose first row is

(a) $(2/\sqrt{13}, 3/\sqrt{13})$, (b) a multiple of $(1, 1-i)$, (c) a multiple of $(1, -i, 1-i)$.

13.33. Prove that the products and inverses of orthogonal matrices are orthogonal. (Thus, the orthogonal matrices form a group under multiplication, called the *orthogonal group*.)

13.34. Prove that the products and inverses of unitary matrices are unitary. (Thus, the unitary matrices form a group under multiplication, called the *unitary group*.)

13.35. Show that if an orthogonal (unitary) matrix is triangular, then it is diagonal.

13.36. Recall that the complex matrices A and B are unitarily equivalent if there exists a unitary matrix P such that $B = P^*AP$. Show that this relation is an equivalence relation.

13.37. Recall that the real matrices A and B are orthogonally equivalent if there exists an orthogonal matrix P such that $B = P^TAP$. Show that this relation is an equivalence relation.

13.38. Let W be a subspace of V. For any $v \in V$, let $v = w + w'$, where $w \in W$, $w' \in W^\perp$. (Such a sum is unique because $V = W \oplus W^\perp$.) Let $T: V \to V$ be defined by $T(v) = w - w'$. Show that T is self-adjoint unitary operator on V.

13.39. Let V be an inner product space, and suppose $U: V \to V$ (not assumed linear) is surjective (onto) and preserves inner products; that is, $\langle U(v), U(w)\rangle = \langle u, w\rangle$ for every $v, w \in V$. Prove that U is linear and hence unitary.

Positive and Positive Definite Operators

13.40. Show that the sum of two positive (positive definite) operators is positive (positive definite).

13.41. Let T be a linear operator on V and let $f: V \times V \to K$ be defined by $f(u, v) = \langle T(u), v\rangle$. Show that f is an inner product on V if and only if T is positive definite.

13.42. Suppose E is an orthogonal projection onto some subspace W of V. Prove that $kI + E$ is positive (positive definite) if $k \geq 0$ $(k > 0)$.

13.43. Consider the operator T defined by $T(u_i) = \sqrt{\lambda_i}u_i$, $i = 1, \ldots, n$, in the proof of Theorem 13.10A. Show that T is positive and that it is the only positive operator for which $T^2 = P$.

13.44. Suppose P is both positive and unitary. Prove that $P = I$.

13.45. Determine which of the following matrices are positive (positive definite):

(i) $\begin{bmatrix} 1 & 1 \\ 1 & 1 \end{bmatrix}$, (ii) $\begin{bmatrix} 0 & i \\ -i & 0 \end{bmatrix}$, (iii) $\begin{bmatrix} 0 & 1 \\ -1 & 0 \end{bmatrix}$, (iv) $\begin{bmatrix} 1 & 1 \\ 0 & 1 \end{bmatrix}$, (v) $\begin{bmatrix} 2 & 1 \\ 1 & 2 \end{bmatrix}$, (vi) $\begin{bmatrix} 1 & 2 \\ 2 & 1 \end{bmatrix}$

13.46. Prove that a 2×2 complex matrix $A = \begin{bmatrix} a & b \\ c & d \end{bmatrix}$ is positive if and only if (i) $A = A^*$, and (ii) a, d and $|A| = ad - bc$ are nonnegative real numbers.

13.47. Prove that a diagonal matrix A is positive (positive definite) if and only if every diagonal entry is a nonnegative (positive) real number.

Self-adjoint and Symmetric Matrices

13.48. For any operator T, show that $T + T^*$ is self-adjoint and $T - T^*$ is skew-adjoint.

13.49. Suppose T is self-adjoint. Show that $T^2(v) = 0$ implies $T(v) = 0$. Using this to prove that $T^n(v) = 0$ also implies that $T(v) = 0$ for $n > 0$.

13.50. Let V be a complex inner product space. Suppose $\langle T(v), v\rangle$ is real for every $v \in V$. Show that T is self-adjoint.

13.51. Suppose T_1 and T_2 are self-adjoint. Show that T_1T_2 is self-adjoint if and only if T_1 and T_1 commute; that is, $T_1T_2 = T_2T_1$.

13.52. For each of the following symmetric matrices A, find an orthogonal matrix P and a diagonal matrix D such that P^TAP is diagonal:

(a) $A = \begin{bmatrix} 1 & 2 \\ 2 & -2 \end{bmatrix}$, (b) $A = \begin{bmatrix} 5 & 4 \\ 4 & -1 \end{bmatrix}$, (c) $A = \begin{bmatrix} 7 & 3 \\ 3 & -1 \end{bmatrix}$

13.53. Find an orthogonal change of coordinates $X = PX'$ that diagonalizes each of the following quadratic forms and find the corresponding diagonal quadratic form $q(x')$:

(a) $q(x, y) = 2x^2 - 6xy + 10y^2$, (b) $q(x, y) = x^2 + 8xy - 5y^2$

(c) $q(x, y, z) = 2x^2 - 4xy + 5y^2 + 2xz - 4yz + 2z^2$

Normal Operators and Matrices

13.54. Let $A = \begin{bmatrix} 2 & i \\ i & 2 \end{bmatrix}$. Verify that A is normal. Find a unitary matrix P such that P^*AP is diagonal. Find P^*AP.

13.55. Show that a triangular matrix is normal if and only if it is diagonal.

13.56. Prove that if T is normal on V, then $\|T(v)\| = \|T^*(v)\|$ for every $v \in V$. Prove that the converse holds in complex inner product spaces.

13.57. Show that self-adjoint, skew-adjoint, and unitary (orthogonal) operators are normal.

13.58. Suppose T is normal. Prove that

(a) T is self-adjoint if and only if its eigenvalues are real.

(b) T is unitary if and only if its eigenvalues have absolute value 1.

(c) T is positive if and only if its eigenvalues are nonnegative real numbers.

13.59. Show that if T is normal, then T and T^* have the same kernel and the same image.

13.60. Suppose T_1 and T_2 are normal and commute. Show that $T_1 + T_2$ and T_1T_2 are also normal.

13.61. Suppose T_1 is normal and commutes with T_2. Show that T_1 also commutes with T_2^*.

13.62. Prove the following: Let T_1 and T_2 be normal operators on a complex finite-dimensional vector space V. Then there exists an orthonormal basis of V consisting of eigenvectors of both T_1 and T_2. (That is, T_1 and T_2 can be simultaneously diagonalized.)

Isomorphism Problems for Inner Product Spaces

13.63. Let $S = \{u_1, \ldots, u_n\}$ be an orthonormal basis of an inner product space V over K. Show that the mapping $v \mapsto [v]_s$ is an (inner product space) isomorphism between V and K^n. (Here $[v]_S$ denotes the coordinate vector of v in the basis S.)

13.64. Show that inner product spaces V and W over K are isomorphic if and only if V and W have the same dimension.

13.65. Suppose $\{u_1, \ldots, u_n\}$ and $\{u'_1, \ldots, u'_n\}$ are orthonormal bases of V and W, respectively. Let $T:V \to W$ be the linear map defined by $T(u_i) = u'_i$ for each i. Show that T is an isomorphism.

13.66. Let V be an inner product space. Recall that each $u \in V$ determines a linear functional $\hat{u}$ in the dual space V^* by the definition $\hat{u}(v) = \langle v, u \rangle$ for every $v \in V$. (See the text immediately preceding Theorem 13.3.) Show that the map $u \mapsto \hat{u}$ is linear and nonsingular, and hence an isomorphism from V onto V^*.

Miscellaneous Problems

13.67. Suppose $\{u_1, \ldots, u_n\}$ is an orthonormal basis of V. Prove

(a) $\langle a_1u_1 + a_2u_2 + \cdots + a_nu_n,\ b_1u_1 + b_2u_2 + \cdots + b_nu_n \rangle = a_1\bar{b}_1 + a_2\bar{b}_2 + \ldots \bar{a}_n\bar{b}_n$

(b) Let $A = [a_{ij}]$ be the matrix representing $T: V \to V$ in the basis $\{u_i\}$. Then $a_{ij} = \langle T(u_i), u_j \rangle$.

13.68. Show that there exists an orthonormal basis $\{u_1, \ldots, u_n\}$ of V consisting of eigenvectors of T if and only if there exist orthogonal projections $E_1, \ldots, E_r$ and scalars $\lambda_1, \ldots, \lambda_r$ such that

(i) $T = \lambda_1E_1 + \cdots + \lambda_rE_r$, (ii) $E_1 + \cdots + E_r = I$, (iii) $E_iE_j = 0$ for $i \neq j$

13.69. Suppose $V = U \oplus W$ and suppose $T_1: U \to V$ and $T_2: W \to V$ are linear. Show that $T = T_1 \oplus T_2$ is also linear. Here T is defined as follows: If $v \in V$ and $v = u + w$ where $u \in U$, $w \in W$, then

$$T(v) = T_1(u) + T_2(w)$$

ANSWERS TO SUPPLEMENTARY PROBLEMS

Notation: $[R_1;\ R_2;\ \ldots;\ R_n]$ denotes a matrix with rows $R_1, R_2, \ldots, R_n$.

13.25. (a) $[5+2i,\ 4+6i;\ 3-7i,\ 8-3i]$, (b) $[3, -i;\ -5i, 2i]$, (c) $[1, 2;\ 1, 3]$

13.26. $T^*(x,y,z) = (x+3y,\ 2x+z,\ -4y)$

13.27. $T^*(x,y,z) = [-ix+3y,\ (2-3i)x+(2+5i)z,\ (3+i)y-iz]$

13.28. (a) $u = (1, 2, -3)$, (b) $u = (-i,\ 2-3i,\ 1+2i)$

13.32. (a) $(1/\sqrt{13})[2, 3;\ 3, -2]$, (b) $1/\sqrt{2}\ (3)[1, 1-i; -1-i, 1]$
(c) $\frac{1}{2}[1,\ -i;\ 1-i;\ \sqrt{2}i,\ -\sqrt{2},\ 0;\ 1,\ -i,\ -1+i]$

13.45. Only (i) and (v) are positive. Only (v) is positive definite.

13.52. (a and b) $P = (1/\sqrt{5})[2, -1;\ 1, 2]$, (c) $P = (1/\sqrt{10})[3, -1;\ 1, 3]$
(a) $D = [2, 0;\ 0, -3]$, (b) $D = [7, 0;\ 0, -3]$, (c) $D = [8, 0;\ 0, -2]$

13.53. (a) $x = (3x' - y')/\sqrt{10}$, $y = (x' + 3y')/\sqrt{10}$; (b) $x = (2x' - y')/\sqrt{5}$, $y = (x' + 2y')/\sqrt{5}$;
(c) $x = x'/\sqrt{3} + y'/\sqrt{2} + z'/\sqrt{6}$, $y = x'/\sqrt{3} - 2z'/\sqrt{6}$, $z = x'/\sqrt{3} - y'/\sqrt{2} + z'/\sqrt{6}$;
(a) $q(x') = \text{diag}(1, 11)$; (b) $q(x') = \text{diag}(3, -7)$; (c) $q(x') = \text{diag}(1, 17)$

13.54. (a) $P = (1/\sqrt{2})[1, -1;\ 1, 1]$, $P^*AP = \text{diag}(2+i,\ 2-i)$

APPENDIX A

Multilinear Products

A.1 Introduction

The material in this appendix is much more abstract than that which has previously appeared. Accordingly, many of the proofs will be omitted. Also, we motivate the material with the following observation.

Let S be a basis of a vector space V. Theorem 5.2 may be restated as follows.

THEOREM 5.2: Let $g: S \to V$ be the inclusion map of the basis S into V. Then, for any vector space U and any mapping $f: S \to U$, there exists a unique linear mapping $f^*: V \to U$ such that $f = f^* \cdot g$.

Another way to state the fact that $f = f^* \cdot g$ is that the diagram in Fig. A-1(a) *commutes*.

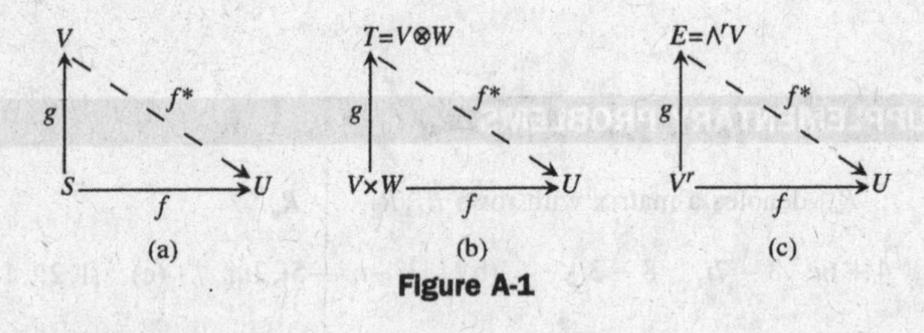

Figure A-1

A.2 Bilinear Mapping and Tensor Products

Let U, V, W be vector spaces over a field $\mathbf{K}$. Consider a map

$$f: V \times W \to U$$

Then f is said to be *bilinear* if, for each $v \in V$, the map $f_v: W \to U$ defined by $f_v(w) = f(v, w)$ is linear; and, for each $w \in W$, the map $f_w: V \to U$ defined by $f_w(v) = f(v, w)$ is linear.

That is, f is linear in each of its two variables. Note that f is similar to a bilinear form except that the values of the map f are in a vector space U rather than the field $\mathbf{K}$.

DEFINITION A.1: Let V and W be vector spaces over the same field $\mathbf{K}$. The *tensor product* of V and W is a vector space T over $\mathbf{K}$ together with a bilinear map $g: V \times W \to T$, denoted by $g(v, w) = v \otimes w$, with the following property: (*) For any vector space U over $\mathbf{K}$ and any bilinear map $f: V \times W \to U$ there exists a unique linear map $f^*: T \to U$ such that $f^* \cdot g = f$.

The tensor product (T, g) [or simply T when g is understood] of V and W is denoted by $V \otimes W$, and the element $v \otimes w$ is called the *tensor* of v and w.

Another way to state condition (*) is that the diagram in Fig. A-1(b) commutes. The fact that such a unique linear map f^* exists is called the ''Universal Mapping Principle'' (UMP). As illustrated in Fig. A-1(b), condition (*) also says that any bilinear map $f: V \times W \to U$ ''factors through'' the tensor product $T = V \otimes W$. The uniqueness in (*) implies that the image of g spans T; that is, span $(\{v \otimes w\}) = T$.

THEOREM A.1: **(Uniqueness of Tensor Products)** Let (T, g) and (T', g') be tensor products of V and W. Then there exists a unique isomorphism $h:T \to T'$ such that $hg = g'$.

Proof. Because T is a tensor product, and $g':V \otimes W \to T'$ is bilinear, there exists a unique linear map $h:T \to T'$ such that $hg = g'$. Similarly, because T' is a tensor product, and $g:V \otimes W \to T'$ is bilinear, there exists a unique linear map $h':T' \to T$ such that $h'g' = g$. Using $hg = g'$, we get $h'hg = g$. Also, because T is a tensor product, and $g:V \otimes W \to T$ is bilinear, there exists a unique linear map $h^*:T \to T$ such that $h^*g = g$. But $1_T g = g$. Thus, $h'h = h^* = 1_T$. Similarly, $hh' = 1_{T'}$. Therefore, h is an isomorphism from T to T'.

THEOREM A.2: **(Existence of Tensor Product)** The tensor product $T = V \otimes W$ of vector spaces V and W over **K** exists. Let $\{v_1, \ldots, v_m\}$ be a basis of V and let $\{w_1, \ldots, w_n\}$ be a basis of W. Then the mn vectors

$$v_i \otimes w_i \quad (i = 1, \ldots, m; j = 1, \ldots, n)$$

form a basis of T. Thus, $\dim T = mn = (\dim V)(\dim W)$.

Outline of Proof. Suppose $\{v_1, \ldots, v_m\}$ is a basis of V, and suppose $\{w_1, \ldots, w_n\}$ is a basis of W. Consider the mn symbols $\{t_{ij} | i = i, \ldots, m, j = 1, \ldots, n\}$. Let T be the vector space generated by the t_{ij}. That is, T consists of all linear combinations of the t_{ij} with coefficients in **K**. [See Problem 4.137.]

Let $v \in V$ and $w \in W$. Say

$$v = a_1 v_1 + a_2 v_2 + \cdots + a_m v_m \quad \text{and} \quad w = b_1 w_1 + b_2 w_2 + \cdots + b_m w_m$$

Let $g:V \times W \to T$ be defined by

$$g(v, w) = \sum_i \sum_j a_i b_j t_{ij}$$

Then g is bilinear. [Proof left to reader.]

Now let $f:V \times W \to U$ be bilinear. Because the t_{ij} form a basis of T, Theorem 5.2 (stated above) tells us that there exists a unique linear map $f^*:T \to U$ such that $f^*(t_{ij}) = f(v_i, w_j)$. Then, for $v = \sum_i a_i v_i$ and $w = \sum_j b_j w_j$, we have

$$f(v, w) = f\left(\sum_i a_i v_i, \sum_j b_j w_j\right) = \sum_i \sum_j a_i b_j f(v_i, w_j) = \sum_i \sum_j a_i b_j t_{ij} = f^*(g(v, w)).$$

Therefore, $f = f^* g$ where f^* is the required map in Definition A.1. Thus, T is a tensor product.

Let $\{v'_1, \ldots, v'_m\}$ be any basis of V and $\{w'_1, \ldots, w'_m\}$ be any basis of W.

Let $v \in V$ and $w \in W$ and say

$$v = a'_1 v'_1 + \cdots + a'_m v'_m \quad \text{and} \quad w = b'_1 w'_1 + \cdots + b'_m w'_m$$

Then

$$v \otimes w = g(v, w) = \sum_i \sum_j a'_i b'_i g(v'_i, w'_i) = \sum_i \sum_j a'_i b'_j (v'_i \otimes w'_j)$$

Thus, the elements $v'_i \otimes w'_j$ span T. There are mn such elements. They cannot be linearly dependent because $\{t_{ij}\}$ is a basis of T, and hence, $\dim T = mn$. Thus, the $v'_i \otimes w'_j$ form a basis of T.

Next we give two concrete examples of tensor products.

EXAMPLE A.1 Let V be the vector space of polynomials $\mathbf{P}_{r-1}(x)$ and let W be the vector space of polynomials $\mathbf{P}_{s-1}(y)$. Thus, the following from bases of V and W, respectively,

$$1, x, x^2, \ldots, x^{r-1} \quad \text{and} \quad 1, y, y^2, \ldots, y^{s-1}$$

In particular, $\dim V = r$ and $\dim W = s$. Let T be the vector space of polynomials in variables x and y with basis

$$\{x^i y^j\} \text{ where } \quad i = 0, 1, \ldots, r-1; j = 0, 1, \ldots, s-1$$

Then T is the tensor product $V \otimes W$ under the mapping

$$x^i \otimes y^j = x^i y^j$$

For example, suppose $v = 2 - 5x + 3x^3$ and $w = 7y + 4y^2$. Then

$$v \otimes w = 14y + 8y^2 - 35xy - 20xy^2 + 21x^3y + 12x^3y^2$$

Note, $\dim T = rs = (\dim V)(\dim W)$.

EXAMPLE A.2
Let V be the vector space of $m \times n$ matrices over a field $\mathbf{K}$ and let W be the vector space of $p \times q$ matrices over $\mathbf{K}$. Suppose $A = [a_{11}]$ belongs to V, and B belongs to W. Let T be the vector space of $mp \times nq$ matrices over $\mathbf{K}$. Then T is the tensor product of V and W where $A \otimes B$ is the block matrix

$$A \otimes B = [a_{ij}B] = \begin{bmatrix} a_{11}B & a_{12}B & \cdots & a_{1n}B \\ a_{21}B & a_{22}B & \cdots & a_{2n}B \\ \cdots & \cdots & \cdots & \cdots \\ a_{m1}B & a_{m2}B & \cdots & a_{mn}B \end{bmatrix}$$

For example, suppose $A = \begin{bmatrix} 1 & 2 \\ 3 & 4 \end{bmatrix}$ and $B = \begin{bmatrix} 1 & 2 & 3 \\ 4 & 5 & 6 \end{bmatrix}$. Then

$$A \otimes B = \begin{bmatrix} 1 & 2 & 3 & 2 & 4 & 6 \\ 4 & 5 & 6 & 8 & 10 & 12 \\ 3 & 6 & 9 & 4 & 8 & 12 \\ 12 & 15 & 18 & 16 & 20 & 24 \end{bmatrix}$$

Isomorphisms of Tensor Products

First we note that tensoring is associative in a cannonical way. Namely,

THEOREM A.3: Let U, V, W be vector spaces over a field $\mathbf{K}$. Then there exists a unique isomorphism

$$(U \otimes V) \otimes W \to U \otimes (V \otimes W)$$

such that, for every $u \in U$, $v \in V$, $w \in W$,

$$(u \otimes v) \otimes w \mapsto u \otimes (v \otimes w)$$

Accordingly, we may omit parenthesis when tensoring any number of factors. Specifically, given vectors spaces $V_1, V_2, \ldots, V_m$ over a field $\mathbf{K}$, we may unambiguously form their tensor product

$$V_1 \otimes V_2 \otimes \ldots \otimes V_m$$

and, for vectors v_j in V_j, we may unambiguously form the tensor product

$$v_1 \otimes v_2 \otimes \ldots \otimes v_m$$

Moreover, given a vector space V over $\mathbf{K}$, we may unambiguously define the following tensor product:

$$\otimes^r V = V \otimes V \otimes \ldots \otimes V \ (r \text{ factors})$$

Also, there is a canonical isomorphism

$$(\otimes^r V) \otimes (\otimes^s V) \to \otimes^{r+s} V$$

Furthermore, viewing $\mathbf{K}$ as a vector space over itself, we have the canonical isomorphism

$$\mathbf{K} \otimes V \to V$$

where we define $a \otimes v = av$.

A.3 Alternating Multilinear Maps

Let $f: V^r \to U$ where V and U are vector spaces over **K**. [Recall $V^r = V \times V \times \ldots \times V$, r factors.]

(1) The mapping f is said to be multilinear or r-linear if $f(v_1, \ldots, v_r)$ is linear as a function of each v_j when the other v_i's are held fixed. That is,

$$f(\ldots, v_j + v_j', \ldots) = f(\ldots, v_j, \ldots) + f(\ldots, v_j', \ldots)$$
$$f(\ldots, kv_j, \ldots) = kf(\ldots, v_j, \ldots)$$

where only the jth position changes.

(2) The mapping f is said to be *alternating* if

$$f(v_1, \ldots, v_r) = 0 \text{ whenever } v_i = v_j \text{ with } i \neq j$$

One can easily show (Prove!) that if f is an alternating multilinear mapping on V^r, then

$$f(\ldots, v_i, \ldots, v_j, \ldots) = -f(\ldots, v_j, \ldots, v_i, \ldots)$$

That is, if two of the vectors are interchanged, then the associated value changes sign.

EXAMPLE A.3 (Determinants)

The determinant function $D: M \to K$ on the space M of $n \times n$ matrices may be viewed as an n-variable function

$$D(A) = D(R_1, R_2, \ldots, R_n)$$

defined on the rows $R_1, R_2, \ldots, R_n$ of A. Recall (Chapter 8) that, in this context, D is both n-linear and alternating.

We now need some additional notation. Let $K = [k_1, k_2, \ldots, k_r]$ denote an r-list (r-tuple) of elements from $I_n = (1, 2, \ldots, n)$. We will then use the following notation where the v_k's denote vectors and the a_{ik}'s denote scalars:

$$v_K = (v_{k_1}, v_{k_2}, \ldots, v_{k_r}) \text{ and } a_K = a_{1k_1} a_{2k_2} \ldots a_{rk_r}$$

Note v_K is a list of r vectors, and a_K is a product of r scalars.

Now suppose the elements in $K = [k_1, k_2, \ldots, k_r]$ are distinct. Then K is a permutation σ_K of an r-list $J = [i_1, i_2, \ldots, i_r]$ in *standard form*, that is, where $i_1 < i_2 < \ldots < i_r$. The number of such standard-form r-lists J from I_n is the binomial coefficient:

$$\binom{n}{r} = \frac{n!}{r!(n-r)!}$$

[Recall $\text{sign}(\sigma_K) = (-1)^{m_K}$ where m_K is the number of interchanges that transforms K into J.]

Now suppose $A = [a_{ij}]$ is an $r \times n$ matrix. For a given ordered r-list J, we define

$$D_J(A) = \begin{vmatrix} a_{1i_1} & a_{1i_2} & \cdots & a_{1i_r} \\ a_{2i_1} & a_{2i_2} & \cdots & a_{2i_r} \\ \cdots & \cdots & \cdots & \cdots \\ a_{ri_1} & a_{ri_2} & \cdots & a_{ri_r} \end{vmatrix}$$

That is, $D_J(A)$ is the determinant of the $r \times r$ submatrix of A whose column subscripts belong to J.

Our main theorem below uses the following "shuffling" lemma.

LEMMA A.4 Let V and U be vector spaces over K, and let $f: V^r \to U$ be an alternating r-linear mapping. Let $v_1, v_2, \ldots, v_n$ be vectors in V and let $A = [a_{ij}]$ be an $r \times n$ matrix over K where $r \leq n$. For $i = 1, 2, \ldots, r$, let

$$u_i = a_{i1} v_1 + a_{i2} v_2 + \cdots + a_{in} v_n$$

Then

$$f(u_1, \ldots, u_r) = \sum_f D_J(A) f(v_{i_1}, v_{i_2}, \ldots, v_{i_r})$$

where the sum is over all standard-form r-lists $J = \{i_1, i_2, \ldots, i_r\}$.

The proof is technical but straightforward. The linearity of f gives us the sum

$$f(u_1, \ldots, u_r) = \sum_K a_K f(v_K)$$

where the sum is over all r-lists K from $\{1, \ldots, n\}$. The alternating property of f tells us that $f(v_K) = 0$ when K does not contain distinct integers. The proof now mainly uses the fact that as we interchange the v_j's to transform

$$f(v_K) = f(v_{k_1}, v_{k_2}, \ldots, v_{k_r}) \quad \text{to} \quad f(v_j) = f(v_{i_1}, v_{i_2}, \ldots, v_{i_r})$$

so that $i_1 < \cdots < i_r$, the associated sign of a_K, will change in the same way as the sign of the corresponding permutation σ_K changes when it is transformed to the identity permutation using transpositions.

We illustrate the lemma below for $r = 2$ and $n = 3$.

EXAMPLE A.4 Suppose $f: V^2 \to U$ is an alternating multilinear function. Let $v_1, v_2, v_3 \in V$ and let $u, w \in V$. Suppose

$$u = a_1 v_1 + a_2 v_2 + a_3 v_3 \text{ and } w = b_1 v_1 + b_2 v_2 + b_3 v_3$$

Consider

$$f(u, w) = f(a_1 v_1 + a_2 v_2 + a_3 v_3,\ b_1 v_1 + b_2 v_2 + b_3 v_3)$$

Using multilinearity, we get nine terms:

$$\begin{aligned} f(u, w) = {} & a_1 b_1 f(v_1, v_r) + a_1 b_2 f(v_1, v_2) + a_1 b_3 f(v_1, v_3) \\ & + a_2 b_1 f(v_2, v_1) + a_2 b_2 f(v_2, v_2) + a_2 b_3 f(v_2, v_3) \\ & + a_3 b_1 f(v_3, v_1) + a_3 b_2 f(v_3, v_2) + a_3 b_3 f(v_3, v_3) \end{aligned}$$

(Note that $J = [1, 2]$, $J' = [1, 3]$ and $J'' = [2, 3]$ are the three standard-form 2-lists of $I = [1, 2, 3]$.) The alternating property of f tells us that each $f(v_i, v_i) = 0$; hence, three of the above nine terms are equal to 0. The alternating property also tells us that $f(v_i, v_f) = -f(v_f, v_r)$. Thus, three of the terms can be transformed so their subscripts form a standard-form 2-list by a single interchange. Finally we obtain

$$\begin{aligned} f(u, w) &= (a_1 b_2 - a_2 b_1) f(v_1, v_2) + (a_1 b_3 - a_3 b_1) f(v_1, v_3) + (a_2 b_3 - a_3 b_2) f(v_2, v_3) \\ &= \begin{vmatrix} a_1 & a_2 \\ b_1 & b_2 \end{vmatrix} f(v_1, v_2) + \begin{vmatrix} a_1 & a_3 \\ b_1 & b_3 \end{vmatrix} f(v_1, v_3) + \begin{vmatrix} a_2 & a_3 \\ b_2 & b_3 \end{vmatrix} f(v_2, v_3) \end{aligned}$$

which is the content of Lemma A.4.

A.4 Exterior Products

The following definition applies.

DEFINITION A.2: Let V be an n-dimensionmal vector space over a field $\mathbf{K}$, and let r be an integer such that $1 \leq r \leq n$. The *r-fold exterior product* (or simply *exterior product* when r is understood) is a vector space $\mathbf{E}$ over $\mathbf{K}$ together with an alternating r-linear mapping $g: V^r \to \mathbf{E}$, denoted by $g(v_1, \ldots, v_r) = v_1 \wedge \ldots \wedge v_r$, with the following property:

(*) For any vector space U over K and any alternating r-linear map $f: V^r \to U$ there exists a unique linear map $f^*: \mathbf{E} \to U$ such that $f^* \cdot g = f$.

The r-fold tensor product $(\mathbf{E}, g)$ (or simply $\mathbf{E}$ when g is understood) of V is denoted by $\wedge^r V$, and the element $v_1 \wedge \cdots \wedge v_r$ is called the *exterior product* or *wedge product* of the v_i's.

Another way to state condition (*) is that the diagram in Fig. A-1(c) commutes. Again, the fact that such a unique linear map f^* exists is called the "Universal Mapping Principle (UMP)". As illustrated in Fig. A-1(c), condition (*) also says that any alternating r-linear map $f: V^r \to U$ "factors through" the exterior product $\mathbf{E} = \wedge^r V$. Again, the uniqueness in (*) implies that the image of g spans $\mathbf{E}$; that is, $\text{span}(v_1 \wedge \cdots \wedge v_r) = \mathbf{E}$.

THEOREM A.5: (Uniqueness of Exterior Products) Let $(\mathbf{E}, g)$ and $(\mathbf{E}', g')$ be r-fold exterior products of V. Then there exists a unique isomorphism $h: \mathbf{E} \to \mathbf{E}'$ such that $hg = g'$.

The proof is the same as the proof of Theorem A.1, which uses the UMP.

THEOREM A.6: (Existence of Exterior Products) Let V be an n-dimensional vector space over $\mathbf{K}$. Then the exterior product $\mathbf{E} = \wedge^r V$ exists. If $r > n$, then $\mathbf{E} = \{0\}$. If $r \leq n$, then $\dim E = \binom{n}{r}$. Moreover, if $[v_1, \ldots, v_n]$ is a basis of V, then the vectors

$$v_{i_1} \wedge v_{i_2} \wedge \cdots \wedge v_{i_r},$$

where $1 \leq i_1 < i_2 < \cdots < i_r \leq n$, form a basis of E.

We give a concrete example of an exterior product.

EXAMPLE A.5 **(Cross Product)**
Consider $V = \mathbf{R}^3$ with the usual basis $(\mathbf{i}, \mathbf{j}, \mathbf{k})$. Let $E = \bigwedge^2 V$. Note $\dim V = 3$. Thus, $\dim E = 3$ with basis $\mathbf{i} \wedge \mathbf{j}, \mathbf{i} \wedge \mathbf{k}, \mathbf{j} \wedge \mathbf{k}$. We identify E with $\mathbf{R}^3$ under the correspondence

$$\mathbf{i} = \mathbf{j} \wedge \mathbf{k}, \ \mathbf{j} = \mathbf{k} \wedge \mathbf{i} = -\mathbf{i} \wedge \mathbf{k}, \ \mathbf{k} = \mathbf{i} \wedge \mathbf{j}$$

Let u and w be arbitrary vectors in $V = \mathbf{R}^3$, say

$$u = (a_1, a_2, a_3) = a_1\mathbf{i} + a_2\mathbf{j} + a_3\mathbf{k} \text{ and } w = (b_1, b_2, b_3) = b_1\mathbf{i} + b_2\mathbf{j} + b_3\mathbf{k}$$

Then, as in Example A.3,

$$u \wedge w = (a_1b_2 - a_2b_1)(\mathbf{i} \wedge \mathbf{j}) + (a_1b_3 - a_3b_1)(\mathbf{i} \wedge \mathbf{k}) + (a_2b_3 - a_3b_2)(\mathbf{j} \wedge \mathbf{k})$$

Using the above identification, we get

$$\begin{aligned} u \wedge w &= (a_2b_3 - a_3b_2)\mathbf{i} - (a_1b_3 - a_3b_1)\mathbf{j} + (a_1b_2 - a_2b_1)\mathbf{k} \\ &= \begin{vmatrix} a_2 & a_3 \\ b_2 & b_3 \end{vmatrix}\mathbf{i} - \begin{vmatrix} a_1 & a_3 \\ b_1 & b_3 \end{vmatrix}\mathbf{j} + \begin{vmatrix} a_1 & a_2 \\ b_1 & b_2 \end{vmatrix}\mathbf{k} \end{aligned}$$

The reader may recognize that the above exterior product is precisely the well-known cross product in $\mathbf{R}^3$.

Our last theorem tells us that we are actually able to "multiply" exterior products, which allows us to form an "exterior algebra" that is illustrated below.

THEOREM A.7: Let V be a vector space over $\mathbf{K}$. Let r and s be positive integers. Then there is a unique bilinear mapping

$$\bigwedge^r V \times \bigwedge^s V \to \bigwedge^{r+s} V$$

such that, for any vectors u_i, w_j in V,

$$(u_1 \wedge \cdots \wedge u_r) \times (w_1 \wedge \cdots \wedge w_s) \mapsto u_1 \wedge \cdots \wedge u_r \wedge w_1 \wedge \cdots \wedge w_s$$

EXAMPLE A.6
We form an exterior algebra A over a field **K** using noncommuting variables x, y, z. Because it is an exterior algebra, our variables satisfy:

$$x \wedge x = 0, \quad y \wedge y = 0, \quad z \wedge z = 0, \quad \text{and} \quad y \wedge x = -x \wedge y, \quad z \wedge x = -x \wedge z, \quad z \wedge y = -y \wedge z$$

Every element of A is a linear combination of the eight elements

$$1, \quad x, \quad y, \quad z, \quad x \wedge y, \quad x \wedge z, \quad y \wedge z, \quad x \wedge y \wedge z$$

We multiply two "polynomials" in A using the usual distributive law, but now we also use the above conditions. For example,

$$[3 + 4y - 5x \wedge y + 6x \wedge z] \wedge [5x - 2y] = 15x - 6y - 20x \wedge y + 12x \wedge y \wedge z$$

Observe we use the fact that

$$[4y] \wedge [5x] = 20y \wedge x = -20x \wedge y \quad \text{and} \quad [6x \wedge z] \wedge [-2y] = -12x \wedge z \wedge y = 12x \wedge y \wedge z$$

APPENDIX B

Algebraic Structures

B.1 Introduction

We define here algebraic structures that occur in almost all branches of mathematics. In particular, we will define a *field* that appears in the definition of a vector space. We begin with the definition of a *group*, which is a relatively simple algebraic structure with only one operation and is used as a building block for many other algebraic systems.

B.2 Groups

Let G be a nonempty set with a binary operation; that is, to each pair of elements $a, b \in G$ there is assigned an element $ab \in G$. Then G is called a *group* if the following axioms hold:

$[G_1]$ For any $a, b, c \in G$, we have $(ab)c = a(bc)$ (the *associative law*).

$[G_2]$ There exists an element $e \in G$, called the *identity* element, such that $ae = ea = a$ for every $a \in G$.

$[G_3]$ For each $a \in G$ there exists an element $a^{-1} \in G$, called the *inverse* of a, such that $aa^{-1} = a^{-1}a = e$.

A group G is said to be *abelian* (or: *commutative*) if the *commutative law* holds—that is, if $ab = ba$ for every $a, b \in G$.

When the binary operation is denoted by juxtaposition as above, the group G is said to be written *multiplicatively*. Sometimes, when G is abelian, the binary operation is denoted by + and G is said to be written *additively*. In such a case, the identity element is denoted by 0 and is called the *zero* element; the inverse is denoted by $-a$ and it is called the *negative* of a.

If A and B are subsets of a group G, then we write

$$AB = \{ab \mid a \in A, b \in B\} \quad \text{or} \quad A + B = \{a + b \mid a \in A, b \in B\}$$

We also write a for $\{a\}$.

A subset H of a group G is called a *subgroup* of G if H forms a group under the operation of G. If H is a subgroup of G and $a \in G$, then the set Ha is called a *right coset* of H and the set aH is called a *left coset* of H.

DEFINITION: A subgroup H of G is called a *normal* subgroup if $a^{-1}Ha \subseteq H$ for every $a \in G$. Equivalently, H is normal if $aH = Ha$ for every $a \in G$—that is, if the right and left cosets of H coincide.

Note that every subgroup of an abelian group is normal.

THEOREM B.1: Let H be a normal subgroup of G. Then the cosets of H in G form a group under coset multiplication. This group is called the *quotient group* and is denoted by G/H.

EXAMPLE B.1 The set **Z** of integers forms an abelian group under addition. (We remark that the even integers form a subgroup of **Z** but the odd integers do not.) Let H denote the set of multiples of 5; that is, $H = \{\ldots, -10, -5, 0, 5, 10, \ldots\}$. Then H is a subgroup (necessarily normal) of **Z**. The cosets of H in **Z** follow:

$$\begin{aligned}
\bar{0} &= 0 + H = H = \{\ldots, -10, -5, 0, 5, 10, \ldots\} \\
\bar{1} &= 1 + H = \{\ldots, -9, -4, 1, 6, 11, \ldots\} \\
\bar{2} &= 2 + H = \{\ldots, -8, -3, 2, 7, 12, \ldots\} \\
\bar{3} &= 3 + H = \{\ldots, -7, -2, 3, 8, 13, \ldots\} \\
\bar{4} &= 4 + H = \{\ldots, -6, -1, 4, 9, 14, \ldots\}
\end{aligned}$$

For any other integer $n \in \mathbf{Z}$, $\bar{n} = n + H$ coincides with one of the above cosets. Thus, by the above theorem, $\mathbf{Z}/H = \{\bar{0}, \bar{1}, \bar{2}, \bar{3}, \bar{4}\}$ forms a group under coset addition; its addition table follows:

+	$\bar{0}$	$\bar{1}$	$\bar{2}$	$\bar{3}$	$\bar{4}$
$\bar{0}$	$\bar{0}$	$\bar{1}$	$\bar{2}$	$\bar{3}$	$\bar{4}$
$\bar{1}$	$\bar{1}$	$\bar{2}$	$\bar{3}$	$\bar{4}$	$\bar{0}$
$\bar{2}$	$\bar{2}$	$\bar{3}$	$\bar{4}$	$\bar{0}$	$\bar{1}$
$\bar{3}$	$\bar{3}$	$\bar{4}$	$\bar{0}$	$\bar{1}$	$\bar{2}$
$\bar{4}$	$\bar{4}$	$\bar{0}$	$\bar{1}$	$\bar{2}$	$\bar{3}$

This quotient group **Z**/*H* is referred to as the integers modulo 5 and is frequently denoted by $\mathbf{Z}_5$. Analogeusly, for any positive integer n, there exists the quotient group $\mathbf{Z}_n$ called the integers modulo n.

EXAMPLE B.2 The permutations of n symbols (see page 267) form a group under composition of mappings; it is called the *symmetric group* of degree n and is denoted by $\mathbf{S}_n$. We investigate $\mathbf{S}_3$ here; its elements are

$$\epsilon = \begin{pmatrix} 1 & 2 & 3 \\ 1 & 2 & 3 \end{pmatrix} \quad \sigma_2 = \begin{pmatrix} 1 & 2 & 3 \\ 3 & 2 & 1 \end{pmatrix} \quad \phi_1 = \begin{pmatrix} 1 & 2 & 3 \\ 2 & 3 & 1 \end{pmatrix}$$

$$\sigma_1 = \begin{pmatrix} 1 & 2 & 3 \\ 1 & 3 & 2 \end{pmatrix} \quad \sigma_3 = \begin{pmatrix} 1 & 2 & 3 \\ 2 & 1 & 3 \end{pmatrix} \quad \phi_2 = \begin{pmatrix} 1 & 2 & 3 \\ 3 & 1 & 2 \end{pmatrix}$$

Here $\begin{pmatrix} 1 & 2 & 3 \\ i & j & k \end{pmatrix}$ is the permutation that maps $1 \mapsto i$, $2 \mapsto j$, $3 \mapsto k$. The multiplication table of $\mathbf{S}_3$ is

	ϵ	σ_1	σ_2	σ_3	ϕ_1	ϕ_2
ϵ	ϵ	σ_1	σ_2	σ_3	ϕ_1	ϕ_2
σ_1	σ_1	ϵ	ϕ_1	ϕ_2	σ_2	σ_3
σ_2	σ_2	ϕ_2	ϵ	ϕ_1	ϕ_3	σ_1
σ_3	σ_3	ϕ_1	ϕ_2	ϵ	σ_1	σ_2
ϕ_1	ϕ_1	σ_3	σ_1	σ_2	ϕ_2	ϵ
ϕ_2	ϕ_2	σ_2	σ_3	σ_1	ϵ	ϕ_1

(The element in the *a*th row and *b*th column is *ab*.) The set $H = \{\epsilon, \sigma_1\}$ is a subgroup of $\mathbf{S}_3$; its right and left cosets are

Right Cosets	Left Cosets
$H = \{\epsilon, \sigma_1\}$	$H = \{\epsilon, \sigma_1\}$
$H_{\phi_1} = \{\phi_1, \sigma_2\}$	$\phi_2 H = \{\phi_1, \sigma_3\}$
$H_{\phi_2} = \{\phi_2, \sigma_3\}$	$\phi_2 H = \{\phi_2, \sigma_2\}$

Observe that the right cosets and the left cosets are distinct; hence, H is not a normal subgroup of $\mathbf{S}_3$.

A mapping f from a group G into a group G' is called a *homomorphism* if $f(ab) = f(a)f(b)$. For every $a, b \in G$. (If f is also bijective, i.e., one-to-one and onto, then f is called an *isomorphism* and G and G' are said to be *isomorphic*.) If $f : G \to G'$ is a homomorphism, then the kernel of f is the set of elements of G that map into the identity element $e' \in G'$:

$$\text{kernel of } f = \{a \in G \mid f(a) = e'\}$$

(As usual, $f(G)$ is called the *image* of the mapping $f : G \to G'$.) The following theorem applies.

THEOREM B.2: Let $f: G \to G$ be a homomorphism with kernel K. Then K is a normal subgroup of G, and the quotient group G/K is isomorphic to the image of f.

EXAMPLE B.3 Let G be the group of real numbers under addition, and let G' be the group of positive real numbers under multiplication. The mapping $f : G \to G'$ defined by $f(a) = 2^a$ is a homomorphism because

$$f(a+b) = 2^{a+b} = 2^a 2^b = f(a)f(b)$$

In particular, f is bijective, hence, G and G' are isomorphic.

EXAMPLE B.4 Let G be the group of nonzero complex numbers under multiplication, and let G' be the group of nonzero real numbers under multiplication. The mapping $f : G \to G'$ defined by $f(z) = |z|$ is a homomorphism because

$$f(z_1 z_2) = |z_1 z_2| = |z_1||z_2| = f(z_1)f(z_2)$$

The kernel K of f consists of those complex numbers z on the unit circle—that is, for which $|z| = 1$. Thus, G/K is isomorphic to the image of f—that is, to the group of positive real numbers under multiplication.

B.3 Rings, Integral Domains, and Fields

Let R be a nonempty set with two binary operations, an operation of addition (denoted by +) and an operation of multiplication (denoted by juxtaposition). Then R is called a *ring* if the following axioms are satisfied:

[R_1] For any $a, b, c \in R$, we have $(a+b)+c = a+(b+c)$.

[R_2] There exists an element $0 \in R$, called the *zero* element, such that $a+0 = 0+a = a$ for every $a \in R$.

[R_3] For each $a \in R$ there exists an element $-a \in R$, called the *negative* of a, such that $a+(-a) = (-a)+a = 0$.

[R_4] For any $a, b \in R$, we have $a+b = b+a$.

[R_5] For any $a, b, c \in R$, we have $(ab)c = a(bc)$.

[R_6] For any $a, b, c \in R$, we have

$$\text{(i) } a(b+c) = ab+ac, \text{ and (ii) } (b+c)a = ba+ca.$$

Observe that the axioms [R_1] through [R_4] may be summarized by saying that R is an abelian group under addition.

Subtraction is defined in R by $a - b \equiv a + (-b)$.

It can be shown (see Problem B.25) that $a \cdot 0 = 0 \cdot a = 0$ for every $a \in R$.

R is called a *commutative ring* if $ab = ba$ for every $a, b \in R$. We also say that R is a *ring with a unit element* if there exists a nonzero element $1 \in R$ such that $a \cdot 1 = 1 \cdot a = a$ for every $a \in R$.

A nonempty subset S of R is called a *subring* of R if S forms a ring under the operations of R. We note that S is a subring of R if and only if $a, b \in S$ implies $a - b \in S$ and $ab \in S$.

A nonempty subset I of R is called a *left ideal* in R if (i) $a - b \in I$ whenever $a, b \in I$, and (ii) $ra \in I$ whenever $r \in R, a \in I$. Note that a left ideal I in R is also a subring of R. Similarly, we can define a *right ideal* and a *two-sided ideal*. Clearly all ideals in commutative rings are two sided. The term *ideal* shall mean two-sided ideal uniess otherwise specified.

THEOREM B.3: Let I be a (two-sided) ideal in a ring R. Then the cosets $\{a+I \mid a \in R\}$ form a ring under coset addition and coset multiplication. This ring is denoted by R/I and is called the *quotient ring*.

Now let R be a commutative ring with a unit element. For any $a \in R$, the set $(a) = \{ra \mid r \in R\}$ is an ideal; it is called the *principal ideal* generated by a. If every ideal in R is a principal ideal, then R is called a *principal ideal ring*.

DEFINITION: A commutative ring R with a unit element is called an *integral domain* if R has no *zero divisors*—that is, if $ab = 0$ implies $a = 0$ or $b = 0$.

DEFINITION: A commutative ring R with a unit element is called a *field* if every nonzero $a \in R$ has a *multiplicative inverse*; that is, there exists an element $a^{-1} \in R$ such that $aa^{-1} = a^{-1}a = 1$.

A field is necessarily an integral domain; for if $ab = 0$ and $a \neq 0$, then

$$b = 1 \cdot b = a^{-1}ab = a^{-1} \cdot 0 = 0$$

We remark that a field may also be viewed as a commutative ring in which the nonzero elements form a group under multiplication.

EXAMPLE B.5 The set $\mathbf{Z}$ of integers with the usual operations of addition and multiplication is the classical example of an integral domain with a unit element. Every ideal I in $\mathbf{Z}$ is a principal ideal; that is, $I = (n)$ for some integer n. The quotient ring $\mathbf{Z}_n = \mathbf{Z}/(n)$ is called the *ring of integers module n*. If n is prime, then $\mathbf{Z}_n$ is a field. On the other hand, if n is not prime then $\mathbf{Z}_n$ has zero divisors. For example, in the ring $\mathbf{Z}_6$, $\bar{2}\bar{3} = \bar{0}$ and $\bar{2} \neq \bar{0}$ and $\bar{3} \neq \bar{0}$.

EXAMPLE B.6 The rational numbers $\mathbf{Q}$ and the real numbers $\mathbf{R}$ each form a field with respect to the usual operations of addition and multiplication.

EXAMPLE B.7 Let $\mathbf{C}$ denote the set of ordered pairs of real numbers with addition and multiplication defined by

$$(a, b) + (c, d) = (a + c, b + d)$$
$$(a, b) \cdot (c, d) = (ac - bd, ad + bc)$$

Then $\mathbf{C}$ satisfies all the required properties of a field. In fact, $\mathbf{C}$ is just the field of complex numbers (see page 4).

EXAMPLE B.8 The set M of all 2×2 matrices with real entries forms a noncommutative ring with zero divisors under the operations of matrix addition and matrix multiplication.

EXAMPLE B.9 Let R be any ring. Then the set $R[x]$ of all polynomials over R forms a ring with respect to the usual operations of addition and multiplication of polynomials. Moreover, if R is an integral domain then $R[x]$ is also an integral domain.

Now let D be an integral domain. We say that b *divides* a in D if $a = bc$ for some $c \in D$. An element $u \in D$ is called a *unit* if u divides 1—that is, if u has a multiplicative inverse. An element $b \in D$ is called an *associate* of $a \in D$ if $b = ua$ for some unit $u \in D$. A nonunit $p \in D$ is said to be *irreducible* if $p = ab$ implies a or b is a unit.

An integral domain D is called a *unique factorization domain* if every nonunit $a \in D$ can be written uniquely (up to associates and order) as a product of irreducible elements.

EXAMPLE B.10 The ring $\mathbf{Z}$ of integers is the classical example of a unique factorization domain. The units of $\mathbf{Z}$ are 1 and -1. The only associates of $n \in \mathbf{Z}$ are n and $-n$. The irreducible elements of $\mathbf{Z}$ are the prime numbers.

EXAMPLE B.11 The set $D = \{a + b\sqrt{13} \mid a, b \text{ integers}\}$ is an integral domain. The units of D are ± 1, $18 \pm 5\sqrt{13}$ and $-18 \pm 5\sqrt{13}$. The elements 2, $3 - \sqrt{13}$ and $-3 - \sqrt{13}$ are irreducible in D. Observe that $4 = 2 \cdot 2 = (3 - \sqrt{13})(-3 - \sqrt{13})$. Thus, D is not a unique factorization domain. (See Problem B.40.)

B.4 Modules

Let M be an additive abelian group and let R be a ring with a unit element. Then M is said to be a (left) R-*module* if there exists a mapping $R \times M \to M$ that satisfies the following axioms:

$[M_1]$ $r(m_1 + m_2) = rm_1 + rm_2$

$[M_2]$ $(r + s)m = rm + sm$

$[M_3]$ $(rs)m = r(sm)$

$[M_4]$ $1 \cdot m = m$

for any $r, s \in R$ and any $m_i \in M$.

We emphasize that an R-module is a generalization of a vector space where we allow the scalars to come from a ring rather than a field.

EXAMPLE B.12 Let G be any additive abelian group. We make G into a module over the ring **Z** of integers by defining

$$ng = \overbrace{g + g + \cdots + g}^{n \text{ times}}, \quad 0g = 0, \quad (-n)g = -ng$$

where n is any positive integer.

EXAMPLE B.13 Let R be a ring and let I be an ideal in R. Then I may be viewed as a module over R.

EXAMPLE B.14 Let V be a vector space over a field K and let $T: V \to V$ be a linear mapping. We make V into a module over the ring $K[x]$ of polynomials over K by defining $f(x)v = f(T)(v)$. The reader should check that a scalar multiplication has been defined.

Let M be a module over R. An additive subgroup N of M is called a *submodule* of M if $u \in N$ and $k \in R$ imply $ku \in N$. (Note that N is then a module over R.)

Let M and M' be R-modules. A mapping $T: M \to M'$ is called a *homomorphism* (or: *R-homomorphism* or *R-linear*) if

(i) $T(u + v) = T(u) + T(v)$ and (ii) $T(ku) = kT(u)$

for every $u, v \in M$ and every $k \in R$.

PROBLEMS

Groups

B.1. Determine whether each of the following systems forms a group G:

(i) G = set of integers, operation subtraction;

(ii) $G = \{1, -1\}$, operation multiplication;

(iii) G = set of nonzero rational numbers, operation division;

(iv) G = set of nonsingular $n \times n$ matrices, operation matrix multiplication;

(v) $G = \{a + bi : a, b \in \mathbf{Z}\}$, operation addition.

B.2. Show that in a group G:

(i) the identity element of G is unique;

(ii) each $a \in G$ has a unique inverse $a^{-1} \in G$;

(iii) $(a^{-1})^{-1} = a$, and $(ab)^{-1} = b^{-1}a^{-1}$;

(iv) $ab = ac$ implies $b = c$, and $ba = ca$ implies $b = c$.

B.3. In a group G, the powers of $a \in G$ are defined by

$$a^0 = e, \quad a^n = aa^{n-1}, \quad a^{-n} = (a^n)^{-1}, \quad \text{where } n \in N$$

Show that the following formulas hold for any integers $r, s, t \in \mathbf{Z}$: (i) $a^r a^s = a^{r+s}$, (ii) $(a^r)^s = a^{rs}$, (iii) $(a^{r+s})^t = a^{rs+st}$.

B.4. Show that if G is an abelian group, then $(ab)^n = a^n b^n$ for any $a, b \in G$ and any integer $n \in \mathbf{Z}$.

B.5. Suppose G is a group such that $(ab)^2 = a^2 b^2$ for every $a, b \in G$. Show that G is abelian.

B.6. Suppose H is a subset of a group G. Show that H is a subgroup of G if and only if (i) H is nonempty, and (ii) $a, b \in H$ implies $ab^{-1} \in H$.

B.7. Prove that the intersection of any number of subgroups of G is also a subgroup of G.

B.8. Show that the set of all powers of $a \in G$ is a subgroup of G; it is called the *cyclic group* generated by a.

B.9. A group G is said to be *cyclic* if G is generated by some $a \in G$; that is, $G = (a^n : n \in Z)$. Show that every subgroup of a cyclic group is cyclic.

B.10. Suppose G is a cyclic subgroup. Show that G is isomorphic to the set $\mathbf{Z}$ of integers under addition or to the set $\mathbf{Z}_n$ (of the integers module n) under addition.

B.11. Let H be a subgroup of G. Show that the right (left) cosets of H partition G into mutually disjoint subsets.

B.12. The *order* of a group G, denoted by $|G|$, is the number of elements of G. Prove Lagrange's theorem: If H is a subgroup of a finite group G, then $|H|$ divides $|G|$.

B.13. Suppose $|G| = p$ where p is prime. Show that G is cyclic.

B.14. Suppose H and N are subgroups of G with N normal. Show that (i) HN is a subgroup of G and (ii) $H \cap N$ is a normal subgroup of G.

B.15. Let H be a subgroup of G with only two right (left) cosets. Show that H is a normal subgroup of G.

B.16. Prove Theorem B.1: Let H be a normal subgroup of G. Then the cosets of H in G form a group G/H under coset multiplication.

B.17. Suppose G is an abelian group. Show that any factor group G/H is also abelian.

B.18. Let $f : G \to G'$ be a group homomorphism. Show that

(i) $f(e) = e'$ where e and e' are the identity elements of G and G', respectively;

(ii) $f(a^{-1}) = f(a)^{-1}$ for any $a \in G$.

B.19. Prove Theorem B.2: Let $f : G \to G'$ be a group homomorphism with kernel K. Then K is a normal subgroup of G, and the quotient group G/K is isomorphic to the image of f.

B.20. Let G be the multiplicative group of complex numbers z such that $|z| = 1$, and let $\mathbf{R}$ be the additive group of real numbers. Prove that G is isomorphic to $\mathbf{R}/\mathbf{Z}$.

B.21. For a fixed $g \in G$, let $\hat{g} : G \to G$ be defined by $\hat{g}(a) = g^{-1}ag$. Show that G is an isomorphism of G onto G.

B.22. Let G be the multiplicative group of $n \times n$ nonsingular matrices over **R**. Show that the mapping $A \mapsto |A|$ is a homomorphism of G into the multiplicative group of nonzero real numbers.

B.23. Let G be an abelian group. For a fixed $n \in \mathbf{Z}$, show that the map $a \mapsto a^n$ is a homomorphism of G into G.

B.24. Suppose H and N are subgroups of G with N normal. Prove that $H \cap N$ is normal in H and $H/(H \cap N)$ is isomorphic to HN/N.

Rings

B.25. Show that in a ring R:

(i) $a \cdot 0 = 0 \cdot a = 0$, (ii) $a(-b) = (-a)b = -ab$, (iii) $(-a)(-b) = ab$.

B.26. Show that in a ring R with a unit element: (i) $(-1)a = -a$, (ii) $(-1)(-1) = 1$.

B.27. Let R be a ring. Suppose $a^2 = a$ for every $a \in R$. Prove that R is a commutative ring. (Such a ring is called a *Boolean ring*.)

B.28. Let R be a ring with a unit element. We make R into another ring $\hat{R}$ by defining $a \oplus b = a + b + 1$ and $a \cdot b = ab + a + b$. (i) Verify that $\hat{R}$ is a ring. (ii) Determine the 0-element and 1-element of $\hat{R}$.

B.29. Let G be any (additive) abelian group. Define a multiplication in G by $a \cdot b = 0$. Show that this makes G into a ring.

B.30. Prove Theorem B.3: Let I be a (two-sided) ideal in a ring R. Then the cosets $(a + I \mid a \in R)$ form a ring under coset addition and coset multiplication.

B.31. Let I_1 and I_2 be ideals in R. Prove that $I_1 + I_2$ and $I_1 \cap I_2$ are also ideals in R.

B.32. Let R and R' be rings. A mapping $f : R \to R'$ is called a *homomorphism* (or: *ring homomorphism*) if

(i) $f(a+b) = f(a) + f(b)$ and (ii) $f(ab) = f(a)f(b)$,

for every $a, b \in R$. Prove that if $f : R \to R'$ is a homomorphism, then the set $K = \{r \in R \mid f(r) = 0\}$ is an ideal in R. (The set K is called the *kernel* of f.)

Integral Domains and Fields

B.33. Prove that in an integral domain D, if $ab = ac$, $a \neq 0$, then $b = c$.

B.34. Prove that $F = \{a + b\sqrt{2} \mid a, b \text{ rational}\}$ is a field.

B.35. Prove that $D = \{a + b\sqrt{2} \mid a, b \text{ integers}\}$ is an integral domain but not a field.

B.36. Prove that a finite integral domain D is a field.

B.37. Show that the only ideals in a field K are $\{0\}$ and K.

B.38. A complex number $a + bi$ where a, b are integers is called a *Gaussian integer*. Show that the set G of Gaussian integers is an integral domain. Also show that the units in G are ± 1 and $\pm i$.

B.39. Let D be an integral domain and let I be an ideal in D. Prove that the factor ring D/I is an integral domain if and only if I is a prime ideal. (An ideal I is *prime* if $ab \in I$ implies $a \in I$ or $b \in I$.)

B.40. Consider the integral domain $D = \{a + b\sqrt{13} \mid a, b \text{ integers}\}$ (see Example B.11). If $\alpha = a + b\sqrt{13}$, we define $N(\alpha) = a^2 - 13b^2$. Prove: (i) $N(\alpha\beta) = N(\alpha)N(\beta)$; (ii) α is a unit if and only if $N(\alpha) = \pm 1$; (iii) the units of D are ± 1, $18 \pm 5\sqrt{13}$ and $-18 \pm 5\sqrt{13}$; (iv) the numbers 2, $3 - \sqrt{13}$ and $-3 - \sqrt{13}$ are irreducible.

Modules

B.41. Let M be an R-module and let A and B be submodules of M. Show that $A + B$ and $A \cap B$ are also submodules of M.

B.42. Let M be an R-module with submodule N. Show that the cosets $\{u + N : u \in M\}$ form an R-module under coset addition and scalar multiplication defined by $r(u + N) = ru + N$. (This module is denoted by M/N and is called the *quotient module*.)

B.43. Let M and M' be R-modules and let $f : M \to M'$ be an R-homomorphism. Show that the set $K = \{u \in M : f(u) = 0\}$ is a submodule of f. (The set K is called the *kernel* of f.)

B.44. Let M be an R-module and let $E(M)$ denote the set of all R-homomorphism of M into itself. Define the appropriate operations of addition and multiplication in $E(M)$ so that $E(M)$ becomes a ring.

APPENDIX C

Polynomials over a Field

C.1 Introduction

We will investigate polynomials over a field K and show that they have many properties that are analogous to properties of the integers. These results play an important role in obtaining canonical forms for a linear operator T on a vector space V over K.

C.2 Ring of Polynomials

Let K be a field. Formally, a polynomial of f over K is an infinite sequence of elements from K in which all except a finite number of them are 0:

$$f = (\ldots, 0, a_n, \ldots, a_1, a_0)$$

(We write the sequence so that it extends to the left instead of to the right.) The entry a_k is called the kth coefficient of f. If n is the largest integer for which $a_n \neq 0$, then we say that the *degree* of f is n, written

$$\deg f = n$$

We also call a_n the *leading coefficient* of f, and if $a_n = 1$ we call f a *monic polynomial*. On the other hand, if every coefficient of f is 0 then f is called the *zero polynomial*, written $f = 0$. The degree of the zero polynomial is not defined.

Now if g is another polynomial over K, say

$$g = (\ldots, 0, b_m, \ldots, b_1, b_0)$$

then the *sum* $f + g$ is the polynomial obtained by adding corresponding coefficients. That is, if $m \leq n$, then

$$f + g = (\ldots, 0, a_n, \ldots, a_m + b_m, \ldots, a_1 + b_1, a_0 + b_0)$$

Furthermore, the *product* fg is the polynomial

$$fg = (\ldots, 0, a_n b_m, \ldots, a_1 b_0 + a_0 b_1, a_0 b_0)$$

that is, the kth coefficient c_k of fg is

$$c_k = \sum_{t=0}^{k} a_1 b_{k-1} = a_0 b_k + a_1 b_{k-1} + \cdots + a_k b_0$$

The following theorem applies.

THEOREM C.1: The set P of polynomials over a field K under the above operations of addition and multiplication forms a commutative ring with a unit element and with no zero divisors—an integral domain. If f and g are nonzero polynomials in P, then $\deg(fg) = (\deg f)(\deg g)$.

Notation

We identify the scalar $a_0 \in K$ with the polynomial

$$a_0 = (\ldots, 0, a_0)$$

We also choose a symbol, say t, to denote the polynomial

$$t = (\ldots, 0, 1, 0)$$

We call the symbol t an *indeterminant*. Multiplying t with itself, we obtain

$$t^2 = (\ldots, 0, 1, 0, 0), \quad t^3 = (\ldots, 0, 1, 0, 0, 0), \quad \ldots$$

Thus, the above polynomial f can be written uniquely in the usual form

$$f = a_n t^n + \cdots + a_s t + a_0$$

When the symbol t is selected as the indeterminant, the ring of polynomials over K is denoted by

$$K[t]$$

and a polynomial f is frequently denoted by $f(t)$.

We also view the field K as a subset of $K[t]$ under the above identification. This is possible because the operations of addition and multiplication of elements of K are preserved under this identification:

$$(\ldots, 0, a_0) + (\ldots, 0, b_0) = (\ldots, 0, a_0 + b_0)$$
$$(\ldots, 0, a_0) \cdot (\ldots, 0, b_0) = (\ldots, 0, a_0 b_0)$$

We remark that the nonzero elements of K are the units of the ring $K[t]$.

We also remark that every nonzero polynomial is an associate of a unique monic polynomial. Hence, if d and d' are monic polynomials for which d divides d' and d' divides d, then $d = d'$. (A polynomial g *divides* a polynomial f if there is a polynomial h such that $f = hg$.)

C.3 Divisibility

The following theorem formalizes the process known as "long division."

THEOREM C.2 (Division Algorithm): Let f and g be polynomials over a field K with $g \neq 0$. Then there exist polynomials q and r such that

$$f = qg + r$$

where either $r = 0$ or $\deg r < \deg g$.

Proof: If $f = 0$ or if $\deg f < \deg g$, then we have the required representation

$$f = 0g + f$$

Now suppose $\deg f \geq \deg g$, say

$$f = a_n t^n + \cdots + a_1 t + a_0 \quad \text{and} \quad g = b_m t^m + \cdots + b_1 t + b_0$$

where $a_n, b_m \neq 0$ and $n \geq m$. We form the polynomial

$$f_1 = f - \frac{a_n}{b_m} t^{n-m} g \tag{1}$$

Then $\deg f_1 < \deg f$. By induction, there exist polynomials q_1 and r such that

$$f_1 = q_1 g + r$$

where either $r = 0$ or $\deg r < \deg g$. Substituting this into (1) and solving for f,

$$f = \left(q_1 + \frac{a_n}{b_m}t^{n-m}\right)g + r$$

which is the desired representation.

THEOREM C.3: The ring $K[t]$ of polynomials over a field K is a principal ideal ring. If I is an ideal in $K[t]$, then there exists a unique monic polynomial d that generates I, such that d divides every polynomial $f \in I$.

Proof. Let d be a polynomial of lowest degree in I. Because we can multiply d by a nonzero scalar and still remain in I, we can assume without loss in generality that d is a monic polynomial. Now suppose $f \in I$. By Theorem C.2 there exist polynomials q and r such that

$$f = qd + r \text{ where either } r = 0 \text{ or } \deg r < \deg d$$

Now $f, d \in I$ implies $qd \in I$, and hence, $r = f - qd \in I$. But d is a polynomial of lowest degree in I. Accordingly, $r = 0$ and $f = qd$; that is, d divides f. It remains to show that d is unique. If d' is another monic polynomial that generates I, then d divides d' and d' divides d. This implies that $d = d'$, because d and d' are monic. Thus, the theorem is proved.

THEOREM C.4: Let f and g be nonzero polynomials in $K[t]$. Then there exists a unique monic polynomial d such that
(i) d divides f and g; and (ii) d' divides f and g, then d' divides d.

DEFINITION: The above polynomial d is called the *greatest common divisor* of f and g. If $d = 1$, then f and g are said to be *relatively prime*.

Proof of Theorem C.4. The set $I = \{mf + ng \mid m, n \in K[t]\}$ is an ideal. Let d be the monic polynomial that generates I. Note $f, g \in I$; hence, d divides f and g. Now suppose d' divides f and g. Let J be the ideal generated by d'. Then $f, g \in J$, and hence, $I \subset J$. Accordingly, $d \in J$ and so d' divides d as claimed. It remains to show that d is unique. If d_1 is another (monic) greatest common divisor of f and g, then d divides d_1 and d_1 divides d. This implies that $d = d_1$ because d and d_1 are monic. Thus, the theorem is proved.

COROLLARY C.5: Let d be the greatest common divisor of the polynomials f and g. Then there exist polynomials m and n such that $d = mf + ng$. In particular, if f and g are relatively prime, then there exist polynomials m and n such that $mf + ng = 1$.

The corollary follows directly from the fact that d generates the ideal

$$I = \{mf + ng \mid m, n \in K[t]\}$$

C.4 Factorization

A polynomial $p \in K[t]$ of positive degree is said to be irreducible if $p = fg$ implies f or g is a scalar.

LEMMA C.6: Suppose $p \in K[t]$ is irreducible. If p divides the product fg of polynomials $f, g \in K[t]$, then p divides f or p divides g. More generally, if p divides the product of n polynomials $f_1 f_2 \ldots f_n$, then p divides one of them.

Proof. Suppose p divides fg but not f. Because p is irreducible, the polynomials f and p must then be relatively prime. Thus, there exist polynomials $m, n \in K[t]$ such that $mf + np = 1$. Multiplying this equation by g, we obtain $mfg + npg = g$. But p divides fg and so mfg, and p divides npg; hence, p divides the sum $g = mfg + npg$.

Now suppose p divides $f_1 f_2 \cdots f_n$. If p divides f_1, then we are through. If not, then by the above result p divides the product $f_2 \cdots f_n$. By induction on n, p divides one of the polynomials $f_2, \ldots f_n$. Thus, the lemma is proved.

THEOREM C.7: (Unique Factorization Theorem) Let f be a nonzero polynomial in $K[t]$. Then f can be written uniquely (except for order) as a product

$$f = kp_1 p_2 \cdots p_n$$

where $k \in K$ and the p_i are monic irreducible polynomials in $K[t]$.

Proof: We prove the existence of such a product first. If f is irreducible or if $f \in K$, then such a product clearly exists. On the other hand, suppose $f = gh$ where f and g are nonscalars. Then g and h have degrees less than that of f. By induction, we can assume

$$g = k_1 g_1 g_2 \cdots g_r \quad \text{and} \quad h = k_2 h_1 h_2 \cdots h_s$$

where $k_1, k_2 \in K$ and the g_i and h_j are monic irreducible polynomials. Accordingly,

$$f = (k_1 k_2) g_1 g_2 \cdots g_r k_1 h_2 \cdots h_s$$

is our desired representation.

We next prove uniqueness (except for order) of such a product for f. Suppose

$$f = kp_1 p_2 \cdots p_n = k' q_1 q_2 \cdots q_m$$

where $k, k' \in K$ and the $p_1, \ldots, p_n, q_1, \ldots, q_m$ are monic irreducible polynomials. Now p_1 divides $k' q_1 \cdots q_m$. Because p_1 is irreducible, it must divide one of the q_i by the above lemma. Say p_1 divides q_1. Because p_1 and q_1 are both irreducible and monic, $p_1 = q_1$. Accordingly,

$$kp_2 \cdots p_n = k' q_2 \cdots q_m$$

By induction, we have that $n = m$ and $p_2 = q_2, \ldots, p_n = q_m$ for some rearrangement of the q_i. We also have that $k = k'$. Thus, the theorem is proved.

If the field K is the complex field **C**, then we have the following result that is known as the fundamental theorem of algebra; its proof lies beyond the scope of this text.

THEOREM C.8: (Fundamental Theorem of Algebra) Let $f(t)$ be a nonzero polynomial over the complex field **C**. Then $f(t)$ can be written uniquely (except for order) as a product

$$f(t) = k(t - r_2)(t - r_2) \cdots (t - r_n)$$

where $k, r_i \in \mathbf{C}$—as a product of linear polynomials.

In the case of the real field **R** we have the following result.

THEOREM C.9: Let $f(t)$ be a nonzero polynomial over the real field **R**. Then $f(t)$ can be written uniquely (except for order) as a product

$$f(t) = kp_1(t) p_2(t) \cdots p_m(t)$$

where $k \in \mathbf{R}$ and the $p_i(t)$ are monic irreducible polynomials of degree one or two.

APPENDIX D

Odds and Ends

D.1 Introduction

This appendix discusses various topics, such as equivalence relations, determinants and block matrices, and the generalized MP (Moore–Penrose) inverse.

D.2 Relations and Equivalence Relations

A *binary relation* or simply *relation* R from a set A to a set B assigns to each ordered pair $(a, b) \in A \times B$ exactly one of the following statements:

(i) "a is related to b," written $a\,R\,b$, (ii) "a is not related to b" written $a \not R\, b$.

A relation from a set A to the same set A is called a relation *on* A.

Observe that any relation R from A to B uniquely defines a subset $\hat{R}$ of $A \times B$ as follows:

$$\hat{R} = \{(a, b) | a R b\}$$

Conversely, any subset $\hat{R}$ of $A \times B$ defines a relation from A to B as follows:

$$a\,R\,b \text{ if and only if } (a, b) \in R$$

In view of the above correspondence between relations from A to B and subsets of $A \times B$, we redefine a relation from A to B as follows:

DEFINITION D.1: A relation R from A to B is a subset of $A \times B$.

Equivalence Relations

Consider a nonempty set S. A relation R on S is called an *equivalence relation* if R is reflexive, symmetric, and transitive; that is, if R satisfied the following three axioms:

[E$_1$] (Reflexivity) Every $a \in A$ is related to itself. That is, for every $a \in A$, $a\,R\,a$.

[E$_2$] (Symmetry) If a is related to b, then b is related to a. That is, if $a\,R\,b$, then $b\,R\,a$.

[E$_3$] (Transitivity) If a is related to b and b is related to c, then a is related to c. That is,

if $a\,R\,b$ and $b\,R\,c$, then $a\,R\,c$.

The general idea behind an equivalence relation is that it is a classification of objects that are in some way "alike." Clearly, the relation of equality is an equivalence relation. For this reason, one frequently uses $\sim$ or $\equiv$ to denote an equivalence relation.

EXAMPLE D.1

(a) In Euclidean geometry, similarity of triangles is an equivalence relation. Specifically, suppose α, β, γ are triangles. Then (i) α is similar to itself. (ii). If α is similar to β, then β is similar to α. (iii) If α is similar to β and β is similar to γ, then α is similar to γ.

(b) The relation $\subseteq$ of set inclusion is not an equivalence relation. It is reflexive and transitive, but it is not symmetric because $A \subseteq B$ does not imply $B \subseteq A$.

Equivalence Relations and Partitions

Let S be a nonempty set. Recall first that a partition P of S is a subdivision of S into nonempty, nonoverlapping subsets; that is, a collection $P = \{A_j\}$ of nonempty subsets of S such that (i) Each $a \in S$ belong to one of the A_j, (ii) The sets $\{A_j\}$ are mutually disjoint.

The subsets in a partition P are called *cells*. Thus, each $a \in S$ belongs to exactly one of the cells. Also, any element $b \in A_j$ is called a *representative* of the cell A_j, and a subset B of S is called a *system of representatives* if B contains exactly one element in each of the cells in $\{A_j\}$.

Now suppose R is an equivalence relation on the nonempty set S. For each $a \in S$, the *equivalence class* of a, denoted by $[a]$, is the set of elements of S to which a is related:

$$[a] = \{x \mid aRx\}.$$

The collection of equivalence classes, denoted by S/R, is called the *quotient* of S by R:

$$S/R = \{[a] \mid a \in S\}$$

The fundamental property of an equivalence relation and its quotient set is contained in the following theorem:

THEOREM D.1: Let R be an equivalence relation on a nonempty set S. Then the quotient set S/R is a partition of S.

EXAMPLE D.2 Let $\equiv$ be the relation on the set $\mathbf{Z}$ of integers defined by

$$x \equiv y (\text{mod } 5)$$

which reads "x is congruent to y modulus 5" and which means that the difference $x - y$ is divisible by 5. Then $\equiv$ is an equivalence relation on $\mathbf{Z}$.

Then there are exactly five equivalence classes in the quotient set $\mathbf{Z}/\equiv$ as follows:

$$A_0 = \{\ldots, -10, -5, 0, 5, 10, \ldots\}$$
$$A_1 = \{\ldots, -9, -4, 1, 6, 11, \ldots\}$$
$$A_2 = \{\ldots, -8, -3, 2, 7, 12, \ldots\}$$
$$A_3 = \{\ldots, -7, -2, 3, 8, 13, \ldots\}$$
$$A_4 = \{\ldots, -6, -1, 4, 9, 14, \ldots\}$$

Note that any integer x, which can be expressed uniquely in the form $x = 5q + r$ where $0 \leq r < 5$, is a member of the equivalence class A_r where r is the remainder. As expected, the equivalence classes are disjoint and their union is $\mathbf{Z}$:

$$\mathbf{Z} = A_0 \cup A_1 \cup A_2 \cup A_3 \cup A_4$$

This quotient set $\mathbf{Z}/\equiv$, called the *integers modulo 5*, is denoted

$$\mathbf{Z}/5\mathbf{Z} \text{ or simply } \mathbf{Z}_5.$$

Usually one chooses $\{0, 1, 2, 3, 4\}$ or $\{-2, -1, 0, 1, 2\}$ as a system of representatives of the equivalence classes.

Analagously, for any positive integer m, there exists the congruence relation $\equiv$ defined by

$$x \equiv y (\text{mod } m)$$

and the quotient set $\mathbf{Z}/\equiv$, denoted by $\mathbf{Z}/m\mathbf{Z}$ or simply $\mathbf{Z}_m$, is called the *integers modulo m*.

D.3 Determinants and Block Matrices

Recall first:

THEOREM 8.12: Suppose M is an upper (lower) triangular block matrix with diagonal blocks $A_j, A_2, \ldots, A_n$. Then $\det(M) = \det(A_j)\det(A_2)\ldots\det(A_n)$.

Accordingly, if $M = \begin{bmatrix} A & B \\ 0 & D \end{bmatrix}$ where A is $r \times r$ and D is $s \times s$. Then $\det(M) = \det(A)\det(D)$.

THEOREM D.2: Consider the block matrix $M = \begin{bmatrix} A & B \\ C & D \end{bmatrix}$ where A is nonsingular, A is $r \times r$ and D is $s \times s$. Then $\det(M) = \det(A)\det(D - CA^{-1}B)$

Proof: Follows from the fact that $M = \begin{bmatrix} I & 0 \\ CA^{-1} & I \end{bmatrix}\begin{bmatrix} A & B \\ 0 & D - CA^{-1}B \end{bmatrix}$ and the above result.

D.4 Full Rank Factorization

A matrix B is said to have *full row rank* r if B has r rows that are linearly independent, and a matrix C is said to have *full column rank* r if C has r columns that are linearly independent.

DEFINITION D.2: Let A be a $m \times n$ matrix of rank r. Then A is said to have the *full rank factorization*

$$A = BC$$

where B has full-column rank r and C has full-row rank r.

THEOREM D.3: Every matrix A with rank $r > 0$ has a full rank factorization.

There are many full rank factorizations of a matrix A. Fig. D-1 gives an algorithm to find one such factorization.

Algorithm D-1: The input is a matrix A of rank $r > 0$. The output is a full rank factorization of A.
Step 1. Find the row cannonical form M of A.
Step 2. Let B be the matrix whose columns are the columns of A corresponding to the columns of M with pivots.
Step 3. Let C be the matrix whose rows are the nonzero rows of M.
Then $A = BC$ is a full rank factorization of A.

Figure D-1

EXAMPLE D.3 Let $A = \begin{bmatrix} 1 & 1 & -1 & 2 \\ 2 & 2 & -1 & 3 \\ -1 & -1 & 2 & -3 \end{bmatrix}$ where $M = \begin{bmatrix} 1 & 1 & 0 & 1 \\ 0 & 0 & 1 & -1 \\ 0 & 0 & 0 & 0 \end{bmatrix}$ is the row cannonical form of A. We set

$$B = \begin{bmatrix} 1 & -1 \\ 2 & -1 \\ -1 & 2 \end{bmatrix} \quad \text{and} \quad C = \begin{bmatrix} 1 & 1 & 0 & 1 \\ 0 & 0 & 1 & -1 \end{bmatrix}$$

Then $A = BC$ is a full rank factorization of A.

D.5 Generalized (Moore–Penrose) Inverse

Here we assume that the field of scalars is the complex field **C** where the matrix A^H is the conjugate transpose of a matrix A. [If A is a real matrix, then $A^H = A^T$.]

DEFINITION D.3: Let A be an $m \times n$ matrix over **C**. A matrix, denoted by A^+, is called the pseudoinverse or Morre–Penrose inverse or MP-inverse of A if A^+ satisfies the following four equations:

$$\textbf{[MP1]}\ AXA = A, \quad \textbf{[MP3]}(AX)^H = AX,$$

$$\textbf{[MP2]}\ XAX = X, \quad \textbf{[MP4]}\ (XA)^H = XA,$$

Clearly, A^+ is an $n \times m$ matrix. Also, $A^+ = A^{-1}$ if A is nonsingular.

LEMMA D.4: A^+ is unique (when it exists).

Proof. Suppose X and Y satisfy the four MP equations. Then

$$AY = (AY)^H = (AXAY)^H = (AY)^H(AX)^H = AYAX = (AYA)X = AX$$

The first and fourth equations use **[MP3]**, and the second and last equations use **[MP1]**. Similarly, $YA = XA$ (which uses **[MP4]** and **[MP1]**). Then,

$$Y = YAY = (YA)Y = (XA)Y = X(AY) = X(AX) = X$$

where the first equation uses **[MP2]**.

LEMMA D.5: A^+ exists for any matrix A.

Fig. D-2 gives an algorithm that finds an MP-inverse for any matrix A.

Algorithm D-2. Input is an $m \times n$ matrix A over **C** of rank r. Output is A^+.

Step 1. Interchange rows and columns of A so that $PAQ = \begin{bmatrix} A_{11} & A_{12} \\ A_{21} & A_{22} \end{bmatrix}$ where A_{11} is a nonsingular $r \times r$ block. [Here P and Q are the products of elementary matrices corresponding to the interchanges of the rows and columns.]

Step 2. Set $B = \begin{bmatrix} A_{11} \\ A_{21} \end{bmatrix}$ and $C = [\mathrm{I_r}, \mathrm{A_{11}^{-1}A_{12}}]$ where I_r is the $r \times r$ identity matrix.

Step 3. Set $A^+ = Q\left[C^H(CC^H)^{-1}(B^HB)^{-1}B^{11}\right]P$.

Figure D-2

Combining the above two lemmas we obtain:

THEOREM D.6: Every matrix A over **C** has a unique Moore–Penrose matrix A^+.

There are special cases when A has full-row rank or full-column rank.

THEOREM D.7: Let A be a matrix over **C**.

(a) If A has full column rank (columns are linearly independent), then $A^+ = (A^HA)^{-1}A^H$.

(b) If A has full row rank (rows are linearly independent), then $A^+ = A^H(AA^H)^{-1}$.

THEOREM D.8: Let A be a matrix over **C**. Suppose $A = BC$ is a full rank factorization of A. Then

$$A^+ = C^+B^+ = C^H(CC^H)^{-1}(B^HB)^{-1}B^H$$

Moreover, $AA^+ = BB^+$ and $A^+A = C^+C$.

EXAMPLE D.4 Consider the full rank factorization $A = BC$ in Example D.1; that is,

$$A = \begin{bmatrix} 1 & 1 & -1 & 2 \\ 2 & 2 & -1 & 3 \\ -1 & -1 & 2 & -3 \end{bmatrix} = \begin{bmatrix} 1 & -1 \\ 2 & -1 \\ -1 & 2 \end{bmatrix} \begin{bmatrix} 1 & 1 & 0 & 1 \\ 0 & 0 & 1 & -1 \end{bmatrix} = BC$$

Then

$$(CC^H)^{-1} = \frac{1}{5}\begin{bmatrix} 2 & 1 \\ 1 & 3 \end{bmatrix}, \quad C(CC^H)^{-1} = \frac{1}{5}\begin{bmatrix} 2 & 1 \\ 2 & 1 \\ 1 & 3 \\ 1 & -2 \end{bmatrix}, \quad (B^H B)^{-1} = \frac{1}{11}\begin{bmatrix} 6 & 5 \\ 5 & 6 \end{bmatrix}, \quad B(B^H B)^{-1} = \frac{1}{11}\begin{bmatrix} 1 & 7 & 4 \\ -1 & 4 & 7 \end{bmatrix}$$

Accordingly, the following is the Moore–Penrose inverse of A:

$$A^+ = \frac{1}{55}\begin{bmatrix} 1 & 18 & 15 \\ 1 & 18 & 15 \\ -2 & 19 & 25 \\ 3 & -1 & -10 \end{bmatrix}$$

D.6 Least-Square Solution

Consider a system $AX = B$ of linear equations. A *least-square solution* of $AX = B$ is the vector of smallest Euclidean norm that minimizes $\|AX - B\|_2$. That vector is

$$X = A^+B$$

[In case A is invertible, so $A^+ = A^{-1}$, then $X = A^{-1}B$, which is the unique solution of the system.]

EXAMPLE D.5 Consider the following system $AX = B$ of linear equations:

$$\begin{aligned} x + y - z + 2t &= 1 \\ 2x + 2y - z + 3t &= 3 \\ -x - y + 2z - 3t &= 2 \end{aligned}$$

Then, using Example D.4,

$$A = \begin{bmatrix} 1 & 1 & -1 & 2 \\ 2 & 2 & -1 & 3 \\ -1 & -1 & 2 & -3 \end{bmatrix}, \quad B = \begin{bmatrix} 1 \\ 3 \\ 2 \end{bmatrix}, \quad A^+ = \frac{1}{55}\begin{bmatrix} 1 & 18 & 15 \\ 1 & 18 & 15 \\ -2 & 19 & 25 \\ 3 & -1 & -10 \end{bmatrix}$$

Accordingly,

$$X = A^+B = (1/55)[85, 85, 105, -20]^T = [17/11, 17/11, 21/11, -4/11]^T$$

is the vector of smallest Euclidean norm which minimizes $\|AX - B\|_2$.

D.7 Properties of AA^T and A^TA

Let A be any m × n matrix. Then A^TA and AA^T are both symmetric since

$$(A^TA)^T = A^TA^{TT} = A^TA \quad \text{and} \quad (AA^T)^T = A^{TT}A^T = AA^T$$

One can also show that A^TA and AA^T have the same nonzero eigenvalues.

Recall [Theorem 9.14] that a symmetric matrix M is orthogonally diagonalizable, that is, there exists an orthogonal matrix P and diagonal matrix $D = [d_i]$ such that

$$P^{-1}MP = D$$

where the columns of P are the eigenvectors of M and the d_i are the eigenvalues of M.

In particular, the symmetric matrix M is called:

(i) *positive definite* if, for every nonzero u, $\langle u, Mu\rangle = u^T Mu > 0$

(ii) *positive semidefinite* if, for every nonzero u, $\langle u, Mu\rangle = u^T Mu \geq 0]$

In such a case, the diagonal elements in $P^{-1}MP = D$ are: (i) positive, (ii) nonnegative.

A^TA is positive definite if A has full column rank since $Au \neq 0$ when $u \neq 0$; so

$$\langle u, A^TAu\rangle = u^TA^TAu = \langle Au, Au\rangle > 0$$

Similarly, AA^T is positive definite if A has full row rank. since $A^Tu \neq 0$ when $u \neq 0$; so

$$\langle u, AA^Tu\rangle = u^T AA^Tu = u^TA^{TT}A^T = \langle A^Tu, A^Tu\rangle > 0$$

On the other hand, A^TA and AA^T are always both positive semidefinite.

D.8 Singular Value Decomposition

Let A be any $m \times n$ matrix of rank r. Then there exists a factorization of A of the form

$$A = U \Sigma V^T$$

where U is an m-square orthogonal matrix, V is an n-square orthogonal matrix, and $\Sigma = [\sigma_{ij}]$ is an $m \times n$ *generalized diagonal matrix*, that is, $\sigma_{ij} = 0$ for $i \neq j$. Such a factorization is called a **singular value decomposition (SVD) of A,** and the diagonal entries $\sigma_i = \sigma_{ii}$ of Σ, usually listed in descending order, are called the ***singular values*** of A.

THEOREM D.9: Every matrix A with rank r>0 has a **singular value decomposition.**

We indicate how the entries U, V and Σ in the SVD of A are obtained.

Recall AA^T is symmetric and positive definite (or positive semidefinite). Accordingly, AA^T is orthogonally diagonalizable, that is, $AA^T = P^{-1}DP = P^TDP$ where the columns of the orthogonal matrix P are eigenvectors of AA^T, and the entries of D are the eigenvalues of AA^T and they are nonnegative.

Assuming $A = U \Sigma V^T$, we have

$$A A^T = U \Sigma V^T (U \Sigma V^T)^T = U \Sigma V^T V \Sigma U^T = U \Sigma^2 U^T$$

Thus the columns of U in the SVD of A are the normalized eigenvectors of AA^T and the entries σ_i in Σ are the square roots of the eigenvalues of AA^T.

Similarly, assuming $A = U \Sigma V^T$, we have

$$A^TA = (U \Sigma V^T)^T U \Sigma V^T = V \Sigma U^T U \Sigma V^T = V\Sigma^2V^T$$

Thus the columns of V and the rows of V^T are the normalized eigenvectors of A^TA and the entries σ_i in Σ are the square roots of the eigenvalues of AA^T (which are the same as the eigenvalues of A^TA).

EXAMPLE D.6. We find the singular value decomposition (SVD) of $A = \begin{bmatrix} 4 & 4 \\ -3 & 3 \end{bmatrix}$. Note

$$AA^T = \begin{bmatrix} 4 & 4 \\ -3 & 3 \end{bmatrix}\begin{bmatrix} 4 & -3 \\ 4 & 3 \end{bmatrix} = \begin{bmatrix} 32 & 0 \\ 0 & 18 \end{bmatrix}$$

The eigenvalues are 32 and 18 with corresponding eigenvectors $[1, 0]^T$ and $[0,1]^T$. Thus

$$U = \begin{bmatrix} 1 & 0 \\ 0 & 1 \end{bmatrix} \text{ and } \Sigma = \begin{bmatrix} \sqrt{32} & 0 \\ 0 & \sqrt{18} \end{bmatrix} = \sqrt{2}\begin{bmatrix} 4 & 0 \\ 0 & 3 \end{bmatrix}$$

Also

$$A^TA = \begin{bmatrix} 4 & -3 \\ 4 & 3 \end{bmatrix}\begin{bmatrix} 4 & 4 \\ -3 & 3 \end{bmatrix} = \begin{bmatrix} 25 & 7 \\ 7 & 25 \end{bmatrix}$$

Hence $\Delta(t) = t^2 - 50t + 576 = (t - 32)(t - 18)$. Thus, again, the eigenvalues are 32 and 18.

The normalized corresponding eigenvectors are $[1/\sqrt{2}, 1/\sqrt{2}]^T$ and $[-1/\sqrt{2}, 1/\sqrt{2}]^T$. Thus

$$V^T = \begin{bmatrix} 1/\sqrt{2} & 1/\sqrt{2} \\ -1/\sqrt{2} & 1/\sqrt{2} \end{bmatrix} = \frac{1}{\sqrt{2}}\begin{bmatrix} 1 & 1 \\ -1 & 1 \end{bmatrix}$$

As expected,

$$U\,\Sigma\,V^T = \begin{bmatrix} 1 & 0 \\ 0 & 1 \end{bmatrix}\sqrt{2}\begin{bmatrix} 4 & 0 \\ 0 & 3 \end{bmatrix}\frac{1}{\sqrt{2}}\begin{bmatrix} 1 & 1 \\ -1 & 1 \end{bmatrix} = \begin{bmatrix} 4 & 4 \\ -3 & 3 \end{bmatrix} = A$$

INDEX